Handbook of Instrumental Techniques for Analytical Chemistry

Frank A. Settle, Editor

To join a Prentice Hall Internet mailing list, point to:
http://www.prenhall.com/mail_lists/

Prentice Hall PTR, Upper Saddle River, NJ 07458

Library of Congress Cataloging-in-Publication Data

Handbook of instrumental techniques for analytical chemistry / Frank
  A. Settle, editor.
     p.   cm.
   Includes bibliographical references (p. –) and index.
   ISBN 0-13-177338-0 (hardcover)
   1. Instrumental analysis.   I. Settle, Frank A.
QD79.I5H36 1997
543'.07--dc21
                                               97-10618
                                               CIP

Editorial Production: *Precision Graphic Services, Inc.*
Acquisitions Editor: *Bernard M. Goodwin*
Manufacturing Manager: *Alexis R. Heydt*
Marketing Manager: *Miles Williams*
Cover Design Director: *Jerry Votta*
Cover Design: *Bruce Kenselaar*

Prentice Hall books are widely used by corporations and government agencies for
training, marketing, and resale.

The publisher offers discounts on this book when ordered in bulk quantities. For more
information, contact: Corporate Sales Department
                      Phone: 800-382-3419
                      FAX: 201-236-7141
                      E-mail: corpsales@prenhall.com

Or write: Prentice Hall PTR
          Corp. Sales Dept.
          One Lake Street
          Upper Saddle River, NJ 07458

Printed in the United States of America

10 9 8 7 6 5 4 3 2 1

ISBN 0-13-177338-0

Prentice-Hall International (UK) Limited, *London*
Printice-Hall of Australia Pty. Limited, *Sydney*
Prentice-Hall Canada Inc., *Toronto*
Printice-Hall Hispanoamericana, S.A., *Mexico*
Prentice-Hall of India Private Limited, *New Delhi*
Prentice-Hall of Japan, Inc., *Tokyo*
Simon & Schuster Asia Pte. Ltd., *Singapore*
Editora Prentice-Hall do Brasil, Ltda., *Rio de Janeiro*

Dedication

This book is dedicated to the memories of three generalists who made major contributions to the development of the discipline of analytical chemistry and had a great influence on the editor as well as on many practitioners of analytical chemistry. Dr. Sidney Sigga, professor of chemistry at the University of Massachusetts, was an outstanding industrial and academic chemist who developed the analytical procedure into an art form. Dr. Lockhart B. "Buck" Rodgers, professor of chemistry at the University of Georgia, Purdue University, and the Massachusetts Institute of Technology, was an outstanding scientist and educator who worked across many areas of analytical chemistry. Dr. John K. Taylor, a scientist for over 50 years at the National Institute of Standards and Technology (the old National Bureau of Standards), was a national treasure of analytical knowledge. In addition to their professional contributions, all three were gentlemen who also contributed much to society through their relationships with family, colleagues, and students.

The book is also dedicated to the memory of Dale Baker, the author of the chapter on capillary electrophoresis, who completed this chapter before his death.

Contents

Chapter 4   Quantitative Measurements 73
Earl Wehry

Chapter 5   Managing Laboratory Information 81
Robert Megargle

Chapter 6   Laboratory Automation 101
Joe Liscouski

Section II   Separation Methods

Chapter 7   Introduction 121
Mary Jane Van Sant

Chapter 8   Gas Chromatography 125
Mary Jane Van Sant

Section III   Qualitative Optical Spectroscopic Methods

Chapter 26 Molecular Fluorescence and Phosphorescence Spectrometry 507

Earl L. Wehry

Chapter 27 Chemiluminescence 541

Timothy A. Nieman

Section V Mass Spectrometry

Chapter 28 Introduction. 563

Charles L. Wilkins

Chapter 29 Mass Spectrometry of Volatile Analytes. 567

J. Throck Watson

Acknowledgments

The editor gratefully acknowledges the contribution of the section editors, who organized the sections, found excellent authors, obtained manuscripts, and edited the resulting chapters. The hard work of the chapter authors is also recognized; their efforts form the heart of the handbook. The editor wishes to thank Betty Sun for having the vision to support the project in its early stages, and for her encouragement throughout the project. The fine efforts of Kirsten Dennison and her colleagues at Precision Graphics were an important factor in the appearance of the handbook.

Preface

Scientists, engineers, and technicians in a variety of fields often need information from chemical analysis in their work. Chemists themselves often require information provided by instrumental techniques outside their areas of specialization. This handbook allows readers to find information about the analytical instrumental techniques that are commonly used to provide information on chemical composition and structure.

The handbook presents each technique in a uniform format, in a style that can be understood by a reader who is not familiar with the particular technique. Each chapter is structured to provide a description of the information the technique can provide, a simple explanation of how it works, examples of its application, and practical information such as names of instrument vendors, relative costs of instruments and materials, training and education of personnel, and references for more detailed information. This format also facilitates comparison of techniques. The use of different authors to cover a broad spectrum of techniques resulted in some differences of style, but overall the handbook achieved its goal.

The techniques are grouped into eight sections, each with a different editor. I worked primarily with the section editors, each an experienced authority, to select the techniques for the sections. The editors were given a template and a sample chapter, which they in turn gave to their chapter authors. The authors were encouraged to provide theory and applications at an introductory level.

The chapters in the introductory section cover topics relevant to all techniques, including sample preparation, quantitative measurements, information management, and laboratory automation. The remaining seven sections address the major areas of chemical analysis: separations, qualitative spectrometry, quantitative spectrometry, mass spectrometry, electrochemistry, surface analysis, and polymer analysis. Each section is introduced by an overview chapter written by the section editor.

It is often the case that several complementary techniques may be required to provide the desired information. Also, as might be expected, there is some overlap among techniques, for example, the use of electrochemical techniques and mass spectrometers as detectors for separation techniques. In these cases, cross references provide the linkage among techniques. An extensive glossary and comprehensive index assist the reader with the terminology associated with each technique and relationships among techniques.

I welcome any suggestions for improvements in the organization, scope, and content of the handbook, as well as notification of any errors.

Frank A. Settle
19 Sixty West Drive
Lexington, VA 24450

Introduction

Chapter 1

How to Use the Handbook

Frank Settle, Editor

Virginia Military Institute
Department of Chemistry

Purpose and Scope

The goal of the *Handbook of Instrumental Techniques for Analytical Chemistry* is to provide scientists and engineers from many disciplines with an easily understood reference to current, established techniques of chemical analysis. The *Handbook* describes each technique in uncomplicated terms, gives common applications, points out its strengths and limitations, and provides references to more detailed discussions. The intent is to assist the reader in identifying the techniques that can best provide information to solve the problem at hand. For some problems, there may be a choice of techniques or the option of using complementary techniques. The *Handbook* is intended to provide the reader with sufficient knowledge to interact with analytical specialists or to pursue more detailed references.

Many of the techniques are used in growing areas such as biotechnology and materials science. However, techniques used predominantly for medical analysis are omitted as well as highly specialized techniques used in fundamental research. Discussions of hyphenated techniques, such as gas chromatography–mass spectrometry, are presented in the chapters on the individual techniques, with appropriate emphasis. For example, the gas chromatography chapter identifies mass spectrometers as detectors and refers the reader to the more detailed presentation in the mass spectrometry chapter. Conversely, the mass spectrometry chapter cites gas chromatography as a useful method for the preparation and introduction of samples and refers to the gas chromatography chapter.

Organization

The *Handbook* is designed for ease of use. This chapter contains general information and tables to assist the reader in selecting techniques for a given problem. The remaining chapters in the first section cover the fundamental principles of sample preparation for both inorganic and organic analyses, quantitative analysis, laboratory automation, and management of laboratory information.

The main part of the *Handbook* consists of seven sections, each addressing a group of related techniques. Techniques in Sections II through VI are organized according to fundamental phenomenon, chromatography, electromagnetic spectroscopies, mass spectrometry, and electrochemistry. The remaining two sections, on surface analysis and macromolecular analysis, include techniques specific to these major areas. Each section opens with an introduction by the section editor, followed by chapters on specific techniques. The identical format of each chapter facilitates comparison and selection of techniques. An introductory summary page assists the reader in deciding whether the technique is applicable to the problem at hand. The organization for each technique is shown in the following outline.

Summary
 General Use
 Common Applications
 Samples
  State, Amount, Preparation
 Analysis Time
 Limitations
 Complementary or Related Techniques
Introduction
 Brief History
 Current Use
How It Works
 Physical and Chemical Principles
 Instrumentation (Modular Approach)
  Description of Each Major Component and Its Operation
What It Does
Analytical Information
 Qualitative
 Quantitative
  Accuracy and Precision
  Detection Limits
Applications
 General Discussion
 Specific Examples
Nuts and Bolts
 Relative Costs
  Capital Outlay
  Maintenance and Operation
 List of Instrument Manufacturers (see also Buyers' Guides, page 16)

Nuts and Bolts (*continued*)
Required Training
Operation
Interpretation of Data
Service and Maintenance Requirements
Suggested Readings (Appropriate Level for Readers)

Analytical Methodology

Experienced analytical chemists use a systematic approach to obtain the required information from samples. This approach consists of a series of steps and accompanying thought processes. Professor Sidney Siggia, a general analytical chemist with extensive experience in industry and academics, captured this process in his book *Survey of Analytical Chemistry* (1). Much of his philosophy is reflected in the following sections.

Understanding and Defining the Problem

The goal of every chemical analysis is to obtain the required information within a period of time acceptable to the customer. This means that the analyst must know what information is needed to solve the current problem. For example, in qualitative analysis, complete identification of analytes is not necessary in many cases. Many problems require only a general classification or partial identification. Siggia cites the example of $C_{13}H_{22}(OCH_2CH_2)_4OSO_3H$, a surface-active agent for detergents, which can be analyzed at various levels depending on the information required. A cursory analysis would show the analyte to be an anionic detergent. Partial identification would indicate the analyte to be a sulfate ester of a fatty alcohol adduct of ethylene oxide. Finally, a complete analysis would identify the compound completely as written.

It is important for the analyst to determine the specific information required by a client. In the case of the detergent, the formulation people may need to know only the general classification, anionic detergent, because they mix components to produce a product with desirable characteristics. The detergent manufacturers may be most interested in the identification of functional groups and type of compound (sulfate ester of fatty alcohol adduct of ethylene oxide) for comparison with their products. When the compound is used in a commercial product, government regulations may require complete characterization of the compound.

Quantitative determinations are approached in the same way. Some problems require high precision and accuracy whereas much lower values are acceptable for others. Because there is considerable difference in the effort (and expense) required to provide different levels of qualitative and quantitative information, analysts must work closely with the clients to define the goals of analyses and determine proper limits for investigations.

History of the Sample and Background of the Problem

Before attempting any laboratory analysis, the analyst should become familiar with the background of the problem and the history of the sample. Background information can originate from many sources.

The client (such as a chemist, engineer, salesperson, or lawyer) who brought the sample for analysis should know the reason for the analysis and the origin of the sample. The more knowledgeable the client, the better the goals of the analysis can be defined. For example, when a client mixes reactants under certain conditions and does not get the expected products, the identities and amounts of the unexpected products can provide useful information. Knowledge of the reactant identities and the reaction conditions narrows the number of possible products for the analyst.

Analyses of competitors' products are commonly requested by sales and legal departments. In these cases, knowledge of materials used for specific applications and the products manufactured by the competitors provide the analyst with useful background information. Sample histories also contain information on how, where, and when the sample was collected, transported, and stored. Every sample has a history. A good analyst obtains and uses the sample history to maximum advantage in solving problems.

Literature Search

Efficient use of the scientific, technical, government, and commercial literature is essential to efficient analysis. References from chemical journals can indicate by-products of the reaction under consideration and thus ease the process of identification. Patents or commercial literature usually contain the composition of industrial materials. Registry numbers from *Chemical Abstracts* provide an easy way to locate information on specific compounds. Analytical methods are available from many sources: texts, review articles, monographs devoted to specific sample types, and standards organizations, such as the American Society for Testing and Materials (ASTM) and the Association of Official Analytical Chemists (AOAC). Many analyses must be performed using protocols approved by the Environmental Protection Agency (EPA), the Food and Drug Administration (FDA), or other government agencies. Even if the search does not locate a method for the problem at hand, it may provide ideas for developing a new analytical method.

Use of electronic media can make literature searches more efficient and effective. Sites available on the Internet include libraries, *Chemical Abstracts*, government documents, instrument vendors, electronic journals including *Analytical Chemistry*, and other special World Wide Web sites such as the Virginia Tech Encyclopedia of Chemical Instrumentation (http://www.chem.vt.edu/chem-ed/analytical/ac-methods.html). In today's electronic environment, the problem is one of filtering the massive amount of information to obtain what is needed for the problem at hand. The more one knows about the problem and the techniques available, the more effective is the search to provide useful information.

Plan of Action and Execution

Once the goals and limits of the analysis are defined and the literature search is completed, a more detailed plan of action must be developed. The analyst selects the method or combination of methods most likely to provide the desired information. A good analyst is always alert to the chemistry involved in the analysis as well as instrumental techniques. Thus, a combination of chemical and instrumental techniques may be used.

Major components to be considered in planning an analysis are shown in Fig. 1.1. Each component is important in obtaining reliable information from the analysis. Field sampling and laboratory subsampling procedures must be designed to ensure integrity of results. Proper procedures must be used to store both samples and standards. All samples must be properly labeled and recorded. Laboratory operations are often performed on samples before measurement. These physical or chemical procedures may reduce or remove interferences, adjust analyte concentrations to a range suitable for measurement, or produce species from the analyte that have quantitatively measurable properties. These procedures include dissolution, separation, dilution, concentration, and chemical derivatization.

Control of the chemical environment is often necessary to ensure that the analyte is measured in the desired form and to minimize the effect of interferences. Environmental parameters include temperature, pH, and atmosphere surrounding the sample. Instrumental parameters such as

Figure 1.1 Components of an analytical method.

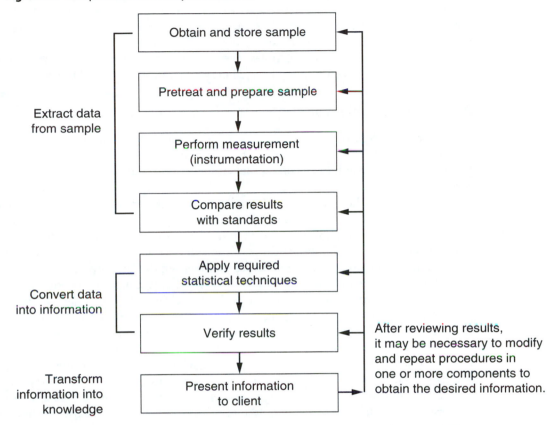

the amplitude and frequency of input signal, detector sensitivity, and sampling rate must be coordinated to obtain optimum conditions for measurements.

Once the measurement has been obtained, steps must be in place to ensure the desired precision and accuracy for the analysis. The methods best suited for standardizing the analysis must be selected. These methods include calibration curves, internal standards, standard additions, blind samples, and control charts. Statistical methods must be incorporated to assess the accuracy and precision of results.

When planning a method, the analyst must be aware that instrumentation is only one component. The analyst should be able to trace the flow of information through the entire process, not just through the instrumentation. Inadequate attention to any component of an analytical method can render the results meaningless, regardless of the power of the instrumental technique.

As the initial data provide the analyst with insight into the situation, it may become necessary to modify the original plan to obtain the desired information; in other cases no changes are required. The analyst should examine the data critically and be ready to modify the original approach.

Critical evaluation of the data requires a knowledge of the limitations of the techniques so as not to reach unjustified conclusions. However, the analyst must gather all information possible from the data and be aware that this information is often known with varying degrees of certainty.

The analyst determines when enough data have been obtained to reach tentative conclusions. The validity of these conclusions depends on the reliability of the data and the judgment of the analyst. The conclusions should be confirmed whenever possible. In qualitative analysis the identity of the unknown can be confirmed by comparing an actual sample of the material in question to the unknown using fingerprinting techniques such as infrared spectroscopy, mass spectrometry, or nuclear magnetic resonance spectroscopy. Matching spectra confirm the identification. Once a method has been developed to determine the composition of a certain sample or mixture, samples of known concentrations are made and analyzed to confirm the validity of the method. The composition of the known samples must match the matrix of the unknown samples and must cover the range of concentrations expected in the unknown samples of the material under study.

Approaches to Unknowns

"The Analytical Approach," a continuing series of articles found in *Analytical Chemistry*, provides insight into the processes described in the previous sections. According to Siggia, the thought processes of understanding, selection, evaluation, and correlation are necessary for the solution of any problem (1).

An example of the solution to an actual problem found in the construction industry illustrates the steps in the analytical approach (2). The problem was to determine why concrete in a large New York City construction project sometimes failed to set. The problem became so bad that the entire project was halted until the cause of the failure could be determined. Samples of the unset concrete from the project hardened before they reached the laboratory.

The plan of analysis was to select a series of standard methods to determine the presence of materials that retard setting of the concrete. Microscopic analysis revealed an unusual thin film on the cement particles. This suggested the presence of excessive amounts of retardant. Because sabotage was a possibility, the concrete was next analyzed for sugar, a known set-retarder. After samples were extracted with appropriate solvents, colorimetric methods indicated the absence of sugar. Other extracts were analyzed by infrared and ultraviolet techniques for the presence of known organic set-retarders such as polysaccharides and lignosulfates. These results also proved negative.

X-ray fluorescence spectrometry was then used to check for the presence of metals; trace amounts of both lead and zinc were found. A more accurate quantitative technique, atomic absorption spectrometry, revealed that both metals were present in the concrete at concentrations high enough to retard cement hydration. Further investigation determined the origin of the metals: dredged river gravel used as an aggregate in the concrete. This gravel was coated with lead and zinc materials found in the river.

Knowledge of the literature, experience with concrete analysis, and interactions with the clients (construction personnel) contributed to the solution. It was first necessary to understand the problem, then select appropriate methods for analysis, collect samples with historical information, perform the analysis, evaluate the results, and finally correlate the results with the situation to resolve the problem.

Thus, chemical analysis is more than just detecting or determining a specific component or the general composition of a sample. It is the resolution or interpretation of a given problem. What component in the fuel caused the engine to fail? Has a competitor infringed on the patent for the composition of our tinted glass? What substance in the feed line poisoned the catalyst? What is the effective concentration of a new drug? The analytical chemist must work closely with clients to provide the information required for the situation.

Technique Selection

The goal of any analysis is to measure a signal related to the concentration of the analyte. This signal usually contains a background or noise component whose magnitude limits the lowest analyte concentration that can be determined (the detection limit). The precision of the measured signal depends on several factors, some associated with fundamental phenomena and others with instrumentation. These may include noise from a variety of sources, instrument drift, and detector sensitivities to specific analytes. Interferences from components in the sample matrix or solvents can adversely affect many techniques. It is usually necessary to validate the accuracy of results of a technique by comparative analysis of standard reference materials representative of the sample matrix. Chapter 4 discusses quantitative analysis in detail.

Using the philosophy outlined in the previous section, the analyst, working closely with the client, examines the sample and determines the information required to solve the problem. The analyst then chooses the techniques available to provide this information. Table 1.2 follows this same approach to assist the reader in locating techniques for the problem at hand. Table 1.1 contains a key to symbols used in Table 1.2. The page numbers in the Reference column of Table 1.2 direct the reader to the summary pages for the techniques. Once an appropriate technique is located the reader should further verify the suitability of the method by checking any endnotes associated with the technique and the information on the technique summary pages. Some techniques provide more definitive information than others; the amount of information varies with the technique. In many cases, Table 1.2 suggests several techniques for a given situation. Consultation with the information contained in the endnotes and technique summary page will often eliminate techniques from further consideration.

Many of the techniques listed for solid and liquid samples can also be used for gases if the gas is adsorbed onto a solid or absorbed into a liquid. In the case of liquid samples, some techniques listed for solids can be used for inorganic residues after the solution is evaporated to dryness or on organic residues after the solution has been ashed. Conversely, many of the techniques recommended for liquids can be used for solids if the solid can be dissolved in an appropriate solvent. Furthermore, if the solid can be vaporized without decomposition, techniques recommended for gas samples can be used.

Single-component samples and simple mixtures can often be analyzed directly by a given technique without prior separation. Complex mixtures often require separation before analysis. There are a variety of combined, or hyphenated, methods available in which the initial technique separates the analytes and the final technique provides the measurement component. Many gases and low-boiling-point organic liquids (boiling points below 300 °C) are relatively easy to determine.

It is often possible to extend the applications of techniques by chemical reactions. Analytes that are not sensitive to a specific technique may be converted to compounds or complexes that have the required sensitivity. Chelation of metal ions has extended the utility of ultraviolet/visible spectroscopy and quantitative conversion of high-boiling compounds to lower-boiling derivatives has increased the usefulness of gas chromatography.

Table 1.1 provides a key to the symbols used with the flow charts and assists in organizing the information desired from the samples. In some cases it is necessary to preserve the original sample. If elemental information is required, is it desirable to know the oxidation (valence) or complexed state of the element? For example, in the case of iron, does the technique determine iron as iron(II), iron(III), iron metal, iron(III) chelated with heme, or total iron in all forms? The ability to detect and measure an element in a specific chemical state, such as iron(III), is called speciation, abbreviated Sp. The ability to determine specific isotopes of elements as well as masses of both elements, compound fragments, and compounds present in samples is abbreviated as Is/Ms for the charts.

The ability to determine compounds is shown as Cp. Other useful information associated with compounds is obtained from functional group and structural analyses, including sterochemical information. Functional groups are specific groupings of atoms appearing in many compounds, such as the carboxyl grouping, COOH, that can give acidic properties to many organic compounds. Sterochemical analysis provides detailed information on the spatial arrangement of atoms in molecules.

Table 1.1 Key to technique selection chart.

Location of analyte

| B | Sample from bulk of material in question |
| S | Sample from or on surface of material in question |

Physical state of sample

G	Gas
L	Liquid (includes liquids dissolved in liquid solvents)
S	Solid (includes solids with gases adsorbed on surface)
Ds	Solid dissolved in appropriate liquid solvents
Dg	Gas dissolved in appropriate liquid solvent

Amount of sample

| L | Macro > 1 milligram |
| S | Micro ≤ 1 milligram |

Estimated sample purity

P	Pure (> 99%) element or compound
Sm	Simple mixture of up to six major components
M	Complex mixture

Fate of sample

| D | Destructive analysis |
| N | Nondestructive analysis |

Elemental information

El	Total analysis (element present in all chemical forms in sample)
Sp	Speciation (determination of individual elemental oxidation or chelation states)
Is	Isotopic and mass analysis

Molecular information

Cp	Compounds present in sample
Io	Polyatomic ionic species present in sample
Fn	Functional group analysis
St	Structural analysis (structural and sterochemical determinations)
Mw	Molecular weight determination
Pp	Physical property

Analysis type

| Ql | Qualitative analysis |
| Qt | Quantitative analysis |

Analyte concentration

Mj	Major component determination (> 10 wt %)
Mn	Minor component determination (10 to 0.1 wt %)
Tr	Trace component determination (1 to 1000 ppm or 0.0001 to 0.1 wt %)
Ul	Ultratrace component determination (< 1 ppm or < 0.0001 wt %)

Table 1.2 Techniques

Technique	Sample					Information Desired				Reference
	Location	Physical State	Amount	Purity	Fate	Elemental	Molecular	Analysis Type	Analyte Concentration	Chapter(s)
Separation techniques										
Gas chromatography	B	G, L, S, Ds	L, S	Sm, M	N, D[1]	El[2]	Cp, Fn[1]	Ql[1], Qt	Mj, Mn, Tr, Ul[3]	8
High performance liquid chromatography	B	L, Ds	L, S	Sm, M	N, D[1]		Cp, Io, St	Ql, Qt	Mj, Mn, Tr, Ul[3]	9
Ion chromatography	B	L, Ds	L, S	Sm, M	N, D[1]	Sp	Cp, Io, Fn	Ql, Qt	Mj, Mn, Tr, Ul[3]	12
Supercritical fluid chromatography	B	L, Ds	L, S	Sm, M	N, D[1]		Cp	Ql[2], Qt	Mj, Mn, Tr, Ul[3]	11
Capillary electrophoresis	B	L, Ds	L, S	Sm, M	N	Sp	Cp, Io, St	Ql, Qt	Mj, Mn, Tr, Ul[3]	10
Planar chromatography	B	L, Ds	L, S	Sm, M	N		Cp, Io	Q[1], Qt	Mn, Tr[3]	13
Optical spectroscopic techniques: Qualitative										
Infrared spectrometry (dispersive and Fourier transform)	B, S	G, L, S, Ds	L, S	P, Sm	N		Cp, Fn, St	Ql, Qt	Mj, Mn, Tr	15
Raman spectrometry	B, S	G, L, S, Ds	L, S	P, Sm	N		Cp, Fn, St	Ql[4], Qt	Mj, Mn, Tr	16
Nuclear magnetic resonance spectrometry	B	L, S, Ds	L, S	P, Sm	N		Cp, Fn, St	Ql, Qt	Mj, Mn	17
X-ray spectrometry	B	S	L	P, Sm	N	El	Cp, St	Ql, Qt	Mj, Mn, Tr	18

Table 1.2 (continued)

Technique	Sample					Information Desired				Reference
	Location	Physical State	Amount	Purity	Fate	Elemental	Molecular	Analysis Type	Analyte Concentration	Chapter(s)
Optical spectroscopic techniques: Quantitative										
Atomic absorption spectrometry	B	L,Ds	L,S[5]	Sm,M	D	El		Qt	Mj,Mn,Tr,Ul	20
Inductively coupled plasma atomic emission spectrometry	B	L,Ds	L	Sm,M	D	El		Ql,Qt	Mj,Mn,Tr,Ul[6]	21
Inductively coupled plasma mass spectrometry	B	G,L,Ds	L	Sm,M	D	El,Is		Ql,Qt	Mj,Mn,Tr,Ul	22
Atomic fluorescence spectrometry	B	L,Ds	L,S	Sm,M	D	El[7]		Qt	Mj,Mn,Tr,Ul	23
Ultraviolet/visible spectrometry	B	G,L,S,Ds	L,S	P,Sm	N	El[8],Sp	Cp,Io	Ql,Qt[9]	Mj,Mn,Tr	25
Molecular fluorescence spectrometry	B	G,L,S,Ds	L,S	Sm	N		Cp[10]	Qt	Mj,Mn,Tr,Ul	26
Chemiluminescence spectrometry	B	G,L,Ds	L,S	Sm,M	D	El[7]	Cp[11]	Qt	Mj,Mn,Tr,Ul	27
X-ray fluorescence spectrometry	B	L,S,Ds	L	Sm,M	N	El		Ql,Qt	Mj,Mn,Tr,Ul	24
Mass spectrometry (MS)										
Electron ionization MS	B	G,L,Ds	L,S	P,Sm[12]	D	El,Is	Cp,Fn,Io,St	Ql,Qt	Mj,Mn,Tr	29,30
Chemical ionization MS	B	G,L,Ds	L,S	P,Sm[12]	D	El,Is	Cp,Fn,St,Mw[13]	Ql,Qt	Mj,Mn,Tr	29,30
High-resolution MS	B	G,L,S,Ds	L,S	P[12]	D	El,Is	Fn,St,Mw[14]	Ql	Mj,Mn,Tr	30
Gas chromatography mass spectrometry	B	G,L,Ds	L,S	Sm,M	D	El,Is	Cp,Fn,St	Ql,Qt	Mj,Mn,Tr,Ul	31
Fast atom bombardment MS	B,S	L,S	L,S	P,Sm	D	El,Is	Cp,Fn,St,Mw	Ql,Qt	Mj,Mn,Tr,Ul	32
High-performance liquid chromatography MS	B	L,Ds	L,S	Sm,M	D	El,Is	Cp,Fn,St,Mw	Ql,Qt	Mj,Mn,Tr,Ul	33
Laser MS	B,S	S,L,Ds	L[15],S	P,Sm	D	El,Is	Cp,Fn,St,Mw	Ql,Qt	Mj,Mn,Tr,Ul	34

Table 1.2 (continued)

Technique	Sample					Information Desired				Reference
	Location	Physical State	Amount	Purity	Fate	Elemental	Molecular	Analysis Type	Analyte Concentration	Chapter(s)
Electrochemical techniques										
Amperometric techniques	B	L, Ds, Dg	L	Sm[16]	N	EI, Sp	Cp, Io, Fn	Qt	Mj, Mn, Tr, Ul	36
Voltammetric techniques	B	L, Ds, Dg	L	Sm[16]	N	EI, Sp	Cp, Io, Fn	Qt, Ql	Mj, Mn, Tr, Ul	37
Potentiometric techniques	B	L, Ds, Dg	L	P, Sm	N	EI, Sp	Cp, Io	Qt	Mj, Mn, Tr	38
Conductiometric techniques	B	L, Ds	L	P, Sm[16]	N	EI, Sp	Cp, Io	Qt	Mj, Mn, Tr	39
Microscopic and surface techniques										
Atomic force microscopy	S	S	L	P, Sm	N	EI	Cp	Ql[17]		41
Scanning tunneling microscopy	S	S	L	P, Sm	N	EI, Sp	Cp, Io	Ql[17]		41
Auger electron spectrometry	S	L[15], S	L	P, Sm	N	EI		Ql, Qt	Mj, Mn, Tr	42
X-ray photon electron spectrometry	S	L[15], S	L	P, Sm	D	EI[18], Sp		Ql, Qt	Mj, Mn, Tr	43
Secondary ion MS	S	L[15], S	L	P, Sm	D	EI, Is	Cp, Io	Ql, Qt	Mj, Mn, Tr	32, 34, 44

Table 1.2 (continued)

| Technique | Sample | | | | | Information Desired | | | | Reference |
	Location	Physical State	Amount	Purity	Fate	Elemental	Molecular	Analysis Type	Analyte Concentration	Chapter(s)
Techniques for polymers[19]										
Size exclusion chromatography	B	L, Ds[20]	L	Sm	D		Mw	Qt	Mj, Mn	46
Low-angle laser light scattering	B	L, Ds[21]	L	P, Sm	N		Mw	Qt	Mj, Mn	47
Light obscuration particle size techniques	B	L, Ds, Dg	L	Sm	D		Pp	Qt	Mj, Mn	47, 48
Pyrolysis techniques	B, S	Ds, S, L	L	Sm, M	D		Cp, Fn, St	Ql, Qt[22]	Mj, Mn, Tr	49
Thermal techniques	B	S, L, Ds	L	P, Sm	D		Cp, Pp	Ql, Qt	Mj, Mn	50
Mechanical property techniques	B	S, L, Ds	L	P, Sm	D, N		Cp, Pp	Ql, Qt	Mj, Mn	51

1. Depends on the characteristics of the detector.
2. Can determine elemental composition with microwave plasma–optical emission detector.
3. Depends on type of detector and standards verification procedure used.
4. Complements infrared spectrometry; can analyze aqueous samples.
5. For selected elements with electrothermal atomizer.
6. For selected elements.
7. Applicable to nontransition metals and some transition metals.
8. Applicable to many transition elements.
9. Dilution with appropriate spectroscopic solvent often required for liquid solutions.
10. Many compounds that do not fluoresce can be derivatized to compounds that do.
11. Limited to compounds that can be made to engage in a reaction that generates chemiluminescence.
12. MS technique can serve as detector for separation technique with appropriate interface between output of separation technique and input of MS.
13. Intensity of parent peak (P) and P+ peaks makes molecular weight determination easier than with electronionization. Chemical ionization spectrum contains fewer peaks than electron ionization spectrum.
14. Can determine molecular mass and formula with high accuracy and precision.
15. High boiling liquids can be handled.
16. Can serve as detectors for ion and liquid chromatographic separation techniques.
17. Provides high resolution images of surface allowing qualitative and semiquantitative analysis.
18. Sensitive to all elements except hydrogen.
19. Many techniques from the above sections may be used to obtain information from polymer and other macromolecular samples. The techniques in this section are used exclusively for polymer and macromolecular samples.
20. Dilution with appropriate chromatographic grade solvent required.
21. Dilution with appropriate solvent required.
22. Chromatography and mass spectrometry are used to separate and analyze pyrolysis products.

Qualitative analysis determines the identity of the analyte whereas quantitative analysis determines the concentration of analyte present. The concentration ranges listed (major, minor trace, and ultratrace) are those in common use.

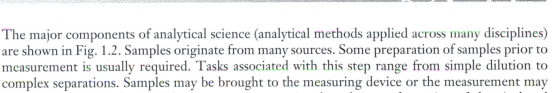

Summary

The major components of analytical science (analytical methods applied across many disciplines) are shown in Fig. 1.2. Samples originate from many sources. Some preparation of samples prior to measurement is usually required. Tasks associated with this step range from simple dilution to complex separations. Samples may be brought to the measuring device or the measurement may be made at the sample site. The measurement step involves the transformation of chemical and physical properties of the sample into data in the form of electronic signals. The conversion of this data into information is the third major component of a method. This final step includes mathematical operations, management of the original data and resulting information, and report generation. These components are generic to any process involving measurements of physical or chemical properties.

The chapters of this section address each of these components:

- Sample preparation (Chapters 2 and 3)
- Quality of measurements (Chapter 4)

Figure 1.2 The components of analytical science.

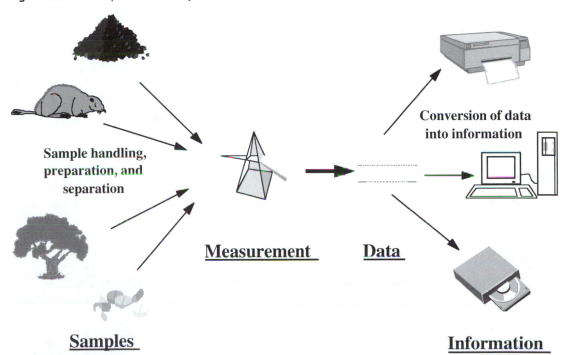

- Acquisition and management of information (Chapter 5)
- Automation of the analytical process (Chapter 6)

Buyers' Guides for Analytical Instrumentation

The following guides are published annually and contain extensive information on the manufacturers and distributors of instrumentation, software, chemical reagents, and laboratory equipment, as well as firms providing analyses and consulting services.

1. "LabGuide" published annually in August by *Analytical Chemistry*.
2. "Buyers' Guide" published annually in February by *American Laboratory*.
3. "Annual Directory" published in January by *Laboratory Equipment*.
4. Product reviews in the "A" pages of *Analytical Chemistry* focusing on a particular technique include general information on the selected technique and information on manufacturers.

Suggested Readings

The editor has found the following references especially useful in understanding the fundamentals of chemical analysis.

ENKE, C., "Chemical Data Domains," *Analytical Chemistry*, 43, no. 1 (1971), 69A. Good paradigm for following transformations of data and information through a chemical analysis.

HUBER, L., *Validation of Computerized Analytical Systems*. Buffalo Grove, IL: Interpharm Press, Inc., 1995. This book is a guide to computer-controlled analytical instruments through the entire validation process from design through implementation, testing, and installation qualification to ongoing calibration and performance qualification.

LAITINEN, H. A., AND W. E. HARRIS, *Chemical Analysis*, 2nd ed. New York: McGraw-Hill, 1975. Emphasizes the noninstrumental aspects underlying instrumental techniques.

MILLER, J. C., AND J. N. MILLER, *Statistics for Analytical Chemistry*, 2nd ed. New York: Ellis Horwood, 1988. An excellent, understandable treatment of important topics in experimental design and statistical analysis of chemical data.

SIGGIA, S., *Survey of Analytical Chemistry*. New York: McGraw-Hill, 1968. Presents a common-sense approach to analytical methodology. Unfortunately, it is out of print.

TAYLOR, J. K., *Quality Assurance of Chemical Measurements*. Chelsea, MI: Lewis Publishers, 1987. Good introduction to analytical process and validation.

August review issue of *Analytical Chemistry*. In even-numbered years, reviews current developments in analytical techniques, and in odd-numbered years, covers application areas, such as air pollution, coatings, and foods.

Articles in the "A" page section of *Analytical Chemistry* provide brief articles on recent advances in instrumentation and applications.

References

1. S. Siggia, *Survey of Analytical Chemistry* (New York: McGraw-Hill, 1968).
2. W. G. Hine, *Analytical Chemistry*, 46, no. 9 (1974), 1230A.

Chapter 2

Sample Preparation in Analytical Chemistry (Organic Analysis)

Ronald E. Majors

Chemical Analysis Group, Hewlett-Packard

Introduction

The major stages of an analytical process are depicted in Fig. 2.1. The proper choice of a measurement technique is only one step in the development of a successful application. All of the steps leading up to the measurement are equally important. The sampling and sample preparation process begins at the point of collection and extends to the measurement step. The proper collection of sample during the sampling process (called primary sampling), the transport of this representative sample from the point of collection to the analytical laboratory, the proper selection of the laboratory sample itself (called secondary sampling), and the sample preparation method used to convert the sample into a form suitable for the measurement step can have a greater effect on the overall accuracy and reliability of the results than the measurement itself. This chapter addresses sample preparation (pretreatment) of organic materials. However, before beginning the discussion of sample pretreatment, we briefly discuss some of the earlier stages of the sampling process.

Sampling

Primary sampling is the process of selecting and collecting the sample to be analyzed. The objective of sampling is a mass or volume reduction from the parent batch, which itself can be homogeneous or heterogeneous. If the wrong sample is collected or the proper sample is collected incorrectly, then all of the further stages in the analysis are meaningless and the resulting data are worthless. Books have been written on the sampling process and how to collect a statistically

Figure 2.1 Steps in the analytical cycle.

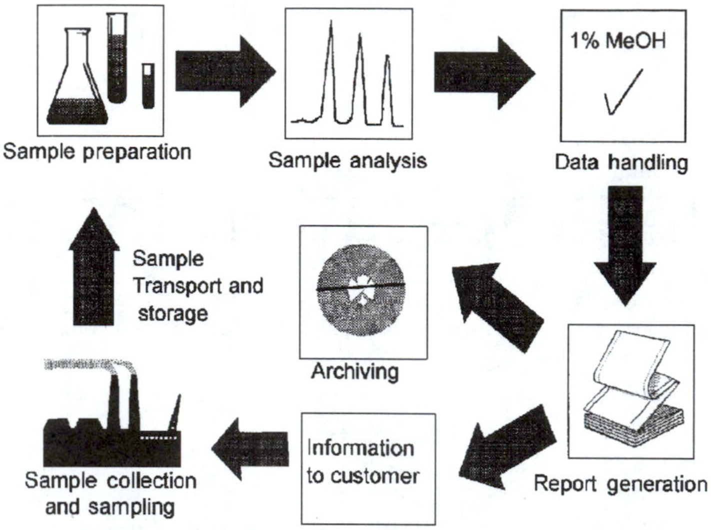

representative sample (1, 2). It is beyond the scope of this chapter to provide primary sampling theory and methodology, but it is important to note that sampling is one of the most overlooked sources of error in analysis. Unfortunately, sampling is sometimes left to people unskilled in sampling methodology and is largely ignored in the education process, especially for nonanalytical chemists.

It is advisable to include a carefully developed sampling plan as part of the overall analysis. Sample information flow must parallel sample flow throughout the analytical process, from collection to report generation. For example, sample tracking begins at the point of collection and can be considered part of the overall analysis process. Proper identification of the collected primary sample by handwritten labels, application of a bar code for automatic reading, writing on the sample container with indelible ink, or other means of documentation must be performed to ensure that later stages of processing can be traced unequivocally to the original primary sample. Likewise, each stage along the analytical process requires proper sample and sub-sample tracking to ensure that Good Laboratory Practices (GLP) are achieved.

Sample Transport and Storage

Once the primary sample is taken, it must be transported to the analytical laboratory without a physical or chemical change in its characteristics. At first glance, this may seem to be a trivial task but when the system under investigation is a dynamic entity, such as samples containing volatile, unstable, or reactive materials, the act of transportation can present a challenge, especially if the laboratory is a long distance from the point of collection (3). Even if a representative primary sample is taken, changes that can occur during transport can present difficulties in the secondary sampling process. Preservation techniques can be used to minimize changes between collection and analysis. Physical changes such as adsorption, diffusion, and volatilization as well as chemical changes such as oxidation and microbiological degradation are minimized by proper preservation. Examples of preservation techniques that can be used between the point of collection and the point of sample preparation in the laboratory are

- Choice of appropriate sampling container
- Addition of chemical stabilizers such as antioxidants and antibacterial agents
- Freezing the sample to avoid thermal degradation
- Adsorption on a solid phase

Once the sample has been brought into the laboratory, storage conditions are equally important to maintain sample integrity before analysis. For thermally labile or volatile samples, the samples should be kept in sealed containers and stored in the refrigerator or freezer. Liquid samples should be kept in a cool, dark area (not exposed to sunlight) until ready for analysis. For more information, Refs. 3 and 4 discuss sampling and sample preparation of volatile samples, Refs. 4 and 5 discuss water, air, and soil samples, and Refs. 6 and 7 discuss biological samples.

Often, prepared laboratory standards, surrogate samples, and blanks are carried through the entire preservation, transport, and storage processes to ensure that sample integrity is maintained. A recent trend in both industrial and environmental analyses is to move the analysis closer to the sample or the process. For example, portable field instruments are becoming more popular in the screening of environmental samples at the site of interest. Likewise, the movement of analytical measurements to at-line or on-line locations may have a profound effect on how samples are collected and analyzed in the future.

Secondary Sampling

Once the sample has made it to the laboratory, a representative subsample must be taken. This process is called secondary sampling. Just as in primary sampling, the size or inhomogeneity of the sample may be a problem in secondary sampling. Once again, statistically appropriate sampling procedures are applied to avoid discrimination, which can further degrade analytical data. (See Refs. 1, 2, and 8 for in-depth coverage of secondary sampling.)

Sample Preparation

The next stage of the sampling process is the preparation of the chosen secondary sample. Sample preparation is the major focus of the remainder of this chapter. Sample preparation is often seen as the last major bottleneck in the analytical process. Over the past several decades, considerable time has been devoted to improving analysis speed, resolution, and automation of analytical measurement techniques and developing and improving data-handling and report-generation software. For example, high-precision, pulseless pumps capable of reproducible isocratic and gradient solvent delivery, together with rapid low-volume, leak-free injectors, microparticulate columns, and greatly improved detection capabilities, have made modern high-performance liquid chromatography (HPLC) a widely used, reproducible, fast, high-resolution analytical technique. The same can be said for gas chromatography (GC) and other instrumental measurement techniques.

In contrast, sample preparation, particularly its automation, has been neglected. Many analytical chemists use time-consuming manual methods that have been around for decades. A GC separation and measurement can require a few minutes; however, preparation of the sample itself can take one or two orders of magnitude longer. The chart in Fig. 2.2 indicates that about two-thirds of a typical chromatographic analysis is spent on sample preparation, requiring more time than collection, analysis, and data management combined. Clearly, speeding up or automating the sample preparation step will reduce analysis time and improve sample throughput.

Figure 2.2 Survey results for distribution of time analytical chemists spend on sample analysis. *(From R. E. Majors, LC/GC 9, pp. 16–20 (1991). Reprinted by permission of Advanstar Communications.)*

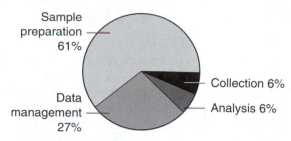

An often overlooked aspect of sample preparation is its effect on error generation. For the most part, sample preparation can be tedious and time-consuming. If one has many similar samples, the repetitive tasks of manually preparing the samples can lead to boredom and inadvertent mistakes can result. Figure 2.3 provides survey data on error propagation for a typical chromatographic method. Sample preparation accounts for almost one-third of the error generated during the performance of an analytical method; operator error is responsible for another 20% or so. Thus, improving and automating sample preparation can decrease error in a typical analytical method by as much as 50%.

Goals and Objectives of Sample Preparation

Successful sample preparation has a threefold objective: namely, to provide the sample component of interest

- In solution
- Free from interfering matrix elements
- At a concentration appropriate for detection and measurement

Two of the major goals of any sample pretreatment procedure are quantitative recovery and a minimum number of steps. Quantitative (99%+) recovery of the analyte is generally desirable.

Figure 2.3 Survey results for the distribution of error generated during sample analysis. *(From R. E. Majors, LC/GC 9, pp. 16–20 (1991). Reprinted by permission of Advanstar Communications.)*

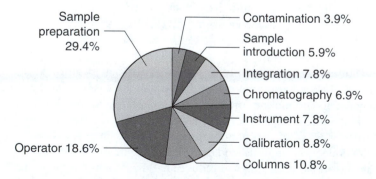

Quantitative recovery may not mean that all of the analyte is included in the final measured sample. For example, if successive sample pretreatment steps are involved, aliquots of intermediate fractions may be used for further processing or injection and the results combined. The smallest number of sample pretreatment steps that can still reach the desired goals will reduce time and effort required, as well as decrease the opportunity for mistakes by the analyst or the accumulation of unavoidable errors. As mentioned above, for further error reduction and less manual effort, automation of sample preparation is generally desirable. Automation can be justified when the sample load becomes significant or where operator exposure to sample or environment must be minimized (as with highly toxic samples or solvents). Automation can often be justified on a cost basis (see Chap. 6).

Separation and Preconcentration Techniques in Sample Preparation

There are three basic approaches in measuring an analyte in the presence of interfering species found in the sample matrix. First, one can consider a selective analytical technique that can measure the analyte in place without the need for isolation. For example, an ion-selective electrode can determine the presence of an ion of interest in a solution containing many other ionic species. A second approach (without analyte isolation) is to convert it in situ into another chemical species. In this approach, a selective derivatization reaction can be used to transform the analyte quantitatively into a species that can be measured more easily. An example is the addition of a chemical reactant that forms a colored complex that can be measured spectrophotometrically. A third approach is the removal of the analyte from the sample matrix by a separation or extraction process. This is the most commonly used approach and is the basis of the remainder of the chapter. On occasion, for extremely complex samples, such as biological fluids or solid waste, an initial separation is carried out to simplify or remove part of the matrix. This step is called preliminary sample cleanup and may involve one or more separation processes.

The most common approach in analyte separation involves a two-phase system where the analyte and interferences are distributed between the two phases. Distribution is an equilibrium process and is reversible. If the sample is distributed between two immiscible liquid phases, the technique is called liquid–liquid extraction (LLE). If the sample is distributed between a solid phase and a liquid phase, the technique is called liquid–solid adsorption. Sometimes, to provide a more favorable distribution, a chemical conversion (by derivatization or ionization changes) is performed before the distribution. In another separation approach, the analyte can be physically or mechanically separated from the matrix. For example, precipitation is a technique where a soluble analyte is converted into an insoluble substance and removed from solution. By filtration, the insoluble precipitate is removed mechanically. Distillation is a sample preparation approach where differential boiling points are used to separate the desired species from undesirable ones: a liquid phase/gas phase distribution.

Often, when analysis involves the measurement of trace amounts of a substance, it is desirable to increase the concentration of the analyte to a level where it can be measured more easily. Concentration of an analyte can be accomplished by transferring it from a large volume of phase to a smaller volume of phase. This preconcentration step is often performed in series or combined with the sample preparation step.

Separations can be carried out in a single batch, in multiple batches, or continuously. A single-batch approach involves a single distribution between two phases, such as a solvent extraction or precipitation. Multiple-batch separations are used when one simple equilibrium is insufficient to separate the desired analyte. Thus, a combination of two or more batch separations could be performed serially. A liquid–liquid extraction performed twice on the same sample using a fresh portion of the extraction solvent is an example of a multiple-batch extraction. Chromatography is an example of a continuous form of separation where one of the phases (moving phase) continuously flows over the other phase (stationary phase). Chromatography is used when the individual analyte properties are so similar that simpler techniques (such as solvent extraction) cannot be used to provide an effective separation. Multidimensional chromatography (Sec. II) involves the use of two or more chromatographic techniques used sequentially. It can be performed either off-line or on-line and is used as a combined prefractionation and separation technique. A technique related to chromatography is trapping or trace enrichment. Here the sample, dissolved in a suitable liquid, is passed through a chromatography column but the conditions are chosen so that the analyte has a very high retention. A large volume of sample is passed through the column to concentrate the analyte on the solid phase. Finally, a strong solvent is used to remove the concentrated analyte.

Types of Samples Encountered

Sample matrices can be classified as organic and inorganic. Biological samples are a subset of organic samples but often require different sample preparation procedures in order to preserve biological integrity and activity. In this chapter, we focus mainly on the preparation of organic samples that are in organic and inorganic matrices. The preparation of biological samples is covered briefly. Inorganic sample preparation is the subject of Chap. 3.

Organic analytes are often classified further as volatile, semivolatile, and nonvolatile. The initial sample can be in various forms: solid, semisolid (including creams, gels, suspensions, colloids), liquid, or gas (Fig. 2.4). Thus, some form of sample pretreatment to isolate the analyte of interest from the matrix is almost always required before sample injection into a chromatograph or measurement by a spectroscopic instrument.

Volatile samples usually are analyzed by gas chromatography (see Chap. 8). There are many sampling techniques for volatile samples, including gas–solid adsorption, headspace (HS) analysis, purge and trap, and vacuum canister sampling of air toxics. Table 2.1 provides a brief description of these technologies. We do not cover these techniques in depth in this chapter. Certain volatile samples are difficult and sometimes impossible to handle by GC or they must be

Figure 2.4 Distribution of sample types. *(From R. E. Majors, LC/GC 9, pp. 16–20 (1991). Reprinted by permission of Advanstar Communications.)*

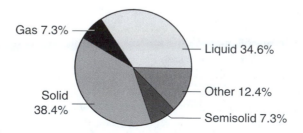

Table 2.1 Typical sampling and sample pretreatment methods for gases.

Method of Sample Pretreatment	Principles of Technique	Comments
Grab sampling	Gaseous sample is pulled into an evacuated glass or metal bulb or canister or by a syringe; gas can also be pumped into plastic bag or other inert container.	Used mostly for volatile compounds in air; samples are returned to laboratory and analytes are isolated and concentrated by cold trapping techniques.
Solid phase trapping	Gaseous sample passed through tube packed with adsorbent (such as silica gel or activated carbon); trapped analytes are eluted with strong solvent.	Used for semivolatile organic compounds in air. Control of gas flow rate is critical for trapping efficiency. Watch for aerosol formation, adsorbent overloading, and irreversible adsorption of reactive analytes. Popular sorbents include silica gel, alumina, porous polymers (Tenax, polyurethane foams), or carbon; chemical or physical complexing reagents may be useful to improve trapping efficiency.
Liquid trapping	Gaseous sample is bubbled through solution that is a good solvent for analytes; analyte has higher affinity for solvent than does gas.	Flow rate should be low enough so as not to create foams or aerosols; complexing agents may be added to solvent to aid trapping. Temperature can be lowered for very volatile species. Process sometimes called impinging.
Headspace sampling	Sample (solid or liquid) is placed in a closed, thermostated glass vial until equilibrium is established. At equilibrium, analytes partition themselves between a gas phase and the solid (or liquid) phase at a constant ratio; gas phase is sampled and injected into GC for analysis.	Used primarily for determination of trace concentrations of volatile substances in samples difficult to handle by conventional GC techniques. Sensitivity can be increased by heating (<100 °C), salting out, adjusting pH, and other means to shift equilibrium. Sometimes water or solvent is added to aid in sample dispersion or to free organics from the matrix, especially for soils and sediments. Can be manual or automated.
Purge and trap (dynamic HS)	Sample (solid or liquid) is placed in closed, thermostated container and the HS vapors are continually removed by means of inert gas flow with subsequent trapping of sample components by solid-phase extraction or cold trapping, then thermally desorbed into GC injection port.	Used when analytes are too low in concentration or have unfavorable partition coefficients in static HS sampling; sometimes called gas phase stripping. Can provide more sensitivity than static HS by accumulating the volatiles until concentration is sufficiently built up for thermal desorption and GC analysis; can be manual or automated.
Thermal extraction	A form of dynamic HS but the sample is heated (controlled) to much higher temperatures, up to 350 °C.	System must be constructed of fused quartz or fused silica so that extracted analytes do not react with hot metal surfaces; system cold spots should be avoided. Used for semivolatile compounds.

derivatized to increase their volatility (see the section on derivatization in chromatography later in this chapter). Most users avoid derivatization whenever possible because a great deal of time and effort go into chemical derivatization processes. Gaseous or volatile samples that are labile, thermally unstable, or prone to stick to metal surfaces can sometimes be better handled by HPLC. To handle a volatile sample by HPLC, trapping is required. To trap volatile compounds, the organic vapor is passed through a solid support then subsequently eluted with a solubilizing liquid; alternatively, the gas can be bubbled into a liquid (impinger) in which the analytes of interest have good solubility.

Compared to volatile compounds or solids, liquid samples are much easier to prepare for analytical measurement because a dissolution or an extraction step may not be involved. Often, dilution in a compatible solvent is all that is required. The major considerations for liquid samples are the matrix interferences, the concentration of analyte, and compatibility with the analytical technique. For example, many HPLC analyses simply involve a "dilute and shoot" sample pretreatment where analyte concentration is reduced by dilution so as not to overload the column or saturate the detector. Popular sample preparation methods for liquids are depicted in Table 2.2.

When a sample is a solid, the sample pretreatment process can be more complex. There are two specific cases: the entire sample is of interest and must be solubilized, or only a part of the solid is of interest and the analytes must be selectively removed. If the solid sample is a soluble salt or drug tablet formulation, the only sample preparation that may be required is finding a suitable solvent that will totally dissolve the sample and the components of interest. If the sample matrix is insoluble in common solvents but the analytes of interest can be removed or leached out, then sample preparation can also be rather straightforward. In these cases, techniques such as filtration, Soxhlet extraction, supercritical fluid extraction (SFE), ultrasonication, or solid–liquid extraction may be useful. Table 2.3 covers the more traditional methods still in widespread use and Table 2.4 covers the most recent methods for the sample preparation of solid samples.

If both the sample matrix and the sample analytes are not soluble in common solvents, then more drastic measures may be needed. For example, a soil sample can be digested in nitric acid to release the analytes of interest. Once the sample is in liquid form, it can be treated like any liquid sample and further sample preparation techniques mentioned above can be applied. For a discussion of dissolution of inorganic matrices, consult Chap. 3.

Sample Preparation for Solid and Semisolid Samples

Sample Size Reduction

Before sample preparation, solid or semisolid substances must be put into a finely divided state. Procedures to perform this operation are usually physical methods, not chemical methods. The reasons for putting the sample into a finely divided state are that finely divided samples are more homogeneous, so secondary sampling may be carried out with greater precision and accuracy, and they are more easily dissolved or extracted because of their large surface-area–volume ratio.

There are many methods for reducing the size of solid samples. Grinding is the method most often used for hard, solid samples. The classical mortar and pestle with plenty of manual effort is recommended for general grinding of solid samples. One must ensure during the grinding process that little heat is generated if the sample contains thermally labile or volatile compounds. With the intense localized heat that may be generated during rigorous grinding, volatile com-

Table 2.2 Typical sample pretreatment methods for liquids and suspensions.

Method of Sample Pretreatment	Principles of Technique	Comments
Solid-phase extraction (SPE)	Liquid is passed through solid phase, which selectively removes analyte (or matrix); analyte is eluted with strong solvent; mechanism is the same as LC/HPLC.	Wide variety of stationary phases are available for selective removal of desired inorganic, organic, and biological analytes; specialty phases are available for drugs of abuse, carbohydrates, and trace enrichment of water.
Liquid–liquid extraction (LLE)	Sample is partitioned between two immiscible phases chosen to maximize differences in solubility; separatory funnel is used for small sample volumes.	Beware of formation of emulsions; break them with heat, addition of salt, filtration through a filter paper; change K_D by different solvent or chemical equilibria-affecting additives (such as buffers for pH adjustment, salts for ionic strength, complexing agents, and ion pairing agents); many published methods are available; continuous extractions are used for low K_D or large volumes.
Dilution	Sample is diluted with solvent that is compatible with analytical measurement technique to avoid chromatographic column overload or to be in linear range of detector or spectrophotometer.	Solvent should be compatible with analytical measurement technique; solvent should not be too strong for HPLC mobile phase conditions so that injection causes unacceptable band broadening. "Dilute and shoot" is a typical sample prep method for simple liquid samples such as pharmaceutical formulations.
Evaporation	Liquid is removed by gentle heating at atmospheric pressure with flowing air or inert gas or under vacuum.	Do not evaporate too quickly; bumping can lose sample; watch for sample loss on wall of container; do not overheat to dryness. Best under inert gas such as N_2; rotary evaporator works best; automated systems (such as Turbovap) are available.
Distillation	Sample is heated to boiling point of solvent and volatile analytes are concentrated in vapor phase, condensed, and collected; steam distillation involves boiling with water or purging with steam and collecting distillate.	Mainly for samples that can be volatilized; sample can decompose if heated too high; vacuum distillation can be used for nonvolatile compounds.
Microdialysis	A semipermeable membrane is placed between two aqueous liquid phases and sample solutes transfer from one liquid to the other based on differential concentration.	Enrichment techniques such as SPE are required to concentrate dialysate; microdialysis is used for examination of extracellular chemical events in living tissue and has been used on-line with microLC columns; dialysis can also be used on-line to deproteinate samples before HPLC because large proteins cannot pass through membranes; ultrafiltration and reverse osmosis can be used in a similar manner.
Lyophilization	Aqueous sample is frozen and water removed by sublimation under vacuum.	Good for nonvolatile organics; large sample volume can be handled; possible loss of volatile analytes; inorganics are concentrated.
Filtration	Liquid is passed through paper or membrane filter to remove suspended particulates.	Highly recommended to prevent backpressure problems in HPLC and to preserve column life in chromatography; make sure membrane filters are compatible with solvent so that they do not dissolve or swell during experiment; use large-porosity (>2 μ) filters for maximum flow or small-porosity filters (<0.2 μ) to get rid of bacteria.
Centrifugation	Sample is placed in tapered centrifuge tube and spun at high force (several G's) and forced to bottom of tube; liquid is decanted.	Quantitatively removing solid sample from tube sometimes presents practical problem; ultracentrifuge normally not used for simple particulate removal.
Sedimentation	Sample is allowed to settle when left undisturbed in a sedimentation tank; settling rate dependent on Stokes's radius.	Extremely slow process; manual recovery of different size particulates at different levels depends on settling rate.

Table 2.3 Traditional extraction methods for solid samples.

Method of Sample Pretreatment	Principles of Technique	Comments
Solid–liquid extraction	Sample is placed in stoppered container and solvent is added that dissolves analyte of interest; solution is separated from solid by filtration (sometimes called shake/filter method).	Solvent is usually boiled or refluxed to improve solubility; sample is in finely divided state to aid leaching process. Sample can be shaken manually or automatically; sample is filtered, decanted, or centrifuged to separate from insoluble solid.
Soxhlet extraction	Sample is placed in disposable porous container (thimble); constantly refluxing solvent flows through the thimble and leaches out analytes, which are continuously collected in boiling pot.	Extraction occurs in pure solvent; sample must be stable at boiling point of solvent. The process is slow but extraction is carried out unattended until complete; process is also inexpensive. Best for freely flowing powders; excellent recoveries.
Homogenization	Sample is placed in a blender, solvent is added, and sample is homogenized to a finely divided state; solvent is removed for further workup.	Used for plant and animal tissue, food, environmental samples. Organic or aqueous solvent can be used; dry ice or diatomaceous earth can be added to make sample flow more freely. Finely dispersed sample promotes more efficient extraction.
Sonication	Finely divided sample is immersed in ultrasonic bath with solvent and subjected to ultrasonic radiation. An ultrasonic probe or ultrasonic cell disrupter can also be used.	Dissolution is aided by ultrasonic process; heat can be added for additional extraction. Process is safe and rapid; best for coarse, granular materials. Multiple samples can be done simultaneously; contact with solvent is efficient.
Dissolution	Sample is treated with solvent and taken directly into solution with or without chemical change.	Inorganic solids may require acid or base to completely dissolve; organic samples can often be dissolved directly in solvent. Heat is required for many samples, especially organic polymers; consult solubility tables for common compounds and salts.

pounds can be lost and thermally unstable compounds degraded. In general, most samples will thermally withstand the rigors of grinding. If the sample is particularly hard and difficult to grind manually, a diamond mortar is recommended. The diamond mortar is a cylinder constructed of hardened steel. A close-fitting steel rod fits inside the cylinder and the sample is pulverized by hammering with the rod. If the material is soft, a ball mill is recommended. The sample is placed into a porcelain cylinder containing porcelain, stainless steel, or hard flint balls. After the cylinder is sealed, it is rotated, shaken, or vibrated until the material inside is ground into a finely divided state. The material may be sieved for a more homogeneous distribution.

In order to grind malleable or elastic samples, such as rubber or plastics, they must be cooled to make them more brittle. Addition of dry ice or liquid nitrogen to the sample will cause it to be brittle enough for grinding. Dry ice can be added directly into a mortar or ball mill. Make sure that the dry ice is prepared from high-quality carbon dioxide; otherwise, trace impurities in the

Table 2.4 Modern extraction methods for solid samples.

Method of Sample Pretreatment	Principles of Technique	Comments
Accelerated (enhanced) solvent extraction (ASE or ESE)	Sample is placed in a sealed container and heated to above its boiling point, causing pressure in vessel to rise; extracted sample is automatically removed and transferred to vial for further treatment.	Greatly increases speed of liquid–solid extraction process and is automated. Vessel must withstand high pressure; extracted sample in diluted form requires further concentration; safety provisions are required.
Automated Soxhlet extraction	A combination of hot solvent leaching and Soxhlet extraction; sample in thimble is first immersed in boiling solvent, then thimble is raised for Soxhlet extraction with solvent refluxing.	Manual and automated versions are available; uses less solvent than traditional Soxhlet and solvent is recovered for possible reuse. Extraction time is decreased due to two-step process.
Supercritical fluid extraction (SFE)	Sample is placed in flow-through container and supercritical fluid (such as CO_2) is passed through sample; after depressurization, extracted analyte is collected in solvent or trapped on adsorbent, followed by desorption by rinsing with solvent.	Automated and manual versions are available; to affect "polarity" of supercritical fluid, density can be varied and solvent modifiers added. Collected sample is usually concentrated and pure because CO_2 is removed after extraction; matrix has an effect on extraction process.
Microwave-assisted extraction (MASE)	Sample is placed in an open or closed container and heated by microwave energy, causing extraction of analyte.	Extraction solvent can range from microwave-absorbing (MA) or non–microwave-absorbing (NMA); in MA case, sample is placed in high-pressure, non–microwave-absorbing container and heated well above its boiling point. Also in the MA case, the sample and solvent can be refluxed at atmospheric pressure, analogous to solid–liquid extraction; in NMA case, container can be open, with no pressure rise; safety provisions are required.

CO_2 may contaminate the sample. Liquid nitrogen grinding is best carried out in a freezer mill or cryogenic mill, where the entire operation is carried out under liquid nitrogen.

Drying of Samples

Samples are often received for analysis in a damp or wet state. The amount of moisture in a sample can have a definite impact on quantitation. Standardization by the drying of solid samples is a prerequisite to good analytical chemistry. The most common method of drying is to heat the sample in an oven to constant weight. Chapter 3 provides a good discussion on the drying of inorganic samples and most of the procedures and precautions cited there are applicable to organic and biological samples as well. Biological samples should not be heated over 100 °C for fear of decomposing or destroying the material. Inorganic samples such as silica gel or soil generally need to be heated well over the boiling point of water to ensure that moisture is removed; however, adsorbed volatile analytes may be lost during such procedures. Hydrophobic organic samples may not require heating at all because water absorption is minimal. However, organic vapors present in the laboratory can be adsorbed by solid organic samples and heating is one method to remove physically sorbed organics. For hydroscopic samples, such as acid anhydrides, drying in a vacuum dessicator is used. Likewise, samples that could be oxidized by heating in the presence of air should be dried by vacuum desiccation or under nitrogen. Freeze drying is a technique for preserving the integrity of heat-sensitive samples (such as biological samples) or where loss of volatile analytes might be a problem. Freeze drying is accomplished by first freezing the sample and then removing moisture from the frozen sample by vacuum.

Extraction of Soluble Components

In many analyses, it is desirable to handle the sample in the liquid state. However, there are cases where we do not want the primary sample to dissolve completely because we are interested only in the soluble components. Some examples are polymer additives, fats in food, and pesticides in soil. In these cases, we want to selectively remove (leach) one or more of the sample components, leaving behind the insoluble fraction. This fraction is removed by decanting or by filtration and the lechate solution containing the desired components is treated further, if necessary, before analysis. Tables 2.3 and 2.4 cover some of the more popular techniques used for the extraction of soluble organic compounds from an insoluble solid matrix. Some of these techniques follow.

Extraction of Solid Samples

In the last section, we discussed methods for the pretreatment of solid samples to ready them for sample preparation. Tables 2.3 and 2.4 provided an introduction to sample preparation methods for solid samples. Table 2.3 listed some traditional methods and Table 2.4 provided a look at newer methods, some of which use the basic principles of the older extraction methods but provide faster results, have better automation potential, and use less organic solvent.

Classical Extraction Technology

The extraction of analytes from sample matrices requires the right combination of solvent and technique. Table 2.3 lists popular traditional methods for the sample preparation of solid samples. Most of these methods (such as Soxhlet extraction and leaching) have been around for over 100 years and are time-tested and provide results that are accepted by most scientists. Regulatory agencies such as the United States Environmental Protection Agency (EPA) and the Food and Drug Administration (FDA) and their equivalents in other countries accept these classical methods for the extraction of solid samples. For the most part these methods use organic solvents, often in copious amounts, although there has been a trend in recent years to miniaturize these systems to minimize sample and solvent requirements.

Solid–liquid extraction takes many forms. The shake-flask method merely involves the addition of a solvent (for example, organic solvent for organic compounds and dilute acid or base for inorganic compounds) to the sample and by agitation allows the analytes to dissolve into the surrounding liquid until they are removed as completely as possible. This method works well when the analyte is very soluble in the extracting solvent and the sample is quite porous. To get more effective solid–liquid contact, samples must first be brought into a finely divided state. Heating or refluxing the sample in hot solvent may be used to speed up the extraction process. The shake-flask method can be performed in batches, which increases overall sample throughput. Once the analytes are removed (determined during method development by making analyte measurements as a function of time), the insoluble substances are removed by filtration or centrifugation.

Sonication can be used to get faster and more complete extraction. The ultrasonic agitation allows more intimate solid–liquid contact and the gentle heating that results during sonication can aid the extraction process. Sonication is also a recommended procedure for the pretreatment of solid environmental samples. For example, EPA Method 3550 for extracting nonvolatile and semivolatile organic compounds from solids such as soils, sludges, and wastes specifies sonication extraction. In this method, different extraction solvents and sonication conditions are recommended depending on the type of pollutants and their concentration in the solid matrix. Homogenization in the presence of solvent is also an effective way to maximize extraction yield.

By far the most widely used method for the sample pretreatment of solids is Soxhlet extraction. In this method, the solid sample is placed in a Soxhlet thimble, a disposable porous container made of stiffened filter paper. The thimble is placed in the Soxhlet apparatus, where refluxing extraction solvent condenses into the thimble and leaches out the soluble components. The Soxhlet apparatus is designed to siphon the solvent with extracted components once the inner chamber holding the thimble fills up with solution to a certain volume. The siphoned solution containing the dissolved analytes then returns to the boiling flask and the process is repeated over and over again until the analyte is successfully removed from the solid sample. Soxhlet extractions are usually slow, often approaching 18 to 24 hr. However, the process takes place unattended, so once the sample is loaded and refluxing begins, there is little operator involvement until the conclusion of the extraction. Each sample requires a dedicated apparatus. Thus, one often sees rows of Soxhlet extractors in the fume hood in laboratories that use this technique. Soxhlet extraction is less expensive than some of the more modern extraction techniques. Soxhlet extraction glassware itself is rather inexpensive. However, the most common extractors use hundreds of milliliters of high purity solvent. Small-volume Soxhlet extractors and thimbles are available for small amounts of sample, down to milligram sizes.

In the Soxhlet process, fresh extraction solvent is always presented to the sample. Because the dissolved analyte is allowed to accumulate in the boiling flask, it must be stable at the boiling point of the extraction solvent. Method development in Soxhlet extraction involves finding a solvent or solvent mixture that has a high affinity for the analyte and a low affinity for the solid sample matrix. The solvent should have a high volatility because it must be removed at the conclusion of the extraction in order to concentrate the analyte of interest. Fortunately, there are many published methods and a quick literature search for the same or similar samples to your own may save time overall. Because this form of extraction is one of the oldest methods, it is the de facto standard and many newer extraction technologies, such as SFE, accelerated solvent extraction, and microwave-assisted extraction, are compared to Soxhlet extraction. However, these mechanisms of extraction, especially matrix effects, may be different and such comparisons are not always relevant.

Modern Technologies for the Extraction of Solids

For many years analysts have been content to perform traditional sample preparation methods. These methods were time-tested and they were familiar with the processes and procedures. However, as the need for increased productivity, faster assays, and more automation arose, newer extraction techniques were developed to meet these requirements. Table 2.4 lists some of these methods. Some of these methods have automated the traditional methods and made them easier to use. Other methods were developed that used new technology. For the most part, these newer approaches, especially those that are automated, are more expensive in terms of the initial purchase price but may cost less on a per-sample basis.

Automated Soxhlet Extraction

Automated Soxhlet extractors have increased the speed of the normal Soxhlet extraction process and decreased the amount of solvent required. Figure 2.5 depicts a new twist on the Soxhlet experiment: three-step extraction as performed in the Soxtec System (Tecator Division, Perstop Analytical Inc., Silver Spring, MD). In the first step, the extraction thimble is actually immersed in boiling solvent for a period of time. In the traditional Soxhlet extraction, the condensed solvent is allowed to drip into the extraction thimble during the entire extraction process. The temperature of this condensing solvent is lower than the solvent's boiling point and thus requires much longer to extract the analytes than boiling solvent. In the modern approach, the boiling solvent provides a quicker initial extraction.

In step two, the sample holder is raised manually to its upper position above the boiling solvent, which now contains most of the extractable material. This step is the rinsing step and resembles the normal Soxhlet extraction process but is much shorter. Finally, in step three, the distilled solvent is collected for possible reuse or disposal. At the same time, this step concentrates the analyte in the boiling flask during the solvent removal process.

Figure 2.5 Three-step Soxhlet extraction process performed by Soxtec Extraction System. Shown are (a) solubilization of extractable matter from sample immersed in boiling solvent; (b) rinsing of extracted solvent (similar to conventional Soxhlet extraction); (c) concentration of extracted sample and collection of distilled solvent for reuse or disposal. *(Courtesy of Perstorp Division of Tecator, Inc.)*

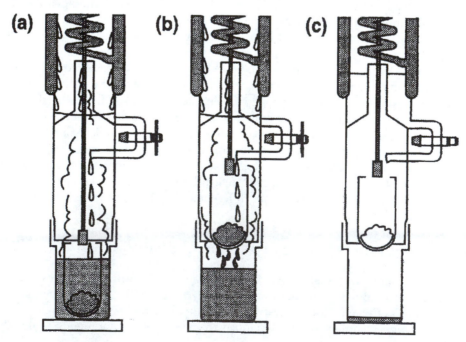

In the Soxtec system, these steps are manual but the Soxtherm Automated Extractor (O.I. Analytical, Sample Preparation Products Group, Columbia, MO) automates them. In step one, the Soxtherm extractor also immerses the sample in boiling solvent. In step two, the solvent level is lowered by shunting off portions of the condensate so that the sample is no longer immersed in the solvent. In this step, the remaining extractable substances are rinsed into the boiling flask. All of these operations can be programmed and six samples can be extracted unattended.

Because of the immersion step, both systems decrease the extraction time by a factor of 4 to 10 compared with the traditional Soxhlet method. With both systems, operator exposure to solvents is minimized and solvent usage is less than half that of traditional Soxhlet extraction. The solvent is collected and recycled. Typical applications include total fat in meat (Association of Official Analytical Chemists, Reference Methods 960.39, 976.21, and 985.15); pesticides, polychlorinated biphenyls (PCBs), and polynuclear aromatic hydrocarbons (PAHs) in soil and plants; oil and grease in soil, sludge, and water; additives in plastics and rubber; pesticides in oatmeal; and fat in potato chips. Recovery of analytes is comparable to traditional Soxhlet extraction but with much faster extraction times. In a study of the recovery of oil and grease in spiked soils using a 1,1,2-trichlorotrifluoroethane extraction of a potting soil and sand mixture, recoveries of 89% with a 2.3% relative standard deviation were found using the Soxtherm system (9). Another manual device that does more conventional Soxhlet extraction combined with a Kuderna–Danish concentration is the One Step Soxhlet System (Corning Inc., Corning, NY).

Supercritical Fluid Extraction

For a detailed description of SFE, see Chap. 11. Table 2.4 briefly covered SFE in relation to other modern extraction technologies. SFE is becoming a very popular technique for the removal of nonpolar to moderately polar analytes from solid matrices. Due to extraction thimble design, liquids are harder to handle but can be successfully extracted (10).

Although SFE is newer than other extraction techniques, it has found widespread application in the areas of environmental, food, and polymer sample preparation. In environmental analysis, the EPA has approved several methods including total petroleum hydrocarbons, PAHs, and organochloropesticides in soils and sludges with several more on the way. Because supercritical fluid (SF) CO_2 is an excellent solvent for fats, it has proven useful in the food industry. By careful adjustment of SF density, cholesterol can be extracted without extracting fats, so SFE is used as a fractionation technique. In the natural product area, it is being investigated as a technique for extraction of compounds that may have medicinal value. In the area of polymers, the penetrating power of SF CO_2 provides extraction of additives in less than an hour; such extractions formerly required many hours by Soxhlet or ultrasonic extraction methods. For example, additives in high-density polyethylene, a difficult polymer to solvent extract, can be successfully extracted by SFE, although the extraction time is over an hour. Pharmaceutical chemists have found SFE useful for extraction of drug materials from tablet formulations and tissue samples.

Microwave-Assisted Extraction

Most of us are familiar with microwave technology when it comes to heating food at home. Recently, there has been a great deal of activity in using microwaves as an alternative to hotplate techniques in the laboratory. Microwave heating is unique in that you heat the sample, which in turn heats the container, rather than the other way around. The use of industrial-grade microwave ovens is now an accepted replacement technique for the acid digestion when determining metals. Organic, inorganic, or biological samples and the digesting acidic solution are placed together in a closed, chemical-resistant, non–microwave-absorbing container. The container is then subjected to microwave radiation for a controlled period of time. Using microwaves, the solution is heated up to a high temperature and the hot acid digests the sample matrix, be it a soil or organic polymer. This destructive digestion process is much faster than the hotplate technique and much safer. In fact, the EPA has already written microwave digestion into several of its procedures for metal determinations. Chapter 3 covers applications for inorganic matrices, so we will not discuss microwave digestion further here.

The use of microwave heating with organic solvents is gaining interest. Microwave-assisted solvent extraction (MASE) is an alternative to conventional solvent extractions (11). Chemical compounds absorb microwave energy roughly in proportion to their dielectric constants: the higher the value of the dielectric constant, the higher the level of absorption of microwave energy. There are two limiting approaches to MASE: use of a microwave-absorbing extraction solvent of high dielectric constant and use of a non–microwave-absorbing extraction solvent of low dielectric constant. In the microwave-absorbing solvent approach, the sample and solvent are placed together in a closed vessel, similar to those used for microwave digestions. Under microwave irradiation, the solvent heats to well over its boiling point and the hot solvent provides a rapid extraction of analyte under moderate pressure (usually a few hundred pounds per square inch). For these higher-pressure extractions, the non–microwave-absorbing sample containers are made of polytetrafluoroethylene (PTFE), quartz, or advanced composite materials that combine optimum chemical and temperature resistance and good mechanical properties. Acid resistance is less of an issue here. In the microwave-absorbing solvent approach, an open (nonpressurized) container can be used and the sample solvent can be refluxed as in solid–liquid extraction. This approach has been successful in the extraction of polymer additives, priority pollutants such as PAHs, pesticides, and PCBs from soils and sediments, and in vitamins from food.

In the non–microwave-absorbing solvent approach, sometimes called microwave-assisted process (MAP) (12), the sample and solvent are placed in an open vessel. In this approach, the solvent does not become hot because it cannot absorb the microwave radiation. However, the sample itself, which usually contains water or other compounds possessing a high dielectric constant, absorbs the microwave radiation and can release the heated analytes into the surrounding cool liquid, chosen for its solubility characteristics. The latter approach is more gentle, is generally carried out under atmospheric or low-pressure conditions, and can be used for heat-sensitive or thermally labile analytes. Examples of the use of the non–microwave-absorbing solvent approach are extraction of essential oils from plant materials (apiolc from sea parsley, peppermint, cedar oil, sulfur-containing components from garlic, perfumery and flavoring volatiles from flowers), lipids from fish (12), and organochloropesticides from sediment samples (13).

In MASE, less solvent is used than in conventional Soxhlet or liquid–liquid extractions. The extraction solvents can be selected based on their microwave-absorbing ability so that there is little or no heating, a great deal of heating, or anything in between. In addition, the extraction solvent power for the analytes of choice can be an important selection criteria. In the non–microwave-absorbing solvent extraction approach, for sample matrices that have little absorption of microwaves such as dehydrated or dry materials, the addition of a solvent of high dielectric constant such as water or methanol to the sample can aid in the extraction process by rehydrating the matrix. In addition to the choice of extraction solvent, MASE has many experimental parameters that can be varied to optimize the extraction, such as heating time, pulsed heating versus continuous heating, stirring versus nonstirring, closed container versus open container, and outside cooling of vessel versus noncooling. Multiple samples can easily be extracted simultaneously, resulting in increased throughput. The MASE techniques are safe because in most cases the laboratory worker is not exposed to the extraction solvents. However, workers should exercise caution when dealing with microwave radiation and pressurized, closed containers.

Accelerated or Enhanced Solvent Extraction

Instead of microwaves as a source of heating closed extraction vessels, the use of a conventional air oven allows the use of stainless steel extraction cells. The technique is called accelerated solvent extraction (ASE, Dionex Corp., Sunnyvale, CA) or enhanced solvent extraction (ESE, Suprex Corp., Pittsburgh, PA) and it uses organic solvents at high temperature under pressure. Many parts of the experimental apparatus are the same as those used in SFE: a pump for pumping solvent into and out of the extraction vessel, extraction vessels with an automated sealing mechanism to withstand the pressures generated, an oven for heating the sample compartment, and collection vials to hold the collected extracts (Fig. 2.6). The process consists of the following steps: (1) sample cell loading, (2) solvent introduction and pressurization, (3) sample cell heating (under constant pressure), (4) static extraction, (5) transfer of extract to sealed vial with fresh vent wash of solid sample, (6) nitrogen purge of cell, and (7) loading of next sample. Once the sample is loaded into the extraction cell, the entire process is automated and time programmable. ASE provides unattended preparation for up to 24 samples.

When compared to conventional Soxhlet extraction, however, the extraction times for solid samples are greatly reduced (10 to 20 min versus 12 to 24 hr), the amount of solvent is far less (10 to 20 ml versus 200 to 400 ml), and the process is more automated. The recoveries and reproducibility of the two techniques are equivalent. When compared to SFE, ASE/ESE is comparable in speed, organic solvent usage, and cost. Its major advantages over SFE are that its principle is easier to understand because it is similar to conventional solvent extraction and matrix effects have less influence on the extraction. Compared to microwave-assisted extractions, the ASE/ESE techniques are more automated. Although the technique is relatively new, a proposed Method 3545 for the extraction of base-neutrals and acids, chlorinated and organophosphorous pesticides, and PCBs has already appeared in the EPA's SW-846 Manual (15). Table 2.5 compares typical experimental parameters for these newer extraction methods. All of these newer methods require less time, labor, and solvents than the older methods.

Figure 2.6 Schematic representation of the accelerated solvent extraction system. Valve pumping, pneumatic, and heating functions are automated; operating conditions are selected in response to a programmed schedule. *(From B. E. Richter, J. L. Ezzell, D. Felix, K. A. Roberts, and D. W. Later, American Laboratory 27 (4), 24–28 (1995). Reprinted with permission.)*

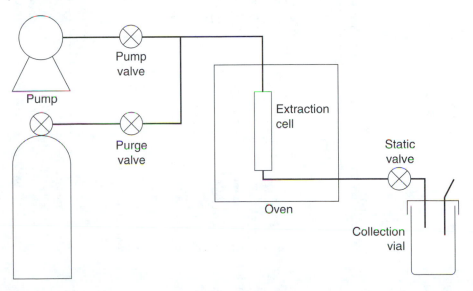

Table 2.5 Comparison of extraction methods for sample preparation of solids.

Parameter	Sonication	Soxhlet (traditional)	Soxhlet (modern)	SFE	ASE (ESE)	Microwave-Assisted (closed container)	Microwave-Assisted (open container)
Sample size, g	20–50	10–20	10–20	5–10	5–15	2–5	2–10
Solvent volume, ml	100–300	200–500	50–100	10–20*	10–15	30	20–30
Temperature, °C	Ambient–40	40–100	40–100	50–150	50–200	100–200	Ambient
Pressure	Atmospheric	Atmospheric	Atmospheric	2000–4000 psi	1500–2000 psi	1500–2000 psi	Atmospheric
Time, hr	0.5–1.0	12–24	1–4	0.5–1.0	0.2–0.3	0.2–0.3	0.1–0.2
Degree of automation[†]	0	0	++	+++	+++	++	++
Number of samples[‡]	High	1	6	44	24	12	12
Cost[§]	Low	Very low	Moderate	High	High	Moderate	Moderate

* When organic modifier is used to effect polarity.

[†] For the most complete commercial instrument; 0 = no automation, + = some automation, ++ = mostly automated, +++ = fully automated.

[‡] Maximum number that can be handled in commercial instruments.

[§] Very low = <$1000, Low = <$10,000, Moderate = $10,000–20,000, High = >$20,000.

Sample Preparation Methods for Liquid Samples

Liquid samples have an advantage over solid samples in that at least one less pretreatment step must be carried out: getting the sample into a liquid form. Table 2.2 provides an introduction to sample preparation methods for liquid samples. Of these methods, most laboratories use only a few for routine sample preparation. Distillation, for example, is mainly used for volatile compounds, although work in the area of vacuum distillation for environmental samples has shown that high boilers can be handled as well (16). Steam distillation, where live steam is continuously blown through a sample mixture or the mixture boiled with water, proves useful for labile, volatile, and low-molecular-weight organic compounds. Lyophilization is mostly used for the concentration and handling of biological samples. In this section we deal with the methods in common use in analytical laboratories.

Solvent (or Liquid–Liquid) Extraction

Solvent extraction is one of the most popular sample preparation methods. When carried out with two immiscible liquid phases, it is commonly called liquid–liquid extraction (LLE). The basic idea behind LLE is the separation of components (solutes) from mixtures by partitioning them between two immiscible solvents (phases). The solutes partition between the two phases according to their relative solubilities and an equilibrium is established. In most LLEs, one of the phases is aqueous and one of the phases is an immiscible organic solvent. The concept "like dissolves like" works well in LLE. A hydrophobic molecule prefers an organic medium whereas an ionic compound prefers to remain in aqueous solution.

Several useful equations can help illustrate the extraction process. The Nernst distribution law states that any neutral species will distribute between two immiscible solvents so that the ratio of the concentrations remains constant.

$$K_D = C_o / C_{aq} \qquad (2.1)$$

where K_D is the distribution constant, C_o is the concentration of the analyte in the organic phase, and C_{aq} is the concentration of the analyte in the aqueous phase. A more useful expression is the fraction of analyte extracted, E, given by Eq. 2.2:

$$E = C_o V_o / (C_o V_o + C_{aq} V_{aq}) = K_D V / (1 + K_D V) \qquad (2.2)$$

where V_o is the volume of organic phase, V_{aq} is the volume of aqueous phase, and V is the phase ratio V_o / V_{aq}.

Because extraction is an equilibrium process, a finite amount of solute may be found in both phases. Therefore, additional extractions or other sample preparation steps may be needed. The application of chemical equilibria can also affect solute behavior in the extraction process. By judicious manipulation of the chemical equilibria to make the solute more organic-solvent–soluble or aqueous-soluble, an additional dimension can be added to the LLE process.

The organic solvent is chosen to have a low solubility in water, possess physical properties that cause analytes to have a greater affinity for it than for the water, have volatility for easy removal and concentration after extraction, and be compatible with the analytical measurement technique. To increase the value of K_D, several approaches may be used:

- The organic solvent can be changed to increase solubility of the analyte.
- If the analyte is ionic or ionizable, its K_D may be increased by suppressing its ionization to make it more soluble in the organic phase. It can also be extracted into the organic

phase by the formation of an ion pair through the addition of a hydrophobic counterion.

- Metal ions can be complexed with hydrophobic complexing agents. For example, a dithizone complex of Pb (II) can be quantitatively extracted into chloroform at pH 10.
- The "salting out" effect can be used to decrease an analyte's concentration in the aqueous phase (by addition of an inert, neutral salt to the aqueous phase).

The mechanics of the LLE process are well known and the reader is referred to textbooks that cover this subject (17, 18). Briefly, most LLEs are performed with simple separatory funnels and, in the classical methods, typically tens of milliliters of volume are used. For simple extractions, K_D should be large because there are practical limits on the volume ratio of the extracting solvents. With most separatory funnel extractions, quantitative extractions (more than 99% solute removal) require two or three extractions. With multiple extractions, the volume of total organic solvent used may be quite large.

In solvent extraction, the organic species of interest can be transferred into either phase, depending on the selected conditions. Solutes extracted into the organic phase are easier to recover by evaporation of solvent, whereas solutes extracted in the aqueous phase may be directly injected into a reversed-phase HPLC column. To illustrate how one can manipulate chemical equilibria to effect better extraction, consider the extraction of an amine from an aqueous solution. If the aqueous phase is buffered at least 1.5 pH units above its ionization constant (K_b), where the amine is in its protonated form, it will remain in the aqueous phase and compounds that are more soluble in the organic phase will be extracted. On the other hand, if the pH of the aqueous solution is adjusted so that the amine is in its un-ionized (neutral) form, it can be more easily extracted into the organic phase. If the amine is already in the organic phase, on the same basis, it can be extracted into an acidic aqueous phase.

If the K_D value is unfavorable, additional extractions may be required for better solute recovery. In this case, a fresh portion of immiscible solvent is added to extract additional solute. Normally, the two extracts are combined. Generally, for a given volume of solvent, multiple extractions are more efficient in removing a solute quantitatively than a single extraction. Sometimes, back extractions can be used to achieve a more complete sample cleanup. In a back extraction, conditions are adjusted so that the solute is transferred from one phase then re-extracted into a fresh portion of another phase that favors its partitioning. For example, if in the above example, the amine were transferred as a neutral compound into the organic phase out of a basic aqueous phase, it could be transferred from the organic back into a fresh portion of acidic aqueous solution, further purifying it from other extractables. In some cases, to improve selectivity, the fresh aqueous phase could contain a type of buffer or reagent different from that of the original aqueous solution.

If K_D is very low or the sample volume (V_{aq}) is high, it becomes nearly impossible to carry out multiple simple extractions in a reasonable volume. Also, if the extraction rate is slow, it may take a long time for equilibrium to be established. In these cases, continuous liquid–liquid extraction is used, where pure solvent is recycled through the aqueous sample. These extraction devices are allowed to run for an extended period of time (12 to 24 hr) and quantitative extractions can be achieved. Small-scale laboratory countercurrent continuous extraction units are commercially available. If larger amounts of material are required for preparative applications, a larger countercurrent extraction apparatus, such as a Craig countercurrent unit, can be used. In such an apparatus, both layers are exchanged numerous times by a series of successive extractions.

In recent years, microextractions have gained popularity. Here extractions are carried out with smaller amounts of organic solvent, where the organic-to-aqueous solvent ratios are in the range of 0.001 to 0.01. Extraction efficiency is lowered and extractions are generally nonquantitative. For quantitative analysis, internal standards must be used and calibration standards must also go through the extraction process. However, analyte levels in the organic phase are raised relative to macro extractions and solvent usage is greatly reduced. These extractions are carried out in a volumetric flask, with the organic extraction solvent being a lower density than water. Thus, the small volume of organic solvent accumulates in the neck of the flask for easier removal. Extractions can also be carried out automatically with modern autosamplers capable of manipulating 2-ml vials and performing microextractions of small volumes of aqueous samples (19, 20).

Solid-Phase Extraction

Solid-phase extraction (SPE) is a widely used selective sample preparation technique that is replacing many of the classical liquid–liquid extraction methods. The objectives of SPE are to reduce the level of interferences, minimize the final sample volume to maximize analyte sensitivity, and provide the analyte fraction in a solvent that is compatible with the analytical measurement technique. As an added benefit, SPE serves as a filter to remove sample particulates. The SPE technique isolates analytes and simplifies sample matrices using the same fundamentals as classical liquid chromatography. Both SPE and LC are differential migration processes in which compounds are retained and eluted as they are swept through a porous medium by mobile phase flow. The main physical differences are in the column shape (length and internal diameter), the particle size of the packing material, and the flow rates and elution solvent profiles. Table 2.6 covers the main uses of SPE.

Table 2.6 Main purposes of solid-phase extraction.

Purpose	How Performed	Estimated Usage	Application Examples
Removal of interferences	Interferences are allowed to pass unretained through the cartridge with analytes remaining sorbed, or analytes pass through the cartridge, with interferences remaining on the cartridge.	50%	Removal of proteins from biological fluids, fats and lipids from food, ionic compounds from aqueous samples, and drugs of abuse from urine, and extractions of dioxans from waste water.
Analyte concentration	Conditions are chosen to achieve strong retention values (retentive stationary phase/weak mobile phase); elution in a small volume of volatile organic solvent.	35	Trace enrichment of ppb of polynuclear aromatics from water, trace pesticides in urine, caffeine from beverages, and therapeutic drugs from plasma.
Phase exchange	Analyte present in emulsion, suspension, or undesirable solvent is sorbed on SPE cartridge, dried, and eluted with desired solvent.	5	Exchange of aqueous solvent for nonaqueous one with intermediate dry nitrogen flush.
Solid-phase derivatization	Specifically coated SPE phases selectively derivatize analytes as they pass through the cartridge.	5	2,4-dinitrophenyl hydrazine-coated cartridges that selectively derivatize carbonyl-containing compounds; amines and polyamines in air; organic acids in water.
Sample storage and transport	Vapor or liquid samples are collected at factory or in field on an SPE cartridge or disk and transported to laboratory.	5	Soil gas analysis; trace organics in water.

In its simplest form, SPE uses a small plastic disposable column or cartridge (Fig. 2.7), often the barrel of a medical syringe, packed with a small amount of sorbent, usually less than a gram, although cartridges with up to 10 g are commercially available. The packing is contained in the barrel by frits, just like an HPLC column. The diameter of the packing material, usually in the 40-μ range, is larger than that used in HPLC. Consequently, due to shorter lengths and larger particles, SPE cartridges provide greatly reduced theoretical plate counts, often 10 to 100 times lower than those of an HPLC column. Because efficiency is relatively unimportant in SPE and cost is more important, irregularly shaped particle packings rather than spherical packings are used. Overall, the principles of separation, phase selection, and method development approaches for SPE are the same as for LC and HPLC.

SPE disks have proved to be popular for large volume sample preparation. The disks resemble the membranes used for filtration but differ in that they have sorbent particles embedded in an expanded PTFE network or fiberglass matrix. The disks are flat, usually 1 mm or less in thickness, with diameters ranging from 4 to 96 mm. The packing material generally makes up 60 to 90% of the total membrane weight. The small length-to-diameter ratio of the disk enables higher flow rates and faster extractions than can be achieved with cartridges. For example, 1 liter of relatively clean water can pass through a 45-mm disk in approximately 20 min but may require 1 to 2 hr when using a 15 mm × 8 mm cartridge bed (21). The disks have proved popular in environmental analysis for the extraction of trace organics in water and several EPA methods for both disks and cartridges are approved as alternatives to LLE methods.

A newer solid-phase microextraction (SPME) format uses solid, fused silica fibers coated with a polymeric GC-like stationary phase (22, 23). The fiber replaces the plunger of a microsyringe. This fiber is dipped into the aqueous solution containing the trace organics to be analyzed. The organics diffuse to and partition into the polymeric coating as a function of their distribution coefficients. Convection speeds up the process. After a predetermined time, the fiber is removed from the solution and placed into the injection port of a gas chromatograph, where the analytes are thermally desorbed. Alternatively, the fiber can be placed into the injection port of an HPLC valve, where analytes are displaced by the use of mobile phase or other solvent. Because SPME is a specialized application of SPE, we do not discuss it further and the remaining coverage refers to cartridges and disks.

Figure 2.7 Typical design of a solid-phase extraction cartridge. I = compound of interest. *(From R. E. Majors, LC/GC 4, pp. 972–984 (1986). Reprinted by permission of Advanstar Communications.)*

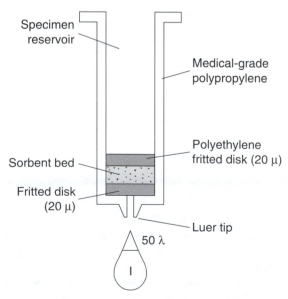

Figure 2.8 Steps in the operation of a solid-phase extraction experiment. * = interference, I = compound of interest. *(From R. E. Majors, LC/GC 4, pp. 972–984 (1986). Reprinted by permission of Advanstar Communications.)*

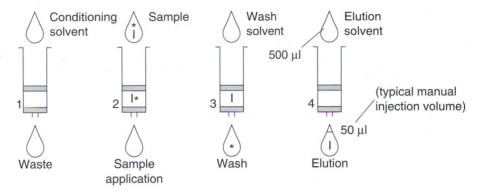

In the conventional SPE experiment (Fig. 2.8), the sample, in a liquid form, is added to the cartridge after first conditioning the sorbent (step 1), which is wetted to place it in a solvated state. Sample solvent (loading) conditions are chosen to either retain the analyte on the sorbent or allow it to pass through the cartridge unretained (step 2). When the analyte (I) is retained, as depicted in Fig. 2.8, interferences (*) are eluted or washed from the cartridge to simplify the original sample (step 3). The analyte is then eluted using a small volume of a stronger solvent (step 4), collected, and further concentrated by solvent evaporation or transferred to an analytical instrument. Depending on the SPE mode, a stronger solvent can be a solubilizing organic solvent (reversed phase or normal phase chromatography), a more concentrated buffer (ion exchange), or an eluting buffer at a differential pH (ion exchange). In practice, SPE not only isolates the analyte but also concentrates it.

The apparatus required to perform the SPE experiment can be very simple. The most basic system uses a syringe to manually push solvent or sample through the cartridge. Sometimes, if the sample is viscous or contains particulates, it is difficult to manually push liquid through the SPE bed. In this case, a vacuum flask that can handle one cartridge at a time can be used. However, if one has many samples, a vacuum manifold system that handles as many as 24 cartridges at a time is recommended (see Fig. 2.9 for a multicartridge system). A vacuum (using an aspirator, vacuum

Figure 2.9 Apparatus for solid-phase extraction. *(From R. E. Majors, LC/GC 4, pp. 972–984 (1986). Reprinted by permission of Advanstar Communications.)*

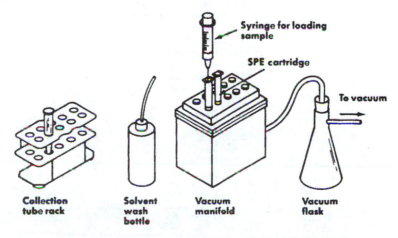

pump, or house vacuum) is applied to pull sample or wash solvent through the cartridges. To hold test tubes for eluent collection, a removable rack is located inside of the vacuum manifold. Finally, a sidearm vacuum flask is placed between the vacuum manifold and the vacuum source to collect rinses and wash solvent.

Because SPE is really a low-efficiency adaptation of HPLC, many phases used in HPLC are also used in SPE versions (Fig. 2.10). The most popular types of SPE phases are the bonded silicas, but inorganic and polymeric materials are available as well. In addition to the phases shown here, there are many specialty phases such as those for isolation of drugs of abuse in urine (24), aldehydes and ketones from air (25), and catecholamines from plasma (26). Certain traditional cleanup normal phases such as Florisil and alumina are used more often in SPE than in HPLC. Like HPLC, the popular long-chain reversed phases such as C18 are used quite often in SPE (Fig. 2.10). These relatively nonselective, hydrophobic phases tend to retain not only solutes of interest but most of the hydrophobic matrix components as well, and offer little selectivity and discrimination. Often, choice of a shorter-chain phase such as C4 or C2 or a phenyl or diphenyl phase provides more selective interactions, allowing better discrimination between solute and matrix.

Generally, SPE cartridge packings tend to be of lower quality than corresponding HPLC packings. Thus, as in HPLC, batch-to-batch reproducibility of SPE phases is a problem with which users have had to contend. However, because SPE is traditionally practiced in its digital form (that is, the solute is "on" or "off" the cartridge), small differences in batch characteristics can be obliterated by swamping out these effects using $k' = \infty$ for solute retention and $k' = 0$ for solute elution. Mixed mechanisms due to secondary interactions occur in SPE. Such mixed mechanisms are true if phase-loading-to-silanol ratios are not well controlled or if the SPE phases are not endcapped. For example, under aqueous conditions the acidic character of residual silanols can cause ionic interactions with cationic groups (such as protonated amines). Likewise, the problems of residual silanols on bonded silica phases can hamper the recovery and quantitation of basic solutes in either organic or aqueous environments, just as in HPLC. Trace metal binding can also occur if the silica gel base material has not been properly acid-washed or prepared from an extremely pure organosilane.

Method development in SPE is similar to method development in HPLC but the presence of the matrix must be taken into account. The matrix complicates method development because, as

Figure 2.10 Solid-phase extraction chemistries currently in use. *(From R. E. Majors, LC/GC 14, pp. 754–766 (1996). Reprinted by permission of Advanstar Communications.)*

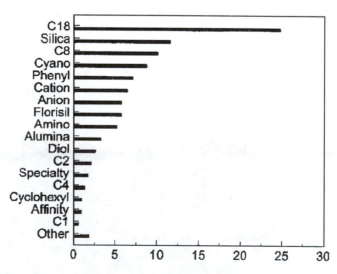

depicted in Fig. 2.11, there are four interactions that may occur, whereas in a typical HPLC method there are only three (analyte, solvent, and stationary phase). For our hypothetical example of Fig. 2.11, conditions should be chosen so that the matrix–solvent interactions and the analyte–stationary phase interactions are maximized and the matrix–stationary phase and analyte–solvent interactions are minimized. As noted in Fig. 2.10, the SPE stationary phases are similar to those of HPLC and may be selected on the same basis. Guidelines for selection of the appropriate SPE phase based on the analyte molecular weight and solubility are found in the literature (27, 28) and are not reproduced here.

Before going into the laboratory and starting method development, one should first characterize the problem as fully and carefully as possible. Questions to be asked include the following:

- What is the real problem? What can SPE accomplish and is it the best choice?
- What are the analytes of interest? What are the known analyte properties (polar, nonpolar, or ionic character, solubility, acid–base properties)? Does the analyte have any functional groups that can be exploited to effect a sample cleanup step? Are they different from the matrix?
- What is the concentration or concentration range of the analyte in the sample?
- What is the composition of the matrix? Does the matrix have any functional groups different from the analyte that might be exploited to effect a separation? What properties of the matrix suggest which mechanisms not to use? What is the typical pH and ionic strength of the matrix? Does the matrix vary from sample to sample?

Other sources of information should be tapped. HPLC retention data on the analytes or interferences can help to suggest some starting points, particularly for phase and solvent selection. Retention data can provide a rough estimate of which loading and eluting solvents you should use, especially if you are using reversed-phase sorbents. If no chromatographic data are available, you can perform a few experiments to observe how organic solvent composition and pH affect the retention time. SPE, unlike HPLC, may provide you with some degree of flexibility in choosing the pH. In HPLC, silica-based sorbents are stable in the pH range of 2 to 7. The sorbents used in SPE also undergo hydrolysis at pH extremes, but the cartridge is typically disposed of after one use and contact times are short. Even so, the presence of a small amount of dissolved silica or hydrolyzed bonded phase may not interfere with the subsequent measurement. However, if this does cause a problem, polymeric SPE devices that are stable over a much greater pH range, often from 1 to 14, can be used.

Figure 2.11 Fourfold interactions in solid-phase extraction.

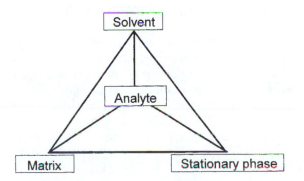

For solubility data, reference works such as the latest editions of the *Merck Index* or the *Handbook of Chemistry and Physics* can be helpful. A literature search on SPE methods for similar analyte–matrix pairs would be useful. SPE cartridge manufacturers have published extensive bibliographies, some in searchable electronic database formats, that can help to locate publications of interest (29–31) or provide initial conditions. Also, manufacturers can often provide application notes on the same or similar compounds. Many offer consulting services on method development.

Numerous SPE cartridges and disks are readily available from a variety of manufacturers. Some manufacturers even assemble method development kits that provide a selection of phases so that you do not have to buy many boxes of cartridges that you do not need. To make phase and solvent selection easier, automated systems are available that can be programmed to step through the many phases and eluting solvents.

Column Chromatography as a Sample Preparation Technique

Before the development of SPE for sample pretreatment, similar separations were carried out using low-pressure or open-column liquid chromatography (LC). Adsorbents such as large-particle silica gel or alumina are packed into large internal diameter (1 to 5-cm) glass columns. The sample is injected with a pipette onto the top of the packed bed and solvent is percolated through the column under gravity flow. Fractions are collected either manually or with a fraction collector and analyte measurement is performed off-line using a spectrophotometer or colorimeter. If analyte concentration is too low in the collected fractions for measurement by the secondary analytical technique, they can be concentrated using other sample preparation techniques, such as evaporation. Relative to HPLC, column efficiency is poor but the technique is simple to carry out.

A new version of LC has become popular in recent years called flash chromatography (32). In this technique, convenient prepacked columns with precleaned adsorbent are used and gentle pressure is applied to aid flow. In addition to the older adsorbents, newer stationary phases such as C8 or C18 bonded silicas are available so that reversed phase chromatography can be performed. Flow-through detectors have been used in flash chromatography.

The use of off-line LC as a cleanup technique for GC and HPLC is well documented (for example, see Refs. 32 and 33). In liquid phase analysis, the cleanup step can be performed in a mode different from the HPLC mode and thus a multidimensional LC–HPLC experiment can provide a clean sample that serves to prolong HPLC column lifetime. Samples can often be fractionated using step gradients and there are many methods for pesticides and drugs in biological fluids, where the techniques have been used successfully. In GC, adsorbents such as Florisil have long been used for sample cleanup before pesticide analysis.

The Role of Membranes in Sample Preparation

Membranes are usually made of synthetic polymeric materials, although natural substances such as cellulose and inorganic materials, such as glass fibers, are also used. The membrane techniques most familiar in sample preparation are in the areas of filtration and SPE. Membrane barrier separation processes are also becoming increasingly popular. Driving forces such as chemical- or electrochemical-potential gradients transport mixtures of chemicals across a restrictive membrane interface (Fig. 2.12). Ultrafiltration, reverse osmosis, dialysis, and electrodialysis are examples of techniques that use barrier processes for concentration, purification, and separation.

Figure 2.12 Typical setup for a flow dialysis experiment.

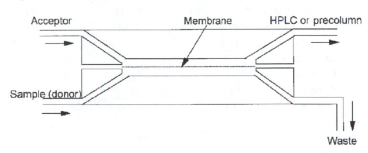

Membranes can be in many shapes but disks, cartridges, and hollow-fiber forms are the most popular. Figure 2.12 shows how a typical flat semipermeable membrane is used in an experimental setup with flowing liquids. Synthetic semipermeable membranes allow passage of certain chemical compounds while preventing or strongly retarding other compounds. Here the driving force is a differential analyte concentration across the membrane. The sample to be purified (called the donor) is placed on one side of the membrane; this sample is in an aqueous solution. On the other side is another liquid, which can also be an aqueous solution (called the acceptor).

A liquid phase can be added to the membrane material to facilitate solute transport. For example, a liquid ion exchanger or a hydrocarbon phase can be imbibed in the membrane and the solubility of an analyte in this supported liquid membrane can be another parameter in the membrane separation process (34). Each side of the membrane in Fig. 2.12 is connected to an independent flow channels. Aqueous solutions are pumped through each of the channels. On the donor side, compounds can be selectively extracted into the supported membrane and then re-extracted on the acceptor side. Using an example of a carboxylic acid we can follow the operation of this purification system. By adjusting the pH of the donor side below the pK_a (the negative log of the acid's ionization constant) of the acid, the ionization can be suppressed, which allows the un-ionized form to be extracted into the organic membrane, which is supported with a long chain hydrocarbon phase. The un-ionized acid diffuses through the membrane to the support hydrocarbon phase and finally to the acceptor side, which has a basic pH, and thus the organic acid is extracted in its ionized form. The analyte of interest can no longer re-extract into the hydrocarbon liquid in the organic membrane. The acceptor side can be static and has a fairly small volume so that the organic acid becomes concentrated during the process. Enrichment factors of several hundred have been demonstrated in samples such as organic herbicides in water (35), amines in plasma (36), and carboxylic acids in soil (37).

If a trap or precolumn is placed between the membrane device and the HPLC column, the analyte concentration can be enhanced further. In this case, the acceptor solution is continuously pumped through the precolumn and once a sufficient concentration of analyte has built up, the precolumn can be switched to the HPLC column and the sample displaced, similar to column switching.

Flow dialysis is a membrane barrier separation process in which the donor (or feed) solution containing the solute of interest and the acceptor solution are both flowing. Ionic compounds and small organic molecules can freely pass through the membrane but molecules larger than 1000 molecular weight (dependent on the molecular weight cutoff of the membrane) cannot penetrate the membrane or do so very slowly. Static dialysis is also used as a purification technique, particularly for biological systems. Microdialysis, a specialized application of dialysis, uses small microprobes (fused silica tubing) with a membrane at the end (38). These probes are placed in living system such as brain tissue or plant tissue and the diffusion of small organic molecules through the membrane can be monitored on-line by HPLC without disturbing the animal or plant. A microsyringe pump is used to pump the sample into a loop injector.

Flow dialysis has also proved effective as an on-line sample preparation technique for the deproteination of biological samples before HPLC analysis. The sample is introduced into the donor chamber and acceptor solvent is pumped through the acceptor chamber to a trace enrichment column placed downstream of the analyzer. Once a sufficient amount of analyte has been isolated for detection, it is backflushed into the HPLC system. These techniques have been automated and are in routine use in many laboratories.

Ultrafiltration (UF) sampling is similar to dialysis but the driving force is not a differential concentration across the membrane, but a pressure differential (10 to 100 psi) draws sample through the probe. The pressure differential physically causes the donor solution to flow through the membrane, carrying along with it small molecules that are collected on the acceptor side. UF cutoff membranes for molecular weight ranges from 300 to about 300,000 are available. Because of the forced flow, large molecules tend to obstruct the UF membranes and they are generally larger than those used for microdialysis. Some examples of the use of UF in sample preparation are the measurement of theophylline in saliva and the in vivo monitoring of acetaminophen in subcutaneous tissues (39).

An advantage over other sample preparation techniques is that the membrane processes require little or no organic solvent. They can provide excellent selectivity factors and, when combined with traps or trace enrichment columns, can yield excellent sensitivity rivaling that of other sample preparation techniques. In addition, membrane techniques are amenable to automation.

Removal of Particulate Matter

The removal of particulate material is important in sample pretreatment. Particulate matter may have originated from the initial sample or been introduced by accident during a multistage sample preparation process. Particulates in the sample are generally undesirable for the analytical measurement technique. For example, in spectroscopy, particulates can cause light scattering, thereby causing erroneous readings. The impact of particulate matter on an HPLC column and the resulting high-column backpressure and short-column lifetimes are well understood. However, an often overlooked consideration is that particulate matter can also be a source of irreversible sample adsorption. For example, it is also known that aged soil particulates can strongly retain polar compounds that may not be removed unless exhaustive solvent extraction is used. Merely filtering an aqueous slurry of a soil sample will not dislodge these strongly held analytes.

The most common methods to remove particulates from the sample are filtration, centrifugation, and sedimentation. For most liquid samples, filtration is recommended, especially as one of the last steps before the analytical measurement. Most chemists use paper filtration, a relatively straightforward technique. To remove fine particulates, filter paper should be of low porosity. Note that the lower the porosity, the longer the sample will take to pass through the filter, but the cleaner the filtrate. Vacuum filtration speeds up the process. Membrane filters can also be used for filtration of samples before measurement. These filters can be purchased loose for placement into filter holders such as the Millipore unit. However, for convenience most users prefer disposable filters, which are available preassembled in inert plastic holders with Luer fittings on the top and bottom. The sample is placed in a syringe and forced through the membrane using gentle pressure. A wide variety of materials, porosities, dimensions, and pore sizes are available. However, for most samples, filters in the range of 0.25-μ porosity to 2-μ porosity are recommended. The 0.25-μ membranes remove the tiniest particulates and large macromolecules but require quite a bit of force to push the liquid sample through the filter, particularly if the sample contains colloidal material or a large amount of fines. Most manufacturers of membrane filters provide detailed information on solvent compatibility. If the wrong solvent is used with these filters, the filter may dissolve, contaminating the sample.

Multidimensional Chromatography

Multidimensional column chromatography (MDCC) (sometimes called column switching, multiphase, multicolumn, coupled column, or boxcar chromatography) is a powerful technique for the separation and cleanup of multicomponent mixtures. In this approach, a fraction from one chromatographic column is selectively transferred to one or more secondary columns for further separation. MDCC is used for

- Trace enrichment of selected analytes.
- Improvement of resolution of part of a complex sample; maximum resolution can be achieved by using different modes, stationary phases, and mobile phases.
- Increasing sample throughput by use of heartcutting, backflushing, front- or end-cutting, and recycle chromatography.

MDCC has been performed by almost any combination of one or more chromatographic techniques imaginable; those that have been successfully coupled both off-line and on-line include all combinations of LC, GC, thin-layer chromatography (TLC), supercritical fluid chromatography (SFC), and capillary electrophoresis (CE). However, the more popular techniques are GC–GC, LC–LC, and LC–GC, with the latter technique seeing limited use.

In concept, MDCC can be performed off-line or on-line. In GC–GC, because of the difficulty of collecting fractions from the vapor phase, most MDCC combinations are performed on-line while in LC–LC and LC–GC can be carried out both ways. In fact, SPE and column chromatography are really off-line MDCC techniques and are not discussed further here. On-line MDCC is achieved through coupling to a second column by means of a high-pressure switching valve that traps a defined volume or collected sample, usually in a loop, and directs it to the second column (heart cutting). Alternatively, the switching valve can be plumbed to divert the mobile phase containing the desired solutes from the first column to the second column for a defined period of time (on-column concentration or trace enrichment).

On-line MDCC techniques are preferable to off-line MDCC because they are more reproducible and more amenable to automation, samples are not exposed to the atmosphere because the experiment is performed in a closed system, overall analysis time is decreased, and sample throughput is increased. However, more complex instrumentation is needed, trace substances in large volumes cannot always be concentrated, and the primary and secondary chromatographic techniques must be compatible. For example, using normal phase HPLC with a hexane mobile phase is difficult to couple on-line to reversed phase HPLC using methanol-water because the solvents are not miscible.

A typical example of on-line MDCC will be illustrated using HPLC–HPLC. To ensure that separations can be achieved on the two HPLC columns, it is often desirable that different mechanisms be used. Two HPLC modes that can be conveniently coupled are aqueous size exclusion chromatography (SEC) as the primary mode and reversed phase chromatography (RPC) as the secondary mode. SEC, a relatively low-resolution chromatographic technique, separates molecules on the basis of their differences in molecular size, whereas RPC separates on differences in hydrophobic character. Water-soluble samples that have both high-molecular-weight components and low-molecular-weight components are prime candidates for on-line MDCC. Proteins are high-molecular-weight compounds that may foul a typical RPC column if not removed from the sample before injection. By selection of the appropriate SEC column, one can partially exclude these higher-molecular-weight compounds and direct the lower-molecular-weight fraction onto the RPC column.

Diverse food products are often fortified with vitamins to enhance their nutritional value. A protein supplement fortified with several B vitamins at levels ranging from 0.001 to 0.04% by weight was investigated using MDCC (40). A method for the determination of the trace amounts of vitamins in this complex mixture was desired. Without sample pretreatment, this problem would be nearly impossible to solve. Initially, SEC experiments were run in order to determine

Figure 2.13(a) Size exclusion chromatogram for the separation of B vitamins in food protein supplement.

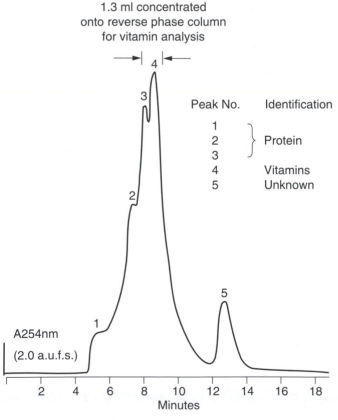

Column: MicroPak TSK 2000 SW (Varian Assoc., Palo Alto, CA) , 30 cm × 7.5 mm i.d.
Mobile phase: Methanol-water (1:9 vol/vol) containing 0.1 M KH_2PO_4 and 0.01 M 1-heptanesulfonic acid.
Flow rate: 1.2 ml/min.
Detector: UV at 254 nm.

the elution profiles of the proteins and the vitamins. Figure 2.13(a) shows the SEC chromatogram run on the mixture. Obviously, low concentrations of the B vitamins could not be determined using this single HPLC mode. However, from this initial experiment, it was determined that the vitamins were actually eluted at the 8.2- to 9.5-ml volume interval (1.3 ml volume). Using a column switching setup shown in Fig. 2.14, this volume was diverted into the RPC column and using ion pair chromatography and gradient elution, the trace amounts of the B vitamins could be separated as depicted in the actual recorder trace of Fig. 2.13(b). Approximately 0.02% of niacin, 0.003% of pyridoxine, 0.009% of riboflavin, and 0.003% of thiamine were found in the food supplement, which compares favorably with other methods for the analysis of vitamins in food matrices (41). The MDCC approach reduced overall sample preparation time and also allowed the determination of trace vitamins in a very complex sample.

Figure 2.13(b) Column switching experiment from size exclusion column to reversed phase HPLC column. 1.3-ml portion of effluent (identified in chromatogram of Fig. 2.13(a)) from Column 1 was directed onto Column 2. After injection, the gradient identified for Column 2 was run at the 10-min mark. The resultant chromatogram of both the partial SEC separation and the reversed-phase chromatography gradient separation is depicted in Fig. 2.13(b). *(Reprinted from J. Chromatogr, 206, J. A. Apfell, T. V. Alfredson, and R. E. Majors, pp. 43–57, copyright 1981 with kind permission of Elsevier Science—NL, Sara Burgerhartstraat 25, 1055 KV Amsterdam, The Netherlands.)*

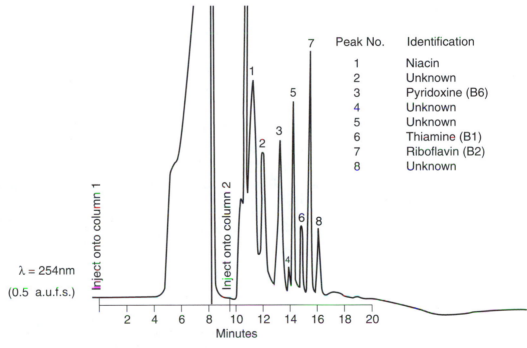

Column 1: MicroPak TSK 2000 SW (Varian Assoc., Palo Alto, CA) , 30 cm × 7.5 mm i.d.
Conditions: Same as in Fig. 2.13(a).
Column 2: MicroPak MCH-10, 30 cm × 4 mm i.d.
Mobile phase: Solvent A, methanol-water (1:9 vol/vol) containing 0.1 M KH_2PO_4 and 0.01 M 1-heptanesulfonic acid. Solvent B, methanol containing 0.1 M KH_2PO_4 and 0.01 M 1-heptanesulfonic acid.
Gradient: A to 80%, B at 15%/min.
Flow rate: 2 ml/min.

Figure 2.14 Valving system used in column switching experiment of Fig. 2.13(b). *(Reprinted from J. Chromatogr, 206, J. A. Apfell, T. V. Alfredson, and R. E. Majors, pp. 43–57, copyright 1981 with kind permission of Elsevier Science—NL, Sara Burgerhartstraat 25, 1055 KV Amsterdam, The Netherlands.)*

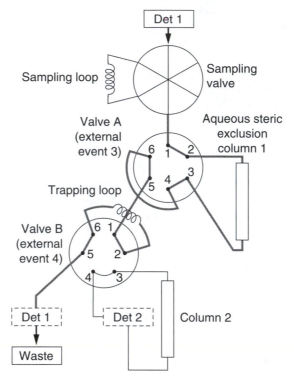

Derivatization in Chromatography

In chromatography it is preferable to separate and detect compounds in their native state. However, certain compounds may be unstable at the temperatures used in GC or do not possess a detectable chromophore for ultraviolet detection in HPLC. In these cases, derivatization may be required. Derivatization involves a chemical reaction between an analyte and a reagent that changes its chemical and physical properties. The main uses of derivatization in chromatography are

- To change the molecular structure or polarity of the analyte for better chromatography
- To stabilize an analyte
- To improve detectability
- To change the matrix for better separation

The derivatization reaction should be rapid and quantitative, with minimal production of by-products. Any excess reagent should not interfere with the analysis or should be easily removed from the reaction mixture.

In GC, derivatization is used primarily to increase the thermal stability of compounds that would otherwise be unsuitable for GC analysis. Compounds that are most often derivatized are those that are of high molecular weight or those that contain polar functional groups. Through derivatization, these polar functional groups are generally replaced with nonpolar functional

Table 2.7 Advantages and disadvantages of precolumn derivatization in HPLC.

Advantages	Disadvantages
Fewer equipment restrictions; simple reaction vessels.	Automation is less straightforward.
Chemical reactions are not constrained by time.	Derivatization reaction must be quantitative and derivatives well behaved (stable).
Wide variety of reactants are available.	Reaction by-products and excess reagent may interfere with chromatography or detection step if not removed.

groups. Peak shapes are often improved because undesirable surface interactions are minimized. Derivatization is also used to enhance detectability by the introduction of organic groups that can be better detected. An example is the incorporation of electron-donating groups, which can be detected by the sensitive and selective electron-capture detector. Virtually all GC derivatization reactions are performed by precolumn derivatization reactions.

In HPLC, most derivatizations are used to enhance analyte sensitivity. For example, amino acids contain weakly UV-absorbing carboxyl and amino groups that do not allow their detection at low levels. By the introduction of UV-absorbing chromophores or by the production of fluorescent compounds, sensitivity can be enhanced by orders of magnitude. In contrast, chiral derivatizations are usually performed to change selectivity for one enantiomeric form over the other. HPLC derivatizations can be performed precolumn or postcolumn. Tables 2.7 and 2.8 list the advantages and disadvantages of each approach.

Most derivatization agents have two key elements: an organic group, which controls the volatility (GC) or detectability (HPLC) and the reactive group, which provides the means of attachment to the analyte. Some criteria in choosing a derivatizing agent are that it must be stable, excess reagent or by-products formed should not be detectable or must be separable from the derivatized analyte, it must be reactive with the analyte under reasonable conditions, it should be nontoxic, and it should be amenable to automation. Because derivatizing reagents and conditions for derivatization are so varied, no attempt is made to list all of them here, but the reader is referred to Refs. 41–44. Table 2.9 lists some of the popular derivatizing reagent types and compounds derivatized for GC; Table 2.10 does the same for HPLC.

For GC, the silylation reagents are the most popular and universal. They are very reactive, can derivatize most functional groups, and, with most compounds, do not form by-products or artifacts. The trimethylsilyl ether derivatives themselves are stable. New silyl homolog or analog re-

Table 2.8 Advantages and disadvantages of postcolumn derivatization in HPLC.

Advantages	Disadvantages
Easy automation.	Reaction must be fast.
Nonquantitative reactions can be used (if reproducible).	Band broadening may occur.
Derivatization does not interfere with chromatography.	Possible incompatibility between the mobile phase and the derivatizing reagents complicates method development.
	Derivatives must respond differently from reagents or by-products or the latter must be separated after derivatization and before detection.
	Fewer chemical reactions are available that meet criteria.

Table 2.9 Typical derivatization reagents and compounds derivatized in GC.

Type of Derivative	Common Examples	Typical Compounds Derivatized
Alkylsilane	TMCS, HMDS, MSTFA, BSTFA*	Alcohols, mercaptans, amines, amides, carboxylic acids
Haloalkylacyl	Haloalkyl acid chlorides and anhydrides (trifluoroacetyl most popular)	Alcohols, amines, amides, mercaptans, carbonyls
Ester	Alcohol + acid catalyst	Carboxylic acids
Alkylates	Alkyl halide + catalyst, diazoalkanes	Carboxylic acids, alcohols, amines, amides, alkylsulphonates, phenols
Oxime	Alkyl- or aryl-hydroxylamine hydrochloride salt	Ketones
Pentafluorophenyl-containing derivative	Various pentafluoro–acid anhydrides and chlorides, aldehydes, hydrazines	Alcohols, amines, phenols, carboxylic acids, ketones

*TMCS = trimethylchlorosilane
HMDS = hexamethyldisilazane
MSTFA = N-methyl-N-(trimethylsilyl)acetamide
BSTFA = N,O-bis(trimethylsilyl)trifluoroacetamide

agents have been developed that have improved hydrolytic stability or better detectability. Silylating reagents containing electron-capturing groups such as chloro, bromo, or iodo (electron capture detector), and cyano groups (thermionic detector) have been developed for improved detectability.

In UV detection for HPLC, a variety of reagents with chromophoric groups are available. Ideally, the chromophoric group should have a large molar extinction coefficient at a convenient detection wavelength and should not be large and polar so that it has little effect on the chromato-

Table 2.10 Fluorescent and UV derivatives for HPLC.

Functional Group Derivatized	Reagents	Detection Principle
Alcohol	Phenylisocyanate	Ultraviolet spectrophotometry
Alcohol	3,5-dinitrobenzoyl chloride	Ultraviolet spectrophotometry
Aldehyde, ketone	2,4-dinitrophenylhydrazine	Visible spectrophotometry
Aldehyde, ketone	Dansyl hydrazine	Fluorescence
Amine, primary	o-phthalaldehyde, fluorescamine	Fluorescence
Amine, secondary	9-fluorenylmethylchloroformate	Fluorescence
Amine, secondary	Dansyl chloride	Fluorescence
Carboxyl	p-bromophenylacyl bromide	Ultraviolet spectrophotometry
Carboxyl	2-naphthacyl bromide	Ultraviolet spectrophotometry
Carboxyl	4-bromomethyl-7-methoxycoumarin	Fluorescence
Phenol	1-naphthyl isocyanate	Ultraviolet spectrophotometry
Thiols	Dabsyl chloride	Ultraviolet spectrophotometry
Thiols	7-chloro-4-nitrobenzo-2-oxa-1,3-diazole	Fluorescence

graphic separation. Because the physical properties of a molecule required for good fluorescence response are special and there are only a few compound types with native fluorescence, fluorescent derivatization reagents are less numerous than chromophoric reagents. With fluorescent derivatives, the mobile phase can have a profound effect on fluorescent yield or can cause quenching and, therefore, sensitivity is affected.

The reagent 9-fluorenylmethylchloroformate (FMOC-Cl) is a popular precolumn derivatizing reagent that reacts in pH 8.5 aqueous solution with primary and secondary biogenic amines and amino acids to give highly fluorescent, stable 9-fluorenylmethyl carbamate derivatives. The excess reagent must be removed because it elutes in the middle of the chromatogram; it can be extracted with hexane but some of the derivatized amines can also be extracted (45). Another approach to remove the excess FMOC-Cl is reaction with heptylamine; the derivative elutes very late in an RPC column and therefore does not interfere with the derivatized amines.

For postcolumn detection, ortho-phthalaldehyde (OPA) is an ideal reagent that reacts with primary amines and amino acids in the presence of a thiol such as 2-mercaptoethanol to give a highly fluorescent derivative. The reagent itself is nonfluorescent, so it does not interfere with the fluorescent measurement. OPA can also be used as a precolumn derivatization reagent but certain derivatives are unstable, so timing between reagent addition and injection into the chromatograph is critical. An example of the successful application of OPA postcolumn detection is typified by EPA Method 531, which is used for the determination of carbamate pesticides by HPLC. Carbamate pesticides have the general structure shown below.

Hydrolysis Step

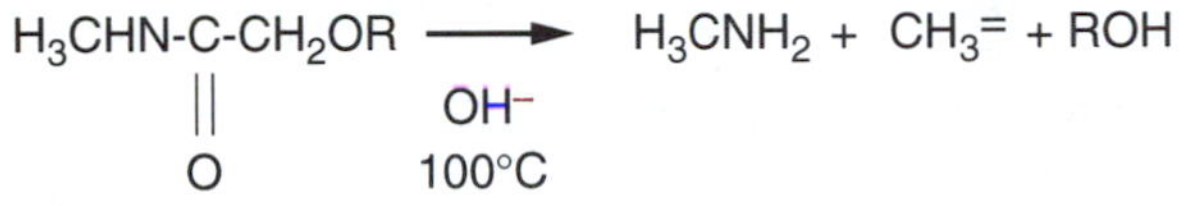

Carbamate structure

Derivatization Step

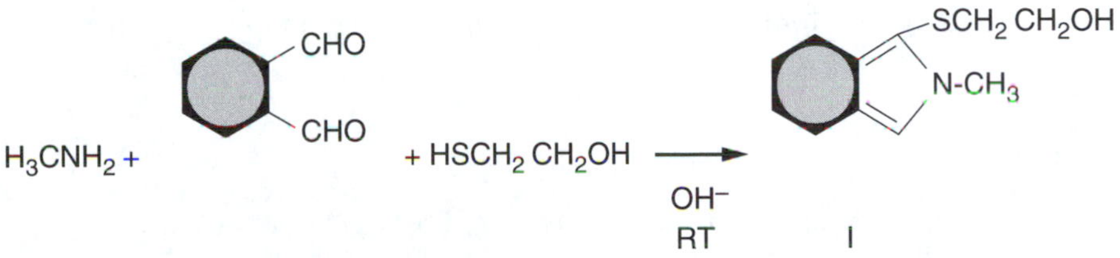

Each R-group represents a different commercial product or its metabolite. The postcolumn derivatization is a two-step process after the separation of the carbamates by RPC. In the first step, the carbamates are hydrolyzed at elevated temperature to generate methylamine. In the second step, the methylamine is derivatized by OPA and 2-mercaptoethanol to give I, which shows optimal excitation at 330 nm and maximum emission at 465 nm. Determination of carbamates in the sub-μg/l concentration range are possible using fluorescence detection. A typical chromatogram is shown in Fig. 2.15.

Figure 2.15 Chromatogram of the HPLC separation of carbamate pesticides by postcolumn derivatization. *(From Pickering Product Bulletin #2, "Carbamate Pesticide Analysis for HPLC," reprinted with permission of Pickering Laboratories, Inc.)*

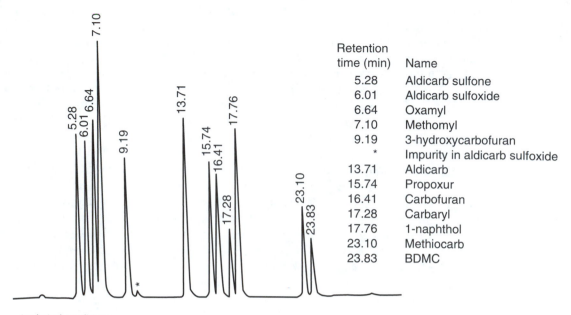

Retention time (min)	Name
5.28	Aldicarb sulfone
6.01	Aldicarb sulfoxide
6.64	Oxamyl
7.10	Methomyl
9.19	3-hydroxycarbofuran
*	Impurity in aldicarb sulfoxide
13.71	Aldicarb
15.74	Propoxur
16.41	Carbofuran
17.28	Carbaryl
17.76	1-naphthol
23.10	Methiocarb
23.83	BDMC

Analytical conditions:
Column: C18, 5µ, 4.6mm × 150 mm.
Mobile phase: 25% methanol in water for 30 sec, followed by 26.5 min linear gradient to 75% methanol in water.
Flow rate: 1.0 ml/min.
Column temp.: 42 °C.
Detector: Fluorescence, excitation wavelength: 330 nm; emission wavelength: 465 nm.
Postcolumn system conditions:
Flow rate: 0.3 ml/min for both reagents.
Reaction dwell times: 1) hydrolysis step 23 sec @ 100 °C.
2) OPA derivatization step 8 sec @ room temperature.

Although organic synthesis reactions can be mastered by the average chemist, it is more convenient to purchase a derivatization kit that contains prepacked reagents, catalysts, and step-by-step directions for specific analytes. Many chromatography and chemical suppliers have such kits in their catalogs.

References

1. P. M. Gy, *Sampling of Particulate Materials, Theory and Practice* (Amsterdam: Elsevier, 1982).

2. R. Smith and G. V. James, *The Sampling of Bulk Materials* (London: The Royal Society of Chemistry, 1981).

3. C. L. P. Thomas, *LC-GC*, 10 (1992), 832.

4. L. H. Keith, ed., *Principles of Environmental Sampling* (Washington, D.C.: American Chemical Society, 1988).

5. J. A. Graham, *Anal. Chem.*, 63 (1991), 613A.

6. C. T. Wehr and R. E. Majors, *LC-GC*, 5 (1987), 548.

7. R. E. Majors and D. Hardy, *LC-GC*, 10 (1992), 356.

8. P. M. Gy, *Sampling of Heterogeneous and Dynamic Material Systems* (Amsterdam: Elsevier, 1992).

9. K. P. Kelly and E. E. Conrad, "Automated Soxhlet Extraction of Solid Matrices for Oil and Grease or Fat Content for Trace Organics Determination," *Application Note No. 24* (Columbia, MO: O.I. Analytical, 1995).

10. D. R. Gere and E. M. Derrico, *LC-GC*, 12 (1994) 432.

11. B. W. Renoe, *Amer. Lab.*, 26, no. 12 (1994), 34.

12. J. R. J. Pare, J. M. R. Belanger, and S. Stafford, *Trends in Anal. Chem.*, 13, no. 4 (1994), 176–84.

13. F. I. Onuska and K. A. Terry, *Chromatographia*, 36 (1993), 191–4.

14. B. E. Richter and others, *Amer. Lab.*, 27, no. 4 (1995), 24–8.

15. J. L. Ezzell and others, *LC-GC Mag.*, 13, no. 5 (1995), 390–8.

16. M. H. Hiatt, *Anal. Chem.*, 67 (1995), 4044.

17. G. D. Christian and J. E. O'Reilly, *Instrumental Analysis*, 2nd ed. (Newton, MA: Allyn & Bacon, 1986).

18. D. C. Harris, *Quantitative Chemical Analysis*, 4th ed. (New York: W.H. Freeman, 1994).

19. R. E. Majors and K. D. Fogelman, *Amer. Lab.*, 25, no. 2 (1993), 40W–40GG.

20. T. Studt, *R&D Magazine*, 32, no. 3 (1992), 90.

21. T. A. Dirksen, S. M. Price, and S. J. St. Mary, *Amer. Lab.*, 25, no. 18 (1993), 24–7.

22. C. L. Arthur and J. Pawliszyn, *Anal. Chem.*, 62 (1990), 2145.

23. C. L. Arthur and others, *J. Environ. Sci. Technol.*, 26 (1992), 979.

24. G. E. Platoff and J. A. Gere, *Forens. Sci. Rev.*, 3 (1991), 117.

25. C. M. Druzik, D. Grosjean, A. Van Neste, and S. Parmar, *Intern. J. Environ. Anal. Chem.*, 38 (1990), 495–512.

26. V. Dixit and V. M. Dixit, *J. Liq. Chrom.*, 14, no. 14 (1991), 2779–800.

27. M. Zief and R. Kiser, *Amer. Lab.*, 22, no. 1 (1990), 70–83.

28. *Solid Phase Extraction for Sample Preparation* (Phillipsburg, N.J.: J.T. Baker Inc., 1994).

29. P. D. MacDonald and E. S. P. Bouvier (eds.), *Solid Phase Extraction Applications Guide and Bibliography* (Milford, MA: Waters, 1995).

30. *Applications Bibliography* (Harbor City, CA: Varian Sample Preparation Products, 1992).

31. *BakerBond spe Bibliography* (Phillipsburg, N.J.: J.T. Baker Inc., 1994).

32. R. E. Majors and T. Enzweiler, *LC-GC*, 6 (1988), 1046–50.

33. U.S. Environmental Protection Agency, *Interim Methods for the Sampling and Analysis of Priority Pollutants in Sediments and Fish Tissue* (Cincinnati: U.S. EPA Monitoring and Support Lab., 1981).

34. J. A. Jonsson and L. Mathiasson, *Trends in Anal. Chem.*, 11, no. 3 (1992), 106–14.

35. G. Nilve and R. Stebbins, *Chromatographia*, 32 (1991), 269–77.

36. H. Lindegard and others, *Anal. Chem.*, 66 (1994), 4490–7.

37. G. Nilve, G. Audunsson, and J. A. Jonsson, *J. Chromatogr.*, 471 (1989), 151.

38. C. E. Lunte, D. O. Scott, and P. T. Kissinger, *Anal. Chem.*, 63 (1991), 773A–779.

39. M. C. Linhares and P. T. Kissinger, *Anal. Chem.*, 64 (1994), 2831–5.

40. J. A. Apffel, T. V. Alfredson, and R. E. Majors, *J. Chromatogr.*, 206 (1981), 43–57.

41. J. F. Gregory, *J. Agr. Food Chem.*, 28 (1980), 486.

42. B. King and G. S. Graham, *Handbook of Derivatives for Chromatography* (Philadelphia: Heyden and Son, Ltd., 1979).

43. S. Ahuja, *Selectivity and Detectability Optimization in HPLC* (New York: Wiley, 1989).

44. C. F. Poole and S. K. Poole, *Chromatography Today* (Amsterdam: Elsevier, 1993).

45. S. Einarsson, B. Josefsson, and S. Lagerkvist, *J. Chromatogr.*, 282 (1983), 609.

46. *Pickering Product Bulletin #2* (Mountain View, CA: Pickering Laboratories Inc., 1989).

47. R. E. Majors, *LC-GC*, 9 (1991), 16–20.

48. R. E. Majors, *LC-GC*, 4 (1986), 972–84.

49. R. E. Majors, *LC-GC*, 14 (1996), 754–66.

Chapter 3

Sample Preparation in Analytical Chemistry (Inorganic Analysis)

John R. Moody and Charles M. Beck II

National Institute of Standards and Technology, Inorganic Analysis Chemistry Division

The vast majority of analyses today are done by instrumental means. Because of this we often overlook or downplay the importance of sample preparation. However, sample preparation is crucial to the outcome of an analysis. The best instrumentation and computer acquisition systems cannot overcome poor sample acquisition or poor sample preparation. It is beyond the scope of this chapter to go into all types of sample preparation in detail. We intend to give the reader some idea of common techniques used for *inorganic* analysis. Most of the principles involved in sample preparation are based on well-established analytical procedures. Despite the importance of instrumentation in modern analytical science, it is always necessary to note that nothing can come out of the instrument that is better than what was presented for analysis. How do we transform a rock or a polymer into a form suitable for analysis? No amount of idealistic thinking can totally remove human intervention from all sample acquisition and preparation steps.

Fundamental laboratory operations such as quantitative sample transfer, volumetric calibration, and accurate and precise weighing techniques have not disappeared in the instrumental age. However, these ideas have been largely ignored or overlooked as the emphasis of modern analytical training focused on the instrumental technique. Nevertheless, a good understanding of classic procedures and the mechanisms by which they work is required to improve sample preparation. Instead of being viewed as a burden, sample preparation should be viewed for what it can do to improve sample analysis. Thus, we first describe classical analysis and then relate it to the context of analytical chemistry.

Classical Chemical Analysis

Classical analysis is the collective term used to describe an analysis whose measurement step is a gravimetric or titrimetric determination. *Instrumental analysis* is the collective term used for an analysis whose measurement step involves an instrument. Although the use of instrumental analysis grew slowly until the 1940s, most analyses before that time were done classically. Since the 1940s, instrumental analysis has become more important because of the increasing demands for sensitivity, speed, and economy, so that today instrumental analysis has assumed the major role in the analytical laboratory.

Gravimetry and titrimetry are based solely on atomic weights (relative atomic masses), experimentally measured masses, and atomic-weight ratios obtained from well-characterized chemical reactions. (When titrations are done on a volume basis instead of a mass basis, the volume must be related to mass through density.) The classical methods of gravimetry and titrimetry, along with coulometry and isotope dilution mass spectrometry (IDMS), are called definitive methods because they have exactly defined errors and under the proper conditions need no external calibrants. The results can be said to include the true value within the statistical uncertainty of the method. Coulometry depends upon Faraday's laws relating the moles of a substance reacted to the fundamental units of current and time. In IDMS the analytical result is based only on isotope ratio measurements. In contrast, all other instrumental methods, including IDMS, rely on responses that possess no absolute quantitative meaning until they are calibrated with known standards.

In both classical and instrumental analysis, the final determination is almost always preceded by one or more preparatory steps. The precision and accuracy of the analytical result are no better than the collective precision and accuracy of all the preparatory steps. Most of the preparatory steps in an instrumental analysis have been adopted directly from classical analysis. Most analytical chemists working today received their education in the instrumental age. Many of the techniques used in the preparatory steps in a classical analysis are not as significantly emphasized as in the preinstrumental era. Therefore, a brief look at some components of classical analysis will be useful.

Gravimetry is the determination of an element or species through the measurement of the mass of the relatively insoluble product of a well-characterized chemical reaction involving that element or species. The insoluble product may be a gas evolved from solution, or it may be a nonvolatile solid residue from a low-temperature volatilization or a high-temperature ignition. Usually, the insoluble product is a precipitate formed in an aqueous solution.

One or more preparatory steps are almost always necessary before doing a gravimetric determination. The element or species being sought (the analyte) must be determined on a homogeneous specimen that represents the entire sample, be it a solid, liquid, or gas. If the specimen is not already a solution, it must be put into solution by a dissolution procedure. These include open-beaker acid dissolution, dry ashing, fusion, microwave-oven digestion, and high-temperature, high-pressure digestions (bomb- and Carius-tube techniques). Appropriate precautions must be taken to avoid mechanical and volatilization losses. Next, interferences must be eliminated through a separation, a series of separations, or masking (complex formation). For gravimetry, an interference is any element or species that would coprecipitate with, or be occluded by, the precipitate being sought. These preparatory steps are required before the element or species of interest can be determined by precipitating the insoluble element or stoichiometric compound.

For high-accuracy work, the systematic errors due to coprecipitation, occlusion, and the solubility-product effect must be evaluated. Despite precautions, all precipitates contain small amounts of coprecipitated and occluded substances. These substances can be determined by various instrumental techniques and appropriate corrections made. Due to the solubility product effect, all filtrates from precipitates contain small amounts of dissolved precipitate, which also can be determined by instrumental techniques.

An example of a gravimetric determination is the assay of a sample for strontium. If all the necessary preparatory steps have been properly carried out, one has an aqueous solution that contains all the strontium in the original sample and is free from all interferences. Strontium is precipitated at pH 8.5 as strontium oxalate by slowly adding a saturated solution of ammonium oxalate and letting the precipitate ripen (age) overnight. The strontium oxalate is filtered and the filtrate reserved for instrumental determination of unprecipitated strontium. The precipitate is ignited in a quartz crucible at 1100 °C to convert it to strontium oxide. The tared crucible is reweighed and the mass of strontium oxide found by difference. After the precipitate is checked for impurities by instrumental means, the amount of strontium in the sample is calculated by summing the strontium in the precipitate and in the filtrate and correcting for any low impurity amounts in the strontium oxide.

The preparation of the sample for the final gravimetric determination has already involved the common analytical operations of weighing, transferring, dissolving, and making one or more separations. The gravimetric determination itself involves the use of several additional common analytical operations. The precipitation technique must minimize coprecipitation and occlusion. In some instances, coprecipitation and occlusion are minimized when the precipitating agent is added slowly to the analyte. In other cases, adding the analyte slowly to the precipitating agent gives the best results. The next step is filtration, which involves the transfer of both the precipitate and the supernatent liquid to the appropriate type of filter paper, followed by the proper washing of the precipitate. Finally, accurate results depend heavily on the use of proper techniques in the drying and ignition of the precipitate. As with the initial weighing of the sample, the inclusion of a buoyancy correction is essential for the accurate weighing of the precipitate. In the example just given, instrumental techniques have been used to provide corrections for errors that were minimized but not corrected for in the past. Gravimetric procedures that are fully optimized by instrumental procedures can have greatly improved accuracy and precision and justify the use of the term *definitive method* because all known errors have been accounted for.

Titrimetry is the determination of an element or species through the measurement of the mass of a chemical necessary to complete a definite chemical reaction in a solution containing that element or species. The mass of the chemical is usually obtained indirectly by measuring the volume of a standard solution (solution of known concentration) of that chemical, but for very accurate work a mass titration should be used. Titrimetric methods are based on acid–base reactions, precipitation reactions, complexation reactions, and oxidation-reduction reactions. Endpoints generally are determined by observing a color change in an indicator. Endpoints also are commonly determined by observations of potential differences at electrodes, changes in conductivity, changes in optical absorbance, and many other physical or chemical properties. A titrimetric determination is usually preceded by one or more preparatory steps, many of which are the same as or similar to those used in gravimetry. Unlike in gravimetry, coprecipitation and occlusion interferences do not exist except in titrimetric methods that are based on precipitation reactions. However, interferences that cause competing side reactions must be eliminated or accounted for in all titrimetric methods. In addition, any systematic error resulting from the indicator's deviation from marking the true equivalence point must be evaluated and a correction made.

An example of a titrimetric determination is the assay of a sample for vanadium. With instrumental methods, it is now possible to know exactly what interfering elements are present in the sample and to design the appropriate procedure to isolate the analyte. As with gravimetry, when all the necessary preparatory steps have been performed, one should have an aqueous solution containing all the vanadium in the sample and no interfering substances. The vanadium is oxidized to V^{5+} with excess potassium permanganate, which is subsequently destroyed by the addition of excess nitrite. The excess nitrite is destroyed with excess urea, which does not interfere with the redox titration for vanadium. The vanadium is titrated to V^{4+} with standardized ferrous ammonium sulfate solution using diphenylamine sulfonate indicator.

Besides the common analytical operations already used in sample preparation, the titrimetric assay also involves the calibration and proper use of volumetric flasks, pipets, and burets, together with any necessary temperature correction. Volumetric flasks, pipets, and burets are calibrated by weighing how much water they contain or deliver and making buoyancy corrections. Gravimetric titrations eliminate most of these requirements and have the advantage that masses are easier to determine accurately than volume. The density of a solution changes with temperature, so all volumetric procedures must be temperature corrected. The mass of solution does not change with temperature, so gravimetric titrations do not need to be corrected for temperature.

Finally, one must also understand the reaction mechanism of the indicator used to be able to evaluate the indicator error. In the example given, the diphenylamine sulfonate indicator is itself a redox system that consumes a tiny amount of the V^{5+} analyte, which requires a correction in the final calculations. There are other ways to determine the endpoint of a titration, including potentiometric, thermometric, amperometric, spectrophotometric, and conductimetric techniques that can eliminate the indicator or improve the ability of the method to determine the endpoint accurately.

This brief overview shows that the gravimetric or titrimetric determination is the last in a series of steps that make up a classical analysis. Because most of the preparatory steps are identical for instrumental techniques, an accurate and precise instrumental analysis depends on someone in the analytical chain having training and experience in classical analysis. This type of training is de-emphasized in modern undergraduate analytical chemistry. This summary is not a substitute for that training, but we try to point the way for self-study through references to well-known texts as we outline general sample preparation steps.

Sampling

All analyses, whether classical or instrumental, inorganic or organic, require the analyst to start with a known amount of a sample. Unless otherwise stated, it is assumed that the analysis is based on the dry mass of the sample. The lack of a well-defined sample is often the source of disagreement between laboratories or between producer and consumer. The problem is less obvious in the university teaching laboratory, where samples usually are relatively dry and well defined. In industry, however, many samples are by nature wet or highly hydrated, so an accurate analysis depends on an appropriate drying procedure. In many cases analysts use drying procedures without considering the sample source or method of sample acquisition.

Analytical results are valid only if the sample represents the material of interest. Consequently, many books have been devoted to the sampling process. If one has a highly heterogeneous material, such as a powder with constituents of two different colors, it is easy to see when the sample is well mixed or blended. However, when sampling other materials, it is not so easy to see when the sample is truly representative. Additionally, one may be confronted with a gross sample size. Sample reduction is beyond the scope of this chapter, but is needed in such instances. Obviously, the procedure used (such as grinding, riffling, or quartering) must produce a final sample that is representative of the original material. Today, the sampling and even the sample reduction procedures are usually done in the field by a technician.

However the sample acquisition is done, the analytical chemist must look into how the sampling is conducted. Even if the sampling is representative, the sample integrity can be altered by many other processes, such as contamination and loss or degradation of analyte between the time of sampling and analysis. The responsibility of the analyst is to ensure that all steps in the method produce a valid, uncompromised sample. A few questions early in an analytical procedure can save much grief later if it is subsequently discovered that the samples are not valid.

Different sample types may need different types of protection before transport to the laboratory. Organic samples may easily lose volatile compounds and liquid samples often lose many trace elements to the container wall during storage and transport. Whether the sample is liquid or solid, the correct container and storage conditions must be chosen to protect the integrity of the sample. Many trade sources give procedures for the handling of particular samples (such as American Society for Testing and Materials manuals).

Sample Drying

Once a representative sample has been obtained, most analyses usually start with sample drying. Many methods are used for drying. These include lyophilization, conventional gravity ovens, forced-air ovens, equilibration in a desiccator over an appropriate desiccant, vacuum ovens, and infrared and microwave ovens. Historically, the conventional gravity oven and equilibration over a desiccant have been used for most inorganic analyses. By careful choice and control of drying conditions, these methods often yield the same sample composition. In other words, the same samples, even with different moisture contents, dried by two different laboratories using the same method will yield the same composition. When this ideal is not achieved, differences in sample composition among different laboratories occur.

The term *drying* itself is misleading because all the different methods can produce different results depending on the sample composition. We might assume that the weight loss during the drying step is due to the loss of water; however, the weight loss is also due to other volatile materials being lost. Thermal methods of drying might be expected to show a greater loss of all volatiles than room temperature methods due to changes in the relative values of the vapor pressure of volatile constituents with temperature. Nevertheless, the differences among procedures are not predictable, especially with new sample matrices. This only serves to emphasize that a drying procedure should be carefully thought out and agreed on by all parties involved in the analysis.

In developing a procedure, the analyst measures the weight loss over a time and repeats the measurements over enough time to establish a stable sample weight. A plot of the data forms a curve that gradually flattens with longer drying times. It is difficult to attain complete dryness for many materials and the analyst must use some judgment in deciding when the material is sufficiently dry to use. Ideally one would choose a drying time that would lead to such a small difference in sample composition from sample to sample that these differences are negligible compared with the analytical uncertainty. In nonideal cases, the analyst must compromise. Some methods may speed up the process, but in other instances the material itself may not dry in a complete or reproducible manner. The goal is not to achieve the same weight loss in all samples, but to achieve the same base percentage of residual water and other volatile materials in each sample. If this is done, all analysts will base their reports on the same sample basis. Methods that can be shown to yield the same base sample composition will therefore yield consistent analytical results.

Because the moisture composition of a sample may change with time, temperature, and elevation, it is important that weight loss not be the sole criterion used in evaluating a sample drying procedure. From this discussion it should also be obvious that no method of drying is more correct than another. Analyte loss by some drying methods may preclude drying the sample taken for analysis. However, a separate sample may be dried and used to correct the other sample to a dry weight basis. Analysts choose a procedure based on convenience, speed, availability, and other factors. Furthermore, drying is not the only choice facing the analyst. It may not be possible to dry some samples without changing composition or causing decomposition. In some circumstances, therefore, it may be better to investigate a procedure where samples can be equilibrated at a given relative humidity. Thus, one can produce a constant composition of water among all samples instead of attempting to dry all samples completely. The result is still the same because all analysts and laboratories may achieve the same base sample composition before analysis.

The only other major topic regarding the choice of a drying method is related to the type of analysis. In ultratrace analysis, the oven may be a substantial source of contamination and the analyst must carefully choose the drying apparatus. Nonmetallic devices such as desiccators (giving due attention to possible contamination from the dessicant) are best; corrosion-resistant devices are a distant second choice. Note that for trace element work, stainless steel devices are neither corrosion-resistant nor noncontaminating. An alternative for ultratrace analysis is to take the sample as is and then dry a separate specimen to provide a correction factor for the as is sample. This may increase one component of uncertainty in the analysis but it does avoid the problem of contamination and for some analytes contamination may be by far the greater cause of uncertainty of the analysis. In addition to contamination, another factor in selecting a drying procedure is analyte loss in the drying procedure. In such cases, samples should be taken as is and corrected to a dry weight basis based on the weight loss of a sample that is dried.

Sample Weighing

The term *weighing* is a misnomer because the mass of the unknown is actually compared with the mass of a known standard using the analytical balance. Historically, most balances were dual pan, and a number of procedures such as substitution weighing were used to overcome its shortcomings (such as unequal arm lengths). Today, most laboratories use single-pan or substitution balances. As the various masses are "added" in by twisting knobs, mass is actually removed from the balance arm. The mass removed is nominally the same as the mass of the object being weighed and the optical scale is used to read the last few digits of the sample weight. These automatic balances (as opposed to dual-pan balances) always have a constant load on the balance arm. Within its capacity, the automatic mechanical balance is one of the finest laboratory instruments ever devised. Load capacities range from kilograms (for large balances) to a few grams or less (for ultramicrobalances).

In the last decade, the electronic balance has become the instrument of choice in most modern laboratories. Electronic balances allow the inexperienced analyst to weigh samples easily with great precision. In addition, balances may be equipped with a digital interface to a laboratory computer. This eliminates the most common error in the use of the balance: improper transcription of the sample weight from the balance to the laboratory notebook. However, the electronic balance introduces a new set of problems. Whereas each individual weight in a mechanical balance has a calibration correction that must be used for careful work, the electronic balance must be carefully checked for linearity. In addition, whereas the mechanical balance changes calibration very slowly over time, the electronic balance calibration may change erratically with temperature.

When making a buoyancy correction, the calculation for an electronic balance is not always obvious. With a two-pan or a one-pan balance one is always comparing the sample with the balance weights. The buoyancy correction is based on the difference in the buoyant effects for the sample and the balance weights. For most balances, the weight calibrations are adjusted to give an apparent density for the stainless steel weights of 8.0. Because the electronic balances are user-adjusted for calibration, for linearity, or for both, the method of buoyancy correction is not a constant. Many electronic balances have a calibration check weight built in. If the masses are adjusted to their weight in the vacuum, there will be no correction for their buoyancy. If not, the buoyancy correction is needed for the weights, although they are not used in the sample weighing because they were used for the balance calibration. Consult the manufacturer for specific advice on the proper use of the internal calibration weight. Even with a properly operating electronic balance it is a good idea to have a calibrated weight set around to check the accuracy of the balance frequently. Most quality assurance schemes require this.

Due to their fragility, more care is given to the placement of mechanical balances in the laboratory. With the small size and portability of the electronic balance, much less care is used and usually it makes little difference. However, for semimicro (reads to 0.01 mg) and micro balances, proper placement of the balance is important. Even for top-loading and other balances that are less sensitive to vibration and other problems, it is still a good idea to become familiar with balance performance under various laboratory conditions. Building vibration is a problem for more sensitive balances. Some electronic balances have circuitry to integrate or average out the effects of vibration. Nevertheless, performance can be improved by eliminating the vibration. The easiest approach is to search for the "quietest" spot in the laboratory. In new laboratory construction, a separate building pillar, mechanically isolated from the rest of the building, is the best way to ensure low vibration for critical measurements. Large-mass balance tables with elastomeric isolators can minimize many vibration problems when physical alterations to the building are not feasible. Vibration can lead to erratic and irreproducible weighings.

Ideally, the balance has a stable temperature and humidity environment and is shielded from air currents, direct sunlight, and the body heat of the operator. Stable temperature and humidity prevent drift due to changing balance arm lengths or inadequate temperature compensation circuitry. The best temperature and humidity are those that stabilize the sample. At low humidity, powders acquire a static charge that makes them difficult to handle. Likewise, hygroscopic materials are difficult to handle at high humidity. A humidity range that we have found desirable for general-purpose laboratories is 45 to 55% relative humidity. Glove boxes with environmental controls should be considered when normal environmental conditions in the laboratory are not compatible with the sample. Balances may be operated easily inside an environmental chamber. In addition, the chambers eliminate any operator exposure to toxic or dangerous samples. Direct sunlight and operator body heat may affect sensitive balances. Obviously, such balances should be removed from heat sources or air-conditioning outlets in the laboratory. Thermal effects are seen in the slow drift of the balance zero point or tare point over time. Sample replicate weighings are less precise under conditions of thermal drift.

Samples are almost never weighed directly on the balance pan. Most sample weighings are done by difference. That is, a weighing bottle, boat, foil, scoop, or other device is first weighed with the sample in it. The sample is then transferred to a beaker and the weighing container is re-weighed. The difference is the amount of a sample transferred, assuming that the operation was quantitative (no spills). Gravimetric dilutions are easily made using a precision top-loading balance to record the weights of the empty bottle, the sample added, any other additions, and the final weight after dilution with solvent. Moderately hygroscopic materials may be weighed in a weighing bottle by quickly adding the sample on a top-loading balance to obtain an approximate weight. When closed, the bottle may be reweighed more accurately. The sample can then be transferred and the weighing bottle reweighed. Except for the most hygroscopic materials, this procedure can be applied to many moisture sensitive materials.

Sample Dissolution

Sample dissolution techniques are divided roughly into extraction and total dissolution techniques. For some analyte analysis, sample preparation often consists of getting the sample into a form that allows easy extraction of the analytes. This may involve macerating, lyophilizing, grinding, or other techniques designed to reduce the particle size of the sample in order for the solvent to come into contact with all parts of the sample. Extraction procedures are simple and rapid compared with total dissolution procedures. For most inorganic analytes, simple solvents cannot leach out trace elements, so most inorganic analysis involves a total sample dissolution. However, this does not mean that solvent extraction or simple dilute acid extraction cannot retrieve a significant amount of the analyte. In fact, such procedures are the basis of many regulatory requirements that require the measurement of leachable elements.

Here, we are concerned with the more quantitative process of a total sample dissolution. In both major and trace element analysis, sample dissolution is the digestion or mineralization of a sample to render it soluble and to destroy organic matter or other matrix material that may interfere with the recovery of the analyte. There are three types of sample dissolution procedures used. The oldest technique is dry ashing. Combustion techniques are closely related except that they are usually conducted in a bomb or combustion tube. Whether dry ashings are done in a furnace or with a more modern plasma source, the technique depends on high temperature and oxygen to convert the sample into more soluble oxides. Various chemicals may be added as ashing aids that either speed up the process or help to fully oxidize the sample. Dry ashing techniques may not always be advisable because there is often a loss of volatile elements such as selenium and mercury. Depending on the temperature there may even be a loss of elements such as sodium or potassium that have some volatility at the temperatures required for complete dry ashing.

Oxidative fusion procedures are the most powerful techniques for dissolving difficult sample types such as refractories. A large amount of the flux material must be mixed with the sample before the fusion reaction. These reactions may be carried out in platinum, zirconium, or other inert containers. Oxidative fusions are generally not advisable for trace analysis because even the highest-purity fusion mixture will yield microgram levels of contaminants per gram of fusion mixture. Other, possibly higher levels of contamination can come from the fusion vessel and the furnace. In addition, the amount of flux material (up to 10 to 20 g) required per gram of sample means that there will be significant additional elements present in the final sample that may produce significant interferences. Because of the analytical blank problems and foreign matrix elements introduced by oxidative fusion, the technique is usually used as a last resort. For some techniques such as X-ray fluorescence, however, this may be an ideal sample preparation technique because it yields a uniform solid solution that can be reproduced from sample to sample.

The last category of dissolution procedures is wet oxidation. Wet oxidation techniques involve the use of the mineral acids (usually hydrochloric, nitric, perchloric, sulfuric, and hydrofluoric acids) and heat. They may be performed in open or closed systems with as much heat as can be tolerated by the container. Wet dissolution techniques are very old and little has been learned about them in the latter half of the twentieth century. Sample digestions are usually performed using combinations of acids to achieve a complete dissolution. For example, the combination of hydrogen peroxide and nitric acid is a useful technique for the quick oxidation of organic samples. Hydrogen peroxide traditionally has been avoided for trace element analysis because of the high impurity levels in commercial hydrogen peroxide. Newer semiconductor grades of hydrogen peroxide, however, eliminate these impurities and, therefore, a mixture of hydrogen peroxide and nitric acid is a useful technique for the rapid oxidation of most organic samples. In most of these procedures, excess nitric acid is required.

Another combination, nitric and sulfuric acid, is a good dehydrating and oxidizing mixture. It is necessary to test for the completeness of oxidation by adding small amounts of nitric acid before copious fumes of sulfuric acid develop. Darkening of the digestion solution indicates the need to add additional nitric acid. One disadvantage of this combination is that it could form insoluble sulfates such as barium sulfate. Another serious disadvantage for many atomic spectroscopic techniques is that a sulfate matrix leads to spectral interferences for some elements. Sulfuric acid provides a very-high-temperature reaction mixture, making it difficult to use in Teflon[1] ware. A high temperature, however, is one advantage of the nitric acid–sulfuric acid combination because it leads to rapid oxidation.

One of the most effective combinations for trace element analysis is nitric and perchloric acids. Use an excess of nitric acid and test for completeness of oxidation before evaporating to fumes of perchloric acid. Dark brown to black solutions are an indication of incomplete oxidation and signal the need for more nitric acid. This can be added dropwise until the solution decolorizes. Despite its hazard potential, perchloric acid is one of the most useful reagents for sample preparation today. The hazard potential of perchloric acid use is greatest for service personnel who must work on inadequate fume exhaust systems. Perchloric acid is both rapid and complete with respect to oxidation and solubilization of the sample. Among acids, perchloric acid produces the smallest quantity of insoluble salts. Potassium perchlorate is an exception because it is only sparingly soluble in cold water. Hydrofluoric acid may be used in combination with any of the above mixtures when glasses or minerals are present in the sample. Obviously, when hydrofluoric acid is used, a plastic vessel made of a material such as Teflon must be used. Both hydrofluoric acid and perchloric acid have safety risks that the analyst should be prepared to handle; however, because both are extremely useful reagents, their analytical advantages are often persuasive.

Wet digestions may be done in open beakers or test tubes under total reflux using an Erlenmeyer flask fitted with a condenser, or in pressure vessels often called bombs. Another variant is the Carius technique, in which the sample is sealed in an ampule. This is placed in a pressure-tight steel jacket (often with dry ice to provide external pressure around the ampule), and then heated in an oven to several hundred degrees. Several commercial apparatuses are currently available that allow the attainment of high pressure during the dissolution. The combination of high temperature and pressure allows very rapid dissolutions, although with greater risk than open digestions.

In ultratrace analysis, most wet digestions should be performed in Teflon or quartz containers to reduce contamination. Teflon ware is available in Teflon pertetrafluoroethylene (PTFE), Teflon fluorinated ethylene-propylene (FEP), and Teflon perfluoroalkoxy (PFA). The PTFE material is white and opaque and is made under pressure and heat by pressing powder into the shape of the article. It is not recommended because it is somewhat porous. It does have the high-

---

1. Certain commercial equipment, instruments, and materials are identified in this report to specify adequately the experimental procedure. Such identification does not imply endorsement by the National Institute of Standards and Technology, nor does it imply that the materials or equipment identified are necessarily the best available for the purpose.

est service temperature of any fluoropolymer at 260 °C. Teflon FEP is a translucent and colorless melt polymer usually used to make laboratory apparatus by injection or blow molding. Physically, Teflon PFA is similar to Teflon FEP except that it has a higher temperature limit of about 250 °C compared with Teflon FEP's limit of about 205 °C. Other fluoropolymer resins are available, such as polyvinylidenedifluoride (PVDF), but they are not commonly available in the form of beakers or flasks. Consult labware catalogs for recommended temperature limits for particular pieces of fluoropolymer labware, as these may vary somewhat from literature values. All fluoropolymers may deform at high temperatures. At excessive temperatures they also melt, char, and release corrosive fluorine vapors.

For perchloric acid digestions for trace analysis, Teflon beakers of 30 ml to 100 ml are used with Teflon tetrafluoroethyl (TFE) covers in place over the beaker. Initially the temperature is increased slowly and, with organic materials in particular, some foaming may occur. However, as the foaming subsides, the temperature can be increased and the cover gradually removed to allow water and oxides of nitrogen to escape. The maximum digestion temperatures are limited by the type and heat dissipation characteristics of the Teflon ware used. A full beaker can withstand a higher temperature than one that is empty or nearly dry. Gradually, evaporation leads to a concentrated perchloric acid mixture and at this point the cover should be replaced for a time to enhance the action of perchloric acid. Finally, it is removed to permit the evaporation of excess perchloric acid. Any darkening of the reaction mixture as it evaporates indicates incomplete oxidation. A few drops of nitric acid should be added as a precaution to ensure complete oxidation of the sample and to prevent rapid oxidation by the perchloric acid that could lead to the loss of the sample. Finally, the Teflon ware should be watched as the sample evaporates to dryness to be sure that the Teflon ware does not stick to the hotplate (temperature setting too high). Remember that in a wet digestion a higher-boiling acid cannot be displaced by a lower-boiling acid. Thus, even if hydrochloric acid were used as the dissolution medium, the sample could be converted to nitrate salts by adding nitric acid and slowly evaporating. However, if sulfuric acid were used for the dissolution, no amount of treatment with nitric acid would convert the salts to nitrates. Higher-temperature reactions require quartz or glass vessels but do not permit the use of hydrofluoric acid.

A good fume hood is necessary for all wet digestions. Typically, fume hoods are the dirtiest part of the laboratory. Laminar flow clean air fume hoods with class 100 or class 10 clean air are available for use and strongly recommended. For trace analysis, all components of the fume hood should be metal-free to avoid contamination from corrosion products. Another effective but less desirable approach for removing acid fumes is to use a fume eradicator supplied with filtered nitrogen and exhaust the system through a water pump. The main disadvantage of the open beaker procedure is slowness. Open digestions do not handle difficult-to-dissolve samples as well as a pressure system. Open digestions are best done overnight, thus neutralizing much of the speed disadvantage. It takes a few hours to a day or more to dissolve more resistant samples completely.

One obvious way to speed up the reaction is to increase the temperature or pressure, or both, and closed digestion systems have been developed for that purpose. These may be the newer Teflon bombs or they may be old techniques such as the Carius tube technique. There are also modern techniques such as the microwave digestion oven and closed pressure systems that do not require an ampule. Regardless, vessels can operate at pressures up to 100 kPa and temperatures up to at least 300 °C. Under these conditions many samples can be completely digested in under an hour using only hydrochloric or nitric acid. The major difficulty is that organic sample loads must be small, typically less than 100 mg, because of the quantity of gas produced during oxidation. The unsealing process can also be dangerous because of the pressure in the device during oxidation. When a commercial high-pressure apparatus is used, the manufacturer's instructions should be followed exactly for purposes of safety.

Microwave digestion systems have become popular as a means of speeding up or automating the digestion procedure. Microwave systems may be either open or closed, and there are also flow-through or flow-injection systems for adaptation to on-line analysis. Microwave-digested samples may not be as fully oxidized and may require additional treatment on a hotplate. Whether digested by microwave systems or by high-pressure systems, most of the excess acid is usually removed by evaporation. The laminar flow clean air fume hood is the best place for this evaporation.

For trace element analysis, the apparatus and digestion conditions are usually chosen to produce the lowest analytical blank or contamination. Therefore, Teflon ware is the preferred apparatus in use today. However, an equally good choice for many chemists is quartz ware. Although fragile and more expensive, quartz apparatus can be heated to a higher temperature than plastics or glass. Quartzware may have lower blanks for some elements and have less risk for sample cross-contamination than plastic labware. A cross contamination problem occurs with plastic vessels when they are used for different types of samples from one analysis to another. Often, there is some carry over of major elements from the prior sample when using plastic labware. The best procedure is to dedicate a group of containers and apparatus for each type of sample to be used in the laboratory.

Many commercial high-purity acids are available for ultratrace analysis. These should be used instead of American Chemical Society (ACS) analytical reagent grades if the blank requirements are extremely low. Besides high-purity acid, the chemist should have scrupulously clean apparatus and clean air facilities available for sample manipulation, digestion, and evaporation. Similarly, for the organic chemist, extremely pure grades of solvent are available today for sample extraction and dilution. Of course, other procedures should be tailored as necessary to avoid contamination. These procedures depend on the type of sample being analyzed.

Extraction, Preconcentration, and Separations

As the focus of analytical chemistry has changed from macro to trace analysis, the role of the sample preparation chemist has changed with it. In addition to, or even instead of a sample dissolution, the analyst may be called on to perform an extraction or preconcentration of the sample. There are two complementary goals for these procedures. The first is to improve sensitivity by increasing the mass of analyte available for analysis. For instance, if a sample had 10 ng of copper per gram of sample and a relatively large amount of sample were available, then an amount as large as 10 g could be dissolved. The sample digest could be selectively extracted to yield the 100 ng of copper concentrated into 1 g (or less) of solution after the extraction. Preconcentration by a factor of 10 to 100 or more is often used in analytical chemistry to improve sensitivity. Preconcentration procedures may involve extraction with a chelating agent in a solvent, ion exchange, precipitation, scavenging, electrodeposition, or other procedures. Often, an additional goal of preconcentration is the elimination of part of the matrix. The sample matrix often limits the achievable detection limit by the instrumental technique (spectral or chemical-combination) and interferes with the determination of the analyte element. Thus the elimination of the matrix is usually desirable.

Extraction and preconcentration are really subsets of the more general technique of separations and so many texts have been written on these techniques that little detailed information is given here. Two competing problems require the analyst's attention. The first is the sample yield or the recovery of the analyte through the various stages of extraction and preconcentration or separation. When the analyte yield drops below 99%, obviously the procedure is not quantitative. The other problem is that of contamination and the analytical blank.

The analytical blank for a method may be defined as the sum of all analyte contributed by the chemical procedures to the sample measurement. The analyte measured is the analyte present in the sample plus the analyte contributed by the sample preparation. Instrumental backgrounds or noise may produce similar problems but they are not the same. High blanks obscure trace measurements and create large uncertainties in the measurement. Therefore, the analyst sometimes has to balance high yields against high blanks. For practical instrumental analytical procedures, therefore, a compromise between these factors is necessary. A good sample preparation procedure has both high recovery of the analyte and low blank levels under the conditions of usage. Consistency may be more important than either a low blank or high recovery if that consistency leads to better statistical confidence in the resulting data.

Despite these advantages, the major disadvantage of separation procedures is that they require considerable analyst time. Few fully automated procedures exist for inorganic separations other than for ion chromatography or flow-injection analysis. Most organic chromatography procedures are available in fully automated capacities. For inorganic samples, however, most of the attention of the analyst has been toward using instrumental techniques to minimize or eliminate deficiencies in sample preparation. Most spectrometric techniques require little sample preparation other than dissolution and dilution. Thus far, we have seen that proper sample preparation for analysis can improve the sensitivity or the selectivity of the analytical procedure. Likewise, it is also possible to improve the precision of the procedure by providing more analyte to the instrument. Precision may also be improved by removing interferences that may lead to erratic positive or negative analytical errors.

The future of sample preparation is not clear, but it is apparent that the trend in laboratory instrumentation is toward fully automated systems including sample preparation. Such automated preparation could improve analytical precision. For others, it may limit choices to those available on the menu. Some manual sample preparation is likely to survive just to handle difficult samples or to provide the analyst with greater options for sample preparation. Of course, designers of automated sample preparation systems need a good understanding of the processes that are being automated.

Dilution and Matrix Matching

After a sample dissolution or a chemical separation, the analyte may not be in the correct chemical or physical state for presentation to the instrument. Each instrument requires a particular type of sample form, solvent, oxidation state, concentration, and so forth. Therefore, the analyst must modify the sample to meet the requirements of the instrumental technique. To change oxidation state, an appropriate oxidizing or reducing agent must be used. Perhaps the pH must be adjusted to some critical value. Both acetic acid and ammonium hydroxide of high purity may be prepared and together they make a useful buffer with a low analytical blank.

For trace metal analysis by spectrometric procedures, it is often desirable to present the sample to the instrument in a dilute solution of nitric acid (for instance, to avoid mass interference problems with chloride solutions by inductively coupled plasma–mass spectrometry, or ICP–MS). Samples dissolved in HCl are evaporated to dryness under a clean environment and then redissolved in nitric acid and diluted to the desired concentration of nitric acid. The analyst might also wish to add an internal standard at this point. The easiest approach is to add the internal standard at the desired concentration to the acid used to redissolve the sample. In addition, there is usually a target range of optimum concentration for the instrumental technique. This ensures that the analyte is within the linear part of the dynamic range of the instrument where it has both adequate sensitivity and precision. Dilution may be used to adjust the sample concentration downward. At very low analyte concentrations, there are problems associated with the instrumental limits of detection. These include a degraded signal-to-noise ratio and, therefore, poorer analytical precision. The only noninstrumental way to avoid this problem is to get more analyte (a larger sample) or increase the concentration by decreasing the sample volume. The lower limit of this approach is determined by the minimum sample volume needed for analysis by the instrument.

Preparation of Standard Solutions

In theory, few things are more simple than the preparation of a standard solution. All that is required is to dissolve a known amount of analyte and dilute to the appropriate volume or mass in a solvent to prepare the desired concentration of the analyte. In practice, however, this is more difficult than it appears. In preparing analytical standard solutions, starting with high-purity materials is desirable. By avoiding impurities we accomplish two things. First, we eliminate the potential for interferences in the analysis. Second, by having an extremely high-purity material we have greater confidence in the amount of the element taken.

There are two ways to determine the purity of a material. The first is to determine all of the impurities and then subtract from 100%. By difference we have the purity factor for the element. The second way is to determine the element concentration analytically, which is a much more difficult proposition. Consequently, especially for metals, the purity of materials is expressed as a number of nines. For instance, *five nines pure* means 99.999% pure. Most commercial spectrometrically determined certificates of impurities do not take into account some elements such as carbon, hydrogen, oxygen, and nitrogen that can exist as water, hydrocarbons, or gases dissolved in the metal. Thus, 1 g of a pure iron rod that contains 10 µg of metallic impurities by spectrometric analysis could have 20 or even 100 µg of oxygen present. Therefore, the estimation of the purity of the element by subtraction of the measured impurities from 100% can provide inaccurate results, but it is normally the only method available to the analyst.

In the case of metals, the physical form is of concern. A sample taken from a metal rod is better than a sample taken from a metal form that has a larger surface area such as a foil, powder, or smaller sizes of shot or flake. The ratio of surface impurities to total mass is much better for the rod than for other physical forms. A rod can be cut to the desired length to produce a specimen of the nominal mass required. Before weighing and dissolution, the metal should be cleaned by etching with acid, rinsed with water, and then quickly dried.

If one does not use a pure metal, then the other source available for obtaining an element in solution is a compound. Most chemical compounds have the disadvantage that they are somewhat nonstoichiometric. Stoichiometry is the chemical composition of the compound—in other words, the elemental formula. Purity statements may refer to the assay of the compound with respect to one element or the purity may be determined by the method described for metals. Very few compounds, including certified reference materials, are assayed for every element present. For instance, the chloride in KCl is usually assayed and the potassium is taken to be equivalent to the chloride concentration, although this is not always necessarily so. The ACS Committee on

Analytical Reagents now specifies the element or property assayed for ACS Reagent Grade chemicals. There is no consistent method used in industry for assays or purities, so the analysts should be aware of this and ask enough questions of the vendor to satisfy their needs.

To prepare an elemental solution, one carefully weighs out an amount of a pure, dry compound or a pure, clean metal and calculates the dilution needed. The balance used must have the necessary precision and accuracy to suit the accuracy goals of the standard solution being prepared. Now the material can be dissolved in the appropriate acid and diluted to the required volume either volumetrically or gravimetrically. Note that these steps require an attention to detail that may not be known to the average analyst. Thus, it is no surprise that for most analysts, the preparation of standard solutions is left to commercial sources. The analyst can ensure quality of analytical measurements by preparing control charts of calibrant solution concentration. These may be prepared by using one old lot as the control and comparing all new lots with the control lot. New lots of calibrant that do not agree with the control lot should be discarded.

Analytical methods may require that the analyte standard be prepared in a sample matrix that closely simulates the sample as far as the major elements are concerned. This can complicate matters considerably because trace amounts of analytes will be present from impurities in the major elements used for the matrix matching solution, especially where the ratio of elements is extreme (10^6:1 or higher). The difficulty of preparing exact matches of matrices by gravimetric or volumetric dilution of known elemental standards becomes complex. Under these conditions the actual concentrations may not be exactly what was calculated from the concentrations of the starting solutions. This may be due to contamination or a partial loss of analyte through precipitation, adsorption, or other processes.

Standard solutions are also used in the analysis of samples by the method of standard addition. We need to know accurately the concentration and quantity of the standard added and the amount of the sample to which the standard is being added. Given these quantities and the instrument response for each solution, we can compute the concentration of the element in the unknown sample. This technique depends on the accuracy with which these additions and solution preparations can be made, and it requires a linear or known response of the analytical system to the interferent.

Once prepared, standard solutions do not have an indefinite lifetime. Organic compounds have many possible degradation paths, including bacterial. Sterile solutions are often used as one way to extend the lifetime of solutions. Low storage temperatures are also used to help stabilize solutions. For simple solutions of most cations, the most effective solution is to ensure that the acidity is at least 1 or 2% acid. Most inorganic species may be held at concentrations of 1 mg/l to 1 µg/l indefinitely if the acidity is high enough. Of course, there are other issues in the storage of solutions such as contamination. Glass containers do not breathe or transpire solvent readily, but glass slowly dissolves in solvents, leading primarily to the contamination of the solution with the most abundant elements in glass (Si, K, Na, Ca, Mg, and Al). How much contamination depends on the storage time. Quartz is a more nearly ideal container because it is essentially pure silica and the only significant impurity leached out over time is silica.

Polyethylene containers are the most commonly used for solution storage for inorganics, whereas glass is more typically used for organic solutions. The major disadvantage of polyethylene is the transpiration loss through the vessel walls, which can be about 0.25% of the total content per year. Teflon and other fluoropolymers reduce the transpiration loss but do not eliminate it. If the solution must be stored for a long time or if the acid concentrations must be particularly high (as much as 30% for concentrated Ti solutions), then the fluoropolymer containers are preferable because they are not attacked by the solutions or acids. Polyethylene containers may show some embrittlement under adverse conditions. Storing solutions in a humidity cabinet at about 70% relative humidity greatly extends the storage life of the solutions. Taring the containers is another way to track solution losses and to decide when to prepare a new solution.

Advantages of Sample Preparation

Sample preparation can be used to achieve a number of ends that usually cannot be reached by instrumental means. For example, there are limits to how much interference may be present before the accuracy of the instrument is adversely affected. Separations or extractions can be used to isolate the analyte and remove the interferent from the sample. Similarly, careful sample preparation may be used to compensate for the inherent limitations of a particular instrument when the more advantageous instrument is not available. The same analyte isolation procedures may be used to preconcentrate the analyte and increase the concentration of analyte measured by the instrument. These approaches may be used to overcome shortcomings of particular instrumentation, but the analyst should also be able to use these procedures as a means of extending the state of the art.

Chemical reactions may also be used to create new analytical methods. Colorimetric methods are quite old but are still commonly used. Relatively few chemicals are strongly colored or fluorescent. However, one can use chromophores, complexation reagents, or other chemicals to produce colored, fluorescent, or UV absorptive compounds that form the basis for an analysis. For example, most rare earth elements are difficult to detect by an ion chromatography or high-pressure liquid chromatography system. When an appropriate reagent is pumped and mixed into the sample stream after the analytical column, a strongly absorptive species is formed in the UV or visible wavelengths. Flow injection analysis systems also may depend on the mixing of an appropriate reagent to produce a detectable analytical signal. Not only are these chemical reactions desirable, but the success of a number of analytical systems depends on them. To work well, the systems must be optimized, bearing many of the principles of quantitative analysis in mind.

Conclusions

We have tried to show, through a series of examples of analytical operations, that the historical basis of analytical chemistry now called wet chemistry has not disappeared in the instrumental age of analysis. Although the examples are not all-inclusive, we hope that the few examples presented will give today's analyst a starting point. We intended to suggest the care and philosophy behind the development of a successful sample preparation procedure. Many of the Suggested Readings that follow may be old or dated in many respects, but this is because there has been little emphasis on these subjects in modern times. Obviously, sample preparation procedures for every chapter are beyond the scope of this book. However, we hope that readers will dissect their analytical procedures and use this chapter as a reference source for evaluating some steps of the procedure and as a guide for thinking critically about the other steps.

Suggested Readings

Classical Analysis, General Analytical

Epstein, M. S., and J. R. Moody, Jr, "Definitive Measurement Methods," *Spectrochim Acta*, 46B (1991), 1571–5.

Furman, N. H. ed., *Standard Methods of Chemical Analysis, The Elements*, 6th ed. New York: Krieger, 1975.

Hamilton, L. F., and S. G. Simpson, *Calculations of Analytical Chemistry*, 6th ed. New York: McGraw-Hill, 1960.

HAMILTON, L. F., AND S. G. SIMPSON, *Quantitative Chemical Analysis*, 12th ed. New York: Macmillan, 1963.

HILLEBRAND, W. F., G. E. F. LUNDELL, H. A. BRIGHT, AND J. I. HOFFMAN, *Applied Inorganic Analysis*, 2nd ed. New York: Wiley, 1953.

KODAMA, K., *Methods of Quantitative Inorganic Analysis*. New York: Interscience, 1963.

KOLTHOFF, I. M., V. A. STENGER, R. BELCHER, AND G. MATSUYAMA, *Volumetric Analysis*, 3 vols. New York: Interscience, 1942, 1947, 1957.

KOLTHOFF, I. M., E. B. SANDELL, E. J. MEEHAN, AND S. BRUCKENSTEIN, *Quantitative Chemical Analysis*, 4th ed. New York: Macmillan, 1969.

LAITENEN, H. A., "History of Trace Analysis," *Journal of Research National Bureau of Standards*, 93 (1988), 175–85.

LAITENEN, H. A., AND G. W. EWING, EDS., *A History of Analytical Chemistry*. Washington, D.C.: The Division of Analytical Chemistry of the American Chemical Society, 1977.

LUNDELL, G. E. F., "The Chemical Analysis of Things as They Are," *Industrial Engineering Chemistry Analytical Edition*, 5 (1933), 221–5.

LUNDELL, G. E. F., AND J. I. HOFFMAN, *Outlines of Methods of Chemical Analysis*. New York: Wiley, 1938.

SZABADVÁRY, F., *History of Analytical Chemistry*. New York: Pergamon Press, 1966.

SZABADVÁRY, F., AND A. ROBINSON, *The History of Analytical Chemistry*, in G. Svehla, ed., *Comprehensive Analytical Chemistry*, vol. X. New York: Elsevier, 1980.

TREADWELL, F. P., AND W. T. HALL, *Analytical Chemistry*, 9th ed., vol II, *Quantitative Analysis*. New York: Wiley, 1942.

Sample Dissolution

BOCK, R., *A Handbook of Decomposition Methods in Analytical Chemistry*, trans. by L. L. Marr. New York: Wiley, 1979.

GORDON, C. L., W. G. SCHLECHT, AND E. WICHERS, "Use of Sealed Tubes for the Preparation of Acid Solutions of Samples for Analysis, or for Small-Scale Refining: Pressures of Acids Heated Above 100 °C," *Journal of Research National Bureau of Standards*, 33 (1944), 457–70.

KINGSTON, H. M., AND L. B. JASSIE, EDS., *Introduction to Microwave Sample Preparation, Theory and Practice*. Washington, D.C.: American Chemical Society, 1988.

MATUSIEWICZ, H., AND STURGEON, R. E., "Present Status of Microwave Sample Dissolution and Decomposition for Elemental Analysis," *Progress in Analytical Spectroscopy*, 12 (1989), 21–39.

SMITH, G. F., *The Wet Chemical Oxidation of Organic Compositions Employing Perchloric Acid*. Columbus, OH: G. Frederick Smith Chemical Co., 1965.

ŠULCEK, Z., AND P. POVANDRA, *Methods of Decomposition in Inorganic Analysis*. Boca Raton, FL: CRC Press, 1989.

ŠULCEK, Z., P. POVANDRA, AND J. DOLEŽAL, *Decomposition Procedures in Inorganic Analysis*, in CRC *Critical Reviews in Analytical Chemistry*, vol. 6, issue 3. Boca Raton, FL: CRC Press, 1977.

WASHINGTON, H. S., *The Chemical Analysis of Rocks*, 4th ed. New York: Wiley, 1930.

Sample Handling, Analytical Theory, and Separations

BERG, E. W., *Physical and Chemical Methods of Separation*. New York: McGraw-Hill, 1963.

MA, T. S., AND V. HORAK, *Microscale Manipulations in Chemistry*, in P. A. Elving and J. A. Winefordner, eds., *Chemical Analysis*, vol. 44. New York: Wiley, 1976.

MINCZEWSKI, J., J. CHWASTOWSKA, AND R. DYBCZYNSKI, *Separation and Preconcentration Methods in Inorganic Trace Analysis*, trans. by M. R. Masson. Chichester, U.K.: Horwood, 1982.

MIZUIKE, A., *Enrichment Techniques for Inorganic Trace Analysis*. Berlin: Springer-Verlag, 1983.

MOODY, J. R., "NBS Clean Laboratories for Trace Element Analysis," *Analytical Chemistry*, 54 (1982), 1358–76A.

MOODY, J. R., R. R. GREENBURG, K. W. PRATT, AND T. C. RAINS, "Recommended Inorganic Chemicals for Calibration," *Analytical Chemistry*, 60 (1988), 1203–18A.

PIERCE, W. C., D. T. SAWYER, AND E. L. HAENISCH, *Quantitative Analysis*, 4th ed. New York: Wiley, 1958.

WALTON, H. F., *Principles and Methods of Chemical Analysis*, 2nd ed. Englewood Cliffs, N.J.: Prentice-Hall, 1964.

Sampling

ASTM E-300, *Standard Recommended Practice for Sampling Industrial Chemicals*. Philadelphia: American Society for Testing and Materials, 1973 (reapproved 1979).

KRATOCHVIL, B. G., AND J. K. TAYLOR, "Sampling for Chemical Analysis," *Analytical Chemistry*, 53 (1981), 924–38A.

KRATOCHIVIL, B. G., AND J. K. TAYLOR, *A Survey of the Recent Literature on Sampling for Chemical Analysis*, NBS Technical Note 1153. Gaithersburg, MD: National Bureau of Standards, January 1982.

MOODY, J. R., "The Sampling, Handling and Storage of Materials for Trace Analysis," *Philosophical Transactions Royal Society, London section A*, 305 (1982), 669–80.

Chapter 4

Quantitative Measurements

Earl Wehry

University of Tennessee Department of Chemistry

To make an informed choice of instrumental method for solving a specific problem in quantitative chemical analysis, certain key issues must be addressed. Three especially important concerns are limits of detection; calibration; and interferences, contamination of samples, and loss of analyte. A quantitative analytical procedure may involve several steps before the actual measurement including some or all of the following: acquisition and storage of the sample, dissolution of the sample, removal of interferences, and preconcentration of the analyte. (See Chaps. 2 and 3.) The performance of the preliminary steps, rather than that of the quantitative measurement itself, may ultimately determine the quality of the results that are obtained.

Limit of Detection, Limit of Quantification, and Sensitivity

A technique is obviously useless for the task at hand if it cannot detect the analyte in the sample. Consequently, a major criterion in the choice of technique is the anticipated ability of each method under consideration to detect the analyte at the levels expected in the sample under study.

The question arises as to how one can be sure that, in any specific measurement, the analyte has actually been detected. This is not always an easy question to answer (1–3). The limit of detection (LOD) of a technique can be conceived of as the smallest concentration, or amount, of analyte that can be established as being different, at a reasonable statistical confidence level, from a blank (a material similar in composition to the sample except that the analyte is absent) (3–5). Several definitions of *limit of detection* (LOD) are used in the literature, and authors do not always clearly specify what definition they use. One common definition of *LOD* is the concentration (or quantity) of analyte that produces a signal that exceeds the signal observed from a blank by an amount equal to three times the standard deviation for the measurement on the blank (5, 6). Because the standard deviation depends on the number of replicate measurements (N), the LOD also depends on N; the International Union of Pure and Applied Chemistry (IUPAC) recommends use of $N = 20$ for the blank, assuming that a true analytical blank is used and the measurement conditions for the sample and blank are as nearly identical as possible (6). For more detail, see Long and Winefordner (7), Boumans (4), and Ingle and Crouch (3).

Due to the statistical nature of a detection limit, there are occasions when an analyte that actually is absent is reported as present (a false positive), and other situations when an analyte that is present is reported as not detected (a false negative) (8). The probability that a false positive or false negative will occur depends on the definition of *limit of detection* that is used (8, 9). To a certain extent, therefore, the appropriate definition of *LOD* in a given situation depends on whether greater harm is likely to be caused by false positives or false negatives.

At analyte concentrations at or near the LOD, the precision of the measured data is usually poor. Thus, it is inadvisable to attempt to quantify the analyte unless its concentration is well above the LOD. A limit of quantification, defined as the analyte concentration for which the signal exceeds that for a realistic analytical blank by 10 times the standard deviation, is often specified as the smallest analyte concentration that one should attempt to quantify (5). The interpretation of analytical data that fall below the limit of quantification and limit of detection is discussed by Helsel (10).

Many chapters in this book give "ballpark" values of LODs for the various techniques. Though useful (especially for comparing the performances of different methods), estimates of LODs must be approached with caution. Most reported LODs are obtained using instrumentation functioning at peak performance levels, operated by experienced personnel working in highly controlled settings, usually with very clean samples.

Moreover, the LOD of an analytical procedure may depend on factors other than the characteristics of the measurement itself, especially if the sample is very complex and multistep pretreatment of the sample is required before the actual measurement is made. A complex sample that exhibits a large blank signal (and therefore presumably contains one or more additive interferences) is usually more problematic than a clean sample. The LOD for an analytical procedure can be compromised severely if either loss of analyte occurs before the actual measurement or the sample is contaminated before the measurement. On the other hand, the use of preconcentration techniques (11) before the measurement step can greatly improve the effective LOD for a procedure. The entire sequence of operations, from procurement of the initial sample up through the measurement step, not just the characteristics of the measurement step itself, ultimately determines the LOD that may realistically be expected.

When these facts are considered along with the intrinsically statistical definition of a detection limit, it becomes evident that the limit of detection "is not a well-defined concentration or amount, but instead reflects a relatively wide region of uncertainty" (2, 12). (The public tends not to be aware of this fact, occasionally with unfortunate consequences.)

The LOD for a method can be stated in terms of either concentration or absolute mass of analyte (12). For methods that require only a very small volume of analyte, the LOD in *absolute mass* units tends to be very small even if the LOD expressed in *concentration units* is relatively large. (Electrothermal atomic absorption spectrometry is a classic example, although—as discussed in Chap. 21—this is a technique that exhibits low LODs in concentration units as well as absolute mass units.) Whether the more relevant units for the LOD are concentration or absolute mass depends on the nature of the application under consideration. In most instances, the LOD in concentration units is the more useful figure of merit, unless the quantity of available sample is severely limited.

Methods that can detect small concentrations of analytes often are called "sensitive" methods. Strictly speaking, sensitivity is the manner in which a change in analyte concentration produces a change in the resulting signal (13). It is possible for a method to be very sensitive but to exhibit relatively poor limits of detection (if, for example, it is very unselective and thus prone to produce large signals for blanks). On the other hand, an insensitive method is very unlikely to exhibit a low LOD. Thus, sensitivity is a necessary but not sufficient condition for the achievement of a low LOD.

Calibration

Calibration is the process of determining the relationship between a measured signal and the concentration of analyte that generated the signal. This is an especially important issue when the species that generates the signal is not the analyte itself, but some other entity produced by chemical transformation of the analyte. For example, in atomic absorption, emission, and fluorescence spectroscopy, the analyte, which is usually an ion in a solution or solid sample, is converted to an atom in the gas phase; the atom is then detected. In inductively coupled plasma–mass spectrometry (ICP–MS), the analyte is converted into a gas-phase ion. Any variable that affects the efficiency of the processes by which the analyte undergoes conversion to gas-phase atoms or ions alters the signal observed for a given concentration of the analyte element.

A similar situation may exist in molecular UV-visible absorption and fluorescence spectrometry (especially the latter). If the analyte does not absorb or fluoresce strongly, it may be "derivatized" (that is, chemically converted to a species that absorbs or fluoresces more strongly). In chemiluminescence, the species that produces the signal is almost never the analyte itself. Instead, the analyte is converted via a chemical reaction to a species that emits light, or else acts as a catalyst of such a reaction.

In X-ray fluorescence (XRF), the analytical signal is virtually always generated by the analyte itself, rather than some other species produced via chemical transformation of the analyte. However, as discussed in Chap. 25, this fact does not ensure that calibration of signals in XRF is a trivial process.

Accurate calibration is a prerequisite to obtaining accurate quantitative results. Automation of instruments has tended to simplify calibration procedures, but has also made it easier for the unwary analyst to use inappropriate calibration methods. Calibration procedures for specific techniques are discussed in the individual chapters; general discussions of the issue are also available (14, 15).

Interferences, Contamination, and Loss of Analyte

An interference (or interferent) is any constituent of a sample that acts in such a way as to complicate the process of quantifying the analyte. Closely related to interferences, in terms of their effects on the quality of quantitative results, are sample contamination and loss of analyte.

Concern about interfering species increases as the analyte concentration decreases. For example, suppose we have a sample that is a 99.999% pure material and the analyte is present in that material as a trace contaminant at a concentration of 1 ppb. A simple calculation shows that in this highly pure material, it is theoretically possible for 10^4 different species to be present, each at 1 ppb. If one of these is our analyte, any or all of the 9,999 other possible ppb-level concomitants could act as interferences in the determination of that analyte.

Interferences can be conveniently divided into two classes (16):

- Multiplicative interferences: species that do not produce signals of their own, but alter the signal produced by a given concentration or quantity of analyte.

- Additive interferences: species that produce signals of their own, which add to the analyte signal and can be misconstrued as having been produced by the analyte. Additive interferences produce a finite blank signal (see the discussion of limits of detection earlier in this chapter).

Most interfering species act as either multiplicative or additive interferences. Occasionally, however, a sample constituent can act as both an additive and multiplicative interference. A multiplicative interference may operate in many different ways. Three common examples are as follows:

It may decrease the available amount of the specific form of the analyte to which the measurement technique responds. For example, in atomic absorption spectroscopy (AAS), inductively coupled plasma–atomic emission spectroscopy (ICP–AES), inductively coupled plasma–mass spectrometry (ICP–MS), and atomic fluorescence spectroscopy (AFS), any sample constituent that forms a refractory compound with the analyte element tends to decrease the efficiency with which the analyte is converted to gas-phase atoms or ions, the species that are actually detected in these techniques. Any such species therefore acts as a multiplicative interference, and the signal for the analyte is smaller than it would have been if the interference were absent.

It may affect the efficiency of the physical process by which the signal is generated. Such a phenomenon, called quenching, is common in molecular fluorescence spectroscopy; the fraction of excited analyte molecules that fluoresce is decreased in the presence of interfering species. Consult Chap. 27 for details.

It may alter the efficiency with which the signal produced by the analyte is transmitted to the detector. For example, in X-ray fluorescence (XRF), molecular UV-visible fluorescence (MFS), and chemiluminescence (CL), radiation emitted by analyte atoms (XRF) or molecules (MFS, CL) may be absorbed by other sample constituents, preventing that radiation from being detected and thus decreasing the signal produced by a given concentration of analyte.

Many techniques often require that the analyte be transformed into some other species that is then detected. Any sample constituent that affects the efficiency of this analyte transformation process acts as a multiplicative interference. In AAS, ICP–AES, AFS, ICP–MS, and XRF, such phenomena often are called matrix effects. Multiplicative interferences often decrease the signal generated by a given concentration of analyte, and thus cause reported analyte concentrations to be low unless proper correction is made for the interference.

It is sometimes possible to ameliorate the effect of a multiplicative interference by diluting the sample to the point at which the interference ceases to occur. Calibration techniques such as standard addition (sometimes called analyte addition) and internal standard (sometimes called internal reference) methods (17–20), in which the sample is spiked with known quantities of the analyte or another substance, may deal effectively with multiplicative interferences. Another way of dealing with multiplicative interferences is to separate the analyte from them before the measurement is made; this can be time-consuming and not always practical, particularly when the analyte concentration is very low.

Additive interferences can cause severe problems. By definition, the signal produced by an additive interference is not easy to distinguish from that generated by the analyte. The presence of an uncompensated additive interference causes the analyte concentration to be overestimated, and leads to a false positive if the analyte is not present in the sample. Additive interferences may be constituents of the sample, or may be introduced in the analytical procedure as by use of contaminated reagents and glassware or contact with contaminated laboratory air (21). Generally it is not possible to compensate for additive interferences by conventional types of calibration methods, such as standard additions or internal standardization (17).

There are several ways to deal with additive interferences. The safest approaches are to choose a selective technique that does not respond to the interferences present in the sample or to separate the analyte from additive interferences before performing the measurement. The former approach is not possible if no very selective method for the analyte exists or the analyst does not have much information about the nature of possible interferences present in the sample. The latter procedure tends to be time-consuming and may introduce significant risk of loss of analyte.

As in the case of multiplicative interference, it may be possible to reduce or eliminate errors due to additive interferences by diluting the sample before the measurement, provided that one does not thereby decrease the analyte concentration to such an extent that the analyte itself cannot be detected with confidence.

Alternatively, if a blank is available, the signal produced by the additive interferences may be measured and then subtracted from the signal observed for the sample. This is possible only if a true blank (that is, a material identical with the analytical sample except for the absence of the analyte) is available. This is seldom the case, although sometimes an adequate synthetic blank can be prepared for a well-defined type of sample such as sea water. Another possibility is to produce an internal blank from the sample itself, by eliminating the analyte or the signal produced by the analyte (22). Alternatively, the use of instrumental techniques to discriminate signals produced by the analyte from those generated by additive interferences may eliminate, or at least decrease, analytical errors caused by the presence of additive interferences. Examples include background correctors in atomic absorption spectroscopy, derivative techniques in molecular absorption or fluorescence spectroscopy, and time resolution in molecular fluorescence spectroscopy. Use of data treatment methods that decrease or eliminate the effect of additive interferences is another possibility (23–25).

Clearly, the selectivity of an analytical technique (its susceptibility, or lack thereof, to additive interference) is a key factor in choosing one technique over another. To select a technique on the basis of its selectivity, the analyst generally must have substantial knowledge regarding the composition of the sample, to allow the most likely additive interferences to be identified ahead of time.

Although it does not, strictly speaking, fit the definition of an additive interference, contamination of a sample by the analyte obviously has the same ultimate effect as an additive interference: reporting an analyte concentration that is too high.

The problems of additive interference and sample contamination are particularly severe in environmental analysis. Analyte concentrations often are very low. Moreover, samples are usually complex, and thus may contain additive interferences. For example, if one examines the analytical data reported between 1981 and 1993, one would infer that the lead and cadmium concentrations in the Quinnipiac River in Connecticut have declined dramatically. However, what actually has taken place is a major improvement in sample-handling techniques, decreasing the extent to which the river water samples are contaminated with Cd, Pb, or additive interferences before quantification (26); the quality of the river water itself may have changed little, or even declined.

Thus, one must be very cautious when interpreting compilations of measured trace pollutant concentrations; the values may be much too high (27). "Rigorous control of contamination during sampling, transport, and storage of samples is a prerequisite for valid analysis" (28). The article from which this quotation is taken describes ultratrace-level metal determinations in Greenland snow, and describes in detail the precautions that must be taken if valid quantitative data are to be obtained for complex samples containing low analyte concentrations.

Another risk is loss of analyte. This is most likely to be a problem when the analyte concentration is low to begin with and the sample workup procedure involves numerous steps. Loss of trace sample constituents from solution by adsorption onto the walls of bottles, flasks, syringes, spectrophotometer cells, and beakers is an ever-present danger (29, 30); proteins and other biomolecules are particularly prone to adsorb on container walls. Loss of volatile or semivolatile analytes by evaporation is also a potential problem, as are dissolution or extraction steps in an analytical procedure that proceed with less than 100% efficiency (21). Chemical degradation of organic analytes during storage or workup of samples is another possibility that must be guarded against (31).

Sampling

Obviously, acquisition of a representative sample is crucial, as is avoidance of contamination or analyte loss in the process of acquiring and storing the sample. These important issues are discussed in detail elsewhere (21, 32, 33).

Verification of Accuracy

Once a quantitative analysis has been performed, how confident can one be in the quality of the results obtained? Precision is easy to measure; accuracy is much harder to assess (34). One way to evaluate the accuracy of a procedure is to apply it to a standard reference material (SRM, a sample in which the analyte concentration is certified by the U.S. National Institute of Standards and Technology, the Institute for Reference Materials and Measurements of the European Commission, or some other authoritative body) (9, 34, 35). A problem here is that unless the SRM is very similar to the unknown samples analyzed by the procedure in question (particularly in terms of the identities of potential interferences present), achievement of high accuracy in determination of the analyte in the SRM does not automatically ensure similarly high accuracy in the determination of that analyte in actual samples.

Although the number of available SRMs has increased dramatically in recent years, one may be confronted by samples for which appropriate SRMs are not readily available. Another way to evaluate the accuracy of an analytical procedure that does not require an SRM is to determine the analyte in the sample using two or (preferably) more independent techniques that have no operations in common (2, 21, 36–38). The "no operations in common" proviso is important if the sample pretreatment, rather than the measurement step itself, is responsible for systematic errors. If the independent methods produce results that do not differ significantly (at a reasonable statistical confidence level), then one can have considerable (albeit less than 100%) confidence that the methods in question achieve satisfactory accuracy for the analyte in the sample type in question (37).

In addition to deficiencies in procedures, apparatus, and the performance of analysts, insidious effects of the laboratory environment itself (such as fluctuations in temperature, humidity, or pressure) may affect analytical results. Some striking and unexpected examples are discussed by Rogers (38). Although analytical scientists often exercise great care in ensuring proper functioning of instrumentation, they may be less careful in taking measures to minimize errors that can be introduced by the environment in which instruments are used.

Finally, one should be aware of the continuing development of chemometrics (the application of statistical and mathematical methods to chemistry). Several of the issues mentioned in this section (calibration, dealing with interferences, and establishing the validity of limits of detection and quantification) are central concerns of current research in chemometrics, which is summarized in the biennial fundamental review issue of *Analytical Chemistry* (39).

References

1. L. A. Currie, "Detection in Analytical Chemistry," ACS Symposium Series, vol. 361 (Washington, D.C.: American Chemical Society, 1988).

2. L. B. Rogers, *J. Chem. Educ.*, 63 (1986), 3.

3. J. D. Ingle, Jr., and S. R. Crouch, *Spectrochemical Analysis* (Englewood Cliffs, N.J.: Prentice Hall, 1988), 172.

4. P. W. J. M. Boumans, *Anal. Chem.*, 66 (1994), 459A.

5. American Chemical Society Committee on Environmental Improvement, "Principles of Environmental Analysis," *Anal. Chem.*, 55 (1983), 2210.

6. IUPAC Commission on Spectrochemical and Other Optical Procedures for Analysis, *Anal. Chem.*, 48 (1976), 2294.

7. G. L. Long and J. D. Winefordner, *Anal. Chem.*, 55 (1983), 712A.

8. Currie, ACS Symposium Series, 1.

9. L. A. Currie, *Pure Appl. Chem.*, 64 (1992), 456.

10. D. R. Helsel, *Environ. Sci. Technol.*, 24 (1990), 1766.

11. Z. B. Alfassi and C. Wai, *Preconcentration Techniques for Trace Elements* (Boca Raton, FL: CRC Press, 1992).

12. J. D. Winefordner and C. Stevenson, *Spectrochim. Acta*, 48B (1993), 757.

13. J. D. Ingle, Jr., *J. Chem. Educ.*, 51 (1974), 100.

14. J. N. Miller, *Analyst*, 116 (1991), 3.

15. Y. Wang, D. J. Veltkamp, and B. R. Kowalski, *Anal. Chem.*, 63 (1991), 2750.

16. T. C. O'Haver, in J. D. Winefordner, ed., *Trace Analysis: Spectroscopic Methods for Elements* (New York: Wiley, 1976), 15.

17. O'Haver, in *Trace Analysis*, p. 41.

18. M. Bader, *J. Chem. Educ.*, 57 (1980), 703.

19. Ingle and Crouch, *Spectrochemical Analysis*, p. 178.

20. B. E. H. Saxberg and B. R. Kowalski, *Anal. Chem.*, 51 (1979), 1031.

21. C. Veillon, *Anal. Chem.*, 58 (1986), 851A.

22. O'Haver, in *Trace Analysis*, 33.

23. B. E. Wilson and B. R. Kowalski, *Anal. Chem.*, 61 (1989), 2277.

24. D. C. Baxter and J. Ohman, *Spectrochim. Acta*, 45B (1990), 481.

25. Z. Zhang, Z. Piao, and X. Zeng, *Spectrochim. Acta*, 48B (1993), 403.

26. G. Benoit, *Environ. Sci. Technol.*, 28 (1994), 1987.

27. H. L. Windom and others, *Environ. Sci. Technol.*, 25 (1991), 1137.

28. R. Lobinski and others, *Environ. Sci. Technol.*, 28 (1994), 1467.

29. J. M. Tramantano, J. R. Scudlark, and T. M. Church, *Environ. Sci. Technol.*, 21 (1987), 749.

30. A. Lopez Garcia and others, *Analyst*, 264 (1992), 241.

31. M. P. Maskarinec and others, *Environ. Sci. Technol.*, 24 (1990), 1665.

32. B. Kratochvil and J. K. Taylor, *Anal. Chem.*, 53 (1981), 924A.

33. L. H. Keith, *Principles of Environmental Sampling* (Washington, D.C.: American Chemical Society, 1988).

34. H. Marchandise, *Fresenius J. Anal. Chem.*, 345 (1993), 82.

35. S. D. Rasberry and T. E. Gills, *Spectrochim. Acta*, 46B (1991), 1577.

36. S. B. Schiller and K. R. Eberhardt, *Spectrochim. Acta*, 46B (1991), 1607.

37. M. S. Epstein, *Spectrochim. Acta*, 46B (1991), 1583.

38. L. B. Rogers, *Anal. Chem.*, 62 (1990), 703A.

39. S. D. Brown, T. B. Blank, S. T. Sum, and L. G. Weyer, *Anal. Chem.*, 66 (1994), 315R.

Chapter 5

Managing Laboratory Information

Robert Megargle

Cleveland State University Department of Chemistry

This chapter recognizes that analytical laboratories are producers of vital information for the organizations they serve. It considers the factors that affect the quality of that information and deals with the way information is managed. Basic concepts are first considered, followed by the application of computers to information management.

Laboratories as Information Producers

If asked to briefly describe their jobs, many chemists working in an industrial or service analytical laboratory would say "I perform tests and measurements on samples." Their focus is on processing samples, performing good measurements, and reporting data. *Data* is defined here as the result of an observation or measurement. There is often a mental barrier between the laboratory and the rest of the organization. Too often, responsibility is accepted only for what happens inside the lab.

On the other hand, the organization sees the laboratory as a place that produces information. *Information* is defined as data in context or combined with other data, sufficient to support decisions and conclusions. *Knowledge* is defined as the interpretation of information. The laboratory test results are important mostly because they answer questions and solve problems. The methods used to obtain the data are of lesser concern. Laboratory techniques and the types of instruments become an issue to the organization only if there is suspicion that the information is incorrect.

The Laboratory Testing Loop

The role of an analytical chemistry laboratory as an information producer is illustrated by the flow chart in Fig. 5.1. There is a process. Tests are performed to monitor the process, or perhaps to carry it out. The process might be a manufacturing operation being sampled for quality control. It might be a research study where the laboratory data will determine the experiment outcome. For a forensic laboratory, the process could be a crime scene. A patient would be the process for a clinical laboratory. Other examples of processes are the environment, a drinking water supply system, or a blood bank. Ultimately, our focus must always be on the process. It is the reason the laboratory exists.

In some cases, analytical testing is done by on-line instruments. In other cases, samples are drawn and sent to a laboratory. In the laboratory, the samples may be treated or prepared in some way, and then subjected to analytical measurement. The results of the measurement, on-line or in the lab, must be captured. They may be recorded in a laboratory notebook, typed or drawn on paper by an instrument, or transmitted directly to a computer system. In many cases, the raw data must be processed. This can include offset and scaling, normalization, curve fitting, data smoothing, Fourier transform, and a variety of other techniques. Then the experimental results are often organized to bring together related information and placed in temporary storage until needed. An example might be to collect all the test results for a given sample.

All laboratory results are eventually saved in permanent archives. Before that, however, they will probably be interpreted. The tests were performed for a reason. Someone reviews the results to see whether the reason has been satisfied. The interpretation is also likely to be archived, although not necessarily within the laboratory.

Interpretation produces knowledge that is often used to make decisions. Those decisions can lead to actions that affect the process, thus completing the laboratory testing loop. Quality control results are used to alter production parameters. Raw material tests may cause the company to reject a delivery. Final inspection tests may result in batches of product being discarded. Research results can determine the next experiment. They might help engineers determine the optimum

Figure 5.1 The information flow loop for an analytical laboratory.

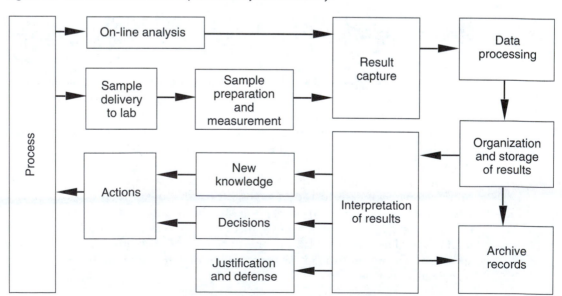

production conditions. Forensic results focus the efforts of detectives and can lead to arrest and prosecution. Clinical results help the physician make a diagnosis and prescribe treatment. Environmental testing can result in site cleanup, litigation, and fines. Water testing results might be followed by more treatment. Infected blood units can be rejected because of blood bank testing.

New knowledge might be a goal of lab testing. Knowledge can also be used to modify or enhance the process, so it too can lead to actions. Lab tests can help answer questions that are generated by research experiments. Such results can identify or help characterize new materials that might have market potential. New knowledge might result in fewer mistakes, which in turn exposes people to fewer risks and usually means less waste. Knowledge can lead to an advantage over competitors and increase profits for the company.

Another reason for testing is justification and defense. Documenting compliance with regulations and protecting against liability are examples. Test results can demonstrate product quality and safety and show that industry standards were met. Laboratory results may be needed to verify that environmental laws have been satisfied or that health and safety standards have been met.

Some companies maintain laboratories to help deal with customer complaints. Here test results might help locate problems at the customer site or result in changes in the manufacturing process. A petroleum company, for example, might test returned samples of gasoline to verify complaints about gas line freeze-up. Paint companies might perform tests to determine why certain batches of paint failed to adhere or did not cure properly. Finally, laboratory test results are a part of the permanent records of the company. They help show what the company did, how well they did it, what they gained, and what they lost.

Good information produced by the laboratory is critically important. In some situations, it can prevent millions of dollars of waste, lost customers, fines, liability, and even bankruptcy of the corporation. Laboratories should therefore adopt a broader mindset. If they view themselves as problem solvers and information providers, analytical laboratories will find ways to better serve the organization. This can lead to higher morale, less resistance to adequate funding of the laboratory, and more efficient corporate operation. In this mode, laboratory personnel are loyal to corporate goals, not just to the laboratory.

Properties of Good Information

Three characteristics of information are important in this context. They are described here as the three Rs. First, the information must be reliable, meaning that it is of good quality. Second, it must be relevant, meaning that it answers the questions and solves the problems at hand. Third, it must be responsive, meaning that it is available in time to be useful.

Reliability of Information

The attributes of laboratory data that establish good quality are as follows:

- Accuracy: how close the result is to the true value.
- Precision: how close repeated measurements are to one another, which also allows an estimation of the number of significant figures that can be justified.
- Limits of detection: how small an amount can be detected.

- Limits of quantification: how small an amount can be reliably measured quantitatively.
- Linearity: the range over which the instrument produces linear response.
- Interferences: other substances that might throw off the test result. Interferences fall into three categories: those that also give a signal and make the result too high, those that tie up the analyte and give a result that is too low, and those that poison the measurement system.
- Selectivity: how well potential interferences are rejected.

The correctness of laboratory test results is established by three separate processes: validation, calibration, and quality assurance. All three processes usually depend on standard materials.

Standard Materials

A standard is a material that has a known value for some measurable parameter. For some analytical measurements, precisely known standards do not exist. In these cases, a reference material that is stable for some period of time is often used instead. It does not precisely establish a known value, but it can serve as a benchmark for evaluation of the analysis results over this time interval. When it is time to replace the reference with a new batch, the old and new materials are both tested for some number of days in order to establish a correlation that can be carried forward. It should be noted that when precise standards are not available, the analytical methods cannot produce exactly known results. However, if consistency can be achieved, the relative test results are usually sufficient to control the process or answer the questions. Reference materials are treated the same as standards in this discussion.

Standards can be used two ways. A calibrator material is a standard used to adjust the instrument. It may include a blank used to set the zero end of the scale and one or more knowns used to establish a calibrated range. A quality assurance (QA) material is a standard used during routine instrument performance checks, or to validate a test method. The distinction is important, even if the actual materials are similar. If the correct answer is not obtained with a calibrator material, the instrument is adjusted until the right answer is produced. Hence, there should be no deviation between the instrument value and the calibrator value. A QA material, however, is not used to adjust the instrument. The instrument result is compared to the known value and any deviation is a measure of the error.

All standard materials must be documented and saved in the permanent records of the laboratory. They are an essential link in the evidence chain that establishes the quality of the analytical results. For purchased commercial standards, one should save the name of the standard, catalog number, lot number, and expiration date. In-house standards should be prepared according to written directions. A record should be kept of the recipe, when the standard was prepared, what materials were used, and who made it. This person effectively accepts responsibility for the quality of the standard. The use of the standards should be documented, probably as part of the records of each work session where they are used.

Validation

Validation is a process we go through to be convinced that what we are doing is correct. No good scientist reports a result if he or she lacks confidence in the reagents, the procedures, the instrument, or the technique. This has always been a part of scientific thinking, but in recent years validation has acquired a more important and formal status. In the past, validation was usually left to each individual technician, section head, or supervisor. Some workers were very careful to test their methods and instruments before using them on unknown samples and some were not. In the latter case, bad data were sometimes generated, only to be detected when the lab results proved to be ineffective in solving problems or answering questions. By this time, corporate inefficiencies and losses had been experienced, and the reputation of the laboratory had suffered.

Low-quality information from poorly validated methods has become an acute issue in several fields where laboratories in the private sector supply data to government agencies. In particular, the U.S. Food and Drug Administration (FDA) and the U.S. Environmental Protection Agency (EPA) depend on information produced by the industries subject to their regulatory oversight to regulate those industries. This potential conflict of interest was resolved by requiring formal validation procedures of all methods and systems that can affect the final reported information and by making the results of that validation subject to government inspection. Thus, these industries must not only provide the government with data needed for regulatory decisions, but also produce documentation that shows that the information was obtained by valid methods.

Although most other industries do not face government mandates, validation is a scientifically sound concept. It should be part of the installation of any new test method, instrument, or computer system that processes laboratory data.

Conceptually the validation process is clear, but sometimes it is difficult to carry out some of the steps. In general, every factor that affects the correctness of a result should be identified. Then a check or test must be selected and performed that demonstrates that each factor is working correctly and under control.

A popular way to validate an analytical method is to apply it to standard materials. Many validation questions are answered if the instrument results agree with the known values. When validation testing is done, a variety of known samples should be tested. They should include samples at the upper and lower limits of the normal range, and some should contain possible interferences. It is tempting to approach validation thinking, "I want to show that this method works." It is better to think, "I want to find all the problems with this method that I can."

A validation test should be done in each laboratory for all new test methods, even if they are taken directly from one of the standardizing organizations such as the American Society for Testing and Materials (ASTM) or the Association of Official Analytical Chemists (AOAC). Methods from standards organizations have been tested at numerous sites and their validity has achieved industry consensus. A local validation, however, will show that the method works for your equipment and sample types and that you are doing it correctly. It is also wise to practice a new method on knowns before trying it on real samples, where the answer matters to the organization.

It is important to keep records of the validation process. They are part of the evidence that the laboratory is producing correct results. If the standard materials are stable, they should also be saved. If not, the recipe used to produce these knowns should be kept. The objective is to be ready to repeat the validation test at a later time if necessary. It is important to record the time and date of the validation, who did the measurements, and who signed off (accepted as correct) the validation results.

Maintaining Validation

There is more to getting reliable results than proving the scientific validity of the testing procedure. The technique also has to be performed correctly in routine operation. It is important to confirm the capabilities of the personnel who will do the test and ensure that the correct procedures are followed. In regulated laboratories, training records of personnel must be saved. Standard operating procedures (SOPs) must be written to define how the test is done. There must be evidence that the SOPs are followed. All of this is subject to government inspection. Although nonregulated laboratories may not have to face inspectors, the issues are still important. All laboratories must have confidence that the validated test method is being performed correctly.

Keeping a test method in a valid state over time is also important. The SOPs should specify schedules for routine maintenance and revalidation tests. Expiration dates of standards, as well as all required reagents, should be respected. Certain comparison tests should be performed when new batches of standards or reagents are put into service.

Another issue is change control. These are the policies to be followed when any changes are made to the instruments, procedures, or computers. Any substantial change should be followed by complete revalidation. For minor changes, it may be acceptable to revalidate only the factors likely to be affected.

Original data should not be anonymously altered. Any change should leave the original value legible. The time and date of the change, the author of the change, and the reason for the changed value should be recorded. This rule does not apply to typing errors that are caught by the typist, but it does apply once the data are officially filed in either an electronic or paper record system.

Security may be important, especially in regulated laboratories. The government agencies want to be sure that test results cannot be lost, distorted, or changed either accidentally or intentionally. Security issues include controlling who has access to the laboratory, keeping backup copies of data, putting primary lab notebooks in a secure location, preventing unauthorized computer use, and keeping viruses and other corrupting programs out of computers.

Calibration

Some instruments must be calibrated at every work session. Others need less frequent calibration. Balances, for example, should probably be checked with standard weights at least once a year. Some parameters of certain instruments have built-in calibration and never need to be checked. The wavelength scale of a Fourier transform infrared spectrometer, for example, depends on the wavelength of a laser, which already has the status of a primary standard. Monochromators of scanning instruments, however, should be adjusted periodically.

Because instrument calibration obviously affects test results, a record should be kept of all such calibrations. It is part of the evidence chain. When calibration is infrequent, this record can be maintained in an instrument log book. It should include the date and responsible person. The identity of any standards used should be noted. When instruments need frequent calibration, records are usually saved with the results of each work session. Normally it is sufficient to document which standards were used as calibration materials that day.

Quality Assurance

Quality assurance (QA) techniques are used routinely to add confidence to the measurement process. The frequency of QA testing depends on the instrument and type of measurements being made. In some cases, QA may be done only a few times a year. In other situations, QA is important for every work session. Quality assurance involves analyzing QA materials as if they were samples, except that the results are treated differently. QA results are collected as a historical record of performance of the measurement system. The remainder of this chapter assumes that QA is performed at least once daily.

QA results are often displayed graphically to reveal instrument drift or worsening precision. Figure 5.2 shows a sample QA chart. There are many variations to this basic scheme. In this graph, the measured values of the QA standard are plotted on the y-axis against the day they were analyzed on the x-axis. If the standard was tested more than once a day, only one result is chosen for the plot (for example, the first one). The center line of the plot is the target value for this standard and instrument. It might be an average result obtained by measurement with this system, or it might be the absolute known value of the standard. Lines at one and two standard deviations on

Figure 5.2 A typical quality assurance chart.

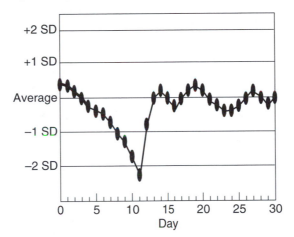

both sides of the target value are also shown. They are based on the measured or expected precision for this measurement.

A typical interpretation of this control chart follows. Values within one standard deviation are considered normal. When results are between one and two standard deviations, such as on days 9 and 10 in Fig. 5.2, there is concern but the test results are still accepted. On day 11, when the result exceeded two standard deviations, the method was judged to be out of specification. Samples measured during this period will not be accepted. Some adjustment or repair was made because the measured QA standard was back in the control range the next day. The instrument stayed in proper operation the rest of the month. The plot shows a continuous drift to low values for days 0 to 11. This indicates a problem with the instrument or the reagents. Normally the plot should vary randomly about the target value, as it does after day 12.

When Fig. 5.2 is viewed as a whole, it is called a retrospective evaluation because it looks back over a 30-day period to judge instrument quality. QA samples can also be used to help verify the customer sample results during a work session. Here one concentrates only on the QA values measured that day. Suppose the instrument produces the correct results for QA samples at the beginning of the work session, and again at the end. If no problems were noticed between these checks, then it is reasonable to assume that the measurement process was in control for the duration of the work session. If this check fails, some or all of the testing may have to be repeated. When one has less confidence in the method, more frequent QA measurements can be made. The entire work session is not lost if the final QA results are off.

In general, a different set of standard materials should be used for QA and instrument adjustment. Many scientists recommend that QA materials be tested in a random pattern to improve the chance of detecting systematic errors. Sometimes blind QA samples are introduced into the job stream by lab supervisors. They are marked as customer samples and the analysts are not told they are knowns. This is done to avoid the natural tendency to be especially careful when you know your work is being evaluated.

Relevance of Information

The second attribute of good information is relevance. The information must answer the questions at hand. Test methods must be selected whose results will satisfy the needs of the customer.

The test must produce a result of sufficient accuracy and precision for the intended purpose. Spending time and money to produce a result to four significant figures is wasted if one significant figure is all that is needed. On the other hand, an imprecise result is worthless if it fails to distinguish between possible answers or actions. The information must be as complete as the problem warrants.

Establishing that results produced by a given test actually answer the questions being asked may be more difficult than showing that the result is correct. Sometimes the relationship is obvious, but often it requires expert knowledge of the process being tested. Because the process can vary so widely depending on the organization, little general guidance can be given. The issue of test relevance must be considered at a high professional level. Laboratory staff should be ready to devise new solutions to serve real needs, be prepared to offer advice, and be active in any decision, planning, or troubleshooting activities in the organization. This requires that they understand the process, their profession, and company goals. Knowledge and experience are important. Sometimes experiments can be devised to prove correlations between lab test results and factors in the process. In more difficult cases, trial and error may be necessary.

Responsiveness

The third attribute of good information is response time. The lab result has little value if it arrives after the decision it was intended to support has already been made. At best it can confirm the decision or prepare people for a failed outcome. At worst, the result is ignored and the test might as well never have been done.

Returning test results in time to be useful depends on the time frame of the process. Sometimes minutes are important, other times days are sufficient. Usually the issues here are organization, staffing, and communications. For very fast time frames, the laboratory instruments may be limiting. If faster techniques cannot be found, on-line analysis and automatic controls may be necessary.

In most analytical laboratories, failure to meet required response time is due to increased workload, inferior equipment, poor communication systems, inefficient organization of the lab, or insufficient staff. Usually it is a management issue. When the problem arises, laboratory procedures should be evaluated. Time-limiting factors should be identified and streamlined, if possible. Faster instruments may be necessary. Electronic communication might be substituted for interoffice mail. Personnel might be reassigned to help critical areas. Additional staff might be hired. Computers can sometimes help resolve logistical and communication problems.

Laboratory Information Management System

The laboratory information management system (LIMS) is a computer system designed to help the analytical laboratory. Many modern instruments contain computers to control measurement parameters, collect data, and perform data processing functions. The LIMS, however, is a computer system intended to apply to the laboratory as a whole. It manages not only the information output of the laboratory (that is, the test results), but also the information needed inside the laboratory to function smoothly. This information includes factors such as

- What samples are in the laboratory?
- What tests must be performed?
- Where should the tests be performed?
- What control samples are needed?
- What is the status of each test?
- What reports must be printed?
- Are any tests overdue?
- What is the backlog of tests and samples?

The LIMS also helps with another important problem: keeping up with the ever-increasing demand for paperwork. Recordkeeping requirements become more extensive as the need to document every facet of the chain of evidence for every test result becomes more critical. As we have seen, this need for documentation is being driven by the need to meet government regulations, protect against liability, assign accountability, produce test results that will stand up in court, and follow increasingly higher standards for good laboratory practice.

A LIMS is a computer database customized for use in the analytical laboratory. It contains a series of data tables that hold information about the laboratory, the samples received, the tests performed, the results obtained, and the reports produced. Information is added to the LIMS during the daily work of the laboratory, and the LIMS is then used to produce the various outputs needed to satisfy customers, laboratory management, upper management, and permanent record-keeping requirements.

Most modern LIMSs are purchased systems because the cost, complexity, and time required to write LIMS software is beyond the capabilities of most laboratories. Some companies have elected to design and write their own systems, usually because they have special needs that commercial systems cannot meet. Almost always, these are companies that already have extensive expert computer skills and enough resources that they can afford the 2 to 5 years it typically takes to develop a system from scratch.

The commercial systems have to be flexible so that one LIMS program can cover many laboratories. Only by using the same software at many sites can the LIMS vendor offer a product at reasonable cost. To achieve this flexibility, the software in most modern LIMSs follows a table-driven strategy. The software is designed so that all the information that might vary from site to site is placed into appropriate data structures in the database. Thus, information such as the names of clients, names of laboratory personnel, names of tests performed by the laboratory, types of reports, and other such site-specific information is put into the LIMS as data. This is called static information because it changes only occasionally over time as the laboratory operates. All static information must be entered into a new LIMS before it can be used in the laboratory.

The LIMS also contains dynamic information that changes daily as the work gets done. This includes identification of samples and tests, test results, changes in test or sample status, and other information related to the ongoing work. The rest of this chapter describes a typical LIMS and its ability to assist the laboratory workflow. The LIMS components are illustrated in Fig. 5.3.

Test Order

For the laboratory, the workflow process begins when there is a request for a lab test. Someone enters the test order into the LIMS, which causes a new entry to be made in a table of test orders. Most LIMSs first require the person placing the order be identified by giving a name, code, or customer number. This is used to identify an entry in a separate customer table inside the LIMS.

Figure 5.3 Functional components of a LIMS.

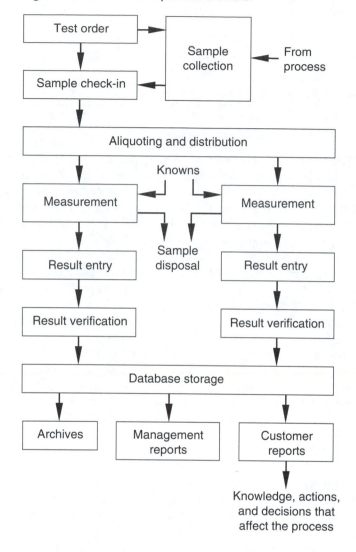

The customer table was created when static information was entered, but it can be changed when new customers are accepted or old customers are deleted. The existence of this entry authorizes this customer to place the test order. The entry has the customer's name, address, telephone number, and other information. Perhaps there is a status, such as a company rank or department. There may be an accounting number to which lab services are billed.

If no customer entry exists, most LIMSs insist that one be created before accepting the test order. Proper identification of the individual who ordered the test is part of the official record of this test. It cannot be omitted if the audit trail is to be maintained. The computer record of this test order includes a reference to the customer table entry of this person. The cross-reference effectively assigns responsibility for the order.

The test order must identify the test to be performed. A name or code for the test is typed. Test descriptions have already been entered into a table as part of the static information. An error is reported if the test order identification does not match an entry in this test description table.

The LIMS will not accept unknown test orders. Each test description entry contains information about a test the laboratory can perform. This table contains everything the LIMS must know to process every test. Items in each entry include sample type needed, minimum sample volume, the workstation where the test is performed, expected completion time, special handling directions, the type of result expected (such as numeric, textual, or spectra), measurement units, calculations needed, billing rates, and much more. There must be information in this entry to allow the LIMS to execute every function concerned with the routing, scheduling, reporting, and disposition of this test. The more functions the LIMS can perform with respect to a test, the more information must be in the test description table.

Additional information may be entered when the test order is placed. Special priority might be ordered, perhaps at extra cost. There may be test options to be determined, such as the instrument to be used or wavelength ranges of interest for spectra. The customer might indicate that only trace levels are of interest. The time and date the order is placed is likely to be recorded automatically by the LIMS.

Sample Collection

Nearly every test order requires a sample. Test orders and sample orders are best treated separately. Often several tests are ordered for the same sample, and on occasion, some tests require several samples to produce one test result. Each test order may require a new sample or it may be assigned to a sample already known to the LIMS. In some cases, the sample is still to be collected; in others the sample already exists in the laboratory. Because there is no one-to-one correlation between sample and test, a separate table is most often used to save information about each sample.

The sample must be identified. The customer probably has some identification scheme of his or her own, and that should be preserved in the LIMS. Retrieval of information is easier for the customer if the customer identification is kept and can be used as a search key. The LIMS, however, must uniquely identify every sample from every source, so it assigns its own sample number. This avoids the problem created if two customers by chance use the same sample number. Sample numbers, called accession numbers in some laboratories, are usually assigned sequentially. The number must have enough digits that the time it takes to roll over is longer than the time any sample will be in the system. Two active samples cannot have the same number. Some systems attach the date to the sample number to make it forever unique.

Samples often arrive in batches, with the same set of tests to be applied to all of them. Occasionally certain samples in the batch may require additional or fewer tests. In some organizations, it is important to maintain the association of each sample with the batch in which it came. In others, no correlation exists. The batch is simply a convenient way to bring unrelated samples to the laboratory. When batches are important, the sample identification system must have a batch number plus a sample number within the batch.

Additional information must be kept about each sample. The time and date the sample was collected are most important. They tell when the process was sampled. The test results to be obtained are associated with that time and date. Often collection time must be supplied by the customer who delivered the sample and is then manually put into the LIMS. The site of sample collection and the sample type may be entered to further identify the nature of the sample. The person who collected the sample is part of the evidence trail, and should be indicated in the sample description table. Container type and the initial volume of sample are useful. Remaining sample volume can be calculated from the latter by subtracting the volume needed for each test ordered on this sample. Remaining sample can be used to determine whether another test can be safely added to the sample. It is useful to note whether the sample presents any particular hazards to personnel. Examples include corrosive, poisonous, radioactive, flammable, or biological hazards. The LIMS can use this to print warnings on sample labels. Final disposition of any remaining sample may be entered. Options include disposal, return to customer, or save as a standard for future tests.

A reference to the sample is saved in the entry of the test order table, and references to the tests to be performed can be saved in the entry of the sample description table. These computer cross-references establish the evidence trail for the way the test orders and samples are used to produce test results.

Most LIMSs print a sample label. This is superior to hand-written labels that might be on the container from the customer. They are not subject to handwriting problems and they contain the unique laboratory number used by the LIMS to track the sample and the test results. The label can include warning codes and disposal information. It might indicate which tests are to be performed. Machine-readable labels are being used more and more to help guard against mix-ups. Bar codes are most popular. If samples are always identified by scanning the bar code, the only place where a sample mix-up can occur is when the label is first applied to the container.

Workflow modes of operation for test ordering and sample collection vary for different companies. The difference profoundly affects the way the LIMS operates. In many companies, test orders and samples arrive at the laboratory together. The laboratory has no prior knowledge of the needed test or the sample. In this case, test ordering, sample identification, and sample check-in are done together at a LIMS terminal inside the laboratory. When the entry is completed, the status of the sample is automatically received.

In other organizations, however, it is appropriate to allow the test orders to be placed in advance of sample delivery or even sample collection. In this mode, most LIMSs automatically create entries in the sample description table as necessary to satisfy the needs of test orders that have been placed. Part of the test order process may be to specify when sample collection is to be performed. Possible responses include specific time and date or next routine collection time. These entries do not correspond to actual samples yet, and would have the status "ordered" or "not yet collected." The LIMS may be programmed to combine test orders as much as possible so that they can be performed on one sample, or the LIMS may be set up to expect separate samples for separate tests. The latter is more convenient for the laboratory that must route samples to different workstations. The former is more convenient for sample collectors. Few LIMSs routinely schedule new test orders for existing samples already in the lab. It might be an option available on special demand. The usual presumption is that a new test order is a request for current information about the process, not the state of the process at some time in the past corresponding to the old sample.

Sometimes test orders may be standing directions. In a quality control laboratory, for example, standard procedure might require the process be sampled at regular time intervals, perhaps every 1 or 2 hours. Here the LIMS can be programmed to automatically generate test orders at the scheduled times. Sometimes test orders may be part of a long-term study, and may be ordered far in advance. In animal research, for example, a testing regimen may be established at the beginning of a study that may last a few months to several years. Different tests can be requested on different schedules. Different animal studies may be in progress in the company at the same time, and each may have a different testing regimen. The LIMSs in these companies may have a very complex set of standing test orders lasting far into the future.

When the LIMS knows it will need samples before they are obtained, it can help the organization in several ways. At a minimum, it can generate overdue sample reports. These alert personnel that an expected sample has not yet been received at the laboratory. Someone can be dispatched to locate and fix the problem. Even better, the LIMS can generate a reminder to collect samples before they are overdue. In this way, the LIMS is helping do the job correctly, before there is an error.

There are sometimes problems with sample collection beyond staying on schedule. It is important to get a homogeneous sample that truly represents the process. It cannot be contaminated during collection or on its trip to the laboratory. It must be properly labeled. Sometimes preservatives are required. Studies in many industries have found that inaccurate test results are due more often to sampling errors than to measurement errors. For these reasons, some laboratories elect to collect the samples themselves. This mode of operation is not suitable in all environments, but is effective where it can be used. The sample collector, trained and based in the laboratory, knows the importance of good sampling technique.

In this mode, the LIMS can provide collection lists because it knows what samples are due to be collected. The list can tell how much sample is needed, calculated by adding the volume requirements for each scheduled test. Sample labels can be printed in advance. The sample collector takes this information on rounds to the sites where samples are needed, collects a sample, and immediately applies a label. Labeling errors are minimized when legible computer labels are applied right after the sample is drawn. Poor sampling technique is reduced because of the individual's special training. Schedules are more likely to be met because of the LIMS prompting.

Sample Check-In

The sample check-in step tells the LIMS that a sample has arrived at the laboratory. The LIMS now knows that tests on this sample can be scheduled. When the laboratory deals only with test orders that come with samples, sample check-in is implicit. The new test order implies that a sample has been received. When test orders are allowed in advance of the sample, however, an entry must be made in the LIMS to record their arrival.

When the samples are delivered by the customer, each one must be checked in manually. The sample collection time is entered from records kept by the customer. The sample delivery time is automatically captured by the LIMS. If the samples have not yet received LIMS-printed labels, they get them during sample check-in. Manual entry might be through the keyboard of a LIMS terminal or by scanning the bar code if the sample already has a laboratory label. This might happen if there are remote label printers connected to the LIMS. Staff in the production area might use these LIMS-printed labels for samples that are then shipped to the laboratory. If labels are not printed until a sample is about to be drawn, the LIMS can automatically capture the sample collection time.

The sample check-in step can be streamlined when laboratory staff collect the samples. The sample collector returns to the laboratory after rounds and reports to the LIMS the collection list number just used. The individual may then indicate whether any samples on the list were missed because of special circumstances. The LIMS then knows that the rest of the samples on that collection list are present in the laboratory. They are all checked-in in one step.

If a precise sample collection time is not needed, the LIMS may assign an average time to all samples on the collection list. A better estimate of collection time might be to prorate the time from start of rounds to end of rounds, assuming samples were obtained in order as they appear on the collection list. If even more accurate sampling time is necessary, the sample collector must make a notation on the collection list next to each sample, and manually enter that data into the LIMS at the end of rounds.

Aliquoting and Distribution

There are different modes of processing samples once they are in the laboratory. In a few laboratories, one sample is never used for more than one test. These are likely to be very specialized cases. Most labs at least have to allow for the possibility of testing one sample at several workstations. In many labs, multiple tests on a sample are routine. In small laboratories, the sample might be passed from workstation to workstation as needed to satisfy all test requirements. With larger numbers of samples, this plan breaks down. It is too likely that samples will get lost or fail to reach all the workstations where they must be tested. Here there are two possible schemes.

All samples might be kept in a sample room. When a sample is needed to perform a test, the technician checks the sample out of that room and takes it to the appropriate workstation. Some sample is extracted to perform the test. Afterward, the remaining sample is returned to the sample room. This approach might be favored in a laboratory that must strictly account for every sample, or where samples must be under high security, such as a forensic laboratory. When this operational mode is used, the LIMS keeps track of all check-outs and check-ins to the sample room. At any time the LIMS can report the location of every sample. This is called sample tracking, and it is an important function of the LIMS in any large laboratory that might be processing hundreds or even thousands of samples at any time. At that sample count, knowing the whereabouts of any given sample is no small achievement.

In addition to knowing the current location, the LIMS might also keep a historical record of the travels of any sample. This is called a chain of custody, and it is important when test results are to be used in legal cases. The chain of custody documents who had possession of and was responsible for the sample at all times. It is evidence that the sample could not have been switched or altered by unknown parties.

A second scheme is to divide the original sample into aliquots and send them separately to the different workstations where the tests are to be performed. The LIMS can provide directions to the aliquoting technician. It knows what tests are needed for each sample and how much sample is required at each workstation. When aliquots are prepared, they must be labeled. The LIMS will probably produce these labels. The sample numbering system must allow each aliquot to be associated with the original sample delivered to the lab. All aliquots share the same source information as the parent. After splitting the sample, however, each aliquot must be treated as a separate sample. From that point, they will have different histories.

Measurement

When the samples are in the lab and ready for testing, the LIMS can schedule them at the appropriate workstations. Each analyst requests that the LIMS produce a worklist. It shows the samples that are to be tested at that person's workstation. This list gives the sample numbers, the tests needed, and any hazard codes. It might show the expected completion time so the analyst can judge which tests are closest to being overdue. High-priority samples might be listed first. It might indicate how the sample is to be disposed of. In short, the worklist helps the analyst organize the day's activity and properly deal with every sample.

Result Entry

There are several ways that test results can be entered into the LIMS. Manual entry is typical in low-volume workstations. The worklist printed by the LIMS may have space to record test results. As the analyst finishes testing each sample, the results are written on the paper. At the end of the work session, the analyst takes the worklist to the nearest LIMS terminal and types in the test results. The LIMS helps this process by displaying a screen on the terminal organized the same as the printed worklist.

Avoiding errors in this mode of operation depends on the diligence of the analyst. There are two basic types of error. The results may be reported for the wrong sample. The mix-up can occur when the sample is chosen for testing, when the test result is copied to the worklist, or when the worklist result is typed into the LIMS. The analyst must be very careful to check and double check sample numbers at these times. Two, the results can be mistakenly copied. This can happen when the instrument readings are written on the worklist or when the worklist entries are typed into the LIMS. The most likely errors are transposing digits, omitting a digit, or adding an extra digit to the test result. Transcription errors that make the result absurd are easily caught. It is more difficult to detect a transcription error that is still a reasonable test answer.

Transcription errors can be virtually eliminated by interfacing the instrument to the LIMS. Now the data transfer is electronic, which is much more reliable than manual transfer. It is rare for any error to occur, and even more rare that an incorrect transfer will be a reasonable test result. Usually failed electronic communication shows up as gibberish. Interfacing also reduces clerical work, freeing the analysts to be more productive with their main area of expertise: analyzing samples.

Sample mix-up errors can be reduced to almost zero if sample labels have bar codes. At every point where sample identity must be entered or confirmed, the bar code is scanned. This also saves the analyst time because scanning a bar code is much quicker than reading and double checking a long sample number printed on a sample label.

Despite the advantages, most low-volume workstations are not connected to the LIMS. The cost of the interface is often not justified by the number of tests performed at the workstation. High-volume workstations, however, are another story. Here the savings in personnel time and the reduced error rates easily justify the added hardware and software needed for interfacing.

Unfortunately, the interfacing problem is complex. Every instrument and LIMS is configured a little differently. Although the electronic connections may be standardized (at least to a degree), the information content of messages passed between instruments and the LIMS is not. The LIMS must not only receive the electronic signals correctly from the instrument, but must also be able to understand the information it receives. Only in this way can the LIMS properly place test results and other data into the right slots in the database. Each interface is likely to be a separate project, sometimes requiring that new software be written. It is therefore expensive.

Some LIMSs attempt to deal with instrument interfacing generically, using database table parameters to account for instrument differences. It helps considerably, but there are still some instruments that defy even this approach. The issues are sufficiently complex that an expert is usually needed to make the right interface table entries.

Standardization of message content to and from instruments should vastly reduce the interface problem. Instrument and LIMS vendors can write their software to a common message format, and expect proper communication to be accomplished. These standards are being developed in the mid-1990s, but are likely to be applied only to new instruments designed after their acceptance. Older instruments will continue to be used for many years, so instrument interfacing is likely to remain a problem for some time.

High-volume workstations are more likely to use automated instruments. Many of these instruments are equipped with sample trays that can be loaded at the beginning of a work session and left unattended to perform the tests. Selected locations in the sample tray are filled with blanks, calibrators, and QA materials. The instrument uses the blanks and calibrators to adjust itself. The QA materials may be processed by either the instrument, the LIMS, or both. The usual QA charts are produced. Some of these instruments perform the same set of tests on every sample; others can be programmed to perform only selected tests so that every sample can be analyzed differently.

When automated instruments are interfaced to the LIMS, close cooperation is needed between the two systems to achieve reliable operation. One major problem is sample identification.

When the LIMS gets a test result from the instrument, it must know which sample was measured so the data can be stored correctly. In many current implementations, this correlation is achieved only by instrument tray position. At the beginning of the work session, the LIMS prints or displays an instrument load list. It tells the technician how to load the instrument, including the unknowns and all calibration and QA materials. It is up to the technician to correctly follow these directions. Mix-up errors occur if the tray is not loaded correctly. Alternatively, the technician might load the instrument and then type the sequence of samples and standard materials into the LIMS. This requires more keyboard time by the technician. It is also subject to sample mix-up errors, plus the additional possibility that the technician may mistype a sample number. A compromise is to let the LIMS print a load list but allow the technician to modify it. Most LIMSs permit all these options.

Bar code labels eliminate sample mix-up errors. Some automated instruments can be equipped with a bar code reader, and in the future almost all will have this option. The instrument reports the sample number found on the sample label to the LIMS along with the test results. Now the sample tray can be loaded in any order by the technician, and the system correctly identifies every sample. This advantage is so significant that laboratories should insist bar code readers be part of any new automated instruments they purchase.

A dialog with the LIMS is needed for instruments that can vary the tests they perform on each sample. In one strategy, the instrument reads the bar code and asks the LIMS what tests are to be run on this sample. The LIMS responds. The instrument then carries out the tests and returns the results to the LIMS. The cycle repeats when the next sample is processed. In an alternative strategy, the LIMS downloads into the instrument computer all the sample numbers to be run at that workstation, along with all tests that have been requested. When the instrument reads the sample bar code, it looks in its own database to find the test orders for that sample. The tests are run and the results are reported to the LIMS. The latter strategy allows the instrument to function for a while even if the LIMS interface is temporarily out of service. The results are transmitted later when the connection is restored. However, there is a problem if the instrument encounters a sample about which it has no local knowledge. Then it has to query the LIMS or report an error to the operator.

As a part of the entry of each result, the LIMS records who performed the test and entered the value. The time and date the result was logged into the LIMS are recorded. These facts are a part of the official record of this test and are needed to assign responsibility.

Result Verification

No test result should be reported to the customer until it has been reviewed by a responsible person. This step is called verification, or sometimes sign-off. It is an official approval that the results are acceptable. The time and date of verification and the person who signed off the result should be recorded as a part of the official record.

At the very least, this process should catch missing, silly, or absurd results. It might involve much more. For manually entered results, for example, the verification step is a chance to catch typing errors. The technician is presented with a screen of results from the last work session. These can be checked against the values recorded on the worklist. The technician is also given one last opportunity to detect unusual or out-of-range results. A technician may elect to repeat tests when the results are suspect.

Verification can be assisted by the LIMS when there are known ranges or accepted values for the test results. Values that are out-of- bounds can be highlighted for special scrutiny. Sometimes delta checks can be used. Here, the result for the current test is compared to the last test result on the same sample source. A delta check is failed if the results differ by more than a predefined limit. This check is valid only when one can assume that the underlying process should not change too rapidly. A failed delta check indicates that the process may be out of control or one of the test results is faulty. Further investigation is warranted.

Verification should include review of the real-time quality control results. QA results may be displayed for approval. If QA is out of specification, some or all of the sample tests may have to be repeated.

When data are transferred to the LIMS through an interface, transcription errors are unlikely. Verification consists of making sure reasonable values are recorded in the LIMS. Limit checks, delta checks, and a review of QA samples should also be done with interfaced instruments, if appropriate.

Although review of QA results over time is not officially needed to verify the test results of a given day, it is useful to help spot problems at the workstation. Drift and worsening precision can be spotted before they become a problem, and corrective action taken. Retrospective quality control charts should be reviewed often.

Database Storage

This block in Fig. 5.3 is intended to remind us that the LIMS is a database application. All of the activities described above involve transactions with this database. Records of test orders, samples delivered to the lab, sample collection lists, work session assignments, test results, standard materials used, and verification have all been created in the database. The arrows in Fig. 5.3 are overly simplistic. Database storage really connects to all other boxes. At this point in the flow diagram, all information about the testing process is available. What remains is to output this information in appropriate reports.

Customer Reports

Test results may be reported to customers after they are verified. When high priority is needed, each result can be transmitted to the client as soon as it is authorized. More often, however, the LIMS waits until most or all of the test results ordered for a sample are ready. Then one report is printed that gives all the results. If samples are submitted in batches, all tests for the entire batch may be reported together. The purpose is to provide the customer with information in the most useful form. Reports can be printed and mailed to customers, or they may be transmitted electronically over local area networks, e-mail, or fax.

Report formats should be designed with care. Most customers do not see all the sample preparation and measurement procedures that are needed to get a test result. They see only the final report. To them, that report is the product of the laboratory. Their judgment of the laboratory depends in large part on their impressions of the lab report. Good report formats also reduce reading errors, improve productivity by transferring information from the page to the mind of the reader more quickly, and result in less need for follow-up explanations or interpretations.

A good report format groups related information together. The report should contain all the information needed to understand the test result and omit all extraneous information. Units of measure should be included, for example, but the name of the analyst who did the test just clutters the report. That name is important for the archives and to preserve the chain of evidence, but it should be available only on demand in case of a problem. Color, bold fonts, large and small type, borders, column lines, underlines, and highlight boxes can be used to improve the appearance and clarity of the report. The visual appearance of the report should direct the reader to the important information, leaving the supporting information less prominent.

When reports are to be printed, limitations of the printing device must be considered. High-speed printing may not allow the use of graphics. When graphics are used, printing speed usually drops. If a complex report format requiring fancy printer features is designed, there will be trouble if that printer fails. Backup printers may not have all the features. Some laboratories design all reports for the least capable printer, but that lowers the quality of reports at all times.

Some laboratories use preprinted forms to improve clarity and make a better impression on the client. Fixed information, such as the name of the laboratory, can be attractively placed on the form. Shading and color can be used to highlight important information. However, preprinted forms are hard to change, the paper is more expensive, and longer lead time is required for ordering more stock. There is extra work if the printer must be loaded with different forms. Printing reports on white paper is less elegant but more flexible. Report formats can be changed easily and different reports can be interspersed with no problems. Paper stock is less expensive and more readily available. Preprinted forms are found more often in contract laboratories whose main income comes from selling analytical services. Making a good impression on the customer is important to stay in business. White paper reports are most often used in laboratories operated by and used internally within a corporation. Here economy and flexibility are more important than image.

Archives

Virtually all laboratories keep permanent records. They are required by government regulation in some industries, and by corporate policy in nearly every other case. Good archive records allow the evidence trail to be examined for any measurement made by the laboratory. The archive record is important because the test result may have influenced an important action or decision. It may be necessary in the future for the company to defend that action, at which time it becomes important to defend the underlying test result.

Without computers, archive records are hard to maintain. They usually involve placing paper reports in file cabinets. If the files are extensive, employees may have to be hired just to maintain these records. The space required becomes large, in some cases occupying large rooms. The volume can be reduced with microfilm. Usually the paper records are photographed. Retrieval of information from the archives is difficult. A piece of paper or microfilm can be filed only by one attribute. For example, suppose test results are filed by date of completion. Then it becomes difficult to find the tests done by a particular analyst, for a given client, or for a particular sample type. The date of completion must be known to find any given record. The search may be very lengthy without that knowledge. Cross-references can be created, but they are very labor-intensive to maintain manually.

The LIMS can provide extensive help managing the archives. The actual records can be kept on paper as before, but the LIMS can maintain all the necessary indexes for efficient retrieval. Alternatively, the archive information may be stored in computer-readable form, such as magnetic tape, magnetic disk, or laser disk. Magnetic tape has limited shelf life and should probably be avoided in new LIMS designs. Computer technology is more expensive than paper, but it greatly facilitates use of the archive records. New records can be added easily and the effort to retrieve existing data is enormously reduced. After a number of years, the archives contain substantial computer-searchable information, and can become an important resource for retrospective studies.

One problem with computerized archives is maintaining compatibility with future technology. Some definite plan must be in place to make sure that records created today can still be used when systems are upgraded. One strategy is to commit to maintaining existing hardware just for the archives. Another is to insist at the time of any system enhancement that older archives be copied to the newer media or translated to the newer format.

Management Reports

The LIMS can periodically review the information in the database and print various management reports. A missing sample report is useful when test orders are accepted in advance of a sample. It lists the samples that are due but have not yet arrived at the lab. An overdue test report informs management about uncompleted test orders that have been in the laboratory too long. These reports tell management about errors and problems so corrective action can be taken in a timely way. A remaining workload report might be printed every day. It tells how much work is in progress, and gives management an early warning if the lab is slipping behind schedule.

Several LIMS reports help laboratory managers prepare reports to higher management. A test count report tells how many of each kind of test was performed by the laboratory over a selected period of time. It documents lab productivity and can help determine what supplies to order. A turnaround report gives the average time needed to complete each kind of test performed by the lab. It is useful to answer critics who complain about service, or to justify additional people or instruments to solve problems. Getting approval for laboratory improvements is easier with the kind of solid documentation a LIMS can provide. Turnaround and test count reports can be tabulated by workstation and by time and date. These reports might reveal periods of overload and periods of slack time. They may help management assign staff and schedules. The LIMS can produce staffing reports and personnel logs. This is usually possible because almost everyone in the laboratory uses the LIMS regularly. The LIMS might be able to prepare task accounting reports. These might be used for billing or internal chargeback accounting, as well as showing how the laboratory resources are being used.

LIMS Enhancements

Additional functions are sometimes added to the core LIMS beyond what is shown in Fig. 5.3. This is usually done to enhance the value of the LIMS to the organization by providing better information. These enhancements are very dependent on the nature of the business the lab supports. Most are aimed at improving reliability, relevance, or responsiveness or increasing the efficiency of the laboratory. They may attempt to aid interpretation of the laboratory data, perform more lab functions with less human input, or direct the flow of information more quickly to where it is needed. A few examples suffice to illustrate the possibilities.

In many laboratories, the concept of a material specification is appropriate. Quality control laboratories, for example, may test raw materials and intermediate or final product for acceptability. Water control laboratories have a similar need. Here the tests are used to determine whether the water is safe to drink or safe to dump into a river. In all these cases, the lab tests are being done to see whether the material meets certain requirements. These requirements can be stored in the LIMS database (if the LIMS is appropriately designed) and be used to automatically check test results. Each material the lab can process is given a name, a list of lab tests to be performed, and a range of acceptable results for each test. When a sample of this material comes in, the LIMS is simply told to run this specification. All the tests in the specification are automatically scheduled. When the results are complete, the values are automatically compared to the acceptable limits. The LIMS can directly issue a certificate of compliance, reject the sample, or refer the results to an authority for a decision. In these environments, the LIMS must have the additional capability of creating, editing, and deleting these material specification profiles.

A clinical laboratory in a hospital will have a patient database in addition to the core LIMS. The patients are the source of the samples, but the laboratory customer is the physician: he or she orders the tests and receives the results. The LIMS for a laboratory supporting a well-structured research program might have a study database. Each individual study undertaken by the company would be entered. Examples include animal studies or agricultural field research. Information in the study table would include the sources of sample, the sampling schedules, the test profiles, the duration of the project, and a host of other research-specific information. This LIMS may be capable of scheduling sample collections and automatically ordering tests for the entire project.

Many LIMSs are customized in the way they process test results. Sometimes sets of results are subjected to various statistical analysis routines. Test results may be automatically reported to larger corporate computer systems. An important concept in most LIMSs is a read-only mode of data sharing. Other computer systems may obtain LIMS data and process it as they wish. The original LIMS data, however, remain intact inside the LIMS as the official record of the laboratory results. The only changes to LIMS data are those that are properly authorized, following good laboratory practice.

Conclusions

A properly designed LIMS that is chosen wisely for the laboratory in which it is installed is a tremendous asset. It is hard to see how analytical testing laboratories with moderate to large sample and test counts can remain competitive without one. The continuing stress for more efficiency, fewer errors, higher workloads, and increasing demands for documentation of performance and accountability are important driving forces. The systems are likely to become more sophisticated as analytical chemistry, like all other sciences, continues to exploit the advantages of the information age.

References

1. R. D. McDowell, ed., *Laboratory Information Management Systems.* (Wilmslow, U.K.: Sigma, 1987).

2. R. R. Mahaffey, *LIMS: Applied Information Technology for the Laboratory* (New York: Van Nostrand Reinhold, 1990).

3. *E1578: Standard Guide for Laboratory Information Management Systems (LIMS)* (West Conshohocken, PA: ASTM, 1993).

4. M. D. Hinton, *Laboratory Information Management Systems* (New York: Marcel Dekker, 1994).

5. A. S. Nakagawa, *LIMS: Implementation and Management* (Cambridge, U.K.: The Royal Society of Chemistry, 1994).

Chapter 6

Laboratory Automation

Joe Liscouski

Laboratory Automation Standards Foundation

What Is Laboratory Automation?

Laboratory automation is the application of computing, information, robotics, electronic, and mechanical engineering technologies to laboratory problems. For the sake of this discussion, it includes situations where a computer is an integral part of the laboratory equipment (for example, Fourier transform techniques in infrared and nuclear magnetic resonance experiments). Topics in the field include

- Sample storage and preparation (this includes robotics); this treatment of sample preparation is light because it is covered in more detail elsewhere in this book.
- Instrument control, data acquisition, data analysis, and reporting.
- Laboratory information management systems; these are covered in depth in Chap. 5.

Automation is used to assist people in carrying out repetitive work (freeing them for other work) and to provide them with capabilities that might not be cost-effective otherwise (such as curve fitting, database searching, and graphic treatments of analytical data). The primary benefits of automated systems include

- Reduced cost of laboratory operations
- Improved quality of data and reproducibility of results
- Better data management
- Better tools for carrying out laboratory work (automated analysis of chromatograms, automated sample preparation, and laboratory information management systems for management and test results organization)

Beyond those points there are added benefits outside the laboratory. Laboratories, particularly analytical ones, do not exist in isolation. They support research, product and process development, and manufacturing operations, as well as other groups. Well-designed systems take into account the requirements of other groups. Integrating the lab's information management system with process control and inventory management can speed the release of products to customers, reducing the time they spend in inventory and the cost of inventory management. Process control groups can benefit from easier access to critical data, not only in speed of access but also in getting it in a form that can be used efficiently with their electronic data systems.

A number of factors are driving the movement to laboratory automation systems, primarily economic and quality factors:

- People's time is valuable and companies are looking to use automation to make better use of employees.

- Well-trained people are becoming increasingly rare, so efforts that can free them from routine work are valuable.

- Increasing workloads drive companies to look for alternatives to people for routine work.

Automation systems provide better-quality data and tighter control over the way routine work is done. Electronic transfers of data and information, once tested and proven, are not prone to transcription errors.

The rest of this chapter covers the development of laboratory automation, the directions in which technology is moving, the considerations that must be reviewed in planning for systems, and the automation options for various phases of laboratory work.

Approaches to Laboratory Automation

There are two ways of looking at laboratory automation: automation focused on solving a particular problem and efforts that try to view laboratory work as a process, with automation as a means of improving that process. In most cases, particularly in industrial settings, the latter is preferred.

Focused Automation Problems

These often occur in research settings, particularly where small numbers of people are involved. The purpose of automation is to make an instrument more useful and relieve bottlenecks. This is how automation in laboratory work began. For the sake of the discussion, we look at gas chromatography as an example; chromatography was one of the first techniques to extensively exploit automation.

From the vendor's viewpoint, everything centers on the instrument; automation is a way of remaining competitive, and the types of automation used should improve the lab's ability to use that instrument (Fig. 6.1).

Before microprocessors became available, automation was mechanical. Autoinjectors were the first instrument add-on, solving the problem of getting samples into the instrument and freeing the analyst from the tedium of manual injections; the samples still needed preparation. Injection volumes were more reproducible, and people were freed from one problem. That shifted the bottleneck from processing samples to processing data. That problem was attacked through a progression of microprocessor-based systems culminating in the single and multi-instrument data

Figure 6.1 Manual methods give way to automated systems. *(© LASF 1992.)*

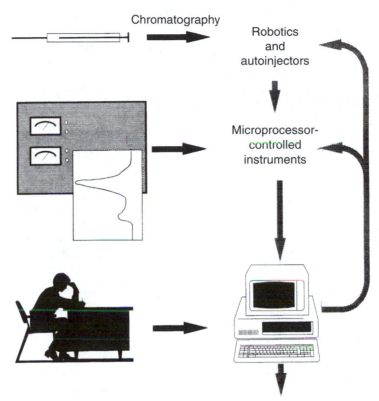

acquisition systems of today. Making all this work today is simple; for the most part, just plug it in and use it. Getting to that point required the solution of a number of problems.

Instrument Interfacing

The central problem is acquiring and digitizing a signal that accurately reflects the output of the instrument's detectors. At their core, all instruments are analog in nature, providing voltages, currents, or mechanical movements as a measure of whatever the instrument was designed to do. In chromatography, that output is a low-level voltage (micro- or millivolts at the detector, or volts if amplifiers are used). Voltage is the most common form of instrument detector output in analytical chemistry. The first of two issues is to capture that signal and represent it as a number in a digital computer. The second is to do it on a regular and repeatable basis.

Reading an alcohol or mercury thermometer is a matter of performing an analog-to-digital (A/D) conversion. The movement of the liquid in the tube is an analog response to changes in temperature. Reading the temperature is a matter of projecting the top of the liquid against a scale and making an estimate of the numerical value—an A/D conversion. Computer systems with A/Ds do essentially the same thing: they compare a voltage against a scale and make a best guess of the actual value, "best" being limited by the resolution (number of bits) of the A/D.

Suppose your thermometer is scaled from 0 to 100 degrees Celsius and it has 4 divisions at 25-degree intervals. That is the equivalent of a 2-bit A/D ($2^2 = 4$); 256 divisions would be an 8-bit A/D ($2^8 = 256$). The more divisions, the more reproducible and reliable the reading. There are three types of A/D converters on the market today; each has its strengths and applications (the details will change as technology improves):

| | **Type of A/D** | | |
	Successive Approximation	**Integrating**	**Flash**
Number of bits	16–18	24	10–12
Conversion rate (conversions/sec)	0 to 100,000	0 to 50	0 to 1,000,000+
Application	General purpose, suitable for most instruments, sensitive to electrical noise.	Slowly changing signals (principally chromatography and slow scanning spectroscopy), insensitive to noise.	Best suited for very fast data acquisition where high resolution is a factor; can digitize noise.

The second requirement for analog data acquisition is that of making readings on a repeatable basis. If you were asked to make temperature readings every 10 minutes, you would use a clock to measure the intervals. Computers do the same using a programmable real-time clock that provides a repeatable time base; each time the interval elapses a reading is taken. If the reading rate were once an hour, you might set an alarm to ring when the interval was up, so that you could interrupt whatever you were doing, make the reading, and return to the interrupted task. Readings at the rate of once per second would find you dedicated to clock watching (or trying to convince someone that this is fun and they should do it). The computer equivalent puts some constraints on the computer operating system: It is either dedicated to clock watching (appropriate for high data acquisition rates) or it requires the ability to respond to an interruption; many commercial operating systems lack this capacity and respond with "Application unexpectedly terminated." Doing it properly requires an understanding of the operating system, the computer hardware, and good programming.

For the most part, analog data acquisition is of interest in special cases in which the experimenter has no alternative to analog data collection (a new type of instrument or experiment, for example). Most commercial instrumentation comes with either a data system for doing data acquisition or an RS-232 port that can be used to control the instrument and transfer data. (RS-232 has been the standard in chemical instrumentation for the last 20 years.)

RS-232 ports are the serial ports used on computers for serial printers, modems, and American Standard Code for Information Interchange (ASCII) terminals. They provide bidirectional communications (subject to vendor implementations) between an instrument and a computer system. The fact that an RS-232 port exists means that the device has an internal processor: Something had to capture the signal, convert it a digital value, encode it in ASCII, and control the data transfer. That reduces instrument interfacing to a communications problem—still not a trivial issue.

In order for communications to take place, there must be agreement on the communication protocol and data formats. These should be covered in the instrument's user manual. Serial data transfer occurs by sending characters between the device and the computer. Characters (printable and nonprintable) are either 7 or 8 bits in length and conform to the ASCII code. The following items must be understood (all settings are dictated by the manufacturer):

- Conformance to RS-232 wiring specifications: vendors may wander from the specification, so checking the manual is essential.

- Baud rate: data transfer rate is usually one of the following: 110, 300, 1200, 2400, 9600 (others are possible, but these are the common values).

- Number of bits per character: the character set can be represented by either 7 bits (128 possible characters) or 8 bits (256 characters).

- Parity: set to even, odd, or none. It is used in error detection, but often ignored.

- Number of stop bits: it is either 11 (if 110 baud is used) or 10. 110 baud is used for electromechanical terminals that need an extra bit time to complete each character's operations.

Next is the data communications protocol. What facilities has the vendor put in to test whether a data transfer has occurred without corruption? Many have none; others provide a means of checking the data for validity and retransmission if an error has been detected. This is a critical point. Those that do not provide error detection are assuming that you have handled all noise-related issues and that data transmission is reliable. They are betting on the probability of transfer and not the guarantee of correct transmissions. If errors occur they usually occur in bunches and instead of "7.35," "#%$#" is received. You are responsible for detecting the problem and taking corrective action. That is not satisfactory in automated systems. Positive control and error detection are essential.

Other interfacing techniques are available to experimenters: parallel and IEEE-488. These methods are not widely used in analytical chemistry.

Parallel interface (also called digital interfacing) is the transfer of digital data between two devices with all data bits being sent at the same time over separate wires, one wire per bit (in parallel, as compared to serial I/O [RS-232], in which the data bits are sent one at a time over the same wire). This technique is expensive and usually used only for short cabling lengths (less than 10 feet; RS-232 can be used with cables over 100 feet long). Its use today is limited to control circuits for switches. No communications protocols are available for instruments using this technique.

The IEEE-488 bus, first introduced in 1975, is a significant advance over RS-232 because it provides better control of data movement, has a standardized communications protocol, and a standardized connector and cable. Its use in analytical chemistry is limited to a few vendors' modular instrument systems, where it serves as a communications and control path. Although it is superior technically, its short cable lengths (maximum of 2 meters per cable), high cost of implementation, and late arrival on the scene limit its role. It is widely used in electronics applications (such as frequency generators and digital voltmeters).

Cable length was an important consideration in the early days of laboratory automation, when one computer (costing $50,000 to $100,000) may have had to serve a number of instruments, some of them in other rooms. The mid-1980s saw the arrival of the IBM-compatible PC as a common laboratory system. Its low cost, permitting one-instrument-one-computer systems, could provide a basis for renewed interest in IEEE-488, or follow-on techniques, if serial communications protocols are not standardized. Cost will still be a factor, particularly because serial ports are a standard part of PC computers. (Note: when the total cost of hardware, software, and implementation is taken into account, the differential is small.)

The National Institute for Standards and Technology has an organization called Computer Automation for Analytical Laboratory Systems (CAALS). That group is developing communications standards for modular instrumentation that hold promise as a successor to current technologies. Their first standard specification was published in 1994.

Instrument Control and Data Processing

Once we were assured of collecting accurate data, the next step in automating chromatographs was the analysis of the data. (Until microprocessor-controlled instrumentation can set instrument operating parameters such as temperatures and flow rates, the data systems will be acquisition devices and provide little control.) This requires data filtering (to reduce noise), peak-picking, and reporting routines, followed by data storage and archiving. These systems may have an interface to the autosampler that allows them to control the timing and sequence of sample injections (Fig. 6.2). This is the point where most instrument vendors stop. Their focus is to sell the instruments (primarily) and the automation as an add-on. End-users may view the automated data capture as essential, depending on their workload. Beyond that, integration and data transfer with other systems is an exercise for the user.

At this point, you have an automated instrument that may be completely satisfactory for a small laboratory or research project, but not a laboratory serving a company in the chemical, pharmaceutical, biotechnology, aerospace, or other industries. If automation is limited to this view, the end result will be islands of automation, not automated laboratories.

Laboratory-Wide Automation Technology

Applying automation to an entire laboratory is another matter entirely. It is not the sum of a number of individual efforts: automating all the instruments does not yield an automated facility. What is needed is the systematic application of automation to laboratory work. The first step is to view the analytical laboratory as a place where a process is going on, rather than a room where a number of discrete, independent analyses are performed. Figure 6.3 shows the data flow model developed by the Laboratory Automation Standards Foundation. This viewpoint shows the relationship between the steps in a lab's work: the output of one becomes the input of another. Those connections must be planned. The vendors of laboratory instruments, data systems, robotics, and software view laboratory automation and computing from their perspective, as shown in the earlier chromatography example. Vendors of laboratory information management systems (LIMSs) view the lab from the point of view of the central LIMS database (as evidenced clearly by the American Society for Testing and Materials *LIMS Guide*), with all lab activity fed from or directed by the LIMS. These

Figure 6.2 Control and data flow in automated systems. (© *LASF 1992.*)

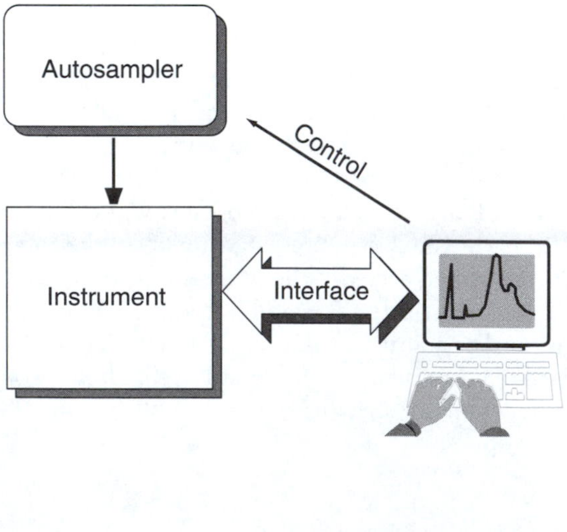

Figure 6.3 LASF laboratory model. (© *LASF 1992*.)

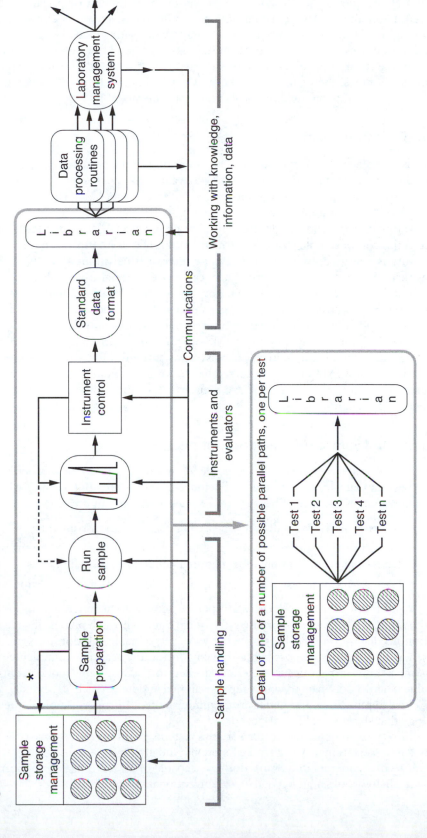

*Sample/specimen return.

points of view do not mesh well unless a considerable amount of consulting and software is involved. Designing a lab automation system with interconnectivity as a criterion, rather than an afterthought, will result in a system that is easier to work with and manage.

Laboratories that perform tests on samples use a set of events that control the sequence of tasks in the laboratory (Fig. 6.4). These events represent the administrative level of lab work that connects samples, their submitters (the lab's customers), and the laboratory work flow. The next few sections of this chapter examine each stage of the model.

Sample Storage Management

This is an area that has only recently received attention from automation vendors. Samples are stored in a variety of places including shelves, vaults, and refrigerators. Most storage areas are designed assuming that the storage/retrieval agent is a human being suitably equipped with basket, large pockets, and a stepladder to reach the top shelf.

Automated Systems Integration Corporation (Camp Hill, PA) has developed a system for the mechanization of sample storage management; Fig. 6.5 is an illustration of one installation. The structure is a large box—room size—whose interior is partitioned in sets of shelves. A conveyor runs through the system and a robot arm can move material from a shelf to a tray on the belt. A computer database maintains an inventory of samples, locations, and related information. The computer can be given a list of samples and the bar-coded containers will be delivered to the receiving station. The box can be temperature and humidity controlled.

This equipment coupled with a LIMS provides some intriguing possibilities, particularly with a LIMS's automated sample login (many systems have the ability to accept samples that have been logged into the LIMS on a prearranged basis). Samples could be delivered to the storage system equipped with a bar-code reader, have their arrival verified and entered into the LIMS, and then scheduled for analysis.

Automated Sample Preparation: Robotics

Sample preparation is covered in more detail in Chaps. 2 and 3 of this book. Laboratory robots have been in place for over a decade. Their acceptance is growing, particularly as the demands on human resources increase. There are three types of robotics in use today:

- Autosamplers, which deliver prepared materials to an instrument
- Fixed-function, programmable systems capable of doing sample preparation for specific types of instrumentation using a limited number of external devices (such as balances)
- General-purpose systems, user-programmable and adaptable to a wide range of applications

The first two have had an easy acceptance because their functions and roles are well defined and bounded. The general-purpose systems have gone through phases of wide interest and anticipation, to disinterest, and back to being viewed as a viable solution for some problems. As with any new technology the possibilities quickly outrun the realities of implementation, then reality sets in. Reality in robotics is the effort needed to make systems work. Companies realized the problem and addressed it with preprogrammed modules (such as Zymark's Pye-technology), packaged applications, and better programming systems. Properly applied, they are a potent technology that should not be overlooked.

Two types of general-purpose robotics systems are in use: fixed-base (the arm moves and the robot base is stationary) and tracked systems (the arm is on a linear track, which can be from a meter to many meters in length). Both are capable of a wide range of applications. The material covered below looks a bit beyond most conventional laboratory robotics.

Figure 6.4 LASF laboratory model with controlling events. *(© LASF 1992.)*

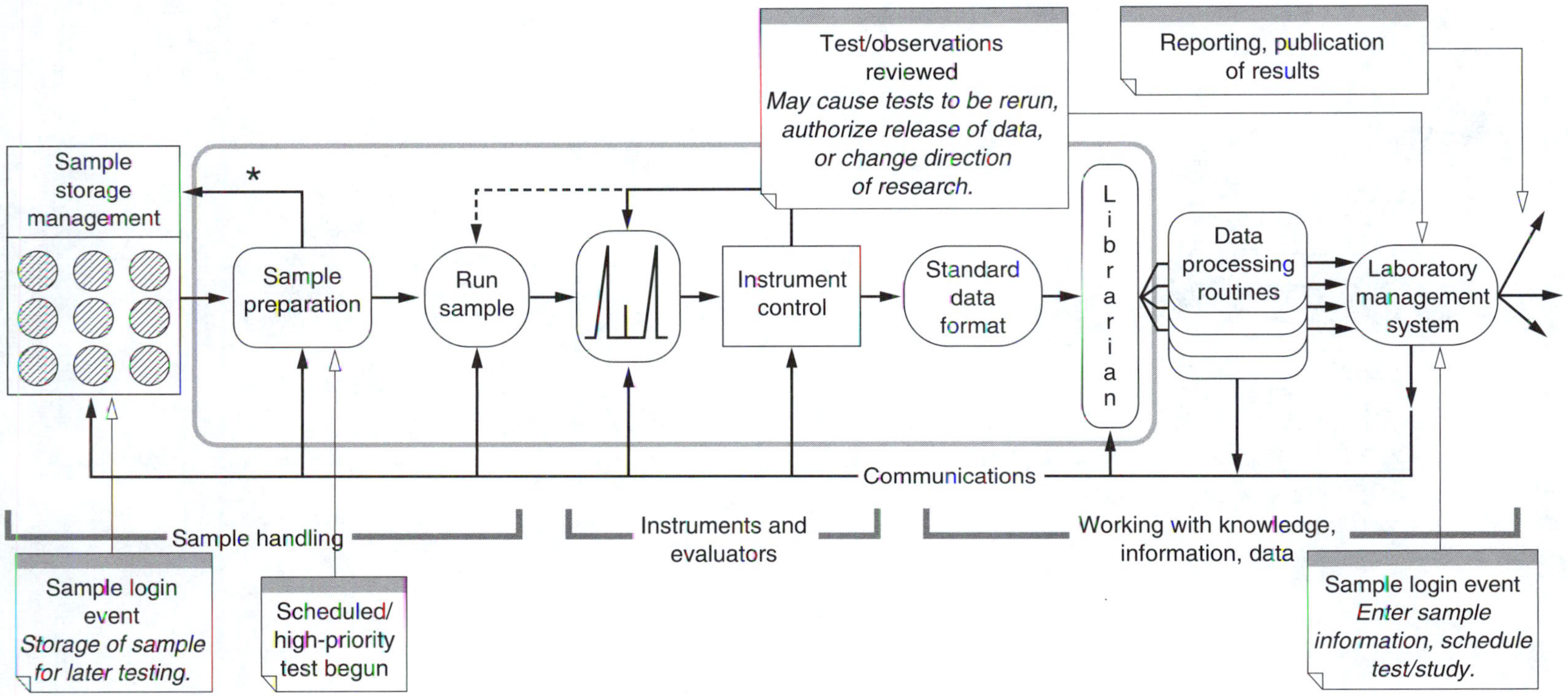

*Sample/specimen return.

Figure 6.5 Automated sample storage system.

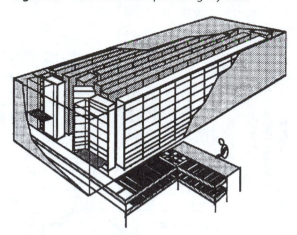

The major application for lab robots is as a replacement for manual activities, primarily sample preparation, performing titrations, and introducing samples into instruments, although these systems are capable of much more. The limitations on their use reflect the boundaries of the views of the vendors rather than their true potential, and center around the control system (Fig. 6.6).

The robot control system is designed to manage all of the actions the robot is capable of taking (such as changing hands, getting a sample tube, and placing something on a balance), actuating control signals (closing a switch or turning on a motor through a switch box), and reading and writing files. Some systems have message transfer capabilities. The programming for most robotics applications revolves around these actions, essentially carrying a sequence of preprogrammed events. Their interaction with an instrument, aside from placing something in one or reading a balance or pH meter, is minimal.

Figure 6.6 Interaction between robotic system and other components of automated system. (© *LASF 1992.*)

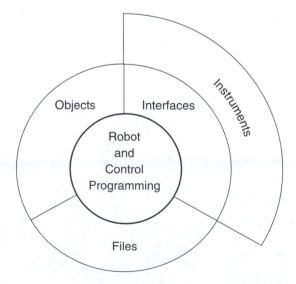

If we take one step further back, our perspective on laboratory robotics can change dramatically. Consider a situation where the robot's programming, rather than being a complete program for a task, is instead a series of control modules that carry out certain functions (such as weigh sample, put sample in instrument, and prepare sample number N). Note that these are higher-level primitives than the typical control programming. The robot controller looks to incoming messages from a supervisory system to tell it which of these primitives should be executed at any point.

The supervisory system's programming allows it to access a list of samples to be processed, test pass/fail/alarm criteria, the sample preparation protocol, and data from the instruments used in the analysis. Data handling from the instrument is one area where this approach is an improvement over conventional robotics. Robots are often used when the instrument output consists of a very small number of fully reduced data points that can be acted on with simple decision algorithms. Full instrument support of the type described here may include data acquisition, processing, and reduction to final values.

The sequence of steps in an automated process would look like this:

Step	Supervisory System	Robot	Instrument* Subsystem
1	Receive list of samples and standards to be processed.		
2	Initiate sample prep by sending a sequence of commands to robot.	Robot executes primitives as received and carries out actions.	
3		Prepared sample is placed in instrument.	
4	Trigger instrument to begin analysis.		Sample is processed, data acquired and processed, results sent to supervisor.
5	Supervisor compares sample results against specifications. If sample is OK, proceed to next; otherwise, initiate backup procedure.		

* May be part of the supervisory system or a separate module on another computer.

The backup procedure could include a check of the instrument's calibration, a check of reference samples, and a repeat of the sample in question. The result is that if a sample is in question, the system automatically generates the support information needed to make informed decisions.

The system should periodically check calibrations, even if all samples are within specification, to show that the entire process is working properly and the results are indeed valid.

This runs counter to traditional approaches to robotics. Many robotics advocates stress the point that for best efficiency, the robot should always be moving and several samples should be in some stage of preparation, much like an assembly line. There are situations where that is perfectly appropriate. The approach should be dictated by the efficiency of the overall process, not necessarily on optimizing subprocesses.

Robotics has yet to see its full potential realized in laboratory work. That is going to take a change in mindset, moving to a systematic analysis perspective and concentrating on the process of lab automation and data flow. It will also require a change in the physical architecture of laboratories.

Laboratories are designed the way they were decades ago. The laboratories' mission must drive automation and information handling, and that in turn must drive the physical layouts.

Instrument Systems

Fundamentally, instrumentation has not changed much in the last decade. The major advances have been in automation, data collection, and processing. From the standpoint of the instrumentation specialist, that device is the center of the universe and everything else is subservient to it (Fig. 6.7). Note that the models used for LIMS, robotics, and instruments, from the point of view of specialists in those areas, do not mesh well—the connections are left as an exercise for the user. The details of each instrument and the handling of data are covered under the various techniques in this book.

In 1992 a fundamental change in the underlying structure of laboratory automation was announced. The Analytical Instrument Association (AIA), a group of the major instrument vendors, announced the release of the first data format/data interchange standard; the initial entry was for chromatography (the AIA can be contacted at 225 Reinekers Lane, Suite 625, Alexandria, VA 22314). When the standard was adopted by the vendor community, a chromatogram could be collected by one data system and exported to another where the file could be read and processed. This was a significant step forward.

Before this development, the data storage format for instrument data was considered a proprietary information and the user community could not gain access to the data in its native format. Some systems permitted export of the data in files, but it could not be used anywhere else—no one knew how to read the files. The AIA's work broke this limitation.

The data in the files are divided into a series of categories as described by the ASTM Analytical Data Interchange Storage Standard (ADISS) model (ASTM subcommittee E49.52):

- Raw data: the data as produced by the instrument's detectors and captured by a data acquisition system
- Final results: the results of processing the raw data

Figure 6.7 Laboratory automation from the instrument vendor's perspective. (© *LASF 1992.*)

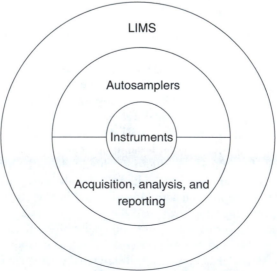

- Full data processing method: a description of all information needed to understand how the data were processed (the software used, model numbers, and versions)
- Full chemical method: all chemical/physical method information needed to repeat the experiment under exactly the same chemical or physical conditions
- Product quality information: sample description and information needed for regulatory compliance

The files—databases—are built on network Common Data Format (netCDF), a public-domain package available via the Internet through Unidata (it is also supplied with the standards kits from the AIA). NetCDF runs on virtually any computer because it is distributed as source files that must be compiled on each machine using the C language compiler.

One of netCDF's unique features is that it is a self-describing data format. There is sufficient information in each data file to allow a utility supplied with the package to find out what fields are available and provide you with their contents. The files can also be translated to ASCII formats that anyone can read.

The standard development work is being extended to include specifications for mass spectrometry, Fourier transform infrared spectroscopy, and the atomic spectroscopy techniques of atomic absorption, inductively coupled plasma emission, directly coupled plasma emission, and other emission techniques. There are considerations for standards developments for other techniques.

All of this yields a singularly important benefit: when the standard is implemented, the user community can export instrument-derived data in a format that can be read by other applications, taking advantage of analysis tools that may not be supported by the instrument data systems vendor. Why is this important to you?

- That data represent an investment in the time and effort to get them and as a result have some inherent value that should be managed.
- The test data may be necessary to support product claims in the future (product liability), provide support for patent applications, or accompany submissions to the Food and Drug Administration.
- The data may be needed to respond to customer requests.
- You may want access to the data for long-term studies (product stability studies, for example).
- There may be more powerful analysis techniques available than were provided with the initial instrument vendor's package. These standardized files can be used as a means of moving data between the original system and those packages.
- The data may be the only description of a sample you have, once all the initial analysis is completed (the actual sample may have been completely used or discarded), and is the only basis for answering questions about an analysis.

Data Storage and Management

This block in the LASF model is called the data librarian; it is a concept rather than something you can purchase, at least at the time this chapter was written.

It addresses a fundamental need in laboratory work. The advent of data format standards gives us the ability to publish large numbers of files (thousands per week in some labs), each containing the data for a particular analysis or series of instrument runs. How can we manage them? Creating subdirectories on a particular computer is not sufficient because it is easy to duplicate names and erase existing data with the same name. File names also force the development of an index that can be searched so that material can be found easily.

Consider the New York Public Library. If you want to find information on a particular topic, you search the card catalog for the titles you need, go to the circulation desk, make the requests, and eventually receive the material you want. You do not know how the books are actually stored, and you do not really care as long as you get what you want when you want it.

Consider your laboratory. You would like all analytical data on a particular product produced between two dates. Naturally, it is needed yesterday. What kind of exercise would that be? Suppose a new baseline correction technique was developed and you wanted to reprocess all data using the existing algorithm for comparison purposes. Suppose a stability program is in process and a degradation product peak begins to appear, and you want to go back over the past two years and check other materials.

These are reasonable requests for data that may be difficult to satisfy without the computer equivalent of the New York Public Library's data storage and retrieval system. It does not exist, but you need to begin thinking about organizing your data in such a way that it can be found and worked with.

Data Processing Libraries

Although the library to house the data files may not exist yet, the ability to create those files in standardized formats does. That opportunity is not being lost on software developers. This change in data handling frees researchers, primarily interested in analysis rather than acquisition, from the burden of having to create a complete system in order to work on only one aspect of it.

What we can expect to see over time—and products have been released—is the development of libraries of data analysis procedures for processing analytical data using the data format/exchange standards as input. This has some interesting possibilities, particularly if coupled with the librarian concept noted above. To compare two different approaches to chromatographic data analysis, we could build up a collection of representative and problem chromatograms, feed them to both algorithms, and compare the results.

The comparison is much more useful than duplicate injections because small variations in actual sample handling can bring the reality of small differences in results into question. Using data files as input means that the data are exactly the same and any differences are significant.

Laboratory Information Management Systems

This subject is examined in detail in Chap. 5. However, a couple of points must be considered here.

LIMSs originally developed as sample tracking systems. Instrument data systems handled the laboratory data and the administrative information about samples was managed by the LIMS. Things have become a bit more complex since then.

There are three major viewpoints concerning the nature of LIMS:

- LIMS is the centerpoint of all laboratory activities; it coordinates them and manages the interfaces and data collection.

- LIMS is not a product, but rather a concept that encompasses the entire laboratory.

- LIMS really means sample tracking and management and is only one component of a laboratory-wide structure.

These choices are not a matter of right or wrong, but of how you want your laboratory to work.

A Laboratory Automation Example

In this section of the chapter we consider two different approaches to a laboratory automation problem: one that would have been used before the development of data format standards and one that uses some of the ideas noted earlier. In both instances the problem includes

- Automated sample storage
- Robotics for sample preparation and insertion into an instrument
- The movement of test results into a LIMS

The instrument generates a large number of data points—thousands per sample—that are reduced to a dozen or so parameters from which a single numerical value is derived. For the sake of space, we concentrate on data flow issues rather than the details of each stage.

Pre–Data-Format Standards

Stage	Comments
Automated sample storage	The worklist produced by the LIMS would be used to select samples from the storage unit; that link could be via a person entering information into the storage units control system, or, if files could be transmitted to the control system, it could parse the input and obtain the needed samples.
Automated sample preparation	The robotics system would take the sample containers, remove, weigh, and extract the samples, and prepare them for introduction into the instrument. The procedure is under control of the robot's CPU and there is no interaction with external computers. If desired, a more sophisticated system could be used.
Instrument and data acquisition and processing system and data entry into the LIMS	The detector output is sampled and processed and the results stored. One of two things can happen: • Sample results can be printed and manually entered into the LIMS. • A communications link can be established between the data system and the LIMS. In the second case, the programming will probably be a customized link, unique for every installation. (Some LIMS vendors have off-the-shelf packages for some data systems.) The data stream, either direct or via file transfer, is parsed on the final result entered into the LIMS sample record. The major issues with this method stem from maintaining a correlation between sample results and the data they are derived from. Initially a link is maintained pairing the raw data in one system and a copy of the final result in the LIMS. However, once the data storage on the data system becomes full, it must be purged and old data put into archival storage. At that point, the pairing of data (coordinating instrument data with that on a LIMS) must be maintained by the archival records of what information was put on what archival storage. Searching of historical data depends on reloading the instrument

	data system from the archives (which may not always be possible), performing a search, extracting the needed data, and purging the system again.
LIMS	The LIMS provides sample tracking, worklist generation, backlog reports, sample results entry, data checking against specifications, and reporting.

Using Industry Standard Data Formats for Exporting Instrument Data

Stage	Comments
Automated sample storage	Same as above.
Automated sample preparation	Same as above.
Instrument and data acquisition and processing system	The detector output is sampled and processed and the results exported in standardized data formats, or just sampled and exported for later processing.
Data librarian (note: this capability does not yet exist as a separate commercial product)	The instrument data is entered into the librarian. The librarian, using a hierarchical storage system, is responsible for maintaining the data and its migration across media; the user is required to know only what he or she wants and the librarian will get it. The linkage between a LIMS and the instrument data is made through the librarian.
Data processing libraries	The development of standardized data format will spur the development of standalone data analysis packages from instrument vendors, third-party vendors, university researchers, and chemists working in industry; some third parties are already developing products that use the AIA standard as a data source. These routines can request one or more files from the librarian, perform the computations, and report the results, including entry into a LIMS (using file transfer or other means, depending on the vendor).
LIMS	The LIMS provides sample tracking, worklist generation, backlog reports, sample results entry, data checking against specifications, and reporting.

The development of data format standards will have a considerable impact on the design, development, and implementation of laboratory systems. This brings us to a key element in laboratory automation.

Planning for Laboratory Automation

Laboratory automation is the application of automation technologies to laboratory work in order to improve people's ability to

- Get work done
- Apply cognitive talents to efforts where intelligence is of value

- Reduce the cost of work
- Improve the quality of work and analytical results
- Enable those working in laboratories to pursue research problems
- Put the results of the company's investment in lab work to better use through better data management and more rigorous analysis

There is no set pattern for laboratory automation. The LASF model deliberately avoids noting what automation tools are being used in each step, because those implementation decisions are a matter of individual discretion. The question you must consider is this: How do you want your lab to work?

That is not a trivial concern. Different people get their work enjoyment from different tasks. Some prefer designing systems, others prefer analyzing data; still others see developing new procedures as rewarding occupation. Automation should not be so thorough as to remove the reasons we got into chemistry from our daily activities.

The sequence of planning events would look like this:

- Determine your goals for automation. What do you want to accomplish as a result of the investment?
- Examine the way your lab works. What should be changed, what preserved?
- Draw a process flow for lab work, perhaps a modification of the LASF model. What steps will be automated and to what degree? How are the steps related to each other? How are data and information moved from one step to the next? Where are they stored? What role do standards play?
- What are your priorities in implementing a system?
- Determine an implementation plan.
- Periodically review progress to ensure that it is meeting the goals that were set. What alterations are necessary?
- Keeps things as flexible as possible; the technology is changing and you do not want to lock yourself out of new products.

The planning process should include training, the development of new standard operating procedures and documentation of how the overall system is supposed to work, and documentation for changes in direction as they occur.

Validation

Many people believe that system validation is required only in FDA- or EPA-regulated industries. That is not true. Validation is enforced in regulated industries but required everywhere. The purpose of a validation program is to ensure that a system is demonstrated to work as expected. It is a matter of proof of concept and implementation. It is also a matter of building confidence in a laboratory's data. Confidence can eliminate the need to repeat work, which adds cost and time delays into problem resolution.

The details of a validation program are up to the discretion of each laboratory. The articles listed below will give you a starting point to work from.

Suggested Readings

Validation Programs

CLARK, A. S., "Computer Systems Validation: An Investigator's View," *Pharmaceutical Technology*, 12, no. 1 (Jan. 1988), 60.

ROSSER, M., "Draft Guidelines on Good Automated Manufacturing Practices: A Conference Report," *Pharmaceutical Technology*, 14, no. 4 (Apr. 1994), 74.

Software Development Activities Referencing Materials and Training Aids for Investigators (1987), an FDA publication.

TETZLAFF, R., F., "GMP Documentation Requirements for Automated Systems, Part 1," *Pharmaceutical Technology*, 16, no. 3 (Mar. 1992), 112.

TETZLAFF, R., F., "GMP Documentation Requirements for Automated Systems, Part II," *Pharmaceutical Technology*, 16, no. 4 (Apr. 1992), 60.

Additional References

CAMPBELL, J., *The RS-232 Solution*. San Francisco: Sybex Press, 1989.

DESSY, R. E., *The Electronic Laboratory: Tutorials & Case Histories in Laboratory Automation*. Ann Arbor, MI: Books on Demand, 1985.

DESSY, R. E., "The Analytical Laboratory as Factory: A Metaphor for Our Times," *Analytical Chemistry*, 65, no. 18 (Sept. 15, 1993), 802A.

HUBER, L., *Validation of Computerized Analytical Systems*. Buffalo Grove, IL: Interpharm Press, Inc., 1995.

LISCOUSKI, J., *Laboratory Automation & Computing: A Strategic Approach*. New York: Wiley, 1994.

LITTLE, J. N., "Advances in Laboratory Robotics for Automated Sample Preparation," *Chemometrics & Intelligent Laboratory Systems: Laboratory Information Management*, 21, nos. 2 and 3 (Dec. 1993), 199.

MAJ, S. P., "Analysis and Design of Laboratory Information Management Systems," *Chemometrics & Intelligent Laboratory Systems: Laboratory Information Management*, 13, no. 2 (Dec. 1991), 157.

MCDOWALL, R. D., "A Matrix for the Development of a Strategic Laboratory Information Management System," *Analytical Chemistry*, 65, no. 20 (Oct. 15, 1993), 897.

NELSON, D. C., "A Status Report on the Analytical Instrument Association Data Communications Standards Program," *Chemometrics & Intelligent Laboratory Systems: Laboratory Information Management*, 26, no. 1 (May 1994), 43.

SETTLE, F. A., P. J. WATER, AND H. M. KINGSTON, "An Expert-Database System for Sample Preparation by Microwave Dissolution. 2. Electronic Transfer and Implementation of Standard Methods," *Journal of Chemical Information and Computer Science*, 32, no. 4 (July 1, 1992), 349.

Separation Methods

Chapter 7

Introduction

Mary Jane Van Sant

Hewlett-Packard
Americas' Technical Center

From its infancy in the late 1800s, with contributions from Runge (1), Day (2), and Tswett (3), separation methods, commonly known as chromatography, have grown in significance and popularity to become a leading type of analysis in instrumental analytical chemistry. Today, chromatographic techniques have little to do with the separation of color (the technique name evolved from the earliest work of separating dyes or plant pigments on paper), but do involve the separation of components in a sample mixture. A number of types of separation methods have developed over the years to accommodate the various physical and chemical states of the many sample mixtures one may be interested in separating and analyzing. This section presents six chapters on gas chromatography (GC), liquid chromatography (LC), and capillary electrophoresis (CE), supercritical fluid chromatography and extraction (SFC), ion chromatography (IC), thin-layer (planar) chromatography (TLC).

This is not a comprehensive list of all available chromatographic techniques. Given the scope and size limitations of this publication, only the most common established chromatographic techniques are introduced and defined. The target audience for this publication is scientists and engineers, some from disciplines other than chemistry, who are seeking introductory information about the techniques listed. The goal is to use a common presentation format to provide easily understood explanations of the techniques in uncomplicated language. The reader is provided with explanations of how the technique works and what it does; types of samples analyzed; economic considerations such as the cost of equipment, maintenance, and training requirements; a summary of the routine operational characteristics; and a list of references for those requiring more in-depth and technical information. From these chapters, the reader should obtain an adequate understanding in order to interact with those more experienced in the field when choosing a technique to perform a desired analysis, pursue further study of the technique, or learn to perform an analysis.

Chromatographic techniques are dynamic processes wherein a mobile phase transports the sample mixture across or through a stationary-phase medium. As the sample comes in contact with the stationary phase, interactions occur. A partitioning or separation of the components in the mixture results from the differential affinity of each component with the stationary phase.

Components with the least affinity for the stationary phase pass through more quickly than the components with greater affinity for the stationary phase. As the separated components emerge or elute, a detector responds with a signal change that is plotted against time, thus producing a chromatogram. In thin-layer chromatography components do not emerge from the stationary-phase bed. Here, they are detached on the surface of the layer after the chromatographic process has been interrupted and the mobile phase is evaporated.

The mobile phase is either a liquid or gas and the stationary phase is a solid or liquid. Table 7.1 lists the various separation techniques and the most common types of mobile and stationary phases used.

The theoretical explanations for the partitioning or separating phenomenon for the various techniques can be fairly complex. Intermolecular forces involving mechanical, physical, or chemical processes are responsible for the differential distribution of the components and the resulting chromatogram. It is beyond the scope of this book to present extended theoretical discussions, but some of the chapters of this book may include some explanation of these forces. More rigorous treatment of partitioning theory can be found in other textbooks, such as those by Willard and others (4), Grob (5), Meyer (6), and Snyder and Kirkland (7).

All of the chromatographic techniques included in this book are suitable for both qualitative and quantitative analyses. At the beginning of each chapter, a Summary provides information that will help the reader match his or her separation needs with the appropriate technique. Additional-

Table 7.1 Common types of phases for separation techniques.

Technique	Mobile Phase	Stationary Phase
Gas chromatography	Gas (helium, nitrogen, or hydrogen).	Viscous liquid such as squalane, polyethylene glycol, or polymethyl siloxane. Adsorbent solid such as silica, molecular sieves, alumina, and porous polymers. Phase is housed in a column of glass or metal.
Liquid chromatography	Liquid water or organic solvents such as methanol, acetonitrile, propanol, or hexane.	Solid silica or a polymer such as polysaccharide or polystyrene housed in a column made of stainless steel.
Capillary electrophoresis	Electrolyte or buffer solution.	Buffer-filled capillary column, subject to applied voltage, creating migration of charges species.
Supercritical fluid chromatography	Carbon dioxide in supercritical state; may have modifier such as methanol. Other choices include pentane, hexane, sulfur hexafluoride, and isopropanol.	Modified silicas or polymers used in packed columns and cross-linked polymethyl siloxanes in capillary columns.
Ion chromatography	Aqueous solutions of acids, bases, or salts. These solutions are sometimes modified with water-miscible organic solvents such as acetonitrile or methanol.	Ion exchange resins, alkyl-bonded porous silica resins, and styrene–divinylbenzene polymers.
Planar chromatography (thin-layer chromatography)	Solvent mixtures such as hexane/acetone, chloroform/ethyl acetate, ethyl acetate–methanol, butanol/acetic acid/water, or methanol/acetonitrile.	Precoated layers of silica gel, aluminum oxide, cellulose, polyamide, or ion exchange material supported by glass, plastic sheets, or aluminum foil.

Table 7.2 Sample types and sensitivity ranges.

Technique	Type of Sample Analyzed	Sensitivity Range
Gas chromatography: very versatile, economic, and widely used.	Solid, liquid, or gaseous volatile organics and inorganic permanent gases	Parts per trillion or nanogram/liter to percent or grams/liter levels
Liquid chromatography: worldwide, a leading analytical technique for components not suitable for volatilization.	Liquid volatile and nonvolatile organic, inorganic, and biological compounds, polymers, chiral compounds, thermally labile compounds, small ions, and macromolecules	Parts per billion or micrograms/liter to percent or grams/liter levels
Capillary electrophoresis: analysis of large biomolecules requiring only nanoliter quantities of sample.	Liquid polar and nonpolar compounds, some elements, nonionic and ionic organics, inorganic anions and cations, macromolecules, and chiral compounds	Parts per billion or micrograms/liter to percent or grams/liter levels.
Supercritical fluid chromatography: can do GC and LC type analyses; instrumentation rather complex.	Solid, liquid, or gaseous, thermally labile and nonvolatile analytes as well as most samples analyzed by GC or LC	Parts per trillion or nanograms/liter to percent or grams/liter levels
Ion chromatography: provides separation of inorganic and organic ions and ionizable species.	Liquid inorganic anions or cations, organic acids, amines, amino acids, carbohydrates, and nucleic acids	Parts per billion or micrograms/liter to percent or grams/liter levels
Planar chromatography: simple, economic alternative to LC with simultaneous analysis of a number of samples.	Same as for liquid chromatography	Parts per billion or micrograms/liter to preparative scale

ly, at the end of each chapter a Nuts and Bolts section provides information concerning cost, training, and maintenance requirements.

In order to determine the best technique for a desired separation one must consider its compatibility with the state of the sample as well as the sensitivity requirements. Table 7.2 lists the techniques with a brief comment and compares the type of sample analyzed and the sensitivity ranges for the various techniques.

References

1. F. F. Runge, *Farbenchemie, I , II*, and *III* (1834, 1843, 1850); *Ann. Phys. Chem.*, 31 (1834), 65; 32 (1834), 78.
2. D. T. Day, *Proc. Am. Phil. Soc.*, 36 (1897), 112; *Science*, 17 (1903), 1007.
3. M. Tswett, *Ber. Dsch. Bot. Ges.*, 24 (1906), 116, 384; 25 (1907), 71–74.
4. H. H. Willard and others, *Instrumental Methods of Analysis*, 7th ed. (New York: D. Van Nostrand, 1988).
5. R. L. Grob, *Modern Practice of Gas Chromatography*, 3rd ed. (New York: Wiley, 1995).
6. V. R. Meyer, *Practical High-Performance Liquid Chromatography* (New York: Wiley, 1988).
7. L. R. Snyder and J. J. Kirkland, *Practical HPLC Method Development* (New York: Wiley, 1988).

Chapter 8

Gas Chromatography

Mary Jane Van Sant

Hewlett-Packard
Americas' Technical Center

Summary

General Uses

- Performs dynamic separation and identification of all types of volatile organic compounds and several inorganic permanent gases
- Performs quantitative and qualitative determination of compounds in mixtures
- May be destructive or nondestructive, depending on the type of detector used
- Can determine molecular conformation (structural isomers) and stereochemistry (geometrical isomers) with appropriate detector
- Can be configured for analysis of specific compounds using a wide choice of options
- Can be automated for analyses of solid, liquid, and gas samples and may include automated steps in sample preparation

Common Applications

- Separation of compounds in a mixture via a column and detection for qualitative and quantitative identification by any of several types of detectors
- Used in most types of manufacturing industry for analyses of raw materials, intermediates, or final product
- Used in environmental, forensic, pharmaceutical, and clinical/medical applications

Samples

State

Solid, liquid, or gas samples can be analyzed.

Solids
- Can be dissolved or extracted into a liquid solution that is injected into the gas chromatograph (GC) for analysis
- May be placed as-is in an auxiliary headspace analyzer and the out-gass126ed volatiles are analyzed by the GC
- May be extracted using supercritical fluid extraction

Liquids
- May be diluted, extracted, concentrated, or analyzed as-is via syringe injection
- May be sampled from a continuous stream via automated analysis using valves

Gases
- May be introduced by manual injection using a gas-tight syringe
- May be sampled from a continuous stream via automated analysis using valves

Amount

The amount of sample required for analysis depends on its initial state and concentration. Sample preparation procedures may be used to adjust the amount such that the following ranges are actually introduced into the GC:

Solids Headspace analysis used for solids and sludges requiring a minimum of a few micrograms to a maximum of a few grams

Liquids
- Syringe injection of sample volumes from 10 nL to 10 μL
- Purge and trap analysis using either a 5- or 25-mL purge vessel
- Valved analysis using sampling loops of volumes from 0.2 μL to 1 mL

Gases
- Syringe injection using gas-tight syringes from 10 μL to 25 mL
- Valved analysis using sampling loops of volumes from 0.250 to 10.00 mL

Preparation

Preparation can range from little or no preparation to sophisticated extractions, derivatizations, cleanup, or isolation procedures and depends on the type of sample analyzed. Automated sample preparation devices are available:

- Purge and trap analyzers
- Headspace analyzers

- Pyrolyzers
- Supercritical fluid extractors
- Robotic sample preparation stations

Analysis Time

Actual GC run time for sample analysis ranges from less than 1 minute to 1 or 2 hours. Many analyses take about 20 to 60 min. Sample preparation time may vary from a few minutes to several hours. Calibration procedures are often performed and require preparation of solutions or mixtures. These procedures may take a few minutes when one-point calibration is used or up to several hours if multilevel calibration is required.

Limitations

General

- It is not conducive for analysis of large molecular compounds such as proteins and polymers.
- The more sophisticated hyphenated systems (GC-FTIR [-Fourier transform infrared], GC-MS [-mass spectroscopy], GC-FTIR-MS, GC-AED [-atomic emissions detection]) are more costly (more than $50K).
- Minimal elemental information is available for most samples (except with GC-AED applications).
- Technique is used primarily for organic rather than inorganic analyses.

Accuracy

- Related to the type of injection, column, and detector used
- Ranges from 0.3% to less than 3% relative standard deviation

Sensitivity and Detection Limits

- Is most dependent on the injection process
- Ranges from parts per trillion or nanograms per liter levels to percent or grams per liter levels
- Is most dependent on the type of detector used

Complementary Techniques

- Liquid chromatography provides analyses of biological, macromolecules, and polymers that are more easily handled in a liquid state rather than a gaseous state.
- Supercritical fluid chromatography provides analyses of volatile, nonvolatile, and thermally labile substances.
- Capillary electrophoresis provides enhanced analyses in the area of bioscience and pharmaceuticals.
- Ion chromatography provides analyses of inorganic ions and organic acids.

Introduction

Gas chromatography is a dynamic method of separation and detection of volatile organic compounds and several inorganic permanent gases in a mixture. GC as an instrumental technique was first introduced in the 1950s and has evolved into a primary tool used in many laboratories. Significant technological advances in the area of electronics, computerization, and column technology have yielded lower and lower detectable limits and more accurate identification of substances through improved resolution and qualitative analysis techniques. GC is a very versatile technique that can be used in most industry areas: environmental, pharmaceutical, petroleum, chemical manufacturing, clinical, forensic, food science, and many more. Several leading manufacturers of gas chromatographs provide fairly extensive resources for training, method development, and operational support services.

The purpose of this chapter is to provide an introduction to the technique of gas chromatography. It is not intended to provide a rigorous treatment of the subject. The references listed at the end of this chapter provide pertinent cited publications for those interested in a more in-depth and academic study of GC.

How It Works

GC involves the partitioning of gaseous solutes between an inert gas mobile phase and a stationary liquid or solid phase. Figure 8.1 is a simplified diagram of the major component parts of a gas chromatograph. These parts can be identified as the gases, the injection port, the column, the detector, and the data acquisition system, consisting of an electrometer and recorder/integrating device.

Gases

The first major component of a GC is the carrier gas. A carrier gas supply is always present and usually consists of helium, nitrogen, hydrogen, or a mixture of argon and methane. The function of the carrier gas is to carry the sample through the system. The gas one chooses depends on the specific application and type of detector used; however, helium is one of the most commonly used gases. Additional gases might include hydrogen and air, which are associated with certain detectors used in the gas chromatograph. For example, the flame ionization detector requires a flame and needs hydrogen and air to support the combustion. The gases are commonly supplied by compressed gas cylinders, but gas generators are also an option as a gas source. Gas purity is a major consideration when obtaining gas supplies and one should consider the sensitivity and selectivity of the detector used when determining the appropriate purity level to use (the more pure, the higher the cost). An assortment of regulators, metal tubing, and connectors are used to interface the gases to the gas chromatograph. It is recommended that moisture traps and perhaps other purity traps be used to minimize the contribution of contaminants from the gas sources. Typical compressed gas cylinders contain pressure between 2000 and 2500 psi, but it is customary to work with supply line pressures to the GC in the range of 20 to 100 psi. Thus appropriate regulators and controllers are used to step down and control the pressure and flows into the instrument.

Figure 8.1 Typical gas chromatograph. *(Courtesy of Hewlett-Packard Company.)*

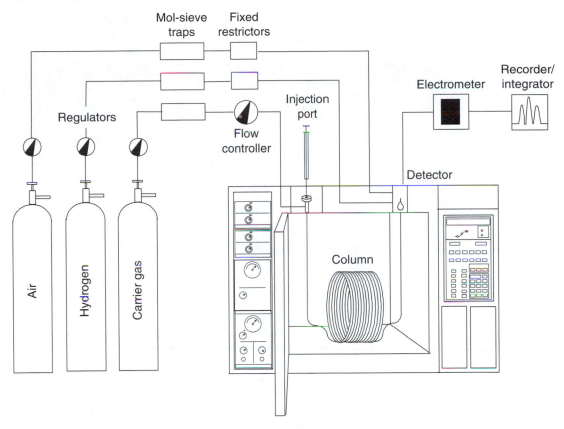

Injection Ports and Sample Introduction

The next major component of a gas chromatograph is the injection port or GC inlet. The purpose of the inlet is to introduce the sample into the carrier gas stream. Several types of inlets and sample introduction techniques are available. The most common type of analysis involves the injection of 1 to 3 µL of a liquid sample into a heated inlet. The injection is accomplished by a manual technique or by an automated injecting device that can accommodate a number of sample vials. The injection port is designed to interface with either packed or capillary columns. A packed injection port allows the entire volume of sample injected to be introduced into a packed or high-capacity capillary column (commonly known as a Megabore™ column) for analysis. In recent years, there has been a significant decline is the use of packed columns because of the significantly improved chromatographic performance offered by capillary columns.

A split injection port is designed to allow only a fraction of the injected volume of sample into the capillary column. Capillary columns have a limited sample capacity, so some of the sample injected is allowed into the column and the remaining sample is vented or split away. Figure 8.2 shows the flow dynamics for the split inlet on the HP5890 GC. A total carrier gas flow into the inlet is divided into three portions as it passes through the injection port. A small flow of 1 to 3 mL/min passes across the face of the septum in order to sweep away contaminants that out-gas from the septum. This septum bleed is vented out through the septum purge vent. Most of the total flow into the inlet passes down through the center of the injection port and the split occurs via an alternate pathway at the bottom of the inlet. A small flow passes down into the column and the

Figure 8.2 Split inlet of the HP5890 GC. *(Courtesy of Hewlett-Packard Company.)*

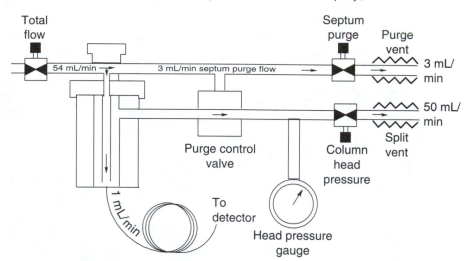

bulk of the flow travels out the split vent. Therefore, when sample is injected, some of the sample is allowed into the column for analysis and most of the sample is thrown away via the split vent. A purge control valve is used to direct the flow out of the injection port through the split vent. For quantitative reproducibility, it is important that the proportion of sample analyzed be known. Thus, one calculates the split ratio using the following formula:

$$\text{Split ratio} = \frac{\text{Split vent flowrate} + \text{Column flowrate}}{\text{Column flowrate}} \tag{8.1}$$

In Fig. 8.2, the split ratio is calculated as (50 + 1)/1, or 51:1. Typical split ratio values range from 50:1 to 400:1 and depend on the diameter of the column used. The more narrow the column, the lower the column's sample capacity and, thus, the higher the split ratio used in order to avoid column overload.

Many split injection ports can also operate in a splitless mode. In the splitless mode most of the sample is allowed to enter the column. This technique is suitable for samples with trace concentration levels. The splitless technique is a rather complex method in which many parameters must be considered. The most common of these parameters include efficiently transfering sample into the column, using chemically deactivated glass liners to minimize reactive sites in the injection port, choosing a solvent appropriately so that it is the first eluting component, performing temperature programming for oven temperature control, and manipulating valve switching times to minimize the amount and size of the solvent peak. Figure 8.3 shows the redirection of flows through the inlet via a purge control valve, such that the sample injected into the inlet passes into the column rather than split between the column and split vent, as seen earlier in the split mode.

In order to achieve the maximum sensitivity, especially of earlier eluting components, it is necessary to switch the position of the purge valve back to the split configuration about 0.5 to 2.0 min into the analysis. This step allows for the removal of some of the excess solvent, minimizing the solvent peak tail and preventing the solvent peak from masking components eluting near the solvent.

The procedures used for setting the flows associated with the split and splitless techniques are viewed as cumbersome by most novice users and generally some degree of instruction is required. The traditional injection ports have manually controlled regulators and flow controllers that are adjusted to deliver required flows in accordance with measurements made by an electronic flow

Figure 8.3 Splitless configuration. *(Courtesy of Hewlett-Packard Company.)*

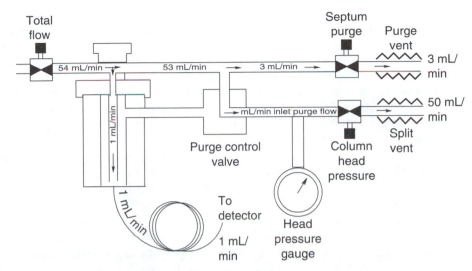

measuring device or a glass bubble flowmeter. In recent years, GC manufacturers have introduced electronic pneumatic-controlled pressure regulation of flows. Hewlett-Packard, Perkin-Elmer, and Hitachi have introduced completely electronic pressure-controlled GCs that do not require the manual measurement of any flows, a welcome development for many neophyte users.

A technique known as cool on-column injection is designed to allow the sample to go directly into the column without vaporization. This approach is suitable for samples that are thermally labile or prone to react with the components in the injection port. This last technique is considered the most accurate and precise mechanism for sample introduction with capillary columns (1). It has not found widespread routine use in the United States but has had some acceptance in research and development efforts.

Liquid samples from a process stream may also be introduced into the injection port via an automated, fixed volume, heated liquid sampling valve interfaced with the gas chromatograph. Gaseous samples may be injected into the inlet by either a gas-tight syringe or by an automated gas sampling valve of fixed volume.

Auxiliary sample introduction devices include headspace analyzers, purge and trap systems, pyrolyzers, and air-monitoring concentrator units. These devices are interfaced with the gas chromatograph in order to perform preparatory steps before sample introduction. The headspace analyzer allows one to place a solid, viscous liquid, sludge, or similar substance in an enclosed vial and expose the contents to controlled temperatures, thus releasing volatile organics in the vapor space contained in the vial. The vapor space is then transported to the injection port via a heated transfer line for subsequent chromatographic analysis. The purge and trap device is commonly used in the analysis of volatile organics in water samples. A sample of water is sparged with carrier gas and the volatiles are transported to an adsorbent trap and thus concentrated. After a specified time the contents of the trap are thermally desorbed into a transport stream that is introduced into the injection port for analysis. The pyrolysis unit is used to thermally degrade high-molecular-weight substances such as polymers into smaller fragments with boiling points more suitable for analysis by gas chromatography. The fingerprint pattern of the fragments is used to identify the parent polymer. The air-monitoring concentration device provides cryogenic concentration followed by thermal desorption of volatile organic compounds (VOCs) in the air at sub–part-per-billion levels. The system is used in compliance with several Environmental Protection Agency (EPA) methods for VOC analysis.

Columns

The third major component of the gas chromatograph is the column, which is responsible for the separation of the components in the sample mixture. The first GC columns used were packed columns developed in 1951 and 1952 (2, 3). The first open tubular or capillary column was introduced in 1958 (4) and because of its significant ability for high-quality separations, the capillary column has become the column of choice. Figure 8.4 shows a cross-section diagram of the porous layer open tubular (PLOT) column and the traditional wall-coated open tubular column.

When describing the columns, one states the length in meters, diameter in millimeters, stationary phase film thickness in micrometers, and type of stationary phase. The inner walls of the capillary tube are coated with either a solid porous material or a viscous liquid material. The capillary column is made of fused silica quartz coated on the outside with a polyimide to give it durability. The inner diameter commonly ranges from 0.05 to 0.53 mm, the length from 10 to 150 m, and the film thickness of stationary phase from 0.05 to 3 μm.

Most experienced chromatographers use the capillary column in performing their required separations; however, many government-regulated methods (such as those governed by the EPA or the Food and Drug Administration) may specify the use of packed columns. Others continue to use packed columns for fixed-gas analyses because of the lower costs compared to PLOT column alternatives. Figure 8.5 is a cross-section diagram of a typical packed column. Descriptions of the columns include length in feet or meters, diameter in inches or millimeters, concentration of liquid stationary phase in percent, type of liquid phase, and size and type of solid support. Packed columns contain either a porous support material packed into the metal or glass tubing for gas–solid chromatography, or a viscous liquid phase coated on the outside of a solid support material packed into the column for gas–liquid chromatography.

There are dozens of types of column stationary phases to choose from when packed columns are used and determining the appropriate phase is a tedious process. Because of the significantly improved separation ability of capillary columns, the process of column selection has been simplified over the years. Often one can determine a suitable column by consulting the application section of the leading chromatography supplier's catalog. The Vendor section of this chapter lists the major suppliers. Figure 8.6 is an application note from Hewlett-Packard's catalog. It specifies all of the pertinent parameters needed for setup and analysis of free organic acids in water.

Figure 8.4 Capillary columns. *(Courtesy of Hewlett-Packard Company.)*

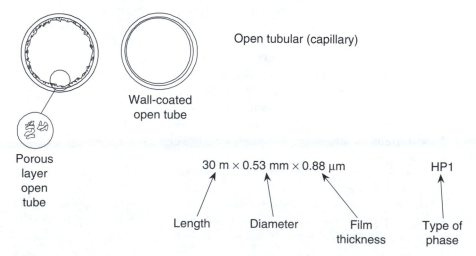

Figure 8.5 Packed column. *(Courtesy of Hewlett-Packard Company.)*

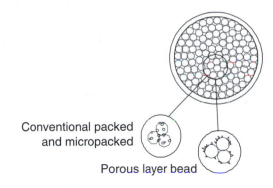

Conventional packed
and micropacked

Porous layer bead

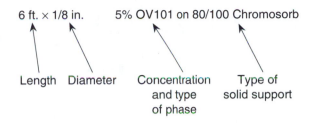

6 ft. × 1/8 in.   5% OV101 on 80/100 Chromosorb

Length   Diameter   Concentration   Type of
                     and type   solid support
                     of phase

One can obtain the starting conditions and suggested column for most common applications by referencing these application notes. If the desired application is not readily available, the technical support chemists affiliated with the chromatography vendors are available to recommend a column.

The more narrow the diameter, the greater the column's separation efficiency or peak sharpness. One generally chooses a column of about 0.2 mm or smaller when analyzing for trace concentrations or complex separations of 50 or more components. The type of stationary phase determines the relative order of elution of the components in the sample mixture. One chooses a stationary phase that will differentially select components in the sample mixture.

A simple analogy can be used to explain the separation phenomenon in columns. One should imagine a large open tubular column with a stationary phase coating of chocolate adhering to the inner walls of the column. Three species are present in a mixture: one that abhors chocolate (component A), one with a moderate liking for chocolate (component B), and one that loves chocolate (component C). If the mixture starts to wander down the column, a distribution begins to develop, with components B and C interacting with the chocolate and component A moving right through. The first to elute would be A, the second eluent would be B, and the third would be C, with the most affinity for the chocolate. The type of stationary phase and its ability to interact with the components in the mixture affect the separation. Let's change the coating in the column to chocolate with almonds. Assume that component C, although a lover of chocolate, is also allergic to almonds. Now when the components elute, the order is quite different, with A and C coeluting and B following later. Thus, the selective properties of the column have changed.

In the selection of stationary phases for a desired separation, it is important that the components in the mixture have differential affinity for the stationary phase; this is called the column's selectivity. This differential affinity is based on physiochemical interactions. The primary mechanism in gas–liquid chromatography is a difference in solubility of the components in the liquid

Figure 8.6 Application note. *(Courtesy of Hewlett-Packard Company.)*

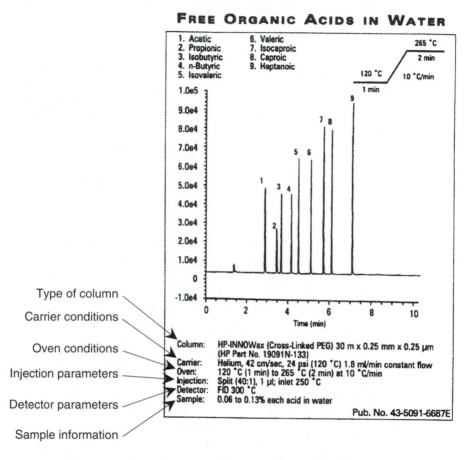

stationary phase and theories about polar and nonpolar behavior have been used to explain the complex interplay of forces that produce the distribution of components. Common liquid stationary phases used for the separation of most organic substances include methyl and phenyl polysiloxanes as well as polyethylene glycols. In gas–solid chromatography components are separated based on differences in adsorption to the stationary phase and "retention of sorbate components is determined by (1) the chemical nature and geometric pore structure of the sorbent, (2) molecular weight of sorbate molecules and their geometric and electronic structures, and (3) temperature of the column"(5). Common solid, adsorbent stationary phases used for the separation of permanent gases and volatile substances include molecular sieve, aluminum oxides, and porous polymers. These latter materials have been used in PLOT columns and have gained increasing popularity in the last 10 years as the manufacturing technology has improved.

The novice chromatographer can easily perform routine analyses without extensive understanding of column separation theory; however, if one's responsibilities include method development and performance optimization, a fundamental understanding of separation theory is recommended. The Suggested Readings by Robert L. Grob and Karen Hyver listed at the end of this chapter cover the theory of separation in great detail, and one should refer to these sources for further explanation.

Figure 8.7 depicts a representative chromatogram that shows the signal output versus time plot of the separated components.

Figure 8.7 Typical capillary column chromatogram. *(Courtesy of Hewlett-Packard Company.)*

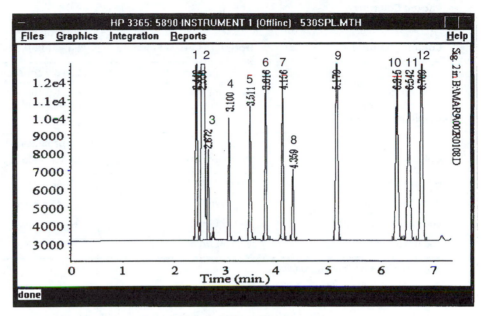

Capillary column analysis: 30 m × 0.32 mm × 0.25 μm HP5
Column flowrate = 1 mL/min   Linear velocity =   20.5 cm/sec
Isothermal 160 degrees   FID
1 μL HP column checkout sample at split ratio = 100:1

===

Area percent report

===

PK #	Ret Time	Area	Height	Type	Width	Area %	Name
1	2.442	62321	45133	BV	0.022	0.2127	Pentane
2	2.556	2.90451E+007	1.87002E+007	VV	0.024	99.1066	Solvent
3	2.672	8303	5036	VB	0.026	0.0283	Solvent impurity
4	3.100	10129	7143	BB	0.023	0.0346	Undecane
5	3.511	14978	7438	BB	0.032	0.0511	4-Chlorophenol
6	3.816	16871	9417	BB	0.028	0.0576	1-Decylamine
7	4.156	17124	9425	BB	0.029	0.0584	Tridecane
8	4.359	7748	3907	BB	0.031	0.0264	Methyl caprate
9	5.179	33486	14578	BB	0.036	0.1143	Tetradecane
10	6.315	27045	9249	BB	0.046	0.0923	Acenaphthylene
11	6.542	29540	8682	BB	0.053	0.1008	1-Dodecanol
12	6.789	34293	11407	BB	0.047	0.1170	Pentadecane

Total area = 2.9307E+007

The retention times in minutes are listed in the report and are labeled on the chromatogram. The data system begins recording detector signal changes when the sample is injected into the injection port and the analysis is started. As the components are separated and reach the detector, the signal increases and the detection or retention time of each peak is recorded. The other information in the report is discussed later in this chapter.

The retention time of a component should be reproducible at constant operating conditions of oven temperature and column flowrate and is thus used to qualitatively identify the component. Therefore, given the conditions in Fig. 8.7, pentadecane, peak 12, should always have a retention time of 6.789 min and the predicted retention time is used to identify the component.

Table 8.1 Guidelines used for finding the optimum flowrate or linear velocity for a particular column.

Packed Columns

Column Diameter	Optimum Carrier Flowrate	Approximate Optimum Carrier Linear Velocity
1/4"	50–60 mL/min	2.6–3.2 cm/sec
1/8"	20–30 mL/min	4.2–6.3 cm/sec

Capillary Columns

Column Diameter	Optimum Carrier Flowrate	Approximate Optimum Carrier Linear Velocity (100 °C oven temp.)
HP Series 530 μm	3–5 mL/min	22–38 cm/sec
320 μm (wide bore)	1–2 mL/min	20–41 cm/sec
200 μm (narrow bore)	.4–.6 mL/min	21–32 cm/sec
100 μm (high speed)	.2–.5 mL/min	19–31 cm/sec

(Courtesy of Hewlett-Packard Company.)

Increasing the flowrate shortens the retention time and decreasing the flowrate results in longer retention times. In optimizing separation, it should be noted that one does not arbitrarily set the flowrate or linear velocity; it is set for optimum separation efficiency. Table 8.1 suggests the ranges of flowrate and linear velocity as related to the column diameters that generally yield the best efficiency. One would adjust the flowrate within these ranges to achieve the desired separation in the minimum amount of time.

The oven temperature parameters are generally determined by experimentation. There are two modes of analysis: isothermal (a constant oven temperature throughout the analysis) and temperature programming (a temperature gradient throughout the analysis). Isothermal analysis is commonly used when the difference in boiling points of the highest- and lowest-boiling components in the mixture is 100 °C or less. For mixtures whose components contain a wide range of boiling points, temperature programming results in shorter analysis runtimes and more efficient separation, especially of the later-eluting components. Increasing the oven temperature shortens the retention time, so one can adjust the oven temperature to provide the desired separation while minimizing the runtime. The column's stationary phase has a maximum temperature limit and as the operating temperature of the oven approaches this limit, the lifetime of the column is compromised and the background signal of the detector is elevated. One can obtain the temperature limits of the columns in GC supply catalogs, and methods should be developed such that the patience of the operator is weighed against the degradation of the column and the needed separation.

In general, capillary columns require little maintenance. They are most susceptible to dirty samples or the accumulation of nonvolatile material from the sample or carrier gas. It may be necessary to condition the column periodically by increasing the temperature of the oven to drive out the contaminants. These contaminants appear as very broad ghost peaks or cause an increased detector background signal. One can also rinse the column with a few milliliters of solvent. The column lifetime is directly related to the nature of the sample and the routine operating oven temperature. The cleaner and more neutral the sample, the longer the lifetime. The higher the oven temperature used, the shorter the lifetime.

Detectors

The fourth major component of the gas chromatograph is the detector. It is a device that senses the presence of components different from the carrier gas and converts that information to an electrical signal. The characteristics influencing one's choice of detector include selectivity and sensitivity. Selectivity is the ability of the detector to recognize and respond to the components of interest. Not all detectors respond to all components. Sensitivity is the concentration level detected and formally is defined as the change in the response with the change in detected quantity. One needs to match the selective properties of the detector to the components of interest in the sample mixture, as well as considering the lowest level of detection of the detector and the concentration level of those components. The following are brief descriptions of the common detectors.

Thermal Conductivity Detector (TCD)

A filament temperature increases as analytes present in the carrier gas pass over it, causing the resistance to increase. It is the oldest type of GC detector used. Although its sensitivity level has dramatically improved through the years, it is still the least sensitive of the detectors. Its popularity has endured because of its ability to respond to any type of constituent that is different from the carrier gas (for example, if helium is the carrier gas, any other component would be detected as long as its concentration was in the nanogram or parts-per-million level).

Flame Ionization Detector (FID)

Organic components burn in a flame, producing ions that are collected and converted into a current. This is the most widely used detector because it offers fairly low detection limits (picogram or parts-per-billion concentrations) and responds to any type of hydrocarbon component.

Electron Capture Detector (ECD)

As electronegative species pass through the detector, they capture low-energy electrons, causing a decrease in cell current. Those interested in very low levels of halogenated compounds (containing Cl, Fl, or Br), such as pesticides, use this detector.

Nitrogen-Phosphorous Detector (NPD)

Nitrogen and phosphorous compounds produce increased currents in a flame enriched by vaporized alkali metal salt. This is a detector of choice for analysis of organophosphorus pesticides and pharmaceuticals.

Flame Photometric Detector (FPD)

Sulfur and phosphorous compounds burn in a flame, producing chemiluminescent species that are monitored at selective wavelengths.

Electrolytic Conductivity Detector (ELCD)

Halogens, sulfur, or nitrogen compounds mix with a reaction gas in a reaction tube. The products are mixed with a suitable liquid, which produces a conductive solution, and the change in conductivity is monitored.

Photoionization Detector (PID)

Molecules are ionized by excitation with photons from an ultraviolet (UV) lamp. The charged particles are then collected, producing current. This detector is used for analyses of aromatic and UV ionized compounds.

Mass Selective Detector (MSD)

Molecules are bombarded with electrons, producing ion fragments that pass into the mass filter. The ions are filtered based on their mass/charge ratio. The very popular (GC-MS) gas chromatography-mass spectroscopy technique yields excellent qualitative identification of components by matching the compound's mass spectrum with spectra included in libraries that are part of the system. This detector is discussed in detail in Chap. 31 of this book.

Infrared Detector (IRD)

Molecules absorb infrared energy, the frequencies of which are characteristic of the bonds within that molecule. Though not very commonly used, this detector is a powerful tool for distinguishing isomers. Fourier transform infrared (FTIR) systems usually include libraries of the IR spectrum that are used for qualitative identification. A very sophisticated GC-FTIR-MS system can obtain 99% confidence levels of identification.

Atomic Emissions Detector (AED)

Molecules are energized by a plasma source and separated into excited atoms. As electrons return to their stable state, they emit light, which is element-specific. This is a relatively new detector option and yields elemental information about the separated components in a mixture.

Many gas chromatographs are limited to two detectors that can be used simultaneously, so one purchases the GC configured with the desired detectors. Table 8.2 lists the common HP5890 GC detectors used and their operating characteristics. Other GC vendors can provide comparable information for their detectors.

Data Acquisition

The last major component of the GC system is the data acquisition system. The data system is used to translate the electrical signal generated by the detector into a peak chromatogram. The area of the peak is representative of the concentration of the component; the larger the concentration, the larger the detector signal generated and the larger the peak area depicted.

The most commonly used devices for generating the chromatogram and report are the integrator and the personal computer (PC). Although the integrator is generally a more user friendly device, in recent years the PC with software used for controlling the GC, as well as for calculating reports, has become a more popular alternative. The PC has distinct advantages for those requiring more customized reporting options and those needing more sophisticated management and archiving of data. Data can also be integrated and reported using larger computerized laboratory automation systems.

The report shown in Fig. 8.7 was generated by a PC data system. The signal output is reported as area and height counts. The Type column in the report defines the point at which the integration or calculation of the peak began and ended. For example, *BB* refers to a peak that began at the baseline and ended at the baseline; *VV* refers to a peak that began at a valley point and ended

Table 8.2 Operating conditions for HP5890 detectors commonly used in capillary GC.

Type	Typical Samples	Detection Limits	Flow rate (mL/min)		
			Carrier + Makeup	H$_2$	Air
Flame ionization detector	Hydrocarbons	10–100 pg 10 ppb–99%	20–60	30–40	200–500
Thermal conductivity detector	General	5–100 ng 10 ppm–100%	15–30	—	—
Electron capture detector	Organohalogens, chlorinated solvents and pesticides	0.05–1 pg 50 ppt–1 ppm	30–60	—	—
Nitrogen–phosphorus detector	Organonitrogen and organophosphorus compounds	0.1–10 pg 100 ppt–0.1%	20–40	1–5	70–100
Flame photometric detector (393 nm)	Sulfur compounds	10–100 pg 100 ppb–100 ppm	20–40	50–70	60–80
Flame photometric detector (526 nm)	Phosphorus compounds	1–10 pg 1 ppb–0.1%	20–40	120–170	100–150
Photoionization detector	Compounds ionized by UV	2 pg C/sec	30–40	*	—
Electrolytic conductivity detector	Halogens, N, S	0.5 pg Cl/sec 2 pg S/sec 4 pg N/sec	20–40	80	—
Fourier transform infrared	Organic compounds	1000 pg of strong absorber	3–10	—	—
Mass selective detector	Tunable for any species	10 pg to 10 ng (depending on selective ion monitoring versus scan)	0.5–30†	—	—
Atomic emissions detector	Tunable for any element	0.1–20 pg/sec (depending on element)	60–70	Preset	Preset

*Refer to Detector Manual.
†Dependent on the type of interface.
(*Courtesy of Hewlett-Packard Company.*)

at a valley point. Ideally, one would want to optimize conditions for the separation so that as many peaks as possible are resolved or separated baseline to baseline. The width of the peak is reported and it is noted that as the retention time gets longer, the peak shape becomes broader for isothermal analyses. This is a common occurrence and is related to the dispersive forces occurring in the column during separation. The more narrow the peak width, the more efficient or sharper the separation. The relative concentration in area percent is also shown in this report. Calculation of quantity is discussed further later in this chapter. The names of the components also appear in the report. Other pertinent information for this analysis is also shown in Fig. 8.7: the type of column used, the column flowrate and linear velocity, the oven temperature, the type of detector, the volume and type of sample injected, and the split ratio used.

What It Does

Gas chromatographic analysis provides the chemist with a chromatogram and report of the composition of a sample containing a mixture of components. Generally GC systems are configured and set up for analyzing primarily routine, quality control type samples or for primarily nonroutine research and development (R&D) analyses. GC systems used for routine analyses can be set up with the latest automation enhancements so that minimum operator assistance is required. These enhancements include robotic sample preparation stations, auto injectors, and programmed data systems. GC systems used for R&D and method development are usually configured for more versatility. One of the major reasons for the widespread use of gas chromatography is the ability to configure the instrument for one's specific needs and budget.

Analytical Information

Qualitative

Qualitative identification can be determined by GC-MS techniques. A mass spectrum of any or all peaks in the chromatogram is compared with spectra contained in libraries housed by the system's computer. The final report contains a list of the most probable identification of each peak based on matches found in libraries containing thousands of compound entries. One of the most powerful qualitative tools is the GC-FTIR-MS system, which provides both an IR and MS spectrum for each separated component.

When two-dimensional detectors such as FIDs and ECDs are used, qualitative identification is a bit more involved. One must rely on matching retention times of known compounds with the retention times of the components in the unknown mixture. Thus, these detectors are most often used when one is performing target compound analysis, where one has a good idea of the compound identifications and uses reference standards for determining retention times. It is important to remember that any changes in oven temperature or column flowrate will affect the retention time, thus affecting the accuracy of identification. Therefore, one must be assured that the same operational parameters are used when analyzing known standards and unknowns. Before the development of capillary columns and the GC-MS technique, a very involved system using calculated retention indices was used to identify true unknowns in a mixture. Today, anyone interested in identification of unknown components generally relies on the GC-MS technique along with matching of retention times.

Although one skilled in mass spectra interpretation techniques may be able to recognize coelution of two or more components, it is better to provide the best chromatographic separation by the column before detection. A technique of dual-column confirmation may be used to better ensure that separation of all components has occurred. One performs two separate analyses using two different columns of different stationary phase polarities, such as a nonpolar methyl silicone stationary phase and a more polar 50% phenyl methyl silicone or polar carbowax phase. Changing the type of stationary phase can affect the fingerprint pattern of peaks by changing the elution order as well as the retention times of components. By using two different types of stationary phases one ensures that all components have been adequately separated before qualitative identification. Many FDA- and EPA-regulated methods require dual-column confirmation.

Quantitative

For the most accurate quantitative results, calibration procedures are required. Calibration involves the correlation between a known concentration of a component and the resultant detector signal generated when that component is detected. Electronic integrators are used to convert the detector signal to either peak area or height. Calibration algorithms are incorporated into the integrator or computer software used for data acquisition and analysis. Thus, one creates calibration tables in the data system through the use of prompted dialogs or menu items. The simplest procedure involves creating a single-level calibration and is achieved by the following process. First one optimizes the chromatographic parameters for the desired separation and identification of components (such as choice of column, detector, injection process, and oven temperature). Then a mixture of the components is prepared with a known amount of each of the desired components is present. One analyzes the mixture and an area percent report is generated. This is a default report that includes the retention times of the components detected and an area percent calculation of the component concentration.

$$\text{Area percent of component A} = \frac{\text{Area of component A}}{\text{Total area of all peaks}} \times 100 \qquad (8.2)$$

In most cases, this report is the least accurate determination of concentration because the detector sensitivity and selectivity can vary significantly between components. One uses this report in order to determine the retention times of the components. Using the calibration dialog of the data system, the retention times, known amounts, and names of the components are typed in, creating the calibration table. The system uses the known concentration amounts and the detected area values to calculate the response factors for each of the components. The response factor is the amount divided by the area. When one analyzes a sample containing an unknown amount of a component, the data system multiplies the response factor for that component by the area detected for the unknown and determines the concentration of the unknown.

$$\text{Response factor} = \text{Amount/Area} \qquad (8.3)$$

$$\text{Unknown amount of component} = \text{Area of the peak} \times \text{Response factor} \qquad (8.4)$$

This simple form of calibration is known as the external standard method. There are additional methods available, including the internal standard and normalized percent methods.

The multilevel calibration procedure is a more accurate calculation of concentration. This procedure involves preparing mixtures of the components at more than one concentration level. After analyzing each mixture of known amounts and entering the information into the calibration dialog, the data system calculates a calibration curve of peak area or height versus concentration. A four-level calibration of a mixture containing three components would involve the preparation and analysis of four solutions containing progressively larger concentrations of each of the three components, A, B, and C.

The top diagram in Fig. 8.8 shows the analysis results for these four solutions. The data system would generate three calibration curves, one for each of the components. Figure 8.8 also illustrates a single-point calibration for component A (there would be two additional curves for the other components) and a multilevel calibration for component C (there would be additional curves for the other components). The data system usually offers a choice of curve-fitting techniques, such as point-to-point, linear, and nonlinear. The appropriate calibration curve is used to calculate the unknown concentrations of the components. Many government-regulated methods require the use of multilevel calibration procedures. One should seriously consider using multilevel calibration when the concentration levels of the components approach the minimum detectable limits of the detector and when using the more sensitive detectors such as ECDs, NPDs, and FPDs, which have limited linear dynamic ranges.

Figure 8.8 Single-level and multilevel calibration. *(Courtesy of Hewlett-Packard Company.)*

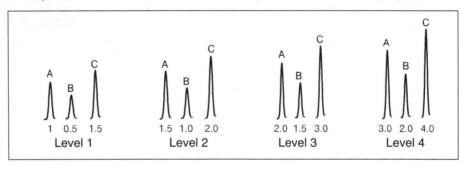

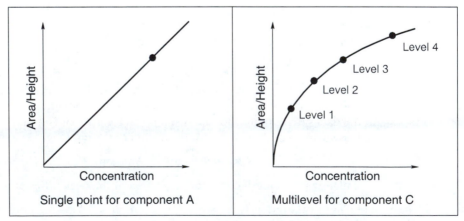

Applications

The list of gas chromatographic applications is very extensive. The easiest way for a novice to become familiar with the capabilities of the GC technique is to peruse the example applications listed in the GC vendors' supply catalogs. Figure 8.6 is a reproduction of an application note from Hewlett-Packard's *Chemical Analysis Columns and Supplies Catalog*. These example notes show the chromatogram and the key instrument parameters used. Hewlett-Packard's catalog and J&W Scientific's *Catalog and Technical Reference* each list over 200 example applications. The vendor catalogs classify the applications by analysis type. The following is an example list of applications offered by J&W Scientific:

- **Environmental:** benzene, toluene, xylene, diesel fuel, chlorinated pesticides, herbicides, organophosphorus insecticides, amines, aromatics, base neutrals, dioxins and furans, polynuclear aromatic hydrocarbons, phenols, other semivolatiles, air and soils, water, and other volatiles

- **Food and flavors:** beverages, essential oils, fames, fats and oils, fragrances, natural products, organic acids, and sterols

- **Industrial chemicals:** alcohols, aldehydes, amines, aromatics, esters, ethers, glycols, halocarbons, ketones, and other solvents

- **Life sciences:** anesthetics, drugs of abuse, over-the-counter drugs, pharmaceuticals, pharmaceutical residual solvents, and steroids

- **National Institute for Occupational Safety and Health methods:** alcohols, aromatics, esters, and ketones

- **Petroleum:** light hydrocarbons and gases, naphtha, gasoline, middle distillates, heavy petroleum products, and wax

Hewlett-Packard Company offers an *HP Environmental Solutions Catalog*, which focuses on environmental applications.

Nuts and Bolts

Relative Costs

The cost of a complete GC system can range from $5,000 to $100,000. The simplest single-injection-port, single-detector unit with electronic integrator can be obtained for less than $10K. The costs increase as one adds injector and detector options and automation features. The cost increases rapidly to the $70–100K range when configuring the system with a mass selective detector and computerized data systems. The cost of capillary columns ranges from about $300 to $1000. Consumable supplies are generally priced under $100.

Vendors for Instruments and Accessories

American Laboratory Buyer's Guide, February 1996, lists over 80 suppliers of gas chromatography systems. The leading vendors for instruments and supplies are listed below with an indication as to whether the vendor is an instrument manufacturer (I), consumables supplier (C), or both (I & C):

Alltech Associates Inc. (C)
2051 Waukegan Rd.
Deerfield, IL 60015
phone: 708-948-8600, 800-255-8324
fax: 708-948-1078
email: 73554.3372@compuserve.com
Internet: http://www.alltechweb.com

Chrompack Inc. (I & C)
1130 Hwy. 202, S.
Raritan, NJ 08869
phone: 908-722-8930, 800-526-3687
fax: 908-722-8365
email: cpinfous@mars.superlink.net
Internet: http://www.chrompack.com

Gow-Mac Instrument Co. (I & C)
277 Brodhead Rd.
Bethlehem, PA 18017-8600
phone: 610-954-9000
fax: 610-954-0599
email: gowmacgs@fast.net

Hewlett-Packard Co. (I & C)
2850 Centerville Rd.
Wilmington, DE 19808
phone: 302-633-8504, 800-227-9770
fax: 302-633-8902
Internet: http://www.hp.com/go/chem

J & W Scientific (C)
91 Blue Ravine Rd.
Folsom, CA 95630-4714
phone: 916-985-7888, 800-223-3424
fax: 916-985-1101
email: 70530,2047@compuserve.com
Internet: http://www.JandW.com

The Perkin-Elmer Corp. (I & C)
761 Main Ave.
Norwalk, CT 06859
phone: 800-762-4000
fax: 203-762-4228
email: info@perkin-elmer.com
Internet: http://www.perkin-elmer.com

Restek Corp. (C)
110 Benner Circle
Bellefonte, PA 16823-8812
phone: 814-353-1300, 800-356-1688
fax: 814-353-1309
Internet: http://www.restekcorp.com

Shimadzu Scientific Instruments, Inc. (I & C)
7102 Riverwood Dr.
Columbia, MD 21046
phone: 800-477-1227
fax: 410-381-1222
Internet: http://www.shimadzu.com

Supelco, Inc. (C)
Supelco Park
Bellefonte, PA 16823-0048
phone: 814-359-3441, 800-247-6628
fax: 814-359-3044, 800-447-3044
email: Supelco@Supelco.sial.com
Internet: http://www.Supelco.sial.com/Supelco.html

Varian Analytical Instruments (I & C)
P.O. Box 9000, Dept. 87
San Fernando, CA 91340
phone: 800-926-3000
fax: 713-240-6752
Internet: http://www.varian.com

VICI Valco Instruments Co. Inc.
P.O. Box 55603
Houston, TX 77255
phone: 713-688-9345
fax: 713-688-3948
Internet: http://www.vici.com

Thermal Separations Technologies
(formerly Carlo-Erba and Fisons)
3661 Interstate Park Rd. North
Riviera Beach, FL 33404
phone: 800-685-9535
fax: 561-845-8819

Required Level of Training

Routine operation of the simpler GC systems can be performed by anyone after an introductory training session of about a day or two. This author has trained many people with less than a high school education (but who had some lab instrument operation experience) on the routine operation of an HP5890. Operation of the more complex systems using the more sophisticated detectors and PC-based data acquisition systems requires considerably more training and depends on the background level of the operator (those with some college chemistry background generally master the operation of these systems more easily). For an understanding of the fundamental principles, in addition to the operational procedures, most novices require about a week of training.

Instrument troubleshooting and method development tasks are mastered through experience as well as through training. Training classes at all levels of expertise, at locations throughout the United States and Canada, and for various durations of time from 1 day to several days, are offered by many instrument and consumable supply vendors, the American Chemical Society, and training institutes associated with some major universities (such as Virginia Tech at Blacksburg). Costs of these training sessions range from about $300 per day to about $2000 per week.

Many vendors offer computer-based and video training options.

Service and Maintenance

The degree of service and maintenance required is related to the complexity of the system. Most systems require routine inspection and cleaning procedures of the injection port and detector. The frequency of maintenance depends on the cleanliness of the sample and the frequency of sample analyses. Many GCs have self-diagnostics available to help identify problems. Most instrument vendors offer service contracts as well as service support options, including telephone assistance. Some vendors are pursuing on-line diagnostic support via modem in the GC for direct communication with a service engineer. Operator error is a common cause of problems associated with systems operated by novice users, so adequate training should be pursued to minimize these occurrences.

Suggested Readings

BRUNO, T., *Chromatographic and Electrophoretic Methods*. Englewood Cliffs, N.J.: Prentice Hall Chemical Instrumentation Series, 1991.

GROB, R. L., ED., *Modern Practice of Gas Chromatography*, 3rd ed. New York: Wiley, 1995.

HYVER, K. J., *High Resolution Gas Chromatography*, 3rd ed. Wilmington, DE: Hewlett-Packard Co., 1989.

J&W SCIENTIFIC, *Catalog and Technical Reference 1996–1997*. Folsom, CA: J & W Scientific, 233–56.

LOFFE, B. V., AND A. G. VITENBERG, *Headspace Analysis and Related Methods in Gas Chromatography*. New York: Wiley Interscience, 1984.

MCNAIR, H. M., AND E. J. BONELLI, *Basic Gas Chromatography*. San Fernando, CA: Varian Instruments, 1969.

References

1. K. Knauss, J. Fulleman, and M. P. Turner, *Journal of High-Resolution Chromatography, Chromatogr. Commun.*, 4 (1981), 641–3.

2. E. Cremer and F. Prior, *Z.Elektrochem.*, 55 (1951), 66; E. Cremer and R. Muller, *Z. Elektrochem.*, 55 (1951), 217.

3. A. T. James and A. J. P. Martin, *Biochem. J.*, 50 (1952), 679–90.

4. M. J. E. Golay, in V. J. Coates, H. J. Noebels, and I. S. Fagerson, eds., *Gas Chromatography* (New York: Academic Press, 1958), 1–13.

5. R. L. Grob, ed., *Modern Practice of Gas Chromatography*, 3rd ed. (New York: Wiley, 1995), 104.

Chapter 9

High-Performance Liquid Chromatography

Phyllis Brown and Kathryn DeAntonis

University of Rhode Island
Department of Chemistry

Summary

General Uses

- Separation of a wide variety of compounds: organic, inorganic, and biological compounds, polymers, chiral compounds, thermally labile compounds, and small ions to macromolecules
- Analysis of impurities
- Analysis of both volatile and nonvolatile compounds
- Determination of neutral, ionic, or zwitterionic molecules
- Isolation and purification of compounds
- Separation of closely related compounds
- Ultratrace to preparative and process-scale separations
- Nondestructive method
- Qualitative and quantitative method

Common Applications

- Measuring levels of certain compounds such as amino acids, nucleic acids, and proteins in physiological samples

- Measuring the levels of active drugs, synthetic byproducts, or degradation products in pharmaceutical dosage forms
- Measuring levels of hazardous compounds such as pesticides or insecticides
- Monitoring environmental samples
- Purifying compounds from mixtures
- Separating polymers and determining the molecular weight distribution of the polymers in a mixture
- Quality control
- Following synthetic reactions

Samples

State

Sample must be in liquid form for injection into the instrument; solid samples must be dissolved in a solvent compatible with the mobile and stationary phases.

Amount

1–100 µL injected (generally 5–10 µL); mass amounts injected vary depending on the sensitivity and dynamic range of the detector for the analyte.

Preparation

Limited or extensive sample prep may be required as defined by the relative complexity of the sample. Sample preparation may include any of the following steps: dilution, preconcentration, filtration, extraction, ultrafiltration, or derivatization.

Analysis Time

Analysis time is in a range from 5 min to 2 hr (generally 10–25 min). Sample preparation differs from sample to sample. Sample preparation may be extensive and require more time than the analysis.

Limitations

- Compound identification may be limited unless high-performance liquid chromatography (HPLC) is interfaced with mass spectrometry.
- Resolution can be difficult to attain with complex samples.
- Only one sample can be analyzed at a time.
- Requires training in order to optimize separations.
- Time analysis can be long (compared with capillary electrophoresis).
- Sample preparation is often required.

Complementary or Related Techniques

- Mass spectrometry (MS) provides structure identification and molecular weight information.
- Nuclear magnetic resonance (NMR) provides detailed information on molecular structure.
- Infrared spectroscopy (IRS) provides information on functional groups.

Introduction

High-performance liquid chromatography (HPLC) was developed in the late 1960s and 1970s. Today, it is a widely accepted separation technique for both sample analysis and purification in a variety of areas including the pharmaceutical, biotechnological, environmental, polymer, and food industries. HPLC is enjoying a steady increase in numbers of both instrumental sales and publications that describe new and innovative applications. Some recent growth areas include miniaturization of HPLC systems, analysis of nucleic acids, intact proteins and protein digests, analysis of carbohydrates, and chiral analyses.

How It Works

Chromatography is a technique in which solutes are resolved by differential rates of elution as they pass through a chromatographic column. Their separation is governed by their distribution between the mobile and the stationary phases. The successful use of liquid chromatography (LC) for a given problem requires the right combination of a variety of operating conditions such as the type of column packing and mobile phase, column length and diameter, mobile phase flow rate, column temperature, and sample size. In order to select the best combination of chromatographic conditions, a basic understanding of the various factors that affect LC separations is necessary.

HPLC instrumentation is made up of eight basic components: mobile phase reservoir, solvent delivery system, sample introduction device, column, detector, waste reservoir, connective tubing, and a computer, integrator, or recorder. A block diagram of a general LC system is shown in Fig. 9.1.

Mobile Phase Reservoir

The mobile phase reservoir can be any clean, inert container such as an empty solvent jug, a laboratory flask, or a commercial reservoir. It usually contains 1 to 2 L of solvent, and it should have a cap that allows the tubing inlet line to pass through.

It is important to degas solvents before use because small gas bubbles present in the mobile phase can collect in other components, particularly in the pump heads and the detector, and ruin the analysis. When preparing the mobile phase it is important to use highly purified buffer

Figure 9.1 Block diagram of general LC system. *(Figure is reproduced from* High Performance Liquid Chromatography, *2nd edition (S. Lindsay) with permission of the University of Greenwich.)*

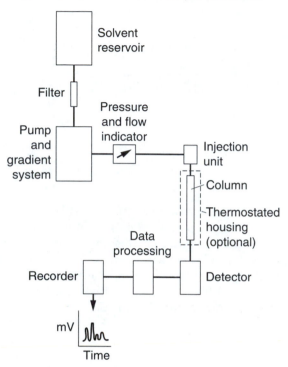

salts and reagents, preferably HPLC grade. Any impurities present in the reagents can lead to chromatographic artifacts, and small particles can collect in the narrow tubing or in the column and cause a void. The mobile phase should be filtered before use in order to eliminate these small particulates.

Solvent Delivery Systems

The purpose of the pump, or solvent delivery system, is to ensure the delivery of a precise, reproducible, constant, and pulse-free flow of mobile phase. There are two classes of HPLC pumps: constant pressure pumps and constant flow pumps, with the latter being by far the most common.

The most common type of HPLC constant flow pump is the reciprocating piston pump, in which a piston is driven in and out of a solvent chamber (internal volume of 10 to 100 μL) by an eccentric cam or gear. On the forward stroke, the inlet checkvalve closes, the outlet checkvalve opens, and the mobile phase is pumped to the column. On the return stroke, the checkvalves reverse and solvent is drawn into the chamber. In the single-head reciprocating pump, 50% of the time the mobile phase flows to the column and 50% of the time the chamber is refilling. With the twin-head reciprocating pump, two pump heads operate simultaneously but 180° out of phase with each other. As a result, mobile phase flows to the column 100% of the time, providing an essentially pulseless flow.

Most separations can be done using isocratic elution, which is the use of a single-solvent system that does not change during the analysis. For more complex analyses, gradient elution is required. Gradient elution is done by gradually strengthening the mobile phase composition throughout the separation. The highly retained compounds are eluted more quickly and the compounds that are eluted earlier remain well resolved. A basic gradient could be 0:100 acetonitrile:water to 60:40 acetonitrile:water in 60 minutes, which leads to an increase in acetonitrile concentration of 1% per minute. A gradient can be linear (as described), convex, or concave, or a complex sequence of each to achieve the desired separation.

Gradient flow can be generated in three ways. In all cases, a computer-controlled pumping mechanism is required. In the first case, controlled amounts of each eluent (up to four solvents can be used) are metered into a mixing chamber before reaching the high-pressure pump, which sends the mixture to the column. In the second case, the amount of each solvent is regulated by a proportioning valve, which is controlled by a microprocessor. The mixed solvent then enters the high-pressure pump and flows to the column. In the third case, the delivery of multiple high-pressure pumps is controlled individually with a programming device, and the mixture is sent to a high-pressure mixing chamber. Because multiple pumps are required, this is the most expensive system.

Low-pressure mixing is less expensive because it requires only one pump, but it is more susceptible to gas bubbles because the solvents are mixed at atmospheric pressure. Many high-pressure systems operate reliably without degassing because the solvents are mixed at sufficiently high pressure to keep the gas in solution. However, bubbles may get into the detector.

Sample Introduction Devices

A variety of sample introduction devices exist, manual and automatic, that use primarily a valve mechanism. They give excellent precision and are easy to use. When the valve is in the load position, the sample loop (generally 10 to 50 μL in volume) is filled. For best results, a two- to five-fold excess of sample should be passed through the loop to ensure that the previous sample has been thoroughly purged. Turning the valve from the load to inject position connects the sample loop to the high-pressure mobile phase stream, whereby the sample is sent to the column.

Column

The column is the heart of the HPLC instrument because the separation occurs here. It is generally made of 316-grade stainless steel, which is relatively inert to chemical corrosion and is packaged with the desired stationary phase. Common dimensions for analytical scale columns are in the range of 10 to 25 cm long and 3 to 9 mm inner diameter.

Detectors

The important role of the HPLC detector is to monitor the solutes as they are eluted from the column. The detector generates an electrical signal that is proportional to the level of some property of the mobile phase or solutes. Detectors can be divided into two categories: bulk property detectors and solute property detectors. A detector that measures a property of both the solute and mobile phase, such as a refractive index detector, is a bulk property detector. Likewise, a detector that measures a property of the solute only, such as a UV detector, is a solute property detector. Solute property detectors are significantly more sensitive than bulk property detectors, on the order of 1000 times or more. Some characteristics of a good HPLC detector are sensitivity, linearity, predictability in response, reliability, nondestructiveness, ease of use, and low dead volume.

UV Absorbance Detectors

Over 70% of all of the HPLC detectors are UV absorbance detectors (Chap. 25). The mobile phase is passed through a small flow cell, where the radiation beam of a UV/visible photometer or spectrophotometer is located. As a UV-absorbing solute passes through the flow cell, a signal is generated that is proportional to the solute concentration. Only UV-absorbing compounds, such as alkenes, aromatics, and compounds that have multiple bonds between C and O, N, or S are detected. The mobile phase components should be selected carefully so that they absorb little or no radiation.

Absorption of radiation as a function of concentration, c, is described by the Beer–Lambert law:

$$A = \varepsilon bc \tag{9.1}$$

where A = absorbance,
ε = molar extinction coefficient, and
b = flow cell path length.

Three types of absorbance detectors are available: fixed-wavelength, variable-wavelength, and photodiode array. A fixed-wavelength detector uses a light source that emits maximum light intensity at one or several discrete wavelengths that are isolated by appropriate filters. For example, with a commonly used mercury lamp, the wavelengths of 254, 280, 313, 334, and 365 nm can be selected. A fixed-wavelength detector is the most sensitive and least expensive of the three, but it is inflexible in wavelength selection. A variable-wavelength detector uses a relatively wide band-pass UV/visible spectrophotometer. It offers an increased number of UV and visible wavelengths, but it is more expensive than the fixed-wavelength detector. In order to generate real-time spectra for each solute as it is eluted, a photodiode array is used. Comparison of spectra generated chromatographically with a known spectrum is useful for solute identification. In addition, software has been designed to evaluate peak purity with diode-array-generated data. The major disadvantage with scanning wavelength technology is a loss in sensitivity.

Fluorescence Detectors

A fluorescence detector (Chap. 26) senses solute fluorescence in much the same way that UV detectors sense absorbance. Instrumentally, fluorescence detectors are similar to UV absorbance detectors except that the excitation and emission radiation in the flow cell are perpendicular to each other.

Selectivity and sensitivity are the major advantages of a fluorescence detector. For the appropriate naturally fluorescent solutes or solutes tagged with a fluorescent derivative (because few solutes fluoresce naturally), these detectors can be much more sensitive than UV absorbance detectors. In addition, because both the excitation and emission wavelengths can be varied, the detector can be made highly selective.

Electrochemical Detectors

Electrochemical detectors (Chap. 36, 37, and 39) measure the current associated with the oxidation or reduction of solutes as they are eluted from the column. The suitability of electrochemical detection depends on the redox characteristics of the solute molecules in the environment of the mobile phase. Electrochemical detectors have the advantages of high sensitivity, high selectivity, and wide linear range, but they can be difficult to work with.

Conductivity Detectors

Conductivity detectors (Chap. 39) are used primarily to detect ions in conjunction with ion chromatography (IC). This type of detector senses the conductivity of the eluent as it is pumped through the detector. Because charged ions are more conductive than the eluent, a signal is generated as they are pumped through.

Refractive Index Detectors

Refractive index detectors sense the difference in refractive index between the column eluent and a reference stream containing mobile phase only. These detectors are the closest to universal detectors in HPLC because any solute can be detected as long as its refractive index is different from that of the mobile phase.

Although they are very versatile, refractive index detectors have many significant drawbacks. First, because the differences in the absolute refractive indices of many substances commonly analyzed by HPLC are small, the sensitivity of these detectors is generally lower than that obtained with UV and fluorescence detectors. Second, because refractive index is dependent on temperature, thermostating both cells is mandatory. Third, because the detector is also sensitive to pressure variations, a pulseless flow is essential. Finally, gradient elution is difficult to use because the detector is affected by variations in the mobile phase. However, a gradient can be used if the detector is configured with two columns and two flowing liquid streams.

Mass Spectrometer

The mass spectrometer (MS) is a very important HPLC detector because of its ability to generate structural and molecular weight information about the eluted solutes. The combination of HPLC and mass spectrometry allows for both separation and identification in the same step, an advantage none of the other detectors provide.

The major difficulty in using mass spectrometry is in designing the interface. The flow rate in HPLC is on the order of 1 mL/min, which is two to three orders of magnitude larger than the flow that can be taken by the conventional mass spectrometer vacuum systems. A second problem with using mass spectrometry is the difficulty of vaporizing nonvolatile and thermally labile molecules without degrading them. HPLC-MS is further discussed in Chap. 33.

Connective Tubing

The connective tubing in an HPLC system should be made of a material that is inert to the solvents in the mobile phase. It is usually made of stainless steel or inert plastic. The connections between the tubing and the different components in the system are fitted with unions that are designed to minimize dead volume. Zero dead volume (ZDV) or limited dead volume (LDV) fittings are used to ensure minimal band broadening due to dead volume. The fittings and thread sizes from different manufacturers are generally not compatible. Special care must be taken when assembling columns and fittings so that they match, and the amount of dead volume is minimized.

Care should be taken when selecting the tubing inner diameter. Beyond the point of injection, the inner diameter of the tubing should be kept to a minimum in order to reduce the effects of band broadening. For analytical scale chromatography, the tubing before injection is generally on the order of 0.030 in. and after injection, the tubing should be no larger than 0.010 in. For preparative scale chromatography, the tubing is larger due to the need for higher flow rates.

Computer, Integrator, or Recorder

A data collection device such as a computer, integrator, or recorder is connected to the detector. It takes the electronic signal produced by the detector and plots it as a chromatogram, which can be evaluated by the user. Recorders are rarely used today because they are unable to integrate the data. Both integrators and computers can integrate the peaks in the chromatograms, and computers have the further advantage that they electronically save chromatograms for later evaluation.

What It Does

Chromatography can be done in a variety of modes. The best mode for a given separation depends on the structural characteristics of the solutes to be separated and the analysis requirements. The most frequently used modes are partition chromatography (including ion-pair chromatography), adsorption chromatography, ion exchange and ion chromatography, size exclusion chromatography, affinity chromatography, and chiral chromatography.

Stationary phases for most modes of chromatography consist of a central core (usually silica or a polymer such as a polysaccharide or polystyrene) with the desired functional groups fixed to the surface. A variety of different functional groups can be used for each mode. Stationary phase particles are usually spherical, with a diameter of 5 to 10 μm, and are tightly packed in the column.

Mobile phase selection is critical for partition, adsorption, and ion-exchange chromatography and less critical for other modes. Table 9.1 lists some important properties of several of the most frequently used chromatographic solvents. The UV cutoff and refractive index are important factors when UV and refractive index detection are used. The polarity index (P') and eluent strength ($\varepsilon°$) are polarity parameters that aid in the choice of mobile phase for partition and adsorption chromatography, respectively (1).

The chart in Fig. 9.2 demonstrates how an HPLC method might be selected based on the molecular mass and the solubility of the sample. For many samples there is a choice of method, and in many cases separation can be achieved by reverse phase chromatography using a bonded silica stationary phase. If it is an option, reverse phase bonded phase HPLC is considered first because it is often faster, cheaper and experimentally easier than the alternatives. It has been estimated that 75% of all chromatographic methods are done in this mode.

Partition Chromatography

Partition chromatography has become the most commonly used chromatographic mode. The terms normal phase and reversed phase are used in partition chromatography to describe the relative polarities of the mobile and stationary phases. The pioneers of chromatography used highly polar stationary phases such as alumina or silica along with a relatively nonpolar mobile phase such as hexane or *i*-propylether. Relative polarities of solvents are determined using the polarity index, P′ (see Table 9.1). Because the original work was done this way, it is described as normal

Table 9.1 Properties of some common solvents used in HPLC.

Solvent	UV Cutoff (nm)	Refractive Index	Boiling Point (°C)	Viscosity (cP, 25 °C)	Solvent Polarity (P')	Solvent Strength (ε°)*
Isooctane	197	1.389	99	0.47	0.1	0.01
n-Hexane	190	1.372	69	0.30	0.1	0.01
Benzene	280	1.498	80	0.60	2.7	0.32
Methylene Chloride	233	1.421	40	0.41	3.1	0.42
n-Propanol	240	1.385	97	1.9	4.0	0.82
Tetrahydrofuran	212	1.405	66	0.46	4.0	0.82
Ethyl acetate	256	1.370	77	0.43	4.4	0.58
Chloroform	245	1.443	61	0.53	4.1	0.40
Dioxane	215	1.420	101	1.2	4.8	0.56
Acetone	330	1.356	56	0.3	5.1	0.56
Ethanol	210	1.359	78	1.08	4.3	0.88
Acetic acid		1.370	118	1.1	6.0	
Acetonitrile	190	1.341	82	0.34	5.8	0.65
Methanol	205	1.326	65	0.54	5.1	0.95
Water		1.333	100	0.89	10.2	

* The solvent strength is listed for alumina stationary phases. Multiplication by 0.8 gives the value for silica.
(*Adapted from* Introduction to Modern Liquid Chromatography, *2nd edition by L. Snyder and J. J. Kirkland, © 1979 by John Wiley & Sons, Inc. Reprinted by permission of John Wiley & Sons, Inc.*)

phase chromatography. Reversed phase is the opposite of normal phase because the stationary phase is nonpolar, often a hydrocarbon, and the mobile phase is relatively polar, such as water or acetonitrile. In normal phase chromatography, the least polar component is eluted first, and increasing the polarity of the mobile phase then decreases the elution time. Conversely in reversed phase chromatography the most polar compound is eluted first and increasing the mobile phase polarity increases the retention time. These relationships are illustrated in Fig. 9.3.

Ion-Pair Chromatography

Partition chromatography in the reversed phase mode is often used for the separation of highly polar, multiply charged, and strongly basic compounds in ion-pair chromatography. An ion-pairing reagent is included in the mobile phase such that it can bind to the ionic solute, generating an uncharged ion pair that can be resolved by reverse phase. Ion-pairing reagents are compounds with polar head groups (such as ammonium or sulfate) and nonpolar tails (such as phenyl or pentyl).

Adsorption Chromatography

Adsorption, or liquid–solid, chromatography is generally used for the separation of small compounds that are soluble in nonpolar solvents. The solutes are physically adsorbed on the stationary phase, and separation results from differences in adsorption strength. Although there is some overlap between adsorption and partition chromatographies, the modes are considered complementary.

Figure 9.2 Choice of HPLC method. *(Figure is reproduced from* High Performance Liquid Chromatography, *2nd edition (S. Lindsay) with permission of the University of Greenwich.)*

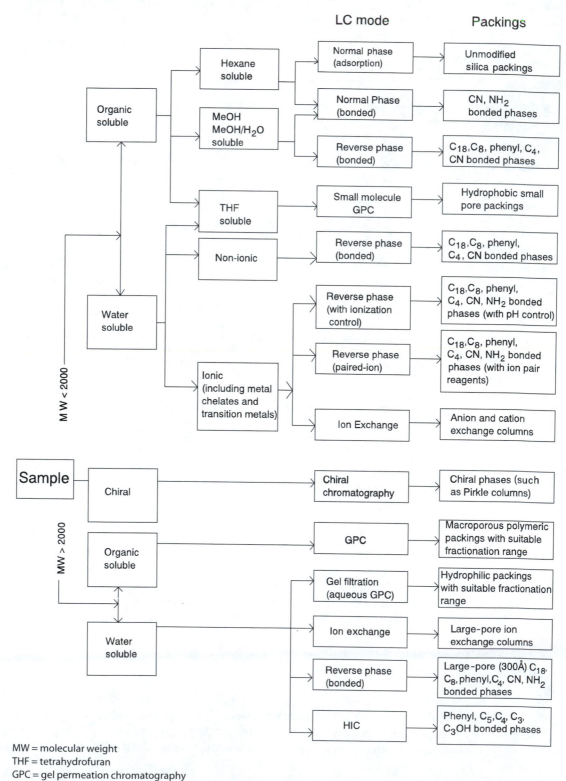

MW = molecular weight
THF = tetrahydrofuran
GPC = gel permeation chromatography
HIC = hydrophobic interaction chromatography

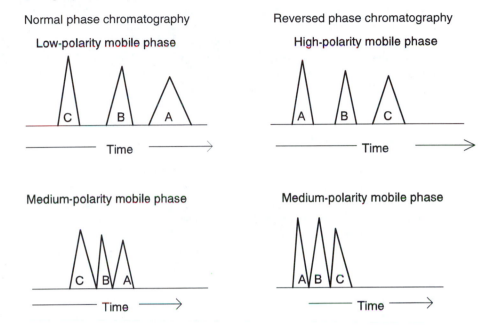

Figure 9.3 Relationship between polarity and elution times for normal phase and reversed phase chromatography. *(From* Principles of Instrumental Analysis, *Third Edition by Douglas A. Skoog, copyright © 1985 by Saunders College Publishing, reproduced by permission of the publisher.)*

Only two stationary phases are used—alumina and silica—with silica being the preferred adsorbent due to its higher sample capacity and wider range of useful applications. Mobile phases in adsorption chromatography are selected in much the same way as they are selected in partition chromatography. Instead of the polarity index, however, the solvent strength index, $\varepsilon°$, is used (see Table 9.1).

Ion Exchange and Ion Chromatography

Ion exchange and ion chromatography are used for the separation of ionic or ionizable species. These modes are covered in detail in Chapter 12.

Size Exclusion Chromatography

Size exclusion or gel permeation chromatography (GPC, Chap. 48) is used for the separation of large compounds such as polymers and proteins with molecular weights greater than 1000 daltons. In GPC, solutes are separated by physical size.

The stationary phase particles have a network of uniform pores into which some solute particles can diffuse. Molecules that are larger than the pore are excluded and unretained, whereas molecules that are smaller than the pores are retained inside. Residence time inside the pore depends on the size of the solute. Of the particles that fit into the pores, the larger ones spend the least amount of time in the pores and are the least retained, and the smaller ones are retained the most. Because retention is based entirely on physical impedence, not chemical interaction, the mobile phase does not play a critical role.

Organic gels and silica-based particles are used in GPC stationary phases. The gels are most often made from polymers of cross-linked styrene divinylbenzene or polyacrylamides. The degree of cross-linking defines the size of the pores. Silica particles have average pore sizes ranging from 40 to 2500 Å. A variety of pore sizes for each type of stationary phase are made to accommodate separations in different molecular weight ranges. A calibration curve is included with each column and defines the molecular weight range for which the column is useful. In the calibration curve, molecular weight is plotted against retention volume Vr ($t_R \times \mu$ where t_R is retention time). The exclusion limit defines the molecular weight of a completely unretained solute. The permeation limit is the molecular weight below which the solutes completely penetrate the pore (2).

Affinity Chromatography

Affinity chromatography is used for the analysis and purification of a variety of biological macromolecules (such as proteins and nucleic acids) and small molecules using the principle of biospecific adsorption. Specific ligands, which can be a variety of proteins, dyes, or nucleic acids, are bound to a solid support material and used as the chromatographic stationary phase. The ligands are selected based on their specific binding affinity for a desired substrate. When the sample is injected onto the column, the substrate is bound to the ligand while the remaining matrix passes through. Later, the bound substrate is washed from the stationary phase.

The stationary phase for affinity chromatography is a solid support, such as agarose (or other polysaccharide or polyamide) or porous glass bead, which is connected to the specific ligand via a spacer arm. The spacer arm is a short alkyl chain that is inserted to eliminate any steric interferences that could prevent the solute from reaching the ligand.

The mobile phase, or elution buffer, in affinity chromatography has two roles. First, it must allow for the strong binding between the ligand and the substrate, and then, once the undesired byproducts are removed, the ligand substrate interaction must be broken. The difference between the initial and final elution buffers can be a number of things; often a change in pH or ionic strength is successful.

Chiral Chromatography

In the last decade, efforts have been made to improve separations of chiral compounds. Effective separations have been developed using both chiral mobile phase additives and chiral stationary phases (CSPs), but more emphasis has been placed on the CSPs. In either case, enantiomeric resolution is the result of preferential complexation between the chiral resolving agent and one of the isomers. In order for it to recognize the chiral character of the solute, the chiral resolving agent must have chiral character itself.

CSPs are solid supports, usually silica gel, with the chiral resolving agents immobilized on the surface. Four basic types of chiral agents are available. In type I CSPs, attractive interactions, such as hydrogen bonding, pi–pi, and dipole interactions, lead to complexation between the solute and the CSP. Some type I chiral resolving agents are D-naphthalene and cellulose triacetate. Type II CSPs form inclusion complexes, which are chiral cavities in which solute fits. The most commonly used type II CSPs are α- and β-cyclodextrin. Type III CSPs involve chiral ligand exchange. Diastereomeric metal complexes (often copper [II] is used) are formed with both the solute and a selector ligand (such as proline or another amino acid), which is bound to the stationary phase. Type IV CSPs are proteins bound to a modified silica support. Complexation between the immobilized proteins, such as bovine serum albumin and chymotrypsin, and the solute result from hydrophobic and polar interactions.

There are three classes of chiral mobile phase additives: chiral ion-pairing agents, chiral ligand exchangers, and chiral inclusion complexes. The latter two are discussed in the stationary phase section. Chiral ion-pairing is like reverse phase ion-pairing except that a chiral ion-pairing agent is used.

Applications

HPLC methods have been developed in a variety of areas including organics, biologicals, inorganics, small molecules (less than 200 g/m), macromolecules, pollutants, polymers, and many others.

1. Bioseparations.

The analysis of biological compounds is an important area of interest for the biotechnology and pharmaceutical industries. The analysis of DNA, carbohydrates, lipids, proteins, peptides, and amino acids are among the major areas of interest in the use of HPLC for bioseparations. Figure 9.4 is a chromatogram of a series of derivatized amino acids using an AccQ-Tag C18 reversed

Figure 9.4 Reversed phase chromatography of amino acid standards at the (a) 50 pmol and (b) 1 pmol levels. Detector full-scale responses were (a) 90 and (b) 4.5 mV. Chromatography was done using a multistep gradient with (a) 140 mM sodium acetate, 17 mM TEA titrated to pH 5.05 with phosphoric acid containing 1 mg/L disodium ethylenediamine tetraacetic acid and (b) 60:40 acetonitrile:water (V:V). A fluorescence detector was used (excitation wavelength 250 nm and emission wavelength 395 nm). The column was a reversed phase AccQ-Tag C18 column (15 cm × 3.9 mm) from Waters Corp. thermostated at 37 °C. The flow rate was 1 mL/min. *(From S. A. Cohen, K. M. DeAntonis, and D. A. Michaud, in R. H. Angeletti, ed., Techniques in Protein Chemistry IV (New York: Academic Press, 1993), pp. 289–306. Copyright © 1993 by Academic Press.)*

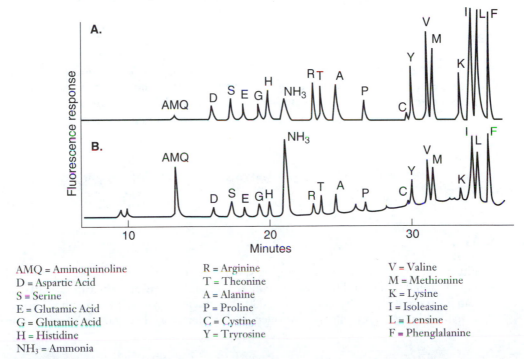

AMQ = Aminoquinoline
D = Aspartic Acid
S = Serine
E = Glutamic Acid
G = Glutamic Acid
H = Histidine
NH₃ = Ammonia

R = Arginine
T = Theonine
A = Alanine
P = Proline
C = Cystine
Y = Tryrosine

V = Valine
M = Methionine
K = Lysine
I = Isoleasine
L = Lensine
F = Phenglalanine

Figure 9.5 GPC separation of several standard proteins. *(Courtesy of* Waters Sourcebook of Chromatography, *1992.)*

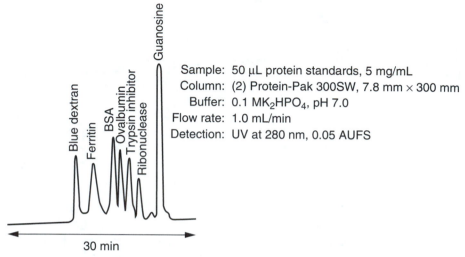

Sample: 50 µL protein standards, 5 mg/mL
Column: (2) Protein-Pak 300SW, 7.8 mm × 300 mm
Buffer: 0.1 MK_2HPO_4, pH 7.0
Flow rate: 1.0 mL/min
Detection: UV at 280 nm, 0.05 AUFS

phase column with gradient elution and fluorescence detection (excitation wavelength 250 nm and emission wavelength 395 nm). The detection limit for this sensitive analysis is as low as 40 fmol for some of the amino acids (3).

Proteins are generally analyzed by one of three approaches. Ion exchange can be used to capitalize on the ionic character of the amino acid residues on the protein. Affinity chromatography can be used to capitalize on the specific binding affinity of proteins. Because proteins are large compounds (molecular weight is larger than 1000 g/m), their analysis can be done using gel permeation chromatography. Figure 9.5 is a chromatogram that demonstrates the GPC separation of a number of standard proteins using a Protein-Pak 300SW column from Waters Corp. (molecular weight range 10,000 to 300,000 g/m) with isocratic elution and UV detection (280 nm)(4).

2. Ion Analysis.

Ion analysis (Chap. 12) is important for a variety of applications in the environmental and industrial areas. Figure 9.6 demonstrates the separation of a number of standard cations at the low parts-per-million level. The analysis is done using an IC-Pak C M/D column (cation exchange column) from Waters Corp., with isocratic elution and conductivity detection (4).

Nuts and Bolts

Relative Costs

Financing a complete chromatographic system requires consideration of two costs: the cost of the initial capital outlay and the cost of maintenance and operation. The capital outlay for a basic system is on the order of $10,000 to $12,000. This includes a dual-piston pump for isocratic elution, a manual injector, and a single-wavelength UV detector. The most expensive systems are com-

Figure 9.6 Ion chromatography separation of several standard cations. *(Courtesy of* Waters Sourcebook of Chromatography, *1992.)*

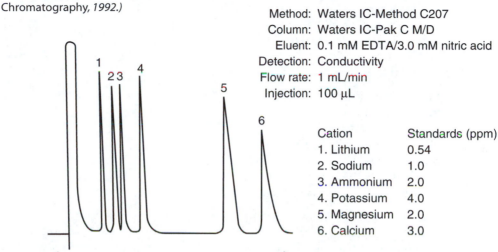

Method: Waters IC-Method C207
Column: Waters IC-Pak C M/D
Eluent: 0.1 mM EDTA/3.0 mM nitric acid
Detection: Conductivity
Flow rate: 1 mL/min
Injection: 100 μL

Cation	Standards (ppm)
1. Lithium	0.54
2. Sodium	1.0
3. Ammonium	2.0
4. Potassium	4.0
5. Magnesium	2.0
6. Calcium	3.0

pletely automated and computerized systems that cost $50,000 and up and include a multisolvent pumping system for both isocratic and gradient elution, an autoinjector, a photodiode array detector, and a computer to run the system and store the chromatographic data. If a mass spectrometry detector is included in the package, the cost may exceed $100,000.

Vendors for Instruments and Accessories

I = instrument manufacturer; C = consumables supplier

Beckman Instruments, Inc. (I & C)
2500 Harbor Blvd.
Fullerton, CA 92634
phone: 714-773-6707, 800-724-2345
fax: 714-773-8186
Internet: http://www.beckman.com

Bio-Rad Laboratories (I & C)
2000 Alfred Nobel Dr.
Hercules, CA 94547
phone: 510-741-1000, 800-424-6723
fax: 800-879-2289
Internet: http://www.bio-rad.com

Dionex Corp. (I & C)
P.O. Box 3603, 1228 Titan Way
Sunnyvale, CA 94088-3603
phone: 408-737-0700, 800-723-1161
fax: 408-730-9403
email: marcom@dionex.com
Internet: http://www.dionex.com

Hewlett-Packard Co. (I & C)
2850 Centerville Rd.
Wilmington, DE 19808
phone: 302-633-8504, 800-227-9770
fax: 302-633-8902
Internet: http://www.hp.com/go/chem

J & W Scientific (C)
91 Blue Ravine Rd.
Folsom, CA 95630-4714
phone: 916-985-7888, 800-223-3424
fax: 916-985-1101
email: 70530,2047@compuserve.com
Internet: http://www.JandW.com

The Perkin-Elmer Corp. (I & C)
761 Main Ave.
Norwalk, CT 06859
phone: 800-762-4000
fax: 203-762-4228
email: info@perkin-elmer.com
Internet: http://www.perkin-elmer.com

Restek Corporation (C)
110 Benner Circle
Bellefonte, PA 16823-8812
phone: 814-353-1300, 800-356-1688
fax: 814-353-1309
Internet: http://www.restekcorp.com

Shimadzu Scientific Instruments, Inc. (I & C)
7102 Riverwood Dr.
Columbia, MD 21046
phone: 800-477-1227
fax: 410-381-1222
Internet: http://www.shimadzu.com

Supelco, Inc. (C)
Supelco Park
Bellefonte, PA 16823-0048
phone: 814-359-3441, 800-247-6628
fax: 814-359-3044, 800-447-3044
email: Supelco@Supelco.sial.com
Internet: http://www.Supelco.sial.com/Supelco.html

Waters Corp. (I & C)
34 Maple St.
Milford, MA 01757
phone: 508-478-2000, 800-254-4752
fax: 508-872-1990
email: info@waters.com
Internet: http://www.waters.com

Required Level of Training

The operation of a chromatographic system must be considered from two perspectives. In order for a complete novice to be able to follow a well-defined protocol, 1 to 5 days of intensive training is sufficient. Many chromatographic instrument companies offer training programs for this type of operator. Training to develop methods, interpret data proficiently, and maintain the instrument requires a more substantial investment of time. Depending on the complexity of the analysis, weeks to months of training, study, and practice are required to be successful.

Service and Maintenance

Maintenance is divided into three categories: routine maintenance, normal maintenance, and major components. Routine maintenance incorporates all of the supplies that must be replenished on a daily basis, such as mobile phase, chart paper, and sample vials. Normal maintenance includes instrument upkeep that must be done on a less frequent basis, including checkvalve, seal, and detector lamp replacement, and injector rebuilding. Major components such as power supplies and circuit boards are things that fail infrequently. All of these types of maintenance must be kept up in order to have an efficient system. Total maintenance costs generally run on the order of 5 to 10% of the list price of the instrument per year for an average system, which is one that runs approximately 10,000 samples per year. Maintenance is on the order of 5% if the operator does it on his or her own and 8 to 10% if a maintenance contract is purchased that includes regular service visits to replace major components.

Special Considerations for Regulated Environments

The Food and Drug Administration (FDA) and the Environmental Protection Agency (EPA) have specific regulations for HPLC analysis, and many industries must follow their regulations. If these regulations must be met, it is wise to purchase an instrument from a company that is ISO 9000 (or higher) approved. In addition, a select group of vendors have service training and documentation designed to help the user meet installation, operation, maintenance, and calibration requirements of the FDA and EPA.

Suggested Readings

Books

BROWN, P. R., *High Pressure Liquid Chromatography: Biochemical and Biomedical Applications.* New York: Academic Press, 1973.

LINDSAY, S., *High Performance Liquid Chromatography.* New York: Wiley, 1992.

Journals

LC-GC and *LC-GC International*

The Journal of Liquid Chromatography

The Journal of Chromatography

American Laboratory

International Laboratory

Advances in Chromatography series

References

1. L. R. Snyder and J. J. Kirkland, *Introduction to Modern Liquid Chromatography*, 3rd ed. (New York: Wiley, 1996).

2. D. A. Skoog, *Principles of Instrumental Analysis* (Orlando, FL: Holt, Rinehart, & Winston, 1985).

3. S. A. Cohen, K. M. De Antonis, and D. A. Michaud, "Compositional Protein Analysis Using 6-amino-quinolyl-N-hydroxysuccinimidyl carbonate, a Novel Derivatization Reagent," in R. H. Angeletti, ed., *Techniques in Protein Chemistry IV* (New York: Academic Press, 1993), pp 289–306.

4. *Waters Sourcebook of Chromatography*, 1993.

5. S. Lindsay, *High Performance Liquid Chromatography* (New York: Wiley, 1992).

Chapter 10

Capillary Electrophoresis

Dale R. Baker

Consultant, Boulder, CO

Summary

General Uses

- Separation and identification of polar and nonpolar compounds and some elements, including nonionic and ionic organic compounds, inorganic anions and cations, macromolecules (such as proteins and oligonucleotides), and chiral compounds
- Quantitative and qualitative determination of compounds and some elements in mixtures
- Determination of molecular weights of large biomolecules and isoelectric points of proteins
- Depending on the type of detector used, can be nondestructive or destructive
- Can be automated for analysis of liquid samples or solid samples dissolved in a liquid

Common Applications

- Applicable to the separation and determination of many of the same types of samples as high-performance liquid chromatography (HPLC), ion chromatography, and slab gel electrophoresis
- Used in biochemical, clinical, environmental, food, forensic, and pharmaceutical applications

Samples

State

- Solid or liquid samples can be analyzed.
- Solids must be dissolved in a liquid, often the same liquid that is used as the electrolyte, which is then injected into the capillary.
- Liquids may be injected directly, concentrated, diluted, or reconstituted.

Amount

Typical injection volumes are 1 to 50 nL, but in some cases, volumes of up to about 1 μL are injected.

Preparation

Preparation usually involves only dissolution of the sample; however, extraction, derivatization, or filtration may be required. The sample should be in a liquid, with no solids present.

Analysis Time

Separation times range from a few seconds to about an hour. Sample preparation time depends on the nature of the sample and ranges from a few minutes to several hours. For quantitative analysis, calibration with standards of known concentrations is required and may take up to a few hours. Most instruments have the capability for automated calibration.

Limitations

General

- Not well suited for determination of nonpolar, low-molecular-weight, volatile compounds, which are best determined by gas chromatography.
- Not well suited for determination of nonionic, high-molecular-weight polymers.
- Not as sensitive as HPLC.

Accuracy

Precision ranges of 1 to 2 relative standard deviation (%)

Sensitivity and Detection Limits

Sensitivities of mg/L (parts per million) to μg/L (parts per billion)

Complementary or Related Techniques

- Gas chromatography provides analyses of low-molecular-weight, volatile compounds.
- Supercritical fluid chromatography provides analyses of volatile compounds.
- HPLC provides analyses of nonionic polymers by gel permeation chromatography.

Introduction

When an electric field is applied across a tube containing a conductive solution, an electrolyte, which contains charged solutes, the solutes migrate through the solution. The rates and directions of their migration depend on the signs and magnitudes of their charges as well as their sizes. This phenomenon is called electrophoresis.

Under the influence of an electric field, electroosmotic flow (EOF) causes movement of the electrolyte through the tube. The charged solutes, the ions, migrate through the tube, with the highly charged ions migrating the fastest and the lesser charged ions the slowest. Neutral molecules are not influenced by the electric field and move through the tube under the influence of just the EOF and are not separated from each other. However, neutral molecules can be separated from each other using the technique of micellar electrokinetic capillary chromatography. The charged solutes move through the conductive medium as zones that, in the absence of any molecular diffusion, do not get wider as they move.

Early electrophoretic separations were performed using glass tubes filled with a conductive medium. These tubes were relatively large and the separation efficiencies were poor, due to zone spreading caused by thermal convective diffusion, and the ions were not well separated from each other.

Passage of an electric current through a conductive medium produces joule heat. The larger the diameter of the tube, the more joule heat is generated, which causes spreading of the zones giving poor separations. Molecules in the center of the tube migrate faster than those near the wall because the viscosity of the electrolyte is lower in the center.

The temperature difference, Δt, between the center of a tube and its wall is given by

$$\Delta t = \frac{0.239\text{W}}{4k}r^2 \tag{10.1}$$

where W is the heat in watts m^{-3}, k is thermal conductivity in cal sec^{-1} cm^{-1} °C^{-1}, and r is the tube radius. It can be seen that a reduction in tube radius will reduce the temperature difference and, consequently, reduce the convective diffusion within the tube. Reducing the tube diameter by a factor of 10 reduces Δt by a factor of 100.

One way to minimize convective diffusion is to add a nonconductive stabilizing medium, such as a gel, to the solution. In this case, a slab of gel is immersed in a conductive medium, the sample is applied in preformed wells in the gel, and an electric field is applied across the gel. The nonconductive gel has a high area-to-volume ratio and dissipates heat better than a free solution of the same dimensions. This technique, slab gel electrophoresis, has been widely used for the separation of large biological molecules such as proteins and oligonucleotides. Media that have been used include paper, cellulose acetate membranes, and agarose or polyacrylamide gels. When polyacrylamide is used, the technique is called polyacrylamide gel electrophoresis (PAGE). This is in contrast to free-solution electrophoresis, which is the technique in which no medium is added to the conductive solution.

Another approach to minimizing convective diffusion is to reduce the tube diameter. Teflon capillaries of 200 μm internal diameter were used for free-solution electrophoresis; these small-diameter tubes reduced convective zone spreading. Later, Pyrex capillaries of 75 μm internal diameter were used. These very-small-diameter capillaries minimized zone spreading due to convective diffusion. Also, any heat that was generated was effectively dissipated due to the high area-to-volume ratio of the small-diameter capillaries. Because the small-diameter capillaries provide excellent heat dissipation, high voltages, up to 30 kV, can be used. The advantages of high voltage are that separation times are reduced and very efficient separations are achieved. Part of the reason for the high efficiency of capillary electrophoresis is that the solutes move through the capillary in a plug flow as opposed to laminar flow, as is the case in HPLC.

One of the most common means of detecting solutes as they pass through a capillary is with an ultraviolet/visible (UV/Vis) absorbance detector, similar to the types that are widely used in HPLC. Glass capillaries cannot be used with UV/Vis detection at wavelengths shorter than about 280 nm. However, fused silica capillaries can be used at shorter wavelengths, down to 190 nm, which allow detection of more compounds. Another advantage of fused silica capillaries is that they are flexible and not easily broken.

How It Works

A capillary electrophoresis system is represented in Fig. 10.1; the main components are the inlet and exit electrolyte reservoirs, sample container, capillary, detector, and high-voltage (HV) power supply.

The negatively charged electrode is the cathode, which attracts cations, and the positive electrode is the anode, which attracts anions.

Electrophoresis is performed by filling the reservoirs and capillary with a buffer solution, the electrolyte. The capillary inlet is placed into a sample vial, the sample is introduced, then the capillary inlet is placed back into the inlet reservoir and an electric field is applied. The electric field causes the solutes to migrate through the capillary and they are detected by the detector and its output is usually displayed at an integrator or computer. This output is a plot of detector response versus time and is called an electropherogram. Each peak in the electropherogram represents one of the components as it migrates through the detector. Because the compounds migrate through the detector at different times, an electropherogram is produced in which the separated compounds appear as peaks with different migration times. The starting time, a time of zero, is the

Figure 10.1 Representation of a capillary electro-phoresis system. The output from the detector is typically sent to an integrator or computer. *(Reprinted courtesy of Hewlett-Packard Company.)*

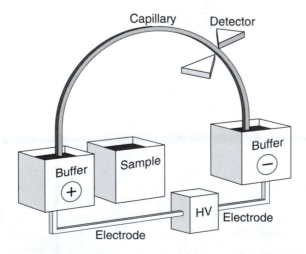

time when the electric field is turned on. Migration times are measured at the apices of the peaks, usually in units of minutes.

Sample Separation

Capillary electrophoresis is typically performed using fused silica capillaries of 50 to 75 μm internal diameter, 375 μm outer diameter, and 50 to 100 cm length. Applied voltages of up to 30 kV are commonly used. Under these conditions, very fast and efficient separations can be attained, as illustrated in Fig. 10.2 which is an electropherogram showing the separation of 16 anions in only 4 minutes.

When an electric field is applied to an electrolyte containing a charged solute, the solute migrates through the electrolyte with a velocity v_{EP}, in cm/sec, given by

$$v_{EP} = \mu_{EP}E \tag{10.2}$$

where μ_{EP} is the electrophoretic mobility, in $cm^2\ V^{-1}\ sec^{-1}$ and E the applied electric field, in V/cm. Solutes are separated because they migrate through the capillary at different velocities. A solute's electrophoretic mobility, μ_{EP}, is given by

$$\mu_{EP} = \frac{q}{6\pi\eta r} \tag{10.3}$$

where q is the charge on the ionic solute, η the viscosity of the electrolyte, and r the radius of the solute. It can be seen from Eq. (10.3) that the greater the charge-to-size ratio (q/r), the higher the electrophoretic mobility and, from Eq. (10.2), the higher the velocity. Highly charged, small molecules have the highest velocity and move through the capillary the fastest and larger molecules with lower charges move slower. Neutral molecules have a charge, q, of zero and are not influ-

Figure 10.2 Electropherogram showing the separation of 16 anions. *(Reprinted from* Journal of Chromatography, *640, M. P. Harrold, M. J. Wajtusik, J. Riviello, and P. Henson, pp. 463–471, copyright 1993 with kind permission of Elsevier Science—NL, Sara Burgerhartstraat 25, 1055 KV Amsterdam, The Netherlands.)*

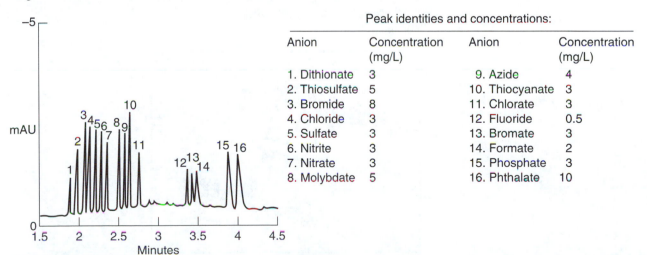

Peak identities and concentrations:

Anion	Concentration (mg/L)	Anion	Concentration (mg/L)
1. Dithionate	3	9. Azide	4
2. Thiosulfate	5	10. Thiocyanate	3
3. Bromide	8	11. Chlorate	3
4. Chloride	3	12. Fluoride	0.5
5. Sulfate	3	13. Bromate	3
6. Nitrite	3	14. Formate	2
7. Nitrate	3	15. Phosphate	3
8. Molybdate	5	16. Phthalate	10

enced by the electric field. An increase in viscosity of the electrolyte causes a reduction in electrophoretic mobility and reduces the electrophoretic mobilities of all ions.

When an electrolyte, also called a run buffer, is placed in a fused silica capillary and an electric field is applied, a volumetric flow of the electrolyte occurs. The flow is due to electroosmosis and is called the electroosmotic flow (EOF). EOF usually goes toward the cathode, but it can be eliminated or even reversed.

Depending on the pH of the electrolyte, the inner wall of a fused silica capillary can have different amounts of Si–O$^-$ groups present at the surface. Surface silanol (Si–OH) groups are ionized to negatively charged Si–O$^-$ groups at pH above about 3. This ionization can be enhanced by first passing a basic solution through the capillary, followed by the electrolyte.

The negatively charged Si–O$^-$ groups attract positively charged cations from the electrolyte. These cations are strongly attracted to the capillary wall and form a fixed layer of positively charged cations. The cations in the fixed layer are not of sufficient density to electrically neutralize all of the Si–O$^-$ groups so a mobile layer of cations forms. The cations are attracted to the cathode and the ones in the mobile layer move in that direction. Because the cations are solvated, they drag the bulk electrolyte solution with them through the capillary, thus creating the EOF, as represented in Fig. 10.3.

Between the fixed and mobile layers is a plane of shear. There is an electrical imbalance at the plane of shear that causes a potential difference between the layers. This is called the zeta potential, ζ.

The electroosmotic flow (EOF) rate varies with pH of the electrolyte. Below about pH 3, the SiOH groups at the capillary wall are fully protonated and the rate of EOF is very low. The rate of EOF increases with increasing pH, reaching a maximum at about pH 10.

The EOF profile is relatively flat, as opposed to a pumped flow profile, as in HPLC. An advantage of a flat flow profile is that all of the solute molecules experience the same velocity component caused by EOF regardless of their cross-sectional positions in the capillary. Therefore, the solutes elute as narrow bands, giving narrow peaks of high efficiency. Under the influence of the

Figure 10.3 Representation of electroosmotic flow. The negatively charged spheres represent SiO$^-$ groups on the surface of the capillary wall. The arrow indicates the direction of the flow and the flat line under the arrow indicates that the flow profile is relatively flat. (*Reprinted courtesy of Hewlett-Packard Company.*)

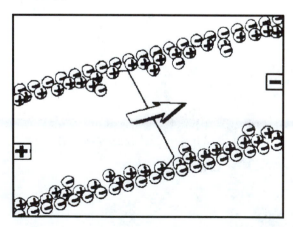

uneven flow profile of pumped flow, the solutes in the center of the tube move faster than those nearer the wall, resulting in relatively broad peaks.

Solute molecules are separated because of differences in their electrophoretic mobilities and are moved through the capillary by the EOF. Anions are attracted to the positively charged anode and, if there were no EOF, would simply migrate to the inlet reservoir without passing through the capillary and detector. However, the EOF is usually greater than the electrophoretic movement of anionic solutes, and can be strong enough to carry even small, triply charged anions toward the negatively charged cathode through the detector.

Anionic solutes are pulled back toward the anode, and move at a rate that is lower than the EOF. Neutral solutes are not influenced by the electric field and move through the capillary at the same rate as the EOF. Cations move toward the cathode under the influence of both the electric field and EOF and move faster than the EOF.

The order in which solutes pass through the capillary is cations, then neutrals, then anions. A representation of the resulting electropherogram is shown in Fig. 10.4.

Charged solutes are separated from each other because of differences in their electrophoretic mobilities. However, neutral solutes are not separated from each other. The solutes with the highest mobility and the first to migrate through the capillary are small, highly charged cations. Small, highly charged anions are the last to migrate through the capillary. Neutral solutes can be separated using a different type of capillary electrophoresis, micellar electrokinetic capillary chromatography.

The velocity of the electroosmotic flow, v_{EOF}, is given by

$$v_{EOF} = \frac{E\varepsilon\zeta}{4\pi\eta} \tag{10.4}$$

where ε is the dielectric constant of the electrolyte, ζ is the zeta potential, E is the applied electric field in V/cm, and η is the viscosity of the electrolyte.

Electroosmotic mobility, μ_{EOF}, is given by

$$\mu_{EOF} = \frac{\varepsilon\zeta}{4\pi\eta} \tag{10.5}$$

It can be seen from Eqs. (10.4) and (10.5) that

$$v_{EOF} = \mu_{EOF}E \tag{10.6}$$

The electric field, E, is equal to V/L where V is the applied voltage and L is the total length of the capillary.

EOF can be calculated by injecting a neutral marker, a nonionic molecule, such as benzyl alcohol, that moves through the capillary at the same rate as the EOF, and measuring its migration time, t_m. Its velocity, v_{EOF}, can be calculated from $v_{EOF} = l/t_m$, where l is the length of the capillary to the detector. The electroosmotic mobility, μ_{EOF}, can then be calculated by rearranging Eq. (10.6).

Figure 10.4 Drawing of an electropherogram indicating the order of elution due to EOF. Neutral molecules are not separated from each other. *(Reprinted courtesy of Hewlett-Packard Company.)*

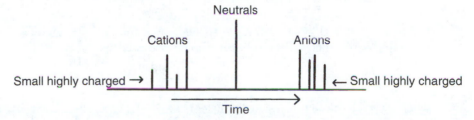

Anything that causes changes in the right side of Eq. (10.4) will cause changes in the EOF. It is important that the EOF be constant during and between analyses because variations in EOF will cause shifts in migration times, which may result in erroneous peak assignments and peak area measurements. Factors that can cause changes in EOF are as follows:

- Increasing voltage increases electroosmotic flow.
- Increasing electrolyte pH increases electroosmotic flow.
- Increasing the concentration or ionic strength of the electrolyte decreases electroosmotic flow.
- Increasing temperature increases electroosmotic flow.
- Adding an organic solvent may increase or decrease electroosmotic flow, depending on how it affects the viscosity, dielectric constant, and zeta potential.
- Modifying the capillary wall can reduce, eliminate, or even reverse the electroosmotic flow.

Usually, the detector side of the capillary has a negative charge and attracts the positively charged cations in the mobile layer so EOF is toward the cathode. The order of elution is cations, neutrals, then anions, as shown in Fig. 10.4.

If analyzing for just anions, the separation time can be reduced by reversing the direction of EOF and the polarity of the applied electric field, such that the anode is near the detector end of the capillary. The order of elution will now be anions, neutrals, then cations.

A method for reversing the direction of EOF is to add a quaternary amine as an alkyl ammonium salt, a flow modifier, to the electrolyte. The positively charged hydrophilic ends of the quaternary amines become fixed to the Si–O⁻ groups on the capillary wall through ionic interactions. The hydrocarbon ends of the fixed quaternary amines associate with the hydrocarbon ends of the free quaternary amines through hydrophobic interactions. The positively charged ends of the associated amines attract anions from the electrolyte. These solvated anions move toward the anode, dragging the electrolyte toward the positive electrode, resulting in reversal in direction of EOF.

Alkyl ammonium salts that have been used as flow modifiers include cetyltrimethylammonium bromide (CTAB), cetyltrimethylammonium chloride (CTAC), tetradecyl-trimethylammonium bromide (TTAB), amines such as diethylenetriamine (DETA), and a diquaternary ammonium salt, 1,6-bis(trimethylammonium)hexane hydroxide (hexamethonium hydroxide).

Migration time, t_m, is the time it takes a solute to migrate through the effective capillary length, l, which is the length of the capillary to the detector and is given by

$$t_m = \frac{lL}{(\mu_{EP} + \mu_{EOF})V} \tag{10.7}$$

where L is the total capillary length, μ_{EP} is the electrophoretic mobility of the solute, μ_{EOF} is the electroosmotic mobility of the electrolyte, and V is the applied voltage. Efficiency is expressed as the number of theoretical plates, N, and can be calculated from

$$N = 16(t_m/w)^2 \tag{10.8}$$

where w is the peak width measured at the base of the peak. It can be seen that narrower peaks give higher efficiencies. Efficiency can also be expressed as

$$N = \frac{(\mu_{EP} + \mu_{EOF})V}{2D} \tag{10.9}$$

where D is the solute's diffusion coefficient.

Resolution, R, is a measure of how well two adjacent peaks in an electropherogram are separated from each other. Resolution can be expressed as

$$R = 0.177(\mu_2 - \mu_1)[V/D(\mu_{AVE} + \mu_{EOF})]^{1/2} \qquad (10.10)$$

where $\mu_2 - \mu_1$ is the difference in electrophoretic mobilities of the solutes and μ_{AVE} is the average of their electrophoretic mobilities .

From Eq. (10.7), (10.9), and (10.10) it can be seen that a high voltage will produce the highest resolution with the highest efficiency in the shortest time. The voltage should not be so high that any joule heat that is produced cannot be dissipated. For a given set of conditions, the maximum voltage that should be used can be determined from an Ohm's law plot, a graph of observed current versus applied voltage. This graph should be linear with a zero intercept. If excessive heat is produced, the resistance goes down, causing an increase in current, which is indicated by an increase in the slope of the plot. The voltage at which nonlinearity occurs is the maximum that should be used. At voltages higher than this, efficiency and resolution may be reduced because of convective diffusion. The maximum voltage depends on capillary dimensions and the electrolyte's composition, pH, and concentration.

When there is a large difference in the electrophoretic mobilities of the solutes, that is, when $\mu_2 - \mu_1$ is large, good resolution is achieved. This difference can often be improved by optimization of the pH of the electrolyte. The best resolution is attained when the pH of the electrolyte is near the pK_a of the solutes.

As capillary length is increased, resolution is increased, assuming the zones are not broadened by diffusion, because two solutes migrating at different velocities have more time to be separated . Resolution is proportional to the square root of the ratio l/L. Because increasing capillary length also increases analysis time, Eq. (10.7), the shortest capillary that gives the desired resolution should be used.

Sample Introduction

In capillary electrophoresis a sample is not injected; rather, it is introduced into the capillary. The term *injection* is a carryover from chromatography, where the sample is usually injected into a flowing mobile phase. Even though the term injection is inappropriate, it is used here to conform to common usage. Sample volumes are very small in capillary electrophoresis, usually nanoliters, compared to HPLC or gas chromatography, in which microliters are injected.

Samples are placed in containers that usually have volumes of a few microliters to a few milliliters. Samples are injected by placing the inlet of the capillary into the sample container, introducing the sample into the capillary, and placing the capillary back into the inlet reservoir. Samples can be injected by either hydrodynamic (or hydrostatic), or electrokinetic (electromigration) injection.

For best resolution and peak shape, the concentration of the injected sample should be approximately one-hundredth the concentration of the electrolyte. The presence of high concentrations of solute ions may distort the electric field in the capillary, causing distortion of peak shapes.

Hydrodynamic injection is done by either pressure or siphoning. Siphon injection, also called gravity injection, is done by raising the sample container, allowing the sample to siphon into the capillary. Injection volume, V_i, can be calculated from

$$V_i = 2.84 \times 10^{-8} \frac{Htd^4}{L} \qquad (10.11)$$

where V_i is the sample volume in nanoliters, H is the height the sample is raised in millimeters, t is the time the sample is raised in seconds, d is the capillary inner diameter in microns, and L is the total capillary length in centimeters. For example, if the sample is raised 50 mm for 10 sec, and the capillary is 50 cm long and 50 μm internal diameter, the injection volume is 1.78 nL. Equation (10.11) assumes that the sample has the same viscosity as water and its temperature is 20 to 22 °C.

Pressure injection can be done by either pressurizing the sample vial or by applying a vacuum to the exit reservoir. The volume injected can be calculated from

$$V_i = \frac{Pd^4\pi t}{128\eta L} \tag{10.12}$$

where P is the pressure and h is the viscosity. For example, if a pressure of 25 mbar is applied for 2 sec with a capillary that is 75 cm long and 50 μm internal diameter, the injection volume is 1 nL.

In electrokinetic injection, an electric field is applied between the sample vial and the exit reservoir, which causes the sample components to migrate into the capillary. Neutral molecules are pulled into the capillary by the EOF, whereas charged solutes move into the capillary because of both EOF and electrophoretic migration. Hydrodynamic injections do not work well for gel-filled capillaries, so electrokinetic injection is usually required.

The quantity injected, Q, is given by

$$Q = V\pi c t r^2 \frac{\mu_{EP} + \mu_{EOF}}{L} \tag{10.13}$$

where V is the voltage, c is the sample concentration, t is the time the voltage is applied, r is the capillary radius, μ_{EP} is the electrophoretic mobility of the solute, and μ_{EOF} is the electroosmotic mobility.

Because the quantity of solute injected depends on μ_{EP}, there exists a sampling bias when using electrokinetic injection. Assuming a sample contains two solutes that have different electrophoretic mobilities and are present in equal concentrations, it can be seen from Eq. (10.13) that different quantities of these solutes would be injected. Larger amounts of early eluting solutes with high mobilities are injected than of later eluting solutes with lower mobilities.

The quantity injected also depends on the EOF and the electrophoretic velocity of nondetected ions in the sample solution. Different sample quantities may be injected from sample to sample if there are different concentrations of electrolytes in the sample solutions. As the electrolyte concentration in the sample solution decreases, its electrical resistance increases, causing the electrophoretic velocity and EOF rate of the sample solution to increase. Hence, more sample is injected from a dilute solution than from a concentrated solution. To help overcome this problem, a high concentration of an electrolyte that is not detected can be added to each sample. Electrokinetic injection is subject to more variables than hydrodynamic injection and, in general, is not as precise. Most commercially available instruments provide both electrokinetic and hydrodynamic injections.

A sample can sometimes be concentrated after it has been injected into the capillary if it is dissolved in a solution, such as water, that has a lower conductivity than the electrolyte. This is called sample stacking. The sample, dissolved in water or dilute electrolyte solution, is introduced into the capillary using hydrodynamic injection. Thus, there is a plug of water at the capillary inlet, with the rest of the capillary filled with electrolyte. When voltage is applied, a higher electric field develops across the plug of water because of its high electrical resistance. Because electrophoretic velocity is proportional to electric field strength, the ions in the sample rapidly migrate toward the electrolyte, forming a concentrated band or stack. When they reach the electrolyte they slow down because the electrical field is lower. The stack of concentrated ions then proceeds through the capillary.

The volume of sample that can be injected by stacking is limited. The EOF in the sample plug is faster than the EOF in the electrolyte. This difference in flows may cause laminar flow within the capillary if a large volume of water is injected. Laminar flow may cause the stacked sample zones to broaden, reducing the resolution. A method for stacking large volumes of sample solution is available. In this method, a large volume of sample dissolved in water is injected and, as the sample is stacked, the sample solution is backed out of the capillary by reversing the polarity of the applied voltage. Because the sample solution is removed, there is no zone broadening due to laminar flow.

Sample Detection

Detection methods that have been used in capillary electrophoresis include UV/Vis absorbance, fluorescence, laser-induced fluorescence, mass spectrometry, conductivity, amperometry, radiometric, and refractive index. Detection limits are approximately 1 mg/L to 1 µg/L.

UV/Vis absorbance and fluorescence detectors are the most widely used. With these detectors the solutes can be detected while they are still in the capillary. A small section of the polyimide outer coating on the capillary is scraped or burned off and that section is placed in the light beam of the detector. The clear section of the capillary serves as the cell window. These detectors are applicable for compounds that sufficiently absorb light or fluoresce.

Indirect absorbance and fluorescence detection can sometimes be used for compounds that do not absorb light or fluoresce. In the case of indirect absorbance detection, a light-absorbing compound is added to the electrolyte. When a nonabsorbing solute elutes it causes a decrease in absorbance, which gives a dip in the baseline, a negative peak. The polarity of the output of the detector is reversed so positive peaks appear at the integrator. Indirect fluorescence detection can be used in the same way, in which case a compound that fluorescences is added to the electrolyte.

Indirect absorbance has been used for analysis of nonabsorbing inorganic anions in which pyromellitic acid was added to the electrolyte. The electrophoretic mobility of the light-absorbing compound that is added to the electrolyte should match that of the solutes in order to maintain peak symmetry.

Absorbance is proportional to the pathlength of the cell in a UV/Vis detector. In HPLC the cell pathlengths of UV/Vis detectors are typically 1 to 10 mm. In capillary electrophoresis the capillary serves as the detector cell and the pathlength is equal to the inner diameter of the capillary, usually 50 to 75 µm. Detection limits in HPLC are approximately 100 times lower than in capillary electrophoresis. Increasing the pathlength of the cell increases absorbance and lowers detection limits.

A Z-shaped cell can increase the pathlength to 1 to 3 mm, with a proportional increase in peak height. Also, capillaries are available that have a bubble in the portion of the capillary that serves as the cell window. The bubble has a diameter of 150 µm and gives approximately three times the peak height of a 50-µm internal diameter capillary.

A list of common detectors and approximate detection limits is given in Table 10.1.

Table 10.1 Capillary electrophoresis detectors and their approximate detection limits.

Detector	Approximate Detection Limit, µg/mL
Absorbance, UV/Vis	10^{-1}
Indirect absorbance, UV/Vis	1
Fluorescence	10^{-3}
Indirect fluorescence	10^{-2}
Laser-induced fluorescence	10^{-6}
Mass spectrometer	10^{-4}
Amperometric	10^{-5}
Conductivity	10^{-3}

Reprinted courtesy of Hewlett-Packard Company.

Small-diameter capillaries reduce the joule heat that is produced and enhance heat dissipation. Depending on the electrolyte, capillary, and voltage used, the heat that is generated may not be effectively dissipated. High temperatures within the capillary may cause nonreproducible migration times, broad peaks due to convective diffusion, bubbles in the capillary (which can interrupt the flow of current and cause the instrument to turn off), or sample decomposition. For example, proteins may be denatured at high temperature. Therefore, the capillary should be cooled to make sure the heat is dissipated.

A summary of the features of capillary electrophoresis is as follows:

- High-efficiency separations are attained.

- High voltages can be used.

- Short separation times are achieved because of the high voltages and the short columns.

- Small sample volumes are required.

- Small amounts of reagents must be purchased and disposed of.

- Different modes of capillary electrophoresis are available, so different types of samples can be analyzed.

- Aqueous media, buffers, are usually used so aqueous samples can be injected directly.

- Ambient temperature for separations minimizes any sample decomposition or denaturation.

- Automation is available for automated sample injection, method development, data display, quantitation, and data storage.

What It Does

The most frequently used modes of capillary electrophoresis are capillary zone electrophoresis (CZE), micellar electrokinetic capillary chromatography (MECC), capillary gel electrophoresis (CGE), capillary isoelectric focusing (CIEF), and capillary isotachophoresis (CITP). The different modes of capillary electrophoresis often complement each other and, in some cases, samples may be separated by two or more modes.

The availability of different modes of capillary electrophoresis greatly enhances its versatility. Because a fused silica capillary can be used for more than one mode, changing modes often involves changing only the electrolyte. For example, changing from CZE to MECC often requires only the addition of a detergent to the electrolyte.

CZE is sometimes called free-solution capillary electrophoresis (FSCE). Although FSCE is the more appropriate term, *CZE* has gained widespread usage and so is used here. In CZE, the capillary and the inlet and exit reservoirs are filled with an electrolyte of constant composition, that is, a free solution with no medium such as a gel. CZE is widely used as it is relatively simple and is applicable to separations of anions and cations in the same run. Because there is just an electrolyte solution in the capillary there are no matrix effects in CZE. However, there may be interaction of the solute with the capillary wall.

In CZE, separation occurs because the solutes move through the capillary at different rates due to differences in their electrophoretic mobilities. Electrophoretic mobility is mainly dependent on the charge-to-size ratio of an ion and, to some extent, on its shape. EOF causes the solutes to move through the capillary from the anode to the cathode with the order of elution being cations, neutrals, then anions. Because the ions are separated on the basis of their charge-to-size ratios, neutral compounds are not separated from each other in CZE. However, neutrals are separated from the ions.

CZE can be used to separate almost any ionic compounds that are soluble in an electrolyte, such as small inorganic anions, large biomolecules, and, using nonaqueous buffers, even water–insoluble compounds.

Micellar electrokinetic capillary chromatography (MECC) is sometimes called micellar electrokinetic chromatography (MEKC). MECC provides an electrophoretic method for separating neutral molecules where the separation is on the basis of differences in distributions of neutral solutes between micelles and the electrolyte.

In MECC, a detergent such as sodium dodecyl sulfate (SDS) is added to the electrolyte at concentrations above its critical micelle concentration (CMC). SDS, $[CH_3-(CH_2)_{11}-O-SO_3]$, has a hydrophilic, negatively charged moiety on one end of the molecule, the sulfate group, with the rest of the molecule being a neutral, hydrophobic, 12-carbon hydrocarbon. At concentrations above its CMC, which in the case of SDS is 8.27 mM, a detergent will form micelles, which are aggregates of individual detergent molecules. Micelles form in a spherical shape with the nonpolar, hydrophobic, hydrocarbon ends of the detergent in the center of the sphere, away from the aqueous electrolyte. The polar, ionic ends of the detergent are on the outside of the sphere. Micelles can be anionic, cationic, zwitterionic, or nonionic.

When a sample is introduced into the capillary it distributes itself between the hydrophobic interiors of the micelles and the aqueous electrolyte. The distribution is determined by the solute's polarity, or water solubility. A water-soluble solute, which is not soluble in the micelles, spends most of its time in the electrolyte. A water-insoluble solute is solubilized by the micelles and spends most of its time in the micelles. Separation is on the basis of differences in the distribution of solutes between micelles and electrolyte.

Anionic micelles are attracted to the anode, but the EOF carries them toward the cathode, so the velocity of the micelles is less than the EOF. Solutes that are insoluble in the micelles move through the capillary at the rate of the EOF and are the first to elute, at a time of t_0. Solutes that are totally solubilized by the micelles are the last to elute, at a time of t_{mc}. Solutes that spend part of the time in the micelles elute between t_0 and t_{mc}, as shown in Fig. 10.5, with retention times that are proportional to the time they spend in the micelles.

Figure 10.5 Representation of a micellar electrokinetic capillary chromatogram. *(Reprinted courtesy of Hewlett-Packard Company.)*

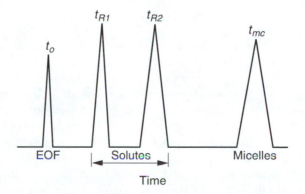

All solutes will elute between t_0 and t_{mc}. t_0 can be determined by injecting a solute that is insoluble in the micelles, such as methanol, and measuring its migration time. Sudan III can be used as a totally solubilized solute to determine t_{mc}. To increase the space between t_0 and t_{mc} a micelle with a higher mobility should be used.

MECC may be considered to be analogous to reversed phase HPLC, where solutes distribute themselves between a polar mobile phase and a nonpolar stationary phase. However, in MECC the micelles are not stationary; rather, they may be considered a nonpolar pseudo-phase.

In addition to separating neutral compounds, MECC can also be used for separation of ionic substances. MECC has been used for separation of samples as diverse as inorganic anions and large peptides.

The capacity factor, k', is the ratio of number of moles of solute in the micelles to the number of moles in the electrolyte or aqueous phase, n_{mc}/n_{aq}, and is given by

$$k' = \frac{t_R - t_0}{t_0(1 - t_R/t_{mc})} \tag{10.14}$$

where t_R is the retention time of the solute of interest. A solute that is completely solubilized by the micelles elutes at t_{mc}, then $t_R = t_{mc}$ and k' is infinite. A solute that is totally insoluble in the micelles elutes at t_0 and k' is zero. The resolution, R_s, between two adjacent peaks in a micellar electrokinetic chromatogram can be expressed as

$$R_s = \frac{N^{1/2}}{4} \times \frac{\alpha - 1}{\alpha} \times \frac{k_2'}{1 + k_2'} \times \frac{1 - t_0/t_{mc}}{1 + (t_0/t_{mc})k_1'} \tag{10.15}$$

where N is the number of theoretical plates, α is the selectivity, k_2'/k_1', and the subscripts 1 and 2 refer to the first and second solute, respectively.

In capillary gel electrophoresis (CGE), the capillary is filled with a gel, usually either a crosslinked polymer such as polyacrylamide/bisacrylamide or a linear polymer such as polyacrylamide. These gels have pores within them and as solutes migrate through the capillary they are separated by a molecular sieving mechanism on the basis of their sizes. Small molecules are able to pass through the pores and elute first, whereas larger molecules are retarded by the gel and elute later.

In CGE, solutes move through the capillary by electrophoresis and can be separated by differences in both size and electrophoretic mobilities. CGE works well for the separation of charged molecules that vary in size but not in their charge-to-size ratios regardless of their chain lengths, such as oligonucleotides or DNA restriction fragments. Large biomolecules, such as proteins and oligonucleotides, have traditionally been separated by slab gel electrophoresis, but CGE provides faster, more efficient separations. Also, unlike slab gel electrophoresis, CGE has on-line detection and quantitation.

Capillary isoelectric focusing (CIEF) is a focusing type of capillary electrophoresis in which the solutes, usually proteins, are separated on the basis of differences in their isoelectric points (pI). Proteins are amphoteric; that is, they can exist as either anions or cations depending on the pH of the solution. The pH at which a protein is neutral is called its isoelectric point. At pH below this, it exists as a cation and at pH above this, it exists as an anion.

In CIEF, a solution of ampholytes and the sample are placed in the capillary and an electric field is applied. The vial that contains the cathode is filled with a catholyte, a basic solution, and the vial that contains the anode is filled with an anolyte, an acidic solution. Ampholytes are compounds that have a range of pI values and they form a pH gradient in the capillary. The protein sample and ampholytes are introduced into the capillary together. Large sample volumes can be introduced as the entire capillary volume is available. When an electric field is applied, proteins and ampholytes migrate through the capillary until they reach a point where they are uncharged, that is, where the pH of the solution is equal to their pIs.

A treated capillary is used in which there is no EOF, ensuring that the movement of solutes is solely on the basis of differences in their pIs. Electroosmotic flow can be eliminated by coating the capillary wall with polyacrylamide.

When the solutes and ampholytes reach areas where they are uncharged the current decreases and becomes stable, indicating that the focusing is complete. The solutes can be detected by pressurizing the capillary or by adding salt to the anolyte or catholyte, forcing the ampholytes and solute zones through the detector. Because the solutes focus into very narrow zones, there may be precipitation of a protein within the zone if a high concentration of protein is injected. Because the zones are very narrow, resolution in CIEF is quite good.

CIEF can be used to determine the pI of a protein by focusing proteins of known pIs, creating a calibration plot of pI versus migration time, focusing the protein, measuring its migration time, and determining its pI from the plot.

Capillary isotachophoresis (CITP) is performed by sandwiching a sample between a leading and terminating, or trailing, buffer in a capillary and applying an electric field in the constant current mode. CITP cannot separate anions and cations in the same run. For purposes of this discussion, it is assumed that the sample is to be analyzed for anions. The leading buffer is selected such that its anions have a higher mobility than any anions in the sample. Conversely, the anions in the terminating buffer have lower mobility than those in the sample. When an electric field is applied, the anions in the buffers and the sample migrate toward the anode. The anions in the leading buffer have the highest mobility. All other anions migrate at the same velocity as the leading buffer anions. The solute molecules arrange themselves in the capillary in order of mobility, then an equilibrium is reached. The sample molecules thus form distinct zones. Because velocity is the product of mobility and electric field, as shown in Eq. (10.2), each zone has a different electric field.

Each zone narrows or spreads until they all have similar concentrations that are approximately the same as that of the leading buffer, meaning that wide zones are made up of sample components that were initially very concentrated and narrow zones are made up of components that were initially more dilute.

The output that is obtained in CITP is different from typical chromatograms or electropherograms, which consist of peaks for each component in the mixture. In CITP, there is no space between the solutes so the output appears as steps, as illustrated in Fig. 10.6, for UV/Vis absorbance and conductivity detectors.

Figure 10.6 Representation of the output from capillary isotachophoresis with absorbance and conductivity detectors. *R* represents increasing resistance. *(Reprinted courtesy of Hewlett-Packard Company.)*

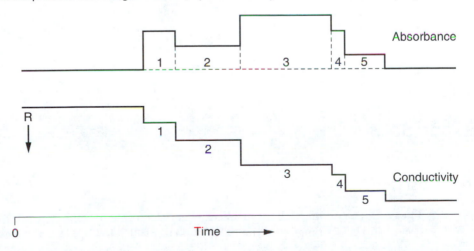

Each zone is of similar concentration, so the amplitude of a step does not reflect the initial concentration of the solute in the sample; instead, the width of each step is proportional to concentration.

Analytical Information

Capillary electrophoresis provides quantitative and qualitative determinations of compounds and some inorganic elements in mixtures. Quantitation is accomplished by injecting standards of known concentrations of the compounds in the mixture that are to be determined, measuring the detector response for each compound, constructing a calibration plot of either peak height, peak area, or peak area divided by migration time versus concentration for each compound, injecting the sample containing unknown concentrations of the components, and determining the concentrations of components in the sample from a calibration plot. Qualitative analysis is performed by comparing migration times of standards to migration times of compounds in a mixture.

Sensitivities of mg/L (parts per million) to µg/L (parts per billion) and precision ranges of 1 to 2 relative standard deviation (%) are typically obtained.

Applications

Capillary electrophoresis (CE) is applicable to the determination of a wide variety of sample types, including polycyclic aromatic hydrocarbons, chiral compounds, water- and fat-soluble vitamins, amino acids, inorganic ions, organic acids, carbohydrates, environmental pollutants, gunshot and explosive constituents, nucleosides and nucleotides, pharmaceuticals, catecholamines, peptides, and proteins.

Instrument and column manufacturers provide catalogs and references for many CE applications. For example, Hewlett-Packard Company offers Volumes 1 (1994) and 2 (1995) of *Applications of the HP³ᴰ Capillary Electrophoresis System*, which presents example applications for over 160 analyses.

Nuts and Bolts

Relative Costs

The cost of a commercially available capillary electropherograph depends on the capability of the system and prices range from about $15, 000 to $100,000. Capillaries cost about $100–$400.

Vendors for Instruments and Accessories

I = instrument manufacturer;   C = consumables supplier

Beckman Instruments, Inc. (I & C)
2500 Harbor Blvd.
Fullerton, CA 92634
phone: 714-773-6707, 800-724-2345
fax: 714-773-8186
Internet: http://www.beckman.com

Bio-Rad Laboratories (I & C)
2000 Alfred Nobel Dr.
Hercules, CA 94547
phone: 510-741-1000, 800-424-6723
fax: 800-879-2289
Internet: http://www.bio-rad.com

Dionex Corp. (I & C)
P.O. Box 3603, 1228 Titan Way
Sunnyvale, CA 94088-3603
phone: 408-737-0700, 800-723-1161
fax: 408-730-9403
email: marcom@dionex.com
Internet: http://www.dionex.com

Hewlett-Packard Co. (I & C)
2850 Centerville Rd.
Wilmington, DE 19808
phone: 302-633-8504, 800-227-9770
fax: 302-633-8902
Internet: http://www.hp.com/go/chem

The Perkin-Elmer Corp. (I & C)
Applied Biosystems Div.
850 Lincoln Center Dr.
Foster City, CA 94404
phone: 415-570-6667, 800-345-5224
fax: 415-572-2743
email: pebio@perkin-elmer.com
Internet: http://www.perkin-elmer.com

Thermo Separation Products (I & C)
P.O. Box 49031
San Jose, CA 95161
phone: 408-526-1100, 800-532-4752
fax: 408-526-1074

Waters Corp. (I & C)
34 Maple St.
Milford, MA 01757
phone: 508-478-2000, 800-254-4752
fax: 508-872-1990
email: info@waters.com
Internet: http://www.waters.com

Required Level of Training

Operation of a capillary electropherograph can be performed after 2 or 3 days of training. Method development and data interpretation require a background in chemistry as well as weeks to months of experience, depending on the complexity of the analysis.

Service and Maintenance

Unlike high-performance liquid chromatographs, a CE operates at low pressures. Unlike a gas chromatograph, a CE operates at ambient or slightly above ambient temperatures. Therefore, a CE usually does not require as much maintenance. The most complex component of a CE is usually the detector. The most common detector, UV/Vis absorbance, requires periodic changing of the lamp, as does a fluorescence detector. The capillary serves as the flow cell in these detectors, so there is no need to clean or replace a flow cell with these detectors.

Most commercially available CEs have a lot of low-pressure plumbing, which requires periodic inspections for leaks.

Suggested Readings

BAKER, D. R., *Capillary Electrophoresis*. New York: Wiley, 1995.

HEIGER, D. N., *High-Performance Capillary Electrophoresis: An Introduction*. Grenoble, France: Hewlett-Packard Co., 1992.

ODA, R. P., T. C. SPELSBERG, AND J. P. LANDERS, "Commercial Capillary Electrophoresis Instrumentation," *LC-GC*, 12 (1994), 50–1.

Chapter 11

Supercritical Fluid Chromatography and Extraction

Larry Taylor

Virginia Polytechnic Institute and State University
Department of Chemistry

Summary

General Uses

- Separation of a wide variety of volatile and nonvolatile compounds: organic, organometallic, biological, polymer, chiral, and thermally labile
- Analysis of impurities
- Isolation and purification of compounds
- Separation of closely related compounds
- Ultratrace to preparative and process-scale separations
- Nondestructive method
- Qualitative and quantitative method
- Fractionation of complex mixtures
- Interfaceable to all types of detectors

Common Applications

- Prepare samples for analysis
- Measure the levels of active drugs, synthetic byproducts, or degradation products in pharmaceutical dosage forms
- Measure levels of hazardous compounds such as pesticides or insecticides
- Monitor environmental samples
- Purify compounds from mixtures such as surfactants, hydrocarbons, and other chemicals
- Separate polymers and polymer additives and determine the molecular weight distribution of the polymers in a mixture
- Quality control
- Follow synthetic reactions

Samples

State

Sample must be dissolved in liquid or supercritical fluid for injection into the supercritical fluid chromatography (SFC) instrument; sample may be liquid, semisolid, or solid for supercritical fluid extraction (SFE) application.

Amount

0.1 to 1.0 µL injected; mass amounts injected vary depending on the sensitivity and dynamic range of the detector for the analyte and the SFC column dimensions; amount taken for SFE depends on the analyte concentration in the matrix (1 to 20 mg extracted analyte).

Preparation

Limited or extensive sample prep may be required as defined by the relative complexity of the sample. Sample preparation may include any of the following steps: dilution, preconcentration, filtration, extraction via liquid or supercritical fluid, or derivatization.

Analysis Time

Analysis time for SFC is in a range from 2 min to 1 hr. Sample preparation differs from sample to sample (generally 10 to 60 min for SFE). Sample preparation may be extensive and require more time than the analysis.

Limitations

- Compound identification is limited unless SFC or SFE is interfaced with mass spectrometry or infrared spectrometry.

- Resolution can be difficult to attain with complex samples.
- Only one sample can be analyzed at a time.
- Requires training in order to optimize separations.
- Time of analysis is relatively short (compared with liquid extraction and high-performance liquid chromatography).

Complementary or Related Techniques

- SFC is compatible with all flame-based detectors (such as flame ionization detectors, electron capture detectors, and flame photometric detectors), spectrometric detectors (such as mass spectrometers and Fourier transform infrared spectrometers), and miscellaneous detectors (such as UV, fluorescence, and light scattering).
- SFE can be coupled directly to gas chromatography, supercritical fluid chromatography, high-performance liquid chromatography, Fourier transform infrared, and mass spectrometry.

Introduction

The physical state of a substance can be described by a phase diagram, as shown in Fig. 11.1, which shows three curves: a sublimation curve, a melting curve, and a boiling curve. These curves define the regions corresponding to the gas, liquid, and solid state. Points along the curve define situations where there is equilibrium between two of the phases. The vapor pressure (boiling) curve ends at T_c, P_c, the critical point. A supercritical fluid (SF) is defined as any substance that is above its critical temperature (T_c) and critical pressure (P_c). T_c is therefore the highest temperature at which a gas can be converted to a liquid by an increase in pressure. P_c is the highest pressure at which a liquid can be converted to a traditional gas by an increase in the liquid temperature. In the critical region there is only one phase and it possesses properties of both gas and liquid. Just above the critical temperature of the substance, liquidlike densities are rapidly approached with modest increases in pressure in the range of 0.7 to 2 times the critical pressure. Higher pressures are required to attain liquidlike densities for temperatures further above critical. Solvent power of an SF, unlike that of a liquid, increases with density at a given temperature and increases with temperature at a given density. The increase in density is not linear with pressure, but rather the rate of increase is much greater in the vicinity of the critical point.

The first workers to demonstrate the solvating power of SFs for solids were Hannay and Hogarth in 1879 (1). They studied the solubilities of cobalt(II) chloride, iron(III) chloride, potassium bromide, and potassium iodide in supercritical ethanol ($T_c = 243$ °C, $P_c = 63$ atm). They found that the concentrations of the metal chlorides in supercritical ethanol were much higher than their vapor pressures alone would predict. They found that increasing the pressure caused the solutes to dissolve and that decreasing the pressure caused the dissolved materials to precipitate as a "snow."

An SF exhibits physicochemical properties between those of liquids and gases. Mass transfer is rapid with SFs. Their dynamic viscosities are nearer to those found in normal gaseous states. The diffusion coefficient is (in the vicinity of critical point) more than 10 times that of a liquid. As was the case for density, values and subsequent changes for viscosity and diffusivity are dependent

Figure 11.1 Phase (pressure–temperature) diagram for CO_2: *CP* = critical point, *TP* = triple point, P_c = critical pressure, T_c = critical temperature. *(From H. Brogle, Chemistry & Industry 385, 1982. Reprinted by permission.)*

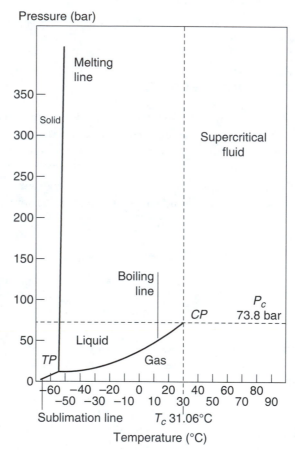

on temperature and pressure. The viscosity and diffusivity of the SF approach those of a liquid as pressure is increased. Diffusivity increases with an increase in temperature, whereas viscosity decreases (unlike gases) with a temperature increase. Changes in viscosity and diffusivity are more pronounced in the region of the critical point. Even at high pressures (300 to 400 atm), viscosity and diffusivity are one to two orders of magnitude different from liquids. Therefore, the properties of gaslike diffusivity, gaslike viscosity, and liquidlike density combined with the pressure-dependent solvating power have provided the impetus for applying SF technology to analytical separation problems.

Although the dissolving capacity of an SF (or dense gas) was known to be determined by its density (pressure–temperature), extraction and separation of mixtures with SFs aroused little interest during the first half of the twentieth century. Since 1980, however, there has been rapid development of supercritical fluid extraction (SFE) (2) in various domains such as the extraction of hops, cholesterol from butter, flavors, residual solvents and monomers from polymers, and eicosapentanoic acid from fish oils. Additional applications appear to be on the horizon. Remediation of soil, demilitarization of propellants, chemical reaction and synthesis of polymers and organic chemicals, impregnation and in situ deposition of chemicals, removal of nicotine from tobacco, nucleation and particle size regulation, and cleaning of electronic and metal parts are but a few of the possible uses for SFs. Supercritical CO_2 will be especially useful in the food and pharmaceuticals industry, where toxicity of the traditional extraction medium, solvent entrapment, and the thermal stability of the materials are often a concern. In truth, SF technology has become an interdisciplinary field and is seriously being pursued by chemical engineers, materials scientists, environmentalists, agronomists, and biologists.

Supercritical fluid chromatography (SFC) affords several unique features relative to GC and HPLC, such as a larger selection of gradient parameters to effect the optimum separation. These parameters include density, pressure, temperature, and mobile phase composition. In addition, selectivity is created by a combination of both analyte–stationary phase and analyte–mobile phase interactions. Other features promise to enhance the popularity of SFC, such as its ability to interface to both GC-like and LC-like detectors and its adaptability to all column dimensions. SFC also offers an environmentally and economically acceptable alternative to liquid mobile phases.

The maturation of SFC has been uneven, and it is only during the last 5 years that one has witnessed a sustained, broad-based research effort. The use of an SF as a chromatographic mobile phase was first demonstrated in 1962 by Klesper, Corwin, and Turner (3). Four years later the first SFC chromatogram appeared, using isobaric conditions. It was not until 1970 that the use of pressure programming in an SFC experiment was demonstrated. The renaissance of SFC is generally recognized to have come in 1981–1982 with Hewlett-Packard's introduction of SFC instrumentation for packed-column SFC at the Pittsburgh Conference. Concurrent with this event was the first report on the use of open tubular capillary columns in SFC. Since 1986, when capillary SFC instrumentation was introduced by several vendors at the Pittsburgh Conference, many new developments have been reported in a rapidly growing number of publications.

Although SFC is generally recommended for thermally labile and nonvolatile analytes, its primary thrusts during the 1980s came from workers in the GC field rather than the LC field. Partly for this reason, a greater emphasis was placed on open tubular capillary columns than on packed columns. No doubt the ability to work with longer columns, which yield much greater theoretical plate numbers, contributed to this scenario. The single most important factor accounting for the early emphasis on open tubular capillary columns was probably their inertness (lack of activity) compared with packed columns. This was a result of their low surface area and much lower porosity than that of silica packing materials. As recently as 1988, these perceptions convinced many workers that there was no future for packed columns in SFC. There now appears to be a resurgence of interest in packed columns, brought on partly by the availability of more deactivated silica-based stationary phases and the ability to use mobile phase gradients during the chromatography. Smaller diameter and more uniform packing materials and the ability to pack narrower fused-silica columns that can be coupled together have also contributed to renewed interest.

How It Works

To understand the principles by which an SF behaves, it is beneficial to first picture a gas as a solvent. As is well known, a gas can be used to perform separations based on vapor pressure. The gas is placed in contact with the sample at elevated temperatures and analytes dissolve into the gas as a function of their volatility. This is routinely done in the cases of thermal desorption and gas chromatography. Increased temperature often results in thermal decomposition of the sample. On the other hand, many materials, especially those of polymeric origin, do not exhibit sufficiently high vapor pressures to be separated by thermal methods. They, like most solids, show no appreciable solubility in a gas. Under such conditions, the enthalpy of dissolution requires a stronger interaction between analyte and solvent in order to overcome solute–solute or solute–matrix interactions. For example, in the dissolution of naphthalene in SF ethylene around the critical point, the partial molar volume of the solute begins to decrease dramatically as ethylene molecules cluster around the naphthalene. The average distance between molecules decreases and nonideal gas behavior begins to govern the interactions between the solvent and sample, accounting for the tremendous enhancement in solubility. It is possible to fine tune the solvating strength of the SF anywhere in the range from the ideal gas to nearly that of the pure liquid. Because of the noncompressibility of liquids, this phenomenon is peculiar to the SF. In this region, the chemist has more means to control solvent behavior than at any other time. It is even possible, by adding small quantities of cosolvents to the SF, to custom design an SF for a specific application. Several manufacturers now offer instruments devoted to both SFC and SFE. The typical benchtop SFE apparatus (Fig. 11.2) consists of a gas supply, a pump and controller used to pressurize the gas, a temperature-controlled oven, an extraction vessel, a backpressure regulator, and a collection device (4). Several of the components of an SFC system (that is, the gas supply, pump, oven, and backpressure regulator) are identical to those of an SFE system.

For analytical SFE or SFC systems the gas supply is simply a laboratory-sized cylinder from any number of manufacturers. Carbon dioxide, by virtue of its moderate critical parameters, high purity, and low cost, is the most common SF used today, but many other potentially useful fluids are available. SFE-grade and SFC-grade CO_2 is available. The SFE grade is purer and more expensive. It is required for analytical extractions if a universal detector is being used because nonvolatile impurities in the CO_2 become trapped along with the analytes that have been separated. Such impurities may interfere with the analyte assay. Some gas suppliers provide premixed fluids for applications where pure SF CO_2 is inadequate. One must recognize, however, that the composition of the mixed fluid is not constant over the entire cylinder volume (that is, cosolvent concentration in the liquid phase increases because of highly different volatilities of CO_2 and polar liquid modifier). Although this observation may not have much impact in SFE, it could be critical in attaining high SFC precision.

From the cylinder that contains both gas and liquid, the latter is drawn from the tank to the pump, where it is pressurized to the desired value. Two major types of pumps are found in SFE and SFC instruments: syringe and piston. Syringe pumps have fixed volumes and any run in progress must be interrupted if the pump is emptied during the run. In order to minimize this problem, dual-syringe pump arrangements are used. Piston pumps are limited only by the liquid volume of the gas–liquid supply cylinder. However, it is necessary to cool the piston heads in some manner so that only the noncompressible liquid phase is pumped. It appears that most new SFE

Figure 11.2 Basic components of an SFE instrument.

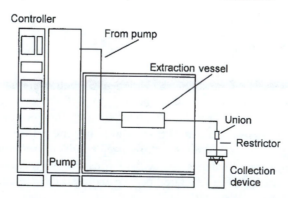

instruments now incorporate piston pumps in their design. The majority of manufacturers actually provide two pumps (usually as an option), with one delivering the primary fluid and the other (usually a micro-LC pump) delivering the desired level of cosolvent. SFE pumps typically have maximum pressures between 6000 and 10,000 psi, and can provide liquid flow rates to approximately 20 mL per minute. The controller on many SFE and SFC instruments allows several gradients, CO_2 pressure or mobile phase, to be programmed into the method. Because SFC requires more uniform flow, the syringe pump is currently more popular, although Hewlett-Packard has introduced a piston pump for SFC that yields essentially pulseless flow.

From the pump the liquid travels to a heated zone, where it becomes supercritical, and then to an extraction vessel for SFE, where the sample is contained. The extraction vessel is housed in an oven so that the desired temperature can be maintained. Ovens provided with analytical SFE instruments usually have maximum temperatures from 100 to 150 °C. The extraction vessels contained in the oven are usually stainless steel cartridges with volumes from a fraction of a milliliter to more than 10 mL, although larger vessels are available. At least one manufacturer (ISCO) provides disposable cartridges of polymeric composition. All are sealed on both sides with frits (usually 0.5 mm) to prevent the entrainment of particulates. In the case of SFC the extraction vessel is replaced by an injector and a column.

Backpressure in analytical SFE and SFC can be provided by a fixed restrictor (usually a narrow-bore fused silica or stainless steel capillary). With a fixed restrictor the flowrate cannot be controlled independently of the pressure. Thus, linear restrictors must be replaced with different diameters or lengths of tubing in order to maintain constant flowrates under different pressure conditions. This is seldom done in actual practice; therefore, the flowrate indeed varies with density programming. The narrow orifices of restrictors can also be prone to plugging if a large amount of borderline soluble material is being extracted or separated. Heating the restrictors helps to alleviate (not eliminate) plugging problems. High-pressure electronically controlled micrometering valves are becoming more popular in SFE and SFC. These backpressure regulators, called variable restrictors, allow flowrates to be adjusted to constant levels at different densities (pressures). Fixed restrictor design, though not important in SFE, is critical to the success of SFC. Currently, flame-based detectors must use a fixed restrictor. Integral fixed restrictors for packed columns and frit-fixed restrictors for open tubular columns are recommended (Fig. 11.3). Variable restriction is now a reality for ultraviolet-visible detectors.

During a typical SFE application the sample is placed in the high-pressure extraction vessel and equilibrated to the desired temperature. The extraction fluid is then allowed to flow into the vessel and is pressurized to the desired value. At this point, either a static, dynamic, or combination static–dynamic extraction can be initiated. As analytes are dissolved in the SF they must be removed by the SF phase into a separate region. Normally, the analytes travel through the restrictor, where the SF decompresses and the analytes deposit in some sort of trapping device. This is often 5 to 15 mL of liquid or several grams of solid support. In the case of solid supports, such as stainless steel spheres or C_{18} chromatographic packing material, the analytes must be rinsed from the support in a separate step. A solvent that exhibits high solvating power should be chosen so that only 1 to 2 mL of rinse liquid is required. To trap the extract in a liquid, the restrictor end is usually immersed in the collection solvent. There the decompressed SF rises rapidly to the surface, leaving the extracted analyte in the liquid. If a liquid trap is used, high SF flow rates can result in violent bubbling following decompression of the SF in the solvent and physical removal of the extract by the aerosol is a possibility. Also, if the extract exhibits poor solubility in the collection solvent, poor recoveries usually result. The best recoveries during liquid collection of the extract usually result from a combination of low to moderate SF flowrates in a solvent known to exhibit an affinity for the compounds of interest. Once the analyte has been either rinsed from the solid phase trap or deposited into the liquid trap, it is ready for chromatographic or spectrometric analysis.

Figure 11.3 Properties and design of most popular fixed restrictors for SFC. *(Reprinted from T. L. Chester, Analytical Instrumentation Handbook, ed. G. W. Ewing, Marcel Dekker, Inc. New York, 1990, courtesy of Marcel Dekker, Inc.)*

Integral restrictor

- 50- to 100-μm internal diameter capillary tube
- Heat-sealed tip
- Tip abraded to 1 to 3 μm internal diameter
- Mechanically robust
- Commercially available

Frit restrictor

- 50- to 100-μm internal diameter capillary tube
- Porous ceramic frit
- Cut to adjust flow
- Rugged
- Prone to clog
- Commercially available

For SFC, CO_2 supply and pump requirements are similar. The oven may also be the same, but it must be large enough to accommodate the chromatographic column. Open tubular fused silica (1- to 50-m) columns and packed (10- to 25-cm) stainless steel columns are routinely used. Stationary phases for open tubular columns should be highly cross-linked because the solvating power of CO_2 at high density and temperature can cause column bleeding. Normal GC columns will not suffice. Open tubular columns for SFC are more narrow than GC columns (50 μm rather than 320 μm) because linear velocities in SFC are 10 to 100 times higher than optimum. A portion of this lost efficiency is regained by using a smaller-inner-diameter column. It should also be noted that with a fixed restrictor, linear velocity continues to increase as density increases with concomitant loss in column efficiency. For a packed column, wherein efficiency is proportional to particle diameter, traditional liquid chromatographic columns can be used. Usually these stationary phases are activated such that in order to elute polar analytes, modified CO_2 is required. On the other hand, polymer-coated particles are available that have been extensively cross-linked. Moderately polar material can be separated with 100% CO_2 on these columns. Although much of early SFC column development concerned the evaluation of alkyl and aryl bonded phases, which are traditionally used for reversed phase chromatography, SFC is more like normal phase chromatography (5). For this reason, more polar columns are required for achieving separation of polar analytes.

The difficulties of sample introduction in SFC can vary depending on the type of column used (6). The issue is of fundamental importance to capillary column users, as rather small injection volumes (20 to 60 nL) are required if solvent overload is to be avoided. An injection of only 200 nL onto a 50-μm internal diameter column results in a 10% loss of resolution. Split injection methods (dynamic split, delayed split, and timed split) provide a simple solution to the problem of injection volume, but they also have detrimental effects. Dynamic split injection, for example, can exhibit nonlinearity and sample discrimination. Although timed split injection has been shown to be an excellent alternative to dynamic split, there remains the problem of sample loss as a result of splitting. Samples must be concentrated if split injection is to be used.

Quantification at the trace level is thus impractical with a split injection system, especially for weakly concentrated analytes or limited quantities of sample (7). Direct injection would be the ideal choice, but for open tubular SFC it is less than easy. Typically, direct injection capillary SFC requires the insertion of a retention gap before the column in order to separate the analytes from the solvent front and focus them at the head of the column. Packed columns certainly pose much less of a sample injection problem because of their greater loadability, and direct injection methods have proven highly effective and reproducible for packed-column use. Interest in sample injection for packed columns has increased as ever-larger injection volumes are desired and efforts are made to perfect SFC as a preparative scale method.

Most recent studies on sample introduction have focused on the optimization of methods developed for elimination of the solvent. As an alternative, samples can be dissolved in the fluid itself, much as is done in LC by mixing samples with the mobile phase. A truly solventless injection technique, on-line SFE-SFC, is commercially marketed and steadily progressing. Improvements in trapping, recovery, and solute focusing continue to be made. Eventual system refinements may well promote SFE-SFC beyond expectations as it combines sample preparation with separation, needs no organic solvent, and enables maximum sample concentration.

Detection using both selective and high information detectors with SFC is an active area of research. Flame ionization with open tubular columns and ultraviolet spectroscopy with 4.6-mm packed columns are routinely used as detectors. Small-inner-diameter packed columns can be easily used with flame ionization detection, whereas ultraviolet detection with open tubular columns has not been perfected. SFC is truly a friendly interface, having been coupled with electron capture, thermionic, flame photometric, sulfur chemiluminescence, and light scattering. Spectrometric detection on-line has also become feasible with SFC. Mass spectroscopy, infrared, atomic emission, and nuclear magnetic resonance have all been used with a variety of columns.

What It Does

SFE

The advent of SFE has dramatically reduced the burden of sample preparation. Compared to traditional techniques such as Soxhlet extraction, SFE cuts extraction times (in many cases by more then 90%) and reduces the number of preparation steps. Unlike most conventional extraction solvents, supercritical CO_2, the fluid used in the overwhelming number of SFEs, is nonpolluting, nontoxic, and relatively inexpensive. A majority of analysis time in the analytical laboratory of today is spent in preparing the sample. This preparation is the major source of error and the major consumer of labor in the laboratory. In many situations it also leads to extensive organic solvent usage and high solvent-disposal costs. Because analytical laboratories are continually called on to increase sample throughout and lower cost, advances in sample preparation are highly desirable. SFE technology offers the sample preparation chemist the opportunity to efficiently and economically improve recovery, increase reproducibility, decrease the use of halogenated solvents, and provide cleaner extracts to the measurement instrument. In order to be truly useful in the analytical laboratory, SFE must be thoroughly understood. A hasty approach to method development can lead to problems and even disillusionment with the technology. A lack of understanding regarding system parameters (such as type of fluid, density, pressure, temperature, flow rate, extraction time, extraction mode, and analyte collection) can result in misinformation and erroneous conclusions. Before considering these parameters one must obviously have evaluated the analyte's solubility in the SF and the nature of the matrix from which the analyte is to be extracted. Even at this point the temptation to look for shortcuts rather than to examine the application systematically may lead to disappointing results, especially for complex samples. For example, fluid purity, extraction system cleanliness, the chemical nature and extent of coextractives, the design and volume of the extraction vessel, and the suspected concentration of analyte in the matrix are all matters for contemplation when setting out to do sample preparation for SFE before analysis. Ideal analytical SFE should be highly reproducible, fast, efficient, quantitative at sub–parts-per-billion levels, and free of foreign contamination, yield a concentrated sample that can be easily manipulated, and preserve the chemical integrity of the analyte to be extracted. In a growing number of instances this is now routine.

SFC

Steroids, drugs of abuse, pharmaceutical products, agricultural and environmental compounds, and food products are polar and nonvolatile, so GC cannot be applied to the analysis of these samples without derivatizing the analytes. HPLC has been a very popular technique for the analysis of polar, nonvolatile, and high-molecular-weight compounds. Both reversed phase and normal phase HPLC methods have been widely used for the analysis of samples such as food products, drugs, and pesticides. Currently SFC offers an attractive alternative for the separation of volatile, nonvolatile, and thermally labile material. The greater diffusivity and lower viscosity afforded by an SF relative to a liquid has been demonstrated to yield faster, more efficient separations, and liquidlike densities enable many thermally inaccessible components to be solubilized. Specific advantages of SFC over HPLC include greater resolution per unit time, shorter total analysis time, much more rapid methods development, and simpler interface to any detector.

Applications

SFE

Single-base gun propellants typically contain nitrocellulose, 2,4-dinitrotoluene, diphenylamine (stabilizer), and plasticizers such as diethyl phthalate and dibutyl phthalate. Because they age, propellant stocks must be continually monitored to ensure that they contain sufficient quantities of stabilizer. SFE and subsequent analysis of these materials is an effective way to monitor propellant stability and to study the chemistry of propellant aging. SFE may also serve as the basis for the design of the practical method to reclaim nitrocellulose from unusable propellant stocks currently in storage. Propellant samples are usually provided in mat or pellet form. To increase the extraction efficiency, samples are cut in order to expose more surface area to the extraction solvent. The sample is loaded into the extraction vessel, pressurized, and brought to a temperature of 50 to 75 °C. (Carbon dioxide becomes supercritical at 31 °C and 72 atm.) Once the extraction is complete, the analytes are trapped, recovered, and analyzed. SFE increases both the speed and the efficiency of extraction. Whereas a typical Soxhlet extraction of a single-base propellant requires 12 to 72 hr, SFE accomplishes the task in less than 2 hr.

Animal studies play an important part in the development and testing of new drugs. Often, such drugs are administered in an admixture with animal feed. One of the problems in the solvent extraction of drugs from feed matrices is the coextraction of other components such as triglycerides. Quantification requires that the analyte of interest be separated from the other extractables, and for this purpose SFE has been used followed by liquid chromatography with UV detection.

An important objective in the development of an analytical method is high recovery with low relative standard deviation (RSD). Because preliminary results indicated the possibility of a lack of uniformity in the distribution of the drug in the feed material, it was decided to prefect the extraction method first under controlled conditions before testing it on feed samples. For this purpose, aliquots of a solution of known concentration of the drug (a polar aromatic carboxylic acid) were transferred to filter paper. After drying, the drug-embedded filter paper was subjected to SFE and recovery.

Once the extraction method had been perfected, it was applied to the original feed samples. The average recovery from these extractions was 68.6%, with an RSD of 20.6%. When the artificially spiked sample of feed was prepared by mixing a solution of the drug in methylene chloride with ground feed and the dried slurry extracted, average recoveries rose to 84.0% and the RSD dropped a full order of magnitude to 2%. These results tend to confirm the suspicion of sample inhomogeneity and indicate to the formulator that the current protocol for mixing drugs into the feed may need to be revised. It also indicates how SFE can play an important part in the quality control of dosage administration in feeding studies.

For a number of years, the paper industry has been interested in the amount and the nature of substances that could be extracted from bleached wood pulp (the concern being that the bleaching process could produce environmentally hazardous compounds). Soxhlet extraction of several bleached wood pulps from northern and southern softwoods with three typical solvents—cyclohexane, ethyl acetate, and water—and the SFE of these same materials using both 100% CO_2 and methanol-modified CO_2 have been compared. Results show that the time saved by SFE is comparable to that of the extraction of propellants; the Soxhlet extraction took 72 hr and the SFE 3 hr. Drying and concentration of sample extracts was also reduced from 6 to 3 hr.

In general, wood pulp is treated either by a chemithermomechanical process (CTMP), which bleaches via ozone or peroxide oxidation, or by the Kraft process (KP), which bleaches through the action of chlorine. Overall, CTMP bleaching produced approximately four times the amount of extractables as did the KP treatment. A GC/MS analysis of the SFE extracts exhibited a large number of compound peaks. When the SFE was performed with CO_2 modified with 2% methanol, the percentage recovery did not change, but some compounds of increased polarity were detected.

SFC

Various classes of polar compounds have been analyzed by both packed and capillary columns. Rapid analysis of several drugs of abuse was carried out on a 2.1-mm internal diameter packed column using methanol modified CO_2 as the mobile phase. Derivatives of oligo- and polysaccharides containing up to 18 glucose units were resolved on a capillary column and a degree of polymerization (DP) was assigned to the sugar derivatives. A mixture of thermally labile azo dyes was separated on a capillary column using a flame ionization detector. A highly deactivated 25% biphenyl capillary column was used for the elution of a derivatized polar penicillin antibiotic. Similar columns were used for analysis of fatty acids, triglycerides, and lipids. Capillary column SFC has been reported for the analysis of surfactants (Fig. 11.4), polyols, and alkaloids. A mixture of an antimalarial drug and related compounds was separated on a packed column using 0.3% formic acid in CO_2 as the mobile phase. Mixtures of steroids, barbiturates, and tetrahydrocannabiol with its metabolites were analyzed on capillary columns. The analysis of several lipids and sterols has been reported by several workers using a N_2O mobile phase under isoconfertic conditions. A packed capillary column with alcohol modified n-hexane mobile phase was used to separate mixtures of both fat-soluble and water-soluble vitamins and MS was used for their detection. A mixture of five pesticides and herbicides was separated on a capillary column with a multichannel UV detector and a full UV spectrum of each solute was achieved. MS has been successfully applied to the SFC

Figure 11.4 SFC of alcohol ethoxylate. *(From the 1988 Workshop on Supercritical Fluid Chromatography held in Park City, UT, January 12–14, 1988, p. 296. Reprinted with permission.)*

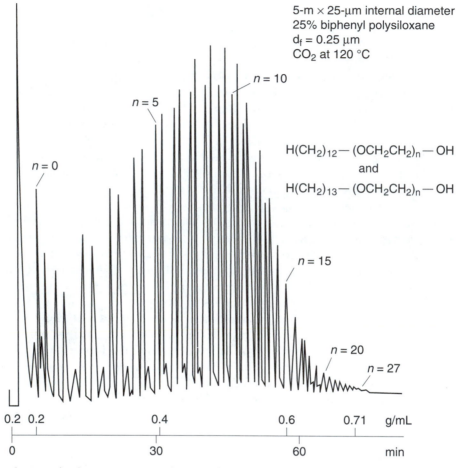

5-m × 25-μm internal diameter
25% biphenyl polysiloxane
$d_f = 0.25$ μm
CO_2 at 120 °C

$$H(CH_2)_{12}-(OCH_2CH_2)_n-OH$$
and
$$H(CH_2)_{13}-(OCH_2CH_2)_n-OH$$

analysis of pesticides because it provides very low detection limits and also leads to positive identification of the analytes. Mixtures of carbamates and organophosphorous pesticides have been analyzed by both packed and capillary columns and the mass spectra are achieved with detection limits in the range of nanograms (8). Metalloporphyrins have been separated on the analytical scale packed column and on a capillary column using alcohol-modified CO_2 as the mobile phase.

Nuts and Bolts

Relative Costs

Several companies offer SFE and SFC systems that span wide ranges of price and sophistication. For SFE, the range is $10,000 to $80,000 depending primarily on the degree of automation, the restrictor type, and the nature of the trapping mechanism. Those beginning to work in this area should purchase a simple manual system and experiment to find out how to successfully extract a specific type of sample. Certain vendors offer modular components for assembling an SFE in-house, which is a more economical alternative. Several vendors provide autosampler capacity up to 44 samples. The number of simultaneous extractions that can be performed may range as high as eight, although sequential extractions (one at a time) appear to be gaining in popularity. Maximum system pressure varies between 6000 and 10,000 psi. Liquid maximum pump flowrates range from 4 mL/min to over 100 mL/min. Extraction vessel capacity is also a variable among the vendors (0.15 to 300 mL).

For SFC, prices range from $26,000 to $64,000 depending on the number of pumps, the type of restrictor, and on-line SFE capability. Certain instruments can be used only with packed columns because of a patent position held by one vendor regarding the use of open tubular columns for SFC. The dual pump arrangement affords one the opportunity to use composition gradients with packed column, as in HPLC, with the pressure and temperature held constant. Open tubular columns often use neat CO_2 with pressure or density programming. Flame ionization detection is standard equipment on all SFCs. Several vendors also supply ultraviolet detectors for packed column applications, wherein modified fluids are used as the mobile phase.

Vendors for Instruments and Accessories

I = instrument manufacturer; C = consumables supplier

Vendors for SFE

Applied Separations Inc. (I & C)
930 Hamilton St.
Allentown, PA 18101
phone: 610-770-0900
fax: 610-740-5520
email: appliedseps@enter.net

Dionex Corp. (I & C)
P.O. Box 3603, 1228 Titan Way
Sunnyvale, CA 94088-3603
phone: 408-737-0700, 800-723-1161
fax: 408-730-9403
email: marcom@dionex.com
Internet: http://www.dionex.com

Hewlett-Packard Co. (I & C)
2850 Centerville Rd.
Wilmington, DE 19808
phone: 302-633-8504, 800-227-9770
fax: 302-633-8902
Internet: http://www.hp.com/go/chem

Isco Inc. (I & C)
P.O. Box 5347
Lincoln, NE 68505
phone: 402-464-0231, 800-228-4250
fax: 402-464-0318
email: info.sid@isco.com

JASCO (I & C)
8649 Commerce Dr.
Easton, MD 21601
phone: 410-822-1220, 800-333-5272
fax: 410-822-7526

Keystone Scientific Inc. (C)
Penn Eagle Industrial Park
320 Rolling Ridge Dr.
Bellefonte, PA 16823
phone: 814-353-2300, 800-437-2999
fax: 814-353-2305
email: 102414.1157@compuserve.com
Internet: http://www.showdex.com

Vendors for SFC

Alltech Associates Inc. (C)
2051 Waukegan Rd.
Deerfield, IL 60015
phone: 708-948-8600, 800-255-8324
fax: 708-948-1078
email: 73554.3372@compuserve.com
Internet: http://www.alltechweb.com

Berger Instruments (formerly Hewlett-Packard) (I & C)
31 Waterloo Ave.
Berwyn, PA 19312
phone: 302-737-4297
fax: 610-296-8952

Dionex Corp. (I & C)
P.O. Box 3603, 1228 Titan Way
Sunnyvale, CA 94088-3603
phone: 408-737-0700, 800-723-1161
fax: 408-730-9403
email: marcom@dionex.com
Internet: http://www.dionex.com

Gilson Inc. (I & C)
3000 W. Beltline Hwy., P.O. Box 620027
Middleton, WI 53562-0027
phone: 608-836-1551, 800-445-7661
fax: 608-831-4451
email: gilson@gilson.com

Required Level of Training

Neither SFE nor SFC has yet reached the black-box stage in which the analyst can put a sample in at one end and get a result out the other end. Neither SFE or SFC is a fully mature technology. At this point SFE is a more user-friendly technique. Knowledge of the chemical properties of both the analyte and the matrix is important for SFE. In addition, one must ensure that the mechanics of SFE have been optimized (such as sufficient supercritical fluid flow, minimal dead volume, efficient trapping, and ample vessel volume sweeps). For SFC, the manipulation of a fragile, small-diameter (about 25 μm), fused silica fixed restrictor attached to the column outlet can be frustrating and tedious because the length and diameter of the restrictor determine the flowrate at a set density. An electronically controlled micrometering valve (such as a variable restrictor) eliminates these problems but this capability exists only with packed columns. Development of analytical methods involving either SFE or SFC requires a college education. For routine operation, an associate college degree should be sufficient. Quantitative determinations at trace levels require at least an introductory analytical course.

Service and Maintenance

Most instruments have diagnostic software that assists the operator in troubleshooting. A high percentage of problems can be solved by trained laboratory personnel, thus eliminating the need for costly visits by the vendor's engineers. Fixed restrictors in SFE and SFC must be replaced periodically because they sometimes plug or become brittle and break. Modified fluids are especially damaging to restrictors. For SFC, restrictors must be either manufactured in-house by tapering a capillary tube or bought (as an integral restrictor or frit restrictor) for approximately $50 each. When performing trace analysis the entire SFE or SFC system (including the valves and transfer tubing) must be scrupulously cleaned on a regular basis.

Suggested Readings

LEE, M. L., AND K. E. MARKIDES, EDS., *Analytical Supercritical Fluid Chromatography and Extraction*. Provo, UT: Chromatography Conferences, Inc., 1990.

McHUGH, M., AND V. KRUKONIS, *Supercritical Fluid Extraction*, 2nd ed. Boston: Butterworths, 1994.

SMITH, R. M., ED., *Supercritical Fluid Chromatography*. London: Royal Society of Chemistry, 1988.

TAYLOR, L., *Supercritical Fluid Extraction*. New York: Wiley, 1996.

WENCLAWIAK, B., ED., *Analysis with Supercritical Fluid: Extraction and Chromatography*. New York: Springer Verlag, 1992.

WHITE, C. M., ED., *Modern Supercritical Fluid Chromatography*. Heidelberg, Germany: Dr. Alfred Huthig Verlag, 1988.

References

1. J. B. Hannay and J. Hogarth, *Proc. Roy. Soc.* (London), 29 (1879), 324; 30 (1880), 178.

2. S. B. Hawthorne, *Anal. Chem.*, 62 (1990), 633A.

3. E. Klesper, A. H. Corwin, and D. A. Turner, *J. Org. Chem.*, 27 (1962), 700; J. L. Hedrick, L. J. Mulcahey, and L. T. Taylor, *Mikrochim Acta*, 108 (1992), 115.

5. L. T. Taylor and H-C. K. Chang, *J. Chromatogr. Sci.*, 28 (1990), 357.

6. C. H. Kirschner and L. T. Taylor, *J. High. Resolv. Chromatogr.*, 16 (1993), 73.

7. T. C. Chester, *J. Chromatogr.*, 299 (1984), 424.

8. R. D. Smith, B. W. Wright, and C. R. Yonker, *Anal. Chem.*, 60 (1988), 1323A.

Chapter 12

Ion Chromatography

John Statler

Dionex Corporation

Summary

General Uses

- Separation and detection of ions and ionizable species
- Profiling, that is, determining the qualitative distribution of mixtures of oligomeric ions
- Simultaneous determination of several ions in mixtures
- Concentration of ionic species in samples of low concentration, often with simultaneous elimination of interfering matrix

Common Applications

- Determination of inorganic anions or cations, organic acids, amines, amino acids, carbohydrates, or nucleic acids in a variety of samples
- Monitoring water quality
- Determination of the composition of industrial wastes
- Monitoring the quality of intermediates in industrial processes
- Determination of ionic composition of biological solutions
- Separation of components of mixtures before mass spectrometry or other spectroscopic techniques
- Identification of ionic impurities
- Purification of components from mixtures

Samples

State

Liquids that are aqueous or water-miscible solvents can be analyzed directly. Water-immiscible liquids, solids, and gases must be extracted into or dissolved in aqueous solution before analysis.

Amount

Samples are usually introduced as 5- to 200-μL volumes, although volumes as large as 100 mL can be introduced using chromatographic preconcentration techniques when additional sensitivity is needed.

Preparation

Dilution or dissolution and filtration are the most common sample preparation procedures. Extraction may be required for nonaqueous samples or preconcentration for dilute samples. The need for precolumn derivatization is rare.

Analysis Time

Excluding sample preparation, analysis times range from less than 3 min to more than 2 hr. Most commonly, however, analysis time is 10 to 15 min.

Limitations

General

- Analyses are performed sequentially.
- Analytes can be misidentified or their quantities incorrectly determined if other components are not well separated.
- The analysis consumes eluent, which must be replenished regularly.

Accuracy

For routine analysis, accuracy is about 3%, but under carefully controlled conditions and with the use of an internal standard, relative standard deviation less than 1% is possible.

Sensitivity and Detection Limits

For a standard sample size of 25 μL, detection limits are about 1 to 5 μg/L (ppb) for most common inorganic ions. However, this can vary a great deal depending on the detector response of the analytes, the nature of the separation method, and interfering components in the sample.

Complementary or Related Techniques

- Atomic absorption, atomic emission, and inductively coupled plasma spectroscopies, used for determining the total amount of a metal rather than the amount of a certain ionic form of that metal
- Mass spectrometry, used to obtain information on chemical structure and molecular mass
- Nuclear magnetic resonance, used to obtain information on chemical structure
- Infrared spectroscopy, used to obtain information on chemical structure, particularly before ion chromatography to identify functional groups that can offer a key to separation or detection
- Capillary electrophoresis, used as a confirmatory technique because it relies on an unrelated separation mechanism

Introduction

Definition

In his comprehensive book on the subject, Small defines ion chromatography (IC) as "the chromatographic separation and measurement of ionic species" (1). Common applications of ion chromatography include the determination of simple anions, such as chloride and sulfate, simple cations, such as sodium and calcium, transition metals, lanthanide and actinide metals, organic acids, amines, amino acids, and carbohydrates.

Early Developments

Modern ion chromatography began with a report by Small, Stevens, and Bauman (2) wherein they described a way to combine an ion exchange chromatographic separation with simultaneous conductometric detection for the determination of anions including chloride, sulfate, nitrate, and phosphate, or of cations including sodium, ammonium, potassium, and calcium. The key element was their development of a device, later known as a suppressor, to lower the background conductometric signal resulting from the liquid mobile phase, or eluent, while enhancing the conductometric signal from the analyte ions. Originally, this pairing of ion exchange chromatography with suppressed conductometric detection was synonymous with IC. Eventually, however, the term expanded to include other detection methods and other chromatographic modes.

The first of the alternative detection techniques, nonsuppressed conductometric detection, emerged in the late 1970s (3). Other routine techniques, including amperometry, optical absorbance, and fluorescence (4), as well as less common techniques such as inductively coupled plasma spectroscopy and mass spectroscopy, have been used in the years since then. Likewise, ion exclusion, ion pairing, and chelation chromatographies have been used in addition to ion exchange for the separation of ions (5).

Nevertheless, even with all these methods of separation and detection falling within the modern definition of ion chromatography, IC as it is most often practiced today is still largely an ion exchange separation with suppressed conductometric detection.

Improvements in Separation

Synthetic ion exchange resins have been available for many decades, but the first resins used in modern IC applications were surface-functionalized styrene divinylbenzene, in many ways very much like those in common use today. Common anions were separated in about 30 min. The alkali metal cations could be separated in about 25 min, but the alkaline earth cations required a separate analysis. These analyses, especially for the cations, seem slow by today's standards, but were vastly superior to the laborious, single ion, wet chemical options. On today's high-efficiency ion exchange resins, common anions can be resolved in less than 3 min, and the common group I and II cations can be determined together in about 10 min.

Much of this improvement in speed has been a result of improvements in resin selectivity and uniformity. Most resins used for IC have a thin surface region of ion exchange sites, minimizing band broadening caused by diffusion of the ions into the resin interior. In some cases, analysis speed can also be increased with the use of gradient ion chromatography (6). This technique allows increasing eluent concentration to separate very weakly and very strongly retained ions in a short period of time.

Selectivity

The ideal ion exchange site is a permanent charge (that is, one whose charge does not change with pH) that does not have secondary (non–ion-exchange) interactions.

Since the beginning of IC, the most common functional group for anion exchange has been some sort of quaternary alkyl ammonium group. This group has a permanent positive charge, regardless of the pH of the surrounding eluent. The alkyl groups on the ammonium and the composition of the surrounding resin can be varied to create different anion exchange environments. This allows the manufacturing of columns to produce resins with different selectivities for specialized applications.

Early cation exchangers usually had sulfonic acid groups as the cation exchange sites. The principal advantage of the sulfonic acid is that even at low pH it remains charged. However, these columns are most often used with column switching techniques or with eluents containing both monovalent and divalent ions to elute the monovalent and divalent analyte ions at similar times. In recent years, carboxylic acids have become popular because the affinities of mono- and divalent analyte cations for the functional group are more similar, so simple, unchanging isocratic eluents are capable of separating all cations in similar times.

Improvements in Detection

Conductivity, the most common detection method in IC, has advanced by reducing noise through improvements in electronics and better isolation and control of cell temperature, by improved response through detector cell design, and, in the case of suppressed conductometric detection, by improved suppressor design to decrease internal volume and dispersion. All of these have helped lower detection limits for many common ions to about 0.1 ng, or less than 10 µg/L (ppb) for a 25-µL injection.

Electrochemical or amperometric detection as it was first used in IC was single-potential or DC amperometry, useful for certain electrochemically active ions such as cyanide, sulfite, and iodide. But the development of pulsed amperometric detection (PAD) for analytes that fouled electrode surfaces when detected eventually helped create a new category of IC for the determination of carbohydrates. Another advancement, known as integrated amperometry, has increased the sensitivity for other electrochemically active species, such as amines and many compounds containing reduced sulfur groups, that are sometimes weakly detected by PAD (7).

Current Use

Types of Samples

IC is most often applied to aqueous samples or to solid samples that will dissolve in or can be extracted into aqueous solutions. For aqueous samples, sample preparation may be unnecessary or may consist of dilution or filtration before introducing the sample into the IC system. Insoluble solid samples, such as soil or air filters, are usually extracted into an aqueous solution for analysis. Analysis of gaseous samples presents some interesting challenges in sample preparation (8) and is currently an active area of research. Nonaqueous liquid samples, such as organic solvents, require either little sample preparation if the solvent is miscible with water, or a liquid–liquid extraction to an aqueous solution if the solvent is immiscible.

Types of Analytes

What sets IC apart from most other techniques for the determination of ions is the ease with which it can determine several ionic analytes simultaneously. In principle, any species that can exist as an ion can be determined by IC. In addition to the common anions, such as chloride, bromide, sulfate, nitrate, and phosphate, and common cations, such as lithium, ammonium, magnesium, and calcium, a multitude of weak acids and bases can be ionized by adjusting pH. Amines are cationic below a pH of about 9. Carboxylic acids are anionic above a pH of about 3. Amino acids may be anionic at high pH or cationic at low pH. Sugars and similar carbohydrates, though not commonly considered ions, are actually very weak acids and can be chromatographed as anions above a pH of 12 or 13. Transition and lanthanide metals, though often considered cations, readily form complexes with chelating anions and are often chromatographed as anionic complexes.

How It Works

IC is the merging of a chromatographic technique for separating ions with a technique for detecting ions and determining concentrations. Although the separation and detection are closely linked in practice, the two processes can be conceptualized independently. Anion exchange and cation exchange are by far the two most common separation techniques, but the alternative chromatographic methods of ion exclusion, ion pairing, and chelation have some advantages in certain cases. Ions can be detected and measured using several methods depending on the sensitivity and specificity needed. Species that are ionic at or near neutrality can usually be detected using conductivity, the most common form of detection in IC. In many cases, however, amperometric, optical, ICP, or mass spectrometric methods may be preferred.

Separation: Ion Exchange

Ion exchange is a process in which a charged analyte (also called a solute or eluite) in a flowing solution competes with an eluent (mobile phase) ion of like charge for sites having the opposite charge on a stationary phase (Fig. 12.1). The sites of opposite charge are often called functional groups. Before introducing analyte ions into the system—that is, before injection—the functional group ions are paired with the eluent ions, maintaining electrical neutrality in the stationary phase. When a sample is injected, new ions compete with the eluent ions at the functional group site. Analyte ions that compete successfully (that is, those that have a high affinity for the ion-exchange site) are retained longer than ions that do not compete well with the eluent ions. As with all chromatographic processes, ion exchange can be thought of as a process involving the rapid movement of the analyte between two phases, in this case a liquid mobile phase and a solid stationary phase. In the stationary phase, the ions are immobilized. Ions travel through the column only in the mobile phase. The more time an ion spends in the stationary phase, the more slowly it moves through the column.

The key to the separation of analyte ions is the differential affinities a functional group has for different analyte ions. For example, if analyte A has a higher affinity for the stationary phase site than does analyte B, A will compete with the eluent ions more successfully for those sites and be retained longer. A will therefore elute from the column after B. The relative affinities of analytes for the sta-

Figure 12.1 Representation of the ion exchange process for an anion with hydroxide eluent.

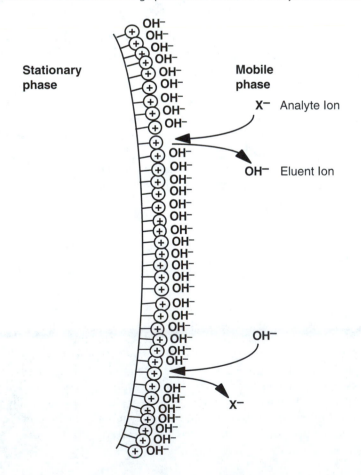

tionary phase are known as the selectivity. Selectivity is determined by many parameters of the separation, including type of functional group, stationary phase environment near the functional group, characteristics of eluent ions, eluent ion concentration, nonionic or oppositely charged eluent additives such as solvents or ion-pairing agents, and temperature. The first two parameters are determined by the design of the ion exchange column and are usually optimized for a given class of analytes, such as common inorganic anions, organic acids, or inorganic cations. The other parameters can be adjusted by the analyst to tailor the separation to specific requirements, but usually ion-exchange columns are designed with a specific set of chromatographic conditions in mind.

The quantity of functional group sites in the stationary phase is known as the capacity. Capacity is usually expressed as the number of equivalents per column or equivalents per gram of resin. A higher capacity results in longer retention of the analyte ions. Capacity is independent of selectivity (that is, capacity can be increased or decreased without altering selectivity) and is determined by the resin manufacturer.

Separation: Ion Exclusion

Ion exclusion, in many respects, is complementary to ion exchange. Like ion exchange, the stationary phase is an ion exchange resin, although it has a very high ion exchange capacity and is the same charge as the analyte ions. Its greatest utility is the separation of weakly ionized species while eluting strongly ionized species in the void volume.

The ion exclusion process (Fig. 12.2) relies on the establishment of an electrical potential between a dilute mobile phase and a stationary phase of high ion-exchange-site concentration. The

Figure 12.2 Representation of the ion exclusion process for a carboxylic acid.

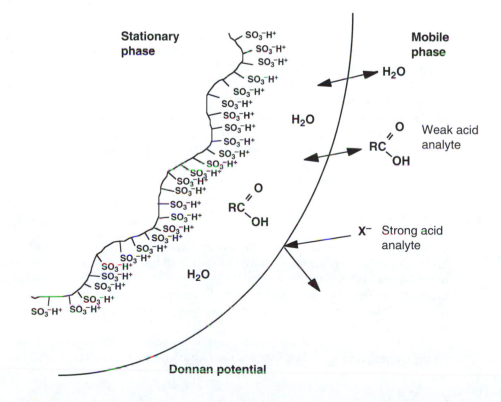

relatively high concentration of ion exchange sites in the resin dictates a high concentration of counterions in the stationary phase to maintain electrical neutrality. However, diffusion forces tend toward equalizing counterion concentrations in the mobile and stationary phases, a situation that leaves the stationary phase charged the same as the analyte ions. This situation of high potential energy is often called the Donnan potential. The Donnan potential permits neutral molecules to enter the stationary phase, whereas analyte ions are repelled or excluded, hence the term *ion exclusion*. Weakly ionized species exhibit intermediate behavior and are separated from one another based largely on their extent of ionization. The most common application of ion exclusion is for the separation of organic acids using a sulfonated macroporous cation exchange resin in the hydronium ion form. The separation of weak bases using a macroporous anion exchange resin is possible, but is much less common.

Separation: Ion Pairing

Ion-pair chromatography, also known as ion-interaction or dynamic ion-exchange chromatography, is a technique using a neutral hydrophobic stationary phase and a mobile phase containing a hydrophobic ion, sometimes called the ion-pairing agent, having a charge opposite to that of the analyte. A common explanation of the ion-pair mechanism is that the ion-pairing agent is associated with the stationary phase and functions as an ion-exchange functional group. For anionic analytes, quaternary ammonium ions, such as tetrabutylammonium, are most commonly used as ion-pairing agents. Alkylsulfonates, such as octanesulfonate, are most commonly used for cationic analytes. The most common stationary phases are alkyl-bonded porous silica resins, commonly used in reversed phase HPLC, and macroporous styrene–divinylbenzene polymers.

Ion-pair chromatography, unlike ion-exchange chromatography, has some flexibility with respect to selectivity and capacity. Selectivity and capacity can be altered by simply changing the ion-pairing agent, and capacity can be increased by increasing the ion-pairing agent's concentration. Historically, ion-pair chromatography has also had the advantage of compatibility with certain organic solvent modifiers, such as methanol or acetonitrile, added to the mobile phase to alter selectivity, but with the availability of solvent-compatible ion exchange resins today, organic modifiers can be used to alter selectivity in ion exchange as well. The presence of ion-pairing agents can complicate the use of conductometric and amperometric detections; nevertheless, both detection techniques are commonly used with ion-pair chromatography. The combination of ion-pair chromatography and suppressed conductivity detection is often called mobile phase ion chromatography (MPIC).

Analytes most suited to ion-pair chromatography are large, hydrophobic ions because the slow mass transfer or secondary hydrophobic interactions that these types of ions exhibit with typical ion-exchange stationary phases leads to poor efficiency, and often poor resolution.

Separation: Chelation

Certain organic groups, such as dicarboxylates, tend to form complexes with metal ions, but resins with such chelating functional groups have not found wide use as stationary phases. Although these resins have a high selectivity for certain metal ions, notably transition and lanthanide metals, the exchange process is too slow for efficient chromatographic separations. Instead, metals are usually chromatographed by anion exchange as anionic complexes of pyridine dicarboxylic acid or similar anionic chelating agents.

Chelating resins are often used for metal concentration or matrix elimination. Many spectroscopic techniques experience interferences from main group metals, such as sodium and calcium, which can be present at concentrations that are several orders of magnitude higher than those of the metal analytes. IC, on the other hand, has the selectivity for interference-free determination of many transition or lanthanide metals.

Detection: Suppressed Conductivity

In the mid-1970s, Small and coworkers (2) recognized that the very nature of ion exchange chromatography required an eluent ion to exchange with the analyte ion. Conductivity, a property shared by all ions, would be the ideal choice as a universal detection method, but the background conductance of the eluent would reduce the sensitivity of the technique. They solved this problem by using a second column, a suppressor column, having the opposite functionality as the analytical column and placing it between the analytical column and the detector. For determining anions, the suppressor column would be a cation exchange column in the hydronium ion form and for determining cations the suppressor would be an anion exchange column in the hydroxide ion form. When the salt of a weak acid or base is used as the eluent, the suppressor reduces the conductometric signal from the eluent.

As an example, consider what would happen during suppression of an eluent containing sodium bicarbonate and sodium carbonate, a common eluent in IC (Fig. 12.3). The carbonate and bicarbonate anions exchange with the analyte anions (such as chloride) during the separation. The effluent then passes through a cation exchange column in the hydronium ion form. Sodium ion is exchanged for hydronium, forming carbonic acid, a weak acid having a very low conductometric signal. An analogous reaction occurs in the case of cations. Another benefit of the suppression reaction is that the response for the analyte ions is often increased. According to Kohlrausch's law, the measured conductivity of an ionic compound is the sum of the equivalent conductances of the anion and cation. The equivalent conductance of hydronium is the highest of all cations, so the measured conductance of each anion increases after suppression because the hydronium counterion contributes more to the total conductance. Similarly, hydroxide ion has the highest equivalent conductance of all anions, so the total conductance for cations increases after suppression. The principal disadvantages to this approach are that the suppressor column causes some dispersion of the analytes before detection and the suppressor must be discarded or regenerated as the hydronium ions are depleted and the column can no longer suppress the eluent.

The suppressor device has changed through its evolution (9). Today it is commercially available in three versions: a packed column similar to the original one used about 20 years ago, a slurry of suspended ion exchange resin that is added to the effluent after the analytical column to accomplish the suppression reaction, and an ion-exchange membrane device that continuously supplies the suppressing ion from a chemical source or from the electrolysis of the water in recycled, suppressed eluent. The packed column suppressor is based on long-tested technology, but it must be regenerated or replaced periodically as it becomes exhausted. The postcolumn reagent approach has the lowest startup costs, but the suppressing resin is a consumable, discarded with the column effluent. The membrane devices have low dead volume, so they lose little efficiency during suppression, and the electrolytic version of the device has no consumable costs and almost no maintenance, but these types of suppressors have higher startup costs.

Analytes detectable by suppressed conductance are those that, when suppressed, are ionic. Generally, this means that anions with pK_as below about 5 and cations with pK_bs below about 5 can usually be detected by this method. In other words, if the analyte is ionic at pH 7, the approximate pH of most eluents after suppression, suppressed conductance is a viable detection method.

Figure 12.3 Suppression reaction in a column-style suppressor.

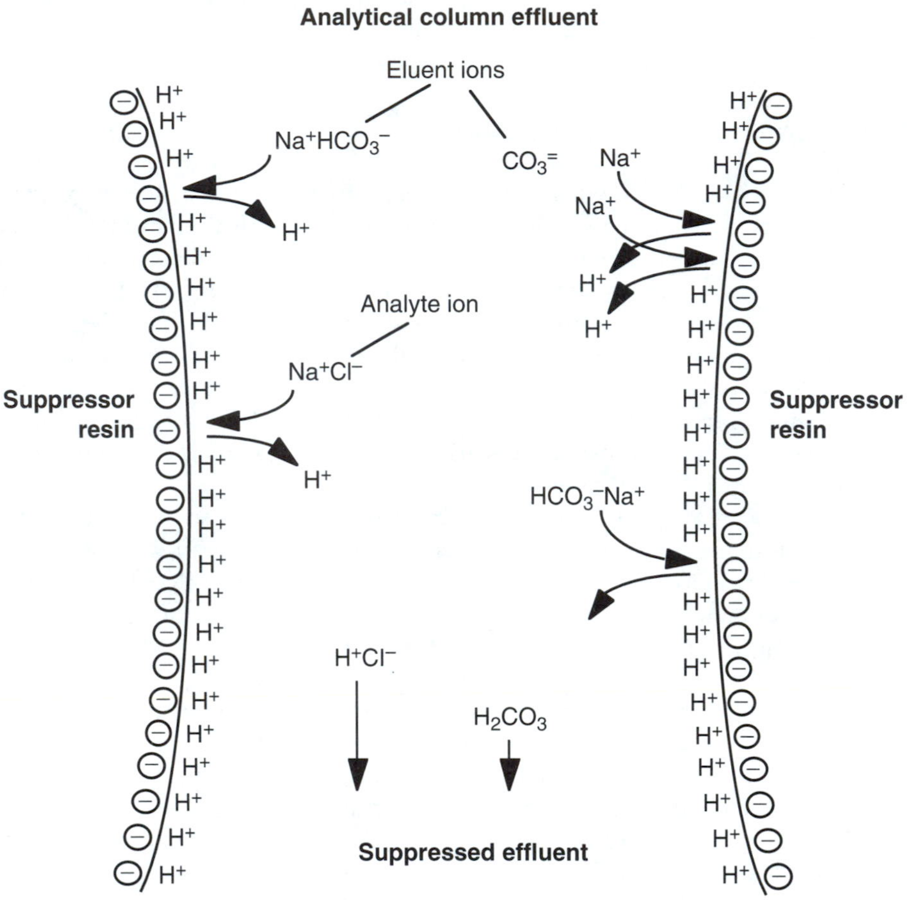

Analytical column effluent

Detection: Nonsuppressed Conductivity

Nonsuppressed IC, sometimes called single-column IC, operates with a higher background conductance. The higher background necessitates careful control of temperature, so conductivity detectors (Chap. 39) designed for nonsuppressed systems generally have thermostated cells and sometimes insulated chambers for the analytical column and the detector cell. Analytical columns designed for nonsuppressed IC are generally of lower capacity so that less concentrated eluents, which have lower background conductance, result in analysis times comparable to the suppressed systems.

The detector must be able to detect a change in conductance as the analytes pass through it. This change may be an increase or a decrease. Most anion methods exhibit an increase because weakly conductive eluents, that is, eluent anions having equivalent conductances lower than the analyte anions, are used. Common nonsuppressed eluent systems include benzoate, phthalate, and borate/gluconate. Although they are much less common, highly conductive eluents such as sodium hydroxide have also been used for anion determination in nonsuppressed systems, producing a decrease in conductance as the analyte elutes. This decrease occurs because common anions have lower equivalent conductances that hydroxide. In these latter cases, the signal from the detector is commonly inverted because chromatographic convention displays analytes as positive peaks.

Most nonsuppressed cation systems are analogous to the second anion system in that they use a highly conductive eluent, usually hydrochloric or nitric acid, and detect analyte cations as a decrease in the measured conductance. In this case, the decrease is due to the lower equivalent conductance of analyte cations compared to the very high equivalent conductance of hydronium.

Small (1) compares nonsuppressed and suppressed conductometric detection on theoretical grounds. He calculates that detection limits for nonsuppressed IC are not as low as those for suppressed IC and the dynamic range is smaller than for suppressed methods. On the other hand, nonsuppressed methods have the advantage that weak acid anions and weak base cations can be detected more readily. For example, a cation with a pK_b of 9 would be only about 1% ionic after suppression (pH of 7), but in 1 mM HCl (pH of 3) it would be about 99% ionic.

Detection: Single-Potential Amperometry

Any analyte that can be oxidized or reduced is a candidate for amperometric detection (Chap. 36). The simplest form of amperometric detection is single-potential, or direct current (DC), amperometry. A voltage (potential) is applied between two electrodes positioned in the column effluent. The measured current changes as an electroactive analyte is oxidized at the anode or reduced at the cathode. Single-potential amperometry has been used to detect weak acid anions, such as cyanide and sulfide, which are problematic by conductometric methods. Another, possibly more important advantage of amperometry over other detection methods for these and other ions, such as iodide, sulfite, and hydrazine, is specificity. The applied potential can be adjusted to maximize the response for the analyte of interest while minimizing the response for interfering analytes.

Detection: Pulsed Amperometry

An extension of single-potential amperometry, (Chap. 36) is pulsed amperometry, most commonly used for analytes that tend to foul electrodes. Analytes that foul electrodes reduce the signal with each analysis and necessitate frequent cleaning of the electrode. In pulsed amperometric detection (PAD), a working potential is applied for a short time (usually a few hundred milliseconds), followed by higher or lower potentials that are used for cleaning the electrode. The current is measured only while the working potential is applied, then sequential current measurements are processed by the detector to produce a smooth output. PAD is most often used for detection of carbohydrates after an anion exchange separation, but further developments of related techniques (7) show promise for amines, reduced sulfur species, and other electroactive compounds.

Detection: Optical

The most commonly used optical detectors are absorbance detectors and fluorescence detectors (Chap. 25 and 26). Fluorescence detectors are rarely used in IC, but absorbance detectors are quite common. Absorbance is used for detection under three circumstances, direct photometric detection, indirect photometric detection, and photometric detection after a postcolumn derivatization.

Some ions, most importantly certain anions, are chromophoric; that is they absorb light. Such is the case with the common anions nitrite and nitrate. These can be detected in the ultraviolet region, by monitoring absorbance at 215 nm, without detecting nonchromophoric anions such as chloride. Absorbance can also be used for detection of many organic ions, such as aryl amines and organic acids, following ion exchange chromatography.

More commonly, however, ions are not chromophoric. Nevertheless, they can be detected indirectly using a chromophoric eluent ion of the same charge, a technique known as indirect photometric detection. During the ion-exchange process, ionic concentration remains constant at the stationary phase functional groups, so at the detector the presence of each equivalent of analyte requires the absence of an equivalent of chromophoric eluent ion. The use of indirect photometric detection is not very common today.

By far, the most common use of absorbance detection in IC is with an ion-exchange separation followed by a derivatization reaction that renders the analytes chromophoric or, in some cases, more chromophoric. Ideally, the postcolumn reaction is fast (complete within seconds) and does not produce interferences (for example, by reaction with eluent components) in the absence of analytes. This technique is commonly used for the determination of transition metals and amino acids. As was mentioned earlier, transition metals are often chromatographed as anion complexes. By adding 4-(2-pyridylazo)resorcinol (PAR) to the column effluent, PAR complexes of the metals rapidly form and are detected by absorbance. Similarly, amino acids can be chromatographed by either anion or cation exchange, derivatized after separation with ninhydrin and detected by absorbance or, if lower detection limits are desired, derivatization with *ortho*-phthalaldehyde allows fluorescence detection.

Detection: Specialized Detectors

Recently, the scientific community has begun coupling ion-exchange separations with specialized techniques for detecting the analytes. The two most common are inductively coupled plasma (ICP) spectroscopy, useful for the determination of metals, and mass spectroscopy (MS).

Coupled to ICP, IC (Chap. 21 and 22) is more a sample preparation step than a chromatographic separation. A column containing a chelating stationary phase is used to selectively concentrate metals from a sample matrix. An intermediate column wash step can selectively remove interfering metals if necessary. Then the column is washed with a strong eluent, delivering the metals of interest to the ICP.

MS (Chap. 33) is not generally compatible with the highly ionic eluents commonly used in IC. However, just as with conductivity detection, eluent ions can be eliminated from the effluent by the use of a suppressor before it enters the MS (10). In these cases, the goal of IC-MS is not the quantification of the analytes, but rather their mass spectral identification and characterization.

What It Does

Instrumentation

Minimally, an IC system consists of an eluent reservoir, an analytical pump to deliver the eluent to the analytical column, an injection valve or other means of introducing the sample, an analytical chromatographic column, a detector, and a data processing device (Fig. 12.4). These components are the same components in an HPLC system. Here, however, we discuss each as it relates to IC.

The eluent reservoir can be as simple as a bottle with a fluid line that leads to the analytical pump. Usually, however, the eluent is kept under a pressure of about 1/3 atm so that eluent is delivered to the analytical pump without fluid interruption. Eluents at high pH, such as sodium hy-

Figure 12.4 The ion chromatographic system.

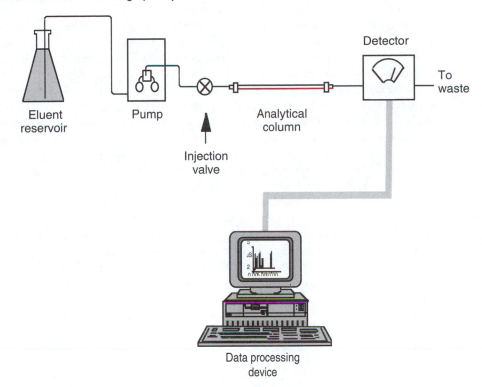

droxide, must be kept under an inert atmosphere to prevent carbon dioxide in the air from forming carbonate in the eluent and thereby altering the eluting ion.

The pump used in IC is usually nonmetallic. This is because many IC methods use strong acids, strong bases, or high concentrations of salts that may corrode metallic systems. Materials such as polyetheretherketone (PEEK) are compatible with the vast majority of IC applications. Metallic pumps may be used if care is taken to maintain them properly. Many manufacturers of metallic pumps for IC recommend regular passivation of the system.

High-sensitivity applications usually require a dual-piston pump for relatively pulse-free operation, although single-piston pumps with pulse dampers are usually sufficient for routine, mg/L-level analysis. The most flexible systems have dual-piston gradient pumps, which allow the analyst to perform gradient methods or to mix eluents isocratically, aiding method development.

The injection valve must introduce a reproducible sample volume into the IC system. The most common means of accomplishing this is by the use of a fixed-loop injection. A length of tubing of known volume is attached to the valve and switched into the eluent stream during injection.

Many injection systems include an autosampler for unattended, high-volume work. The autosampler may have an injection valve built in, or it may deliver the sample to a remote injection valve. The former type is usually chosen if sample sizes are small because having an injection valve in the autosampler shortens the distance that the sample travels between sample vial and injection valve.

Certain low-level IC methods use a concentration column in place of the injection loop on the injection valve. This concentration column may be a guard column of the type used for the analytical separation or it may be a column specifically designed for concentration in a particular application. If the concentration column has a high backpressure, an auxiliary pump is usually required to deliver the sample to that column.

The chromatographic separation takes place on the analytical column. Usually, a guard column is placed before the analytical column to extend its lifetime. Most commonly, the guard column is simply a short, inexpensive version of the analytical column that is replaced as needed. Typically, the guard column also adds about 20% to the capacity of the analytical system.

The detector ordinarily is one of the options discussed previously, namely conductometric, amperometric, optical, or some other method of measuring ions, along with any system used to facilitate detection. These systems include suppressors and postcolumn reaction devices.

The simplest data processing device is a chart recorder that converts a detector output, usually a voltage within a predetermined range, into a time–voltage profile, called a chromatogram. However, this device relies on the analyst to convert the area or height of a peak drawn on the paper to a quantity of analyte, a labor-intensive step.

A recording integrator does much the same thing as a chart recorder, but calculates the peak area automatically and may even calculate analyte quantities based on standards that were injected earlier. Some integrators can also send signals to the instruments, thereby controlling simple functions such as injections and detector range changes.

Data processing through a computer offers the greatest flexibility and automation. As with the integrator, analyte quantities can be determined automatically, but the raw data can also be stored and reprocessed using different parameters at a later time. Computer-based systems usually have complete instrument control as well, so that complete operating conditions can be stored on computer and used later to reproduce the analytical method.

Analytical Information

Qualitative

Like other forms of liquid chromatography, IC can indicate sample composition. The components present elute at nearly unique retention times, determined either by separate injections of known standards of each ion, or more accurately by adding a small amount of a known standard to the sample and identifying the component that experiences an increase in peak size. An estimate of relative concentrations can also be made, but accurate concentrations can be determined only after calibrating the detector response.

One can also obtain information about the distribution of components that differ in the number of repeating units. Examples are samples containing polyphosphates, such as polyphosphoric acid (Figure 12.5), linear oligosaccharides, such as hydrolyzed starch, oligonucleotides, and ionic surfactants, such as linear alkyl benzenesulfonic acid. Although each component may be completely resolved from other components, individual pure standards of each component are difficult if not impossible to obtain, so estimates of concentrations are only qualitative.

Quantitative

IC is first and foremost a quantitative technique. The purity of standards determines the accuracy of the technique, and because pure standards of most ions, usually in the form of salts, typically are easy to obtain, accuracy can be 1 to 2%. Analyte concentrations are determined by establishing a

Figure 12.5 Anion profile of polyphosphoric acid.

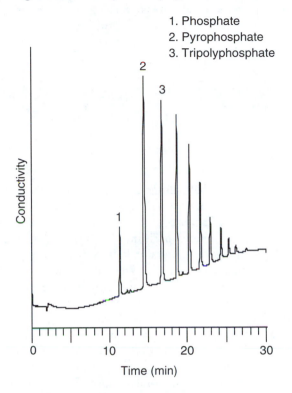

1. Phosphate
2. Pyrophosphate
3. Tripolyphosphate

relationship between known concentrations of standards and their responses, in terms of either peak height or peak area. This is commonly known as the calibration curve. Such a calibration using independently chromatographed standards and samples, an external calibration, usually exhibits precision of less than 3% relative standard deviation or coefficient of variation (CV). The CV often can be reduced to below 1% by the use of an internal standard, a fully resolved component added to the injection of sample or standard. Peak responses are then normalized to the internal standard.

Depending on the exact conditions used, and especially on how well the analyte and detection method are matched, detection limits can vary from below 1 µg/L for a 25-µL sample to above 1 mg/L. Detection limits are sometimes lowered by the use of a concentrator column so that large volumes (as much as 100 mL) can be injected. For the most common applications, inorganic anions and cations using suppressed conductivity detection, detection limits are about 1 to 5 µg/L for a 25-µL sample.

It is usually desirable, though not necessary, to have a linear relationship between analyte concentration and response. The linearity is usually expressed by the coefficient of determination (r^2) over the calibration range. Beer's law for absorbance detection, and the high degree of dissociation of many species in dilute solutions for conductometric detection, are reasons why calibration curves are usually linear ($r^2 > 0.999$) over three, and sometimes four, orders of magnitude. Similar linearities are common for amperometric and fluorimetric detection, but because exact conditions for detection can vary, so can linearity.

For IC the dynamic range—that is, the concentration range over which analytes can be determined—is a function not only of detection but also of separation. Although it is true that better detection limits can extend the dynamic range at low concentrations, it is column capacity that extends the range at high concentrations. The more analyte that can be injected into a system without overloading the column, the greater is the dynamic range. With most detection methods, there is little reason to use low-capacity columns; however, nonsuppressed conductometric detection often uses a low-capacity column to allow the use of low eluent concentrations.

Applications

Conductivity

The determination of most anions and cations, both inorganic and organic, is the mainstay of IC. It is used to analyze drinking water, rain water, soil, foods, chemicals, and countless other samples. Many municipal water facilities monitor ionic content of drinking water to ensure quality and safety (Fig. 12.6). The anion composition of rain and other forms of precipitation tells us a lot about our air quality (Fig. 12.7). Generally, high concentrations of anions in rain indicate high acid content. The ion composition of soil can help determine its suitability for agriculture. Proper reporting of the ionic content of foods and beverages is crucial to helping us maintain healthy diets. Several industries rely on high-purity reagents for their processes and IC is often used to monitor that purity. If ionic impurities are too high, poor-quality products may be produced. IC has also been used for more exotic applications such as determining the composition of moon rocks or the ionic content of ice found in Antarctica that is thousands of years old.

Thousands of publications cover applications of IC with conductivity detection (Chap. 39). The analyst may want to consult one or more of the books listed in the References for a more complete discussion of the topic.

Figure 12.6 Cations in drinking water.

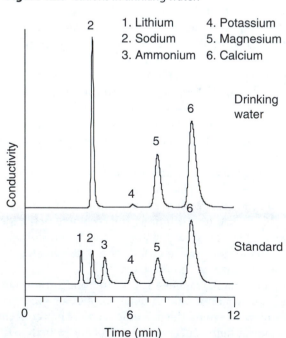

Figure 12.7 Anions in rain water.

1. Fluoride	5. Nitrate
2. Chloride	6. Phosphate
3. Nitrite	7. Sulfate
4. Bromide	

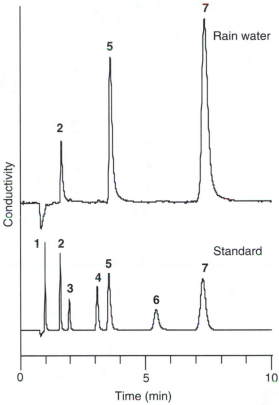

Amperometry

Although it is a relatively new technique, the determination of carbohydrates is one of the most common applications of IC with amperometric detection (Chap. 36). It is the pairing of the most selective separation technique with the most specific and sensitive detection technique for this class of compounds. It has been applied to the analysis of foods (Fig. 12.8), oligosaccharide and polysaccharide profiling, and monosaccharide compositional analysis and oligosaccharide analysis of carbohydrates from glycoproteins.

Amperometry is also used with more traditional ions, such as iodide and sulfite. Each of these ions can be detected by conductivity, but detection limits are much lower using single-potential or pulsed amperometry and, depending on the sample matrix, amperometry may be more specific for these ions than for other ions that are present. Examples of sulfite in beer and wine and iodide in foods have been reported.

An interesting application of amperometric detection in IC is the detection of cyanide or sulfide using a silver working electrode. In these cases, it is not the analyte that undergoes oxidation but the working electrode. A very low potential is applied (0.05 or 0.00 V) and the current measured from the formation of silver complexes according to the reactions

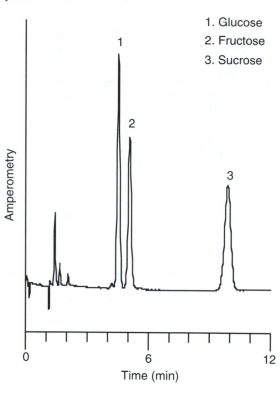

Figure 12.8 Carbohydrate components of orange juice.

1. Glucose
2. Fructose
3. Sucrose

$$2\,CN^- + Ag^0 \rightarrow Ag(CN)_2^- + e^-$$
$$S^= + 2\,Ag^0 \rightarrow Ag_2S + 2\,e^-$$

Because these are weak anions, they can be chromatographed by either anion exchange or ion exclusion. In cases where high concentrations of chloride or other ions that react with silver may be present, ion exclusion is usually the better choice.

Optical

Certain inorganic ions, such as nitrate and nitrite, are chromophoric and can therefore be specifically detected using UV absorbance (Chap. 25) in complex samples such as waste water. Usually, however, the analyst is also interested in the nonchromophoric ions in the sample, so conductivity may be the better choice. Of course, there are a wide variety of applications for chromophoric amines and carboxylic acids. Aromatic amines and carboxylates can be detected by either UV absorbance or conductivity but, in general, as their molecular weights increase their conductometric responses decrease and their UV responses remain the same or increase.

Metals, especially transition metals and lanthanide metals, are most commonly determined with IC by chromatographing them as anionic complexes, of oxalate or pyridinedicarboxylate for instance, and detecting them by absorbance after a postcolumn reaction (Fig. 12.9). In theory, this approach has the advantage over spectroscopic methods of being able to separately determine different oxidation states of a given metal. In practice, this works very well for Cr(III)/Cr(VI) and with some care can work for Fe(II)/Fe(III). However, because many eluents are prone

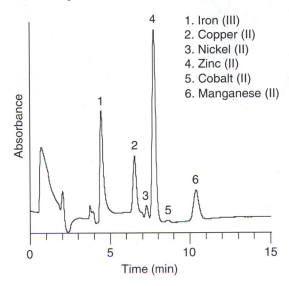

Figure 12.9 Transition metals concentrated from sea water using chelation concentration followed by IC.

1. Iron (III)
2. Copper (II)
3. Nickel (II)
4. Zinc (II)
5. Cobalt (II)
6. Manganese (II)

to upset any redox equilibrium that may exist in the sample and because of the care needed in preventing standards from oxidizing or reducing, IC does not have wide application for oxidation state speciation.

The determination of amino acids was perhaps the first widely used IC application. HPLC methods using precolumn derivatization and a reversed phase separation have replaced IC for amino acid analysis of purified proteins, but IC is still commonly used for physiological and other complex samples that are prone to interferences by reversed phase methods.

Hyphenated Techniques

IC Inductively Coupled Plasma Optical Emission Spectrometry

The coupling of IC, or more accurately chelation concentration, to ICP spectroscopy is a young technique and not widely used, but it has advantages over simple ICP for difficult samples. One of the most challenging types of samples for ICP is one with metals of interest at low concentration (5 to 10 µg/L) and interfering metals (iron, aluminum, calcium) at much higher concentrations of about 50 to 100 mg/L. Chelation concentration can be an automated sample preparation technique that concentrates the metals at low concentration, eliminates the interfering metal, and delivers the sample to the ICP in the column effluent. Standards are concentrated by the same process and delivered to the ICP in the same effluent. The technique has been used for mg/kg-level lanthanide metals in acid digested rock, µg/L-level transition metals in sea water, and a variety of other samples. See Chap. 21 and 22 for further discussion.

IC Mass Spectrometry

IC has been interfaced with MS not for quantitative analysis, but for analyte identification. An ion chromatographic separation, with either a volatile salt eluent or one that can be suppressed before entering the MS, is used to resolve the components. The technique has been applied to oligosaccharides and sulfonic acids (10). There are no commercial instruments dedicated to this technique, so the analyst wanting to work in this research area should be knowledgeable about both IC and MS. See also Chap. 33.

Nuts and Bolts

Relative Costs

Complete system	$$ to $$$
Components	
Single-piston pump	$
Dual-piston pump	$$
Conductivity detector	$ to $$
Amperometric detector	$ to $$
Absorbance detector	$ to $$
Fluorescence detector	$ to $$
Autosamplers	$ to $$
Data systems	
Recorders and integrators	$ to $$
Computer-based	$$
Columns	<$
Suppressors	<$
Eluent reagents	<$

$= 1 to 5K, $$= 5 to 15K, $$$= 15 to 50K

The cost of an IC system is determined by required flexibility and, to a lesser extent, required sensitivity. A dedicated, low-cost system with a single-piston pump and a conductivity detector is reasonably priced and is capable of most routine anion and cation determinations. However, these systems do not allow the analyst to perform gradient separations or to mix eluents isocratically. Single-piston pumps are not as pulse-free as dual-piston pumps, so noise and detection limits are often higher. Complete systems are usually modular, so that the correct component (pump, detector, autosampler, data system) needed for an application can be configured within the system. A modular system is usually the best option if needs may change, requiring a different detector, or if several analyses are to be performed on the same system.

If the system is to perform a new analysis, a new analytical column with optimum selectivity is usually needed. Periodically, columns may have to be replaced due to wear and tear, but the frequency of replacement depends largely on the types of samples and on how the samples are prepared before analysis. A multitude of sample preparation methods have been developed to eliminate sample components that may deteriorate analytical columns. However, many modern analytical columns are robust enough to allow injection of crude samples with regular cleaning of the columns, thus reducing the need for labor-intensive steps before analysis.

Vendors for Instruments and Accessories

I = instrument manufacturer; C = consumables supplier

Alltech Associates Inc. (C)
2051 Waukegan Rd.
Deerfield, IL 60015
phone: 708-948-8600, 800-255-8324
fax: 708-948-1078
email: 73554.3372@compuserve.com
Internet: http://www.alltechweb.com/

Bio-Rad Laboratories,
Life Science Group (C)
2000 Alfred Nobel Dr.
Hercules, CA 94547
phone: 510-741-1000, 800-424-6723
fax: 800-879-2289
Internet: http://www.biorad.com

Dionex Corp. (I & C)
P.O. Box 3603, 1228 Titan Way
Sunnyvale, CA 94088-3603
phone: 408-737-0700, 800-723-1161
fax: 408-730-9403
email: marcom@dionex.com
Internet: http://www.dionex.com

EM Separation Technology (I & C)
480 S. Democrat Rd.
Gibbstown, NJ 08027-1297
phone: 609-224-0742, 800-922-1084
fax: 609-423-4389

Interaction Chromatography (I & C)
2032 Concourse Dr.
San Jose, CA 95131
phone: 408-894-9200
fax: 408-894-0405

SaraSep, Inc. (I & C)
2032 Concourse Dr.
San Jose, CA 95131
phone: 408-432-8536
fax: 408-432-8713
email: sarasep@sarasep.com
Internet: http://www.sarasep.com

Waters Corp. (I & C)
34 Maple Street
Milford, MA 01757
phone: 508-478-2000, 800-254-4752
fax: 508-872-1990
email: info@waters.com
Internet: http://www.waters.com

Required Level of Training

Operation of the IC system can be performed by anyone with a basic, high-school level understanding of chemistry. Instrument troubleshooting is a skill that is usually gained through experience. Many IC manufacturers offer basic courses on the operation and maintenance of their instruments for those who want some additional training.

Integrators and computer-based data systems for IC are usually sufficient for routine quantitative analysis. Only minimal training is necessary to obtain accurate quantitative data, but a sound understanding of analytical chemistry and chromatography is needed to develop a method and to verify initially that the data system is programmed properly.

Service and Maintenance

Many modern IC components have internal diagnostics that can alert the analyst to the source of problems. Some systems even keep records of when maintenance tasks were last performed. The most common maintenance tasks are replacing the piston seals in the analytical pump, replacing lamps in the optical detectors, and replacing the reference electrode in the amperometric detector cell. Regular cleaning of the analytical column may also be necessary if sample matrices that may foul the column are injected. The operator manual for the column usually has instructions from the manufacturer on how best to clean the column.

Suggested Readings

SMALL, H., *Ion Chromatography*, New York: Plenum Press, 1990.

WALTON, H. F., and R. D. ROCKLIN, *Ion Exchange in Analytical Chemistry*, Boca Raton, FL: CRC Press, 1990.

WEISS, J., *Ion Chromatography*, 2nd ed. Weinheim, Germany: VCH. Verlagsgesellschaft mbH, 1995 (English translation).

References

1. H. Small, *Ion Chromatography* (New York: Plenum Press, 1989).

2. H. Small, T. S. Stevens, and W. C. Bauman, *Analytical Chemistry*, 47 (1975), 1801–9.

3. J. S. Fritz, *Analytical Chemistry*, 59 (1987), 335A–44A.

4. R. D. Rocklin, *Journal of Chromatography*, 546 (1991), 175–87.

5. H. F. Walton and R. D. Rocklin, *Ion Exchange in Analytical Chemistry* (Boca Raton, FL: CRC Press, 1990); J. T. Gjerde and J. S. Fritz, *Ion Chromatography*, 2nd ed. (New York: Heuthig, 1987).

6. R. D. Rocklin, C. A. Pohl, and J. A. Schibler, *Journal of Chromatography*, 411 (1987), 107–19; W. R. Jones, P. Jandik, and A. L. Heckenberg, *Analytical Chemistry*, 60 (1988), 1977–9.

7. D. C. Johnson and W. R. LaCourse, *Analytical Chemistry*, 62 (1990), 589A–97A.

8. P. K. Dasgupta, *Analytical Chemistry*, 64 (1992), 775A–83A.

9. S. Rabin and others, *Journal of Chromatography*, 640 (1993), 97–109.

10. R. A. M. van der Hoeven and others, *Journal of Chromatography*, 627 (1992), 63–73; J. Hsu, *Analytical Chemistry*, 64 (1992), 434–43.

Chapter 13

Thin-Layer (Planar) Chromatography

Dieter E. Jaenchen

Camag Scientific Inc.

Summary

General Uses

- Separation of a great variety of compounds (organic, inorganic, biologicals, polymers, chirals)
- Processing or screening many samples quickly
- Analysis in cases where sample preparation is difficult, undesirable, or impossible
- Determination of analytes in complex matrices
- Postchromatographic derivatization
- Flexible detection, that is, detection in the absence of the mobile phase and with changed parameters, if required

Common Applications

- Environmental applications from water analysis to plant residues
- Pharmaceutical applications from stability and impurity studies to drug monitoring in biological fluids
- Biomedical compounds (organic acids, lipids, carbohydrates, amino acids, and steroids)
- Food analysis from carcinogens, drug residues, and preservatives to spices and flavors

Samples

State

The analyte must be dissolved in a suitable liquid for transfer to the adsorbing stationary phase. Viscosity, volatility, wettability, and potential for unintended predevelopment are all factors in solute selection.

Amount

The amount of sample applied (spotted) can vary greatly depending on many factors, with detection and resolution being paramount.

Preparation

Due to the one-time use of the stationary phase, sample preparation is often simpler than for other chromatographic techniques.

Analysis Time

Because many samples can be chromatographed simultaneously, total time must be divided by the number of samples on each plate. Although time for chromatogram development varies greatly with both stationary and mobile phases selected, the influence on time per sample is little. Typically, analysis time for thin-layer chromatography (TLC) or high-performance thin-layer chromatography (HPTLC) is 3 to 15 minutes depending on samples per plate and solvent running distance.

Limitations

General

- Limited to nonvolatile compounds
- Limited to separation numbers (peak capacities) of 10 to 50.

Accuracy

If each step of the technique is automated, precision ranges from 1 to 3% relative standard deviation (RSD). For noninstrumental TLC, RSDs are greater.

Sensitivity and Detection Limits

Very analyte-dependent, with strength of inherent or derivatized visualization/detection being decisive factors. Generally sensitivity is in the nanogram range for absorbance and in the picogram range for measurements by fluorescence.

Complementary or Related Techniques

Coupling or combining TLC with high-performance liquid chromatography (HPLC), mass spectrometry (MS), Fourier transfer infrared (FTIR), or Raman spectroscopy provides interesting possibilities and advantages.

Introduction

The TLC of today is far different from that first described by Kirchner and colleagues in 1951 (1), later standardized by Stahl and colleagues (2). The introduction of high-performance stationary phases (HPTLC) in 1975 (3), followed by the evolution of instrumental techniques, has brought TLC into the modern era.

How It Works

Comparison with High-Performance Liquid Chromatography (HPLC)

One way to explain how TLC works is to draw a comparison between TLC and column liquid chromatography. Column liquid is an on-line process. Samples are chromatographed sequentially. When the next sample is injected, it is essential that the column be in exactly the same condition as it was for the previous sample or calibration standard. This requires careful sample preparation in order to avoid irreversible contamination of the stationary phase or carryover of fractions from one sample to another.

In the closed chromatographic system, all fractions pass through the detectors, where their physical properties are measured. From these, qualitative and quantitative results can be derived. Leaving aside setups where fractions or hard-cuts are collected, no measurement can be added or repeated without repeating chromatography.

In column liquid chromatography, detection takes place in the presence of the mobile phase. Therefore, the mobile phase must be selected with a view to not interfering with the detection wavelength. These considerations, as well as column life considerations, have led analysts to conduct the vast majority of column chromatographic separations in the reversed phase (RP) mode. Although reversed stationary phases are also available for TLC, the application of RP chromatography in TLC is justified only when it yields a better separation (resolution) than normal phase TLC.

The TLC process is an off-line process. A number of samples are chromatographed simultaneously, side by side. The stationary phase is used only once. Sample preparation must be performed only to the extent necessary to ensure that extraneous material in the samples applied to the plate does not interfere with the separation or retain any of the analyte. Carryover of material from one sample to another is not a problem.

In the open (off-line) TLC process all fractions are stored on the plate. Their optical properties are measured by densitometric evaluation, including the recording of in-situ spectra. All such measurements can be repeated without the need to repeat chromatography. The decision to measure with different parameters, or to interpose a derivatization step, can be made upon reviewing first results.

The limitation of TLC is its restricted separation efficiency. Separation efficiency of a chromatographic system can be expressed as the number of theoretical plates (N). HPLC yields a maximum of 10,000 to 15,000 plates, whereas the maximum N numbers reported for modern planar chromatography (HPTLC) are around 5000 (4). The reason for the lower performance of TLC is the fact that, because of its capillary flow behavior, the length of a TLC bed cannot be extended at will, whereas the permissible length of a column is limited only by the pressure available in the system.

Because in TLC it appears more appropriate to define separation in space rather than in time, as is usual in HPLC, the term *separation number* (SN) was introduced (5). SN is defined as the number of fractions that can be accommodated over the available separation distance under the assumption that all fractions are regularly spaced and there is one available for each position.

$$SN = \frac{\log(b_0/b_1)}{\log(1 - b_1 + b_0)/(1 + b_1 - b_0)} \qquad (13.1)$$

or, simplified

$$SN = \frac{z_f}{b_0 - b_1} - 1 \qquad (13.2)$$

where b_0 = half-height width of spot of origin, b_1 = half-height width of spot at R_F = 1 (extrapolated), z_f = migration distance of solvent front, and R_F = retardation factor; $R_F = b/a$, where b = distance traveled by the center of the spot and a = distance simultaneously traveled by the mobile phase.

Using isocratic one-step development over the optimum separation distance of a modern HPTLC plate, one can expect SN to be around 15. It should be noted that the word *isocratic* in context with TLC does not have the same strict meaning that it has in a well-equilibrated HPLC system because in TLC, equilibration between the mobile and the stationary phase progresses gradually while chromatography takes place. Therefore, the word *isocratic* in TLC means that the developing solvent fed to the layer is not intentionally changed during chromatography.

An apparent advantage of HPLC over TLC is its easy access to automation. There are several approaches to automating the TLC process; however, changing the off-line principle of TLC to an on-line closed system in order to achieve complete automation would combine most of the limitations of TLC with those of HPLC. Automation of the individual steps of the TLC process, as set forth in this chapter, appears to be the better choice. One should not overlook that with a stepwise automated TLC workstation, more samples can be processed in an 8-hr workday than with a fully automated HPLC system within 24 hours.

Separation Media

Conventional TLC and HPTLC separation material is available in the form of precoated layers supported by glass, plastic sheets, or aluminum foil. The vast majority of TLC separations are carried out on normal phase silica gel, although other layers such as aluminum oxide, cellulose and reversed phase (alkyl bonded) are available. Recently introduced plates modified with amino, cyano, and diol functional groups bonded to the silica can extend normal phase or reversed phase chromatographic properties. These layers may affect the predominant role of normal phase silica gel in TLC to a certain extent. Also, chiral layers for the separation of optical isomers are available and present some interesting possibilities.

Figure 13.1 Capillary flow diagram.

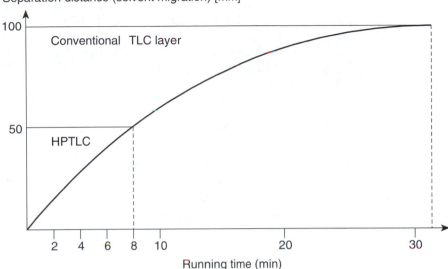

The expression *HPTLC* refers to layers that have a smaller particle size (6 µm instead of 12 to 20 µm), which are slightly thinner (100 or 200 µm instead of 250 µm) and, particularly, whose particle size distribution is significantly closer than that of conventional TLC sorbents. Because of these specifications the optimum separation distance of an HPTLC layer is around 50 mm instead of 100 to 120 mm on a conventional TLC plate.

The flow function of a solvent migrating in a layer by capillary action is depicted in Fig. 13.1. The scale must be seen as relative, as the absolute migration speed depends on the kappa value of the solvent. With all solvents the migration speed decreases with the square of time elapsed and thus with the square of the solvent migration distance.

The optimum migration distance can be defined as the point beyond which the directional migration of the analyte caused by capillary flow of the mobile phase, resulting in resolution, starts to become overrated by the nondirectional migration caused by diffusion. This explains why the separation distance in TLC cannot be extended deliberately, as is possible in column chromatography.

In Fig. 13.1 it can be seen that in HPTLC layers capillary flow chromatography takes place in the fast flow range of the respective developing solvent. From this derive practically all the advantages of HPTLC over conventional TLC (6):

- Reduced diffusion (that is, better separation efficiency)
- Absolute detection (determination) limits reduced by a factor of 10 to 15
- Better economy (because plate costs and solvent consumption per sample are lower)

The advantages of HPTLC separation material have not eliminated the use of conventional TLC layers. Many analysts insist on their use, usually because existing filed methods prescribe them.

Sample Application

The selection of a sample application technique and the device to be used depends on several factors: nature of the analytical task, qualitative or quantitative analysis, workload and time constraints, type of separation layer, conventional TLC or HPTLC, and sample volumes to be applied, which is often dictated by the detection limits.

In order to fully use the separation power of the layer, it is important to restrict the dimension of the sample origin in the direction of chromatography to 3 to 4 mm on conventional layers and 1 to 1.5 mm on HPTLC layers. Sample volumes that can be applied as spots delivered in one stroke are 0.5 to 5 µL on conventional layers and 100 nL to 1 µL on HPTLC layers. Larger volumes may be applied either spotwise with a device with controllable delivery speed or sprayed on in the form of narrow bands.

Sample Application as Spots

The simplest way to apply samples spotwise is to use a fixed-volume pipette that fills by capillary action and delivers its content when it touches the layer. In order to ensure that the layer is not damaged and the spots are precisely positioned, it is advisable to guide the pipette with respect to the lateral position and with reproducible constant pressure. This is possible with a Nanomat, a mechanized spotting device. Disposable glass capillaries suitable for this technique have a volume precision of less than 1%.

Sample Application as Bands

An advantageous alternative to sample application as spots is sample application in the form of narrow bands. This provides the highest resolution attainable with each TLC system (7). In addition, certain systematic measuring errors that occur with densitometric evaluation are drastically reduced by the aliquot scanning technique (scanning with a slit of one-half to two-thirds of the band originally applied).

The Linomat allows sample application in narrow bands by a spray-on technique. This combines sample transfer with sample concentration. The instrument is suitable for applying sample volumes of 2 to 100 µL onto HPTLC layers. This technique is suitable to dramatically lower the determination limits with respect to the concentration in the solution, which is very useful in trace analysis. The reduction of sample volumes by factor 10 from conventional to HPTLC material is not necessary when the spray-on technique is used. Another benefit of the spray-on technique for quantitative analysis is the ability to apply different volumes of the same calibration standard solution instead of the same volume of different concentrations. The spray-on technique also allows unknowns to be simply oversprayed with spiking solutions for the standard-addition method.

By using the Linomat spray-on technique, the gain in precision of quantitative TLC analysis can be expected to be in the range of 30% (that is, from 1.5% RSD to 1.0%) when chromatographic resolution is not a problem. If samples are complex, dirty, tend to tailing, or are otherwise difficult, the choice between band application or spot application can determine whether a meaningful quantitative result can be obtained at all.

Automated Sample Application

Sample application in quantitative TLC analysis can take up as much as one-third of the time needed for the entire analysis. Therefore, attempts to automate or otherwise rationalize this step in the procedure have been made since the early days of TLC.

Devices using a series of capillaries (8) or syringes (9) were used with partial success in conventional TLC. However, none of the multicapillary or multisyringe devices proved to be suitable

Figure 13.2 Automatic TLC Sampler III, application module.

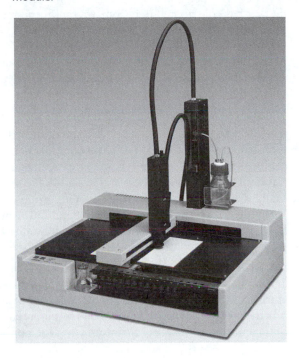

for the smaller dimensions of HPTLC. Applying samples sequentially proved to be the more appropriate approach to solve the problem of automating this step without sacrificing precision.

The Automatic TLC Sampler III is the latest version of this kind (Fig. 13.2). It applies samples automatically from a rack of sample vials. The samples are dispensed from a steel capillary, spotwise by contact transfer or bandwise using the spray-on technique. The sample dispensing speed is selectable between 10 and 1000 nL/s. The application module is controlled by an IBM-compatible PC, normally the same computer that controls densitometric chromatogram evaluation, so that calibration data are entered only once, which contributes to good management practice/good lab practice (GMP/GLP) conformity.

Chromatogram Development

The classical way to develop a thin-layer chromatogram is to immerse the plate with its lower edge in the developing solvent contained in a tank. Flow of the solvent by capillary action starts when the plate is inserted.

While ascending of the solvent progresses, the layer interacts with the vapor phase in the tank. The part of the layer that is still dry adsorbs molecules from the gas phase. From the already wetted part solvent molecules evaporate to an extent depending on the degree of gas phase saturation in the tank with solvent vapors.

It is customary to line the inside of the tank with filter paper soaked with the developing solvent. In some cases this is advantageous, in some cases it is detrimental.

Solvent molecules adsorbed in the dry layer do not mix with liquid solvent ascending in the layer. Unless these are extremely polar molecules, such as water, methanol, acids, and bases, the preadsorbed solvent molecules are pushed ahead by the front. Thus, the apparent solvent front (the boundary between the wet and the dry area) moves ahead of the real front, which is formed by the liquid that has contributed to chromatographic migration of the analyte fractions. The difference between the apparent front, to which observed Rf's relate, and the real front, relating to corrected retention factors (Rf'), can easily amount up to 25% depending on the type of solvent, the shape of the developing device, and the duration of pre-equilibration.

Interaction between layers, dry or wetted, and the gas phase can be largely suppressed by a counterplate arranged at a small distance opposite the layer. This is called sandwich configuration.

Developing solvents can be divided basically into three groups:

Group A solvents, for which pre-equilibration of the dry layer is undesirable. These are neat solvents and mixtures behaving like a one-component solvent. Examples are hexane–acetone 85:15, chloroform–ethyl acetate 1:1, and ethyl acetate–methanol 95:5. With these solvents, chromatograms are developed best in the sandwich configuration. Development in an unsaturated tank is an acceptable compromise. With group A solvents of very low polarity (such as aliphatic hydrocarbons and mixtures of aliphatic and aromatic hydrocarbons), close control of the relative humidity can be essential, whereas humidity has little influence with more polar solvents.

Group B solvents, which require pre-equilibration of the dry layer with the vapors of the solvent. These are solvents containing a volatile acid or base, or solvent mixtures containing a comparatively large amount of a polar component such as methanol or acetonitrile. For these the sandwich configuration is not suitable. Development in a paper-lined tank is an acceptable compromise. It is preferable, however, to develop these chromatograms after controlled (standardized) pre-equilibration of the dry layer.

Group C solvents, which need room for breathing. These chromatograms are best developed in the tank configuration without any equilibration with vapors. Group C solvents are sometimes difficult to distinguish from group B other than by trial and error. Solvent mixtures containing nonpolar and polar components in a proportion close to phase separation belong to this group. Typical examples are butanol–acetic acid–water mixtures.

Two-dimensional linear development with two different solvents at a right angle is suitable to increase peak capacity; however, only one sample can be chromatographed per plate and cochromatography with calibration or identification standards is restricted. Therefore, increasing peak capacity according to the automated multiple development method is the preferred way.

Chromatogram Development in a Tank

A classic flat-bottom tank permits the chromatogram to develop under conditions of partial or complete saturation of the tank atmosphere with solvent vapors. The degree of layer presaturation cannot be controlled unless additional expedients are used. Sandwich configuration can be established by means of a cover plate. Preconditioning the layer at a certain relative humidity, if required, has to be done in a separate device before inserting the plate.

A twin-trough chamber (Fig. 13.3) offers a variety of possibilities to control the vapor phase inside the chamber. It is therefore suitable for all developing techniques discussed.

In preparation for chromatography in the horizontal developing chamber (see Fig. 13.4), the samples are applied parallel to both opposing edges of the plate, which is then developed from both sides toward the middle. This way the number of samples per plate can be doubled. As can be seen from Fig. 13.4, development is possible in all configurations including conditioning at a desired relative humidity, which is established with the appropriate sulfuric acid–water mixture contained in the tray.

Figure 13.3 Twin-trough chamber. Only the trough in which the plate is placed must be filled with solvent when no equilibrium is intended. Standardized pre-equilibrium is possible with the plate placed in the empty trough, while solvent or any other conditioning liquid is contained in the other. Development is started by adding solvent to the trough with the plate.

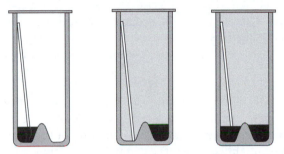

Automatic Chromatogram Development

The automatic developing chamber (ADC) provides instrumental control of chromatogram development. Preconditioning, tank or sandwich configuration, solvent migration distance, and the drying conditions are selectable.

Gradient TLC with Automated Multiple Development

The concept of automated multiple development (AMD) was derived by Burger (10). As with its predecessor technique, programmed multiple development (PMD) introduced by Perry and colleagues (11), the chromatogram is developed repeatedly in the same direction over increasing migration distances. Other than in the PMD process, the developing solvent for each successive run differs from the one used before, so that a stepwise gradient is obtained. Between runs, solvent is completely removed from the chamber and the layer is dried under vacuum.

Unlike a gradient in column liquid chromatography, an AMD gradient starts with the solvent having the strongest elution power. In the successive runs the solvent is varied toward decreasing elution power. In normal phase AMD it is the most polar solvent that is used over the shortest mi-

Figure 13.4 Horizontal developing chamber. (1) HPTLC plate with layer facing downward; (2) counterplate (inserted for development in sandwich configuration, removed for development with group B and C solvents and for humidity control); (3) troughs for developing solvent; (4) glass strips for solvent transfer by capillary action; (5) cover glass plate.

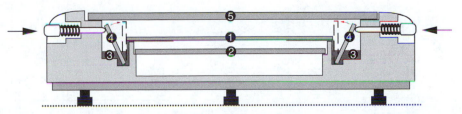

gration distance, the most nonpolar over the longest. Typical distance increments are 3 mm or less for a 20- or 25-step development.

Universal gradient is the term for an AMD gradient that starts with a very polar solvent and is varied via a solvent of medium polarity to a nonpolar solvent. Methanol or acetonitrile are typically used as the most polar component. The choice between these two depends on solubility consider-ations. The central or basis solvent, and to a certain extent the nonpolar solvent, determine selec-tivity. Solvents such as dichloromethane, diisopropyl ether, and t-butylmethyl ether are used as basis solvents in most normal phase AMD applications. n-Hexane is the most often used nonpolar component. There are two requirements for solvents used for the AMD process: They must be suitable for being dried off by vacuum, and the component solvents of a system should have similar migration characteristics (kappa values), so that migration increments can be controlled by time. (See Fig. 13.5.)

Provided all parameters and the solvent migration increments are properly maintained, which is possible only with a fully automatic system (12), the densitogram of a chromatogram track can be superimposed with a matched scale diagram of the gradient. In this way one can identify the sections of the gradient that caused resolution of the respective fractions, as well as the sections that must be modified to achieve complete resolution.

AMD chromatography using a universal gradient is suitable for the separation of complex samples differing widely in the polarity of their components. It is a remarkable feature of the AMD technique that the migration distance of the individual component is largely independent of the sample matrix. (See Fig 13.6.)

During the AMD procedure fractions are focused into narrow bands with a typical peak width of about 1 mm, so that peak capacities more than 50 over the usable separation distance of 80 mm can be achieved. AMD chromatography allows thin-layer chromatography to be used for tasks that could not be solved by TLC in the past.

Postchromatographic Derivatization

Substances not responding to visible or UV light after chromatography must be reacted with chro-mogenic or fluorogenic reagents. The same need exists when certain compounds or classes of compounds are to be identified by specific reactions. In these cases suitable reagents must be ap-plied to the chromatogram as evenly as possible.

Figure 13.5 Typical AMD universal gradient over 25 steps.

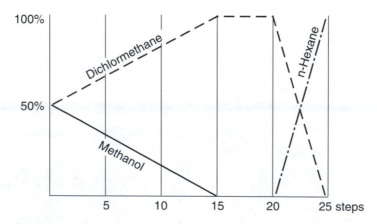

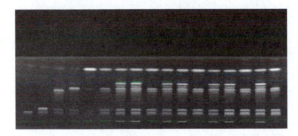

Figure 13.6 Typical AMD chromatogram. Phospholipids from cerebral extracts chromatographed with 25-step universal gradient; photographed under 366-nm UV after postchromatographic derivatization.

Reagent transfer via the gas phase is intrinsically uniform, but only few reagents such as HCl, sulfuryl chloride, and iodine are suitable. Liquid reagents can be applied by spraying or dipping. Of the two, dipping is generally preferred, with a few exceptions, but it must be mechanically controlled to ensure reproducible results. Reagent transfer by spraying cannot be avoided when two or more aqueous reagent solutions are to be used in sequence without intermediate drying, as in diazotation followed by coupling.

Any heating of the chromatogram necessary in the derivatization procedure is preferably performed on a flat, evenly heated surface (TLC plate heater) rather than in a laboratory oven.

Postchromatographic derivatization is the subject of a monograph by Jork and colleagues (13).

Densitometric Evaluation

General Methodological Considerations

For densitometric evaluation of a thin-layer chromatogram, its separation tracks are scanned with a light beam in the form of a slit selectable in length and width. Diffusely reflected light is measured by the photosensor. The difference between the optical signal from the sample-free background and that from a sample zone (fraction) is correlated with the amount of the respective fractions of calibration standards chromatographed on the same plate. Densitometric measurements can be made by absorbance or by fluorescence.

The majority of densitometric measurements of thin-layer chromatograms is carried out in the absorbance mode. The low UV range, under 300 nm down to 190 nm, is the most useful.

Due to light scattering at the particles of the layer, a simple, mathematically well-defined relationship between light signal and amount (concentration) of substance in the layer has not been found. A fair approximation for measurements on a particulate surface by absorbance gives the Kubelka–Munk equation (14), which can be derived for TLC. However, because at least one coefficient has to be determined empirically by trial and error, and today any integrator or software program can handle nonlinear functions, there is no need to linearize the relationship between concentration and optical response.

For scanning by fluorescence, the measured light intensity is directly proportional to the amount of the fluorescing substance. Measurements by fluorescence are therefore more sensitive by a factor of 10 to 1000, and calibration functions are often linear over a comparatively wide concentration range.

For these reasons, substances with inherent fluorescence are always scanned in this mode. For nonfluorescent substances one can perform a pre- or postchromatographic derivatization step to render them fluorescent, if such a reaction is available.

Instead of scanning a chromatogram track with a fixed slit, it is possible to move a light spot in a zigzag or meandering over the sample zones, whereby the swing corresponds to the length of the slit. Disadvantages of this type of scanning are the lower spatial resolution, particularly with HPTLC separation material, and the unfavorable error propagation, when data of sampling points from different positions are averaged.

Features of a Modern TLC Densitometer

Typical features are discussed with a view to the latest TLC densitometer, Scanner 3 (Fig. 13.7).

Two continuum lamps, deuterium and tungsten halogen, cover the spectral range of 190 to 800 nm. A third, high-pressure mercury vapor lamp provides sufficient energy for scanning by fluorescence. Lamps are selected and positioned automatically.

The monochromator has selectable bandwidth of 5 nm or 20 nm. For scanning at lower wavelengths (below 220 nm), the monochromator can be flushed with nitrogen to eliminate signal fluctuations caused by the formation of ozone from oxygen under the influence of short-wavelength UV.

The scanning speed is variable, to a maximum of 100 mm/s. Step resolution can be selected between 25 and 100 μm.

Figure 13.7 Light path diagram of TLC Scanner 3.

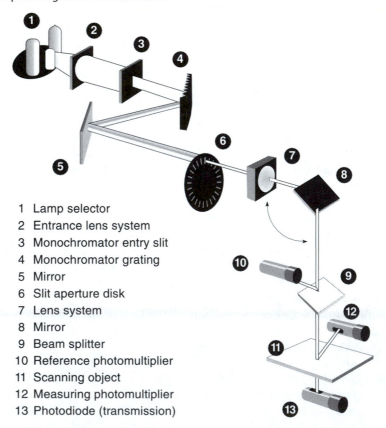

1 Lamp selector
2 Entrance lens system
3 Monochromator entry slit
4 Monochromator grating
5 Mirror
6 Slit aperture disk
7 Lens system
8 Mirror
9 Beam splitter
10 Reference photomultiplier
11 Scanning object
12 Measuring photomultiplier
13 Photodiode (transmission)

All functions of the scanner are controlled by a personal computer with RS 232 interface. The scanner transmits all measurement data in digital form to the computer for processing with the TLC-specific *CA*MAG *TLC* Software, or *CATS*.

Densitometric Chromatogram Evaluation

The functions are explained with reference to the TLC Scanner 3 with CATS software.

The sequence in the quantitative evaluation of a chromatogram is (1) raw data acquisition, (2) integration, (3) calibration and calculation of results, and (4) generation of the analysis report.

Raw data are sampled by scanning the chromatogram plate in the direction of chromatography, track by track. Integration is performed postrun from the raw data gathered during scanning, after all tracks of a chromatogram plate have been measured. The system automatically defines and corrects the baseline and sets fraction limits. The operator can accept these or override the automatic process by video integration.

In the calibration routine, peak data of the unknowns are related to those of the calibration standards. From several calibration functions the most suitable for the task is chosen. The selection is single-level calibration or multi-level calibration with linear or polynomial regression.

The software automatically stores and retrieves all data files. Each report generated contains the date and time of the last change, as well as a unique system-generated identification number for GMP/GLP recognition.

A modern TLC scanner with its dedicated TLC software can do more than quantifying substances on the plate. The following is a selection of capabilities.

Multiwavelength Scanning This term is often misinterpreted as scanning at selectable wavelength, which is a very basic feature of any densitometer fitted with a monochromator. Multiwavelength scanning means that raw data of up to 10 different wavelengths are acquired initially, and can be processed postrun in the integration and calibration routine separately (for example, each substance at the wavelength of its maximum absorbance). Another possibility is to recognize unknowns and identification standards by a multiwavelength plot. Dual-wavelength scanning, which may be used to eliminate matrix effects, is a simple variation of multiwavelength scanning.

Recording In-Situ Spectra Absorption spectra and, within certain restrictions, fluorescence excitation spectra can be measured. The stored spectral data are processed postrun for various purposes the operator selects:

- To determine the optimum wavelength for quantitative scanning
- To identify individual fractions by comparison with spectra of authentic standards cochromatographed on the same plate or stored in a spectrum library
- To check identity by superimposing the spectra of all equidistantly migrated fractions of a chromatogram
- To check the purity of fractions by superimposing the spectra from different positions within a spot

Quantification by Image Processing Quantitative chromatogram evaluation via image processing has been postulated for over 15 years. Several home-built and commercial systems have been described (15–17).

Although favorable results have been reported, it is not expected that image processing will seriously rival scanning densitometers of the kind described above with respect to versatility and accuracy. The strength of densitometric TLC evaluation lies in the selective use of ultraviolet light down to 190 nm, with excellent monochromaticity and including the possibility of recording in-situ spectra. The main limitation of image processing is that no solution has been found for illuminating a TLC plate uniformly with monochromatic light of selectable wavelength comparable to that of a densitometer.

Nevertheless, quantitative TLC evaluation via image processing will have its place because results should be as accurate as required, not as accurate as possible. Whether results are comparable with those obtained with a densitometer depends on how suitable the kind of illumination available in the image device is for the given substances.

What It Does

TLC provides a simple, straightforward means of separating compounds in a mixture either quantitatively or qualitatively with little or no sample preparation. All fractions are stored on the plate for evaluations or postchromatographic derivatization.

Analytical Information

Qualitative

TLC provides a quick, low-cost means of getting a result or answer about a mixture or single compound's integrity. Product uniformity, impurity checks, and reaction progress are often qualitative determinations but with proper technique and instrumentation, can produce excellent quantitative results. Screening for the presence of compounds is a traditional qualitative TLC procedure.

Qualitative results are usually photodocumented with the current trend toward video use.

A different category of qualitative information is substance identification by in-situ spectroscopy. This is accessible via densitometry and discussed in the How It Works section.

Quantitative

One of the basic themes of this chapter is that TLC can yield reproducible qualitative results (1 to 3% RSD) comparable to HPLC. However, a good method, good technique, and proper instrumentation are all required.

Of course, for the technique to be quantitative there must be some means to compare known amounts to unknowns directly from the chromatogram. For reproducible results a densitometer is required, preferably a very versatile one because TLC has such a broad range of applications.

Applications

Applications of planar chromatography range from biochemistry/life sciences, environmental analysis, food analysis, pharmaceuticals, and natural products to toxicology, forensic analysis, and doping control. Substances successfully identified or quantified include amino acids, peptides, pharmaceuticals, phenols, indoles, purines, steroids, synthetic and natural dyes, and vitamins. A particular asset of TLC is its suitability for trace and ultratrace analyses.

Instrument manufacturers provide useful application references. For example, CAMAG Scientific Inc. provides the CAMAG Bibliography Service (CBS), which abstracts articles on TLC. It is produced twice a year and is divided into 38 classifications. Every 5 years past issues are combined into a comprehensive reference book called the Cumulative CAMAG Bibliography Service (CCBS). Additionally, one may consult a book such as the *Handbook of TLC* (18) for application information.

Forced Flow Development.

Several techniques have been reported to accelerate the flow of the mobile phase through a planar bed. Tyihàk et al. (19) described a system in which a planar layer, sealed at its edges by impregnation, is sandwiched between a rigid plate and a flexible membrane under external pressure up to 25 bar. The developing solvent is pumped through the layer. A state-of-the-art overview of the technique, now called overpressure layer chromatography (OPLC) is given by Ferenczi-Fodor, Mincsovics, and Tyihàk in a chapter of the *Handbook of TLC* (18). OPLC is the only forced-flow planar chromatography technique that has gained recognizable but limited acceptance in practice.

Under the term high-pressure planar liquid chromatography (HPPLC), Kaiser (20) describes a technique using up to 200 bar pressure, operating in the circular developing mode. It allows in-flow sample injection and, accordingly, simulation of HPLC techniques as well as coupling HPLC and planar chromatography.

Accelerating the flow of the mobile phase through a planar bed by centrifugal forces is described by Nyiredy and others (21). Although analytical applications have been discussed as well, the main merit of centrifugal planar chromatography seems to lie in its application for preparative purposes.

Thin-layer Radiochromatography.

The separation and measurement of materials labeled with radioisotopes on TLC plates is a widely used method for determining their distribution in a reaction mixture (for example, for following the metabolic pathway of drugs or pesticides).

Radioactive separation zones on a TLC plate can be detected and to a certain extent quantified by autoradiography or by radioscanners. Radioscanners are based on various principles, spark chambers, mechanically driven gas-flow Geiger counters, or imaging-proportional counter scanners. A state-of-the-art overview of thin-layer radiochromatography is given by Clark and Klein in a chapter of the *Handbook of TLC* (18).

Photo and Video Documentation.

Photo documentation of thin-layer chromatograms under visible light, short-wave UV (254 nm), and long-wave UV (366 nm) has been common technique for about 30 years. Standard as well as instant cameras have been used for this purpose. Currently, photography is more and more replaced by video technology. A state-of-the-art overview of photo and video documentation of thin-layer chromatograms is given by Kovar and Morlock in a chapter of the *Handbook of TLC* (18).

Direct Coupling of HPLC and TLC.

Combining different separation techniques governed by different mechanisms into a multidimensional method can increase the potential of the individual techniques by an order of magnitude (22). HPLC is one of the most powerful separation techniques available today for nonvolatile substances. For reasons mentioned above, HPLC most often uses the reversed phase separation mode. On-line coupling of HPLC with AMD using normal phase chromatography results in peak capacities around 500.

A device for mass transfer HPLC/TLC basically consists of a sample spray-on device (modified Linomat or modified Automatic TLC Sampler), which is connected to the column outlet. The maximum quantity of liquid that can be sprayed on a silica layer without washing it away is 10 to 60 μL/min depending on the mobile phase. Therefore, a microbore HPLC unit is the most appropriate for this technique because it avoids the need for a splitter, which would cause complications and other drawbacks such as problems with detection limits.

FTIR Evaluation of Thin-layer Chromatograms.

The combination of TLC with FTIR in-situ evaluation is a useful method, particularly for the identification of complex mixtures and their constituents. Although determination limits are higher than those for UV spectroscopy, the method can be useful for the quantification of substances with no suitable UV response or no access to derivatization. An overview of TLC/FTIR evaluation is given by Kovar (23).

Combination of TLC and Raman Spectroscopy.

Regular Raman spectroscopy can be used for the identification of substances on a TLC plate (24). Koglin (25) has shown that surface-enhanced Raman spectroscopy (SERS) is suitable for the in-situ measurement of compounds in the picogram range. Merits and limitations of regular Raman spectroscopy versus SERS are that in-situ Raman generates spectra that are identical to published Raman spectra measured on solids. Detection limits are comparatively high (in the range of 0.5 to 5 μg per fraction). This means that in-situ Raman spectroscopy is suitable for the identification of totally unknown substances, by reference to a Raman atlas. For SERS evaluation, the layer must be treated with a colloidal silver solution. It is then excited with Ar-ion or He–Ne laser. Spectra are compared to user-measured SERS spectra from a similar layer. The method is very sensitive and is suitable for the identification of a selected number of substances suspected to be present.

Combination of TLC with Mass Spectroscopy.

For the in-situ identification of compounds separated on a TLC plate, molecules must be desorbed from the layer, then ionized and transferred to the mass spectroscopy (MS) cell. Ionization methods used are fast atom bombardment (FAB), liquid secondary ion mass spectrometry (liquid SIMS), time-of-flight SIMS (TOF-SIMS), or laser ionization (LI). Limitation of the interesting TLC/MS combination is that sheets or plates must be cut into small strips or pieces, as there is no MS instrument available that takes 10- × 10-cm or 20- × 10-cm plates. An overview of TLC/MS is given by Busch in a chapter of the *Handbook of TLC* (18).

Nuts and Bolts

Relative Costs

The cost of a complete TLC system ranges from $30,000 to $100,000. The basic equipment needed for modern (quantitative) TLC is a densitometer with computerized data system, which costs about $30K. However, one densitometer workstation can serve several TLC wet stations. Equipment for sample application, chromatogram development, and optional procedures can range from $3K to $70K, depending on desired sophistication and degree of automation. Consumables are inexpensive on a per sample basis.

Vendors for Instruments and Accessories

I = instrument manufacturer; C = consumables supplier.

Alltech Associates Inc. (C)
2051 Waukegan Rd.
Deerfield, IL 60015
phone: 708-948-8600, 800-255-8324
fax: 708-948-1078
email: 73554.3372@compuserve.com
Internet: http://www.alltechweb.com/

Analtech Inc. (I & C)
75 Blue Hen Dr., P.O. Box 7558
Newark, DE 19714
phone: 302-737-6960, 800-441-7540
fax: 302-737-7115
email: analtech@ analtech.com
Internet: http://www. analtech.com/

CAMAG Scientific, Inc. (I & C)
515 Cornelius Harnett Drive
Wilmington, NC 28401
phone: 800-334-3909
fax: 919-343-1834

CAMAG/Switzerland (I & C)
Sonnenmattstrasse 11 CH-4132
Muttenz 1 Switzerland
phone: ++41 61 467 3434
fax: ++41 61 461 0702

Shimadzu Scientific Instruments, Inc. (I & C)
7102 Riverwood Dr.
Columbia, MD 21046
phone: 800-477-1227
fax: 410-381-1222
Internet: http://www.shimadzu.com

Supelco, Inc. (C)
Supelco Park
Bellefonte, PA 16823-0048
phone: 814-359-3441, 800-247-6628
fax: 814-359-3044, 800-447-3044
email: Supelco@Supelco.sial.com
Internet: http://www.Supelco.sial.com/Supelco.html

Whatman Inc. (C)
6 Just Rd.
Fairfield, NJ 07004
phone: 201-882-9277, 800-922-0361
fax: 201-882-5134
Internet: http://www.whatman.com

Required Level of Training

Basic TLC work can be performed by lab technicians after a brief introduction and by following
the instructions supplied with the equipment. The leading manufacturers and distributors of sophisticated instrumentation usually offer training at the time of installation as well as customer
training courses. These courses can be at the manufacturer's or at neutral sites. They usually vary
in both level of instruction and duration. Laboratory personnel with a good background in any
kind of chromatography acquire skill and knowledge of modern TLC quickly.

Service and Maintenance

Generally for the less complex equipment maintenance is very simple and service is usually handled on an exchange or loan basis. For the more complex instruments such as the ones discussed
in this chapter, preventive maintenance agreements are available as well as on-site and off-site service. These instruments, under PC/software control, have self-validating programs. However,
each supplier should be consulted regarding policy.

Suggested Readings

Fundamentals of Thin Layer Chromatography (planar chromatography). Heidelberg, Basel, New York: Dr. A.
    Hüthig Verlag, 1987.

Handbook of TLC, 2nd ed. New York: Marcel Dekker, 1995.

References

1. J. G. Kirchner, J. M. Miller, and G. J. Keller, *Anal. Chem.*, 23 (1951), 420.

2. E. Stahl and others, *Pharmazie*, 11 (1956), 633.

3. H. Halpaap and J. Ripphahn, *Merck Kontakte, Darmstadt, Heft*, 3 (1976), 16.

4. U. A. T. Brinkman, in H. Traitler, R. Studer, and R. E. Kaiser, eds., *Proceedings of the Fourth International Symposium on Instrumental HPTLC (Selvino)* (Bad Duerkheim, Germany: Institute for Chromatography, 1987), 81–92.

5. R. E. Kaiser, *Einführung in die Hochleistungs-Dünnschicht-Chromatographie* (Bad Duerkheim, Germany: Institute for Chromatography, 1976).

6. D. E. Jaenchen, *International Analyst* (May 1987), 36–40.

7. D. E. Jaenchen, *GIT Suppl. Chromatogr.*, (April, 1985), 19–21.

8. M. E. Morgan, *J. Chromatogr.*, 9 (1962), 379–81.

9. R. Mueller and H. Krueger, *GIT Fachz. Lab.*, 17 (1973), 197–203.

10. K. Burger, *Fresenius Z. Anal. Chem.*, 318 (1984), 228–33.

11. J. A. Perry, K. W. Haag, and L. J. Glunz, *J. Chromatog. Sci.*, 11 (1973), 447.

12. D. E. Jaenchen, in R. E. Kaiser, ed., *Proceedings of the Third International Symposium on Instrumental HPTLC (Wuerzburg)* (Bad Duerkheim, Germany: Institute for Chromatography, 1985), 71–82.

13. H. Jork and others, *Thin-Layer Chromatography, Reagents and Detection Methods* (New York: VCH Publishers, 1990 (Volume 1a) and 1994 (Volume 1b)).

14. P. Kubelka and F. Munk, *Z. Techn. Phys.*, 12 (1931), 593.

15. M. Prosek and others, *Computer Anw. Lab.*, 4 (1984), 249.

16. R. M. Belchamber, H. Read, and J. D. M. Roberts, in R. E. Kaiser, ed., *Planar Chromatography*, vol. 1 (New York: Huethig Verlag, 1986), 207–20.

17. Analtech Inc., Newark, DE 19417, Product Bulletin (1990).

18. J. Sherma and B. Fried, eds., *Handbook of TLC*, 2nd ed. (New York: Marcel Dekker, 1995).

19. E. Tyihàk, E. Mincsovics, and H. Kalàsz, *J. Chromatogr.*, 174 (1979), 75–82.

20. R. E. Kaiser, *Einführung in die HPPLC* (New York: Huethig Verlag, 1987).

21. S. Nyiredy and others, *GIT Suppl. Chromatogr.*, (April, 1985), 24.

22. K. Burger, in R. E. Kaiser, ed., *Proceedings of the Seventh International Symposium on Planar Chromatography (Brighton)* (Bad Duerkheim, Germany: Institute for Chromatography, 1989) 33–44.

23. O. R. Frey, K-A. Kovar, V. Hoffmann, *JPC*, 6 (1993), 93–9.

24. C. Petty, *Spectrochemical Acta*, 49 (1993), 645–55.

25. E. Koglin, *GIT Fachz. Lab.*, 38 (1994), 627–32.

Qualitative Optical Spectroscopic Methods

Chapter 14

Introduction

Paul Bouis

Mallinckrodt, Inc.
Mallinckrodt Baker Division
Research and Product Development

Welcome to the world of qualitative spectroscopic instrumental analysis. The term *qualitative* makes it sound like a world of second-rate techniques. This section demonstrates that this is not the case, and that these techniques are indispensable tools in the arsenal of any person involved with analyzing samples. The techniques covered range from infrared (IR), Raman, and nuclear magnetic resonance (NMR) spectroscopy, most often used in studying organic molecules, to X-ray diffraction, a technique used in the identification of asbestos fibers in the environment, for example. A modern laboratory is usually equipped with an IR, and might even have an NMR, but few labs own Raman or X-ray diffraction systems. Infrared analysis is probably the most widely used industrial technique for identification of raw materials, normally done by comparison to a known spectrum of the material. Raman, NMR, and X-ray diffraction are still expert techniques, usually requiring skilled practitioners for interpretation of results. Infrared analysis is just as likely to be used on-line in a process-monitoring role as off-line in a more traditional laboratory environment. Although you will read about on-line process applications for some of the other techniques in the chapters ahead, their use is limited due to their cost and complexity. Despite their label as qualitative analysis tools, all four of these techniques are routinely used in quantitative analysis. Usually this is because in special circumstances, they are the best technique available for a very specific analytical requirement. The use of infrared analysis to quantify the amount of oil and grease in environmental samples is a classic example of this type of application.

The relentless improvement both in quantitative and qualitative analytical chemistry has been fueled by the computer. Experimental techniques that in the past were rarely used by the average practitioner are now used routinely thanks to the affordability of computing power. NMR and IR were both revitalized and revolutionized by the advent of the widespread use of Fourier transform techniques made possible by this computing power. Not only has software influenced improvements in analytical instrumentation, but new hardware has also played a large role. If Raman spectroscopy is to become more available to the analytical chemist, for example, it will be due in large part to the advent of highly sensitive yet affordable semiconductor detectors.

This section begins with a chapter on infrared spectroscopy and is followed by a chapter on its complimentary technique, Raman spectroscopy. Both methods are based on the fact that within any molecule the atoms vibrate with a few definite, sharply defined frequencies characteristic of that molecule. When a sample is placed in a beam of such radiation it absorbs energy at frequencies characteristic of the molecule and transmits or scatters all other frequencies. The resulting spectra are probably the most characteristic physical properties of an organic compound. Raman shifts, as the bands were traditionally called, are a discrete spectrum of lines scattered toward longer wavelengths. In general, no two compounds have identical infrared or Raman spectra. Infrared spectroscopy is very often the surest and most rapid method for identification of organic compounds, in the case of both single compounds and mixtures. It has today almost replaced classic chemical methods in this application. Infrared spectroscopy is often used to follow new and previously unknown reactions. Functional groups can nearly always be identified, even in the abscence of known standards. The infrared spectrum not only identifies a functional group, but also indicates its chemical environment, and in some cases can even be used to predict its chemical behavior. Impurities can often be detected and identified when present in amounts as low as 1%. It is not generally known that Raman instruments actually preceded infrared, but Raman faded from use until the development of the continuous-wave gas laser as a Raman source. Although Raman is inherently less sensitive than infrared, it does not suffer from the interferences caused by water in a sample and offers a different yet complementary view. The similarities and differences between the techniques are covered in detail in the chapter on Raman spectroscopy, along with special applications that take advantage of this unique method.

In Chapter 17, NMR spectroscopy is covered. Structures of most new organic compounds are verified routinely by high-resolution proton and carbon magnetic resonance spectroscopy. Of the nuclei observed, it is often possible to deduce completely their number and distribution by chemical type. Reactions can be monitored for the disappearance of starting materials and the appearance of intermediates and products. In complex problems the combination of NMR with other techniques—especially IR and mass spectroscopy—is used routinely to more efficiently obtain the maximum amount of useful information. NMR was thought to be a mature technique just 20 years ago. But as you will see in Chapter 17, advances such as the ability to run two-dimensional experiments, use magnetic fields of 300 Mhz and greater, and even examine the structure of solids, have led to new and more astonishing applications and techniques. As a result NMR will certainly play an increasingly vital part in the changing role of analytical chemistry in such rapidly evolving areas as the elucidation of the complex structures and behaviors of biomolecules in medicine and biochemistry. The complex hardware, software, and experimental techniques necessary to make such information available to the scientist are explained in detail in Chapter 17.

The Noble prize in chemistry was awarded in 1985 for work on the structural analysis of biomolecules carried out in the 1950s using a technique discovered in the early 1900s: X-ray diffraction. When a randomly oriented aggregate of small crystal fragments (or even single crystals) is irradiated with a monochromatic beam of X-rays, the various planes of atoms diffract the X-ray beam at angles determined by the spacing between the planes. The resulting diffraction pattern is characteristic of the material irradiated. The technique is used to identify crystaline compounds in the pure state or even in mixtures. The several polymorphic forms of a compound can be distinguished, such as the three forms of silica, quartz, tridymite, and cristobalite. Each distinct hydrate of a compound has a different crystal structure and therefore gives a different X-ray pattern. The method is particularly applicable to the determination of the extent to which a polymer has crystallized. Single-crystal diffraction gives information such as unequivocal proof of structure, bond distances and angles, thermal vibration parameters, intermolecular contacts, molecular packing, and absolute configuration of optically active molecules. Today, well-equipped laboratories use this technique as a standalone method or as a complementary technique to other structural analysis tools such as high-field NMR. In Chapter 18 you will learn about some of the spectacular advances that have made this technique an essential tool for the scientist faced with ever more difficult molecular structural analysis.

When combined with the advances in computers and the continued miniaturization of hardware that will inevitably occur, the qualitative analytical techniques covered in this section could evolve into the techniques that launch the next essential applications in analytical chemistry. Envision this possibility as you read the chapters in this section.

Chapter 15

Infrared Spectroscopy

C.-P. Sherman Hsu, Ph.D.

Separation Sciences
Research and Product Development
Mallinckrodt, Inc.
Mallinckrodt Baker Division

Summary

General Uses

- Identification of all types of organic and many types of inorganic compounds
- Determination of functional groups in organic materials
- Determination of the molecular composition of surfaces
- Identification of chromatographic effluents
- Quantitative determination of compounds in mixtures
- Nondestructive method
- Determination of molecular conformation (structural isomers) and stereochemistry (geometrical isomers)
- Determination of molecular orientation (polymers and solutions)

Common Applications

- Identification of compounds by matching spectrum of unknown compound with reference spectrum (fingerprinting)
- Identification of functional groups in unknown substances

- Identification of reaction components and kinetic studies of reactions
- Identification of molecular orientation in polymer films
- Detection of molecular impurities or additives present in amounts of 1% and in some cases as low as 0.01%
- Identification of polymers, plastics, and resins
- Analysis of formulations such as insecticides and copolymers

Samples

State

Almost any solid, liquid or gas sample can be analyzed. Many sampling accessories are available.

Amount

Solids 50 to 200 mg is desirable, but 10 µg ground with transparent matrix (such as KBr) is the minimum for qualitative determinations; 1 to 10 µg minimum is required if solid is soluble in suitable solvent.

Liquids 0.5 µL is needed if neat, less if pure.

Gases 50 ppb is needed.

Preparation

Little or no preparation is required; may have to grind solid into KBr matrix or dissolve sample in a suitable solvent (CCl_4 and CS_2 are preferred). Many types of sample holders and cells are available. Water should be removed from sample if possible.

Analysis Time

Estimated time to obtain spectrum from a routine sample varies from 1 to 10 min depending on the type of instrument and the resolution required. Most samples can be prepared for infrared (IR) analysis in approximately 1 to 5 min.

Limitations

General

- Minimal elemental information is given for most samples.
- Background solvent or solid matrix must be relatively transparent in the spectral region of interest.
- Molecule must be active in the IR region. (When exposed to IR radiation, a minimum of one vibrational motion must alter the net dipole moment of the molecule in order for absorption to be observed.)

Accuracy

In analysis of mixtures under favorable conditions, accuracy is greater than 1%. In routine analyses, it is ± 5%.

Sensitivity and Detection Limits

Routine is 2%; under most favorable conditions and special techniques, it is 0.01%.

Complementary or Related Techniques

- Nuclear magnetic resonance provides additional information on detailed molecular structure
- Mass spectrometry provides molecular mass information and additional structural information
- Raman spectroscopy provides complementary information on molecular vibration. (Some vibrational modes of motion are IR-inactive but Raman-active and vice versa.) It also facilitates analysis of aqueous samples. Cell window material may be regular glass.

Introduction

Infrared (IR) spectroscopy is one of the most common spectroscopic techniques used by organic and inorganic chemists. Simply, it is the absorption measurement of different IR frequencies by a sample positioned in the path of an IR beam. The main goal of IR spectroscopic analysis is to determine the chemical functional groups in the sample. Different functional groups absorb characteristic frequencies of IR radiation. Using various sampling accessories, IR spectrometers can accept a wide range of sample types such as gases, liquids, and solids. Thus, IR spectroscopy is an important and popular tool for structural elucidation and compound identification.

IR Frequency Range and Spectrum Presentation

Infrared radiation spans a section of the electromagnetic spectrum having wavenumbers from roughly 13,000 to 10 cm^{-1}, or wavelengths from 0.78 to 1000 μm. It is bound by the red end of the visible region at high frequencies and the microwave region at low frequencies.

IR absorption positions are generally presented as either wavenumbers ($\bar{v}$) or wavelengths (λ). Wavenumber defines the number of waves per unit length. Thus, wavenumbers are directly proportional to frequency, as well as the energy of the IR absorption. The wavenumber unit (cm^{-1}, reciprocal centimeter) is more commonly used in modern IR instruments that are linear in the cm^{-1} scale. In the contrast, wavelengths are inversely proportional to frequencies and their associated energy. At present, the recommended unit of wavelength is μm (micrometers), but μ (micron) is used in some older literature. Wavenumbers and wavelengths can be interconverted using the following equation:

$$\bar{v} \text{ (in cm}^{-1}) = \frac{1}{\lambda \text{ (in μm)}} \times 10^4 \tag{15.1}$$

IR absorption information is generally presented in the form of a spectrum with wavelength or wavenumber as the x-axis and absorption intensity or percent transmittance as the y-axis (Fig. 15.1). Transmittance, T, is the ratio of radiant power transmitted by the sample (I) to the radiant power incident on the sample (I_0). Absorbance (A) is the logarithm to the base 10 of the reciprocal of the transmittance (T).

$$A = \log_{10}(1/T) = -\log_{10}T = -\log_{10}I/I_0 \tag{15.2}$$

The transmittance spectra provide better contrast between intensities of strong and weak bands because transmittance ranges from 0 to 100% T whereas absorbance ranges from infinity to zero. The analyst should be aware that the same sample will give quite different profiles for the IR spectrum, which is linear in wavenumber, and the IR plot, which is linear in wavelength. It will appear as if some IR bands have been contracted or expanded.

The IR region is commonly divided into three smaller areas: near IR, mid IR, and far IR.

Figure 15.1 IR spectra of polystyrene film with different x-axis units. (a) Linear in wavenumber (cm⁻¹), (b) linear in wavelength (μm). *(Reprinted from R. M. Silverstein, G. C. Bassler, and T. C. Morrill, Spectrometric Identification of Organic Compounds, 4th edition. New York: John Wiley & Sons, 1981, p. 166, by permission of John Wiley & Sons, Inc., copyright © 1981.)*

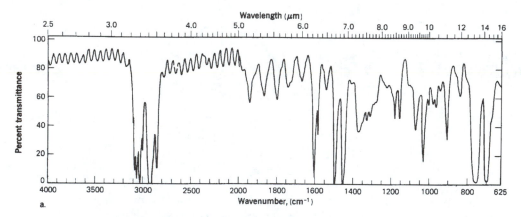

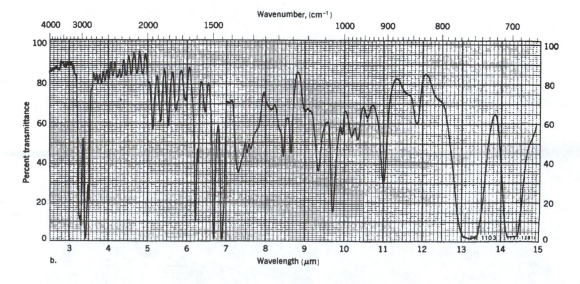

	Near IR	Mid IR	Far IR
Wavenumber	13,000–4,000 cm^{-1}	4,000–200 cm^{-1}	200–10 cm^{-1}
Wavelength	0.78–2.5 µm	2.5–50 µm	50–1,000 µm

This chapter focuses on the most frequently used mid IR region, between 4000 and 400 cm^{-1} (2.5 to 25 µm). The far IR requires the use of specialized optical materials and sources. It is used for analysis of organic, inorganic, and organometallic compounds involving heavy atoms (mass number over 19). It provides useful information to structural studies such as conformation and lattice dynamics of samples. Near IR spectroscopy needs minimal or no sample preparation. It offers high-speed quantitative analysis without consumption or destruction of the sample. Its instruments can often be combined with UV-visible spectrometer and coupled with fiberoptic devices for remote analysis. Near IR spectroscopy has gained increased interest, especially in process control applications.

Theory of Infrared Absorption

At temperatures above absolute zero, all the atoms in molecules are in continuous vibration with respect to each other. When the frequency of a specific vibration is equal to the frequency of the IR radiation directed on the molecule, the molecule absorbs the radiation.

Each atom has three degrees of freedom, corresponding to motions along any of the three Cartesian coordinate axes (x, y, z). A polyatomic molecule of n atoms has $3n$ total degrees of freedom. However, 3 degrees of freedom are required to describe translation, the motion of the entire molecule through space. Additionally, 3 degrees of freedom correspond to rotation of the entire molecule. Therefore, the remaining $3n - 6$ degrees of freedom are true, fundamental vibrations for nonlinear molecules. Linear molecules possess $3n - 5$ fundamental vibrational modes because only 2 degrees of freedom are sufficient to describe rotation. Among the $3n - 6$ or $3n - 5$ fundamental vibrations (also known as normal modes of vibration), those that produce a net change in the dipole moment may result in an IR activity and those that give polarizability changes may give rise to Raman activity. Naturally, some vibrations can be both IR- and Raman-active.

The total number of observed absorption bands is generally different from the total number of fundamental vibrations. It is reduced because some modes are not IR active and a single frequency can cause more than one mode of motion to occur. Conversely, additional bands are generated by the appearance of overtones (integral multiples of the fundamental absorption frequencies), combinations of fundamental frequencies, differences of fundamental frequencies, coupling interactions of two fundamental absorption frequencies, and coupling interactions between fundamental vibrations and overtones or combination bands (Fermi resonance). The intensities of overtone, combination, and difference bands are less than those of the fundamental bands. The combination and blending of all the factors thus create a unique IR spectrum for each compound.

The major types of molecular vibrations are stretching and bending. The various types of vibrations are illustrated in Fig. 15.2. Infrared radiation is absorbed and the associated energy is converted into these type of motions. The absorption involves discrete, quantized energy levels. However, the individual vibrational motion is usually accompanied by other rotational motions. These combinations lead to the absorption bands, not the discrete lines, commonly observed in the mid IR region.

Figure 15.2 Major vibrational modes for a nonlinear group, CH_2. (+ indicates motion from the plane of page toward reader; – indicates motion from the plane of page away from reader.) *(Reprinted from R. M. Silverstein, G. C. Bassler, and T. C. Morrill, Spectrometric Identification of Organic Compounds, 4th edition. New York: John Wiley & Sons, 1981, p. 166, by permission of John Wiley & Sons, Inc., copyright © 1981.)*

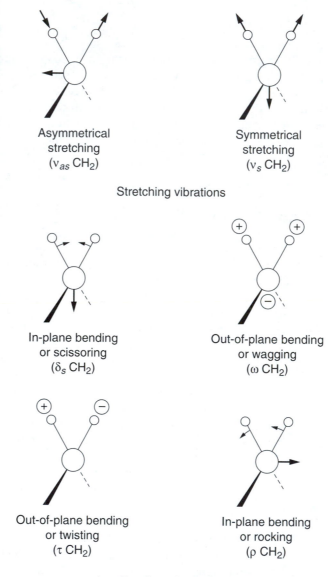

Asymmetrical
stretching
(ν_{as} CH_2)

Symmetrical
stretching
(ν_s CH_2)

Stretching vibrations

In-plane bending
or scissoring
(δ_s CH_2)

Out-of-plane bending
or wagging
(ω CH_2)

Out-of-plane bending
or twisting
(τ CH_2)

In-plane bending
or rocking
(ρ CH_2)

Bending vibrations

How It Works

In simple terms, IR spectra are obtained by detecting changes in transmittance (or absorption) intensity as a function of frequency. Most commercial instruments separate and measure IR radiation using dispersive spectrometers or Fourier transform spectrometers.

Dispersive Spectrometers

Dispersive spectrometers, introduced in the mid-1940s and widely used since, provided the robust instrumentation required for the extensive application of this technique.

Spectrometer Components

An IR spectrometer consists of three basic components: radiation source, monochromator, and detector. A schematic diagram of a typical dispersive spectrometer is shown in Fig. 15.3.

The common radiation source for the IR spectrometer is an inert solid heated electrically to 1000 to 1800 °C. Three popular types of sources are Nernst glower (constructed of rare-earth oxides), Globar (constructed of silicon carbide), and Nichrome coil. They all produce continuous radiations, but with different radiation energy profiles.

The monochromator is a device used to disperse a broad spectrum of radiation and provide a continuous calibrated series of electromagnetic energy bands of determinable wavelength or frequency range. Prisms or gratings are the dispersive components used in conjunction with variable-slit mechanisms, mirrors, and filters. For example, a grating rotates to focus a narrow band of frequencies on a mechanical slit. Narrower slits enable the instrument to better distinguish more closely spaced frequencies of radiation, resulting in better resolution. Wider slits allow more light to reach the detector and provide better system sensitivity. Thus, certain compromise is exercised in setting the desired slit width.

Most detectors used in dispersive IR spectrometers can be categorized into two classes: thermal detectors and photon detectors. Thermal detectors include thermocouples, thermistors, and

Figure 15.3 Schematic diagram of a commercial dispersive IR instrument, the Perkin-Elmer Model 237B Infrared Spectrometer. *(Reprinted by permission of the Perkin-Elmer Corporation.)*

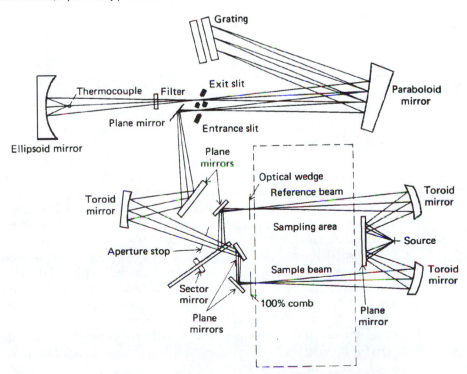

pneumatic devices (Golay detectors). They measure the heating effect produced by infrared radiation. A variety of physical property changes are quantitatively determined: expansion of a nonabsorbing gas (Golay detector), electrical resistance (thermistor), and voltage at junction of dissimilar metals (thermocouple). Photon detectors rely on the interaction of IR radiation and a semiconductor material. Nonconducting electrons are excited to a conducting state. Thus, a small current or voltage can be generated. Thermal detectors provide a linear response over a wide range of frequencies but exhibit slower response times and lower sensitivities than photon detectors.

Spectrometer Design

In a typical dispersive IR spectrometer, radiation from a broad-band source passes through the sample and is dispersed by a monochromator into component frequencies (Fig. 15.3). Then the beams fall on the detector, which generates an electrical signal and results in a recorder response.

Most dispersive spectrometers have a double-beam design. Two equivalent beams from the same source pass through the sample and reference chambers respectively. Using an optical chopper (such as a sector mirror), the reference and sample beams are alternately focused on the detector. Commonly, the change of IR radiation intensity due to absorption by the sample is detected as an off-null signal that is translated into the recorder response through the actions of synchronous motors.

Fourier Transform Spectrometers

Fourier transform spectrometers have recently replaced dispersive instruments for most applications due to their superior speed and sensitivity. They have greatly extended the capabilities of infrared spectroscopy and have been applied to many areas that are very difficult or nearly impossible to analyze by dispersive instruments. Instead of viewing each component frequency sequentially, as in a dispersive IR spectrometer, all frequencies are examined simultaneously in Fourier transform infrared (FTIR) spectroscopy.

Spectrometer Components

There are three basic spectrometer components in an FT system: radiation source, interferometer, and detector. A simplified optical layout of a typical FTIR spectrometer is illustrated in Fig. 15.4.

The same types of radiation sources are used for both dispersive and Fourier transform spectrometers. However, the source is more often water-cooled in FTIR instruments to provide better power and stability.

In contrast, a completely different approach is taken in an FTIR spectrometer to differentiate and measure the absorption at component frequencies. The monochromator is replaced by an interferometer, which divides radiant beams, generates an optical path difference between the beams, then recombines them in order to produce repetitive interference signals measured as a function of optical path difference by a detector. As its name implies, the interferometer produces interference signals, which contain infrared spectral information generated after passing through a sample.

The most commonly used interferometer is a Michelson interferometer. It consists of three active components: a moving mirror, a fixed mirror, and a beamsplitter (Fig. 15.4). The two mirrors are perpendicular to each other. The beamsplitter is a semireflecting device and is often made by depositing a thin film of germanium onto a flat KBr substrate. Radiation from the broadband IR source is collimated and directed into the interferometer, and impinges on the beamsplitter. At the beamsplitter, half the IR beam is transmitted to the fixed mirror and the remaining half is reflected to the moving mirror. After the divided beams are reflected from the two mirrors, they are

Figure 15.4 Simplified optical layout of a typical FTIR spectrometer. *(Reprinted by permission of Nicolet Instrument Corporation.)*

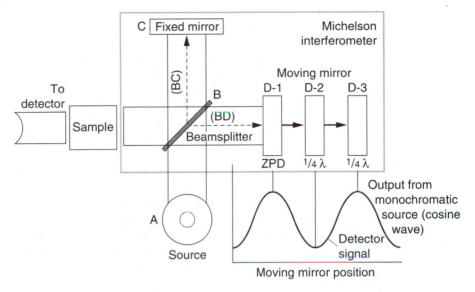

recombined at the beamsplitter. Due to changes in the relative position of the moving mirror to the fixed mirror, an interference pattern is generated. The resulting beam then passes through the sample and is eventually focused on the detector.

For an easier explanation, the detector response for a single-frequency component from the IR source is first considered. This simulates an idealized situation where the source is monochromatic, such as a laser source. As previously described, differences in the optical paths between the two split beams are created by varying the relative position of moving mirror to the fixed mirror. If the two arms of the interferometer are of equal length, the two split beams travel through the exact same path length. The two beams are totally in phase with each other; thus, they interfere constructively and lead to a maximum in the detector response. This position of the moving mirror is called the point of zero path difference (ZPD). When the moving mirror travels in either direction by the distance $\lambda/4$, the optical path (beamsplitter–mirror–beamsplitter) is changed by 2 $(\lambda/4)$, or $\lambda/2$. The two beams are 180° out of phase with each other, and thus interfere destructively. As the moving mirror travels another $\lambda/4$, the optical path difference is now 2 $(\lambda/2)$, or λ. The two beams are again in phase with each other and result in another constructive interference.

When the mirror is moved at a constant velocity, the intensity of radiation reaching the detector varies in a sinusoidal manner to produce the interferogram output shown in Fig. 15.4. The interferogram is the record of the interference signal. It is actually a time domain spectrum and records the detector response changes versus time within the mirror scan. If the sample happens to absorb at this frequency, the amplitude of the sinusoidal wave is reduced by an amount proportional to the amount of sample in the beam.

Extension of the same process to three component frequencies results in a more complex interferogram, which is the summation of three individual modulated waves, as shown in Fig. 15.5. In contrast to this simple, symmetric interferogram, the interferogram produced with a broadband IR source displays extensive interference patterns. It is a complex summation of superimposed sinusoidal waves, each wave corresponding to a single frequency. When this IR beam is directed through the sample, the amplitudes of a set of waves are reduced by absorption if the frequency of this set of waves is the same as one of the characteristic frequencies of the sample (Fig. 15.6).

Figure 15.5 Interferogram consisting of three modulated cosine waves. The greatest amplitude occurs at the point of zero path difference (ZPD). *(Reprinted by permission of Nicolet Instrument Corporation.)*

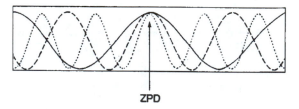

ZPD

The interferogram contains information over the entire IR region to which the detector is responsive. A mathematical operation known as Fourier transformation converts the interferogram (a time domain spectrum displaying intensity versus time within the mirror scan) to the final IR spectrum, which is the familiar frequency domain spectrum showing intensity versus frequency. This also explains how the term *Fourier transform infrared* spectrometry is created.

The detector signal is sampled at small, precise intervals during the mirror scan. The sampling rate is controlled by an internal, independent reference, a modulated monochromatic beam from a helium neon (HeNe) laser focused on a separate detector.

The two most popular detectors for a FTIR spectrometer are deuterated triglycine sulfate (DTGS) and mercury cadmium telluride (MCT). The response times of many detectors (for example, thermocouple and thermistor) used in dispersive IR instruments are too slow for the rapid scan times (1 sec or less) of the interferometer. The DTGS detector is a pyroelectric detector that delivers rapid responses because it measures the changes in temperature rather than the value of temperature. The MCT detector is a photon (or quantum) detector that depends on the quantum nature of radiation and also exhibits very fast responses. Whereas DTGS detectors operate at

Figure 15.6 A typical interferogram produced with a broadband IR source.

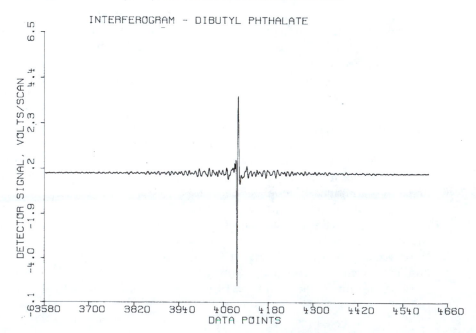

room temperature, MCT detectors must be maintained at liquid nitrogen temperature (77 °K) to be effective. In general, the MCT detector is faster and more sensitive than the DTGS detector.

Spectrometer Design

The basic instrument design is quite simple. Figure 15.7 illustrates the design of a typical FTIR spectrometer. The IR radiation from a broadband source is first directed into an interferometer, where it is divided and then recombined after the split beams travel different optical paths to generate constructive and destructive interference. Next, the resulting beam passes through the sample compartment and reaches to the detector.

Most benchtop FTIR spectrometers are single-beam instruments. Unlike double-beam grating spectrometers, single-beam FTIR does not obtain transmittance or absorbance IR spectra in real time. A typical operating procedure is described as follows:

1. A background spectrum (Fig. 15.8) is first obtained by collecting an interferogram (raw data), followed by processing the data by Fourier transform conversion. This is a response curve of the spectrometer and takes account of the combined performance of source, interferometer, and detector. The background spectrum also includes the contribution from any ambient water (two irregular groups of lines at about 3600 cm^{-1} and about 1600 cm^{-1}) and carbon dioxide (doublet at 2360 cm^{-1} and sharp spike at 667 cm^{-1}) present in the optical bench.

Figure 15.7 Schematic diagram of the Nicolet Magna-IR® 750 FTIR Spectrometer. *(Reprinted by permission of Nicolet Instrument Corporation.)*

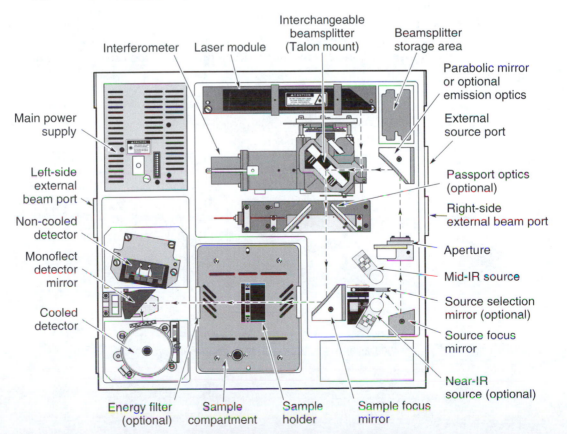

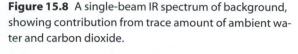

Figure 15.8 A single-beam IR spectrum of background, showing contribution from trace amount of ambient water and carbon dioxide.

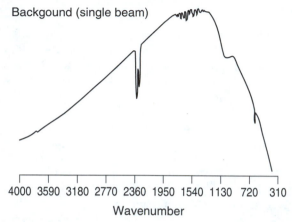

2. Next, a single-beam sample spectrum is collected (Fig. 15.9). It contains absorption bands from the sample and the background (air or solvent).

3. The ratio of the single-beam sample spectrum in Fig. 15.9 against the single beam background spectrum in Fig. 15.8 results in a "double-beam" spectrum of the sample (Fig. 15.10).

To reduce the strong background absorption from water and carbon dioxide in the atmosphere, the optical bench is usually purged with an inert gas or with dry, carbon dioxide–scrubbed air (from a commercial purge gas generator). Spectrometer alignment, which includes optimization of the beamsplitter angle, is recommended as part of a periodic maintenance or when a sample accessory is changed.

Figure 15.9 A single-beam IR spectrum of dibutyl phthalate (a liquid sample).

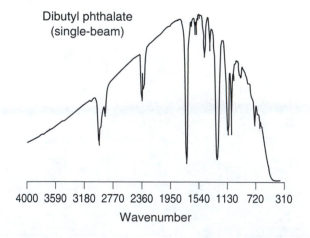

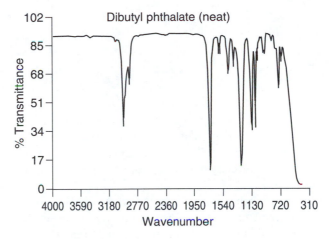

Figure 15.10 The "double-beam" IR spectrum of dibutyl phthalate, produced by ratio of the corresponding single-beam sample spectrum against the single-beam background spectrum.

FTIR Advantages

FTIR instruments have distinct advantages over dispersive spectrometers:

- Better speed and sensitivity (Felgett advantage). A complete spectrum can be obtained during a single scan of the moving mirror, while the detector observes all frequencies simultaneously. An FTIR instrument can achieve the same signal-to-noise (S/N) ratio of a dispersive spectrometer in a fraction of the time (1 sec or less versus 10 to 15 min). The S/N ratio is proportional to the square root of the total number of measurements. Because multiple spectra can be readily collected in 1 min or less, sensitivity can be greatly improved by increasing S/N through coaddition of many repeated scans.

- Increased optical throughput (Jaquinot advantage). Energy-wasting slits are not required in the interferometer because dispersion or filtering is not needed. Instead, a circular optical aperture is commonly used in FTIR systems. The beam area of an FT instrument is usually 75 to 100 times larger than the slit width of a dispersive spectrometer. Thus, more radiation energy is made available. This constitutes a major advantage for many samples or sampling techniques that are energy-limited.

- Internal laser reference (Connes advantage). The use of a helium neon laser as the internal reference in many FTIR systems provides an automatic calibration in an accuracy of better than 0.01 cm^{-1}. This eliminates the need for external calibrations.

- Simpler mechanical design. There is only one moving part, the moving mirror, resulting in less wear and better reliability.

- Elimination of stray light and emission contributions. The interferometer in FTIR modulates all the frequencies. The unmodulated stray light and sample emissions (if any) are not detected.

- Powerful data station. Modern FTIR spectrometers are usually equipped with a powerful, computerized data system. It can perform a wide variety of data processing tasks such as Fourier transformation, interactive spectral subtraction, baseline correction, smoothing, integration, and library searching.

Although the spectra of many samples can be satisfactorily run on either FTIR or dispersive instruments, FTIR spectrometers are the preferred choice for samples that are energy-limited or when increased sensitivity is desired. A wide range of sampling accessories is available to take advantage of the capabilities of FTIR instruments.

What It Does

It is possible to obtain an IR spectrum from samples in many different forms, such as liquid, solid, and gas. However, many materials are opaque to IR radiation and must be dissolved or diluted in a transparent matrix in order to obtain spectra. Alternatively, it is possible to obtain reflectance or emission spectra directly from opaque samples. Some popular sampling techniques and accessories are discussed here.

Liquid cells are used for dilute solutions of solid and liquid samples that are dissolved in relatively IR-transparent solvents. Sampling in solution results in enhanced reproducibility and is often the preferred choice. Unfortunately, no single solvent is transparent through the entire mid IR region. The analyst usually chooses solvents that have transparent windows in the region of interest. The conventional popular solvents are carbon tetrachloride for the region between 4000 and 1330 cm^{-1} and carbon disulfide for the region between 1330 and 625 cm^{-1}. Both solvents are quite toxic, and thus must be handled carefully. One may replace carbon tetrachloride with the less-toxic tetrachloroethylene or methylene chloride and substitute carbon disulfide with n-hexane or n-heptane. Polar solvents such as water and alcohols are seldom used because they absorb strongly in the mid IR range and react with alkali-metal halides, such as NaCl, commonly used for cell windows.

Acquiring acceptable IR spectra of aqueous samples requires use of special types of liquid cells such as thin cells of BaF$_2$, AgCl, or KRS-5 (a mixed thallium bromide–thallium iodide). Aqueous solution measurements can also be accomplished with attenuated total reflectance (ATR) accessories, which are discussed later in this chapter.

Typically, solutions of 0.05 to 10% in concentration are handled in IR cells of 0.1 to 1 mm in thickness. Concentration of 10% and cell path length of 0.1 mm represent one practical combination. In a double-beam spectrometer, a compensating cell is filled with pure solvent and placed in the reference beam. In the single-beam FT instrument, the solvent bands are mostly removed by obtaining the difference spectra through subtraction of solvent spectra from sample spectra. Both fixed-thickness and variable-thickness liquid cells are available commercially. They normally consist of metal frame plates, IR-transmitting windows, and gaskets that determine the path length of the cells.

Salt plates of IR-transmitting materials can be used for semivolatile and nonvolatile liquid samples. Sodium chloride disks are the most popular and economical choice for nonaqueous liquids. Silver chloride or barium fluoride plates may be used for samples that dissolve or react with NaCl plates. A drop of the neat sample is squeezed between two salt plates to form a film of approximately 0.01 mm in thickness. The plates can be held together by capillary attraction, or they may be clamped in a screw-tightened holder or pressed to form a good contact in a press fit O-ring supported holder. It is also possible to place a film of samples on salt plates by melting a relatively low-melting solid and squeezing it between two plates. Sodium chloride salt plates can usually be cleaned with dry methylene chloride or acetone. This smear technique is one of the simplest ways to obtain IR spectra.

Thin films of nonvolatile liquids or solids can be deposited on an IR-transmitting salt plate by solvent evaporation. The sample is first dissolved in a reasonably volatile solvent. A few drops of the resulting solution are placed on the plate. After evaporating off the solvent, a thin film of sample is obtained for subsequent spectra acquisition.

Disposable IR cards have been developed recently by 3M to accommodate samples that are liquids, are soluble in reasonably volatile solvents, or can be smeared on flat surfaces. The cards are made up of a cardboard holder containing a circular IR-transmitting window made of a microporous substrate (polytetrafluoroethylene substrate for 4000 to 1300 cm^{-1} or polyethylene substrate for 1600 to 400 cm^{-1}). Samples are generally applied to the cards by the techniques used for salt plates. The substrate bands can be subtracted from the sample spectra. Besides the convenience, the disposable IR cards are nonhygroscopic, and thus can handle water-containing samples.

Pellets are used for solid samples that are difficult to melt or dissolve in any suitable IR-transmitting solvents. The sample (0.5 to 1.0 mg) is finely ground and intimately mixed with approximately 100 mg of dry potassium bromide (or other alkali halide) powder. Grinding and mixing can be done with an agate mortar and pestle, a vibrating ball mill (Wig-L-Bug from Crescent Dental Manufacturing), or lyophilization. The mixture is then pressed into a transparent disk in an evacuable die at sufficiently high pressure. Suitable KBr disks or pellets can often be made using a simpler device such as a Mini-Press. To minimize band distortion due to scattering of radiation, the sample should be ground to particles of 2 μm (the low end of the radiation wavelength) or less in size. The IR spectra produced by the pellet technique often exhibit bands at 3450 cm^{-1} and 1640 cm^{-1} due to absorbed moisture.

Mulls are used as alternatives for pellets. The sample (1 to 5 mg) is ground with a mulling agent (1 to 2 drops) to give a two-phase mixture that has a consistency similar to toothpaste. This mull is pressed between two IR-transmitting plates to form a thin film. The common mulling agents include mineral oil or Nujol (a high-boiling hydrocarbon oil), Fluorolube (a chlorofluorocarbon polymer), and hexachlorobutadiene. To obtain a full IR spectrum that is free of mulling agent bands, the use of multiple mulls (such as Nujol and Fluorolube) is generally required. Thorough mixing and reduction of sample particles of 2 μm or less in size are very important in obtaining a satisfactory spectrum.

Gas cells can be used to examine gases or low-boiling liquids. These cells consist of a glass or metal body, two IR-transparent end windows, and valves for filling gas from external sources. They provide vacuum-tight light paths from a few centimeters to 120 m. Longer path lengths are obtained by reflecting the IR beam repeatedly through the sample using internal mirrors located at the ends of the cell. Sample gas pressure required to produce reasonable spectra depends on the sample absorbance and the cell's path length. Typically, a good spectrum can be acquired at a partial pressure of 50 torr in a 10-cm cell. Analysis of multicomponent gas samples at parts-per-billion levels can be successfully performed.

Microsampling accessories such as microcells, microcavity cells, and micropellet dies are used to examine microquantities of liquids (down to 0.5 μL) and solids (down to 10 μg). Beam-condensing devices are often used to reduce the beam size at the sampling point. Extra practice is recommended for performing this type of microanalysis.

Attenuated total reflectance (ATR) accessories are especially useful for obtaining IR spectra of difficult samples that cannot be readily examined by the normal transmission method. They are suitable for studying thick or highly absorbing solid and liquid materials, including films, coatings, powders, threads, adhesives, polymers, and aqueous samples. ATR requires little or no sample preparation for most samples and is one of the most versatile sampling techniques.

ATR occurs when a beam of radiation enters from a more-dense (with a higher refractive index) into a less-dense medium (with a lower refractive index). The fraction of the incident beam reflected increases when the angle of incidence increases. All incident radiation is completely reflected at the interface when the angle of incidence is greater than the critical angle (a function of refractive index). The beam penetrates a very short distance beyond the interface and into the less-dense medium before the complete reflection occurs. This penetration is called the evanescent wave and typically is at a depth of a few micrometers (μm). Its intensity is reduced (attenuated) by the sample in regions of the IR spectrum where the sample absorbs. Figure 15.11 illustrates the basic ATR principles.

The sample is normally placed in close contact with a more-dense, high-refractive-index crystal such as zinc selenide, thallium bromide–thallium iodide (KRS-5), or germanium. The IR beam is directed onto the beveled edge of the ATR crystal and internally reflected through the crystal with a single or multiple reflections. Both the number of reflections and the penetration depth decrease with increasing angle of incidence. For a given angle, the higher length-to-thickness ratio of the ATR crystal gives higher numbers of reflections. A variety of types of ATR accessories are available, such as 25 to 75° vertical variable-angle ATR, horizontal ATR, and Spectra-Tech Cylindrical Internal Reflectance Cell for Liquid Evaluation (CIRCLE®) cell.

The resulting ATR-IR spectrum resembles the conventional IR spectrum, but with some differences: The absorption band positions are identical in the two spectra, but the relative intensities of corresponding bands are different. Although ATR spectra can be obtained using either dispersive or FT instruments, FTIR spectrometers permit higher-quality spectra to be obtained in this energy-limited situation.

Specular reflectance provides a nondestructive method for measuring thin coatings on selective, smooth substrates without sample preparation. It basically involves a mirrorlike reflection and produces reflection measurements for a reflective material, or a reflection–absorption spectrum for the surface film on a reflective surface. Thin surface coatings in the range from nanometers to micrometers can be routinely examined with a grazing angle (typically 70 to 85°) or 30° angle of incidence, respectively. For example, lubricant thickness on magnetic media or computer disks is conveniently measured using this technique.

Diffuse reflectance technique is mainly used for acquiring IR spectra of powders and rough surface solids such as coal, paper, and cloth. It can be used as an alternative to pressed-pellet or mull techniques. IR radiation is focused onto the surface of a solid sample in a cup and results in two types of reflections: specular reflectance, which directly reflects off the surface and has equal angles of incidence and reflectance, and diffuse reflectance, which penetrates into the sample, then scatters in all directions. Special reflection accessories are designed to collect and refocus the resulting diffusely scattered light by large ellipsoidal mirrors, while minimizing or eliminating the specular reflectance, which complicates and distorts the IR spectra. This energy-limited technique

Figure 15.11 Schematic representation of multiple internal reflection effect in Attenuated Total Reflectance (ATR). *(Reprinted from 1988 Annual Book of ASTM Standards by permission of American Society for Testing and Materials.)*

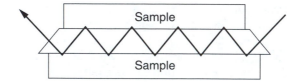

was not popular until the advent of FTIR instruments. This technique is often called diffuse reflectance infrared Fourier transform spectroscopy (DRIFTS).

The sample can be analyzed either directly in bulk form or as dispersions in IR-transparent matrices such as KBr and KCl. Dilution of analyte in a nonabsorbing matrix increases the proportion of diffuse reflectance in all the light reflected. Typically the solid sample is diluted homogeneously to 5 to 10% by weight in KBr. The spectra of diluted samples are similar to those obtained from pellets when plotted in units such as log $1/R$ (R is the reflectance) or Kubelka–Munk units. The Kubelka–Munk format relates sample concentration to diffuse reflectance and applies a scattering factor.

Photoacoustic spectroscopy (PAS) is a useful extension of IR spectroscopy and suitable for examining highly absorbing samples that are difficult to analyze by conventional IR techniques. The size and shape of the sample are not critical. PAS spectra can be obtained with minimal sample preparation and without physical alteration from a wide variety of samples such as powders, polymer pellets, viscous glues, single crystals, and single fibers.

Typically, the modulated IR radiation from an FTIR interferometer is focused on a sample placed in a small cup inside a small chamber containing an IR-transparent gas such as helium or nitrogen. IR radiation absorbed by the sample converts into heat inside the sample. The heat diffuses to the sample surface, then into the surrounding gas atmosphere, and causes expansion of a boundary layer of gas next to the sample surface. Thus, the modulated IR radiation produces intermittent thermal expansion of the boundary layer and generates pressure waves. A sensitive microphone is used to detect the resulting photoacoustic signal.

PAS spectra are generally similar to conventional IR spectra except for some minor differences: Absorbance peaks appear at the same frequency locations, but truncation of strong absorbance bands due to photoacoustic signal saturation is often observed. However, the presence of such truncated bands does not limit the practical use of PAS. Spectral search against standard commercial spectral libraries can be satisfactorily performed. FTIR PAS technique also offers a unique capability for examining samples at various depths from 1 to 20 μm. The acoustic frequencies depend on the modulated frequency of source: The slower the modulation frequency, the greater depth of penetration. Thus, samples such as multilayer polymers can be studied at various depths by simply varying the scan speed of the FTIR spectrometer.

Emission spectroscopy is another technique used with difficult samples such as thin coatings and opaque materials. The sample is normally heated to an elevated temperature, emitting enough energy to be detected. The sample acts as the radiation source, so the normal IR source is turned off. The ability of FTIR instruments to obtain spectra from weak signals makes it possible to study emisssion in the infrared region, even when the sample is at low temperatures such as 50 to 100 °C. Emission spectral bands occur at the same frequencies as absorption bands. The spectra from thick samples can be complicated when radiation from the interior of the sample is self-absorbed by the outer part of the sample.

Infrared microspectroscopy has become a popular technique for analyzing difficult or small samples such as trace contaminants in semiconductor processing, multilayer laminates, surface defects, and forensic samples. Infrared microscopes are energy-inefficient accessories that require the signal-to-noise advantages of FTIR to obtain spectra from submilligram samples. Using a liquid nitrogen cooled mercury cadmium telluride (MCT) detector, samples in the size range of 10 μm can be examined on IR microscopes.

The primary advantages of the IR microscope relate not only to its improved optical and mechanical design, but also to its manipulative capability. In many cases, the major problem in microsampling is focusing the spectrometer beam on the sample. The computerized/motorized control of microscope functions of IR microscope instruments permit these extremely small samples to be moved in the field of view to isolate the portion from which spectra are obtained.

Fiberoptic accessories deliver unique flexibility and diversity in sampling. They are particularly useful in acquiring IR spectra when samples are situated in a remote location or when the unusual size or shape of samples prevents them from fitting well in a standard sample compartment. Many analyses in hazardous or process environments used these devices.

Fiberoptic sample probes or flow cells are coupled to standard FTIR spectrometers with two fiberoptic cables and an optic interface that transfers IR radiation from spectrometer to fiberoptic cables. A variety of probes are available for ATR, specular reflectance, diffuse reflectance, and transmittance measurements. Chalcogenide (GeAsSeTe), a mid IR–transmitting material in the range of 4000 to 900 cm^{-1}, was recently developed by Spectra-Tech and used to make the fiberoptic cables.

Hyphenated Methods Involving Infrared

Gas chromatography/Fourier transform infrared (GC/FTIR) spectroscopy is a technique that uses a gas chromatograph to separate the components of sample mixtures and an FTIR spectrometer to provide identification or structural information on these components. The real potential of GC-IR instrumentation was not widely used until the fast-scanning, sensitive FTIR spectrometers became available commercially.

The most commonly used GC/FTIR interface is a light pipe flow cell. The light pipe is typically a piece of glass tubing 10 to 20 cm long, approximately 1 mm inside diameter, gold coated on the inside, with IR-transmitting windows on each end. This design provides a long path length and low dead volume (90 to 300 μL), resulting in high IR absorbance with minimal peak broadening. The light pipe is connected to the effluent port of the gas chromatograph by a heated transfer line. The gas flow assembly can be heated up to 350 °C to prevent sample components from condensing onto the light pipe and transfer line. Figure 15.12 illustrates the optical design of a GC/FTIR interface.

Figure 15.12 Schematic diagram of a GC/FTIR interface.
(Reprinted by permission of Nicolet Instrument Corporation.)

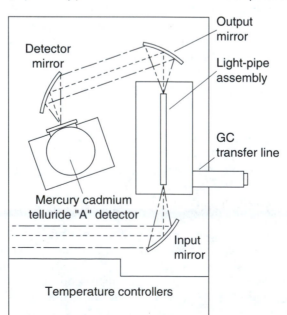

Eluents from a capillary gas chromatograph flow through the transfer line into the light pipe, where the IR spectra are acquired in real time with a rate up to 20 spectra per second. The light-pipe GC/FTIR offers nanogram-level sensitivity. Typically, a usable spectrum can be obtained from 5 to 20 ng of component compound. The flow emerging from the chromatograph is often split between the light pipe and a conventional GC detector (flame ionization, thermal conductivity, or mass spectrometer). This permits the simultaneous generation of a normal chromatogram and the corresponding IR spectra for each chromatographic peak. Alternatively, the total flow after the light pipe can be routed into a conventional detector to provide in-line detection by a flame ionization or mass spectrometer detector. In fact, the combination of a gas chromatograph with an FTIR and mass spectrometer (GC/FTIR/MS) is available commercially.

Although common GC/FTIR spectroscopy is not as sensitive as gas chromatography/mass spectrometry (GC/MS), GC/FTIR offers a major advantage over GC/MS: the ability to identify structural isomers. In addition, the sensitivity of GC/FTIR can be further improved by matrix isolation or direct deposition techniques.

Gas chromatography/matrix isolation/Fourier transform infrared (GC/MI/FTIR) spectroscopy provides subnanogram sensitivity, but is a very expensive technique. The helium carrier gas of a gas chromatograph is mixed with a small amount of argon. While argon is condensed in a track of 300 μm width on a rotating circular gold-coated metal disk cooled at 12 °K, the helium gas is evacuated by pumping. The components separated by the chromatograph are dissolved and trapped in the argon matrix. After the GC run is completed, the argon track is rotated into the IR beam and the reflection–absorption IR spectra are obtained for each component on the cooled surface. Cryogenic temperatures are maintained while the spectra are acquired.

In GC/MI/FTIR, the components are isolated in small areas. Because IR measurements are not made in real time, these components can be held in the IR beam for longer periods, allowing improved signal-to-noise ratios through averaging of multiple scans. Thus, GC/MI/FTIR offers significant sensitivity improvement over light-pipe GC/FTIR.

Gas chromatography/direct deposition/Fourier transform infrared (GC/DD/FTIR) spectroscopy is another sensitive technique that permits a usable spectrum to be obtained with 100 pg of component compound. The separated components are directly deposited in a track of 100 μm width on a liquid-nitrogen–cooled, IR-transmitting disk such as zinc selenide. Transmission IR spectra can be taken in real time. Alternatively, the isolated components can be repositioned in the IR beam after the run is completed to gain the multiscan signal-averaging advantages.

High-performance liquid chromatography/Fourier transform infrared (HPLC/FTIR) spectroscopy uses the same approach as the GC/DD/FTIR to eliminate the mobile phase and gain satisfactory sensitivity.

Conventional flow cells for HPLC chromatograph generally do not provide adequate sensitivity, due to the IR absorption of all HPLC mobile phases. Instead, an HPLC/FTIR interface using the direct deposition technique has been designed and delivers subnanogram sensitivity (1). The interface consists of two concentric fused silica tubes of different internal diameters. While the eluate stream from an HPLC column flows through the inner tube, a sheath of heated gas (helium or air) passes through the outer tube. The nebulized spray is directed to a rotating sample collection disk onto which the component compounds are deposited. During the process the mobile phase is evaporated. The resulting depositions of sample components can then be positioned in the IR beam and their IR spectra collected.

Other techniques involving IR spectrometers, including supercritical fluid chromatography/Fourier transform infrared (SFC/FTIR) spectroscopy, thermogravimetry/Fourier transform infrared (TGA/FTIR) spectroscopy and, gas chromatography/Fourier transform infrared/mass spectrometry (GC/FTIR/MS) have also become available commercially. They generally use more cost-effective flow-through interfaces.

Analytical Information

Qualitative

The combination of the fundamental vibrations or rotations of various functional groups and the subtle interactions of these functional groups with other atoms of the molecule results in the unique, generally complex IR spectrum for each individual compound. IR spectroscopy is mainly used in two ways: structural elucidation and compound identification.

Structural Elucidation

Because of complex interactions of atoms within the molecule, IR absorption of the functional groups may vary over a wide range. However, it has been found that many functional groups give characteristic IR absorption at specific, narrow frequency ranges regardless of their relationship with the rest of the molecule. Generalized tables of the positions and relative intensities of absorption bands (Fig. 15.13) have been established and used to determine the functional groups present or absent in a molecule. The *CRC Handbook of Chemistry and Physics* (2), Silverstein, Bassler, and Morrill's book (3), and a number of other publications all contain useful correlation charts.

Multiple functional groups may absorb at one particular frequency range, but a functional group often gives rise to multiple-characteristic absorption. Thus, the spectral interpretations should not be confined to one or two bands and the whole spectrum should be examined. To confirm or better elucidate the structure of an unknown substance, other analytical information provided by nuclear magnetic resonance (NMR), mass spectrometry (MS), or other chemical analysis should also be used where possible. For systematic evaluation, the IR spectrum is commonly divided into three regions.

The Functional Group Region, 4000 to 1300 cm^{-1} The appearance of strong absorption bands in the region of 4000 to 2500 cm^{-1} usually comes from stretching vibrations between hydrogen and some other atoms with a mass of 19 or less. The O-H and N-H stretching frequencies fall in the 3700 to 2500 cm^{-1} region, with various intensities. Hydrogen bonding has a significant influence on the peak shape and intensity, generally causing peak broadening and shifts in absorption to lower frequencies. The C-H stretching bands occur in the region of 3300 to 2800 cm^{-1}. The acetylenic C-H exhibits strong absorption at about 3300 cm^{-1}. Alkene and aromatic C-H stretch vibrations absorb at 3100 to 3000 cm^{-1}. Most aliphatic (saturated) C-H stretching bands occur at 3000 to 2850 cm^{-1}, with generally prominent intensities that are proportional to the number of C-H bonds. Aldehydes often show two sharp aldehydic C-H stretching absorption bands at 2900 to 2700 cm^{-1}.

The absorption bands at the 2700 to 1850 cm^{-1} region usually come only from triple bonds and other limited types of functional groups, such as C≡C at 2260 to 2100 cm^{-1}, C≡N at 2260 to 2220 cm^{-1}, diazonium salts –N$^+$≡N at approximately 2260 cm^{-1}, allenes C=C=C at 2000 to 1900 cm^{-1}, S-H at 2600 to 2550 cm^{-1}, P-H at 2440 to 2275 cm^{-1}, Si-H at 2250 to 2100 cm^{-1}.

The 1950 to 1450 cm^{-1} region exhibits IR absorption from a wide variety of double-bonded functional groups. Almost all the carbonyl C=O stretching bands are strong and occur at 1870 to 1550 cm^{-1}. Acid chlorides and acid anhydrides give rise to IR bands at 1850 to 1750 cm^{-1}. Whereas ketones, aldehydes, carboxylic acids, amides, and esters generally show IR absorption at 1750 to 1650 cm^{-1}, carboxylate ions usually display stretching bands at 1610 to 1550 and 1420 to 1300 cm^{-1}. Conjugation, ring size, hydrogen bonding, and steric and electronic effects often result in

Figure 15.13 Infrared correlation chart of major functional groups. *(Reprinted with permission from D. R. Lide, Handbook of Chemistry and Physics, 75th edition, Boca Raton, FL: CRC Press. Copyright © 1994 CRC Press, Boca Raton, Florida.)*

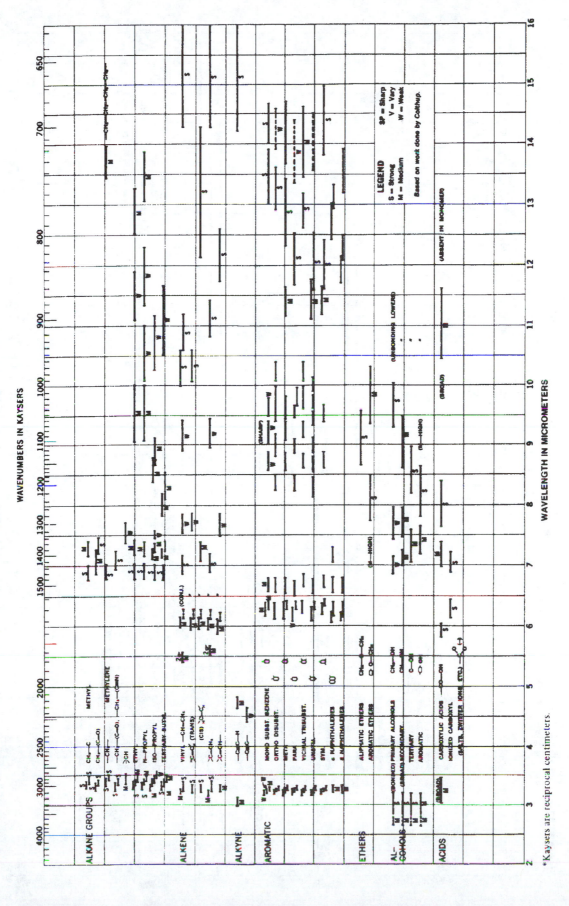

*Kaysers are reciprocal centimeters.

Figure 15.13 *(continued)*

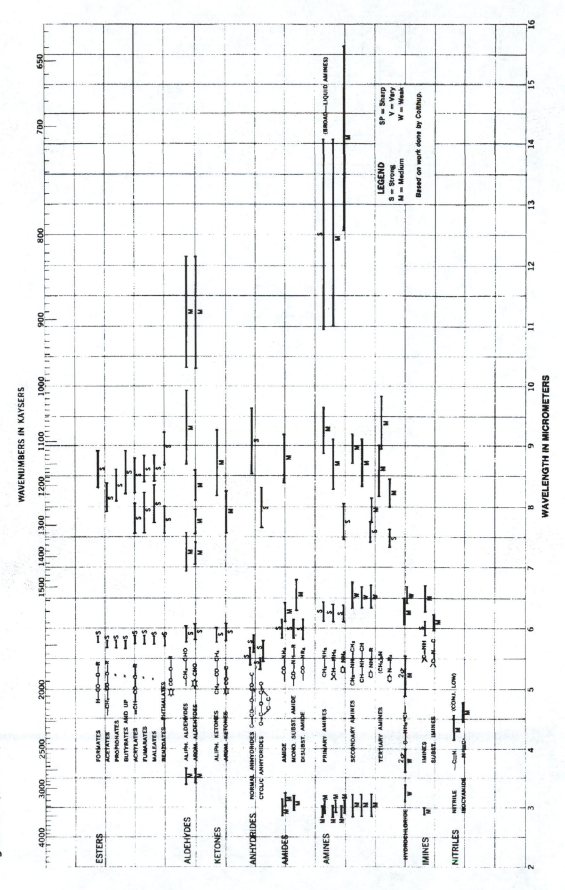

Figure 15.13 (continued)

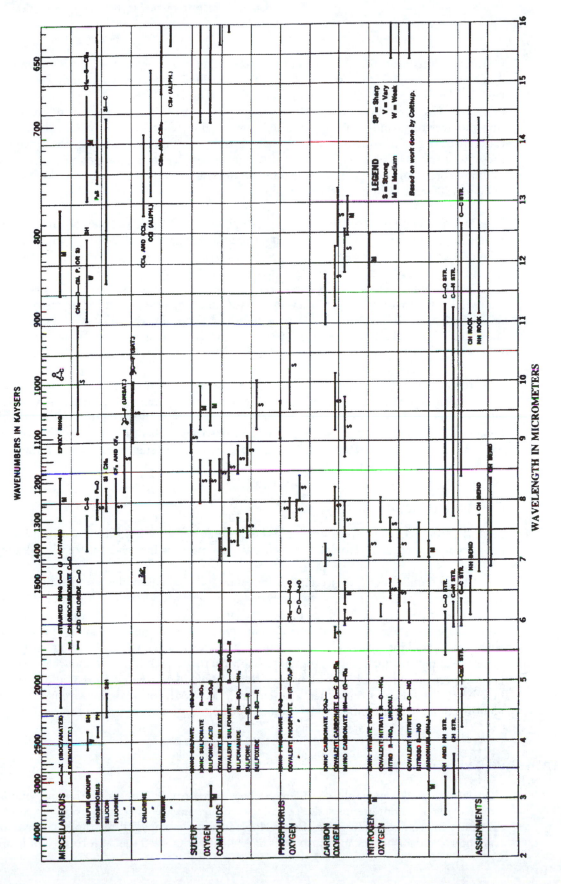

significant shifts in absorption frequencies. Nonconjugated aliphatic C=C and C=N have absorption bands at 1690 to 1620 cm^{-1}, with variable intensities. Aromatic compounds contain delocalized π electrons from the resonance-stabilized double bonds, showing skeletal vibrations (including C-C stretchings within the ring) in the 1650 to 1400 cm^{-1} region and weak combination and overtone bands in the 2000 to 1650 cm^{-1} region. Valuable information about the substitution pattern on an aromatic ring can be obtained by careful examination of absorption bands in these two regions. Molecules containing NO_2 groups, such as nitro compounds, nitrates, and nitramines, commonly exhibit asymmetric and symmetric stretching vibrations of the NO_2 group at 1660 to 1500 and 1390 to 1260 cm^{-1} regions.

The Fingerprint Region, 1300 to 910 cm^{-1} Absorptions in this region include the contributions from complex interacting vibrations, giving rise to the generally unique fingerprint for each compound. A good match between the IR spectra of two compounds in all frequency ranges, particularly in the fingerprint region, strongly indicates that they have the same molecular structures.

Detailed interpretation of IR bands in this region is difficult. However, some assignments of bands in the fingerprint region to a few important vibrational frequencies of functional groups can be done when IR absorptions in other regions are correlated together. For example, esters not only show their carbonyl C=O stretch at 1750 to 1735 cm^{-1}, but also exhibit their characteristic absorption at 1300 to 1000 cm^{-1} from the couplings of C-O and C-C stretches.

The Aromatic Region, 910 to 650 cm^{-1} The IR bands in this region do not necessarily come from the aromatic compounds, but the absence of strong absorption in the 910 to 650 cm^{-1} region usually indicates the lack of aromatic characters. The out-of-plane bending of ring C-H bonds of aromatic and heteroaromatic compounds gives rise to strong IR bands in the range between 910 and 650 cm^{-1}. As previously stated, certain nonaromatic molecules such as amines and amides can also contribute absorption in this region.

Compound Identification

Since the IR spectrum of every molecule is unique, one of the most positive identification methods of an organic compound is to find a reference IR spectrum that matches that of the unknown compound.

A large number of reference spectra for vapor and condensed phases are available in printed and electronic formats. The spectral libraries compiled by Sadtler and Aldrich are some of the most popular collections. In addition, spectral databases are often compiled according to application areas such as forensics, biochemicals, and polymers. Computerized search programs can facilitate the matching process. In many cases where exact match to the spectrum of an unknown material cannot be found, these programs usually list the reference compounds that match the unknown spectrum most closely. This information is useful in narrowing the search. When it is combined with the data from other analysis such as NMR or mass spectrometry, a positive identification or high-confidence level tentative identification can often be achieved.

Quantitative

IR spectroscopy was generally considered to be able to provide only qualitative and semiquantitative analyses of common samples, especially when the data were acquired using the conventional dispersive instruments. However, the development of reliable FTIR instrumentation and strong computerized data-processing capabilities have greatly improved the performance of quantitative

IR work. Thus, modern infrared spectroscopy has gained acceptance as a reliable tool for quantitative analysis.

The basis for quantitative analysis of absorption spectrometry is the Bouguer–Beer–Lambert law, commonly called Beer's law. For a single compound in a homogeneous medium, the absorbance at any frequency is expressed as

$$A = abc \tag{15.3}$$

where A is the measured sample absorbance at the given frequency, a is the molecular absorptivity at the frequency, b is the path length of source beam in the sample, and c is the concentration of the sample. This law basically states that the intensities of absorption bands are linearly proportional to the concentration of each component in a homogeneous mixture or solution.

Deviations from Beer's law occur more often in infrared spectroscopy than in UV/visible spectroscopy. These deviations stem from both instrumental and sample effects. Instrumental effects include insufficient resolution and stray radiation. Resolution is closely related to the slit width in dispersive IR instruments or the optical path difference between two beams in the interferometer of FTIR spectrometers. Stray light levels in FT instruments are usually negligible. Sample effects include chemical reactions and molecular interactions such as hydrogen bonding. The Beer's law deviations result in a nonlinear relationship for plots of absorbance (A) against concentration (c). It is therefore a good practice to obtain calibration curves that are determined empirically from known standards.

Instead of the transmittance scale, absorbance is generally used in quantitative analysis. Absorbance (A) is defined as the negative logarithm of the transmittance (T). According to Beer's law, a linear relationship exists only between the sample concentration and absorbance, not between the sample concentration and transmittance. The linearity of Beer's law plots usually holds better when the absorbance is limited to less than 0.7 absorbance units, although in some cases good linearity has been achieved over more than 2 absorbance units. A number of quantification parameters, which include peak height, peak area, and derivatives, can be used in quantitative analysis. The integration limits for peak area determinations should be carefully chosen to ensure maximum accuracy.

In multicomponent quantitative analysis, the determination of the composition of mixtures involves the use of software packages. These analyses usually assume that Beer's law is additive for a mixture of compounds at a specified frequency. For a simple two-component mixture, the total absorbance, A_T, of the mixture at a given frequency is the sum of the absorbance of two component compounds, x and y, at the specified frequency:

$$A_T = A_x + A_y = a_x b c_x + a_y b c_y \tag{15.4}$$

It is necessary to determine a_x and a_y from absorption measurements of mixtures containing known amounts of compounds x and y at two different frequencies, n and m. Using these values, $a_{x,n}$, $a_{x,m}$, $a_{y,n}$, and $a_{y,m}$, it is possible to use two absorbance measurements from the mixture of unknown composition to determine the concentrations of compounds x and y, c_x and c_y.

$$A_{T,n} = A_{x,n} + A_{y,n} = a_{x,n} b c_x + a_{y,n} b c_y \tag{15.5}$$

$$A_{T,m} = A_{x,m} + A_{y,m} = a_{x,m} b c_x + a_{y,m} b c_y \tag{15.6}$$

Using matrix algebra it is possible to extend this technique to mixtures containing more than two components. The absorbance of a mixture of n independently absorbing components at a particular frequency ν may be expressed in the following equation:

$$A_\nu = a_1 b c_1 + a_2 b c_2 + \ldots + a_n b c_n \tag{15.7}$$

where A_ν = total absorbance of the sample at the frequency, ν, a_j = absorptivity of component j at the frequency ν ($j = 1, 2, \ldots n$), c_j = concentration of component j, and b = sample path length.

Software packages containing matrix methods available with computerized spectrometers simplify the operations associated with multicomponent analysis. If deviations of Beer's law occur, but the law of additivity still holds, sophisticated correlation or statistical evaluation software programs such as least-squares regression, partial least-squares regression, and principal component regression analysis facilitate satisfactory curve-fitting and data-processing tasks.

The broad absorption bands, larger values of absorptivity and sample path length, higher-intensity sources, and more sensitive detectors make the ultraviolet, visible, and near IR regions better suited for quantitative determinations than the mid IR and far IR regions. However, coupling of the advancement of computerized FTIR instrumentation and meticulous attention to detail can make FTIR a viable option for reliable quantitative analysis.

Applications

1. Analysis of Petroleum Hydrocarbons, Oil, and Grease Contents by EPA Methods.

The Environmental Protection Agency (EPA) has established Methods 413.2 and 418.1 for the measurement of fluorocarbon-113 extractable matter from surface and saline waters and industrial and domestic wastes (4). These methods provide semiquantiative determination of petroleum hydrocarbons, oil, and grease by comparison of the infrared absorption of the sample extract with standards.

Petroleum hydrocarbons, oil, and grease include biodegradable animal greases and vegetable oils along with the relative nonbiodegradable mineral oils. They all contain carbon–hydrogen bonds, thus giving rise to C-H stretching absorption in the 3100 to 2700 cm^{-1} region of the IR spectrum. Fluorocarbon-113 (1,1,2-trichloro-1,2,2-trifluoroethane) is one of the chlorofluorocarbons commonly called freons. It contains no C-H bonds, and thus does not absorb IR radiation in the aforementioned 3100 to 2700 cm^{-1} region. The quantity of hydrocarbons, oil, and grease in freon extracts can be estimated by measuring the intensity of C-H absorption band at 2930 cm^{-1}.

The sample is acidified to a low pH (less than 2) and extracted with fluorocarbon-113. Interference is usually removed with silica adsorbent. Depending on the sample concentration, cells of pathlength from 10 to 100 mm can be used to acquire the normal transmission IR spectrum. The concentration of hydrocarbon, oil, and grease in the extract is determined by comparing the absorbance against the calibration plot prepared from standard calibration mixtures. Figure 15.14 shows the different FTIR profiles of three calibration standards. The contributions from the solvent and cell are eliminated by subtracting the reference spectrum of freon from the sample spectrum.

In the standard EPA methods, peak height at a single frequency, 2930 cm^{-1}, is used as the basis for quantification. In the author's laboratory, peak area integration from 3150 to 2700 cm^{-1} is used to quantify the contents of hydrocarbons, oil, and grease. The modified methods have been found to provide significantly improved results in quantitative analysis. Aromatic hydrocarbons have relatively lower absorption intensity in this C-H stretching region, thus giving lower response factors when compared to the IR absorption of oil and grease standards. Using an FTIR instrument, oil and grease at low parts-per-million levels can be readily determined.

Figure 15.14 FTIR spectra of three calibration standards, at 1.5, 3.1, and 6.2 ppm, for oil and grease analysis.

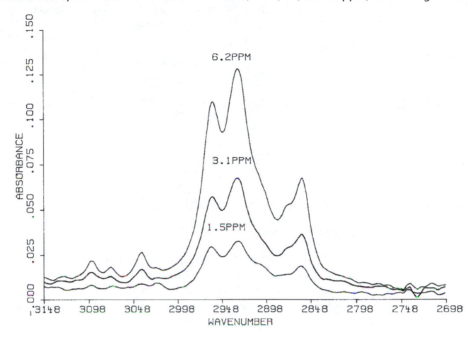

2. Quantitative Analysis of Multicomponent Mixtures of Sulfur Oxygen Anions by Attenuated Total Reflectance Spectroscopy.

Characterization of complex mixtures of sulfur oxygen anions is encountered in studies such as investigating the decomposition of the dithionite anion in acidic aqueous solution. Many techniques such as conventional IR, UV/visible, Raman spectroscopy, and titrimetric and electrochemical analysis all have drawbacks that limit their effectiveness in this challenging analysis. The multicomponent analysis by FTIR attenuated total reflectance (ATR) spectroscopy successfully provides the accurate quantification of multiple sulfur–oxygen anion concentrations in aqueous solution (5).

Sulfur–oxygen compounds have relatively intense S-O stretching absorption bands in the 1350 to 750 cm^{-1} region of the IR spectrum. FTIR/ATR spectroscopy is well suited for quantitative determination of sulfur oxygen anions in strong IR-absorbing aqueous medium. ATR not only uses water-resistant cell material, but also has a very short and reproducible effective path length that goes beyond the sample/crystal interface and into the sample medium. A micro CIRCLE cell from Spectra-Tech incorporating a ZnSe crystal is used in the study. Its basic optics is illustrated in Fig. 15.15. The representative FTIR/ATR spectra of sulfur oxide and nitrate anions are shown in Fig. 15.16.

Aqueous decomposition of sodium dithionite under anaerobic conditions is investigated. Systematic baseline error is characterized and taken into account. Computerized data processing involving least-squares regression is used in the multivariate analysis. Figure 15.17 illustrates the simultaneous measurements of concentrations of seven anions and continuous monitoring of total sulfur and average oxidation states over a 30-min reaction period. This analytical technique successfully determines the reaction orders and reaction stoichiometry shown below:

$$2S_2O_4^{2-} + H_2O \rightarrow S_2O_3^{2-} + 2HSO_3^{2-} \tag{15.8}$$

Figure 15.15 Optic diagram of the Cylindrical Internal Reflectance Cell for Liquid Evaluation (CIRCLE). *(Reprinted by permission of Spectra-Tech, Inc.)*

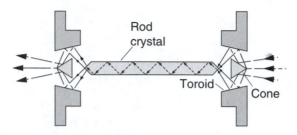

3. Characterization of Heterogeneous Catalysts by Diffuse Reflectance Spectroscopy.

Vibrational spectroscopy has been long established as one of the preferred techniques in obtaining important information on the nature of molecules that are attached to the catalyst surfaces. Diffuse reflectance (DR) spectroscopy has been shown to provide additional advantages over the conventional transmission methods. Detailed characterization of an olefin polymerization catalyst is described here to demonstrate one of the practical applications (6).

Heterogeneous catalysts often consist of active molecules in a distribution of valence states supported on high-surface-area oxides, such as silica and alumina. To better understand Zigler–

Figure 15.16 Representation of FTIR/ATR spectra of nitrate and four sulfur oxide anions. *(Reprinted with permission from D. A. Holman, A. W. Thompson, D. W. Bennett and J. D. Otvos, Analytical Chemistry, Vol. 66, No. 9, 1378–1384. Copyright 1994 American Chemical Society.)*

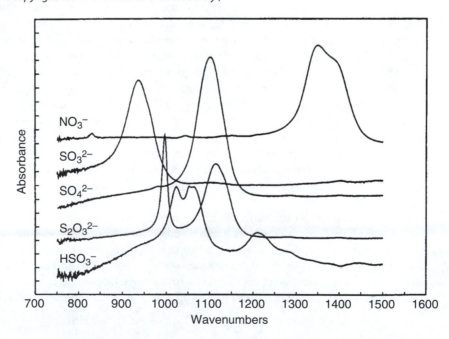

Figure 15.17 Multicomponent analysis of sulfur oxygen anions over a 30-min reaction period of the anaerobic aqueous decomposition of sodium dithionite. (a) $[HSO_3^-]$, (b) $[S_2O_3^{2-}]$, (c) $[S_2O_4^{2-}]$, (d) $[S_3O_6^{2-}]$, (e) $[SO_4^{2-}]$, (f) $[S_2O_5^{2-}]$, (g) $[SO_3^{2-}]$, $\langle OX \rangle$ = average oxidation state, and $[S]_T$ = total sulfur. *(Reprinted with permission from D. A. Holman, A. W. Thompson, D. W. Bennett and J. D. Otvos, Analytical Chemistry, Vol. 66, No. 9, 1378–1384. Copyright 1994 American Chemical Society.)*

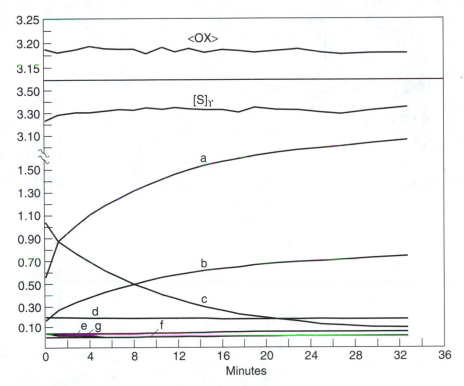

Natta catalysts (important catalysts for olefin polymerization processes), the reactions of attaching titanium chloride ($TiCl_4$) to modified silica surfaces are studied. A diffuse reflectance accessory from Harrick Scientific (Fig. 15.18) and a controlled-atmosphere, high-temperature cell are used in the experiments.

Catalyst samples are diluted by making a 10% w/w dispersion in dry KCl power. Diffuse reflectance spectra of 200 °C pretreated silica gel before and after reaction with hexamethyldisilazane (HMDS) and 600 °C pretreated silica gels are shown in Fig. 15.19. The chemical reactions on modified silica gel surfaces are illustrated in Fig. 15.20. The 200 °C pretreated silica gel exhibits three surface hydroxyl absorption bands at 3740, 3660, and 3540 cm^{-1}, which arise from relatively free, non–hydrogen-bonded, and hydrogen-bonded silanols (Si-OH), respectively. After surface modification with HMDS treatment, the 3740 cm^{-1} peak disappears while the broader bands from hydrogen-bonded silanols remain relatively unperturbed. Baking of the silica gel to 600 °C results in the disappearance of peaks at 3660 and 3540 cm^{-1}, but does not significantly affect the non–hydrogen-bonded silanols' absorption at 3740 cm^{-1}. Diffuse reflectance spectra of 200 °C pretreated silica gel before and after $TiCl_4$ reaction, along with the spectrum resulting from spectral subtraction are shown in Fig. 15.21 (7). The IR absorption bands at 990 and 920 cm^{-1} in the difference spectrum are assigned to the Si-O stretchings of Si-O-$TiCl_3$ and (Si-O)$_2$-$TiCl_2$, respectively. Similar analyses support the other reactions illustrated in Fig. 15.20.

Figure 15.18 Optical diagram of a diffuse reflectance accessory. *(Reprinted by permission of Harrick Scientific Corporation.)*

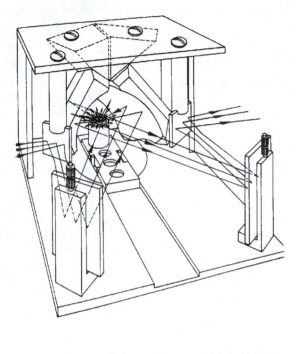

Figure 15.19 FTIR diffuse reflectance spectra of silica gel samples. (a) 200 °C pretreated silica, (b) HMDS pretreated silica, and (c) 600 °C pretreated silica. *(Reprinted by permission from* Spectroscopy, Vol. 9, No. 8 *by J. P. Blitz and S. M. Augustine, © 1994 by Advanstar Communications.)*

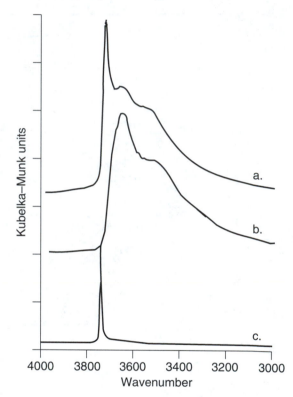

4. Qualitative Analysis of Multilayered Polymeric Films using FTIR Microspectroscopy.

IR analysis can be used to determine the identities of polymer materials in a multilayered film. Using FTIR microspectoscopy, various layers in the polymeric film can be quickly characterized. The qualitative analysis of a three-layer, 20-μm-thick film is described below to demonstrate such an application (8).

The layered thin film is cut as a 2×20-mm sheet. Cross-sections of the film are obtained using a fiber microtome. The individual section is transferred to a NaCl window on a slide positioned on the microscope stage. FTIR spectra are recorded in the transmission mode.

This particular sample has a three-layer composition of Primacor (8 μm), nylon (10 μm), and Primacor (8 μm). Primacor is a copolymer of ethylene and acrylic acid. Nylon is a polyamide polymer. Using an FTIR microscope, infrared transmittance spectra of the multilayered film are obtained (Fig. 15.22). An IR spectrum of pure Primacor can be obtained on the exposed outside layer. The IR spectrum of the center layer exhibits the contributions from both nylon and Primacor. This probably results from spatial contamination, which occurs when a specific layer of 8 to 10 μm or thinner is not masked properly for IR spectrum acquisition due to the poor contrast between the layers or the limitations of the aperture sizes. Functional group mapping can be performed to enhance the spatial resolution. This is accomplished by driving microscope stages across the film

Figure 15.20 Chemical reactions of modified silica gel surfaces. (a) Before TiCl$_4$ reaction and (b) after TiCl$_4$ reaction. *(Reprinted by permission from Spectroscopy, Vol. 9, No. 8 by J. P. Blitz and S. M. Augustine, © 1994 by Advanstar Communications.)*

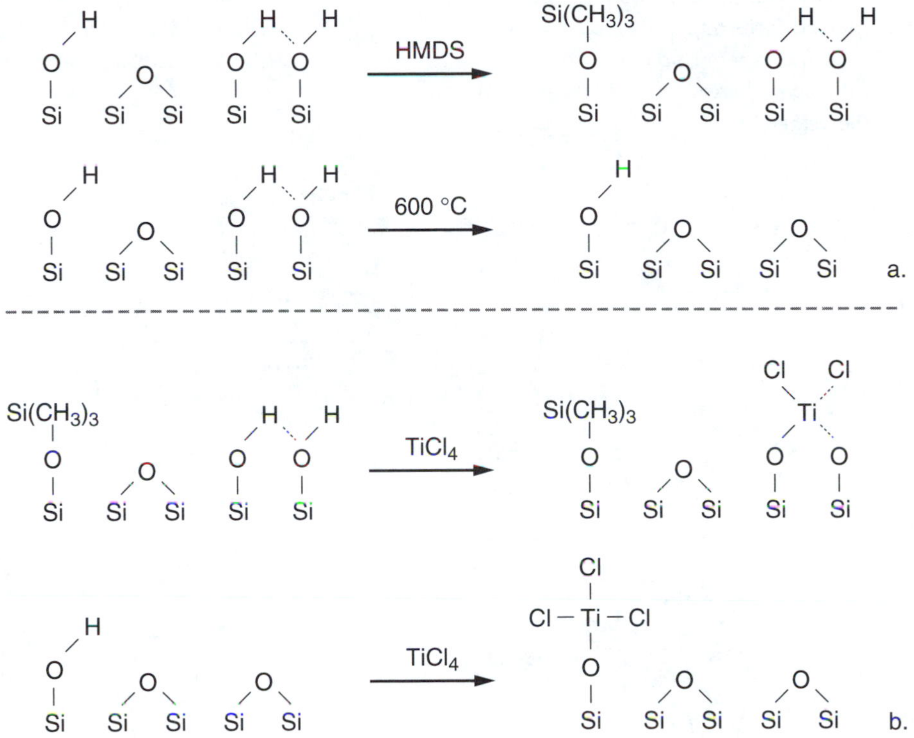

cross-section while mapping the concentration of functional groups that represent characteristics of each component. This type of infrared imaging technique can be more effectively carried out with a computer-controlled, two-dimensional motorized stage and provides a systematic, nondestructive evaluation of sample composition on a microscopic scale.

Nuts and Bolts

Relative Costs

Dispersive	$$ to $$$
FT	$$$ to $$$$
GCIR	$$$
PAS	$$$
IR microscope	$$ to $$$
Reference spectra	$ to $$
Maintenance	10% of purchase price per year
Supplies	<$

$=1 to 5K, $$=5 to 15K, $$$= 15 to 50K, $$$$=50 to 100K, $$$$$=>100K.

Figure 15.21 FTIR diffuse reflectance spectra of silica samples. Spectra before and after TiCl$_4$ treatment of 200 °C pretreated silica and the difference spectrum resulting from spectral subtraction are shown. *(Reprinted from* Colloids and Surfaces, *63, J. P. Blitz, pp. 11–19, copyright 1992, with kind permission from Elsevier Science—NL, Sara Burgerhartstraat 25, 1055 KV Amsterdam, The Netherlands.)*

Figure 15.22 FTIR transmittance spectra of a film with a three-layer composition (Primacor–nylon–Primacor). (a) Outer layer, (b) center layer, and (c) Primacor standard. *(Reprinted by permission from* Spectroscopy, *Vol. 8, No. 8 by T. I. Shaw, F. S. Karl, A. Krishen and L. E. Porter, ©1993 by Advanstar Communications.)*

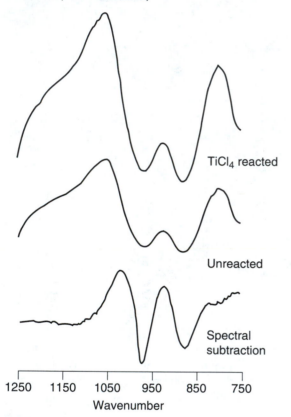

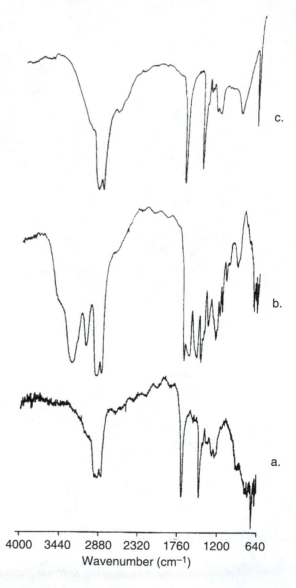

   The major factor affecting the cost of an IR spectrometer is its resolution. Resolution can range from 16 cm^{-1} to 0.1 cm^{-1}. The more expensive ($$$$), higher-resolution (0.1 cm^{-1}) instruments are used to resolve the vibrational bands into their rotational components. These instruments are used in research laboratories to obtain detailed structural information and physical constants for compounds. The lower-priced ($$$), medium-resolution (4 to 2 cm^{-1}) instruments are adequate for most analyses described here.

Vendors for Instruments and Accessories

Instruments

Nicolet Instrument Corp.
5225 Verona Rd.
Madison, WI 53711-4495
phone: 800-232-1472, 608-276-6100
fax: 608-273-5046
email: nicinfo@nicolet.com
Internet: http://www.nicolet.com

The Perkin-Elmer Corp.
761 Main Ave.
Norwalk, CT 06859-0012
phone: 800-762-4000
fax: 203-761-2882
email: info@perkin-elmer.com
Internet: http://www.perkin-elmer.com

Bio-Rad Labs., Digilab Division
237 Putnam Ave.
Cambridge, MA 02139
phone: 800-225-1248
fax: 617-234-7045
email: sales.digilab@bio-rad.com

Bruker Instruments Inc.
19 Fortune Dr.
Billerica, MA 01821
phone: 508-667-9580, 888-427-8537
fax: 508-663-9177
email: optics@bruker.com
Internet: http://www.bruker.com

Mattson Instruments
5225 Verona Rd.
Madison, WI 53717-4495
phone: 800-423-6641, 608-276-6300
fax: 608-273-6818
email: info@mattsonir.com
Internet: http://www.mattsonir.com

Bomen Inc.
450 St. Jean Baptiste
Quebec, G2E 5S5, Canada
phone: 418-877-2944
fax: 418-877-2834

Most instrument manufacturers listed above also provide sampling accessories for their IR spectrometers.

Accessories

Spectra-Tech Inc.
2 Research Dr.
P.O. Box 869
Shelton, CT 06484-0869
phone: 203-926-8998
fax: 203-926-8909

The Perkin-Elmer Corp.
761 Main Ave.
Norwalk, CT 06859-0012
phone: 800-762-4002
fax: 203-761-9645

Harrick Scientific Corp.
88 Broadway
Box 1288
Ossining, NY 10562
phone: 914-762-0020, 800-248-3847
fax: 914-762-0914

International Crystal Laboratories
11 Erie St.
Garfield, NJ 07026
phone: 201-478-8944
fax: 201-478-4201

The Foxboro Company
600 N. Bedford St.
P.O. Box 500
East Bridgewater, MA 02333
phone: 508-378-5400
fax: 508-378-5202
email: tblom@foxboro.com
Internet: http://www.foxboro.com

Axiom Analytical Inc.
17751 Sky Park Circle, #ABC
Irvine, CA 92714
phone: 714-757-9300, 800-Go-Axiom
fax: 714-757-9306
email: goaxiom@aol.com
Internet: http://www.goaxiom.com/axiom

Graseby Specac Inc.
500 Technology Ct.
Smyrna, GA 30082-5210
phone: 770-319-9999, 800-447-2558
fax: 770-319-2488
email: specacusa@aol.com

Wilmad Glass Company, Inc.
Rt. 40 & Oak Rd.
Buena, NJ 08310
phone: 609-697-3000, 800-220-5171
fax: 609-697-0536
email: cs@wilmad.com
Internet: http://www.wilmad.com

Buck Scientific, Inc.
58 Fort Point St.
East Norwalk, CT 06855-1097
phone: 203-853-9444, 800-562-5566
fax: 203-853-0569
email: 102456.1243@compuserve.com
Internet: http://ourworld.compuserve.com/homepages/Buck_Scientific

Required Level of Training

Operation of Instrument

Routine analyses can be performed by analysts with a high school education or an associate college degree. A knowledge of the chemistry of the sample material is useful. Preparation of more difficult samples is an art gained through experience.

Processing Qualitative and Quantitative Data

Qualitative interpretation of spectra requires a minimum of a college organic chemistry course. Quantitative determinations require a minimum of an introductory analytical chemistry course.

Service and Maintenance

Most current spectrometers have diagnostic software that checks the instrument periodically and assists the operator in troubleshooting. The identification and replacement of faulty components are facilitated by the modular design of instrument systems. A high percentage of problems can be solved by trained laboratory personnel, thus eliminating the need for expensive visits by the vendors' engineers. The major components requiring replacement are sources and detectors. The sources for dispersive instruments (Nernst glowers and globar heaters) are much more expensive than the nichrome coil source commonly used with FTIR. However, the replacement of the helium neon laser source used for timing operations in an FTIR spectrometer is expensive (over $600). The beamsplitter of a FT instrument is quite costly (over $3,000). All types of detectors are expensive (over $1000).

Suggested Readings

Books

Annual Book of ASTM Standards, vol. 03.06. Philadelphia: American Society for Testing and Materials, 1995.

GRIFFITHS, P. R., AND J. A. DE HASETH, *Fourier Transform Infrared Spectrometry*. New York: Wiley, 1986.

SILVERSTEIN, R. M., G. C. BASSLER, AND T. C. MORRILL, *Spectrometric Identification of Organic Compounds*, 5th ed. New York: Wiley, 1988.

SMITH, A. L., *Applied Infrared Spectroscopy*. New York: Wiley, 1979.

WILLARD, H. H., AND OTHERS, *Instrumental Methods of Analysis*, 7th ed. Belmont, CA: Wadsworth, 1987.

Articles

BERGLUND, R. A., P. B. GRAHAM, AND R. S. MILLER, "Applications of In-situ FT-IR in Pharmaceutical Process R&D," *Spectroscopy*, 8, no. 8 (1993), 31.

COATES, J. P., J. M. D'AGOSTINO, AND C. R. FRIEDMAN, "Quality Control Analysis by Infrared Spectroscopy, Part 1: Sampling," *American Laboratory*, 18, no. 11 (1986), 82.

COATES, J. P., J. M. D'AGOSTINO, AND C. R. FRIEDMAN, "Quality Control Analysis by Infrared Spectroscopy, Part 2: Practical Applications," *American Laboratory*, 18, no. 12 (1986), 40.

CROOKS, R. M., AND OTHERS, "The Characterization of Organic Monolayers by FT-IR External Reflectance Spectroscopy," *Spectroscopy*, 8, no. 7 (1993), 28.

FUJIMOTO, C., AND K. JINNO, "Chromatography/FT-IR Spectrometry Approaches to Analysis," *Analytical Chemistry*, 64 (1992), 476A.

HARIS, P. I., AND D. CHAPMAN, "Does Fourier Transform Infrared Spectroscopy Provide Useful Information on Protein Structures?" *Trends Biochemical Sciences*, 17, no. 9 (1992), 328.

JONES, R. W., AND J. F. McCLELLAND, "Transient IR Spectroscopy: On-line Analysis of Solid Materials," *Spectroscopy*, 7, no. 4 (1992), 54.

KATON, J. E., AND A. J. SOMMER, "IR Microspectroscopy: Routine IR Sampling Methods Extended to Microscopic Domain," *Analytical Chemistry*, 64 (1992), 931A.

KOENIG, J. L., "Industrial Problem Solving with Molecular Spectroscopy," *Analytical Chemistry*, 66 (1994), 515A.

SPELLICY, R. L., AND OTHERS, "Spectroscopic Remote Sensing: Addressing Requirements of the Clean Air Act", *Spectroscopy*, 6, no. 9 (1991), 24.

WARR, W. A., "Computer-Assisted Structure Elucidation, Part 1: Library Search and Spectral Data Collections," *Analytical Chemistry*, 65 (1993), 1045A.

Training Aids

Audio/video courses: Both audio and video training courses of IR spectroscopy are available from American Chemical Society.

Programmed learning book: GEORGE, W. O., AND P. S. McINTYRE, *Infrared Spectroscopy*, Analytical Chemistry by Open Learning Project, D. J. Mowthorpe, ed. New York: Wiley, 1987.

Short courses: A number of short courses and workshops are generally held in conjunction with American Chemical Society National Meeting, Pittsburgh Conference and Eastern Analytical Symposium. The subjects covered include interpretation of IR spectra, sampling techniques, and accessories. Chemistry short courses of IR spectroscopy are also offered by the Center for Professional Advancement (East Brunswick, N.J.), Chemistry Department of Miami University (Oxford, OH), and Spectros Associates (Northbridge, MA).

Software: IR Mentor, a software aid to spectral interpretation, is available from Sadtler Division, Bio-Rad Laboratories, Inc.

Reference Spectra

Aldrich Chemical Company, Inc.: 17,000 Aldrich-Nicolet FTIR spectra in hard-copy reference books or electronic databases and 12,000 IR spectra in hard-copy books.

Sadtler Division of Bio-Rad Laboratories: More than 150,000 spectra in over 50 different electronic databases, or 89,000 spectra in 119 volumes of hard-copy handbooks.

Sigma Chemical Company: 10,400 FTIR spectra of biochemicals and related organics in hardbound books.

The U.S. Environmental Protection Agency (EPA) vapor phase database (3240 FTIR spectra), Canadian forensic database (3490 spectra), Georgia State Crime Lab database (1760 spectra), and other spectral collections are available through instrument manufacturers.

References

1. An HPLC/FTIR interface is available commercially from Lab Connections, Inc., Marlborough, MA.

2. D. R. Lide, ed., *CRC Handbook of Chemistry and Physics*, 75th ed. (Boca Raton, FL: CRC Press, 1994), 9–79.

3. R. M. Silverstein, G. C. Bassler, and T. C. Morrill, *Spectrometric Identification of Organic Compounds*, 4th ed. (New York: Wiley, 1981), 166.

4. U.S. Environmental Protection Agency, *Methods for Chemical Analysis of Water and Wastes*, 3rd ed., Report No. EPA-600/4-79-020 (Springfield, VA: National Technical Information Service, 1983), 413.2-1, 418.1-1.

5. D. A. Holman and others, *Analytical Chemistry*, 66 (1994), 1378.

6. J. P. Blitz and S. M. Augustine, *Spectroscopy*, 9, no. 8 (1994), 28.

7. J. P. Blitz, *Colloids and Surfaces*, 63 (1992), 11.

8. T. I. Shaw and others, *Spectroscopy*, 8, no. 8 (1993), 45.

Acknowledgment

The author is grateful to Prof. Frank A. Settle, Jr. for providing the detailed handbook template based on infrared spectroscopy. The author would also like to thank Dr. Paul A. Bouis, Prof. Roland W. Lovejoy, and Barbara K. Barr for helpful suggestions and critical reviews of the manuscript.

Chapter 16

Raman Spectroscopy

Dennis P. Strommen

Idaho State University
Department of Chemistry

Summary

General Uses

- Identification of all types of organic compounds
- Determination of functional groups in organic materials
- Identification of all types of inorganic compounds in solid and aqueous solutions
- Determination of the molecular composition of surfaces
- Quantitative determination of compounds in mixtures
- Determination of ground-state structures; especially powerful when used in conjunction with IR data
- Information on excited-state structures

Common Applications

- Identification of compounds by matching spectrum of unknown compound with a reference spectrum (fingerprinting)
- Identification of functional groups in unknown substances
- Identification of reaction components; kinetic studies of reactions
- Detection of molecular impurities or additives

- Identification of polymers, plastics, and resins
- Studies of inorganic compounds
- Molecular structure determination using IR and polarization information
- Nondestructive analysis of biochemical systems in vivo and in vitro
- Surface studies by using surface-enhanced Raman spectroscopy (SERS)
- Studies of electronically excited states of molecules

Samples

State

Solids, liquids, and gases can all be analyzed. Sampling accessories are simple and relatively inexpensive.

Amount

Solids As little as 10 mg can be easily studied and in some cases as little as 10 to 50 ng will give satisfactory spectra. For colored compounds, slightly larger masses may be required to avoid local heating effects.

Liquids Samples can be run in aqueous solution; using resonance Raman, 0.5 mL of mM solutions can give high-quality spectra. Although the volume necessary for normal Raman is even lower, on the order of a few microliters, the concentration must be much higher (about 0.5 M).

Gases Ten torr gives adequate spectra when using a multipass system.

Preparation

Because of the relative weakness of the Raman effect, sample purity is not as critical as in IR unless the impurity fluoresces. If a fluorescent impurity is present, it usually must be removed. Numerous sample holders are available on the open market and in-house devices for special purposes are readily fashioned. In contrast to IR, water need not be removed before the sample is mounted.

Analysis Time

The time required for analysis depends on how much of a spectrum is required, the resolution needed, and the type of instrument being used. The time required for analysis on an FT-Raman or an instrument equipped with an array detector is normally much less than 1 min once the sample is aligned. For a normal scanning instrument, 15 to 20 min is required for a 1000 cm^{-1} scan at usual levels of resolution.

Limitations

General

- Fluorescent impurities can overwhelm the Raman signal.
- Colored samples can absorb incident radiation and either burn or photodecompose.
- Concentrations of solutions must be high unless resonance Raman scattering is used.

Accuracy

In routine analyses, $\pm 5\%$ can be attained.

Sensitivity and Detection Limits

In routine analyses, $\pm 0.5\%$ can be obtained.

Complementary or Related Techniques

- Nuclear magnetic techniques provide additional information on detailed molecular structure.
- Mass spectrometry provides molecular mass information and additional structure information.
- Infrared spectroscopy provides complementary information on molecular vibrations. (Some vibrational modes are IR-inactive but Raman-active and vice versa.)
- In incoherent inelastic neutron scattering, all vibrational modes, including those that are inactive in both IR and Raman, are active.

Introduction

The origins of blue colors in the sky and oceans have long piqued our curiosity. Leonardo da Vinci attributed the color of the sky to scattering of light by suspended particles in the air (1). Maxwell later concluded that the particles were of molecular origin (2). It was as a result of studies such as these that a new effect was discovered. This new effect is observed when light scattered from molecular centers exits from the interaction site with an altered wavelength. This is the Raman effect.

Raman spectroscopy, which began in the early twentieth century, was named in honor of its discoverer, C. V. Raman, who, along with K. S. Krishnan, published the first paper on this technique (3). In turn, Raman's experiments were inspired by the theoretical work of A. Smekal (4); indeed, for many years, the phenomenon was called the Smekal–Raman effect. Raman scattering is one of two spectroscopic methods normally used to probe molecular vibrations. It was the tool of choice for this purpose until the 1940s. It was at that time that commercial IR spectrometers were first introduced and IR became the dominant tool for vibrational analysis. Between then and the late 1960s, Raman spectroscopy was virtually lost in obscurity. Its major limitation was the inherent weakness of the effect, reflected in the necessity for rather large volumes (about 7 mL) and high concentrations (over 1 M) in order to obtain high-quality spectra. The source of radiation during that era was typically the 435.8-nm line of a coiled low-pressure Hg arc lamp. During the late 1960s, however, laser power sources became available. These lasers, with their ability to focus large numbers of photons into small volumes, gave rise to the renaissance of Raman spectroscopy.

How It Works

Raman spectroscopy is similar to infrared (IR) spectroscopy in that they probe molecular vibrations. IR spectroscopy is an absorption phenomenon in which photons matching the energy difference of vibrational levels are absorbed and result in a transition from the ground to an excited vibrational level, whereas the Raman effect is a scattering phenomenon where the energy of the incident photons is much larger than vibrational transition energies. Most of these high-energy photons are scattered without change (Rayleigh scattering). However, a small fraction are scattered from molecular centers with less energy than they had before the interaction. These photons give rise to Raman–Stokes lines. Another series of scattered photons has greater energy than those of the exciting radiation. These are called anti-Stokes photons. They are not discussed further in this chapter. The differences in the energies of the scattered and incident photons correspond to vibrational transitions and thus it is concluded that molecules are promoted to an excited vibrational state, just as they are in IR. The process is depicted schematically in Fig. 16.1, where it is labeled NR (normal Raman). The dotted line is called a virtual state and is little more than a men-

Figure 16.1 Energy-level diagram showing normal Raman (NR), resonance Raman (RR), and infrared (IR) energy transfer processes.

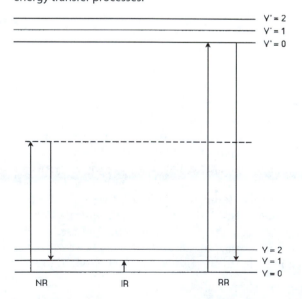

tal crutch that allows us to visualize the process. It must be emphasized that this state does not exist. The information obtained in a Raman experiment can be quite different from that in an IR experiment because the quantum mechanical selection rules are different. In IR, an absorption can take place only if there is a change in dipole moment during the vibration; on the other hand, a change in polarizability (the ease with which the electron cloud is distorted by an external electric field) is required in order for Raman scattering to occur. This gives rise to some rather interesting and useful differences between IR and Raman spectroscopies. Homonuclear diatomic compounds such as H_2, Cl_2, or (in general) X_2 are IR-inactive but active in Raman. These are specific instances of the general rule that molecules with a center of symmetry exhibit a mutual exclusion effect. In this case, if a molecular vibration is active in IR, it is not in Raman and vice versa. Using IR and Raman together allows one to establish the existence of a center of symmetry quite easily. In fact, there is a loose rule that states that if a band is strong in IR it will be weak in Raman and if it is strong in Raman it will be weak in IR. Figure 16.2 shows an example of this behavior. Note

Figure 16.2 Raman (lower trace) and IR (upper trace) spectra of alpha-chloroacetonitrile, Raman-excited with 488-m radiation.

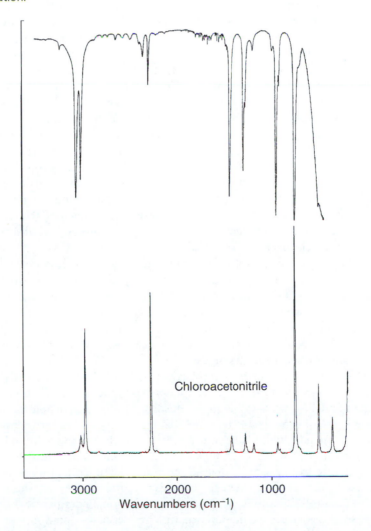

Chloroacetonitrile

particularly the band due to the C–N stretching mode at about 2200 cm^{-1}. All of this occurs because the IR activity of a vibrational mode requires a change in dipole moment during the motion, but Raman activity requires a change in the polarizability. Thus, in a centrosymmetric molecule, the totally symmetric vibration does not give rise to a change in dipole and is IR-inactive. However, it can readily be shown that it does give rise to a polarizability change and consequently is Raman-active.

Raman scattering information is presented as an intensity-versus-wavelength shift in cm^{-1} and can be conveniently compared to its IR counterparts. It is important to realize that the position of a given IR band in cm^{-1} is identical to the wavelength shift of the Raman band because the same transition is being excited by both processes. Thus, fundamental group frequencies from IR tables can be conveniently used in Raman studies to obtain structural information. The number of reference spectra available for Raman is limited as compared to IR. However, several tabulations are available.

As is true for IR, the number of fundamental vibrational modes is $3n - 6$ for nonlinear molecules and $3n - 5$ for linear molecules. The actual number of bands observed in the spectrum may be diminished in the same fashion as in IR. That is, some modes are not active and degeneracies may occur. However, in normal Raman overtones, combination and difference bands are rare. This results in spectra that are significantly simpler than their IR counterparts (see Fig. 16.3).

A limitation of normal Raman spectroscopy is its inherent weakness. Solutions that are on the order of 0.5 M or greater are needed to obtain satisfactory spectra. However, if the energy of the incident photons falls within an excited state manifold (see Fig. 16.1), then the signal due to excitation of particular vibrations may be enhanced as much as one-thousand-fold. This is the resonance Raman effect. It should be noted that the arrowheads corresponding to the energy of the incident photons fall between vibrational energy levels of the excited state. This was done to emphasize the difference between resonance Raman scattering and resonance fluorescence. In the latter, an actual absorption and re-emission takes place that results in depolarized radiation. This is not so for Raman scattering, where the scattered radiation may be highly polarized. Because the signal intensity is 10^3 times stronger than in normal Raman scattering, solutions on the mM scale may thus be used to produce satisfactory spectra. The spectra become even simpler than those obtained with normal Raman because only totally symmetric vibrations show this large signal enhancement. Figure 16.4 dramatically shows this effect using various laser lines to excite the spectrum of Fe(phen)$_3^{2+}$. When the low-energy laser line is used, the spectrum of the compound is virtually unobservable. When energies within the region of the absorption maximum (510 nm) are used, strong clean spectra are obtained.

Another limitation of normal Raman scattering is that it is much weaker than fluorescence emission. The latter may originate from the sample itself or some small impurity in the sample. Because the region of the electromagnetic spectrum where fluorescence occurs is always the same, but the Raman wavelength depends on the exciting line, it is possible to avoid the fluorescence problem by using an alternate exciting line. Recently, the advent of FT-Raman offered another solution to this problem. By exciting in the near IR, 1.06 μ with a Nd–YAG laser, we can avoid exciting the fluorescence in the first place.

Surface-enhanced Raman spectroscopy (SERS) is another recently developed technique for obtaining strong signals from small numbers of organic molecules. It was first reported in 1974 by Fleishman and colleagues (5). Samples adsorbed onto microscopically roughened metal surfaces or colloids give spectra as much as 10^6 stronger than expected from normal Raman. Metals that are effective include gold, silver, and copper, although zinc and gallium have also shown the effect.

Raman spectra are obtained by measuring the intensities and frequencies of scattered radiation originating from a sample that has been irradiated with a monochromatic source. The units on the abscissa are Δ cm^{-1}, where Δ indicates the difference in cm^{-1} between the energy of the incident and scattered photons. These are the vibrational energies of the sample molecules. Typically the source is a laser; a common experimental setup for obtaining spectra is shown in Fig. 16.5.

Figure 16.3 Raman (lower trace) and IR (upper trace) spectra of mesitylene.

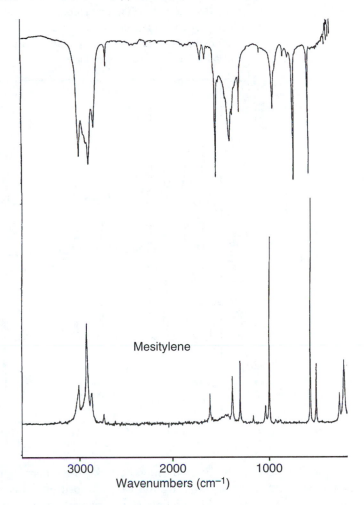

Mesitylene

3000   2000   1000

Wavenumbers (cm⁻¹)

Dispersive Spectrophotometers

These have traditionally used a double-grating monochromator (sometimes even a triple) in order to eliminate the intense Rayleigh scattering and the gratings normally have 1200 to 1800 grooves/ mm. The Czerny–Turner optical arrangement used by Spex Industries is shown in Fig. 16.6. These instruments have four slits that are manually controlled to select resolution and throughput. Recently developed filters, called super notch filters, have eliminated the need for double monochromators in many applications. They can reduce Rayleigh scattering to a point where spectra may be obtained to within 150 cm⁻¹ of the exciting line. However, studies requiring closer approach to the Rayleigh line still require the use of double monochromators. Some of the scattered radiation may be highly polarized and because the efficiency of the gratings depends on polarization, it should be randomized (scrambled) by passing it through a quartz wedge. Failure to do this can result in spectra in which the intensities of certain bands are anomalously low. It is desirable to pass the laser radiation used to excite the Raman scattering through a premonochromator in order to remove the plasma lines from the beam. These plasma lines originate from electronic transitions that, though not lasing, are strong enough to interfere with the spectra and even to damage the detection system.

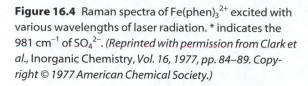

Figure 16.4 Raman spectra of Fe(phen)$_3^{2+}$ excited with various wavelengths of laser radiation. * indicates the 981 cm^{-1} of SO$_4^{2-}$. *(Reprinted with permission from Clark et al., Inorganic Chemistry, Vol. 16, 1977, pp. 84–89. Copyright © 1977 American Chemical Society.)*

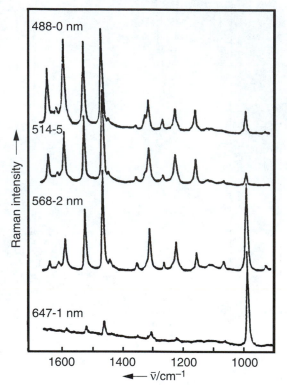

For example, if a spectrum is obtained using the 488.0-nm radiation of an Ar ion laser, a strong plasma line is observed that is the nonlasing 506.19-nm transition. This manifests itself as a band at 737 cm^{-1} in the Raman spectrum. Plasma lines are particularly prevalent when solid samples are being studied. A more thorough listing of these lines and where they appear in spectra may be found in the book *Laboratory Raman Spectroscopy* by Strommen and Nakamoto. It is important to note that fluorescent room lights give rise to at least one spurious line for each excitation wavelength. Thus,

Figure 16.5 Standard 90° arrangement for the detection of Raman scattered radiation. The sample is located at the origin.

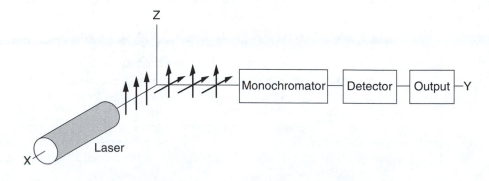

Figure 16.6 The optical arrangement used by SPEX Industries in their 1400 series double monochromators. *(Reprinted with permission from SPEX Industries.)*

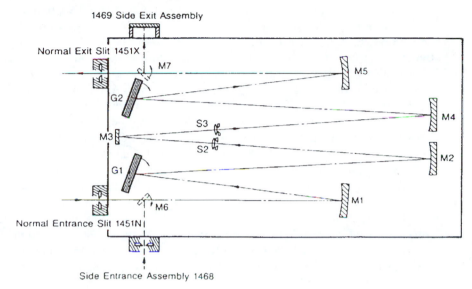

when using 488.0-nm excitation, a line is observed at 2180 cm^{-1} and when using 514.5-nm excitation, the same line is observed at 1125 cm^{-1}.

Detection is most often accomplished by use of a thermoelectrically cooled photomultiplier tube (PMT). Dispersive spectra may also be obtained with photodiode array (PDA) detectors and charge transfer devices (CTDs). The PDA is the simplest and consists of an array of photodetectors. A photodetector can be classified as either a photoconductive or a photovoltaic detector. When operated photovoltaically, the photodetector responds to light and generates an electrical current proportional to the number of photons that strike its surface. When the photodetector is operated as a photoconductor, its electrical resistance is reduced proportionally to the number of photons that strike its surface. Reverse-biasing a photodiode (applying a positive voltage to the photocathode) is generally used on PDAs to increase response time at the expense of optical sensitivity. A PDA is normally used together with an image intensifier to increase the weak Raman signal.

CTDs can generally be divided into two types: charge-coupled devices (CCDs) and the charge-injection devices (CIDs). In both the CCD and the CID arrays, a photosite converts the incoming optical signal to charge, which is stored in the device (integrated). The CCD is a metal-insulator semiconductor that can store charge in localized potential wells. A CCD array can be visualized by picturing the potential wells as a row of buckets (pixels) filled with charge. To read the information from a CCD array, the charge is transferred from the end pixel (bucket) to the readout electronics. The pixel that is adjacent to the end pixel then transfers its charge to the end pixel, which, in turn, transfers it to the readout electronics. This transfer of charge continues sequentially from pixel to pixel until all of the pixels have transferred their charge to the readout electronics. The differences in types of CCD arrays are due mainly to matrix size and the method of reading out the pixels (column readout, row readout, or a combination of the two). There are two main disadvantages to CCD arrays. One is the limitation to sequential readout of the pixels, which causes these arrays to be slower than the PDA. The other is the possibility of blooming. Blooming is an undesirable effect that can be thought of as charge overfilling a pixel (bucket) and spilling into the adjacent pixel. Both of these limitations are addressed in the CID array. The CID array is a hybrid between discrete detectors and monolithic arrays that includes some signal processing on the photodetector substrate. CID arrays can be manufactured from materials that cannot be used in CCDs. It has the advantage of random access to pixels, and is relatively insensitive to blooming.

When selecting the type of array to use (PDA, CCD, or CID), one should consider the complexity of optics necessary to collect and focus the light signal onto the array, the speed of light detection necessary, and the sensitivity required. As a rule, a PDA requires more sophisticated optics and filtering to ensure that the appropriate portion of the light signal strikes only the desired photodetector with no optical crosstalk between elements. With a CCD or CID, this effect is less pronounced because a larger area is available on which to direct the optical signal. The speed of light detection necessary (the system bandwidth) depends on how fast the rise time of the optical signal is. A CCD is the slowest of the three arrays and the PDA has the potential to be the fastest. Sensitivity of the array is an issue if very weak signals are to be detected. Generally, a PDA is more sensitive than a CCD or a CID because the CTD has an associated charge-transfer efficiency associated with it. These efficiencies can be high, but are still less than 1, making these devices less sensitive than a PDA. For overall ease of use, the CCD is probably best because it can be purchased relatively inexpensively, but it is the least flexible of the three arrays.

The choice of photosensitive material for both the PDA and CTD is dictated by the light wavelengths of interest. Materials are available from the visible region to the IR region. However, it should be noted that PDAs and CTDs require cryogenic cooling and are difficult to manufacture with any uniformity between photodetector elements. Two excellent articles on this topic may be found in the literature (6, 7).

Fourier Transform Spectrometers

These instruments use Michelson interferometers for wavelength analysis and Nd–YAG lasers as sources. The principles of Fourier transform (FT) spectroscopy are discussed in Chapter 15. The optical arrangement for the Perkin–Elmer System 2000R Near-Infrared FT-Raman Spectrometer is shown in Fig. 16.7. The radiation source emits at 1.06 μ and the scattered radiation may have wavelengths as long as 1.7 μ, a region that is unsuitable for PMTs, PDAs, and CTDs. In this region photoconductive devices function nicely as detectors. InGaAs and Ge are commonly used materials. They are often used at cryogenic temperatures to lower the noise level and thus raise the signal-to-noise ratio. However, this also reduces the range over which the detector is effective. For example, the cutoff for InGaAs at room temperature is approximately 3500 cm^{-1}, but at liquid nitrogen temperature it drops to 2900 cm^{-1}. The 1.06-μ radiation that is often used in FT-Raman systems is supplied by a continuous-wave (CW) Nd–YAG laser with output power ranging up to 3 W.

It is worth noting again that fluorescent room lights must be rigorously excluded from the optical bench because they give rise to numerous spurious bands between 1700 cm^{-1} and 3000 cm^{-1}.

What It Does

Sampling techniques used in obtaining Raman spectra range from the extremely simple to the rather complicated. In contrast to IR, where water is an impediment due to its strong absorptions across the vibrational energy region, Raman scattering due to water is quite weak. Thus aqueous solutions may be easily examined and glass containers may be used. Low-frequency spectra, down to 50 cm^{-1}, are readily obtained and there is no need to use specialized instruments to change beamsplitters and sources or to maintain the sample in a vacuum, as is the case when probing low-frequency

Figure 16.7 Optical design of the Perkin–Elmer 2000R Near Infrared FT-Raman Spectrometer. *(Reproduced by permission of the Perkin-Elmer Corporation.)*

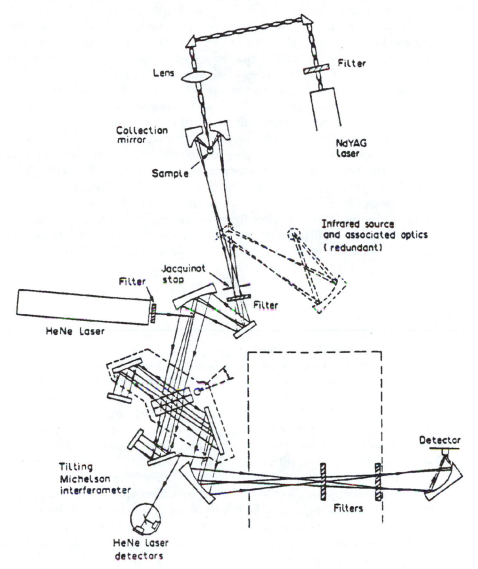

vibrational modes with IR instrumentation. The types of sampling devices used for resonance Raman differ from those used for normal Raman. The following are descriptions of a few of these sampling procedures as well as discussions of SERS and Raman microprobe experiments. More detailed descriptions and additional devices may be found in the monograph by Strommen and Nakamoto.

Normal Raman

Liquid samples may be contained in glass capillary tubes, NMR tubes, cuvettes, and even beakers. In fact, if the bottle containing the sample is clear glass, the sample need not be removed in order to obtain a satisfactory Raman spectrum. When using glass capillary tubes only a few microliters are needed. This is because the scattering volume is limited by the diameter of the laser beam at the focal point. This diameter typically is less than 0.01 mm.

Solid samples are equally simple to deal with. Powders and microcrystalline or amorphous materials may be packed into melting-point capillaries or NMR tubes. Translucent solids may be mounted as-is at the focal point of the laser beam. For example, the spectrum of the handle of a screwdriver can be obtained by placing the handle in the beam. The location of the beam within the handle is not critical to the experiment.

Specialized sampling devices are well-documented in the literature. Temperature studies of Raman band positions and intensities can be done with a Harney–Miller cell (8). This cell holds the sample in a glass capillary that is bathed in a stream of nitrogen whose temperature is controlled and monitored. The lower limit of this cell is determined by the boiling point of nitrogen. Temperatures below this value may be reached with the aid of liquid He transfer cells or, more conveniently, with a closed-cycle He refrigerator, which cools a Cu block on which the sample is deposited. Samples may be vaporized and condensed on the cold surface or they may be trapped (matrix isolation) in an inert gas matrix that is cocondensed with the sample. Spectra obtained in this fashion are free from intermolecular interactions and as a result are very sharp, often revealing details unavailable when the sample is restricted by other environments. High-pressure diamond anvil cells similar to those used in IR experiments are also used in Raman spectroscopy (9).

Resonance Raman

Liquid samples must be handled differently because one must use radiation whose energy matches that of an absorption of the molecule. Given the high photon flux of the laser and the tightness of focus of the beam, local heating becomes a potential problem. In order to prevent both thermal lensing (which distorts the focus) and thermal decomposition, the laser beam must be prevented from irradiating the same volume for any significant length of time. Two general methods have been devised that minimize the local heating effect. In one of these the sample is placed in a round glass container with an optically flat bottom. The container is then rotated rapidly, forcing the sample to the sides. The difficulty with this approach is that the laser beam is brought up from the bottom and it is difficult to keep the laser and the container wall perfectly vertical and at the same time at the exact position of focus for the collection optics. A technique that avoids this problem uses a spinner that holds an NMR tube containing the sample. The beam is brought in on an angle and the scattering is observed at 135°. Spinning devices such as these may be conveniently constructed from the spinner mechanisms of old NMR systems such as the Varian A60.

An alternative approach to the heating problem involves flowing the sample through a closed loop using a peristaltic pump. The loop contains a section of glass capillary tube that is positioned in the laser beam. The advantage of this method is that alignment is rather simple. One can align a capillary containing some strong scatterer such as toluene and then merely replace it with the capillary in the closed loop. The final focal point will be very close. The disadvantage of this method is that it generally requires a much larger amount of the sample than does the rotating cell technique.

Solid samples must also be rotated in order to reduce the possibility of localized heating. This can be done by using the NMR spinner technique described previously. An alternative method requires the preparation of a pellet that is rotated while the beam is focused on or near the edge of the pellet. The pellets are prepared in a fashion similar to KBr pellets for IR spectra. Approximately 200 mg of KBr is ground and placed in a pellet press. Then 5000 to 7000 lb/in.^{2} pressure is applied for approximately 3 min. At this point the press is opened and 100 mg of the ground sample is spread on top as evenly as possible. The sample may also be mixed with an internal standard such as Na_2SO_4 in order that comparisons may be made of relative intensity changes as a function of excitation wavelength. The press is reassembled and the pressure is again raised to 7000 lb/in.² for an additional 3 min. The pellet is then removed and mounted on the rotator at a 45° angle with respect to the laser beam. It should be noted that none of these techniques prevent photodecomposition, which occurs on a time scale that is much faster than rotation times.

FT-Raman

This technique, which is virtually free of interference from fluorescence, can be used to examine a wide variety of samples. Dyes are conveniently examined and may be determined at extremely low levels as described in the following section on SERS. Its main drawback is the fact that water absorbs in the 1.0 μ region and the commonly used source, a CW Nd–YAG laser, radiates energy at 1.06 μ. Thus, the advantage of being able to study aqueous solutions is eliminated. Because of the freedom from fluorescence interference and ease of sampling, FT-Raman spectroscopy has application as an industrial tool. It has been shown to be useful in the study of the morphology of polyethylene and is commonly used to obtain data on polyolefins (10). However, it must be noted that the great advances made in CCD detector technology have somewhat diminished the potential of FT-Raman. Currently available CCD detectors are useful out as far as 1 μ.

SERS

For the SERS effect to be observed, the molecules must be approximately 50 Å or less from the surface. Two of the techniques by which this is accomplished are adsorption onto metal films and adsorption onto colloids. The latter has been put to rather ingenious use in combination with FT-Raman. Samples run on TLC plates are subsequently sprayed with a Ag colloid solution. The resulting admixture evokes the SERS effect. Using this technique, extremely small amounts of material may be studied. In Fig. 16.8, a Raman spectrum of a sample of rhodamine B was obtained in situ after a 40-ng spot on a TLC plate was sprayed with a silver colloid. A recent development in this field involves the use of laser ablation of metals to prepare SERS-active colloids (11). The pure metal is suspended in a solvent and is irradiated for 10 to 15 min with 1064-nm radiation from a Nd–YAG laser. The resulting colloidal suspension is mixed with the substrate to be studied. In this fashion, a Raman spectrum of 5.0×10^{-6} M cytochrome C was obtained with 3 mW of 514.5-nm excitation.

Raman Microprobe

A monochromatic laser beam may be tightly focused through the use of a microscope objective, allowing the operator to examine regions on or within a sample, where the region of interest may be as small as 2.0 μ. Thus, areas of a corroded surface may be microscopically analyzed. Inclusions in minerals and meteorites may also be studied. The latter application is of particular interest to geochemists, who can determine the gas pressure within inclusions by monitoring the position of the 2917-cm^{-1} Raman band, which is pressure-sensitive (12). With this information, the processes involved in the formation of the mineral may be ascertained. When examining surfaces, the instrument may be tuned to a given vibrational mode. This results in an image of the regions on the surface where the particular vibrational moiety exists.

Figure 16.8 Near infrared FT-Raman spectrum of a 40-ng spot of Rhodamine 6G enhanced via the SERS effect. *(Reproduced by permission of the Perkin-Elmer Corporation.)*

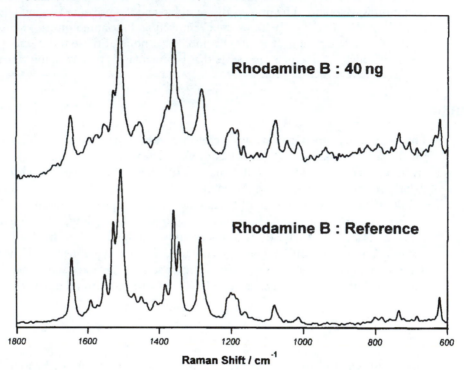

Remote Raman Analysis

Here two techniques are commonly used, one involving the use of light detection and ranging (LIDAR) and the other using optical fibers. LIDAR is discussed in the Applications section of this chapter.

Optical fibers may be used in a fashion similar to that used when obtaining UV or IR spectra. The optical fiber probe is coupled to a laser and to the monochromator. Figure 16.9 shows one such arrangement which was used to monitor three different fuel samples simultaneously (13). The spectra obtained are shown in Fig. 16.10. Industrial uses for a setup such as this are evident.

Raman Depolarization Ratios

No discussion of Raman spectroscopy would be complete without a discussion of depolarization ratios. The depolarization ratio may be defined as

$$\rho = I_\perp / I_{//} \qquad (16.1)$$

where $I_\perp$ = the intensity of scattered radiation whose plane of polarization is perpendicular to that of the incident radiation and $I_{//}$ = the intensity of scattered radiation whose plane of polarization is parallel to that of the incident radiation. Determination of ρ is a rather simple matter. One merely records the spectrum with a piece of polarizing material (the analyzer) placed between the sample

Figure 16.9 Schematic of the sampling setup for simultaneous determination of fuel samples. *(From R. Grayzel, F. Purcell, and F. Adar, "Remote Raman Analysis for Process Monitoring Applications," SPIE Vol. 1681, pp. 148–158, 1992.)*

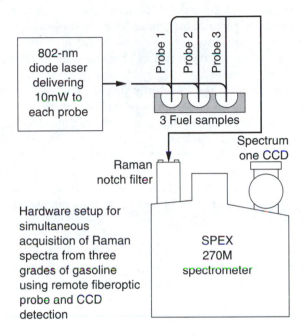

Figure 16.10 Raman spectra obtained using the experimental setup shown in Figure16.9. *(Reprinted with permission.)*

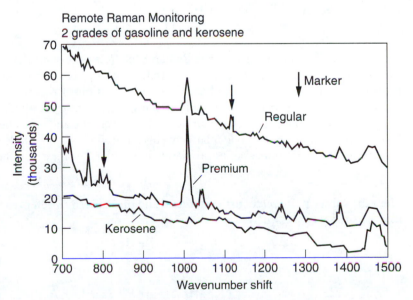

and the monochromator. The analyzer is positioned to pass one type of radiation and then rotated to pass the other.

If the vibrational motion preserves the symmetry of the molecule, then the value of ρ will range from 0 to just less than 3/4. If the symmetry of the molecule is reduced by the vibration, the ratio will be 3/4 within experimental error. In resonance Raman spectra certain values of ρ take on special significance. Thus, a value of 1/3 for a molecule possessing at least one threefold or higher axis of symmetry indicates that the state in resonance with the incident radiation is nondegenerate.

ρ finds its greatest use in determining the shapes of molecules and in associating observed bands with particular vibrational motions. However, there have been other, rather unusual uses of ρ. For example, the change in state from trans to gauche forms causes a concomitant change in the polarization of Raman scattered light. Thus, poly(dimethylsiloxane) networks were investigated (14, 15) using Raman depolarization in order to probe the molecular basis of rubber elasticity. The rotomeric population ($x = n_g/n_t$), where n_g and n_t represent the number of gauche and trans molecules, respectively, was obtained from the depolarization ratios of the 2907 cm^{-1}, CH$_3$ stretching vibration, and the 491 cm^{-1} Si–O backbone stretching mode using the relationship

$$\rho = \rho_g \frac{x}{1+x} \qquad (16.2)$$

Here again, the subscript g refers to the gauche form of the molecule. A more complete discussion of the origins and uses of ρ is available (16).

Analytical Information

It should be noted that Raman spectroscopy is a single-beam technique where internal standards are normally but not always used if quantitative results are required. The intensity of a Raman signal may be represented by Eq. (16.3).

$$I = K\nu^4 \mathcal{J} C \qquad (16.3)$$

Here I represents the intensity of the Raman band, K is a constant that includes the laser power at the sample, ν is the frequency of the scattered radiation, $\mathcal{J}$ is a scattering coefficient that is characteristic of each Raman band, and C is the concentration of the analyte under consideration. A working curve is constructed by obtaining the spectra of a series of concentrations of the desired analyte, each containing the same amount of an internal standard. The internal standard should be chosen so that it has a mode relatively close in frequency to a mode of the analyte under consideration. If this is done, a series of relative intensities, I_{rel}, may be obtained as follows:

$$I_{rel} = [(I_{std}/c_{std})(\nu_a/\nu_{std})^4(\mathcal{J}_a/\mathcal{J}_{std})]C \qquad (16.4)$$

Here the subscript std refers to the standard and the subscript a refers to the analyte. This equation may be simply expressed as

$$I = K'C \qquad (16.5)$$

where

$$K' = [(I_{std}/c_{std})(\nu_a/\nu_{std})^4(\mathcal{J}_a/\mathcal{J}_{std})] \qquad (16.6)$$

As mentioned above, internal standards are not always used and, indeed, were not used in two recent studies (17,18). In the former, a series of inorganic ions were investigated in the millimole range and correlation coefficients were obtained that were greater than 0.99. The latter dealt with the quantitative determination of benzene, toluene, ethylbenzene, and xylene in water. Detection limits as low as 1.4 ppm were obtained in this study but unfortunately correlation coefficients were not reported.

One of the difficulties that is certainly encountered in the absence of internal standards concerns the K of Eq. (16.3). Thus, the signal observed at the detector is a function of the laser power as well as the alignment of the sample. This means that the laser power must be carefully monitored and sample positioning must be carefully controlled.

Applications

1. Analysis of PCl$_3$ Reactor Material.

On-line analysis of hazardous materials in chemical processes is clearly desirable. PCl$_3$, which is a starting material in the production of phosphate esters, is itself manufactured from P$_4$ and PCl$_5$. Knowledge of the P$_4$ and PCl$_5$ concentrations at various times during the process is critical to safe operation. J. J. Freeman and colleagues (19) used FT-Raman spectroscopy to monitor these PCl$_3$ reactor materials. By coupling an FT-Raman spectrophotometer to the sample window of a reactor with a bifurcated optical bundle, they were able to obtain Raman spectra and thus to quantitatively monitor the concentrations of both of these species.

2. Analysis of Aviation Turbine Fuel.

Raman spectra obtained with a conventional system (double monochromator, Ar ion laser, and PMT) for aviation fuels give characteristic patterns (20). These may be used to ascertain the general hydrocarbon make-up of the fuels. As can be seen in Fig. 16.11, the spectrum of JP-4 is dominated by rather narrow bands, whereas the bands in JP-5 are broad. This indicates that JP-4 has a greater amount of aromatic hydrocarbons and JP-5 contains more aliphatic compounds. Once the peaks are classified as belonging to aromatic (narrow bands at 1386 and 1007 cm^{-1}) or aliphatic (broad bands at about 1450 cm^{-1}) constituents, the associated areas can be ratioed to give a measure of the aliphatic/aromatic character of the mixture. Fuel additives such as antioxidants (phenolic in nature) and icing inhibitors (ethylene glycol derivatives) can be seen readily by virtue of their characteristic Raman bands.

3. Qualitative Analysis of Pharmaceutical Products Inside Packaging.

Because of product tampering, pharmaceutical products are now regularly packaged in polymer containers; blister packages such as those used for antihistamine tablets are very common. Although the safety factor for the consumer is greatly increased, analysis of the product is made significantly more difficult and generally requires removal from the package. Thus, only small

Figure 16.11 Raman spectrum. (a) JP-4 using 514-nm laser excitation and (b) JP-5 using 514-nm laser excitation. *(From W. M. Chung et al., "Analysis of Aviation Turbine Fuel Components by Laser Raman Spectroscopy," Applied Spectroscopy, 45, 1991, p. 1528. Reprinted with permission of Applied Spectroscopy.)*

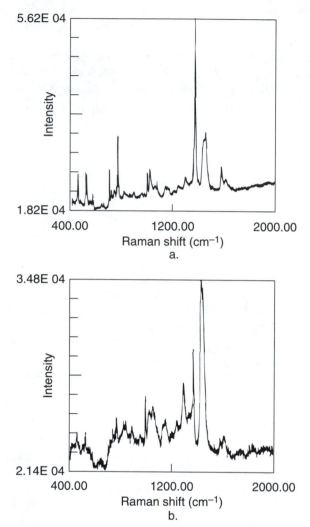

numbers of samples may be examined in a reasonable amount of time and on-line sampling is not a viable option. FT-Raman lends itself to this problem because the scattering due to the polymer coating is generally weak and the fluorescence of the product is not excited by the long-wavelength sources used. A recent investigation of pharmaceuticals contained in blister packages clearly demonstrated the efficacy of this technique (21). Fig. 16.12 shows the FT-Raman spectra obtained for antihistamine tablets both inside and outside the blister package as well as a difference spectrum. The packaging in this case was a poly(vinyl)-based polymer. The source was a CW Nd–YAG laser whose output power was reduced to 375 mW through the use of a series of neutral-density filters.

Figure 16.12 FT-Raman spectra of an antihistamine tablet inside its blister pack (top), removed from its blister pack (middle), and difference spectrum (bottom). *(From D. A. Compton and S. V. Compton, "Examination of Packaged Consumer Goods by Using FT-Raman Spectrometry,"Applied Spectroscopy, 45, 1991, p. 1587. Reprinted with permission of Applied Spectroscopy.)*

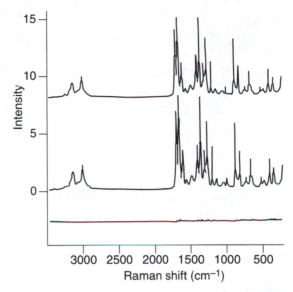

Using this technique, the authors were also able to obtain reasonable spectra from antihistamine microspheres inside gel capsules that were in turn contained in blister packs.

4. Remote Raman Sensing.

A light scattering technique called differential absorption LIDAR (DIAL) has become a useful technique for measurement of stratospheric molecules. LIDAR is a spectroscopic method whereby the Rayleigh (wavelength of incident light is not changed) scattering from an intense laser beam is collected by a telescope. An eximer laser supplies pulses of radiation and the return radiation is examined with respect to the time elapsed and the intensity in order to produce a concentration-versus-height profile of the analyte. In the DIAL modification, two wavelengths are used, one of which is less strongly absorbed. Typically, lasers emitting at 308 and 353 nm are used for this process. DIAL works well in a clean atmosphere where almost all the scattering is Rayleigh scattering. The scattering of aerosols is ignored or accounted for with a small correction factor. However, introduction of volcanic SO_2, such as occurred with the eruption of Mount Pinatubo, causes rapid changes in particle size and aerosol concentration. In order to circumvent this difficulty, McGee and colleagues (22) modified the LIDAR to detect Raman signals from the N_2 molecules. The scattering from the N_2 is shifted by 2331 cm^{-1} with respect to the incident wavelength and return depends on the molecular density. Aerosols do not contribute directly to the Raman return and thus backscattering from these is absent. Thus the extinction due to the aerosols can be calculated from the nonabsorbed beam. Using this technique, McGee and colleagues were able to produce ozone profiles that are valid from 15 to 50 km into the atmosphere.

Nuts And Bolts

Relative Costs

Dispersive	$$$$
FT-Raman	$$$$
PDA or CTD systems	$$$$
Raman-Microprobe	$$$$
Reference spectra	$$
Software	$$
Maintenance	$$
Supplies	$

$= 1 to 5K, $$= 5 to 15K, $$$= 15 to 50K, $$$$= 50 to 100K, $$$$$= > 100K.

Vendors for Instruments and Accessories

Instruments SA, Inc.
3880 Park Ave.
Edison, NJ 08820
phone: 908-494-8660, 800-438-7739
fax: 908-549-5125
Internet: http://www.isaine.com

Perkin-Elmer Corp., Analytical Instruments
761 Main Ave.
Norwalk, CT 06859-0012
fax: 203-761-2882
email: info@perkin-elmer.com
Internet: http://www.perkin-elmer.com

Bruker Instruments, Inc.
19 Fortune Dr.
Billerica, MA 01821
phone: 508-667-9580, 888-427-8537
fax: 508-663-9177
email: optics@bruker.com
Internet: http://www.bruker.com

Kaiser Optical Systems, Inc.
371 Parkland Plaza
P.O. Box 983
Ann Arbor, MI 48106
phone: 313-665-8083
fax: 313-665-8199
Internet: http://www.kosi.com

Nicolet Instrument Corp.
5225 Verona Rd.
Madison, WI 53711
phone: 800-232-1472, 608-276-6100
fax: 608-273-5046
email: nicinfo@nicolet.com
Internet: http://www.nicolet.com

Required Level of Training

Operation of Instrument

For routine operation, a person with 3 years of university education in chemistry or physics is desirable. There are some obvious hazards, electrical as well as visual, involved in working with lasers, so familiarity with these devices is useful. Detector systems can be easily damaged and great care should be exercised when exposing them to radiation. Sample handling is generally simpler than in the case of IR spectroscopy and is quickly mastered.

Processing Qualitative and Quantitative Data

As in the case of IR interpretation, a minimum of 1 year of college organic chemistry is required. Quantitative analysis of samples may be adequately performed by anyone who has completed a college course in analytical chemistry.

Service and Maintenance

Service contracts are available for all components of the instrument. Realignment of the incoming laser beam can be done on-site by resident personnel after minimal instruction. Internal realignment is more difficult and should generally be done by the vendor. If proper care is exercised, this should not be necessary unless the instrument is moved from one site to another. Replacement of the laser tube generally is the most costly maintenance item. A 5-W argon ion tube may cost as much as $10,000. Users who generate more than 5000 hours per year would be well advised to purchase extended tube warranties. Although these warranties are quite expensive, they cost significantly less than the tube itself. The monochromators used in dispersive instruments are generally very reliable and can be expected to give many years of trouble-free service. Computer controllers and detectors are expected to have similar lifetimes as those found with other instruments. Users who employ CTD detectors will need a source of liquid nitrogen. Sample containers are expendable items that range from inexpensive capillary tubes to more expensive specialized devices. The latter can cost up to several hundred dollars each.

Suggested Readings

Books

GARDINER, D. J. AND P. R. GRAVES, EDS., *Practical Raman Spectroscopy*. Heidelberg: Springer-Verlag, 1989.

GRASSELLI, J. G., M. K. SNAVELY, AND B. J. BULKIN, *Chemical Applications of Raman Spectroscopy*. New York: Wiley, 1971.

LONG, D. A., *Raman Spectroscopy*. New York: McGraw-Hill, 1977.

NAKAMOTO, K., *Infrared and Raman Spectra of Inorganic and Coordination Compounds*, 4th ed. New York: Wiley, 1986.

STROMMEN, D. P., AND K. NAKAMOTO, *Laboratory Raman Spectroscopy*. New York: Wiley, 1984.

Articles

ASHER, S. A., "UV Resonance Raman Spectroscopy for Analytical, Physical, and Biophysical Chemistry Part 1," *Analytical Chemistry*, 65 (1993), 59A.

ASHER, S. A., "UV Resonance Raman Spectroscopy for Analytical, Physical, and Biophysical Chemistry Part 2," *Analytical Chemistry*, 65 (1993), 201A.

CUTLER, D. J., "Fourier Transform Raman Instrumentation," *Spectrochim. Acta*, 46A (1990), 131.

GERRARD, D. L., and J. BIRNIE, "Raman Spectroscopy," *Analytical Chemistry*, 64 (1992), 502R.

HENDRA, P. J., "Industrial Value of Fourier Transform Spectroscopy," *Vibrational Spectroscopy*, 5 (1993), 25.

HENDRA, P., and H. MOULD, "FT-Raman Spectroscopy as a Routine Analytical Tool," *International Laboratory* (Sept. 1988), 34.

LEVIN, I. W., and E. N. LEWIS, "Fourier Transform Raman Spectroscopy of Biological Molecules," *Analytical Chemistry*, 62 (1990), 1101A.

STROMMEN, D. P., AND K. NAKAMOTO, "Resonance Raman Spectroscopy," *Journal of Chemistry Education*, 54 (1977), 474.

WAYNE, R. P., "Fourier Transformed," *Chemistry in Britain*, 23, no. 5 (1987), 440.

Short Courses

"Infrared Spectroscopy II. Instrumentation, Raman Spectra, Polymer Spectra, Sample Handling, and Computer Assisted Spectroscopy," Bowdoin College, Brunswick, ME (held in summer).

Reference Spectra

DOLLISH, F. R., W. G. FATELEY, and F. F. BENTLEY, *Characteristic Raman Frequencies of Organic Compounds*. New York: Wiley, 1974.

LIN-VIEN, D., AND OTHERS, *Infrared and Raman Characteristic Frequencies of Organic Molecules*. San Diego: Academic Press, 1991.

The Sadtler Standard Raman Spectra Collection, Bio-Rad Laboratories Inc., Sadtler Division, 3316 Spring Garden Street, Philadelphia, PA 19405.

SCHRADER, B., *Raman/Infrared Atlas of Organic Compounds*, 2nd ed. Weinkeim, Germany: VCH-Verlag-Ges., 1989.

References

1. Leonardo da Vinci, *Trattato della Pittura*.

2. Lord Rayleigh, *Phil. Mag*, 46 (1899), 375, Coll. papers IV, 397.

3. C. V. Raman and K. S. Krishnan, *Nature*, 121 (1928), 501.

4. A. Smekal, *Naturwiss.*, 11 (1923), 873.

5. M. Fleishman, P. J. Hendra, and A. J. McQuillan, *Chem. Phys. Lett.*, 26 (1974), 163.

6. R. B. Bilhorn and others, *Applied Spectroscopy*, 41 (1987), 1114.

7. R. B. Bilhorn and others, *Applied Spectroscopy*, 41 (1987), 1125.

8. F. A. Miller and B. M. Harney, *Applied Spectroscopy*, 2 (1970), 291.

9. W. F. Sherman and G. R. Wilkinson, in R. J. H. Clark and R. E. Hester, eds., *Advances in Infrared and Raman Spectroscopy*, vol. VI. (1980), 158.

10. P. J. Hendra, in J. V. Dawkins, ed., *Applied Science*. (1980).

11. J. Nedderson, G. Chumanov, and T. M. Cotton, *Applied Spectroscopy*, 47 (1993), 1959.

12. D. Fabre and C. R. Couty, *C. R. Acad. Sci. Paris*, 303, no. 4 (1986), 1305.

13. F. Adar, R. Grayzel, and F. Purcell, *S.P.I.E.*, 1681 (1992), 149.

14. J. Maxfield and I. W. Shepherd, *Chem. Phys. Lett.*, 19 (1973), 541.

15. J. Maxfield and I. W. Shepherd, *Chem. Phys.*, 2 (1973), 433.

16. D. P. Strommen, *J. Chem. Educ.*, 69 (1992), 803.

17. D. R. Lombardi and others, *Applied Spectroscopy*, 48 (1994), 875.

18. B. L. Wittcamp and D. C. Tilotta, *Analytical Chemistry*, 67 (1995), 600.

19. J. J. Freeman, D. O. Fisher, and G. J. Gervasio, *Applied Spectroscopy*, 47 (1993), 1115.

20. W. M. Chung and others, *Applied Spectroscopy*, 45 (1991), 1527.

21. D. A. C. Compton and S. V. Compton, *Applied Spectroscopy*, 45 (1991), 1587.

22. T. J. McGee and others, *Geophys. Res. Lett.*, 20, no. 10 (1993), 955.

Acknowledgements

The author wishes to express his appreciation to Dr. René Rodriguez and Ms. Kris Campbell for their helpful comments regarding detector systems.

Chapter 17

Nuclear Magnetic Resonance Spectroscopy

Michael McGregor

University of Rhode Island
Department of Chemistry

Summary

General Uses

- Identification and proof of structure of chemical compounds
- Quantitative determination of sample components
- Detailed information on the spatial orientation of nuclei in a molecule
- Studies of dynamic systems including chemical equilibria, molecular motion, and intermolecular interactions

Common Applications

- Widely applied for structure proof in synthetic chemistry, natural products chemistry, and biochemistry
- Determining stereochemistry and higher-order structures in molecules of all sizes
- Conformational analysis, including assessment of mixtures in conformational equilibria
- Following the course of chemical reactions by identifying and quantifying the starting materials and products

- Intermolecular interactions such as ion-pair formation or enzyme–substrate interactions
- The study of chemical exchange, such as valence bond tautomerism

Samples

State

Samples may be liquid or solid. Solid samples are commonly studied in solution, but may also be used in the solid state

Amount

About 50 μM is required for efficient analysis. Signal averaging can reduce this quantity to 50 nM or less, depending on the complexity of the analysis.

Preparation

Samples are commonly dissolved in deuterated solvents. Solid samples are usually powdered, but single-crystal samples also may be analyzed.

Analysis Time

- Depends on the amount of sample available, magnetic field strength of the spectrometer, sensitivity of the observed nucleus, and complexity of the analysis.
- Routine ^{1}H spectra require about 5 min, routine ^{13}C spectra require 5 to 30 min.
- Dilute samples or complex analyses usually require several hours, up to a realistic maximum of 48 hr.

Limitations

General

- Pure compounds are usually required; mixtures are much more difficult to deal with.
- Background interferences become a limitation with very dilute (ng) samples.
- The method cannot distinguish among magnetically equivalent monomers and dimers.

Accuracy

Accuracy is routinely $\pm 5\%$; under rigorous conditions it is $\pm 0.1\%$.

Sensitivity

Low sensitivity is the principal limitation of the method, due to limited sample amount, observation of an insensitive nucleus, or performance of a complex analysis.

Complementary or Related Techniques

- Mass spectrometry or any method of molecular weight determination
- Electron spin resonance (ESR), a magnetic resonance technique used primarily for structural studies of free radicals

Introduction

In the 50 years since the first observation of nuclear magnetic resonance (NMR), the technique has become an indispensable tool for the chemist. From the time of the first commercial spectrometers NMR has enjoyed an impressive success in solving chemical problems (1). In the early years, applications of proton NMR were quickly extended to structural studies of all major types of organic compounds. The utility of NMR in the study of dynamic chemical systems and in conformational analysis had also become clear (2). Since then, the field has witnessed a vigorous evolution, punctuated by several major revolutionary advances in instrumentation and techniques.

The development of powerful superconducting magnets and the introduction of the pulsed Fourier transform (FT) technique (3) vastly increased the sensitivity and resolving power of the method. The sensitivity improvement meant that NMR could be extended to nearly all elements in the periodic table. Two-dimensional techniques (4) were discovered and rapidly evolved to provide the chemist with the ability to map nuclear connectivities in great detail. Techniques for high-resolution spectra of solids (5) and magnetic resonance imaging emerged as major branches of the field.

Today's NMR spectrometer incorporates the advances of the last 50 years in a versatile, easy-to-use instrument. Automated data acquisition and processing provide access to the most useful experiments on a routine basis.

Technological improvements continue to propel advances in NMR applications. The recent introduction of pulsed-field gradient techniques in high-resolution spectrometers as well as three-dimensional and higher-dimensional techniques promises that applications of NMR will continue to diversify and grow.

How It Works

Physical and Chemical Principles

The following brief discussion of NMR uses a minimum of mathematics for simplicity. Rigorous descriptions of the physical basis for NMR can be found in References 6–9.

Magnetically active nuclei have a property known as spin. As a consequence of having spin, a magnetically active nucleus also has a magnetic moment and an angular momentum. The magnetogyric ratio is the ratio of these two properties. Each magnetically active isotope has a characteristic magnetogyric ratio.

When a nucleus is placed in an external magnetic field, B_0, two aspects of the same physical phenomenon can be observed: nuclear Zeeman splitting and nuclear precession.

Zeeman splitting results in the creation of $2I + 1$ magnetic energy states, where I is the spin quantum number. Nuclei with an atomic mass and an atomic number that are both even numbers have $I = 0$. Such nuclei cannot produce an NMR signal. For spin 1/2 nuclei such as ^1H, ^{13}C, ^{15}N, ^{19}F, and ^{31}P, Zeeman splitting results in two energy levels, shown in Fig. 17.1. The energy separation of these levels is proportional to the strength of B_0. Table 17.1 shows some nuclear properties of the most commonly observed nuclei.

Nuclear precession is illustrated in Fig. 17.2. Precession is the motion of a spinning body whose axis of rotation is constantly changing orientation. The spinning axis describes a cone around the external magnetic field axis. The precessional frequency for any nucleus is equal to the magnetic field strength (B_0) times the magnetogyric ratio.

In the magnetic field B_0, the precessing nucleus may have its magnetic moment aligned with the field, which is the low-energy state, or aligned against the field in the high-energy state. This is somewhat analogous to a magnetic compass, which is in a low-energy state when it aligns with the earth's magnetic field and points north. It would require the input of some energy to put the compass into a high-energy state pointing south.

In a NMR sample there is a large collection of nuclei with a certain amount of kinetic energy. The distribution of nuclei between the two energy states depends on the temperature, the magnetic field strength, and the magnetogyric ratio. Near absolute zero, almost all the nuclei would be in the low-energy state. At room temperature, however, the nuclei have enough kinetic energy to be almost evenly distributed between the energy states. There is only a small excess of nuclei in the low-energy state, on the order of 1 part in 10^5. This is one reason for the low sensitivity of the NMR technique.

The Vector Model

The basic principles of NMR can be described using a vector model. A set of Cartesian coordinates is shown in Fig. 17.3, with the axis of the magnetic field, B_0, oriented in the positive z-direction. The sum of the magnetic moments of the excess of nuclei in the low-energy state may be represented by a net magnetization vector M. This vector may be visualized as a magnet, which will be surrounded with a coil of wire. This magnet-and-coil arrangement can be made to act as if it were an electric motor or an electric generator. In a motor, an electric current applied to the coil induces

Figure 17.1 The nuclear Zeeman splitting for spin 1/2 nuclei.

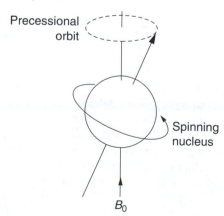

Precessional orbit

Spinning nucleus

B_0

Table 17.1 Nuclear properties of some commonly observed nuclei.

Isotope	% Natural Abundance	Spin	Frequency (MHz) with $B_0 = 2.35$ Tesla
^1H	99.985	1/2	100.00
^{11}B	80.42	3/2	32.08
^{13}C	1.108	1/2	25.14
^{15}N	0.37	1/2	10.13
^{17}O	0.037	5/2	13.56
^{19}F	100.0	1/2	94.08
^{23}Na	100.0	3/2	26.45
^{27}Al	100.0	5/2	26.06
^{29}Si	4.7	1/2	19.86
^{31}P	100.0	1/2	40.48
^{33}S	0.76	3/2	7.67
^{57}Fe	2.19	1/2	3.23
^{119}Sn	7.61	−1/2	37.27
^{195}Pt	33.8	1/2	21.5
^{205}Tl	70.5	1/2	57.69

a rotation of the magnet. This is analogous to excitation in NMR spectroscopy. In a generator, a rotating magnet induces an electrical current in the coil. This corresponds to the detection phase of NMR spectroscopy. This coil-and-sample arrangement, called a probe, is shown in Fig. 17.4.

Pulsed NMR

In order to observe the net magnetization vector, a second magnetic field, B_1, is applied to the sample at a right angle to B_0. The B_1 field is designed to rotate at the same frequency as the nuclei are precessing. B_1 is then static with respect to the precessing nuclei. The net magnetization vector precesses around B_1, moving away from the z-axis. This matching of the B_1 field rotation frequency and the nuclear precession frequency is called resonance.

Figure 17.2 Nuclear precession in a magnetic field B_0.

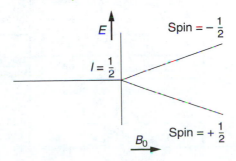

Figure 17.3 The net magnetization vector (M). (a) At equilibrium; (b) after a pulse.

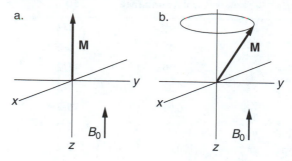

Figure 17.4 An NMR spectrometer including the superconducting solenoid magnet. The top section of the console block is where the excitation pulse is produced. The bottom section is the NMR signal reception system. The expanded view shows a sample within the coil, surrounded by the solenoid magnet.

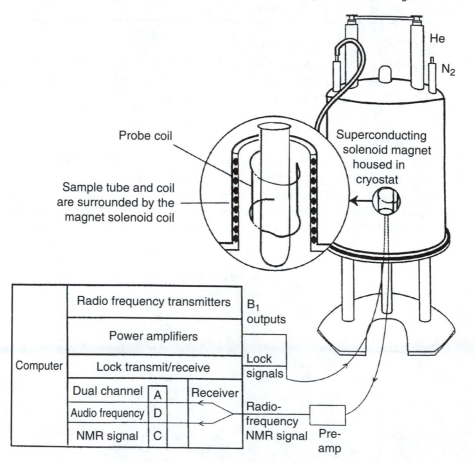

B_1 may be applied for a sufficient amount of time to rotate M into the x–y plane. This is called a 90° rotation, or a 90° pulse. The B_1 field is then turned off, and M lies in the x–y plane, rotating at the nuclear precession frequency relative to the coil. This rotating magnet induces a current in the coil. This is the NMR signal. The signal is called a free induction decay (FID). This is because it is produced by a freely precessing net magnetization vector, inducing a current in the coil, and decays with time as the system returns to equilibrium.

Relaxation

After an excitation pulse, the spins lose their excess energy via interactions with the surroundings and M returns to its equilibrium state. This exponential process is called spin–lattice relaxation. It is characterized by an exponential time constant T_1. The component of the magnetization in the x–y plane also relaxes via additional mechanisms. This is called spin–spin relaxation. This decay is characterized by an exponential time constant T_2. The observed decay of the FID is often driven by the inhomogeneity of B_0. The observed decay time constant is T_2^*. The linewidth at half height of an NMR line is equal to $1/\pi\, T_2^*$.

Chemical Shift

The nucleus experiences the effects of small magnetic fields in its immediate environment. Circulating electrons produce a magnetic field at the nucleus that may either oppose or reinforce the much-larger external field. When the local magnetic field opposes the external field, the nucleus is said to be shielded. A shielded nucleus experiences a lower effective field strength and resonates at a lower frequency.

As a result, each type of nucleus in a molecule may have a slightly different resonant frequency. This difference is called chemical shift. The NMR spectrometer accurately measures all resonant frequencies of the nuclei in a sample in units of hertz, or cycles per second. Chemical shift is defined as the difference between the resonant frequency of a nucleus in one type of chemical environment and that of a reference nucleus, divided by the spectrometer frequency. Because the spectrometer frequency is usually in the megahertz range, whereas the chemical shift range is in the hertz or kilohertz range, this ratio is expressed in parts per million.

Table 17.2 gives the typical chemical shift ranges of two of the most popular spin 1/2 nuclei. Reference compounds such as tetramethyl silane (TMS) provide agreed-upon chemical shift scales for all spectrometers. An excellent discussion of chemical shift can be found in Reference 10.

Spin–Spin Coupling

The small magnetic field of the nucleus exerts an effect on neighboring nuclei called spin–spin coupling. The effect is transmitted through bonds via bonding electrons. If a nucleus without coupling produces a single resonance line, this line is split into two lines by coupling with a neighboring nucleus. The J-coupling pattern for a nucleus contains $2nI + 1$ lines, where n is the number of neighboring nuclei and I is the spin quantum number. Smaller couplings may come from more distant nuclei, sometimes several bonds away. Coupling patterns may be more complicated due to an effect known as strong coupling, if the chemical shift difference between the coupled nuclei is small.

The coupling constant J is measured in hertz and is independent of the external magnetic field strength. The magnitude of the coupling constant J can give information about the stereochemical relationship between coupled nuclei. Three-bond couplings depend on the cosine of the dihedral angle between the coupled nuclei. This dependence is known as the Karplus relationship and it has

Table 17.2 General regions of chemical shifts for ^1H and ^{13}C.

Compound Type	^1H	^{13}C
Alkanes	0.5–1.3	5–35
Monosubstituted alkanes	2–5	25–65
Disubstituted alkanes	3–7	20–75
Cyclopropyl	−0.5–0.5	0–10
$R–CH_2–NR_2$	2–3	42–70
$R–CH_2–SR$	2–3	20–40
$R–CH_2–PR_3$	2.2–3.2	50–75
$R–CH_2–OH$	3.5–4.5	50–75
$R–CH_2–NO_2$	4–4.6	70–85
$R–CH_2–F$	4.2–5	70–80
$R–CH_2–Cl$	3–4	25–50
$R–CH_2–Br$	2.5–4	10–30
$R–CH_2–I$	2–4	−20–0
Epoxides	2.2–2.7	35–45
Nitriles	—	100–120
Alkenes	4.5–7.5	100–150
Allylic	1.6–2.1	18–30
Alkynes	2–3	75–95
Aromatic	6–9	110–145
Benzylic	2.2–2.8	18–30
Acids	10–13	160–180
Esters	—	160–175
Amides	5–9	150–180
Aldehydes	9–11	185–205
Ketones	—	190–220
Hydroxyl	4–6	—

Chemical shifts are expressed in ppm with TMS as a reference.

been used (and abused (11)) in stereochemical analysis for 40 years. Spin–spin coupling is a powerful tool both for establishing connectivity through bonds between atoms in a molecule and for establishing stereochemical relationships.

Nuclear Overhauser Effect

The nuclear Overhauser effect (NOE) is a complex subject that is discussed in a recent definitive monograph (12). The effect may be used to provide information on the spatial relationships of atoms in a molecule or to enhance the response of insensitive nuclei. The effect occurs only between nuclei that share a dipolar coupling, meaning that the nuclei are so close to one another that their magnetic dipoles interact. The most common applications of the NOE are for the sensitivity enhancement of ^{13}C spectra and for determining spatial relationships among protons in a molecule. Examples of the use of the NOE can be found in the Applications section of this chapter.

What It Does

Instrumentation

The modern NMR spectrometer is one of the most sophisticated analytical instruments available. Continuous improvements in magnet technology, electronic circuitry, computer power, and experimental techniques have resulted in an instrument with prodigious problem-solving power. Improvements in the sensitivity and stability of the instruments have increased the range of problems where NMR can be used. Automated user interfaces make instruments easy to use, even for some of the most advanced techniques (13).

A diagram of an NMR spectrometer is shown in Fig. 17.4. The main components are a superconducting solenoid magnet, radiofrequency transmitters and power amplifiers for generating B_1 fields, a probe for delivering B_1 fields to the sample and receiving the NMR signals, a radiofrequency reception system, and a computer for data processing.

The superconducting magnet is a solenoid coil wound with superconducting wire. Once charged with direct current, the superconducting design produces a strong, homogeneous magnetic field that is stable for years, provided that the coil is maintained below its critical temperature. This is achieved by immersing the coil in liquid helium. The magnet is housed in a cryostat, which is a system of concentric dewars enclosed in a vacuum vessel. The liquid helium dewar is surrounded by a liquid nitrogen dewar, and the vacuum vessel surrounds the liquid nitrogen dewar. Boil-off gasses pass out the top of the cryostat through vent/fill stacks. Cryogen boil-off rates are monitored closely, and the liquids refilled as necessary. The magnet should be housed in a temperature-controlled room, away from traffic flow and free from vibration. Clean, dry compressed air is required for the pneumatic system used for spinning samples and lifting samples in and out of the probe.

Built into the magnet are a set of superconducting cryoshims, which are small electromagnets that produce corrective magnetic fields for homogeneity adjustment. These are adjusted at the time of magnet energizing. Another set of shim coils, called the room-temperature shims, surround the probe. These are adjusted for maximum field homogeneity for each sample, either manually or under computer control. Samples may be rotated using an air-powered spinner for improved magnetic field homogeneity in the x–y plane. Recently introduced shim systems can maintain such good homogeneity in the x–y plane that no spinning is required.

Magnetic field strength is maintained at a constant value for each sample by a field-frequency lock circuit. This system monitors the resonant frequency of deuterium atoms in the solvent. The lock system compensates for any change in this frequency by adjusting the magnetic field strength to maintain a stable lock frequency. This stability is essential for efficient signal averaging or long-term experiments.

The probe consists of a coil of wire surrounding the sample space and arranged to produce a magnetic field perpendicular to the external field. The coil is part of a tuned circuit that, when properly tuned for the resonant frequency of choice and impedance-matched to the source transmitter, delivers the B_1 field to the sample. Many probes are capable of variable temperature operation.

After a brief (microseconds) delay to dissipate the power of the pulsed field, the coil functions as the NMR signal detector. The magnetization vector tipped by the pulsed magnetic field precesses around the B_0 field axis and induces a current in the coil. This current is the NMR signal. A sensitive preamplifier, located as near as practical to the coil, picks up and amplifies this weak signal.

The amplified signal is passed to a receiving system that has three functions: amplification, mixing, and quadrature phase detection. The process of mixing, or heterodyning, produces different frequencies from the radiofrequency NMR signal. Ultimately, an audio (kHz) range signal results.

Quadrature phase detection is accomplished by splitting the NMR signal into two channels, whose phases are 90° apart. Once these two channels have been created, proper operation requires that each stage of amplification be equal in both channels and that a phase difference of exactly 90° be maintained. Proper quad-channel balance minimizes spectral artifacts. This technique improves the signal-to-noise ratio of a single acquisition by 40% over a single-channel instrument.

The conversion of the analog NMR signal to digital form imposes some limitations on the receiving system. The analog signal must be at least strong enough to activate the lowest bit of the analog-to-digital (A/D) converter, yet not stronger than can be handled by activating all the bits of the A/D. If the analog signal exceeds the range of the A/D, the free induction decay (FID) is said to be clipped. Baseline distortions and spurious signals may then appear in the spectrum. If the analog signal is too weak, resolution may be lost. Most spectrometers are equipped to automatically adjust the receiver gain appropriately.

If a sample presents weak signals in the presence of strong signals (such as solvent lines), the weak signals may be lost. A 12-bit A/D can accommodate a maximum signal of 2048 units and a minimum signal of 1 unit. In practice, with a 12-bit A/D, this dynamic range problem begins at the point where the strong signals are roughly one thousand times as intense as the weak signals. Various techniques for suppressing the solvent lines are available. Such solvent suppression allows increased amplification of the FID, which makes the weak lines easier to detect.

A final limitation of the A/D system concerns the rate at which the audio frequency is sampled. According to the Nyquist theorem, the sampling rate must be twice as fast as the highest frequency to be detected. Signals that have a higher frequency than the sampling rate can handle will still appear in the spectrum, but at an incorrect or "alias" frequency. Careful attention to the width of the spectral window and the center of the observed frequency range can minimize this problem.

Once in the computer, all the frequencies in the time domain signal are converted to a frequency spectrum by Fourier transformation. The data system then displays and plots the NMR spectrum as directed.

The High-Resolution NMR Spectrum

Resolution in an NMR spectrum is defined as the linewidth at half-height of a suitably sharp NMR signal. In terms of being able to separate different NMR lines, resolution is controlled by three main variables: magnetic field strength, magnetic field homogeneity, and acquisition time. With a given instrument, only the last two are subject to operator control.

The effect of magnetic field strength is illustrated in Fig. 17.5(a). Higher magnetic field strengths produce higher spectral dispersion. This means that two nuclei with different chemical shifts have a larger frequency separation at higher fields. A chemical shift difference of 1 ppm includes a frequency range of 60 Hz in a magnetic field of 60 MHz, but 1 ppm in a 600-MHz magnet includes a 600 Hz frequency range. The increased spectral dispersion available at higher field strengths is one of the primary reasons that high fields are so useful.

The effect of magnetic field homogeneity on resolution is illustrated in Fig. 17.5(b). Good field homogeneity produces good lineshape because all the nuclei contributing to a resonance line

Figure 17.5 Simulated spectra. (a) The effect of magnetic field strength on resolution at constant linewidth; (b) the effect of linewidth on resolution at constant magnetic field strength.

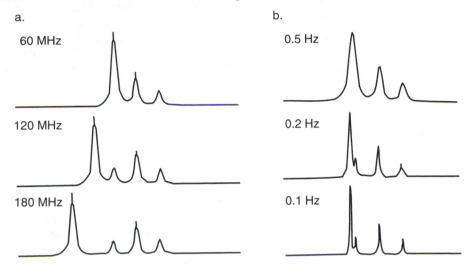

are precessing at the same speed regardless of their location in the sample. Magnetic field gradients result in broadened or distorted lineshapes.

The effect of acquisition time on resolution is defined below. Longer acquisition time means that better resolution is possible. A resolution of 0.5 Hz requires an acquisition time of 2 sec. In practice, the analyst decides what resolution is required and sets the acquisition time accordingly.

$$\text{Resolution (Hz)} = 1/\text{Acquisition time (sec)} \tag{17.1}$$

A high-resolution spectrum requires proper data processing, which may involve the techniques of zero-filling and resolution enhancement.

There are other variables that may degrade the resolution in an NMR spectrum. Vibration of the magnet or sample must be avoided. Vibration damping magnet supports are available for controlling this effect. Temperature fluctuations or gradients can cause resonances to be broadened or shifted. Spectrometers are usually fitted with temperature controllers that can maintain the desired temperature in the probe.

Gradient-Enhanced Spectroscopy

The introduction of actively shielded magnetic field gradients has made the use of pulsed field gradients possible in high-resolution NMR spectroscopy. The active shielding and other techniques reduce local field disruptions caused by eddy currents, which previously made the use of gradients impractical. The use of pulsed field gradients reduces experiment time, minimizes artifacts, and allows new levels of solvent suppression. For example, gradients can be used to select coherence transfer pathways, eliminating the need for extensive phase-cycling schemes. Gradients can minimize subtraction errors associated with difference experiments such as inverse detection. Gradients offer linewidth- and lineshape-independent solvent suppression and multiple solvent suppression and eliminate magnetization transfer from preirradiated solvent peaks. These advantages are especially important in studies of dilute biomolecules in aqueous solutions. Gradients also offer an ideal method for studying diffusion properties. Gradient options can be purchased with new spectrometers or added to existing instruments.

Solids

Solid-state samples present challenging problems. NMR lines in solids tend to be quite broad due to the effects of dipolar couplings between nuclei and the effects of chemical shift anisotropy (CSA). Both of these effects are greatly or completely reduced in solution by rapid molecular tumbling. Ingenious solutions to these two problems have made solid-state NMR spectroscopy one of the most interesting areas in the field.

The effects of CSA can produce line broadenings of many kilohertz. CSA refers to the directional dependence of electronic shielding in a molecule. It is possible to overcome CSA effects by spinning the sample rapidly at an angle to the magnetic field, known as the magic angle. Solids probes use spinning rotors to contain the sample and routinely reach magic angle spinning (MAS) speeds of 10 kHz, fast enough to remove CSA effects.

Decoupling the strong dipolar coupling between protons and carbons in an organic solid can be accomplished with very-high-power decouplers or by multiple pulse line narrowing. This technique uses a series of pulses to average the dipolar interactions by reorienting the spins. Dipolar coupling is exploited for increasing the sensitivity of less-sensitive nuclei using a technique known as cross-polarization. Cross-polarization and magic angle spinning (CPMAS) can produce high-resolution spectra from solids. Figure 17.6 shows a CPMAS ^{13}C spectrum of a solid organic compound. The very broad lines in the static spectrum are sharpened dramatically by the MAS technique.

Equipment requirements for solid-state NMR spectroscopy include high-power transmitters and probes designed to spin the samples at high speeds. Both solids spectrometers and multipurpose spectrometers for solutions and solids are available. Existing spectrometers can usually accommodate solids accessories.

Analytical Information

Qualitative

Among the many applications of NMR, the principal one continues to be the identification and proof of structure of chemical compounds. Today NMR spectrometers are equipped to gather structural information with unprecedented speed and convenience. Although each chemical problem is unique, almost all are approached by identifying the nuclei present in a molecule and defining the coupling relationships among them.

A basic set of experiments for obtaining this information efficiently is routinely available (13). As coupling relationships are identified, certain structural fragments become evident. Further experiments are performed to detect longer-range couplings and through-space couplings in order to connect the structural fragments. Both one-dimensional and two-dimensional experiments are used. Using an example case, the experiments listed in Table 17.3 are illustrated in this section of the chapter.

Sample Preparation

Soluble samples are dissolved in deuterated solvents, which are ordinary solvents that have had their protons replaced by deuterium atoms. For high-quality work, samples may be filtered and degassed to remove paramagnetic impurities such as iron particles and molecular oxygen.

Figure 17.6 Static (above) and MAS (below) ^{13}C spectra of l-Dopa in the solid state. *(Spectra courtesy of Dr. Jim Roberts, Lehigh University.)*

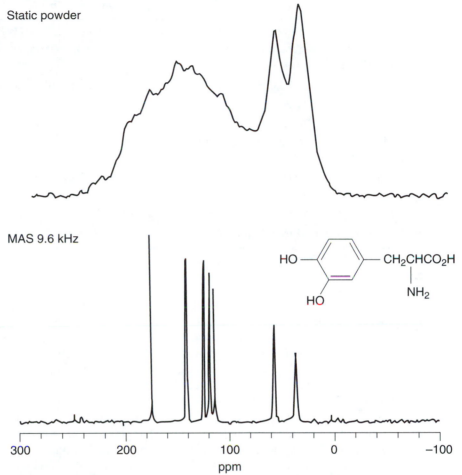

One-Dimensional Spectra

The usual starting point of a structural study is a proton spectrum. There are four main types of information present in the spectrum:

- The number of different proton resonances
- The chemical shifts of these resonances
- The intensity of the proton lines
- The multiplicity or coupling pattern

Figure 17.7(a) shows the proton spectrum of n-butyl acetate. There are five different proton resonances that integrate to a total of twelve protons. The two-proton triplet at 4.2 ppm is due to a CH_2–O group, which usually resonates between 3.5 and 4.5 ppm. The seven remaining protons between 1 and 2 ppm represent the two methylene groups and a methyl group at the end of the n-butyl alkyl chain. The coupling patterns in this spectrum can be understood using the $2nI + 1$ rule. For example, the triplet at 4.1 ppm is due to coupling between the CH_2–O protons and an adjacent methylene group.

Table 17.3 Experiments for obtaining chemical shifts, J-couplings, and dipolar couplings for structure proof.

Experiment	Information
One-dimensional proton and carbon spectra	Chemical shifts, proton coupling constants
Distortionless enhancement by polarization transfer (DEPT) spectra	Identifies methines, methylenes, and methyls
COSY	Identifies proton J-coupling relationships
XHCORR (correlation of hydrogen and any other nucleus), heteronuclear multibond correlation (HMBC)	Identifies proton–carbon couplings
Relayed coherence transfer (RELAY), total correlation spectroscopy (TOCSY)	Identifies long range proton–proton couplings
Correlation by long-range coupling (COLOC), HMBC	Identifies proton–carbon long-range couplings
J-resolved	Obtain proton–proton or proton–carbon coupling constants
NOE difference, NOE spectroscopy (NOESY)	Identify through-space dipolar couplings

Coupling in this molecule can be investigated using homonuclear decoupling. This technique is illustrated in Fig. 17.7(b). Using an additional magnetic field, focused on the protons of interest, it is possible to saturate any resonance in the spectrum. This removes the coupling effect of the decoupled protons and simplifies the spectrum. In this case, the triplet at 4.1 ppm simplifies to a singlet and the sextet at 1.4 ppm simplifies to a quartet. This demonstrates that the decoupled methylene group is located between the simplified methylenes. By stepwise decoupling, the complete proton coupling network of a molecule can be discovered.

Figure 17.7(c) shows the proton-decoupled ^{13}C spectrum of n-butyl acetate. There are six carbon resonances. The chemical shift of these ^{13}C lines allows them to be assigned. The carbonyl carbon signal is at 167 ppm, typical for ester carbonyls. The CH_2–O carbon is located at 64 ppm and the three remaining alkyl carbons are between 35 and 10 ppm.

Figure 17.7(d) shows the DEPT-135 results. This is called an edited carbon spectrum. Only protonated carbons respond so the solvent and quaternary peaks are absent. The number of protons attached to a carbon atom determines the phase of the carbon peak. CH_3 and CH carbons give a positive phase signal, whereas CH_2 carbons give negative phase peaks

This experiment is much more sensitive than the standard ^{13}C experiment due to an effect called polarization transfer. A variation of the experiment, called DEPT-90, gives peaks for C–H carbons only. With these three experiments, the number and type of carbon atom in a molecule can be determined.

Two-Dimensional Experiments

Two-dimensional experiments provide an efficient means of determining the coupling relationships among atoms in a molecule. A two-dimensional experiment involves the creation and observation of a second time domain, which becomes a second frequency domain after Fourier transformation. This is done by collecting a series of FIDs, as shown in Fig. 17.8. These FIDs are acquired after a pair of pulses are separated by a delay time. More complicated two-dimensional experiments may involve more than two pulses and more than one delay time. The delay time between the pulses is called the evolution time. The evolution time is systematically increased in each successive FID. Thus, each new FID represents a new picture of the evolution of the couplings in

Figure 17.7 Analysis of n-butyl acetate. (a) The proton spectrum. Integrated peak intensities appear below each peak. TMS is the reference. Lowercase letters mark peak assignments. (b) The proton spectrum of n-butyl acetate with homonuclear decoupling applied at 1.6 ppm. *(Figure continues on next page.)*

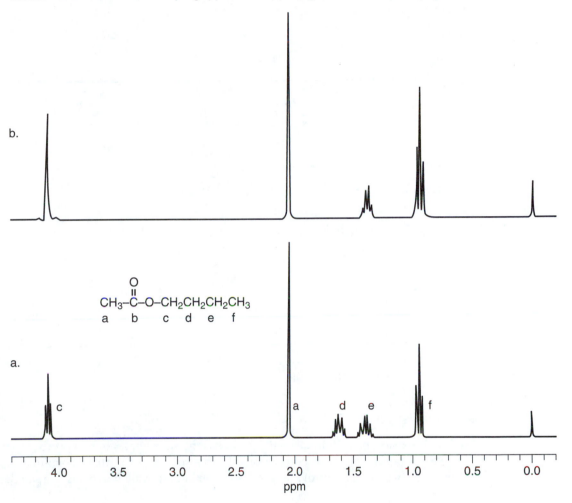

the molecule. After all the FIDs have been Fourier transformed, the data consist of a series of spectra in which the amplitude and phase of a peak change depending on the amount of time that coupling has been allowed to evolve. A new series of FIDs is now constructed by taking one point from each of the spectra, as shown in Fig. 17.8. Fourier transform of these FIDs produces a two-dimensional spectrum, shown in Fig. 17.8 as a contour plot. Nuclei that share a J-coupling produce a correlation peak.

This process can be used to observe several different phenomena that may occur during the evolution time. Chemical exchange, J-coupling, dipolar (through space) coupling, and recently, molecular diffusion (14) have all been used to generate two-dimensional spectra.

Figure 17.9(a) shows a correlation spectroscopy (COSY) spectrum of the aliphatic region of n-butyl acetate. The COSY results are displayed in a contour plot, which consists of an x–y coordinate system where the x- and y-axes represent chemical shift. Diagonal peaks (where x = y) simply represent the data of a one-dimensional spectrum and contain no coupling information.

Figure 17.7 (c) The ^{13}C spectrum of n-butyl acetate in $CDCl_3$. Broadband proton decoupling produces a single carbon line for each atom. TMS is the reference. The $CDCl_3$ peak is at 77 ppm. Lowercase letters mark peak assignments. (d) The DEPT-135 spectrum of n-butyl acetate. CH_3 and CH carbons produce positive peaks and CH_2 carbons negative peaks.

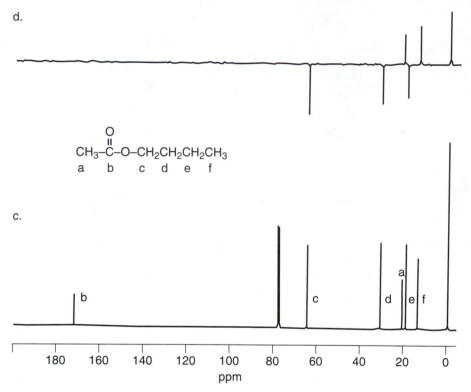

Each symmetrical pair of nondiagonal peaks has the coordinates of two protons that share a coupling. Such peaks are called correlations. In this spectrum the three couplings along the alkyl chain are shown. Variations of the COSY experiment are available to observe smaller, long-range couplings.

A second routine two-dimensional spectrum is shown in Fig. 17.9(b). This is a proton–carbon chemical shift correlation spectrum, known as heteronuclear correlation spectrum (HETCORR). In this spectrum, each correlation peak has the x-coordinate of the chemical shift of a carbon atom and the y-coordinate of the chemical shift of the protons that are directly bonded to the carbon.

At this point in the structure determination of an unknown, all of the carbon atoms and protons in the molecule are known. Major fragments of the molecule are also known. Using the COSY spectrum, for example, coupling in the n-butyl chain has been mapped clearly. The remaining two molecular fragments can now be connected using long-range coupling experiments. Figure 17.9(c) shows a portion of the long range proton–carbon correlation spectrum (COLOC). Correlation peaks in this spectrum are due to long-range proton–carbon coupling. The figure is labeled to show long-range coupling from the acetate methyl protons to the carbonyl carbon and long-range coupling from the –CH_2–O protons to the carbonyl carbon. These data firmly link both alkyl portions of the molecule to each other through the carbonyl carbon. Long-range coupling data such as this are often the key to solving structural problems.

Figure 17.8 The generation of a two-dimensional spectrum. (a) A second time domain is created by observing a series of FIDs. (b) When these FIDs are Fourier transformed, a series of spectra result. (c) The second time domain FIDs are created by taking one point from each of the spectra. (d) When all of the second time domain FIDs are transformed, the two-dimensional spectrum results. A contour plot is shown.

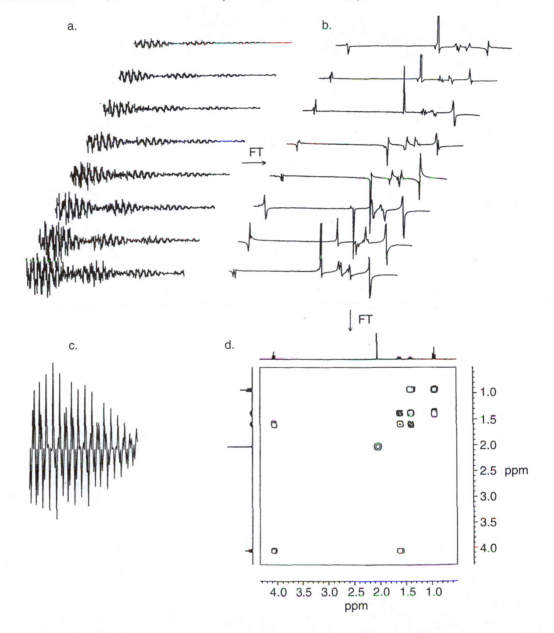

Modern spectrometers are equipped to perform proton–carbon correlation experiments in the inverse mode (15). The method uses proton detection to obtain correlations to the less-sensitive nuclei in a molecule, such as carbon or nitrogen. High-sensitivity makes this the method of choice when very small amounts of sample are available, as is the case with many biological compounds.

Figure 17.9 Analysis of n-butyl acetate. (a) Proton–proton chemical shift correlation spectrum. (b) Proton–carbon chemical shift correlation spectrum. (c) Proton–carbon long-range chemical shift correlation spectrum. Only the carbonyl region is shown.

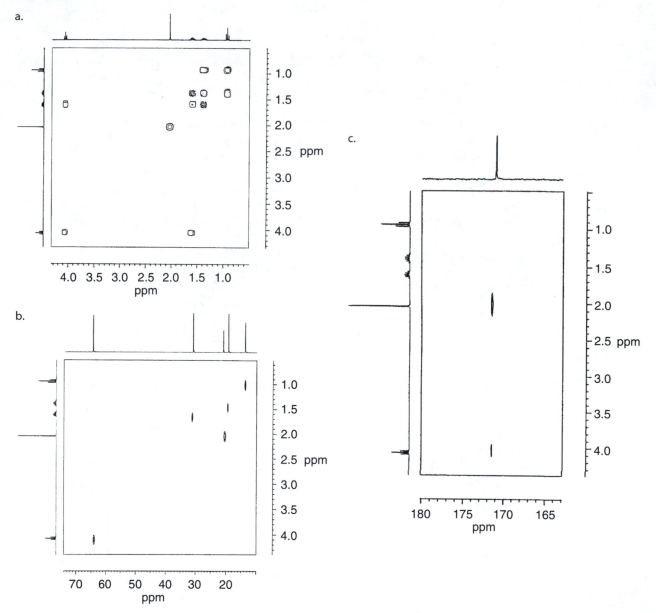

Nuclear Overhauser Effect Spectroscopy

It is possible to detect atoms that are close to each other in space, although the J-coupling is small or nonexistent. A dipole–dipole interaction between two atoms near to each other can be used to produce one-dimensional or two-dimensional correlation peaks. The effect is limited to very short interatomic distances, usually 4 Å or less.

An example of the use of the NOE is given in Fig. 17.10, which shows a NOE difference spectrum of a trisubstituted naphthalene. In this molecule there are two doublets in the proton spectrum. Although it is clear that the two doublets may be assigned to protons 1 and 4, it is not obvious which is which. Only proton 4 is close enough to proton 5 to show an NOE. In the NOE difference experiment, proton 5 is irradiated with a low-power field immediately before the acquisition. Any proton dipolar-coupled to this proton will show an enhanced signal intensity. When a control spectrum is subtracted from the preirradiated spectrum most peaks are canceled and only enhanced peaks remain.

In most cases the observed signal enhancement is less than 20%, possibly less than a few tenths of a percent. The difference spectrum is considerably simplified, and is largely independent of the number of different protons with similar chemical shifts. Thus, the NOE difference experiment is a powerful tool for making proton assignments, simplifying complicated spectra, and differentiating among possible solutions to structural problems. An excellent discussion of the technique can be found in Reference 7.

Figure 17.10 Trisubstituted naphthalene. (a) Control spectrum showing all proton resonances. (b) NOE difference spectrum. The experiment involves preirradiation of proton 5 at 8.5 ppm. The difference spectrum allows the unambiguous distinction of proton 4 from proton 1.

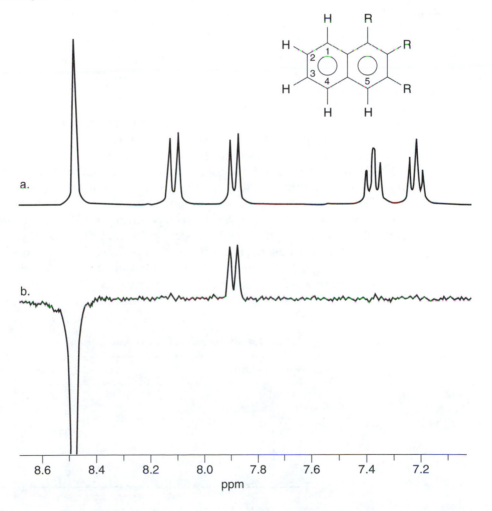

Quantitative

In addition to its role in structure determination, NMR has been widely applied in analytical chemistry. The technique is nondestructive, quantitative, and highly selective for each nuclear isotope, as the precessional frequency of the different isotopes are separated by many megahertz. The method works best with pure compounds, but mixtures can also be used. Chemical shift dispersion often makes it possible to use only one atom or group of atoms in a molecule for quantitative purposes.

The Quantitative NMR Spectrum

The intensity of the NMR signal is directly proportional to the number of nuclei in the probe, making quantitative analysis straightforward. Like any analytical technique, quantification has more stringent requirements and demands closer controls than qualitative analysis. These include sample-dependent and instrument-dependent considerations. The low sensitivity of the technique almost always requires signal averaging, which entails its own requirements.

A robust quantitative method must take into account the longitudinal relaxation time t_1. t_1 is a measure of the time required for the net magnetization vector to return to its equilibrium value after a pulse. This is an exponential process. Table 17.4 shows the percent recovery of magnetization from a pulse as a function of t_1.

Thus, after five t_1 periods, the net magnetization vector has recovered 99.3% of its equilibrium value. However, if two 90° pulses are separated by only one t_1, the second will generate a signal with only 63.2% of the intensity of the first. To avoid this clearly nonquantitative situation, the standard quantitative technique uses a waiting period or recycle time of five times the longest t_1 value among the nuclei being measured.

t_1 can be measured by a variety of techniques. Estimates of t_1 are also available in the literature. It is sometimes possible to add paramagnetic relaxation reagents such as Cr^{+++} to a sample to decrease t_1 (often by an order of magnitude) and shorten analysis time.

The NOE previously discussed must be suppressed in analyses where decoupling is involved. The NOE occurs only during the decoupling period, increasing to a maximum after five t_1 periods. A simple technique called inverse-gated decoupling suppresses the NOE by using a waiting time of five t_1 periods with the decoupler off and then acquiring an FID with the decoupler on.

Standardization of analytical results is accomplished in the same way as other instrumental techniques. In addition to the requirement for high purity, NMR standards must produce a signal that can be distinguished from that of the sample and impurities. Standards may be internal, dissolved in the same solution as the analyte, or external, usually contained in an axially concentric tube. A calibration curve may be prepared with a series of concentrations of an external stan-

Table 17.4 The recovery of magnetization from a pulse.

# of t_1 Periods	Recovery
1	63.2%
2	86.5
3	95.0
4	98.2
5	99.3

dard. The concentration of an analyte may be determined by adding a known amount of a pure standard to a measured volume of a solution with unknown analyte concentration. The number of equivalent nuclei represented by each sample peak and standard peak must be taken into account.

The method of standard addition works well in quantitative NMR. Accurately known amounts of the analyte are added to a solution with unknown analyte concentration. A graphic representation of the data confirms the linear response of the analyte in the concentration range used and allows extrapolation to the original concentration of the analyte.

The purity of a known compound can be determined by NMR, provided that a pure standard with nonoverlapping signal is available. Accurately weighed amounts of sample and standard are placed in solution and peak areas are measured.

With replicate samples and several measurements, the method can determine purities to 0.1% relative standard deviation.

The composition of mixtures may be determined in a straightforward way if each component of the mixture has a recognizable NMR peak.

Applications

NMR spectroscopy has been applied to an extremely wide variety of chemical problems. A typical year sees over 4000 publications on NMR techniques and applications, along with over 100 books and review articles. Biannual reviews of the literature have appeared in *Analytical Chemistry*, providing a convenient starting point for learning about particular applications. Constant improvements in the technique make it certain that NMR will continue to be adapted to new problems in the structure, reactivity, and dynamics of organic and inorganic compounds.

The general topic of proof of structure includes such applications as confirming the structures of reaction products, differentiating among isomers, determining stereochemistry, and conformational analysis.

The outstanding achievement in this area has been the development of techniques for determining the structure of high-molecular-weight biomolecules such as proteins and DNA in solution (16). These complex molecules, with hundreds of atoms and highly overlapped spectra, can now be studied using a combination of two-dimensional and three-dimensional experiments on isotopically labeled samples. Interproton distance constraints obtained through NOE experiments are used by molecular modeling programs to create a three-dimensional structure. Such structures are comparable to structures derived from X-ray diffraction studies on single crystals.

Some of the most interesting discoveries of polymer structure have come from studies where NMR has been used to identify and quantify structural features such as monomer sequence distributions, extent of branching, cross-linking, crystallinity, molecular motions, conformation, tacticity, and copolymer composition (17). Each of these properties may then be related to the bulk properties of the polymer.

As an example, Fig. 17.11 shows a quantitative carbon spectrum of a mixed meta- and para-cresol–formaldehyde novolak copolymer used in photoresists for photolithographic semiconductor fabrication (18). Analysis of the methylene bridging region of the spectrum allows calculation of the types and amounts of bridging in the resin. These data can be directly related to important physical properties such as resist photospeed.

Figure 17.11 Quantitative carbon spectrum of the methylene bridging region of a mixed meta- and para-cresol–formaldehyde novolak. The structure above indicates some of the potential bridged structures in the polymer. The DMSO-d_6 peaks are at left.

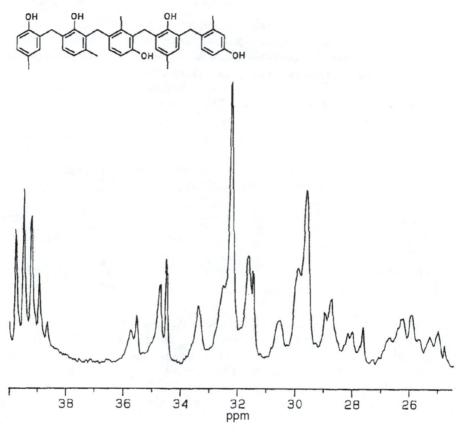

Much of what is known in physical organic chemistry has been made possible by NMR studies (19). The existence of the ring current in aromatic structures is demonstrated by the change in chemical shift of ring protons. The structures of stabilized carbocations have been demonstrated by strong deshielding effects in ^{13}C spectra. The formation of contact ion pairs and equilibria between them and the solvent-separated ion pairs can be studied by NMR spectra of the alkali metal cations. Free-radical interactions in organic molecules have been studied by chemically induced dynamic nuclear polarization (CIDNIP).

Both components of a racemic mixture produce identical NMR spectra. It is possible to produce different spectra from each enantiomer through the creation of diastereomeric species. Three experimental methods are available (20). Diastereomers may be produced by derivatization with a chiral reagent. Diastereomeric adducts may be formed through the use of chiral shift reagents. Diastereomeric solvation complexes may be formed through the use of chiral solvents. As an example, the stepwise addition of a chiral lanthanide shift reagent to a racemic mixture is shown in Fig. 17.12. The enantiomeric purity of the mixture may be calculated from such data.

NMR in Equilibrium Systems

NMR offers a unique method for studying chemical systems in dynamic equilibrium. Reactions and processes with rate constants ranging from 10^{-2} sec^{-1} up to 10^8 sec^{-1} can be studied. Among the many possible applications are following chemical reactions, molecular rearrangements, valence bond tautomerism, conformational changes, rotations about bonds, inversions of pyramidal species, formation of solvated species, ligand binding interactions, enzyme–substrate binding, and protein denaturation.

Figure 17.12 Stepwise addition of a chiral shift reagent to a racemic mixture. One enantiomer reacts with the adduct to form a diastereomeric adduct.

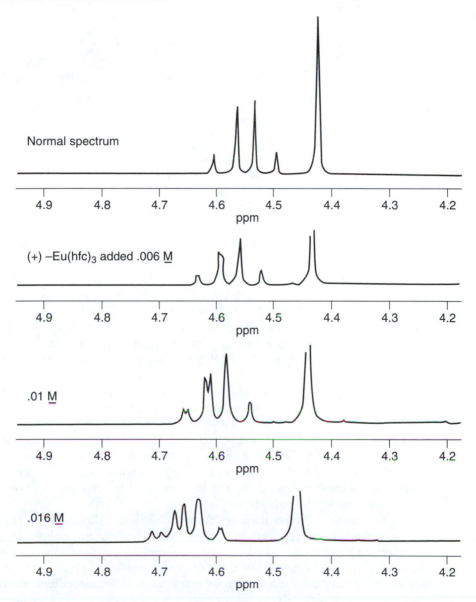

Figure 17.13 Series of spectra following the course of a reaction. Structures and ^{13}C lines of the starting material and product are given at the left. The reaction is the enzymatic epimerization of ibuprofen in a mithchondrial suspension.

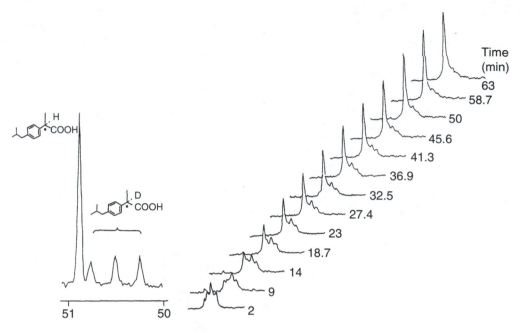

An example of the simplest use of NMR in dynamic systems is illustrated in Fig. 17.13. The rate of a chemical reaction can be followed by measuring the concentrations of reactants or products as a function of time. In this case, (21) the enzymatic removal of a deuterium atom and its replacement by a hydrogen atom on a ^{13}C-labeled site in ibuprofen is followed by ^{13}C NMR. This study illustrates three useful phenomena: site-specific labeling, in which an NMR-active nucleus is synthetically incorporated into a molecule, rendering only that site visible to the spectrometer; a deuterium isotope effect, which shifts the C–D resonance away from the C–H resonance; and C–D coupling. The 1:1:1 triplet is a result of coupling between ^{13}C and ^{2}H, with spin 1.

The binding of ligands such as peptides with the active sites of large macromolecules such as proteins, enzymes, membranes, and cell receptors can be studied by NMR. Exchange rates and binding constants can be determined by studies of relaxation rates and chemical shift changes of the ligands. The structures of the bound ligand and the binding site can be studied through the transferred NOE (22). The NOE that builds up between protons that are close to one another in the bound state is transferred to the free ligands upon dissociation of the complex. The sharp NMR lines of the free ligands allow indirect study of the geometry of bound complexes that are too large for conventional NMR studies.

One of the most interesting applications of NMR is in the study of systems undergoing chemical exchange. The nuclei involved need not be participating in a chemical reaction or even a rearrangement, they need only be capable of existing in more than one magnetically distinct environment. The appearance of the spectrum depends on the concentration of the exchanging species, rate of exchange, and relaxation rate at each exchanging site. For rapidly exchanging species, the spectrum consists of a single resonance at the average of the two chemical shifts, suitably weighted if the concentrations of the two species are not equal. Slowly exchanging species produce two distinct signals. As this slow exchange rate increases, the resonances become closer and closer until they coalesce. At the coalescence point the rate of exchange is 2.22 times the chemical shift difference between sites in the absence of exchange.

This phenomenon is subject to a quantitative analysis, called complete line shape analyses. The linewidth of an exchange-broadened peak is inversely proportional to the lifetime of the nuclei at the exchange site. A number of spectra are obtained at different temperatures, especially near the coalescence temperature and, if possible, at a temperature where the exchange is slow. The free enthalpy of activation for the process is relatively easy to calculate. The Arrhenius energy of activation can also be obtained, although the work is more time-consuming.

Exchanging systems can often be identified using the saturation transfer experiment. Protons at a particular chemical shift in one chemical form are partially saturated by irradiation with a low-power RF field. Upon exchange to a new chemical form the protons retain this saturation. When a nonirradiated control spectrum is subtracted from the irradiated spectrum, all nonexchanging signals are canceled out. Protons that have exchanged produce negative peaks. Figure 17.14 illustrates the use of the saturation transfer experiment to demonstrate the tautomeric equilibrium between keto and hemiacetal forms of an ortho-keto benzoic acid.

Figure 17.14 Demonstration of a tautomeric equilibrium using the saturation transfer experiment. A control spectrum showing all resonances is at top. The bottom spectrum is a difference spectrum obtained by subtracting the control spectrum from a preirradiated spectrum. Saturation is transferred from the ortho protons in the halogenated ring of one of the tautomers (7.05 ppm) to the same protons in the other tautomer (7.3 ppm).

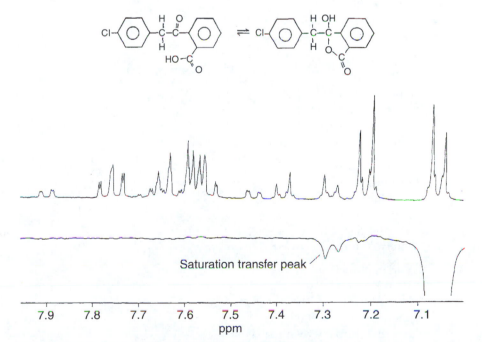

Nuts and Bolts

Relative Costs

NMR spectrometers can be purchased for $50,000 to $1,000,000. The main factor controlling the purchase price is the magnetic field strength. A rough guide to the purchase price is $750 to $1000 per MHz; that is, a 400-MHz system would cost between $300,000 and $400,000.

Solids	$100K
Imaging	$100K
Gradients	$50K
Probes	$10–$30K
Data station	$5–$50K
Automatic sample changer	$40K
Maintenance	5–10% of the purchase price per year
Cryogens	$1–$3K per year

Vendors for Instruments and Accessories

Bruker Instruments Inc.
19 Fortune Dr.
Billerica, MA 01821
phone: 508-667-9580
email: support@bruker.com
Internet: http://www.bruker.com

Hitachi Instruments, Inc.
3100 N. 1st St.
San Jose, CA 95134
phone: 800-548-9001
fax: 408-432-0704
Internet: http://www.hii.hitachi.com

JEOL USA Inc.
11 Dearborn Rd.
Peabody, MA 01960
phone: 508-535-5900
fax: 508-536-2206
Internet: http://www.jeol.com

Varian Associates
3050 Hansen Way
Palo Alto, CA 94304
phone: 415-493-4000
Internet: http://www.varian.com

Required Level of Training

Operation of Instrument

Routine analyses can be done by someone with the training typical of the average lab technician. Instrument setup, calibration, and performance of more complicated analyses usually require a college degree in chemistry and a few years of experience.

Processing Qualitative and Quantitative Data

Qualitative interpretation of spectra requires a college degree in chemistry and several years of experience. Quantitative determinations require the minimum of a course in analytical chemistry and some experience.

Service and Maintenance

The levels of liquid cryogens in the cryostat must be maintained. Liquid nitrogen is usually filled weekly and liquid helium must be filled every 1 to 6 mo, depending on the hold time of the cryostat. Extra-long hold-time cryostats are available for a small additional cost.

Electronic troubleshooting and repair are usually limited to the modular level. Replacement modules can usually be obtained from the manufacturer in 1 day. Calibrations of pulse lengths and sensitivity are generally stable for months.

Suggested Readings

Books

ABRAHAM, R. J., J. FISHER, AND P. LOFTUS, *Introduction to NMR Spectroscopy*. New York: Wiley, 1988.

FRIEBOLIN, H., *Basic One- and Two-Dimensional NMR Spectroscopy*. Weinheim, Germany: VCH, 1991.

SANDERS, J., AND B. HUNTER, *Modern NMR Spectroscopy: A Guide for Chemists*, 2nd ed. Oxford: Oxford University Press, 1993.

SILVERSTEIN, R. M., G. C. BASSLER, and T. C. MORRILL, *Spectrometric Identification of Organic Compounds*, 5th ed. New York: Wiley, 1991.

Articles

FARRAR, T. C., "Selective Sensitivity Enhancement in FT-NMR," *Analytical Chemistry*, 59, no. 11 (May 1987), 679A–90A.

FARRAR, T. C., "Two-Dimensional NMR Spectrometry," *Analytical Chemistry*, 59, no. 11 (June 1987), 748A–81A.

FREEMAN, R., "The Fourier Transform Revolution in NMR Spectroscopy," *Analytical Chemistry*, 65, no. 17 (1993), 743A–53A.

GRONENBORN, A. M., AND G. M. CLORE, "Protein Structure and Determination in Solution by 2D and 3D NMR Spectroscopy," *Analytical Chemistry*, 62 (1990), 2–15.

MARTIN, G. E., AND R. C. CROUCH, "Inverse-Detected Two-Dimensional NMR Methods: Applications in Natural Product Chemistry," *Journal of Natural Products*, 54, no. 1 (1991), 1–70.

WADE, C. G., AND OTHERS, "Automated High Resolution NMR with a Sample Changer," *Analytical Chemistry*, 61 (1989), 107A–11A.

Training Aids

Short courses are available from ACS and NMR Concepts, University of Rhode Island, Kingston, RI 02881.

Reference Spectra

BREITMAIER, E., AND W. VOELTER, *Carbon-13 NMR Spectroscopy*, 3rd ed. Weinheim, Germany: VCH, 1990.

BREMSER, W., AND OTHERS, *Carbon-13 NMR Spectral Data*, 4th ed. (microfiche). New York: VCH Publishers, 1987.

POUCHERT, C. J., AND J. BEHNKE, *Aldrich Library of ^{13}C and ^{1}H FT-NMR Spectra*. Aldrich Chemical Co., 1992.

SADTLER RESEARCH LAB, *^{13}C NMR Spectra*. Philadelphia: Sadtler Research Laboratories.

Additional References

CROASMUN, W. R., AND R. M. K. CARLSON, EDS., *Two-Dimensional NMR Spectroscopy*, 2nd ed. Weinheim, Germany: VCH, 1994.

FREEMAN, R., *A Handbook of NMR*. Essex, UK: Longman Scientific and Technical, 1987.

MARTIN, G. E., AND A. S. ZEKTZER, *Two-Dimensional NMR Methods for Establishing Molecular Connectivity*. Weinheim, Germany: VCH, 1988.

WARR, W. A., "Computer-Assisted Structure Elucidation," *Analytical Chemistry*, 65 (1993), 1045A–50A.

References

1. J. W. Schoolery, *Analytical Chemistry*, 65, no. 17 (1993), 731A–41A.

2. L. M. Jackman and S. Sternhell, *Application of Nuclear Magnetic Resonance Spectroscopy in Organic Chemistry*, 2nd ed. (Oxford: Pergamon Press, 1969).

3. R. Freeman, *Analytical Chemistry*, 65, no. 17 (1993), 743A–53A.

4. T. C. Farrar, *Analytical Chemistry*, 59, no. 11 (June 1987), 748–81.

5. J. S. Waugh, *Analytical Chemistry*, 65, no. 17 (1993), 725A–9A.

6. D. D. Traficante, in C. Dybowski and R. L. Lichter, eds., *NMR Spectroscopy Techniques*, 2nd ed. (New York: Marcel Dekker, 1995), 1–45.

7. J. Sanders and B. Hunter, *Modern NMR Spectroscopy: A Guide for Chemists*, 2nd ed. (Oxford: Oxford University Press, 1993).

8. H. Friebolin, *Basic One- and Two-Dimensional NMR Spectroscopy* (Weinheim, Germany: VCH, 1991).

9. C. P. Slichter, *Principles of Magnetic Resonance*, 3rd ed. (New York: Springer Verlag, 1990).

10. R. J. Abraham, J. Fisher, and P. Loftus, *Introduction to NMR Spectroscopy* (New York: Wiley, 1988).

11. M. Minch, *Concepts in Magnetic Resonance*, 6 (1994), 41–56.

12. D. Neuhaus and M. Williamson, *The Nuclear Overhauser Effect in Structural and Conformational Analysis* (New York: VCH, 1989).

13. C. G. Wade and others, *Analytical Chemistry*, 61 (1989), 107A–11A.

14. H. Barjat and others, *Journal of Magnetic Resonance*, 108 (1995), 170–2.

15. G. E. Martin and R. C. Crouch, *Journal of Natural Products*, 54, no. 1 (1991), 1–70.

16. A. M. Gronenborn and G. M. Clore, *Analytical Chemistry*, 62 (1990), 2–15.

17. A. E. Tonelli, *NMR Spectroscopy and Polymer Microstructure: The Conformational Connection* (New York: VCH, 1989).

18. E. A. Fitzgerald, *Journal of Applied Polymer Science*, 41 (1990), 1809–14.

19. T. H. Lowry and K. S. Richardson, *Mechanism and Theory in Organic Chemistry* (New York: HarperCollins, 1987).

20. D. Parker, *Chem. Rev.*, 91 (1991), 1441–57.

21. W. Shieh and others, *Analytical Biochemistry*, 212 (1993), 143–9.

22. F. Ni and H. A. Scheraga, *Acc. Chem. Res.*, 27, no. 9 (1994), 257–64.

Chapter 18

X-Ray Diffraction

Joseph Formica

*Siemens Industrial Automatic
Analytical Instrumentation*

Summary

General Uses

- Crystal structure determination
- Phase identification
- Quantitative phase analysis
- Texture and stress analysis
- High- and low-temperature and pressure studies

Common Applications

- Identification of unknowns
- Quality control: qualitative and quantitative analysis
- Characterization of polycrystalline and epitaxial thin films
- Quantification of texture and orientation in metals and polymers
- Variable-temperature studies to determine thermal expansion, stability, and phase diagrams

Samples

State

Crystalline solids as powders, single crystals, sheets, foils, or fibers

Amount

- Single-crystal: an uncracked, untwinned crystal 0.1 to 0.5 mm on a side
- Powder: typically several hundred milligrams (or fractions of a milligram if time is not critical)

Preparation

- Single-crystal: mounting on a fiber or in a glass capillary
- Powder: possibly none or grinding solids to fine powder

Analysis Time

Analysis time can range from minutes to days depending on the instrument configuration, sample properties, and analysis type. High-power generators and position-sensitive detectors reduce analysis time, whereas small or poorly diffracting samples increase it. Quantitative analysis usually requires long count times to attain statistical precision. Nonambient diffraction experiments may require long times for equilibration.

Limitations

General

- Crystalline phases only; amorphous materials do not typically yield useful diffraction data.
- Peak overlap may hinder phase identification and quantitative analysis.
- Matrix effects: Strongly diffracting materials may obscure weakly diffracting ones.
- Preferred orientation affects diffracted intensity and is detrimental to quantitative analysis.
- Sample fluorescence may raise the background in the diffraction pattern or may cause saturation in certain detector types.

Accuracy

Quantitative analysis of powder mixtures with less than 1% accuracy and higher precision is achieved under favorable conditions with suitable standards; highly reproducible results are achieved by standardless methods, although they may not be highly accurate ($\pm$ 5 to 20%).

Sensitivity and Detection Limits

In practice, detection limits vary with the system under investigation; in principle, the limiting factor is the available analysis time.

Complementary or Related Techniques

- X-ray fluorescence provides elemental chemical information that may be useful for phase identification.
- X-ray scattering provides information on amorphous materials including solutions and can be used to determine distance distributions, particle molecular weight, particle shape, and radius of gyration.
- Neutron diffraction usually provides better information about the locations of light atoms in a crystal structure.
- Thermogravimetric analysis (TGA), differential thermal analysis (DTA), and differential scanning calorimetry (DSC) can be used for phase identification and quality control and to study solid-state reactions.

Introduction

X-ray diffraction is a versatile analytical technique for examining crystalline solids, which include ceramics, metals, electronic materials, geological materials, organics, and polymers. These materials may be powders, single crystals, multilayer thin films, sheets, fibers, or irregular shapes, depending on the desired measurement. X-ray diffractometers fall broadly into two classes: single-crystal and powder. Single-crystal diffractometers are most often used to determine the molecular structure of new materials. Powder diffractometers are routinely used for phase identification and quantitative phase analysis but can be configured for many applications, including variable-temperature studies, texture and stress analysis, grazing incidence diffraction, and reflectometry.

In 1895, William Röntgen discovered X rays, which are electromagnetic radiation approximately 1 Å (10^{-10} m) in wavelength. In 1912, Max von Laue examined the thesis work of Paul Ewald, who modeled a crystal as a periodic lattice of oscillators with a spacing of the order of 1 Å. If Ewald's model were correct, Laue reasoned that crystals should behave like three-dimensional diffraction gratings for X rays, a hypothesis borne out by experiments later that year (1). Film techniques for powder X-ray diffraction were independently developed between 1915 and 1917 by Peter Debye and Paul Scherrer in Germany and by Albert Hull in the United States. Powder diffractometers with counter detectors appeared in the 1940s, with modern instrument designs credited to William Parrish (2). Single-crystal diffractometers appeared in the mid-1950s with Eulerian cradle designs by Thomas Furnas and David Harker (3).

The operative equation in X-ray diffraction is the Bragg equation (4):

$$n\lambda = 2d \sin\theta \tag{18.1}$$

Figure 18.1 Illustration of Bragg's Law.

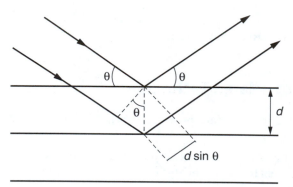

where n is the order of a reflection ($n \in \{1,2,3 \ldots\}$), λ the wavelength, d the distance between parallel lattice planes, and θ the angle between the incident beam and a lattice plane, known as the Bragg angle (Fig. 18.1). When the pathlength in the crystal ($2d \sin \theta$) is a multiple of the wavelength, constructive interference occurs and diffracted intensity is obtained. In general, the d-spacing is a function of the lattice parameters (a, b, c) and angles (α, β, γ) defining the unit cell, and the Miller indices (h, k, l) denoting a particular reflection. As such, it is the geometry of the crystal lattice that determines the positions of the peaks in an X-ray diffraction pattern. In general, the more symmetrical the material, the fewer peaks in its diffraction pattern. The diffracted intensities associated with those peaks are determined by the type and arrangement of atoms within the crystal lattice.

The radiation used in a typical diffraction measurement contains several wavelengths, denoted $K_{\alpha 1}$, $K_{\alpha 2}$, and K_β, which are characteristic of the material producing the X rays (Table 18.1). The smaller the wavelength, the more energetic and penetrating the radiation. Longer-wavelength radiation spreads out the peaks in a diffraction pattern, which may overcome line overlap problems or enhance small peak shifts due to stress. The choice of radiation also depends on the sample characteristics. For example, X-ray fluorescence is excited in iron-bearing materials by copper radiation but not by chromium or cobalt radiation. Sample fluorescence is undesirable because it raises the background intensity and can obscure the useful diffraction information.

Table 18.1 X-ray wavelengths (in Å) and K_β filters for common anode materials.

Anode	$K_{\alpha 1}$	$K_{\alpha 2}$	K_β	K_β **Filter**
Chromium (Cr)	2.28970	2.29361	2.08487	Vanadium (V)
Cobalt (Co)	1.78897	1.79285	1.62079	Iron (Fe)
Copper (Cu)	1.54056	1.54439	1.39222	Nickel (Ni)
Molybdenum (Mo)	0.70930	0.71359	0.63229	Zirconium (Zr)

How It Works

A basic powder diffraction system (Fig. 18.2) consists of an X-ray source, two-circle goniometer (θ and 2θ circles), sample stage, detector, and computer for instrument control and data analysis. A basic four-circle single-crystal diffraction system (Fig. 18.3) consists of an X-ray source, two-circle goniometer (ω and 2θ circles), closed Eulerian cradle (π and χ circles), detector, and computer for instrument control and data analysis. The major components of a diffraction system are described below along with several specialized instruments.

Generator

Modern generators are compact, no longer requiring large oil-filled tanks. Typical laboratory X-ray sources use water-cooled, sealed X-ray tubes with 1- to 3-kW output. The power limit is governed by the ability to remove heat from the anode. Rotating anode generators provide up to 18 kW because the same area on the anode is not continuously heated. The increased power comes at a higher cost and overall complexity than a sealed-tube generator.

Figure 18.2 Siemens D 5000 powder diffractometer shown in the θ/θ configuration with a sealed X-ray tube, fixed slits, sample spinner, and scintillation detector.

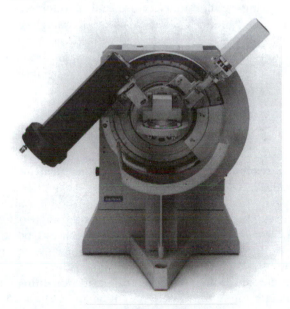

Figure 18.3 Siemens P4 four-circle, single-crystal diffractometer shown with a sealed X-ray tube, closed Eulerian cradle, and scintillation detector.

Goniometer

This device comprises two independent drives that provide accurate and precise motion of the sample stage and detector ($\theta/2\theta$ configuration) or the tube and detector (θ/θ configuration). In either configuration, the goniometer may be oriented to keep the sample horizontal or vertical. The θ/θ configuration maintains the sample in a fixed (and typically horizontal) position, which is often desirable during variable-temperature studies, when examining granular solids, or when using a sample changer. Eulerian cradle systems (such as single-crystal diffractometers) are invariably configured with a fixed X-ray source and with the goniometer lying flat. Dual goniometer systems share a common X-ray source. They provide high throughput (with dual sample changers) or the ability to run routine and nonroutine (such as variable-temperature) samples without reconfiguring the instrument.

Sample Stages

Common attachments for powder diffractometers include sample changers, sample spinners, Eulerian cradles, and environmental chambers. Sample changers accommodate multiple samples (typically 8 to 60), enabling batch processing and high throughput. Sample spinners rotate about an axis normal to the sample surface through the angle ϕ. Complete rotation in ϕ may bring more randomly oriented crystallites in a powder sample into position to diffract, improving the counting statistics of the measurement. Data may also be collected at fixed ϕ values to determine whether sample orientation influences diffracted intensities.

An Eulerian cradle is a mechanical device for tilting and rotating samples. It turns a powder diffractometer into a four-circle system (ω, or θ, 2θ, ϕ, and χ circles) for examination of sample orientation, texture, or residual stress. Closed Eulerian cradles have complete χ circles and are used on single-crystal diffractometers and traditional powder diffractometers configured for texture measurements. Texture measurements obtain pole figures in which intensity is measured for a given reflection (fixed θ and 2θ) while varying ϕ and χ. This information represents the distribution of orientations of the crystallites. For stress (and texture) measurements, open Eulerian cradles enable the detector to pass through a gap in the χ circle to reach high 2θ angles, where relatively large peak shifts can be observed.

Environmental chambers enable experiments at reduced or elevated temperatures and pressures under controlled atmospheres (such as oxidizing, reducing, inert, evacuated, and humid atmospheres). Chambers exist for performing experiments down to liquid helium temperatures (4.2 K) and up to 3000 K.

Detectors

The most common detectors are point detectors: scintillation and solid-state detectors. They collect diffracted intensity from one angle at a time and are scanned through the angular range of interest. Scintillation detectors may be used with K_β filters or with diffracted-beam monochromators. Solid-state detectors provide the energy resolution of a diffracted-beam monochromator without the associated intensity loss and offer a three- to fourfold speed advantage over a scintillation detector and monochromator.

Linear position-sensitive detectors (PSDs) use a fine wire to collect intensity over an angular range simultaneously, offering speed advantages up to one hundred times that of a point detector. These detectors may be held fixed over a 5 to 10° 2θ range or scanned like a point detector over a larger range. The speed of this detector is advantageous for high sample throughput, time-resolved studies, weakly diffracting materials, and the detection of minor phases.

Linear response is an important characteristic of detector systems. Scintillation detectors handle 10^5 to 10^7 counts per second (cps) without saturation, making them ideal for strong single-crystal reflections (such as those from electronic substrate materials). Solid-state detectors handle 25,000 to 50,000 cps linearly, which is more than adequate for most powder diffraction applications. Linear PSDs typically handle 10,000 cps across the length of the detector wire, which again is adequate for most powder diffraction applications, especially those involving small or weakly diffracting samples.

Multiwire area detectors are two-dimensional PSDs that provide in real time the two-dimensional information afforded by classic Debye–Scherrer film techniques (Fig. 18.4). These detectors cover the χ angle, thereby reducing the need for an Eulerian cradle, while providing an order of magnitude better χ-resolution than can be obtained with a cradle system. Area detectors can collect pole figures in minutes rather than the hours or days required with scintillation detectors (5). Two-dimensional area detector diffraction data can be reduced to one-dimensional powder patterns by integration through or along the Debye rings. Integration along the Debye rings gives better intensities for phase identification and quantitative analysis in samples having strong preferred orientation (such as clay minerals).

Charge-coupled device (CCD) area detectors are the latest innovation in area detector technology for examining small molecules. New crystals may be screened in 60 sec without film measurements. High-quality, complete data sets may be obtained with either sealed-tube or rotating-anode sources in 2 to 6 hr, enabling experiments that are too lengthy or impossible with standard four-circle single-crystal systems. Such applications include electron density studies, superlattice discoveries, phase transitions, twinning problems, and data collection for compounds that decay rapidly. This detector offers near real-time data readout and display (less than 2 sec), high spatial resolution (less than 50 μm), and high photon counting efficiency to best handle weak reflections.

Figure 18.4 Debye rings of aluminum oxide (corundum) collected with a multiwire area detector.

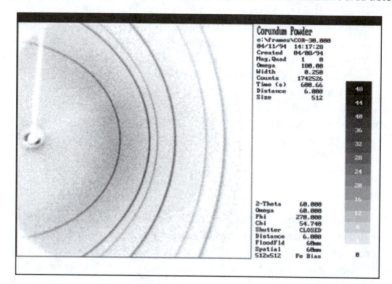

Optics

Modern powder diffractometers include Bragg–Brentano parafocusing geometry (Fig. 18.5). The elements of the beam path include a line source of X-rays, fixed slits to limit radial divergence of the beam and remove scatter, Soller slits (parallel, equally spaced foils) to limit the axial divergence of the beam, a detector slit, and a K_β filter or monochromator. Variable divergence and antiscatter slits are coupled to Bragg angle to maintain a constant irradiated area on the sample. They are especially valuable below 10° 2θ (as for clay mineral analysis). By closing down at low angles, scatter from the main beam is attenuated, and by opening up at higher angles more diffracted intensity reaches the detector than with small fixed slits.

Single-crystal instruments and area detector systems include Debye optics (Fig. 18.6). Conventional single-crystal diffractometers, area detector systems, and powder diffractometers configured with Eulerian cradles include a spot source of X-rays and pinhole collimation. In this way, small specimens can be examined (such as single crystals) or larger ones analyzed with high spatial resolution (as for sample mapping or microdiffraction), and high resolution in χ is achieved.

Other beam conditioners include metal foils to filter K_β radiation (Table 18.1), and crystal monochromators that filter K_β radiation and sample fluorescence and improve peak resolution. On powder diffractometers, curved quartz or germanium incident-beam monochromators select the $K_{\alpha 1}$ wavelength, whereas curved graphite diffracted-beam monochromators pass $K_{\alpha 1}$ and $K_{\alpha 2}$. Strictly monochromatic radiation may be desirable for certain applications, such as structure refinement via Rietveld analysis, but the intensity loss through the monochromator may be as high as 90%. Curved graphite monochromators typically result in a 70 to 75% intensity loss. Flat graphite incident-beam monochromators are often used in single-crystal and area detector systems.

Computers

Powerful, inexpensive personal computers have largely supplanted minicomputers for data collection and analysis. Networking enables data transfer between computers for further processing or storage. Modems and appropriate software enable remote control and diagnostic evaluation of instruments.

Figure 18.5 Powder diffractometer optics: Bragg–Brentano parafocusing geometry.

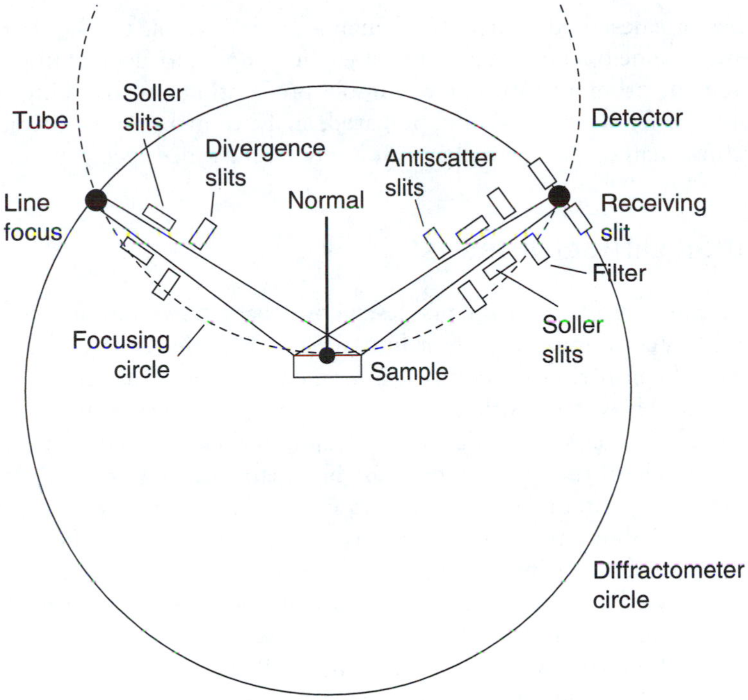

Tube

Soller
slits

Divergence
slits

Normal

Antiscatter
slits

Detector

Line
focus

Receiving
slit

Filter

Soller
slits

Focusing
circle

Sample

Diffractometer
circle

Figure 18.6 Single-crystal diffractometer optics: Debye optics.

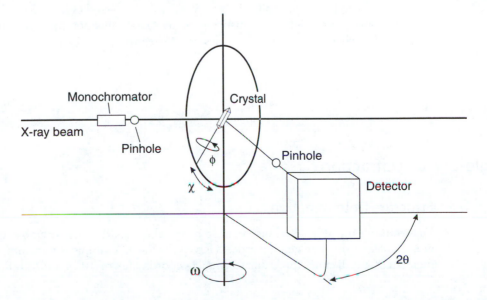

Monochromator

Crystal

X-ray beam

Pinhole

ϕ

χ

Pinhole

Detector

2θ

ω

Process Diffractometers

These are designed to meet the demands of high sample throughput with minimal operator intervention. The operator simply prepares the samples, loads the sample changer, and starts the measurement program. Automated sample preparation and handling systems enable samples to be collected from a process unit, prepared, analyzed in a process diffractometer, then transferred for further analysis (for example, to an X-ray fluorescence spectrometer).

High-Resolution Diffractometers

These are designed for the nondestructive characterization of epitaxial thin films common to the semiconductor industry. These films are nearly perfect single crystals that have extremely narrow peaks. Typical rocking curve measurements (scans of ω with 2θ, ϕ, and χ fixed for a selected reflection) have full widths at half maximum below 0.005°. The X-ray optics must provide a highly collimated, monochromatic incident beam, an Eulerian cradle is required to orient the sample, and the ω-circle of the goniometer must be capable of steps as small as 0.0002°. Channel-cut, four-bounce germanium monochromators are suitable beam conditioners for such applications, providing the flexibility to examine a variety of materials. For specific applications, it is also possible to match the sample with incident- and diffracted-beam conditioners made of the same material. Applications of high-resolution diffraction include determination of layer thickness, lattice mismatch, and chemical composition of epitaxial layers, assessment of substrate misorientation, studies of interdiffusion and growth in multilayer structures, accurate lattice parameter determination, and sample surface mapping for quality control.

Microdiffractometers

Microdiffractometers (Fig. 18.7) include highly collimated X-ray beams to achieve high spatial resolution for examining regions as small as 10 μm in diameter. The spot to be analyzed is pinpointed with a laser beam on a color video image of the sample. This image may be saved along with the diffraction data for a complete record of the measurement. High-precision xyz positioning enables sample mapping and analysis of irregularly shaped objects. Rotating anode generators and position-sensitive detectors drastically reduce data-collection times, making this technique attractive for such applications as forensics, microelectronics, and failure analysis.

What It Does

Single-Crystal Diffraction

Structure Determination

Determining the atomic arrangement in a crystalline solid is the foremost application of single-crystal diffraction. The general process is as follows (6). A sample is positioned in the center of the incident X-ray beam. Approximate angular coordinates of 10 or more reflections are obtained

Figure 18.7 Siemens microdiffraction system shown with multiwire area detector, xyz stage, and video camera.

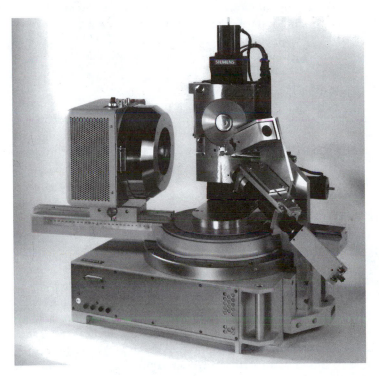

from preliminary measurements made with a film cassette in place of a point detector, with an automatic search routine and a point detector, or with an area detector. These starting values are optimized by an iterative procedure. An autoindexing routine then determines the appropriate unit cell, lattice parameters, and an accurate orientation matrix for the crystal. Once the orientation matrix is determined, the angular locations of all reflections may be calculated. Intensity data are collected for typically several thousand unique reflections, with the data collection strategy depending on the symmetry of the crystal and the detector type (point or area). Corrections are applied for instrumental factors and for the effects of polarization, absorption, and possibly sample decomposition. From integrated intensities, structure factor amplitudes are obtained. The electron density in a crystal depends on both the structure factor amplitude and the phase angle of each reflection. Phase angles cannot be measured, but self-consistent sets of angles may be determined by various analysis techniques (7). When these are combined with the measured structure factor amplitudes, a trial structure is obtained. The parameters in the model of this trial structure are then refined to minimize the difference between the calculated and observed intensities. The results of this procedure are accurate atomic positions from which bond lengths and angles may be obtained.

Crystallographic simulation software has advanced so that molecular models can be built on a computer display, and atoms can be manipulated in the structure with a near-real-time display of the effects on the diffraction pattern. This capability is extremely useful to designers of pharmaceuticals, catalysts (zeolites), and other specialty chemicals.

Absolute Configuration

The absolute configuration of optically active molecules may be determined from the effects of anomalous X-ray scattering (6). These effects are observed by collecting data from Friedel pairs of reflections, that is, (h,k,l) and (-h,-k,-l), the intensities of which are nonequivalent for noncentrosymmetric crystals. Single-crystal X-ray diffraction is the definitive technique for obtaining this information.

Phase Identification

Lattice parameters can be rapidly determined and compared with those reported in such databases as the NIST Crystal Data Identification File or the Cambridge Crystallographic Data File for phase identification. If the phase has not been reported, search results may identify structurally similar materials. Such rapid screening of crystals prevents wasted effort solving published structures.

Powder Diffraction

Crystallite Size and Microstrain

The smallest diffracting domain in a specimen is known as a crystallite. Crystallites from one to several hundred nanometers have broadened peak profiles resulting from incomplete destructive interference at angles near the Bragg angle. From this broadening, crystallite size and size distribution may be determined (8) by a single-line analysis (Scherrer method) or by a more theoretically rigorous multiple-line technique (Warren–Averbach method). Note that crystallite size is not to be confused with particle size determined by other techniques: A particle may contain many crystallites. Crystallite size analysis is used for quality control and to study the effects of processing on grain growth. Microstrain is the degree to which crystallites are deformed relative to unstrained materials. This phenomenon also manifests itself as a broadened peak profile, although the profile shape associated with this type of broadening differs from that of crystallite size broadening. Detectable microstrains range from 10^{-4} to the elastic limit of the material.

Grazing Incidence Diffraction

This technique is used to examine polycrystalline thin films with layers typically no thinner than 500 Å. In such an experiment, the X-ray source is positioned at a glancing angle with respect to the sample. Parallel-beam optics replace the standard Bragg–Brentano optics of a powder diffractometer (9). They consist of long parallel plate collimators preceding the detector. A flat diffracted-beam monochromator (such as LiF or Ge) may be used to improve resolution. The glancing angle affects the depth to which the X-ray beam penetrates the sample, as do the thicknesses and X-ray absorption characteristics of the layers. Typical experiments consist of multiple detector scans at different glancing angles, providing a depth profile of the sample.

Indexing and Lattice Parameter Refinement

Powder diffraction patterns from single-phase materials can be analyzed to determine the crystal lattice type (that is, cubic, tetragonal, orthorhombic, hexagonal, monoclinic, or triclinic), lattice parameters, and the associated Miller indices of the reflections. Well-known indexing routines are those of Visser (analytical approach) and Werner (exhaustive approach); the best known lattice parameter refinement routine is that of Appleman (10). Versions of these and other crystallographic routines are available in the public domain and as part of commercial software packages.

Nonambient Diffraction

Material properties may be studied by diffraction at elevated or reduced temperatures and pressures under a variety of atmospheric conditions (11, 12). Applications include thermal expansion and stability studies, lattice parameter determination, structure identification, phase diagram determination, studies of dynamic structure changes, chemical reactions, reaction kinetics, and catalysis.

Mass spectroscopy, gas chromatography, or infrared spectroscopy can be coupled with high-temperature diffraction experiments to provide evolved gas analysis of a reacting or catalytic system. Position-sensitive detectors enable time-resolved studies of reacting systems over time scales as short as 1 min. They are also advantageous for high-temperature studies because increased atomic motion has a damping effect on diffracted intensity. In situ studies are more representative of actual processing conditions than analysis of treated samples outside the reaction environment. Geoscientists perform high-pressure diffraction studies to simulate conditions deep within the earth. Pharmaceutical companies examine the stability of their products with respect to heat, cold, humidity, and processing such as tableting, which could induce undesirable pressure-induced phase changes.

Single-crystal experiments may be run in a cold nitrogen gas stream to reduce thermal motion and improve data quality (important for compounds with melting points near ambient temperature), to prevent degradation of the sample (such as proteins) in the X-ray beam, and to examine air-sensitive compounds outside capillaries, thereby reducing absorption problems.

Phase Identification

Crystalline materials produce distinct X-ray diffraction patterns that can identify the phases in a material. The International Centre for Diffraction Data (ICDD) maintains the Powder Diffraction File (PDF), a database of single-phase powder diffraction patterns currently numbering over 63,000. Today it is supplied on CD-ROM, and computer search routines can scan the entire database in under 20 sec. For materials not found in the PDF database but for which single-crystal information is available, powder diffraction patterns can also be simulated.

First-generation search/match software required a list of d-spacings and intensities from the measured pattern as input. Current approaches do not require peak picking and use the entire background-subtracted spectrum (13). Elemental analysis of a sample (by AA, ICP, or X-ray fluorescence) enables chemical restrictions to be placed on the search, narrowing the list of potential matches. Users may also create searchable databases of their own proprietary materials.

Quality Control

X-ray diffraction is routinely used to screen materials for impurities. For example, the pharmaceutical industry often deals with polymorphs of a given drug (same chemical composition, different crystal structure), only one of which is biologically active. The ceramic industry analyzes heat-treated products to determine whether the desired phases are present. Other quality control applications include quantitative phase analysis, crystallite size determination in powders, and layer thickness determination in multilayer thin films via reflectometry.

Quantitative Analysis

This is the relationship of diffracted intensity to phase concentration (14). It may be used as a quality or process control technique (such as determination of rutile and anatase in the pigment industry, or α-, β-, and γ-$LiAlO_2$ phases in the ceramics industry). Established applications of quantitative X-ray diffraction include free silica analysis, retained austenite determination in steels, and aluminum bath analysis. Free silica analysis quantifies quartz, cristobalite, and tridymite deposited on silver membrane filters per Occupational Safety and Health Administration and National Institute of Occupational Safety and Health protocols (15). The hardening of steels involves a phase transformation from austenite (face-centered cubic) to martensite (body-centered tetragonal). Retained austenite determination quantifies the austenite content by direct comparison of the integrated intensity of an austenite peak with that of a martensite peak (4). Aluminum bath analysis determines the weight ratio of NaF to AlF_3, the so-called bath ratio, as well as the content of $Na_5Al_3F_{14}$ (chiolite), Na_3AlF_6 (cryolite), and the total lithium and calcium content of an electrolytic aluminum bath. This analysis uses dual detectors, the second of which is used as a lithium or calcium fluorescence channel.

Reflectometry

This thin-film characterization technique is based not on the diffraction of X-rays but on their reflection, which obeys the same physics as the reflection of visible light (16). The value of the critical angle for total external reflection determines the density of a deposited film. Above the critical angle, some radiation penetrates into the film and interference fringes result whose spacing is related to the film thickness. The technique applies to single and multilayer films up to several hundred nanometers thick and can be used to determine surface and interface roughness. Applications include nondestructive characterization of microelectronic materials and coatings.

Residual Stress

When processed materials (metals in particular) are subjected to applied forces, elastic deformation may occur. When the applied forces are removed, there may remain in the material what is known as residual stress (17). Macrostrain is strain uniform over a large number of grains in a material and manifests itself in a diffraction pattern as a peak shift. Residual stress is calculated from measured macrostrain. Depending on the nature of the sample and the experiment, stress may be characterized in one, two, or three dimensions (uniaxial, biaxial, or triaxial stress). Triaxial stress measurements require an Eulerian cradle to orient the sample or a two-dimensional detector to obtain information out of the sample plane. Macrostrain may also be measured by mechanical methods, but unlike diffraction measurements, those techniques are usually destructive. Applications of residual stress measurements include analysis of processed metals (heat-treated, cut, ground, bent, peened, or welded), plastics, polymers, and composite materials.

Texture

Diffraction patterns from samples containing a random orientation of crystallites have predictable relative peak intensities. In samples exhibiting preferred orientation or texture, peaks from certain reflections are enhanced relative to those in a randomly oriented sample. The preferred orientation of crystallites may result from the nature of the material (such as clay minerals with a platy habit) or from processing (such as drawing of wire or rolling of metals). Fibers, sheets, certain polymers and minerals, and many metals exhibit texture, which has a pronounced effect on the physical and mechanical properties of the material. A pole figure is a representation of the distribution of orientations among the crystallites, which is measured with an Eulerian cradle or an area detector system. Orientation distribution function (ODF) calculations to quantify the texture of a material require multiple pole figures. Area detectors can collect pole figures in minutes for rapid and routine quality control in the polymer and metals industries (Fig. 18.8).

Figure 18.8 (111) pole figure of aluminum collected with a multiwire area detector.

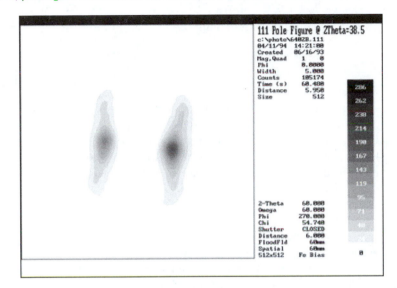

Analytical Information

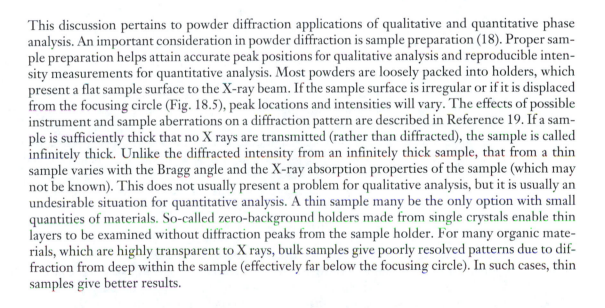

This discussion pertains to powder diffraction applications of qualitative and quantitative phase analysis. An important consideration in powder diffraction is sample preparation (18). Proper sample preparation helps attain accurate peak positions for qualitative analysis and reproducible intensity measurements for quantitative analysis. Most powders are loosely packed into holders, which present a flat sample surface to the X-ray beam. If the sample surface is irregular or if it is displaced from the focusing circle (Fig. 18.5), peak locations and intensities will vary. The effects of possible instrument and sample aberrations on a diffraction pattern are described in Reference 19. If a sample is sufficiently thick that no X rays are transmitted (rather than diffracted), the sample is called infinitely thick. Unlike the diffracted intensity from an infinitely thick sample, that from a thin sample varies with the Bragg angle and the X-ray absorption properties of the sample (which may not be known). This does not usually present a problem for qualitative analysis, but it is usually an undesirable situation for quantitative analysis. A thin sample many be the only option with small quantities of materials. So-called zero-background holders made from single crystals enable thin layers to be examined without diffraction peaks from the sample holder. For many organic materials, which are highly transparent to X rays, bulk samples give poorly resolved patterns due to diffraction from deep within the sample (effectively far below the focusing circle). In such cases, thin samples give better results.

Qualitative

The limits of detection of a given phase depend on several factors, including whether the phase is known to be in the sample, the absorption contrast and relative diffracting ability of the phases, peak overlap, and counting statistics. The statistical precision of detector measurements is related to the square root of the number of counts above background in a diffraction peak (19). In principle, if a phase exists in a mixture, given long enough count times, its peaks will appear above background in the diffraction pattern. Over the same period of time, PSDs offer greatly improved counting statistics compared with point detectors.

Quantitative

Quantitative analysis is the relationship between the diffracted intensity from a phase and its concentration. The intensity diffracted by a single-phase powder specimen depends on many factors, including the multiplicity of the reflection, Lorentz-polarization (geometrical) factor, Debye–Waller (temperature) factor, structure factor, absorption factor, preferred orientation effects, and extinction (if present), as well as instrumental effects. In a multiphase mixture, the diffracted intensity $I_{i\alpha}$ from line i of phase α may be written as

$$I_{i\alpha} = \frac{K_{i\alpha}X_{\alpha}}{\rho_{\alpha}(\mu/\rho)_m} \tag{18.2}$$

where $K_{i\alpha}$ is constant for a given crystal structure and instrument configuration, ρ_{α} is the density of phase α, and $(\mu/\rho)_m$ is the mass absorption coefficient of the mixture (4, 14, 19). For accurate results, the intensities used in quantitative analysis should be integrated intensities (that is, peak areas) rather than peak heights. A curve-fitting technique known as profile fitting may be used to obtain accurate areas, especially from overlapped peaks. Common approaches to X-ray quantitative analysis include the internal standard method, reference intensity ratio (RIR) or Chung method, and Rietveld refinement.

The most widely used and generally most accurate method is the internal standard method. A known weight of a standard material is homogeneously mixed with the sample of unknown composition. Matrix absorption effects, $(\mu/\rho)_m$, are eliminated by ratioing the intensities of unknown peaks to standard peaks, and a linear relationship to the weight fraction of the unknown is obtained. A calibration curve (line) is determined from a set of mixtures of known composition containing a fixed weight of standard.

The reference intensity ratio method or Chung method (20) can be used for standardless quantification. The RIR has historically been defined as the ratio of the integrated intensities of the most intense lines of a 50/50 weight percent mixture of a sample and corundum. Many RIR values have been published over the years (for example, in the PDF database), but many are of questionable accuracy. The recommended practice is to prepare appropriate mixtures and measure RIR values for the materials of interest. A generalized RIR can be defined by substituting a general internal standard for corundum:

$$\mathrm{RIR}_{\alpha,s} = (I_{i\alpha}/I_{js})(I_{js}^{\mathrm{rel}}/I_{i\alpha}^{\mathrm{rel}})(X_s/X_{\alpha}) \tag{18.3}$$

where i and j denote diffraction lines, α and s phase and standard, X weight fraction, and I^{rel} relative intensity of a diffraction line. In the Chung method, one of the phases in the mixture is arbitrarily designated as the standard, giving $N-1$ weight fraction ratios for an N component system. The Nth equation exploits the fact that the mass fractions sum to unity when all phases in a mixture are identified. Thus, analysis can be made without addition of a standard to the unknown. The presence of amorphous materials or unidentified phases invalidates this method.

Rietveld refinement can also be used for standardless quantification with high accuracy (below 1%) when the crystal structures of all phases in a mixture are known. This technique involves calculating the diffraction pattern for the mixture based on a detailed crystallographic model and adjusting the model parameters to minimize the difference between the calculated and measured patterns (21). The mass fractions of the phases are obtained from the intensity scaling factors in the model.

Applications

Figure 18.9 shows the diffraction patterns from three binary mixtures of corundum (Al_2O_3) and boehmite (AlOOH). The phases are identified by their entries from the JCPDS Powder Diffraction File (qualitative analysis). The amounts of each phase present may be determined by one or more of the quantitative analysis methods mentioned earlier. Profile fitting may be used to obtain accurate areas, especially from the overlapped peaks near 38° 2θ.

Figure 18.10 shows grazing incidence data from an amorphous silicon oxide layer (approximately 5000 Å) implanted with germanium on a silicon wafer. During the implantation process, which occurs at elevated temperatures, chromium and iron impurities from a stainless steel clip were codeposited as $CrSi_2$ and $FeSi_2$. Lines from these phases were matched from the JCPDS PDF database and are overlaid on the diffraction patterns. Note that the smaller the grazing angle, the shallower the penetration of the X-ray beam into the sample.

Figure 18.11 shows the phase transition between orthorhombic and rhombohedral forms of potassium nitrate. Data were collected at 1 °C intervals from 100 to 150 °C to observe the reversible phase change near 130 °C in air. A linear PSD scanned the 20 to 35° 2θ range with 0.02° steps in about 4 min. The entire data set was collected in under 3.5 hr. A comparable experiment using a point detector would require at least 64 hr. Before the phase change, certain diffraction peaks are observed to shift to lower angles. This may be understood as a consequence of the thermal expansion of the material. As the sample is heated, the material expands and its lattice parameters increase. Consequently, the d-spacing increases, which decreases the Bragg angle computed from Bragg's Law.

Figure 18.9 Powder diffraction patterns from 20%/80%, 50%/50%, and 80%/20% by weight corundum/boehmite mixtures.

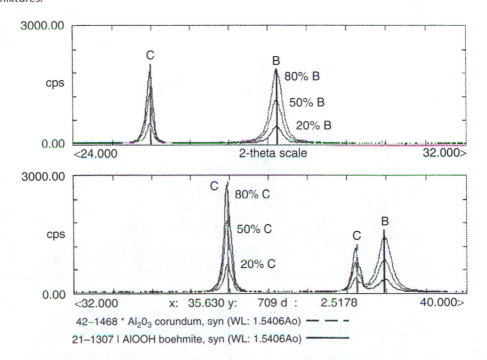

Figure 18.10 Grazing incidence diffraction data from Ge-implanted SiO_2 on Si showing $CrSi_2$ and $FeSi_2$ impurities.

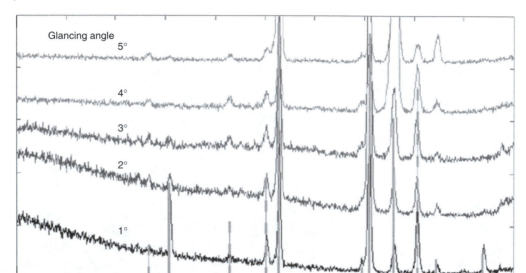

Figure 18.11 Orthorhombic to rhombohedral phase transition of KNO_3, which occurs near 130 °C in air. Data were collected at 1 °C intervals from 100 to 150 °C.

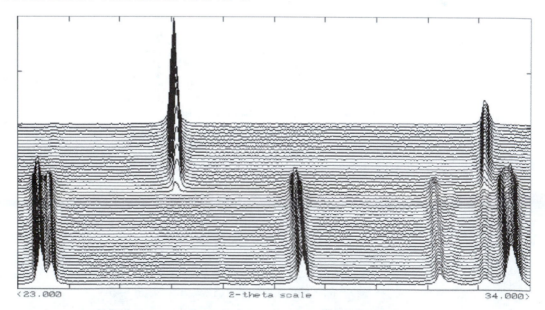

Figure 18.12 Measured and simulated reflectivity curves of a diffusion barrier (TiN/Ti/SiO₂/Si).

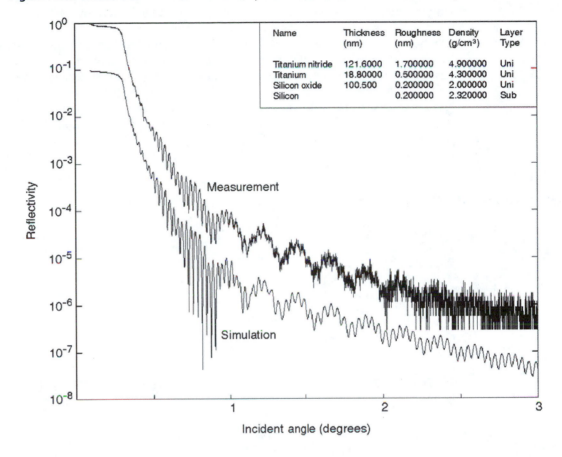

Name	Thickness (nm)	Roughness (nm)	Density (g/cm³)	Layer Type
Titanium nitride	121.6000	1.700000	4.900000	Uni
Titanium	18.80000	0.500000	4.300000	Uni
Silicon oxide	100.500	0.200000	2.000000	Uni
Silicon		0.200000	2.320000	Sub

Figure 18.12 shows the measured and simulated reflectivity curves from a TiN/Ti diffusion barrier (TiN/Ti/SiO₂/Si). The dynamic range of this measurement is six orders of magnitude, but up to eight orders can be covered. The reported thickness, roughness, and density values are the results of an initial simulation and subsequent refinement.

Nuts and Bolts

Relative Costs

A basic powder diffraction system including a goniometer, sealed-tube generator, point detector, computer, and control and analysis software can be purchased for under $100,000. Additional software, detectors, stages, and attachments can drive the price above $150,000. Variable-temperature attachments add $25,000 to $100,000. Eulerian cradles add $30,000 to $60,000.

An area detector diffraction system for general diffraction applications including a goniometer, sample stage allowing complete ϕ rotation, sealed-tube generator, multiwire area detector, computer, and control and analysis software can be purchased for under $200,000.

A basic four-circle single-crystal diffraction system including a goniometer, closed Eulerian cradle, sealed-tube generator, point detector, computer, and control and analysis software can be purchased for under $100,000. Additional software, detectors, and attachments can drive the price above $150,000. Low-temperature attachments add $15,000 to $30,000.

An area detector diffraction system for single-crystal diffraction applications including a goniometer, sample stage allowing complete ϕ rotation, sealed-tube generator, CCD area detector, computer, and control and analysis software can be purchased for under $300,000.

Rotating anode generators add $100,000 to $150,000 to the cost of a diffraction system.

Vendors for Instruments and Accessories

Enraf-Nonius Co.
390 Central Ave.
Bohemia, NY 11716
phone: 516-589-2885
fax: 516-589-2068

Philips Electronic Instruments Inc.
85 McKee Dr.
Mahwah, NJ 07430
phone: 201-529-3800
fax: 201-529-2252

Rigaku/USA, Inc.
200 Rosewood Dr.
Danvers, MA 01923
phone: 508-777-2446
fax: 508-777-3594

Scintag, Inc.
707 Kifer Rd.
Sunnyvale, CA 94086
phone: 408-737-7200
fax: 408-737-9841

Siemens Industrial Automation
Analytical Instrument Business Unit
6300 Enterprise Ln.
Madison, WI 53719-1173
phone: 608-276-3000
fax: 608-276-3014

Required Level of Training

For many applications, X-ray diffraction instruments can operate as turnkey systems, with data processing automatically following data collection. This is true both of single-crystal and powder diffractometers. Most manufacturers offer application and service training courses. Application courses help the user solve particular analytical problems. Service courses are tailored to the technician who keeps the instrument running. The amount of training required to operate a diffractometer is typically a day or two. The relative difficulty of data interpretation depends on the application. A college-level crystallography course may be helpful, but is usually unnecessary. Application-oriented clinics are periodically offered by such organizations as the American Crystallographic Association (ACA) and the International Centre for Diffraction Data (ICDD). The ICDD and ACA also offer educational materials for X-ray diffraction.

Service and Maintenance

Modern X-ray diffraction instruments have comparatively low maintenance requirements. Routine maintenance of sealed-tube diffraction systems involves keeping clean water in the cooling loop for the X-ray tube and infrequently replacing X-ray tubes. Sealed X-ray tubes cost approximately $4000. Their useful life is 1 to 4 yr, depending on the frequency of use, power settings, and intensity decrease one is willing to tolerate as the tube ages. Rotating anode systems also require maintenance of a high-vacuum system. Service contracts cost $7,000 to $25,000 per year depending on the level of support and the accessories covered.

With the advent of ISO 9000 certification, many users periodically run reference standards to maintain a record of instrument performance as part of standard operating procedures. The National Institute of Standards and Technology (NIST) offers a number of standard reference materials for X-ray powder diffraction (Table 18.2). Likewise, most single-crystal instruments are supplied with reference crystals (such as ruby or sulphonium ylid) to verify alignment.

Suggested Readings

Advances in X-Ray Analysis: Proceedings of the Annual Conference on Applications of X-Ray Analysis. New York: Plenum. (published annually)

BISH, D. L., AND J. E. POST, EDS. *Modern Powder Diffraction.* Reviews in Modern Mineralogy, vol. 20. Washington, D.C.: Mineralogical Society of America, 1989.

Table 18.2 NIST Standard Reference Materials for X-ray powder diffraction.

Standard Reference Material Number	Material	Description
1976	α-Al_2O_3 plate	Instrument sensitivity
1878	α-SiO_2	Respirable alpha quartz
675	Fluorphlogopite mica	Low 2θ (large d-spacing)
674a	α-Al_2O_3, ZnO, TiO_2 (rutile), Cr_2O_3, and CeO_2 powders	Powder diffraction intensity
660	LaB_6	2θ X-ray profile
640b	Si powder	2θ/d-spacing
493	Fe_3C	Iron carbide in ferrite
485a, 487, 488		Austenite in ferrite

HAHN, THED, ED. *International Tables for Crystallography* (four volumes). Boston: D. Reidel, 1983.

Powder Diffraction: An International Journal of Materials Characterization. D. K. Smith, ed.

Computer Databases

Cambridge Crystallographic Data File: Includes the references, formulas, space groups, cell dimensions, and atomic positions of more than 60,000 published organic structures. Available from Cambridge Crystallographic Data Centre, University Chemical Laboratory, Lensfield Road, Cambridge CB2 1EW, England.

Inorganic Crystal Structure Database: Includes the references, formulas, space groups, cell dimensions, atomic coordinates, site occupancies, and thermal parameters on more than 35,000 entries. Produced by FIZ Karlsruhe and the GMELIN Institute. Available from Scientific Information Service Inc., 7 Woodland Ave., Larchmont, NY 10538.

JCPDS Powder Diffraction File: Includes *d*-spacing and relative intensities from over 63,000 single-phase X-ray powder diffraction patterns as well as Miller indices, references, formulas, and cell data. Available from the ICDD.

NIST Crystal Data Identification File: Includes the references, formulas, space groups, reduced cell, and volume on more than 182,500 entries. Available from the ICDD.

Resources

American Crystallographic Association (ACA)
P.O. Box 96, Ellicott Station
Buffalo, NY 14205-0096
phone: 716-856-9600
fax: 716-852-4846

International Centre for Diffraction Data (ICDD)
12 Campus Blvd.
Newtown Square, PA 19073-3273
phone: 215-325-9815
fax: 215-325-9823

National Institute of Standards and Technology (NIST)
Standard Reference Materials Program
Room 204, Building 202
Gaithersburg, MD 20899
phone: 301-975-6776
fax: 301-948-3730

Polycrystal Book Service
(specializes in texts on crystallography, diffraction, mineralogy, and microscopy)
P.O. Box 3439
Dayton, OH 45401-3439
phone: 513-223-9070
fax: 513-223-9070

References

1. L. V. Azároff, *Elements of X-Ray Crystallography* (New York: McGraw-Hill, 1968).

2. W. Parrish, in D. McLachlan and J. P. Glusker, eds., *Crystallography in North America* (New York: American Crystallographic Association, 1983).

3. T. C. Furnas, in D. McLachlan and J. P. Glusker, eds., *Crystallography in North America* (New York: American Crystallographic Association, 1983).

4. B. D. Cullity, *Elements of X-Ray Diffraction*, 2nd ed. (Menlo Park, CA: Addison-Wesley, 1978).

5. K. L. Smith and R. B. Ortega, *Adv. X-ray Anal.*, 36 (1993), 641–7.

6. J. P. Glusker and K. N. Trueblood, *Crystal Structure Analysis: A Primer*, 2nd ed. (New York: Oxford University Press, 1985).

7. G. H. Stout and L. H. Jensen, *X-Ray Structure Determination: A Practical Guide*, 2nd ed. (New York: Wiley, 1989).

8. R. Delhez, T. H. de Keijser, and E. Mittemeijer, *Fresenins Zeit. Analyt. Chemie*, 312 (1982), 1–16.

9. R. P. Goehner and M. O. Eatough, *Powder Diffraction*, 7, no. 1 (1992), 2–5.

10. D. K. Smith, in D. L. Bish and J. E. Post, eds., *Modern Powder Diffraction*. Reviews in Modern Mineralogy, vol. 20 (Washington, D.C.: Mineralogical Society of America, 1989).

11. D. D. L. Chung and others, *X-Ray Diffraction at Elevated Temperatures: A Method for In Situ Process Analysis* (New York: VCH, 1993).

12. R. Rudman, *Low-Temperature X-Ray Diffraction: Apparatus and Techniques* (New York: Plenum, 1976).

13. P. Caussin, J. Nusinovici, and D. W. Beard, *Adv. X-ray Anal.*, 32 (1989), 531–8.

14. R. L. Snyder and D. L. Bish, in D. L. Bish and J. E. Post, eds., *Modern Powder Diffraction*. Reviews in Modern Mineralogy, vol. 20 (Washington, D.C.: Mineralogical Society of America, 1989).

15. D. K. Smith, *Power Diffraction* (1997), in press.

16. B. Lengeler, *Adv. X-ray Anal.*, 35 (1992), 127–35.

17. I. C. Noyan and J. B. Cohen. *Residual Stress: Measurement by Diffraction and Interpretation* (New York: Springer-Verlag, 1987).

18. R. Jenkins and others, *Powder Diffraction*, 1, no. 2 (1986), 51–63.

19. H. P. Klug and L. E. Alexander, *X-Ray Diffraction Procedures for Polycrystalline and Amorphous Materials* (New York: Wiley, 1974).

20. C. R. Hubbard and R. L. Snyder, *Powder Diffraction*, 3, no. 2 (1988), 74–8.

21. R. A. Young, *The Rietveld Method* (New York: Oxford University Press, 1993).

Quantitative Optical Spectroscopic Methods

Chapter 19

Introduction

Earl L. Wehry

University of Tennessee
Department of Chemistry

In this section of the handbook, eight techniques that are used primarily for quantitative analyses are discussed:

- Atomic absorption spectroscopy (AAS)
- Inductively coupled plasma atomic emission spectroscopy (ICP-AES)
- Inductively coupled plasma mass spectrometry (ICP-MS)
- Atomic fluorescence spectroscopy (AFS)
- X-ray fluorescence (XRF) spectroscopy
- Molecular ultraviolet/visible (UV-Vis) absorption spectroscopy
- Molecular fluorescence spectroscopy (MFS)
- Chemiluminescence (CL)

These techniques have two key characteristics in common. First, they are based on spectroscopy, or interactions of electromagnetic radiation with matter. At some stage in each of these methods (except ICP-MS), electromagnetic radiation is absorbed or emitted by the analyte. Second, each of these techniques is used primarily for the quantification of specific analytes in samples. This does not mean that they are useless for identification of sample constituents. For example, XRF, ICP-AES, and ICP-MS can be used to identify specific elements present in samples. However, the primary application of the methods described in this section is determination of the concentration or amount of specific elements or compounds in samples.

Spectroscopic methods discussed elsewhere in this handbook also can be used for quantitative purposes. For example, vibrational spectroscopic methods (infrared and Raman spectroscopies) are often used for quantification of specific molecular constituents of materials. However, those methods are considered in another section of the handbook because vibrational spectroscopy is used more often for molecular identification than for molecular quantification.

The reader might also initially think that ICP-MS has strayed from its proper location in the section on mass spectrometry. ICP-MS is the one technique described in this section that is not based on some form of spectroscopic measurement. However, when we consider applications, we find that ICP-MS has little in common with the molecular mass spectrometric methods discussed in the section on mass spectrometry; rather, ICP-MS is used for the same types of analytical purposes as the atomic spectroscopic methods (AAS, ICP-AES, and AFS). Hence, ICP-MS should be considered complementary to (and competitive with) atomic absorption, emission, and fluorescence techniques, rather than with other mass spectrometric techniques.

Five of the techniques discussed in this section—atomic absorption, emission, and fluorescence, ICP-mass spectrometry, and X-ray fluorescence—are elemental analytical techniques. They are used to determine the concentrations or quantities of specific elements present in samples, but generally provide no information regarding the chemical forms of elements in samples. Indeed, in each of the elemental techniques except XRF, samples are subjected to destructive treatment that is specifically intended to convert an analyte element (regardless of its chemical form in the sample) to gas-phase atoms (AAS, ICP-AES, AFS) or ions (ICP-MS). Speciation of an element (that is, identification of the oxidation state and the nature of any complexes present in the original sample) under these conditions is impossible unless appropriate operations (such as chromatographic separations) are undertaken before the sample is atomized (1).

In AAS, ICP-AES, and AFS, gas-phase atoms or ions absorb or emit light in the UV, visible, or near infrared spectral regions. Atomic spectra are line spectra (linewidths of 0.001 nm or less), but the spectra may be very complex (hundreds of lines may be observable for one element). AAS and AFS are primarily single-element techniques (that is, they are used primarily for determination of one, or a few, elements in a sample). In contrast, ICP-AES is a multielement technique.

In ICP-MS, another multielement technique, a sample is converted into ions in the gas phase, and the resulting ions are sorted by their mass-to-charge ratios.

Another multielement technique, X-ray fluorescence, may initially appear to have little in common with AAS, ICP-AES, AFS, or ICP-MS, apart from its use in elemental analysis. In XRF, X-ray photons are emitted by atoms in a sample that have been subjected to illumination with a beam of X rays or (occasionally) protons or other subatomic particles. The experimental techniques involved in generating and measuring X-ray photons are, in principle, similar to those encountered in working with UV or visible photons, but are different in detail. A major attraction of XRF over the other quantitative elemental techniques discussed in this section is that XRF can be nondestructive and applicable directly to solid samples, without dissolution or other major pretreatment. The other atomic spectroscopic techniques are destructive, and can generally deal only with samples that are in the form of liquid solutions (or perhaps slurries). Experimentally, XRF has more in common with certain surface spectroscopic techniques discussed elsewhere in this handbook (especially photoelectron spectroscopy) than with the other techniques discussed in this section. In terms of applications, however, XRF often is complementary to, and competitive with, AAS, ICP-AES, ICP-MS, and AFS.

The remaining three techniques discussed in this section (UV/Vis, MFS, and CL) rely on the measurement of absorption or emission of light in the UV, visible, or near infrared spectral regions by molecules. In each of these techniques, an electronic absorption or emission spectrum can be measured. In principle, therefore, these techniques can be used for molecular identification as well as quantification. However, molecular electronic spectra (especially for solution samples, but often for gases and solids as well) tend to be broad, relatively featureless, and not responsive in an obvious way to the details of molecular structure, such as the presence of specific functional groups. For instance, a reasonably experienced analyst, presented with the UV absorption spectrum of 3-methylphenol in liquid solution, would be able to identify the compound that produced the spectrum as a polar derivative of benzene, but might not be able to proceed much further, even if a library of pure compound spectra were available. Techniques based on molecular electronic spectra are much less broadly useful for identification of molecules (but more widely used for quantification of compounds) than infrared and Raman spectrometry, mass spectrometry, or nuclear magnetic resonance.

The reader should also be aware of a group of molecular fluorescence experiments known as fluorescent probe techniques. The objective of such measurements is to use the spectral characteristics of a fluorescent molecule to make inferences about the nature of the environment in which the fluorescent molecule is found. This may be seen as a qualitative analytical application of molecular fluorescence spectroscopy. However, the information obtained does not pertain to the fluorescent molecule itself, but rather to its surroundings.

Many special issues and concerns arise regarding the proper application of instrumental techniques to quantitative analysis. These issues are discussed in Chapter 4.

Capsule Descriptions of the Techniques Discussed in This Section

Below are brief descriptions of the techniques discussed in this section of the handbook.

Techniques for Quantitative Determination of Elements

Atomic Absorption Spectrometry (AAS)

Procedure Sample solution or slurry is aspirated continuously into a flame, or a small volume of sample is placed in a graphite furnace. Solvent and volatile sample components are driven off, and analyte is converted to gas-phase atoms. Absorption of light by these atoms in their ground states is measured.

Analytes Virtually all metals and many metalloids are analyzed.

Main Advantages
- Instrumentation is relatively simple and inexpensive.
- Low limits of detection (LODs) are possible, especially when electrothermal atomization is used.
- Rapid analyses are possible.

Main Disadvantages
- It is often a single-element technique.
- It is susceptible to multiplicative interferences.
- Solids normally must be dissolved before analysis.

Inductively Coupled Plasma Atomic Emission Spectrometry (ICP-AES)

Procedure Sample solution or slurry is aspirated continuously into an inductively coupled argon plasma discharge. Solvent and volatile sample constituents are driven off and analyte is converted to gas-phase atoms or ions in excited states. Emission of light by the excited atoms or ions is measured.

Analytes Virtually all metals and metalloids are analyzed; some nonmetals that emit in the vacuum-ultraviolet spectral region can also be determined.

Main Advantages
- Simultaneous multielement determinations are possible.
- Most multiplicative interferences are eliminated due to high temperature of plasma discharge.
- Low limits of detection are possible.

Main Disadvantages
- Cost of instrumentation is relatively high.
- Additive interferences (overlaps of emission lines of different atoms) are possible.
- Solids normally must be dissolved before analysis.

Inductively Coupled Plasma Mass Spectrometry (ICP-MS)

Procedure Sample solution or slurry is aspirated continuously into an inductively coupled argon plasma discharge. Solvent and volatile sample constituents are driven off, and analyte is converted to (usually) monopositive ions. Ions are sorted according to their mass/charge ratio and counted.

Analytes Virtually all metals and metalloids and some nonmetals are analyzed.

Main Advantages
- Simultaneous multielement determinations are possible.
- Very low limits of detection are possible.
- There are fewer additive interferences than in ICP-AES.
- Determination of specific isotopes and isotope ratio measurement is possible.

Main Disadvantages
- Cost of equipment is relatively high.
- Additive interferences caused by ionized molecular species may occur.
- Solids must normally be dissolved before analysis.
- It is more susceptible to multiplicative interferences than ICP-AES.

Atomic Fluorescence Spectrometry (AFS)

Procedure Sample or solution is aspirated continuously in a flame, or a small volume of sample is placed in a graphite furnace. Solvent and volatile sample components are driven off and analyte is converted to gas-phase atoms in their ground states. Atoms are excited by an intense light source, and light emitted by the resulting excited atoms is measured.

Main Advantages
- Extremely low limits of detection (down to the single-atom level) are possible.
- Susceptibility to additive interferences is relatively low.

Main Disadvantages
- Suitable commercial instrumentation may not be available; capital cost is usually high if a laser source is used.
- It is often a single-element technique.
- It is susceptible to multiplicative interferences.
- Solids normally must be dissolved before analysis.

X-Ray Fluorescence (XRF) Spectrometry

Procedure Sample is illuminated with a beam of X rays or particles (usually protons) from a suitable source. X rays emitted by atoms excited by the incident energy are measured.

Analytes Any element heavier than boron can be analyzed, although sensitivity is highest for elements heavier than sodium.

Main Advantages
- It is applicable to both solid and liquid (solution) samples.
- It can be nondestructive.
- It is applicable to very small samples.
- Simultaneous multielement determinations are possible.
- It has minimal susceptibility to additive interferences.
- Surface, thin-film, and spatially resolved analyses are possible.
- Rapid analyses are possible.

Main Disadvantages
- Cost of instrumentation is relatively high.
- Limits of detection may be higher than for other elemental techniques (AAS, ICP-AES, ICP-MS, and AFS).
- Susceptibility to multiplicative interferences (matrix effects) is relatively high.

Techniques for Quantitative Determination of Compounds

Molecular Ultraviolet/Visible (UV/Vis) Absorption Spectrometry

Procedure Molecules in liquid, solid, or gaseous sample are excited by light from lamp source; absorption of light is measured.

Analytes Virtually any organic compound containing unsaturation can be analyzed, as well as many inorganic compounds. Chemical derivatization procedures can be used to convert many nonabsorbing analytes to absorbing species.

Main Advantages
- Precision and accuracy are high.
- Rapid analyses are possible.
- It is amenable to process and field applications.
- Susceptibility to multiplicative interference is minimal.
- Cost of instrumentation is relatively low.

Main Disadvantages
- Low limits of detection are not generally possible.
- It is highly susceptible to additive interferences (spectral overlaps).

Molecular Fluorescence Spectrometry (MFS)

Procedure Molecules in liquid, solid, or gaseous sample are illuminated by intense lamp or laser source; emission of light by excited molecules is measured.

Analytes Analysis is limited to aromatic or highly unsaturated molecules, although chemical derivatization procedures to convert many nonfluorescent molecules to fluorescent derivatives have been developed.

Main Advantages
- Very low limits of detection (down to the single-molecule level) are possible.
- Rapid analyses are possible.
- It is amenable to process, field, and remote sensing applications.
- Information content of measurement (polarization and decay time as well as spectral data) is high.

Main Disadvantages
- Few compounds exhibit intense native fluorescence.
- It is highly susceptible to multiplicative and additive interferences unless specialized techniques are used.
- Accuracy and precision are generally lower than those of UV/Vis.

Chemiluminescence (CL)

Procedure Chemical reaction in gas or liquid phase generates electronically excited product that emits light. Analyte is either a reactant or a catalyst of the reaction. The emitted light is measured.

Analytes Analysis is limited to the relatively few species known to engage in, or catalyze, a chemiluminescence reaction (often strong oxidizing or reducing agents).

Main Advantages
- Cost of instrumentation is very low.
- Portable instrumentation is available for field use.
- It can exhibit very low limits of detection.
- Very rapid analyses are possible.

Main Disadvantages
- It is limited to relatively few analytes.
- Lack of selectivity can be a major problem.

References

1. S. C. K. Shum, R. Neddersen, and R. S. Houk, *Analyst*, 117 (1992), 577.

Chapter 20

Atomic Absorption Spectrometry

Joseph Sneddon

McNeese State University
Department of Chemistry

Summary

General Uses

- Quantification of nearly 70 metals in virtually any sample type
- Concentrations as low as parts per billion to as high as weight% with minimal or no sample preparation
- Microliter volumes or microgram masses required
- Well-established and accepted technique

Common Applications

- Biological, medical, and clinical samples (blood, urine, and other body fluids, tissue, hair, teeth, nails)
- Environmental samples (waters, solids, sediments, biota)
- Steel and metal industry
- Pharmaceutical industry
- Food industry
- Air
- Industrial hygiene
- Pollution studies

Samples

Almost any solid, liquid, or gaseous sample can be analyzed. Most samples are converted into homogeneous solutions prior to analysis.

Amount

As low as microliter volumes and milligram masses are used in furnace AAS. Flame AAS can use microliter volumes but is generally performed with milliliter volumes.

Preparation

Aqueous samples can be determined directly with no sample preparation. A complex solution such as blood can be diluted with ultrapure water, which reduces or minimizes interference in the analysis. If the level to be determined is beyond the capability of the technique, then solvent extraction or other concentration techniques must be employed. There are numerous procedures in the literature on these methods. Solid samples need to be dissolved or digested. Ultrapure chemicals are a must in this case. Dissolution can range from a few minutes to up to 24 hr. It may involve classical wet techniques or, more recently, microwave digestion.

Analysis Time

The time to perform a complete analysis is linked to sample preparation. Preparing the sample can range from 0 sec to 24 hr. The actual determination (after calibration setup) ranges from approximately 10 sec for flame AAS to approximately 2 min for furnace AAS.

Limitations

- Provides no information on the chemical form of the metal
- Sample preparation is tedious and time-consuming
- Limited to metals and metalloids
- Destructive technique

Accuracy

Depends on the complexity of matrix. A homogeneous solution at an analyte level 5–10 times above the detection limit will give an accuracy better than ±1%. Accuracy is reduced at or around the detection limit of (typically) ±1–3%. Direct solid determination will give accuracy (typically) ±5–10%.

Linear Range

In a single metal mode, the linear range is typically two to three orders of magnitude above the detection limit. This can be extended using computer software but is generally not advisable. Simple dilution is much easier. In a simultaneous metal mode, the linear range will depend on the metals to be determined and could be less than two orders of magnitude above the detection limit.

Complementary or Related Techniques

- Atomic fluorescence spectrometry
- Atomic emission spectrometry, particularly inductively coupled plasma atomic emission spectrometry
- X-ray fluorescence spectrometry
- Inductively coupled plasma mass spectroscopy
- Neutron activation analysis

Introduction

Atomic absorption spectrometry (AAS) is a widely used and accepted technique capable of determining trace (μg/mL) and ultratrace (sub-μg/mL) levels of elements or metals in a wide variety of samples, including biological, clinical, environmental, food, and geological samples, with good accuracy and acceptable precision. It is arguably the predominant technique in elemental analysis, although it does have some limitations. This chapter describes the theory and basic principles, instrumentation, practice of the technique, and selected applications. The chapter concentrates primarily on general flame and graphite furnace AAS. Certain elements can be determined in AAS by methods particular to their unique properties; for example, mercury is often determined using cold-vapor AAS and covalent hydrides such as selenium and arsenic by hydride generation AAS methods (1). These specific methods are not dealt with in this chapter but can be found in References 2–4. Finally, a brief discussion on some recent developments in AAS is included.

How It Works

Basic Principles

Atomic absorption spectrometry (AAS) was discovered independently by Walsh and Alkemade and Melatz in the early to mid-1950s. It involves the impingement of light of a specific wavelength onto previously generated ground-state atoms. The atoms absorb this light and a transition to a higher energy level occurs. The intensity of this transition is related to the original concentration of the ground state atoms. This can be represented as follows:

$$T = P/P_o \qquad (20.1)$$

where T is the transmittance, P is the power of the light source passing through the sample zone, and P_o is the power of the light source before it passes through the sample zone. The sample zone of path length, b, is relatively long to maximize the amount of light absorbed by the atoms. The amount of light absorbed depends on the atomic absorption coefficient, k. This value is related to the number of atoms per cm^3 in the atom cell, n; the Einstein probability for the absorption process; and the energy difference between the two levels of the transition. In practice these are all constants that are combined to give one constant, called the absorptivity, *a*. *k* is related exponentially to the transmittance as follows:

$$T = P/P_o = e^{-kb} \tag{20.2}$$

In practice, the absorbance, *A*, is used in AAS and is related logarithmically to the transmittance as follows:

$$A = -\log T = -\log P/P_o = \log P_o/P = \log 1/T = kb \log e = 0.43\, kb \tag{20.3}$$

The Beer–Lambert law relates *A* to the concentration of the element in the atom cell, *c*, as follows:

$$A = abc \quad \text{or} \quad A = \varepsilon_o bc \tag{20.4}$$

where *a* is the absorptivity in g/L-cm, ε_o is the molar absorptivity in g/mol-cm, and *b* is the atom cell width in cm. AAS involves the measurement of the drop in light intensity of P_o to P (depending on the concentration of the element). Current and modern instrumentation automatically converts the logarithmic value into *A*. Absorbance is a unitless number, typically 0.01 to 2.0. In practice, it is better to work in the middle of this range (recommended 0.1 to 0.3 *A*) as the precision is poorer at the extremes due to instrumental noise. The most intense transition from the ground state to the first excited state (resonance transition) is the most widely used transition because it is the most sensitive.

Instrumentation

Flame Atomic Absorption Spectrometry

A schematic diagram of a flame atomic absorption spectometer is shown in Fig. 20.1. It essentially consists of six major components: a radiation (light) source, flame atom cell, sample introduction system, monochromator, detection system, and readout. In current commercial instrumentation, all six parts are efficiently combined in a compact, bench-type unit, and often connected to a computer for control, sample preparation station, data reduction, and hard-copy printout.

Radiation Source   The most widely used and accepted radiation source in AAS is the line source, in particular the hollow cathode lamp (HCL) and to a lesser extent the radiofrequency electrode-

Figure 20.1 Schematic diagram of an atomic absorption spectrometer.

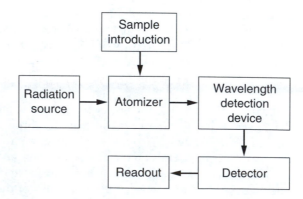

less discharge lamp (EDL). For the most part, the HCL is a line source, with each new metal or element to be determined requiring a separate lamp. Some multielement lamps are available such as Ca-Mg and Cr-Fe-Ni, where the cathode is made of two or three metals with similar properties. In general, these lamps produce a poorer performance than single-element lamps because the difference in volatilities necessitates a compromise in conditions, but do offer the possibility of limited multielement determination. The costs of multielement lamps are lower (per element) than the cost of combined individual lamps.

Hollow Cathode Lamp A schematic diagram of a typical single-element HCL is shown in Fig. 20.2. It is a glass envelope filled with inert gas, usually neon, argon, or occasionally helium, at a pressure of 1 to 5 torr and the hollow cathode made of the pure metal of interest. A voltage is applied between the anode (positive) and the cathode (negative) of about 500 V with a current of 2 to 30 mA and the discharge concentrates into the hollow cathode. The filler gas is ionized (becomes charged) at the anode, and the positive ions produced ($Ar + e^- = Ar^+ + e^- + e^-$) are accelerated by the charge toward the negative cathode. The ions impinge or strike the cathode, causing the metal ions to be removed or sputtered out of the cathode. Further collisions excite the metal atoms and the excited metal ions produce an intense characteristic spectrum of the metal of interest when they return to the ground state. These atoms diffuse back to the cathode or onto the glass walls of the lamp. Modern lamps are constructed of glass but require a silica window for use in the ultraviolet region and have a molded plastic base. The seal between the window and the envelope must be gas tight to prevent loss of pressure and fill gas. HCLs have a finite lifetime, typically 2 to 5 amp-hours, and can deteriorate if not used. This is particularly true for elements with high volatility, such as arsenic and selenium. A short warmup period of a few minutes is usually adequate before use. Current HCLs are very stable over several hours of continuous use.

Electrodeless Discharge Lamp The electrodeless discharge lamp (EDL) was originally developed for atomic fluorescence spectrometry (AFS) and was primarily microwave-excited. These were much more intense than the HCL but somewhat less stable. Radiofrequency-excited EDLs (RF-EDLs) can be used in AAS. They are more intense than HCLs, typically 5 to 100 times, and operate in the 100 KHz to 100 MHz region. A commercially available RF-EDL is shown in Fig. 20.3. Commercial RF-EDLs have a built-in starter, are run at 27.12 MHz from a simple power supply, pretuned, and enclosed to stabilize the temperature. An EDL is run at room temperature. In AAS, the increased intensity does not lead to greatly improved sensitivity; thus, RF-EDLs are recommended primarily for metals that do not make very good HCLs. These metals are highly volatile and their resonance lines are in the low ultraviolet region of the electromagnetic spectrum; they include arsenic (193.7 nm), selenium (196.0 nm), zinc (213.9 nm), and cadmium (228.8 nm). They

Figure 20.2 Schematic diagram of a typical hollow cathode lamp.

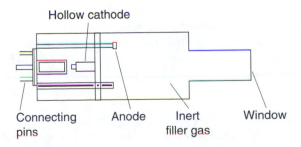

Figure 20.3 Schematic diagram of a radiofrequency electrodeless discharge lamp.

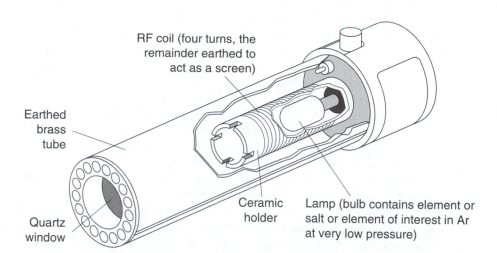

can improve AAS sensitivity by a factor two to three over HCLs but they are more expensive and require the increased cost of the power supply.

Other Sources Various other sources have been proposed for AAS but for the most part have not been widely adopted. They include those used primarily for research or specific tasks, such as continuum and semiconductor diode laser sources for multielement AAS. Other sources include low-pressure discharge lamps, thermal gradient lamps, and plasmas.

Flame Atom Cell In flame AAS, the sample is usually introduced into the flame as a fine mist or aerosol. Flames consist of an oxidant and a fuel. The most widely used flames in AAS are air–acetylene (air is the oxidant and acetylene the fuel) and nitrous oxide–acetylene (nitrous oxide is the oxidant and acetylene the fuel). These flames are called combustion flames. Other flames, called diffusion flames, have been proposed but are not widely used. The primary object of the flame is to dissociate the molecules into atoms. Air–acetylene (2500 °K) does this readily and efficiently for 40 to 50 elements in the periodic table. The other 10 to 20 elements in the periodic table require the hotter nitrous oxide–acetylene flame (3200 °K). A long, thin flame is desirable in AAS for maximum sensitivity.

Sample Introduction System The sample introduction system should reproducibly and efficiently transfer a sample to the atomizer. It should produce no interferences, be reproducible over extended times, have no memory or carryover effects, be independent of the sample type, and be universal for all types of atomizers or atomic spectroscopic techniques. Unfortunately, all these desirable properties cannot be simultaneously obtained and a compromise is often used. Other areas that can be considered include the amount of sample available (which may dictate a discrete or continuous mode), the physical form of the sample (solid, liquid, slurry, or gas), the analytical performance characteristics desirable (such as precision, accuracy, and detection limit), throughput of samples, and the type of atomizer (flame, furnace, or plasma). These factors can influence the type of sample introduction system used. The sample introduction system should be viewed in conjunction with the atomization process. A book describing various sample introduction methods used not only for AAS but atomic spectrometry in general is available (5).

Pneumatic Nebulizers Pneumatic nebulizers (PNs) are the most widely used method of introducing a solution. A jet of compressed air, the nebulization gas, aspirates and nebulizes the solution when the sample is sucked through a capillary tube. In the concentric nebulizer, the sample is surrounded by the oxidant gas as it emerges from the capillary tube, causing a reduced pressure at the tip and thus suction of the sample solution from the container (Bernoulli effect). In the angular or cross-flow nebulizer, a flow of compressed gas over the same capillary at right angles produces the same Bernoulli effect and aspirates the sample. The high velocity of this sample as it exits from the capillary tube creates a pressure drop and shatters the solution into tiny droplets. This aerosol then passes along a plastic expansion chamber, where large droplets may be removed by impactors, spoilers, or baffles. This chamber also allows for the mixing of gases and has a drain tube to remove excess aerosol. An example of the chamber in combination with the pneumatic nebulizer and burner is shown in Fig. 20.4. The object of this combination is to obtain a reproducible and small-range aerosol diameter, ideally 2 to 6 μm in size, before transport to the burner. In general, the pneumatic nebulizer does not have a very high nebulization efficiency, typically 5 to 10%, and is not recommended for high-solid or salt solutions due to clogging and blockage of the capillary tubes.

Other Nebulizers Several other nebulization systems have been developed and characterized for the introduction of solutions to atomizers. They have been proposed to improve aerosol production (nebulization efficiency) and more efficiently handle high-solid dissolved solutions.

Figure 20.4 Diagram of combined burner head, expansion chamber, and pneumatic nebulizer. *(Reproduced by permission of the Perkin-Elmer Corporation.)*

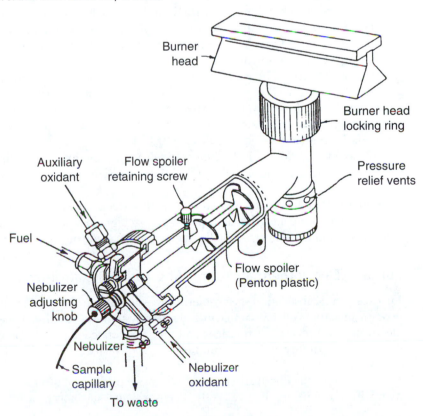

The ultrasonic nebulizer generates a homogeneous aerosol of fine (smaller than 5 μm) droplets, has a high efficiency (30%+), and can allow the carrier gas flow rate and aerosol generation rate to be independently variable. The basic principle involves the use of high-frequency sound waves (typically 0.1 to 10 MHz) to create discrete droplets from a solution, which are then swept to the atomizer through a chamber. Numerous workers have investigated the ultrasonic nebulizer including the principles, piezoelectric transducer, and the nebulizer design as well as its application to difficult samples such as seawater and other high-salt solutions.

The Babington nebulizer was used for solutions with a wide viscosity range of high-salt solutions and suspended particles. The basic principle involves flow of the solution to be nebulized over a glass sphere containing a small hole, slot, or series of very fine holes. Gas forced through these holes disrupts the solution flowing over the sphere and produces an aerosol. Denton and co-workers (6) have developed this system and applied it to tomato sauce, evaporated milk, and untreated whole blood.

Various other nebulizers proposed are the thermospray, frit nebulizer, and high-solids nebulizer (6). In general, these nebulizers are recommended for specific samples and are not universally used or widely accepted.

Monochromator The principal function of a monochromator in AAS is to isolate the wavelength (radiation) of interest from the other wavelengths from the radiation source and light emitted by other elements in the flame. As AAS is a very selective method and has virtually no spectral interferences, the monochromator does not need to have high resolution; typically 0.02 to 2 nm resolution is adequate.

Detection System The most widely used detection system is the photomultiplier tube (PMT). A PMT consists of a photoemissive cathode and several dynodes in a vacuum. The dynodes provide electron multiplication (gain). Typically, a PMT tube is biased negative at 400 to 2500 V with respect to the anode. The photoelectron ejected by the photocathode strikes the first dynode and releases two to five secondary electrons. These are accelerated toward the next dynode and soon give a cascade effect. The gain can be 10^4 to 10^7. This gain is very useful for detecting low levels of light. The cathode is coated with an easily ionized material such as alloys of alkali metals with antimony, bismuth, or silver. The spectral sensitivity of a PMT depends on the material coating of the cathode. A versatile coating is GaAs, which responds from around 190 to around 900 nm with relative high and constant efficiency.

Although the PMT is still the most widely used detection system, the photodiode array (PDA) detector and charge transfer device (CTD) detector have attracted interest in AAS.

Readout Devices Readout devices in early AAS systems were meters with calibrated scales. Current instrumentation includes digital displays and often graphic presentations on video display units or external computers. Alternatively, hard copy can be made on chart recorders, printers, or plotters.

Furnace (Electrothermal) Atomic Absorption Spectrometry

Furnace atomic absorption spectrometry has essentially the same instrumentation as flame AAS except for the atom cell and sample introduction system. An additional need in furnace AAS is faster electronics to process the transient and faster-generated signal. In practice AAS usually has fast electronics capability and most commercial systems have an interchangeable flame and furnace.

Furnace Atom Cell The furnace atom cell or electrothermal atomizer (ETA) for AAS was commercially developed in the late 1960s. Their principal advantages over flame atomizers are the improvement in sensitivity (typically 10 to 100×), the ability to use microvolume (2 to 200 µL) and micromass (few mg) sampling, and in-situ pretreatment of the sample. However, ETAs are prone to interferences, particularly from alkali and alkaline earth halides and require a more complex (and consequently more expensive) system.

The use of an electrically heated tubular furnace was first reported by King in 1905 but for analytical chemistry, the work and system developed by L'Vov around 1960 is regarded as the forerunner of present-day ETAs. It consisted of a carbon electrode in which the sample was applied and a carbon tube that could be heated by electrical resistance. The initial design used a supplementary electrode for preheating the furnace, lined the carbon tube with tungsten or tantulum foil to minimize vapor diffusion, and purged the system with argon to prevent oxidation of the carbon. Later work involved direct heating of the sampling electrode by resistance heating and the tube was made of pyrolytic carbon. After the tube was heated to an elevated temperature, the sample electrode was inserted into the underside of the tube, and vaporization of the sample was confined to the tube, where AAS measurements were made. The system was difficult to operate and the reproducibility could be poor.

In 1967, Massman described a heated graphite atomizer (HGA), which was commercially developed by the Perkin–Elmer Corporation and proved to be the forerunner for all current commercial ETAs. An isothermal furnace system proposed by Woodriff at around the same time was considered more difficult to commercialize, although recent work has shown the advantage of atomization under isothermal conditions. The Massman system was a graphite tube, typically 50 mm long and 10 mm in diameter, that was heated by electrical resistance, typically 7 to 10 V at 400 A. An inert gas, usually argon or nitrogen, flowed at a constant rate of around 1.5 L/min and the entire system was enclosed in a water jacket. A microliter sample was deposited through an entry or injection port in the center of the tube and could be heated in three stages by applying variable current to the system: drying to remove the solvent, ashing or pyrolysis to remove the matrix, and finally atomization of the element. Careful temperature control was required in order to obtain good reproduciblity.

In 1969 West and coworkers developed a rod or filament atomizer. It consisted of a graphite filament 40 mm long and 2 mm in diameter, supported by water-cooled electrodes and heated very quickly by the use of current of 70 A at 10 to 12 V. Shielding from the air was achieved by a flow of inert gas around the filament. Although it was developed primarily for AAS, West and coworkers showed the potential of the system for AFS.

The West filament was the forerunner for the mini-Massman atomizer developed commercially by Varian Associates. A commercial system was called the carbon rod atomizer (CRA 63). Its main advantages were the somewhat simpler and less complex design compared to the HGA, low power requirements (2 to 3 kilowatts), and fast (about 2 sec) heating rate. There were differences between this system and the West filament, principally a hole drilled in a solid cylindrical graphite tube and later a small cup or crucible between two spring loaded graphite rods. The system was proposed for small volumes, typically 1 to 20 µL. In general, higher detection limits and increased interferences were found using the CRA system compared to HGA; manufacture of this type of system was discontinued in the mid-1980s and is not currently commercially available.

A current, top-of-the-line furnace is shown in Fig. 20.5. Most current commercial furnaces are similar to that shown in Fig. 20.5.

Sample Introduction Typical volumes used with a furnace are from 1 to 200 µL. This volume can be introduced to the furnace manually using a micropipette. There are various dedicated micropipettes available for this range as well as adjustable micropipettes. In the late 1970s it was suggested that the precision of furnace AAS could be improved by automatic introduction systems. This led to systems that could be added to furnace AAS for automatic sample introduction. It was shown

Figure 20.5 Current commercial furnace used for AAS. *(Reproduced by permission of the Perkin-Elmer Corporation.)*

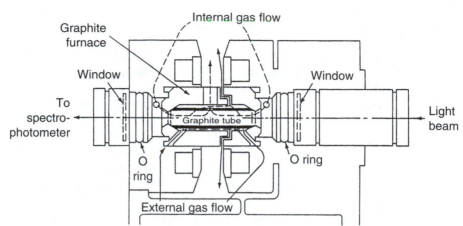

that the precision was not significantly improved by automatic sample introduction but neverthe-less it is now an accepted part of furnace AAS, particularly for unattended operation and when nu-merous samples are to be analyzed. These systems can be incorporated into sample preparation stations. Most systems use a discrete volume for sample introduction, although a continuous sys-tem based on nebulization has been successfully used by Thermo Jarrell Ash for a number of years.

What It Does

Techniques for Quantification of Elements

AAS, in common with many analytical techniques, is not an absolute method of analysis. A com-parison with standards (usually aqueous) is the most common method of performing quantitative analysis. A calibration curve is established in AAS by introducing a known concentration of solu-tion to the system followed by the measurement. The measurement can be compared to the signal generated from an unknown concentration. In practice it is usual to use at least four standards and a blank to construct a linear calibration curve of absorbance versus concentration of analyte to per-form the analyses. It is recommended that the unknown concentration to be determined be no higher than that of the highest standard and no lower than that of the lowest standard. If the signal from the unknown falls outside these two areas, dilution or concentration is recommended. Linear extrapolation of an AAS calibration curve is not recommended due to the lack of linearity, typically 2 to 3 orders of magnitude above the detection limit. Above this level the calibration curve gener-ally exhibits negative deviations from linearity.

Most analyses are performed on samples that are not identical to the aqueous standards. In this case, it may be possible to match the matrix of the unknown with the matrix of the standards. If the matrix of the sample to be analyzed is unknown or varies from sample to sample, then the method of standard additions (known additions) is often used. This method is used to compensate

Figure 20.6 Multiple standard-additions plot. Curve (a) is obtained with a proper blank measurement; in curve (b) the blank measurement does not compensate for the blank interference.

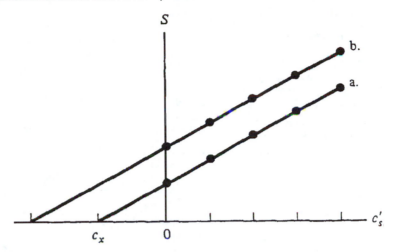

for interferences, both spectral and chemical. This procedure involves measuring the analytical AAS signal from the sample, S; then a small volume of a concentrated standard of known analyte concentration, Sx, is added to a large volume of a sample of S and the analytical AAS of this sample, $S + Sx$, is measured. These steps are repeated using increasing or different concentrations of Sx (Sx_1, Sx_2, etc.) and the total analytical signals are measured. A graph similar to that shown in Fig. 20.6 is established and the unknown concentration obtained by back-extrapolation. See Chap. 4 for further detail.

Interferences

Flame AAS

An interference is a chemical or physical effect that causes the signal to be reduced or increased compared to the signal from the calibrated solution. In order to understand an interference, an understanding of the processes involved in producing atoms from a solution is required. This is shown schematically in Fig. 20.7 and illustrated using magnesium sulfate heptahydrate in solution. The solution containing the magnesium sulfate heptahydrate, $MgSO_4 \cdot 7H_2O$, is introduced or nebulized into the chamber, where it is mixed with the oxidant and fuel gases. Large droplets are removed and a fine mist or aerosol containing uniform and small-diameter (a few microns) particles are carried to the burner. At the flame the solvent (water) is evaporated and salt particles, $MgSO_4 \cdot 7H_2O$ (s), remain.

The sample is then converted to a gaseous state, $MgSO_4 \cdot 7H_2O$ (g). The waters of hydration are removed in several steps, to leave $MgSO_4$ (g). As the temperature increases, the magnesium sulfate is reduced to magnesium oxide (oxides are thermodynamically stable compounds). The final step is the production of the gaseous magnesium atoms from the molecular gaseous magnesium oxide. These atoms are then used for the AAS measurement.

However, this process represents an ideal situation and there are several chemical interferences that prevent or inhibit the formation of neutral gaseous magnesium atoms. These include the

Figure 20.7 Schematic representation of the atomization process in a flame for magnesium sulfate heptahydrate solution.

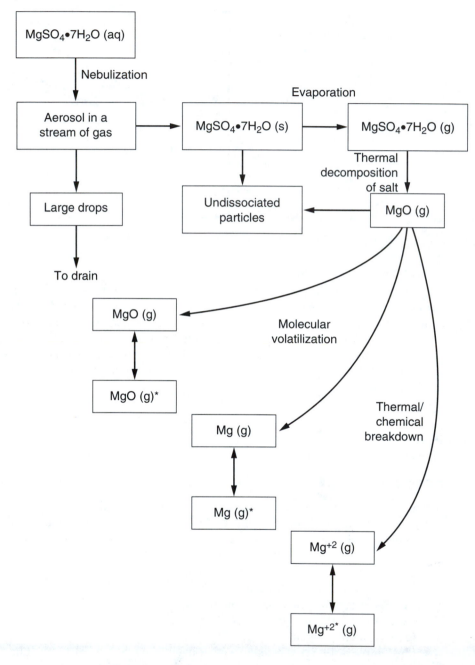

inability of the flame or atomizer to dissociate the molecule into atoms, enhancements caused by the formation and occlusion of more volatile compounds, depressions due to occlusion into refractory compounds, and the fact that ions may be formed instead of atoms.

The ability to form atoms from the molecules, the atomization efficiency, can be maximized using a hotter atomizer because the air–acetylene flame has a temperature of approximately 2500 °K, which is inadequate for complete atomization of many molecules, particularly the refractory oxide and other high-boiling-point molecules. The use of the nitrous oxide–acetylene flame, with a temperature of approximately 3200 °K, improves atomization efficiency (note that this improvement in atomization efficiency may come at the expense of decreased sensitivity). The most well-known case of this type of interference is that of phosphate (and sulfate and silicate) on calcium. In this case, a pronounced "knee" occurs at a certain ratio of P:Ca. This strongly suggests formation of a compound, probably calcium phosphate. The compound is less volatile than calcium chloride, so the formation of calcium atoms is hindered.

The formation of more volatile compounds that may cause an enhancement is not common in flame AAS. An example is the enhancement of selected refractory elements by fluoride. The use of protective reagents, such as EDTA for calcium and 8-hydroxyquinoline for aluminum, minimizes this interference.

Occlusion into refractory compounds can occur when the matrix is refractory (such as rare-earth elements) and the small amount of element is physically trapped in the matrix. This can be minimized using a hotter flame.

The use of an ionization suppressor such as La^{3+} or Cs^+ usually minimizes the formation of ions.

The use of releasing agents can reduce or minimize interferences caused by incomplete or slow breakdown of a molecule into atoms. Examples of releasing agents are lanthanum or strontium, which release calcium from a phosphate interference.

Furnace AAS

Furnace AAS is more prone to interferences, particularly from molecular absorption, particulate (scatter), and atomic (spectral) background. The use of background correction methods is universal in furnace AAS methods, particularly for complex matrices and where resonance (most sensitive) wavelengths in the ultraviolet spectrum are to be used. This is the area where most transition elements have their resonance wavelengths. For a complex matrix, the use of platform atomization (particularly for the more volatile elements), chemical modifiers, and a background correction system are recommended for accurate analyses of trace elements. Flame AAS produces a steady-state signal that depends on how rapidly the sample is introduced. On the other hand, furnace AAS produces a transient signal, not unlike a chromatographic peak. This allows peak area as well as peak height absorbance to be used. Current instrumentation usually gives both pieces of information. In furnace AAS, it is generally recommended that peak area absorbance be used to produce a more accurate result.

Platform Atomization In furnace AAS, the atomizer is usually a graphite tube, typically 2.5 cm long and 1 cm in diameter. The solution is deposited through an entry port and the atomization process starts. It was found that the during atomization cycle (as opposed to the drying or ashing cycle) the signal starts to appear while the temperature in the furnace is still increasing and can lead to error. The use of a platform prevents this by delaying the atomization until the temperature is constant (or not changing as rapidly) in the furnace. This occurs because the sample is not heated directly from the tube furnace but by radiation from the wall. The platform is in minimal contact with the furnace. An example of a furnace with a platform is shown in Fig. 20.8. The platform is particularly effective with highly volatile elements.

Chemical Modifiers Chemical modifiers may be used to increase the volatility of the matrix components and allow their removal at a lower pyrolysis temperature; for example, ammonium acetate, NH_4OAc, is often added to a solution containing sodium chloride, NaCl (which is a difficult

Figure 20.8 Diagram of a graphite tube and platform.

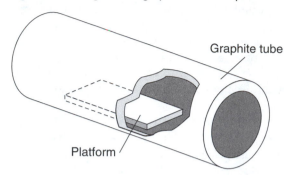

matrix and causes a severe interference in furnace AAS) and the resulting sodium acetate, NaOAc, and ammonium chloride, NH_4Cl, have boiling points below 600 °C, compared to the sodium chloride boiling point of 1400 °C. Thus, the sodium chloride can be removed at the ashing stage of 600 °C. Alternatively, the element to be determined can be made less volatile and allow a higher ashing temperature to be used; for example, cadmium is volatilized at an ashing temperature of 250 °C. The addition of ammonium dihydrogen phosphate, $NH_4H_2PO_4$, as a chemical modifier allows an ashing temperature of 600 °C to be used before loss of cadmium occurs. It is also possible to make the element more volatile. Several chemical modifiers have been proposed for various elements or matrices, with palladium considered the most versatile. Tsalov and colleagues (7) have conducted an extensive review of the literature that includes over 600 publications relevant to chemical modification.

Background Correction Major interferences in graphite furnace AAS are due to particulate matter (scatter), molecular background, and atomic background. These seldom contribute equally to an interference but are rarely discussed separately. They are generally called the background. Early work on correcting for background used the nearby line method but never gained widespread use due to the unavailability of good nearby lines (this method involved moving away from the atomic line, where the signal would be only from the background, and subtracting from the atomic line position, which would have the signal from the element plus the background). The most widely used technique is continuum source (usually deuterium arc) correction first introduced by Koirty-ohann and Pickett in 1965. The basic principle involves light from the continuum source and hollow cathode lamp (HCL) being alternated rapidly through the atom cell. The light from the HCL is absorbed by both the element being determined and the background. The light from the continuum source essentially is absorbed by only the background, not the element of interest. The difference between the hollow cathode lamp and the continuum source signals is the background-corrected signal. This is done automatically in modern instrumentation.

From the mid-1960s to the mid-1970s, continuum source background correction was used almost exclusively. It still enjoys widespread use in flame AAS and is adequate for many applications in furnace AAS. However, since the mid-1970s, an increasing awareness in the limitations of continuum source background correction, particularly in furnace AAS, has led to the development of new background correction systems. Specific concerns about continuum source background correction included the inability to correct when the background exceeds 0.5 absorbance units, and incorrect results in the presence of structured background absorption.

The Zeeman effect background correction technique makes use of the fact that a spectral line splits into its magnetic components in the presence of a magnet. Initial investigations involved placing the magnet around the hollow cathode lamp (direct Zeeman configuration) or atomizer (inverse Zeeman configuration). Inverse longitudinal AC Zeeman is available from Perkin–Elmer, inverse transverse AC Zeeman is available from Perkin–Elmer and Varian, and inverse transverse DC Zeeman is used by Hitachi. A comparison and full discussion of these methods is available elsewhere (8).

The Zeeman effect background correction works on the normal Zeeman splitting effect in which a single transition or resonance line is split into three components. When the magnetic field is applied perpendicular to the radiation (transverse Zeeman), the line is split into three components: a central unshifted π component and two σ components shifted the same distance in positive and negative directions from the π component. The distance is controlled by the magnetic field strength applied; for example, a magnetic field applied strength of 10 kG applied to the magnesium resonance line at 285.2 nm results in the σ components at 3.8×10^{-3} nm from the π component. Furthermore, these components are polarized, with the π component being polarized in the parallel direction and the two σ components in the perpendicular direction. In the parallel mode (longitudinal), the central (π) component is missing. Essentially, longitudinal Zeeman correction works as follows: When the polarized emission beam is parallel to the magnetic field it is absorbed by the atomic vapor of the sample. When the emission light is perpendicular to the magnetic field, there is no absorbance by the atomic vapor, and light scattering and molecular broadband absorption (the background) are generally equally independent of both parallel and perpendicular emission lines. Therefore, if parallel emission lines are used as the sample beam and the perpendicular emission lines are used as the reference beam, electronic subtraction of the two absorbances produces the background-corrected signal.

The advantages of the Zeeman system are that it is unaffected by structured background and spectral interferences can be overcome if lines are 0.02 nm apart. The potential disadvantages are that not all atomic lines produce the normal Zeeman effect, direct overlap of lines cannot be resolved, sensitivity is reduced, and analytical growth curves may be double-valued (that is, two widely different concentrations may give the same absorbance).

The Smith–Hieftje background correction technique involves pulsing the hollow cathode lamp at low current for several milliseconds and at high current for several microseconds (9). The time is precisely controlled so that the same number of photons reach the detector in the same cycle. At high currents the light emitted from the lamp broadens and dips in the middle (ideally until it touches the baseline). This self-reversal is usually to be avoided in AAS. Operating the lamp at low or normal current produces the AAS signal plus the background. Operating the lamp at high currents reduces or eliminates the signal from the sample and produces self-reversal, but the background has not been changed. The signals can be electronically subtracted to produce the Smith–Hieftje background-corrected signal. This background correction method produces an accurate signal for structured background but gives reduced sensitivity and not every hollow cathode lamp produces a usable self-reversal effect.

Recent Developments

AAS is a widely established technique and most recent developments have been in refinements of the technique to meet the needs of its users.

AAS is generally regarded as a single-element technique. However, there is an increasing need to perform analyses of many elements (consider the rise of inductively coupled plasma–atomic emission spectrometry) and AAS can now be performed on a limited multielement atomic absorption spectrometric basis. Several commercial companies, including Thermo Jarrell Ash and Hitachi, have systems capable of detecting from four to eight elements simultaneously. This usually involves a more complex optical system, although Thermo Jarrell Ash have a system based on a scanning galvanometer. This area is reviewed by Farah et al. (10).

Flow-injection analysis (FIA) combined with AAS as a means of sample introduction is attractive because of its potential for a high throughput of samples. It has been claimed that several hundred samples per hour can be analyzed, although a realistic number is 20 to 25 samples per hour. Many companies provide an FIA system as an accessory to AAS. This area is reviewed by Tyson (11).

Combining AAS with chromatographic techniques has been proposed to study various species of element, such as Cr^{3+} versus Cr^{6+}. The chromatographic method separates the different species of element that are eluted from the column at different times, to be detected by the AAS system. The system has not been commercialized, although a system based on gas chromatography–microwave induced plasma is commercially available from Hewlett-Packard. Its main aim is for speciation of volatile organometallic compounds.

One potential area that will attract future interest is direct solid sampling for AAS. Few commercial systems are available (other than the direct introduction of mg masses to the furnace).

Although furnace AAS is constantly being refined, with better-designed furnaces, surfaces, detection systems, and control and data-reduction software, there appear to be no major new developments on the immediate horizon for flame and furnace AAS. This can be attributed to the fact that the technique, as it currently exists, is highly suitable for many analytical problems.

Applications

There is virtually no limit to the applications of AAS. A general guideline is to work or detect at levels no lower than five times above the detection limit for sufficient precision and accuracy. The advantages and disadvantages of using the flame and furnace are given in Tables 20.1 and 20.2. A list of selected detection limits for flame AAS and furnace AAS is shown in Table 20.3. Detection limits from different systems are similar to those shown in Table 20.3. Generally speaking, flame AAS detection limits are 0.1 to 1.0 µg/mL and furnace AAS detection limits are 10 to 100 times lower. It is much easier to work with a flame AAS system. Approximately 67 elements can be detected by AAS.

The list of samples types is unlimited, as described in the introduction and in a recent publication edited by Haswell (12).

Table 20.1 Advantages and disadvantages of flame atomic absorption spectrometry.

Advantages	Disadvantages
Technique is robust.	Toxic and flammable gases used generally prevent unattended operation.
Equipment is easy to use.	
Analysis takes only a few seconds per sample.	Detection limits are in the 0.1 to 1.0 µg/mL range.
Modest cost of system is coupled with modest cost per sample.	Hotter flame required for refractory elements leads to reduced detection limits.

Table 20.2 Advantages and disadvantages of furnace atomic absorption spectrometry.

Advantages	Disadvantages
Microgram masses and microliter volumes may be used.	Analyses take (typically) 2 min.
Detection limits are typically 10 to 100 times lower than those of flame AAS.	Vulnerability to physical/chemical interferences often results in a more complex method involving chemical modification, platform atomization, and background correction.
Technique does not use toxic or flammable gases.	More complex instrumentation requires operating skill and more maintenance.
Can operate unattended.	Precision should be comparable to flame but this is not often the case; it is typically 1 to 5% for homogeneous solutions.
In-situ sample pretreatment can be performed.	

Nuts and Bolts

Vendors for Instruments and Accessories

Buck Scientific, Inc.
58 Fort Point St.
East Norwalk, CT 06855-1097
phone: 800-562-5566
fax: 203-853-0569
email: 102456.1243@compuserve.com
Internet: http://ourworld.compuserve.com/homepages/Buck_Scientific

GBC Scientific Equipment
3930 Venture Dr.
Arlington Heights, IL 60004
phone: 800-445-1902
fax: 708-506-1901

Hitachi Instruments, Inc.
3100 N. First St.
San Jose, CA 95134
phone: 800-455-4440
fax: 408-432-0704
email: info@hii.hitachi.com
Internet: http://www.hii.hitachi.com

Leeman Labs, Inc.
55 Technology Dr.
Lowell, MA 01851
phone: 508-454-4442
fax: 508-452-7429

Table 20.3 Detection limits for selected elements by flame and furnace atomic absorption spectrometry.

Element	Wavelength (nm)	Detection Limit		Furnace (Absolute) (pg)
		Flame (µg/mL)	Furnace* (ng/mL)	
Ag	328.1	0.05	0.01	0.2
Al	309.3	0.50	0.10	2
As	193.7	1.00	0.50	10
Au	242.8	0.20	0.50	10
B	249.8	5.00	10.0	200
Ba	553.6	0.05	0.25	5
Be	234.9	0.02	0.04	0.8
Bi	223.1	0.50	0.10	2
Ca	422.7	0.01	0.20	4
Cd	228.8	0.01	0.005	0.08
Co	240.7	0.03	0.30	6
Cr	357.9	0.03	0.01	0.2
Cs	852.1	0.06	0.05	1
Cu	324.7	0.01	0.04	0.8
Eu	459.4	0.10	0.10	2
Fe	248.3	0.02	0.20	4
Ga	287.4	0.10	1.0	20
Ge	265.2	1.00	3.0	60
Hg	253.6	2.00	5.0	100
K	766.5	0.02	0.02	0.4
Li	670.8	0.01	0.20	4.0
Mg	285.2	0.002	0.001	0.02
Mn	279.5	0.02	0.01	0.2
Mo	313.3	0.06	1.0	20
Na	589.0	0.002	0.005	0.1
Ni	232.0	0.05	0.5	10
Pb	217.0	0.10	0.1	2
Pd	244.8	0.20	1.0	20
Pt	266.0	0.50	2.0	40
Rb	780.0	0.10	0.6	12
Sb	217.6	0.20	0.5	10
Se	196.0	1.00	0.5	10
Si	251.6	0.10	1.0	20
Sn	235.5	0.10	3.0	30
Sr	460.7	0.05	0.4	8
Ti	276.8	0.10	1.5	30
V	318.4	0.15	1.5	30
Zn	213.9	0.01	0.002	0.04

*Based on a volume of 20 µL.
Data taken from J. D. Ingle, Jr., and S. R. Crouch, Spectrochemical Analysis, *Table 10.3, page 300 (1988). Reprinted by permission of Prentice-Hall, Inc., Upper Saddle River, N.J.*

Mattson, Unicam, and Cahn
1001 Fourier Dr.
Madison, WI 53717
phone: 800-423-6641
fax: 608-831-2093
email: info@mattsonir.com
Internet: http://www.netopia.com/ati_inst

Perkin-Elmer Corporation
761 Main Ave.
Norwalk, CT 06859
phone: 800-762-4000
fax: 203-762-4228
email: info@perkin-elmer.com
Internet: http://www.perkin-elmer.com

Shimadzu
7102 Riverwood Dr.
Columbia, MD 21046
phone: 800-477-1227
fax: 410-381-1222
Internet: http://www.shimadzu.com

Thermo Jarrell Ash Corp.
27 Forge Pkwy.
Franklin, MA 02038
phone: 508-520-1880
fax: 508-520-1732

Varian Instruments
P.O. Box 9000
San Fernando, CA 91340
phone: 800-926-3000

There are several other manufacturers of AAS instrumentation and a comprehensive list is available in *American (International) Laboratory Buyers Guide* (available every year from International-al Scientific Communications, Inc., Shelton, Connecticut). Costs range from a simple flame AAS system containing no background correction and simple electronics for data processing ($10K) to a top-of-the-line system with the ability for flame/furnace multielement AAS, two different background corrections systems, sophisticated data-handing systems, sample preparation stations, automatic sample introduction, and computer control (almost $100K). All major manufacturers of AAS instrumentation have extensive worldwide operations with ready availability and service in most countries in the world. All manufacturers of AAS instrumentation provide accessories. There are numerous companies that provide accessories but not instruments or service. They usually provide accessories at a lower cost than the manufacturer does, but generally the quality does not reach the standard and specifications of the accessories provided by the manufacturer.

Consumables

In most cases a specific hollow cathode lamp, costing $150 to $400, is required for each element to be determined. It is possible to have multielement lamps, such as Ca, K, and Na, which cost around $300 to $400, compared to the cost of $600 to $700 for three individual lamps. However, the performance (sensitivity) of the element in a multielement lamp is poorer by a factor of 2 to 3 compared to the single lamp. Therefore, a laboratory that is required to determine many elements requires a large inventory of lamps. This takes up significant space in a laboratory. A further factor is that the lamps do not have an infinite lifetime or shelf-life. Typically, lamps are guaranteed for 2000 mA hours. If they are used at 10 mA, then they are guaranteed for 200 hours of use. Of course, some lamps appear to last forever, whereas others appear to be replaced on a regular basis. Lamps of volatile elements such as cadmium, selenium, and zinc must be replaced frequently. Also, lamps often do not work if they have not been used (shelf lifetime is limited). Electrodeless discharge lamps are recommended for elements whose resonance wavelengths are in the low ultraviolet region, such as arsenic, selenium, and cadmium. They cost slightly more than a hollow cathode lamp but require an external source of power, typically costing over $3000.

For flame AAS, the cost of gases, particularly acetylene or nitrous oxide, must be considered. A tank of acetylene costs around $30 and lasts through about 2 to 3 months of frequent use. Nitrous oxide is a little more expensive and must be replaced more often. Occasional replacement of nebulizer parts is recommended, depending on whether high-solid solutions are being nebulized.

For furnace AAS, a graphite tube lasts around 200 to 500 firings depending on the atomization temperature and complexity of the sample. Graphite tubes cost around $20 to $50 each, depending on the quantity purchased and the vendor. Platforms have lifetime similar to that of graphite tubes and cost $10 to $15 depending on the quantity purchased and the vendor. Other replacement parts for a furnace, such as cones and sensors, should be changed every 3 to 12 months depending on use and complexity of sample, typically costing $1000 to $2000 per year. A tank of argon or nitrogen, which is used to prevent degradation of the furnace by oxygen from the air, lasts 6 to 12 months depending on the use of the instrument.

Required Level of Training

Most commercial vendors of AAS instrumentation offer training courses on the use of the instrumentation from 1 to 3 days, depending on the complexity of the system. Often, there is no charge, as it is included in the cost of the instrument. A disadvantage is that the training is only in the company's major locale or in a major city. It is probably held only every few months. The cost of hotel and meals is not included in the course. However, during the installation, the service engineer usually spends from several hours to a day in providing some training. All major companies have very good communications through telephone, faxes, and e-mail to respond to questions from the buyers. All instrumentation comes with a troubleshooting and help book that can usually locate any problems. Useful courses are offered on AAS at national meetings, typically 4-hour to full-day sessions costing $150 to $500. These provide useful background but are generally not specific to one company's instrumentation.

Most companies supply a "cookbook" of method development. This allows the user to obtain the approximate experimental conditions for a particular element, although it is advisable to optimize to your own particular needs.

In the opinion of the author, satisfactory training in flame AAS takes from a few hours to 1 week. Training in graphite furnace AAS takes from about 1 week to 1 month. This includes familiarity with background correction, changing tubes and furnace parts, chemical modification, autosampler, and data station operation.

Service and Maintenance

All vendors of AAS instrumentation provide service contracts beyond the usual 1-year comprehensive service from purchase and installation of instrumentation. For example, a full-service contract from Thermo Jarrell Ash Corporation (Franklin, Massachusetts), called the Service Assurance Program, is a 1-year agreement providing service to a particular TJA atomic absorption instrument by a TJA service engineer. It is available throughout the United States (except Alaska and Hawaii) and provides the following advantages:

- Predictable and budgetable service costs.
- Positive assurance against large service costs.
- Prompt repairs and minimal downtime.

The 1-year agreement includes all necessary service calls during the period of the agreement. The price covers service labor, travel costs, and materials, unless the instrument has been subjected to misuse, an unsuitable environment, or "acts of god." It does not include items that are normally considered expendable. Customer training is not covered under the agreement.

The cost varies according to the instrument and vendor, whether it is a 1-year or several-years service contract, whether it involves several instruments, or whether a reduced service contract (providing reduced benefits) is required. A full service contract for a top-of-the-line flame/furnace multielement AAS including data station typically costs $6000.

References

1. T. Nakahara, "Hydride Generation Atomic Spectrometry," in J. Sneddon, ed., *Advances in Atomic Spectroscopy*, vol. II (Greenwich, CT: JAI Press, 1994).

2. L. H. J. Lajunen, *Spectrochemical Analysis by Atomic Absorption and Emission* (Cambridge, U.K.: The Royal Society of Chemistry, 1992).

3. J. Sneddon, ed., *Advances in Atomic Spectroscopy*, vols. I and II (Greenwich, CT: JAI Press, 1992, 1994).

4. J. D. Ingle, Jr., and S. R. Crouch, *Spectrochemical Analysis* (Englewood Cliffs, N.J.: Prentice Hall, 1988).

5. J. Sneddon, ed., *Sample Introduction in Atomic Spectroscopy* (Amsterdam: Elsevier, 1990).

6. M. B. Denton, J. M. Freelin, and T. R. Smith, "Ultrasonic Babinton, and Thermospray Nebulization," in J. Sneddon, ed., *Sample Introduction in Atomic Spectroscopy* (Amsterdam: Elsevier, 1990).

7. D. L. Tsalev, V. I. Slaveykova, and P. B. Mandjukov, *Spectrochimica Acta Reviews*, 13 (1990), 225–74.

8. J. Sneddon, *Spectroscopy*, 2, no. 5 (1986), 38–45.

9. S. B. Smith, Jr., and G. M. Hieftje, *Applied Spectroscopy*, 37, no. 5 (1983), 419–25.

10. J. Sneddon, B. D. Farah, and K. S. Farah, *Microchemical Journal*, 49, no. 3 (1993), 318–25.

11. J. F. Tyson, "Flow Injection Techniques for Atomic Spectrometry," in J. Sneddon, ed., *Advances in Atomic Spectroscopy*, vol. I (Greenwich, CT: JAI Press, 1992).

12. S. J. Haswell, ed., *Atomic Absorption Spectrometry* (Amsterdam: Elsevier, 1992).

Inductively Coupled Plasma Atomic Emission Spectroscopy

Arthur W. Varnes

ICP and ICP-MS Services

Summary

General Uses

- Simultaneous determination of over 70 elements in virtually any sample in less than 2 min
- Concentrations from parts per billion to weight% without preconcentration or dilution
- Stoichiometry of chromatographic effluents
- Process control
- Destructive method, but only a few milligrams or milliliters required

Common Applications

- Determination of Environmental Protection Agency (EPA) priority pollutant metals in water, soils, solid wastes, or air samples
- Determination of wear metals in used lubricants
- Quality control to ensure correct elemental composition of raw materials, intermediates, and finished products
- Determination of catalyst poisoning elements
- Elemental composition of unknown materials

Samples

State

Samples are most commonly presented as liquids; however, most solids may be quantitatively dissolved to permit analysis as solutions. Airborne materials may be trapped on appropriate filters and dissolved before analysis. Some research has demonstrated the feasibility of laser ablation or slurry nebulization for solid samples, but virtually all commercial analyses are performed on solutions.

Amount

Solids Typically 50 mg of an inorganic material is dissolved in 50 mL of solution. Smaller samples may be used but sensitivity is degraded and there are concerns about homogeneity. Organic materials may be dry ashed in crucibles or low-temperature plasma ashers, or wet ashed in a high-pressure asher, a microwave oven, or an open beaker. The sample size should be chosen to yield about 10 to 50 mg of inorganic constituents, but may not exceed the safety limits of the technique.

Liquids Typically, 2 mL are required but specialized techniques allow analysis of a few microliters of sample. If sensitivity is not an issue, samples may be diluted to obtain the required volume.

Gases Although in principle one could introduce a small (about 100 mL/min) flow of a gaseous sample into the plasma for analysis, most such samples are analyzed by passing a measured or known volume of sample through a scrubbing solution and analyzing that solution.

Preparation

Although the inductively coupled plasma (ICP) source is very robust, it is still desirable for maximum accuracy to minimize the differences in composition between samples and standards. As noted above, solids are almost always dissolved before analysis. A substantial body of literature is available to guide the analyst. Aqueous samples are acidified to a pH less than 2 with redistilled nitric acid to minimize analyte precipitation. Samples should be prepared and stored in Teflon, polyethylene, polypropylene, or similar containers. Glass vessels should not be used for aqueous solutions because they are very likely to release contaminants or adsorb analytes. Viscous organic samples are diluted with an appropriate solvent (such as xylene, methyl isobutyl ketone, or chloroform) before analysis. Organic vapors diffuse through plastic containers, so it is preferable to use glass for long-term storage of organic samples.

Analysis Time

Instrument warmup and calibration typically require 30 to 60 min; EPA protocols require slightly longer. Aspirating a sample solution, allowing the system to come to steady-state, acquiring data, and washing out the system generally require about 2 min for a simultaneous instrument. Sequential instruments require somewhat longer, depending on the number of elements to be determined, the need for background correction, and the wavelength interval required. Autosampling devices allow unattended operation and most modern instruments include software to perform QC check routines and even shut down the instrument at the end of the run. Sample preparation depends on the type of sample and the equipment available. Often, samples are dissolved overnight and analyzed the following day; samples are diluted to volume while the instrument is warming up.

Limitations

General

- No information on oxidation states of elements or structures of compounds is obtained.
- Element to be determined must have one or more atomic or ionic emission lines in the 160- to 900-nm region; the gaseous elements, halogens, and carbon are generally not determined.
- Volatile organic solvents may extinguish the plasma unless special sample introduction methods are used.
- Major components may emit lines that overlap those of trace components, causing severe degradation of detection limits.
- Differences in composition between samples and standards may alter intensity of inherent plasma background emission or analyte response; care must be taken to avoid errors.

Accuracy

In analysis of simple solutions with known composition, accuracy better than 1% relative for analytes with concentrations greater than 50 times the detection limit may be achieved. Except for unknown samples in which matrix matching is not possible, accuracy better than 2% relative may be routinely achieved for analytes present at 50 or more times detection limit.

Sensitivity and Detection Limits

Most elements may be determined routinely at 10 ppb or less in solution using radial viewing. In favorable cases, sub–parts-per-billion detection limits may be achieved for some elements and low–parts-per-billion for most elements. Recent hardware developments enabling the analyst to view the plasma axially, rather than radially, improve sensitivity by about an order of magnitude; axial viewing, however, is more prone to matrix effects.

Linear Range

For a single emission line, a linear range of four orders of magnitude may be easily attained and as many as six orders may be observed in favorable cases. One may often extend the linear range by using two emission lines for an element, one with high sensitivity for trace concentrations and one with low sensitivity for major concentrations.

Complementary or Related Techniques

- X-ray fluorescence may allow solids to be analyzed with less sample preparation, may afford better accuracy and precision, and, with proper equipment, permits comparison of small regions of samples (that is, inclusion versus bulk or corroded versus normal region).

- Several techniques (such as electron spectroscopy for chemical analysis, Auger, and secondary ion mass spectrometry) allow determination of surface composition, whereas ICP measures bulk composition.

- Electron microscopy with energy dispersive X-ray fluorescence allows determination of elemental composition of extremely small regions of a solid.

- Neutron activation analysis allows determination of element composition with almost no sample preparation or matrix effects; technique requires access to a nuclear reactor and may require several weeks to obtain accurate data on some analytes.

- Ion chromatography and ion-selective electrode allow determination of halides and other anions in solution.

- Combustion techniques are available for determining C, H, O, and N (S may be determined by ICP with a purged or vacuum spectrometer or by combustion).

- Spark and arc sources may be effectively used for atomic emission spectrometry of solids.

- Direct current plasma (DCP) is somewhat less expensive and more tolerant of organic solvents and dirty samples. It may be more prone to chemical interferences because the source is somewhat cooler than the ICP source.

- Flame atomic absorption spectrophotometry (AAS) is more rapid for single-element determinations and has virtually no spectral interference. Graphite furnace AAS has one to three orders of magnitude better sensitivity for most analytes.

- Flame atomic emission spectrometry (flame photometry) provides better sensitivity for group I elements.

- ICP mass spectrometry has two or three orders of magnitude better sensitivity and also yields information on isotopic abundance, but greater care must be taken to recognize and correct for matrix effects arising from differences in composition between samples and standards.

Introduction

Inductively coupled plasma (ICP) emission spectrometry was developed independently by Fassel (1) in the United States and Greenfield (2) in the United Kingdom in the 1970s. At that time the principal analytical atomic spectroscopic techniques were atomic absorption and arc/spark emission. (Flame atomic emission spectroscopy, or flame photometry, was also commonly used for alkaline elements, but generally not for transition or refractory elements.) Flame atomic absorption (AA) affords detection limits on the order of 100 ppb for many elements and is rapid and inexpensive. Graphite furnace AA extends detection limits to sub–parts-per-billion levels for many elements, but is much more time-consuming. Furthermore, at that time furnace AA interferences were not as well understood, so considerable effort was required to overcome matrix effects.

The major limitation of AA, however, was the need to have a lamp that emitted the line spectrum of the analyte. Not only was it expensive to purchase the large number of lamps required, it was also tedious and time-consuming to replace the lamp and prepare a new set of calibration solutions for each analyte of interest. In addition, calibration curves were useful over two orders of magnitude at best.

The author recalls his initial assignment with a former employer. The task was to determine 14 elements in a set of 30 solutions. To obtain the required data, it was necessary to sort the samples into low concentrations requiring furnace AA, normal concentrations for flame AA, and high concentrations requiring dilution before flame AA analysis. It took 10 days to complete the assignment. Two years later, we had commissioned an ICP emission spectrometer. A similar set of samples was completed in one morning!

Other techniques—arc and spark atomic emission spectrometry—were capable of simultaneous multielement analysis. However, those techniques were not well-suited to measuring analytes in solution. Furthermore, the arc source tends to operate at a much lower temperature than the ICP source. Consequently, arc emission spectrometry is much more prone to matrix effects and calibration curves are usually nonlinear. Spark emission spectrometry uses a hotter source, but does not afford good precision. It is necessary to match the sample and standard matrices extremely closely to obtain accurate results with either of these techniques. Even the best spectroscopist and equipment, however, are not capable of routinely attaining precision or accuracy better than 5% relative with DC arc excitation.

ICP offered the best features of those techniques. Detection limits were generally intermediate between those of flame and furnace AA. ICP also afforded simultaneous multielement capability similar to arc and spark techniques. In fact, the first ICP the author purchased (serial no. 20) actually included a DC arc stand. The plasma source was placed in a compartment that was attached to the side of the original spectrometer cabinet and the arc source was removed when ICP determinations were to be performed.

Although the original direct-reading ICP polychromators were accurate and precise and achieved a substantial increase in sample throughput compared to AA or arc/spark emission, they were limited in many respects. The first instruments were not equipped with background correction systems, so it was necessary to match the sample and standard matrices very closely. Also, line selection was heavily influenced by sample matrix, especially major concomitant elements. If an unexpected matrix was encountered, it was often necessary to add another line for one or more analytes to avoid spectral overlaps. Finally, the basic instruments included only a limited number of analytical lines (typically about 10); adding an additional line increased the price by 2 or 3%, so it was necessary to restrict the acquired instrument to 25 detectors to stay within budget.

Sequential scanning instruments were designed to overcome these difficulties (3). Although they offered the promise of determining the entire set of elements with useful emission lines, there were some serious limitations. The most obvious was the need to make a separate measurement for each element. This situation was especially tedious in cases requiring background correction. Furthermore, method development was tedious because it was necessary to check for every possible concomitant element and then select appropriate emission lines to minimize spectral overlap. Reproducibility of wavelengths was less than excellent, so precision and accuracy suffered. On the other hand, sequential instruments were much less expensive than direct-reading polychromators, allowing many firms with budget constraints to enter the market.

In the early 1980s combined simultaneous–sequential instruments were introduced (4, 5). Any elements not available on the simultaneous instrument, or for which the simultaneous line suffered severe interference, could be determined with the sequential spectrometer. A drawback of the early combined simultaneous–sequential systems was that the software did not allow multitasking. Thus, it was necessary to analyze samples first with the simultaneous spectrometer and then to repeat them on the sequential spectrometer. Fortunately this limitation has been corrected in modern simultaneous–sequential combination instruments.

In the mid-1980s another advance by Hausler and Taylor allowed the instrument to serve as a detector for liquid chromatography (6, 7). The author of this chapter built on work of Hausler and Taylor by using time-resolved data acquisition software, originally designed to improve sensitivity and accuracy by exploiting selective volatility of the elements. The software enabled an analyst to specify the number and duration of integrated photomultiplier signals for up to nine analytes of interest. Typically, one would collect 100 signals of 1-sec duration for each analyte element. For highly volatile elements, the signal would increase from baseline shortly after the arc formed and would return to baseline fairly quickly. For refractory elements, the signal might take as long as a minute to appear. Based on data from these studies, the acquisition interval for each analyte could be optimized. Using the same time-resolved data acquisition software, the analyst could pass the sample through a chromatographic column before analysis and record the data from successive time slices. The software could then generate a plot of emission intensity versus time for each analyte of interest, thereby allowing one to determine the separate chemical forms in which the analyte was found in the sample.

Flow injection techniques were also applied in the mid-1980s (8). This capability allowed the analyst to increase sample throughput, work with smaller samples, and improve stability when working with difficult matrices. The promise of increased throughput was somewhat illusory, however, in that replicate analyses required washout and re-equilibration of the plasma before analyses. If a triplicate determination were desired, flow injection was no more rapid than conventional analysis. On the other hand, flow injection was very effective for small samples if the concentration of analytes was sufficient. Because only a few microliters of samples rather than 1 to 2 mL were needed, flow injection was a powerful tool for working with brines or similar difficult sample matrices.

For the first 15 years of commercial instrumentation, analysts still lamented the fact that, although thousands of analytically useful lines were emitted by analytes in the ICP source, even fully fitted polychromators measured only about 1% of them. There would be several advantages to be gained by accessing all useful lines. First, one could minimize potential errors arising from spectral overlaps. Because most elements emit several lines of approximately equal sensitivity, the ability to measure many lines per element would greatly increase the possibility of making a measurement at a line with no significant spectral overlap. Second, by comparing relative intensities of "hard" lines (arising from ionized atoms or elements with high excitation energies) and soft lines (arising from atoms with low excitation energies) (7) for several elements, one could detect mismatch between samples and standards. In addition, one could use internal standard elements with greater accuracy and effectiveness. Third, rapid qualitative and semiquantitative determinations of over 70 elements could be performed easily in 60 sec.

In the mid-1980s a commercial instrument was introduced that incorporated a linear photodiode array detector (9, 10). This instrument did allow access to more lines than polychromators based on photomultiplier tubes. To avoid interference and blooming (spreading of an intense signal on one diode to several adjacent diodes), a mask was used to remove unwanted radiation. A separate mask was required for each matrix, but an automatic mask-changing feature changed masks very quickly with little operator effort. Another concern was that sensitivity was not as high as for photomultiplier tubes. Finally, the arrays had relatively poor resolution and limited wavelength coverage. Despite these limitations, the instrument was commercially successful and did afford greater capabilities than conventional spectrometers.

In the early 1990s, however, new spectrometers incorporating solid-state detectors such as charge-injection devices (CIDs) (11) and segmented charge-coupled devices (CCDs) (12, 13) were introduced. These devices included Echelle gratings to disperse the spectrum in two dimensions and afforded resolution and sensitivity equal to or better than those of conventional direct-reading polychromators and flexibility equivalent to or better than that of sequential systems. A further advantage is that these systems perform true simultaneous background correction (by making measurements at appropriate detector elements near the analytical wavelength to be corrected), thereby improving precision and accuracy when compared to conventional systems in which the background is measured before or after the analyte. The costs of these instruments are indeed much higher than those of sequential instruments but are comparable to or less than a fully equipped direct-reading polychromator. These instruments promise significant improvements in accuracy, linear range, and sample throughput over conventional spectrometers.

As with any instrumental technique, improved detection limits are always desirable. Thus, axial viewing was reduced to commercial practice in the mid-1990s. The torch is oriented parallel to the floor, so the plasma is viewed down the axis rather than across a radius.

The two advantages of this mode are a longer optical path length and very little background emission. Axial viewing generally improves detection limits by an order of magnitude relative to radial detection limits. On the other hand, axial viewing is more prone to matrix effects, so greater operator care is required to ensure accurate results. A major application of this technique is that it allows the analyst to achieve EPA-required detection limits for four elements—arsenic, lead, selenium, and thallium—that traditionally required graphite furnace atomic absorption spectrometry (GFAAS) because radial viewing of the ICP did not achieve the mandated limit for those analytes.

Future directions for instrument development might include simultaneous ICP atomic emission and mass spectrometry. Indeed, such an instrument has already been exhibited by one vendor and several academic publications describe use of fiberoptics to collect optical emissions from an ICP mass spectrometer (ICP-MS) and direct them to a monochromator (14, 15). There are two potential benefits: a linear range of as many as 11 orders of magnitude and interference mechanisms differ between the techniques, so the two techniques should yield different results if an interference is occurring.

The author of this chapter devised a different approach. He used an ultrasonic nebulizer to convert samples into dry aerosols, which were split with a simple glass T-fitting. One portion was conducted to an optical ICP while the remainder was directed to an ICP mass spectrometer. As pointed out by Hieftje (16), the two techniques are somewhat complementary in that the ICP-MS instrument suffers from interferences at low masses, whereas the optical ICP is affected more by high-mass interferants. For many elements one obtains useful data from both instruments; if results do not agree, the analyst is alerted to an analytical error. Furthermore, the two instruments work together to extend linear range and expand the list of analytes determined to over 80 elements.

How It Works

Physical and Chemical Principles

ICP optical emission spectrometry is based on the fact that atoms are "promoted" to higher electronic energy levels when heated to high temperatures. In fact, the plasma temperature is sufficient to ionize most atoms. For about three-quarters of the elements amenable to the technique, the most sensitive line arises from an ion rather than an atom. As the excited species leave the high-temperature region, the absorbed energy is released as ultraviolet and visible photons when the excited atoms decay to lower energy levels or the ground electronic state. Useful emission lines generally occur in the region between 160 and 900 nm. Atomic and ionic emission lines are very narrow, typically less than 5 pm (17), and their wavelengths follow well-understood selection rules.

If one were to introduce a pure, dilute solution of a diatomic salt, such as sodium chloride, into the plasma, resolve the emissions with a diffraction grating and photograph the spectrum, one would observe mostly a dark background with a few lines arising from sodium, chlorine, argon (from the source gas), and hydrogen atoms (from the solvent). (There would also be a few bands arising from polyatomic species, principally OH bands.)

For a more complex solution, each constituent element emits a unique collection of spectral lines. Thus, in principle, a qualitative analysis may be performed by introducing a sample into the plasma and determining whether emission occurs at wavelengths typical of elements of interest. If a signal greater than background is detected, the element may be present; if not, it is at best present in a concentration below the detection limit. The recent introduction of commercial instruments using Echelle gratings and solid-state detectors (charge-injection devices (CID) and charge-coupled devices (CCD), for example) has made this capability feasible. Quantitative analysis may be performed because the intensity of analyte emission is proportional to concentration.

The ICP source is extremely hot. The ICP is estimated to produce a maximum temperature of at least 6500 °K. This extremely high temperature is sufficient to break virtually all chemical bonds in a sample. Consequently, the emitting atoms and ions are virtually independent of one another. A very large fraction of most elements is excited and little self-absorption occurs. As a result, the technique exhibits high sensitivity, a linear range of four or more orders of magnitude, and much reduced chemical interference relative to AAS or arc/spark emission techniques.

Samples are most commonly introduced into the plasma as aerosols. A wide variety of devices are available for sample introduction. Pneumatic nebulizers are the least expensive and most commonly used in commercial devices. The aerosol produced by the nebulizer is generally passed through a spray chamber to remove large droplets and produce a more homogeneous aerosol. While passing though the plasma, the aerosol is vaporized, atomized, perhaps ionized, and then electronically excited. After leaving the extremely hot plasma, the sample emits photons, which are sampled through a narrow entrance slit and dispersed with a grating monochromator. The resolved radiation is measured with a photomultiplier tube or array detector (such as a CCD or CID), which converts the optical signal into an electric signal. An electronic interface converts the signal into an appropriate form for measurement and storage by a dedicated computer. Before analyzing samples, the analyst introduces a blank solution and one or more standard reference solutions. The computer uses these data to generate a set of calibration curves. Emission intensities arising from samples are compared with those arising from the blank and standards to allow calculation of the concentrations of analytes of interest.

If information is desired about the various chemical forms in which an element may be found in a sample, one may pass the sample through a chromatograph before analysis and record the emission intensity as a function of time (6, 7). Qualitative analysis is performed by measuring retention time and concentration by measuring either peak intensity or peak area.

Figure 21.1 is a block diagram of a typical ICP optical emission spectrometer system. Power for the excitation source is provided by the radiofrequency generator. Typical frequencies are 27.5 and 40 MHz; these frequencies are used in diathermy devices, so they do not interfere with commercial communications. Incident power is typically 1.1 to 1.3 kW for aqueous solutions, although some special torches operate with lower power. For organic solutions the power is increased to 1.6 to 2.0 kW because more energy is absorbed by most organic solvents than by water.

Figure 21.2 shows a typical plasma torch, although other designs may be used. Three argon streams are supplied to the torch. The largest flow, normally called the plasma or cooling gas, is directed through the upper arm of the torch at about 15 L/min. A second flow, called the auxiliary gas, passes through the lower arm. It is possible to operate in the aqueous mode with no auxiliary flow, but a small flow (about 1 L/min) is desirable to extend torch life and reduce the silicon background. The third argon stream, called the sample gas, passes through the nebulizer at about 0.8 L/min and carries the sample aerosol into the plasma.

Figure 21.1 Block diagram of a typical ICP optical emission spectrometer system. *(Reprinted with permission from the* Annual Book of ASTM Standards, *copyright American Society for Testing and Materials, 1916 Race Street, Philadelphia, PA 19103.)*

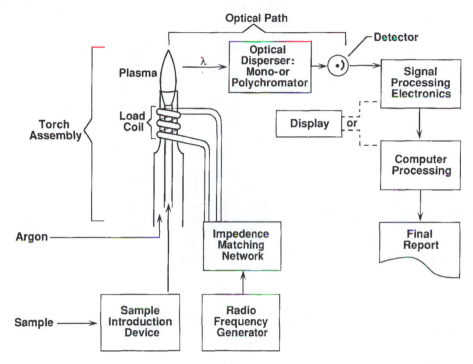

In operation, the plasma torch is secured within the electromagnetic field arising from a conducting helical coil (made from copper or a similar conducting metal), which transfers power from the radiofrequency generator. Because the coil acts as an inductor, the source is called an inductively coupled plasma.

Sample is introduced into the plasma with the aid of a nebulizer. Several types of nebulizers are shown in Figs. 21.3, 21.4, and 21.5. The concentric glass (Meinhard) nebulizer (Fig. 21.3) is very commonly used. It consists of a fine glass capillary concentric with a larger-diameter glass tube drawn to an internal diameter slightly larger than the outer diameter of the inner capillary. The sample is drawn through the central capillary and converted to an aerosol by an argon stream passing through the glass tubing surrounding the sample-introduction capillary. Although the Venturi effect allows the device to draw sample unaided, typically a peristaltic pump is used to improve precision.

Another commonly used nebulizer is of the cross-flow design (Fig. 21.4). Two capillaries are oriented at a 90° angle, one in the horizontal plane and the other in the vertical plane. Typically sample is pumped through the vertical capillary and converted into an aerosol by an argon stream passing through the horizontal capillary.

The Hildebrand grid nebulizer (Fig. 21.5) uses a peristaltic pump to transport liquid sample into a region between two fine-mesh platinum screens. Argon gas at high pressure flows through the screens, converting the liquid to an aerosol.

The three devices described above work very well for solutions with low-viscosity dissolved and suspended solids. For solutions of higher viscosity or dirty solutions, a Babington nebulizer (Fig. 21.6) is usually superior. Typically, the sample is introduced into a V-groove oriented in the

Figure 21.2 Typical plasma torch. *(Reprinted with permission from the* Annual Book of ASTM Standards, *copyright American Society for Testing and Materials, 1916 Race Street, Philadelphia, PA 19103.)*

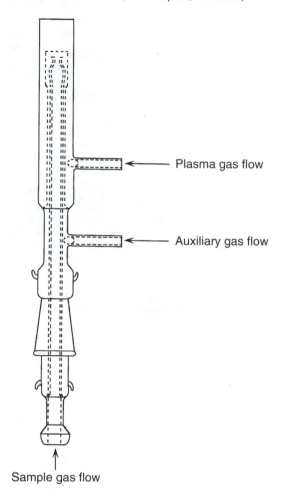

← Plasma gas flow

← Auxiliary gas flow

Sample gas flow

Figure 21.3 Meinhard concentric glass nebulizer. *(Reprinted with permission from the* Annual Book of ASTM Standards, *copyright American Society for Testing and Materials, 1916 Race Street, Philadelphia, PA 19103.)*

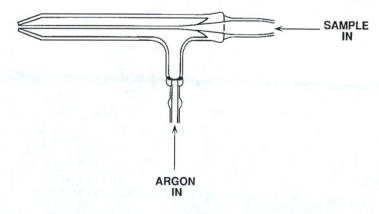

SAMPLE
IN

ARGON
IN

Figure 21.4 Cross-flow nebulizer. *(Reprinted with permission from the* Annual Book of ASTM Standards, *copyright American Society for Testing and Materials, 1916 Race Street, Philadelphia, PA 19103.)*

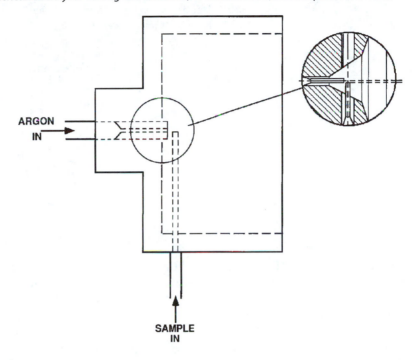

Figure 21.5 Hildebrand grid nebulizer. *(Reprinted with permission from the* Annual Book of ASTM Standards, *copyright American Society for Testing and Materials, 1916 Race Street, Philadelphia, PA 19103.)*

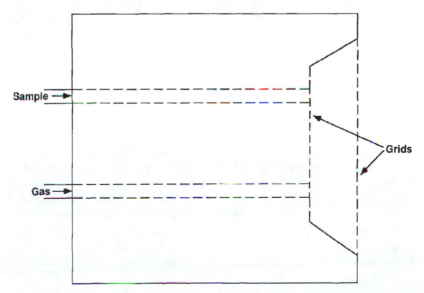

Figure 21.6 Babington nebulizer. *(Reprinted with permission from the* Annual Book of ASTM Standards, *copyright American Society for Testing and Materials, 1916 Race Street, Philadelphia, PA 19103.)*

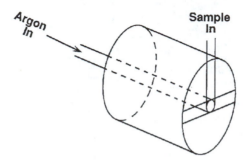

vertical direction. At the bottom rear of the groove, a sapphire or similar resistant material with a small bore allows a stream of high-pressure argon to convert the sample to an aerosol.

All of these devices produce a relatively large range of primary aerosol droplet sizes, typically from submicrometer to more than 100 μm in diameter (18). Furthermore, during their passage from the nebulizer to the plasma, droplets may collide and form even larger secondary and tertiary droplets. Such large droplets may cool or even extinguish the plasma. Consequently, the aerosol is passed through a spray chamber to remove large droplets. The resulting aerosol consists of smaller and more uniform droplets than the primary aerosol but only 1 or 2% of the original aerosol reaches the plasma.

To improve sensitivity, one may use an ultrasonic nebulizer (Fig. 21.7) to increase transport efficiency. In this device, the sample is pumped onto the face of a transducer powered by a radio-frequency generator. The device produces a fine, uniform aerosol. Unfortunately, if transported directly to the plasma, the large amount of solvent would extinguish the plasma. To overcome this experimental difficulty, the wet aerosol is passed through a hot (typically 140 °C) region to vaporize the solvent. The aerosol is then passed through a condenser (typically cooled to 0 °C), which converts it to a dry aerosol. Transport efficiency is typically 10 times higher than for conventional pneumatic nebulizers. Furthermore, because much less solvent reaches the plasma than with conventional nebulizers, detection limits and sensitivities for many elements and ionic species are improved by more than a factor of 10. One concern is that ultrasonic nebulizers may be more prone to carry over from one sample to the next, so care must be taken to avoid errors arising from this memory effect. Another limitation is that samples with high total dissolved solids content may clog the torch.

When aerosol enters the plasma, the solvent is evaporated to yield a dry aerosol. The particles produced by evaporation are vaporized, atomized, and, for most elements, at least partially ionized. The free atoms and ions are thermally excited to more energetic electronic states. As the samples leave the plasma, the excited species are cooled. The excited atoms and ions decay to less energetic states, including the ground state. In the process, ultraviolet, visible, and near infrared photons are emitted. These photons impinge on an entrance slit and pass through it to a diffraction grating, Echelle grating, or similar wavelength-dispersing element. (Some work has been performed using Fourier transform systems (19), which achieve excellent resolution. Two limitations are that detection limits are at least an order of magnitude higher than for conventional systems

Figure 21.7 Ultrasonic nebulizer. *(Reprinted with permission from CETAC Technologies.)*

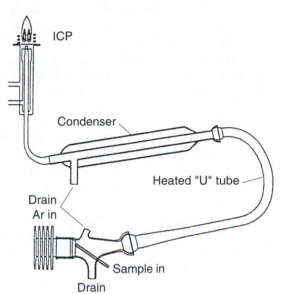

and emissions from major elements increase background noise. Thus, Fourier transform ICP is unlikely to find extensive commercial applications unless these limitations are overcome.)

Spectrometers

Historically, two principal types of spectrometers have dominated the marketplace. Sequential spectrometers are generally less expensive and more flexible, but usually require a higher degree of operator skill and experience. Direct readers are more precise and accurate for well-characterized analyses and afford greater sample throughput, but require a larger capital outlay. Recently, instruments have been introduced using solid-state array detectors, which combine the sensitivity, precision, and accuracy of a photomultiplier tube with the greater wavelength coverage of a photographic plate. Instruments equipped with such array devices are similar in price to direct-reading instruments.

Sequential Spectrometers

These instruments are the least expensive class of ICP emission spectrometers. Most such instruments include a single photomultiplier tube and a moveable grating to select wavelengths in sequential order. Some sequential devices include two or three photomultiplier tubes to decrease analysis time and to optimize response for the wavelengths of interest (solar-blind or red-sensitive tubes, in addition to a broad-range, general-purpose tube). Rather than moving the grating, at least one commercial design uses a fixed grating and translates the photomultiplier tube.

Because there are few detectors, sequential spectrometers are relatively inexpensive. In addition to low price, there is another major advantage for this type of instrument: flexibility. For analysts who cannot predict the analytes to be determined, the approximate concentration range to be covered, and the matrix or matrices to be encountered, the sequential instrument offers the flexibility needed to cope with an ever-changing set of analytical requirements.

Most elements exhibit a large number of spectral lines with sensitivities within a factor of two or three of each other. Thus, if the most sensitive line for an analyte is found to be subject to severe spectral overlaps or background, one may usually identify an alternative line with similar sensitivity and no (or at least much reduced) interferences. Furthermore, one may optimize parameters such as photomultiplier tube voltage, incident power, viewing height, and gas flows for each element.

The major disadvantages of sequential instruments are longer analysis times, greater sample volume, and possibly poorer accuracy and precision. A sequential instrument measures one element at a time (of course, instruments with more than one detector may make a measurement with each detector simultaneously). Thus, each element to be determined increases the analysis time. Furthermore, if plasma parameters are changed, additional equilibration time is required.

Another concern with sequential spectrometers is reproducibility of wavelength selection. Two approaches are used to attempt to return to the identical wavelength for each analyte element. One alternative is to assume that the analyte of interest will produce the largest peak in the vicinity of the analyte wavelength selected. Certainly if the analyte is present at a concentration greater than the instrumental detection limit and no spectral overlap occurs, this assumption will be valid and a peak search routine will indeed allow the instrument to locate the proper wavelength. On the other hand, if the analyte is absent or an interfering element is present at relatively high concentrations, the peak search routine will find the wrong peak and a concentration higher than the actual concentration will be reported. An alternative to peak searching in the window is to use a reference line, for example the carbon line at 193 nm. The instrument performs a peak search routine on the reference line and then moves a predefined number of steps, which ideally will reproduce the wavelength of interest. This method is superior in avoiding selection of the wrong peak or when the analyte is absent, but increases analysis time and may also be less accurate if multiple wavelengths must be determined.

Simultaneous Spectrometers

Conventional direct-reading polychromators use one or more photomultiplier tubes for each analyte of interest. They are faster, more precise, and more accurate for a predictable suite of analytes and interferants. The buyer and vendor carefully review the analytes and matrices to select lines with optimal sensitivity and freedom from interferences. All elements are determined simultaneously, thereby increasing analytical speed. This capability is especially important when dealing with samples of varying composition, because background correction may be accomplished much more quickly than for a sequential system.

Typically, either the entrance slit is translated or a quartz plate just behind the entrance slit is rotated to focus wavelengths near the analyte wavelengths onto the detectors. Sometimes a single translation or rotation allows accurate background compensation for the entire suite of analytes; in any event, it is still much faster than performing a separate background correction measurement for each analyte, as is required in sequential analysis.

Furthermore, because the detectors and grating are stationary, there is no need for a peak search routine. It is possible to overcome matrix-dependent interferences by installing multiple detectors for an element for which the most sensitive, interference-free lines for most analyses suffers a severe interference for a few other matrices. On the other hand, the physical size of the photomultiplier tubes often requires some compromises in line selection. One solution to the problem is to use a mirror to allow a photomultiplier tube to be oriented horizontally, avoiding steric hindrance arising from the need to measure two very close wavelengths. Another alternative is to measure one of the lines in second order, often using a solar-blind photomultiplier tube to avoid interferences. Typically, the maximum number of detectors installed in a polychromator is about 60, although the author is aware of a system in which two gratings have been installed at a 180° angle, allowing use of two Rowland circles and a total of 122 detectors.

The major drawbacks of direct readers are both related to lack of flexibility. First, there must be a detector available for each analyte of interest. Although it is feasible to install additional detectors in the field after acquisition, it is very expensive ($5,000 to $10,000 each, depending on whether other detectors must be moved to allow determination of the additional analyte). Thus, it is desirable to specify a complete list of analytes to the manufacturer before acquisition. Second, if a new matrix is encountered, one or more of the analytical lines may be subject to severe spectral overlaps or background shifts, which degrades precision and accuracy.

Solid-State Array Detector Spectrometers

Recently, solid-state detectors have been introduced to overcome the limitations of sequential and conventional direct-reading spectrometers. The first commercial design used photodiode array detectors. To minimize interference from spectral overlap, masking was used to remove interfering spectral emissions from the light emitted from the plasma when the sample of interest was being aspirated. A separate mask was required for each matrix or set of major elements, but masks could be changed rapidly and automatically with little effort.

In the 1990s two-dimensional detectors incorporating charge-injection devices (CIDs) (11) or charge-coupled devices (CCDs) (12, 13) have been incorporated into commercial devices. Using Echelle gratings and very small individual detecting elements, resolution and sensitivity are virtually identical to sequential or direct-reading polychromators using photomultiplier tubes. These devices allow simultaneous measurement of more than 250,000 lines for the CID device and 5000 lines for the segmented CCD instrument. Furthermore, the cost is virtually the same as for a direct reader equipped with about 25 detectors. It appears that just as Fourier transform IR instruments have replaced grating instruments for IR, solid-state detector devices may eliminate photomultiplier-based direct-reading ICP emission spectrophotometers in the next few years.

Detection, Calculation, and Output

As noted above, detection is typically accomplished by photomultiplier tubes or array detectors that convert optical emission into electrical current. The electrical current consists of at least three components: dark current arising from the fact that the detector is operated at a high electrical potential, so that even in the absence of any light striking the surface, some current flows; nonanalyte emission, including bands from polyatomic species, stray light, and overlap from concomitant element emission; and analyte emission.

To distinguish analyte emission from the other two sources of detector response, a matrix blank solution is aspirated into the plasma and the output current measured. Typically, the output current charges a condenser and the computer measures the voltage to determine the current produced during the sampling interval. If there are substantial differences in composition between standards and samples, background correction must be used. If the background is relatively uniform across the sampling wavelength interval, a single point may suffice for background correction. On the other hand, if the background is sloping, it may be necessary to measure one point on either side of the peak and interpolate to estimate the background. Care must be taken to avoid selecting background points at emission wavelengths of concomitant elements; to do so would result in erroneously large background corrections and low analytical results. One must also be aware that as analytical concentrations increase, lines broaden. Consequently, if background correction wavelengths too close to the analytical wavelength are selected, a portion of the analytical signal may be inadvertently subtracted at high concentrations. If it is known that an analyte is present at relatively high concentrations in all samples, background correction should not be used.

One must be aware of the possibility of errors arising from spectral overlaps, especially for samples containing high concentrations of one or more concomitant elements. Ideally, one selects analyte wavelengths that avoid spectral interferences. In many cases, however, it is not possible to do so. In those circumstances, one must prepare pure solutions of each individual potential interferant. After calibration, one aspirates those solutions and determines the apparent concentration of each analyte arising from spectral emission from the concomitant element. One may then enter spectral overlap correction factors into the software to allow automatic correction of results for spectral overlap.

After the blank is aspirated, one or more standard solutions of known analyte concentration are aspirated to determine the sensitivity for each line of interest. Care must be taken to avoid combining elements that have severe spectral overlaps or are chemically incompatible. For greater sample throughput, however, it is desirable to combine as many elements as possible into each standardization solution. Thus, the analyst must make a tradeoff between throughput and avoiding errors due to spectral overlap or incompatibility.

For many purposes, a blank and a single standard are adequate to define the calibration curve. However, some protocols require the use of four or more standards. Furthermore, if most samples are known to contain analyte concentrations near the detection limit, it is desirable to use a standard of concentration about five times the detection limit to improve accuracy at low concentrations. On the other hand, the precision of measuring the signal arising from a dilute standard is relatively poor, so at least one additional standard with concentration about 100 times the detection limit must also be used.

Before analyzing samples, one should perform an initial calibration blank, initial calibration verification, and an interference check. The initial calibration blank solution should be identical in composition to the original solution used to calibrate the instrument, but obtained from an independent source. Similarly, the initial calibration verification solution should be prepared from another source of analyte elements. The interference check standard contains all major elements and verifies that errors arising from possible spectral overlap have been avoided either by proper line selection or application of a software routine to correct for spectral overlap. It is also desirable to incorporate a quality control check standard—a material similar in composition to the sample, with known concentrations of analytes and prepared in parallel with the samples—to verify adequacy of the sample preparation method.

After calibration and verification, samples are analyzed by a process similar to that used to generate the calibration curve and results are generally printed by a dedicated printer. To assess accuracy and precision, it is desirable to prepare at least 10% of the samples in duplicate. Furthermore, to ensure that matrix effects are not affecting accuracy, it is desirable to spike at least 10% of samples. There are two types of spikes: predigestion spikes, which verify accuracy of the entire sample preparation and analysis process, and postdigestion spikes, which focus entirely on matrix effects. If both check, one can have high confidence in the analytical results. On the other hand, if the postdigestion spike passes but the predigestion fails, that observation points to an inadequacy in the sample preparation process that must be addressed. The EPA has included a helpful section on quality control for ICP emission spectrometry in SW 846 (20).

What It Does

The ICP instrument detection limit (IDL) is typically determined by calibrating the instrument, then aspirating a solution containing an analyte concentration of three to five times the anticipated detection limit. Five to ten successive determinations of the apparent concentration are performed and the standard deviation of the measurements is calculated. The IDL is obtained by multiplying that standard deviation by three. The IDL varies from element to element, depending on the efficiency with which the plasma atomizes or ionizes the element, the probability of populating the excited state from which emission occurs, and the probability of the transition from the upper excited state to the lower excited state or ground state giving rise to the emission. Because of the plasma's high temperature (above 6500 °K), virtually every molecule reaching the plasma is converted into free atoms. Furthermore, many elements are almost totally ionized. Consequently, the vast majority of ICP emission lines used for analyses are ion lines.

Most elements in simple aqueous solution exhibit a detection limit of less than 10 ppb for a conventional commercial instrument using pneumatic nebulization and a 10-sec signal integration time and radial viewing. For favorable elements, such as beryllium and manganese, sub–part-per-billion detection limits may be achieved. Group I elements exhibit relatively poor detection limits (20 to 100 ppb) for direct readers or sequential systems, which do not allow variation in viewing height because their optimal observation zone is about 25 mm above the load coil, whereas most other elements afford optimal sensitivity and freedom from matrix effects at about 16 mm. Rare-earth elements and metalloids also generally exhibit detection limits in the 10- to 50-ppb range. Some elements, such as sulfur, have useful emission lines only below 190 nm; these elements require a vacuum spectrometer or purging with nitrogen or argon to remove oxygen absorption. Axial viewing generally lowers detection limits by about an order of magnitude.

The presence of an interfering element in relatively high concentration severely degrades detection limits. For example, copper exhibits a strong interference on the phosphorus emission line at 213.6 nm. In the absence of copper, the detection limit for phosphorus is about 25 ppb. A 1% copper solution yields a signal equivalent to 1000 ppm phosphorus. Assuming that the precision with which the interference may be determined is 1%, the standard deviation of the interfering signal from copper would be equivalent to 10 ppm of phosphorus. Thus, the presence of copper would increase the detection limit for phosphorus by more than three orders of magnitude! Although this case is an extreme example, the analyst or user of ICP data must be aware of the influence of concomitant elements on detection limits. A table of many of the common interferences is found in Reference 21, but the reader is advised that the magnitude of the interference coefficient varies from instrument to instrument and also varies if experimental parameters are altered. In practice, it is essential to identify the major elements found in the sample, prepare single-element solutions of each, and aspirate them successively to check for possible additional interferences.

The plasma background emission also influences the detection limit. In general, background emission intensity decreases with wavelength, so most analytical lines are found in the ultraviolet region. Furthermore, some molecular bands are present in spectra obtained at 16 mm above the coil; lines overlapped by such bands should not be selected because the molecular emission intensity of the bands varies as difference matrices are aspirated into the plasma. Keep in mind that the detection limit is a function of signal-to-background fluctuations, so a very intense line in the visible region may be more sensitive than an ultraviolet line with lower background but less intensity. Fortunately, most manufacturers are very well acquainted with line and background intensities and spectral overlap problems, so they are very helpful in recommending appropriate lines if provided with thorough information about sample types and analytical requirements.

Analytical Information

Qualitative

Although it is possible in principle to obtain qualitative information from ICP emission spectrometry, that application has not been widely used. The ancestors of ICP emission spectrometry—arc and spark emission spectrometry—allowed the entire spectrum of a sample to be recorded on photographic film or plates, allowing a skilled spectroscopist to determine approximate concentrations of 70 elements from a single exposure. Although it is possible to use photographic plates to record ICP emission spectra, very few commercial analyses have been performed in this manner. In principle, one could perform a qualitative analysis with a sequential spectrometer, but in practice the amount of sample and time required are prohibitive. The expense and physical impossibility of installing individual photomultiplier tubes to monitor multiple lines for each potential element of interest preclude use of direct-reading spectrometers for qualitative analysis. However, at least one commercial instrument developed in the 1980s used an Echelle spectrometer and photographic film to produce a two-dimensional image that allowed qualitative analysis. With the appearance of solid-state detector-based devices it is likely that the use of ICP atomic emission spectrometry for qualitative analysis will become more common.

Quantitative

ICP emission spectrometry's greater strength lies in quantitative analysis. As noted above, calibration of the instrument is rapid and straightforward once the matrix is understood and background correction and spectral overlap correction protocols have been developed. It is tedious to prepare calibration solutions containing 70 or more elements, but high-purity multielement plasma calibration solutions are readily available from several vendors at a modest cost. Using no special calibration or sample introduction techniques, accuracy and precision are generally better than 2% relative and it is not uncommon to achieve better than 1% relative in favorable cases.

A variation on that technique is to spike the sample with several elements not found in the sample. A computer program could then compare relative emission intensities of several lines (hard and soft lines) of the spiking elements to the magnitude of other elements to obtain a reasonably precise and accurate determination of the concentrations of all other elements in the sample.

As an aid to identifying and overcoming spectral overlaps and background shifts, most modern instruments are equipped with PCs and monitors to allow acquisition and display of the spectral regions in the vicinity of analytical lines. Typically, one scans a region of about 1 nm on either side of the analytical wavelength. One should obtain and overlay scans of deionized water, a single-element analyte solution of concentration of 100 to 1000 times the detection limit, the matrix blank, and a representative sample. An examination of the baselines of the various scans allows one to determine the need for background correction. If there is a significant difference in background emission next to the analytical line, one may select background correction points. If the background is flat, a single-point correction is adequate. On the other hand, if the background slopes, one must select points on either side of the peak and interpolate to estimate the background emission at the analytical line. Care must be taken to ensure that no atomic emission from a potential concomitant element occurs at a background point, or an erroneously large correction will be made. The consequence is that the apparent concentration is less than the actual amount present.

If the matrix contains one or more major concomitant elements, it is necessary to overlay spectra obtained from single-element solutions of any concomitant elements to check for spectral overlaps. One must be aware, however, that the single-element concomitant solution may be contaminated with the analyte of interest. Several compendia of analyte lines and potential concomitant interferences are available (22–24). If one observes either a direct spectral overlap (centroid of the apparent concomitant interference is at the identical wavelength of the analyte) or an emission that cannot be found in these tables, the analyst should suspect contamination. If an alternative technique such as graphite furnace AA or ICP mass spectrometry is available, one may use such a method to confirm the presence of a contamination. Finally, in conducting analyses, one may find that many samples yield apparent concentrations less than zero after correction if a contaminant has been mistaken for a spectral overlap.

With a sequential spectrometer or an instrument using a solid-state detector, it is often possible to select an alternative line to avoid the spectral overlaps. For a direct reader or for a sequential instrument, in case an adequate, interference-free line cannot be found, one must calibrate the instrument for each such concomitant element and also determine the interference coefficient. Then, one can correct for the spectral overlap by determining the concomitant concentration in the sample, multiplying by the interference coefficient, and subtracting that amount from the apparent concentration.

Another strength of ICP emission spectrometry is its excellent sensitivity and large linear range. The high temperature of the plasma makes it an extremely efficient excitation source. Furthermore, most elements have one or more useful emission lines in the ultraviolet region, in which the plasma background is very low. Consequently, excellent signal-to-background ratios are exhibited for most elements, thereby affording low detection limits. Furthermore, the relative absence of unexcited species in the plasma reduces reabsorption of emitted photons by ground-state atoms. The source is optically thin, yielding linear response over four to six orders of magnitude. The high sensitivity often allows one to bypass the need for graphite furnace atomic absorption or sample preconcentration. The large linear range reduces the necessity for diluting concentrated solutions, as is often required in flame atomic absorption. As noted above, one can analyze virtually any solution from parts per billion to several percent by using two wavelengths with differing sensitivities to avoid the need for dilution. Finally, the high temperature and electron density of the ICP source greatly reduce "chemical" interferences encountered using other atomization and excitation sources.

Applications

As already noted, ICP emission spectrometry may be applied to determine the concentration of more than 70 elements from low–parts-per-billion to percent levels in a single analysis requiring less than 2 min after calibration. A wide array of techniques have been developed to dissolve solid samples for analysis with virtually no contamination or loss.

The EPA has approved methods for most priority pollutant metals in a variety of matrices (20). Commercial instruments using radial viewing usually cannot achieve the detection limits required for arsenic, lead, selenium, and thallium, so they have usually been performed by graphite furnace AA. However, axial viewing enables the analyst to attain the required detection limits for most samples.

ICP emission spectrometry is a major tool for geochemistry (25, 26). Its speed, excellent sensitivity, accuracy, and long linear range allow rapid, precise analysis of a wide variety of minerals for major, minor, and trace components.

ICP is also used to good advantage in biological and medical research. Many techniques have been developed for rapid dissolution or digestion of tissues or fluids. The speed, accuracy, and large number of elements that can be determined in a single analysis make ICP a very cost-effective technique (27, 28).

Preventive maintenance programs for aircraft, motor vehicles, and machinery also rely on ICP (29). Samples of lubricating fluids are analyzed at specified mileage or hours of equipment operation. A baseline concentration of each element of concern is established. Any excursion from the normal range indicates a potential problem. Preventive maintenance may then be performed.

ICP also is very useful in evaluating catalysts (30). Although major components are generally more readily determined by X-ray fluorescence (XRF), ICP is very valuable for determining trace components present near the XRF detection limit. ICP also is very useful for determining composition of used catalysts; XRF matrix effects may degrade accuracy if there is a significant mismatch between samples and standards. By dissolving the sample and matching the matrix between standards and samples, ICP minimizes matrix effects in analyzing used catalysts. The technique may be used to determine concentrations of impurities ("poisons") as well as to detect changes in concentrations of active elements.

Another application is determination of trace elements in polymers (31, 32). A relatively large sample (5 to 10 g) may be ashed conveniently in a platinum crucible. The residue may be either dissolved in dilute nitric acid or fused with lithium borate. Information about trace element content may be used to evaluate possible contamination during manufacture or processing or to determine whether trace element contamination may be a possible cause of a performance problem.

ICP is also used to advantage in the petroleum industry (33). Refinery feedstocks may be ashed and the residue dissolved for analysis. The pretreatment of units designed to remove catalyst poisoning elements such as iron, nickel, and vanadium may be estimated by comparing the concentrations of the elements in the material being fed to the pretreater with the concentrations in the output from the pretreater. Another application is the determination of trace elements in petroleum coke. Trace elements must not be present above specified levels to permit use of the material in electrodes for metal refining.

Process control ICP automatically maintains operating conditions to make sure that products are within specifications. Samples from various locations in the process streams are analyzed on-line and a computer interprets changes in concentration to alter process temperatures, pressures, or flowrates if concentrations are found to be outside the control limits.

Nuts and Bolts

Relative Costs

An ICP optical emission spectrometer is generally intermediate in capabilities between an AA spectrophotometer and ICP mass spectrometer, so its price is intermediate between the two types of instruments. In early 1996 one could still obtain a serviceable sequential ICP emission spectrometer for about $50,000. If one wished to purchase a combined simultaneous–sequential system with autosampler, one could spend $250,000. With the appearance of the solid-state array detector instruments, however, a fully loaded system will cost about $125,000. Thus, it is likely that few combined simultaneous–sequential systems will be purchased in the future. It is even possible that direct-reading polychromators based on discrete photomultiplier tubes will no longer be produced if the solid-state array detectors live up to expectations.

Vendors for Instruments and Accessories

Instruments S.A., Inc.
3880 Park Ave.
Edison, NJ 08820
phone: 800-438-7739
fax: 908-549-5125
email: john@isainc.com

Leeman Labs, Inc.
55 Technology Dr.
Lowell, MA 08151
phone: 508-454-4442
fax: 508-452-7429

Perkin-Elmer Corporation
761 Main Ave.
Norwalk, CT 06859
phone: 800-762-4000
fax: 203-762-4228
email: info@perkin-elmer.com
Internet: http://www.perkin-elmer.com

Spectro Analytical Instruments
160 Authority Dr.
Fitchburg, MA 01420
phone: 508-342-3400
fax: 508-342-8695

Thermo Jarrell Ash/Baird Corp.
27 Forge Pkwy.
Franklin, MA 02038
phone: 508-520-1880
fax: 508-520-1732

Varian Instruments
P.O. Box 9000
San Fernando, CA 91340
phone: 800-926-3000

VG Elemental/Fisons Instruments
55 Cherry Hill Dr.
Beverly, MA 01915
phone: 800-999-5011
fax: 508-524-1100

Required Level of Training

Instrument Operation

A bachelor's degree in chemistry is sufficient to operate an ICP emission spectrometer. Manufacturers typically present a brief orientation at the time of installation and encourage the new user to operate the instrument for several weeks. The user then attends a training course, typically 5 days, at the vendor's application facility to review fundamentals of the technique and specific operating procedures for the exact instrument in question.

Interpretation of Data

An experienced spectroscopist is needed to ensure that accurate data are obtained. Pitfalls include background shifts arising from mismatch between samples and standards and spectral interferences from major elements. Especially for sequential analysis, it is important to select appropriate lines to minimize errors arising from spectral overlap. Axial viewing and use of solid-state detectors requires a skilled spectroscopist to avoid potential errors and maximize benefits.

Service and Maintenance

Routine ICP maintenance is not time-consuming or expensive. The principal consumable item is pump tubing, which may require replacement daily. The sample tubes may be obtained for less than $2 each, so they represent no major outlay.

The instrument typically consumes about 40 ft^3 of argon per hour; if one is using the instrument regularly, it is desirable to use liquid argon rather than gaseous argon from cylinders. A typical liquid argon cylinder will operate the instrument for 2 weeks if it is operated about 40 hr per week. When purchased as a liquid, the argon required to operate the plasma costs less than $5 per hour.

Typical power consumption is 1 to 2 kW, so cost for electrical power is generally far less than $1 per hour. Similarly, water for cooling the load coil requires only a few liters per minute; if water is expensive or not readily available, one can use a recirculating water bath to minimize water consumption.

If the instrument is located in a friendly environment from which acid fumes are excluded and temperature and humidity are well controlled, the optical and electrical systems should require little maintenance. Depending on the amount and severity of use, the main power tube in the RF generator may have to be replaced about every 2 yr at a cost of $1000 to $2000.

Modern ICP emission spectrometers are very rugged, so little service or maintenance is needed. If concentrated solutions are being analyzed, it is desirable to clean the torch by boiling for several minutes in dilute nitric acid to dissolve any possible residue. For direct-reading instruments using photomultiplier tube detectors, it is desirable to check the alignment of each channel about once a month. Photomultiplier tubes are very durable, but it is desirable to check for stability and sensitivity about once a year. If working with organic solvents, it is desirable to place the torch in a cool muffle furnace, increase temperature to 650 °C, and heat overnight to remove carbon residue.

Suggested Readings

Books

BOUMANS, P. W. J. M., ed., *Inductively Coupled Plasma Emission Spectroscopy*, Parts 1 and 2. New York: Wiley, 1987.

BUSCH, K. W., and M. A. BUSCH, *Multielement Detection Systems for Spectrochemical Analysis*. New York: Wiley, 1990.

MONTASER, A., and D. W. GOLIGHTLY, eds., *Inductively Coupled Plasmas in Analytical Atomic Spectrometry*, 2nd ed. New York: VCH, 1992.

VARMA, A., *CRC Handbook of Inductively Coupled Plasma Atomic Emission Spectrometry*. Boca Raton, FL: CRC Press, 1991.

Review Articles

Recent progress in inductively coupled plasma atomic emission spectrometry is surveyed biannually in the *Analytical Chemistry* Fundamental Reviews issue. The most recent such review is by D. Beauchemin and others, *Analytical Chemistry*, 66 (1994), 462R.

Periodicals

Atomic Spectroscopy

ICP Information Newsletter

Journal of Analytical Atomic Spectroscopy

Spectrochimica Acta, Part B

References

1. R. H. Wendt and V. A. Fassel, *Analytical Chemistry*, 37 (1965), 920.

2. S. Greenfield, I. L. Jones, and C. T. Berry, *Analyst*, 89 (1964), 713.

3. I. B. Brenner and others, *Spectrochim. Acta*, 36B (1981), 785.

4. C. G. Fisher, R. D. Ediger, and J. E. Delany, *Amer. Lab.*, 13 (1981), 115.

5. I. B. Brenner and others, *Amer. Lab.*, 19 (1987), 17.

6. D. W. Hausler and L. T. Taylor, *Analytical Chemistry*, 53 (1981), 1223.

7. D. W. Hausler and L. T. Taylor, *Analytical Chemistry*, 53 (1981), 1227.

8. S. Greenfield, *Spectrochim. Acta*, 38B (1983), 93.

9. G. M. Levy and others, *Spectrochim. Acta*, 42B (1987), 341.

10. S. W. McGeorge and E. D. Salin, *Analytical Chemistry*, 57 (1985), 2740.

11. M. J. Pilon and others, *Appl. Spectrosc.*, 44 (1990), 1613.

12. T. W. Barnard and others, *Analytical Chemistry*, 65 (1993), 1225.

13. T. W. Barnard and others, *Analytical Chemistry*, 65 (1993), 1231.

14. K. Lepla, M. A. Vaughan, and G. Horlick, *Spectrochim. Acta*, 46B (1991), 967.

15. J. R. Garbarino, H. E. Taylor, and W. C. Batie, *Analytical Chemistry*, 61 (1989), 793.

16. G. M. Hieftje and G. H. Vickers, *Anal. Chim. Acta*, 216 (1989), 1.

17. P. W. J. M. Boumans and J. J. A. M. Vrokiing, *Spectrochim. Acta*, 41B (1986), 1235.

18. R. F. Browner, in P. W. J. M. Boumans, ed., *Inductively Coupled Plasma Emission Spectroscopy* (New York: Wiley, 1987), p. 244.

19. A. P. Thorne, *Analytical Chemistry*, 63 (1991), 57A.

20. U.S. EPA, *Test Methods for Evaluating Solid Wastes* (U.S. EPA Statement of Work 846) (Washington, D.C.: U.S. Government Printing Office, 1986).

21. U.S. EPA Method 6010A: *Determination of Elemental Composition by Inductively Coupled Plasma Optical Emission Spectrometry* (Washington, D.C.: U.S. Government Printing Office, 1993).

22. R. K. Winge and others, *Inductively Coupled Plasma Atomic Emission Spectroscopy: An Atlas of Spectral Information* (New York: Elsevier, 1984).

23. P. W. J. M. Boumans, *Line Coincidence Tables for Inductively Coupled Plasma Atomic Emission Spectrometry*, 2nd ed., vols. 1 and 2 (New York: Pergamon Press, 1984).

24. M. L. Parsons and A. Forster, *An Atlas of Spectral Interferences in ICP Spectroscopy* (New York: Plenum Press, 1980).

25. I. Jarvis and K. E. Jarvis, *J. Geochem. Expl.*, 44 (1992), 139.

26. M. S. Cresser and others, *J. Analyt. Atom. Spec.*, 8 (1993), 1R.

27. N. W. Alcock, *Analytical Chemistry*, 65 (1993), 85R.

28. A. Taylor and others, *J. Anal. Atomic. Spec.*, 9 (1994), 87R.

29. K. J. Eisentraut and others, *Analytical Chemistry*, 56 (1984), 1086A.

30. K. J. Chao, S. H. Chen, and M. H. Yang, *Fresenius' Z. Anal. Chem.*, 331 (1988), 418.

31. J. Marshall and others, *J. Anal. Atom. Spec.*, 9 (1993), 349R.

32. D. G. Anderson, *Analytical Chemistry*, 65 (1993), 1R.

Chapter 22

Inductively Coupled Plasma Mass Spectrometry

Arthur W. Varnes

ICP and ICP-MS Services

Summary

General Uses

- Simultaneous determination of over 70 elements in virtually any sample in less than 3 min
- Concentrations from parts per trillion to parts per thousand without preconcentration or dilution in favorable cases
- Empirical formulae of chromatographic effluents
- Determination of isotopic ratios
- Destructive method, but only a few milligrams or milliliters of sample are required

Common Applications

- Determination of trace impurities in semiconductor raw materials, intermediates, and finished products
- Determination of ultratrace elements in geochemistry
- Determination of elemental composition of unknown materials
- Use of stable isotopic tracers as an alternative to radioisotopic tracers
- Determination of trace elements for environmental compliance
- Determination of radionuclides

Samples

State

Samples are most commonly presented to the instruments in liquid form; however, most solids may be quantitatively dissolved to permit analysis as solutions. Airborne materials may be trapped on appropriate filters and dissolved before analysis. Laser ablation accessories are readily available for most commercial inductively coupled plasma mass spectrometry (ICP-MS) instruments, so direct solids analysis is more commonly used with ICP-MS than ICP-AES.

Amount

Solids Typically 50 mg of an inorganic material is dissolved in 50 mL of solution. Smaller sample masses may be used but sensitivity will be degraded; furthermore, for inhomogeneous materials, it may be difficult to prepare a representative sample. Because the sampling orifice diameter is on the order of 1 mm, the orifice is prone to clogging if solutions containing more than 1000 ppm of total dissolved solids are introduced into the plasma. Thus, it is desirable to reduce total solids as much as possible within the constraints of required detection limits. Flow injection techniques may be used to reduce the total mass of material introduced into the mass spectrometer.

Organic materials may be dry ashed in crucibles or low-temperature plasma ashers, or wet ashed in a high-pressure asher, a microwave oven, or an open beaker. The sample should be chosen to yield about 10 to 50 mg of inorganic constituents, but may not exceed the safety limits of the technique.

Liquids Typically 3 mL are required for a complete mass scan but specialized techniques allow analysis of a few microliters of sample. Alternatively, if sensitivity is not of overriding concern, samples may be diluted to obtain the required volume.

Because samples must be transferred from a high-temperature plasma at ambient pressure into a vacuum region through a small orifice, organic matter gives rise to complications with ICP-MS that are not encountered in ICP atomic emission spectroscopy (ICP-AES). Organic matter is converted to elemental carbon in the reducing atmosphere of a conventional argon plasma; this material may plug the orifice very quickly if substantial amounts of organic matter are introduced. This problem may be overcome by adding oxygen to the argon plasma to oxidize elemental carbon to form CO and CO_2. Care must be taken in selecting and controlling oxygen flow rate because excess oxygen will attack the sampling orifice and an inadequate oxygen flowrate will not overcome carbon deposition. Another solution to the problem of carbon deposition is to use a membrane desolvation device to remove volatile organic matter from the sample aerosol before its presentation to the mass spectrometer.

Gases Because gaseous raw materials are used in the semiconductor and other industries, methods and accessories have been developed to facilitate direct analysis of gaseous samples (1).

Preparation

Although the ICP source is very robust, the analyst should minimize the differences in composition between samples and standards to achieve maximum accuracy. As noted above, solids are usually dissolved before analysis. A substantial body of literature exists to guide the analyst. Aqueous samples are acidified to a pH of less than 2 with redistilled nitric acid to minimize analyte precipitation. Samples should be prepared and stored in Teflon, polyethylene, or similar vessels. Glass vessels should not be used for aqueous solutions because they tend to release contaminants or adsorb analytes. Common acids, such as hydrochloric acid and sulfuric acid, give rise to polyatomic spectral interferences not commonly encountered in ICP-AES. Nitric acid does not increase background intensities significantly because the amount of atmospheric nitrogen entrained with the sample aerosol is much larger than the amount of nitrogen arising from nitric acid in the original sample solution. Horlick and associates (2–4) have published definitive papers on polyatomic interferences. Polyatomic spectral interferences for hydrochloric and sulfuric acids may be virtually eliminated by addition of nitrogen (5), xenon (6), or trace amounts of organic compounds (7); however, sensitivity is usually reduced by this approach.

Because matrix effects are much more severe in ICP-MS than in ICP-AES, virtually all quantitative (and most qualitative) analyses are performed with the aid of internal standard elements to compensate for differences in composition between samples and standards (8). Internal standard elements should match the analyte element to be corrected as closely as practical with respect to ionization potential and mass. It is prudent to perform a qualitative scan of unknown samples before selection of internal standard elements to avoid the possibility of spectral overlap with sample components. The author had occasion to determine germanium in a large set of aqueous solutions. The author selected gallium-69, the most abundant isotope of gallium (an element with mass and ionization potential very similar to germanium) as the internal standard. Extremely erratic results were obtained. Further examination revealed that the samples varied by several orders of magnitude in barium concentration, giving rise to differing signals from Ba^{2+} at a mass-to-charge ratio of 69. Because the signal from doubly charged barium was not resolved from that of singly charged gallium, an erroneously high internal standard correction was applied. When the gallium isotope of mass 71 was used, excellent results were obtained.

Commercial instruments are equipped with software that automatically corrects analyte data for the internal standard response. The author has found that a set of five internal standards—scandium-45, yttrium-89, indium-115, terbium-159, and thorium-232—is very effective in correcting for the entire mass range. Internal standard elements may be incorporated into the diluent if the samples are to be processed before analysis or a multichannel peristaltic pump may be used to combine the analyte stream with the internal standard stream immediately before analysis.

Analysis Time

Instrument warmup and calibration typically require 30 to 60 min. During this time, however, the analyst may check mass calibration and optimize lens voltage, torch alignment, and gas flows, either manually or under computer control. After the instrument is ready for operation, aspirating a sample, allowing the system to come to steady-state, acquiring data for the entire mass range, and washing out the system requires about 3 min for a quadrupole-based instrument. High-resolution magnetic sector instruments require considerably more time; 10 min is typical for optimal sensitivity. Isotope ratio experiments typically require about 75 sec per mass; the analysis time depends on the concentration of total analyte and the relative abundances of the isotopes of interest. To correct for mass discrimination, one typically adds an internal standard element of known invariant isotopic composition (9). To determine two isotopes of analytical interest and to obtain data for two internal standard isotopes typically requires about 5 min. If more isotopes are to be determined, the analysis requires more time. Instruments with multiple detectors reduce analysis time and improve accuracy and precision, but are considerably more expensive than instruments with a single detector.

Almost all commercial ICP-MS instruments are very fast sequential-scanning devices. Quadrupole-based instruments typically scan the entire useful mass range (6 for the light isotope of lithium to 254 for U-238-O-16, uranium oxide) in about 100 ms. Obtaining acceptable accuracy and precision for a full-spectrum analytical scan typically requires about 2 min. Equilibration and washout times are slightly longer for ICP-MS than for ICP-AES, so 15 to 20 determinations per hour for the accessible list of elements is typical. Commercial instruments, however, afford the analyst great flexibility in selecting the mass ranges of interest. If only a few isotopes are required, throughput rates are very similar to those of simultaneous ICP-AES instruments.

Unlike atomic absorption spectroscopy (AAS), X-ray fluorescence (XRF), and ICP-AES, ICP-MS can perform isotopic ratio measurements. One impediment to accurate isotope ratio determinations, however, is mass discrimination: Ions with differing mass-to-charge ratios may exhibit slightly different sensitivities. Furthermore, conventional single-element detectors suffer from random fluctuations in ion flux with time; precision and accuracy are typically no better than 1% relative. Thermal ionization mass spectrometry can provide precision and accuracy of about 0.1% relative, but sample throughput is typically an order of magnitude or more poorer than that of ICP-MS.

Autosampling devices allow unattended operation and most modern instruments include software to perform QC check routines and even shut down the instrument at the end of a run. Sample preparation depends on the type of sample, the analytes to be determined, their concentrations, and the equipment and reagents available in the laboratory. Often samples are dissolved overnight and analyzed the following day.

Limitations

General

- Matrix effects are more common and more severe than for ICP-AES.
- Solutions with high solids concentrations may plug orifice.
- Detector lifetime may be severely reduced by exposure to more than a few parts per million of any isotope; however, most commercial instruments include detector protection in the hardware and software.
- Organic solutions require addition of oxygen to the plasma or sample pretreatment to avoid plugging orifice with elemental carbon.
- Polyatomic interferences arising from solvent species and argon cause severe degradation of detection limits for several important elements (silicon, sulfur, potassium, calcium, iron, and possibly arsenic and selenium) for conventional quadrupole-based instruments.
- Oxide, argide (species of the general formula Ar-X, in which X may be any element present in relatively high concentration), and doubly charged species give rise to spectral interferences.
- Initial capital cost, routine maintenance, and time and effort required are all greater than for ICP-AES.
- High-purity reagents and extreme caution to avoid contamination and loss are required to optimize sensitivity and accuracy.
- No information is generated concerning oxidation states or structures of compounds.
- Analyte elements must have ionization potential less than that of argon: 15.8 eV.

Accuracy

In analysis of simple solutions with known composition, accuracy better than 2 percent relative for analytes present at 50 times the detection limit may be achieved using appropriate internal standard elements. Accuracy may be improved if only a small suite of isotopes is to be determined by using peak-hopping for only the analytes of interest and using multiple internal standards.

An option available in ICP-MS that is not available in ICP-AES or AAS is the use of isotope dilution analysis (10). For elements with more than one stable isotope, one may spike the sample with material enriched in a minor isotope for which no isobaric interference occurs. For example, zinc has a minor isotope of mass 67 that is not subject to isobaric interference from either common monatomic or polyatomic interferences. By adding a known amount of zinc-67 to a sample and determining the response in the actual sample, one can then calculate the concentration of zinc in the original sample with accuracy and precision of a few tenths of a percent relative.

Another advantage of ICP-MS over ICP-AES is the opportunity to perform rapid analyses using a single internal standard element. The first step in this procedure is to prepare a solution containing several elements relatively evenly distributed across the entire mass range. The author typically uses a solution containing known concentrations of Be, Al, Sc, Co, Y, In, La, Tb, Pb, and Th for this procedure. Another approach is to use a standard reference material, such as National Institute of Standards and Technology (NIST) 1643c (trace elements in drinking water). The mass spectrum of this solution is recorded and the computer generates a response surface for all analytes based on the mass, relative abundance, and ionization potential of each element. If a sample contains no elements in concentration more than a few parts per million and is similar in bulk composition to the standard solution, accuracy better than 20% relative can be achieved for a 60-sec scan of the entire mass range ($Z = 6$ to $Z = 254$) by adding a mid-mass element (typically indium) as an internal standard.

Isotope ratios may be used to detect possible errors arising from spectral interferences. Most commercial instruments include video monitors that allow the analyst to display and examine each mass spectrum. The instruments also have software that allows peak identification. One places the cursor on the peak of concern and the software displays sequentially all the monatomic and polyatomic species of that mass-to-charge ratio. In addition, for elements with more than one isotope, identification bars of length proportional to relative isotopic abundance are superimposed on all isotopic peaks associated with the monatomic isotope of the selected peak. If the isotopic ratios are consistent with the displayed pattern, it is strong evidence that the peak in question is indeed the suspected element and that it is free from significant spectral interferences.

In cases in which polyatomic interferences are detected, it is often possible to obtain a reliable estimate of the interference and correct the affected signal. For example, if one anticipates an $^{40}Ar^{35}Cl$ interference on ^{75}As, one may measure the responses for selenium and masses 77 and 82. Using the relative isotopic abundances, one may calculate the anticipated response for ^{77}Se from the measured response at mass 82. Subtraction of the calculated ^{77}Se response from the measured ^{77}Se response yields the apparent contribution of $^{40}Ar^{37}Cl$. From the relative abundances of ^{35}Cl and ^{37}Cl, one can then calculate the corresponding response arising from $^{40}Ar^{35}Cl$ and use that value to correct the apparent As concentration for the spectral interference arising from the presence of chlorine in the sample.

Sensitivity and Detection Limits

Sensitivity of a given element depends on three factors: ionization potential, mass, and relative isotopic abundance. The Saha equation (11) relates the degree of ionization to several factors; the most important of these factors is ionization potential. For elements with ionization potential less than 7 eV, ionization of greater than 99% is usually achieved. Mass transmission through the lenses depends on voltage applied to the various members of the interface and focusing lens assemblies. One typically adjusts these voltages to yield the largest response for a mid-mass element such as indium or rhodium. Sensitivity typically decreases for isotopes of lower or higher mass. Finally, sensitivity is really a measure of the number of ions of a given mass-to-charge ratio reaching the detector. The relative abundance of the various isotopes of an element determines the relative sensitivity for isotopes of that element. Using a conventional pneumatic nebulizer (typically a concentric glass nebulizer) with a conventional interface and detector, the response for a quadrupole-based system is typically several million counts per second per part per million for a mid-mass element of modest ionization potential and greater than 90% abundance, such as indium-115. Ultrasonic nebulization typically improves sensitivity by a factor of about 20; one order of magnitude is gained by greater analyte transport efficiency and an additional factor of as much as two arises from desolvation. Specially designed interfaces and other detectors also yield sensitivity increases of an order of magnitude. In assessing an instrument, however, one must keep in mind that detection limits are a function of the signal-to-background ratio, not just the absolute signal intensity.

Horlick and associates (12) have extensively investigated the effect of incident power, sampling depth, and gas flows on sensitivity for several potential analytes. For a fixed incident power and sampling depth, a relatively sharp peak was observed for a plot of response versus nebulizer gas flow; plasma and auxiliary flowrates had virtually no effect. Similarly, for fixed sampling depth and nebulizer flowrate, a sharp peak was observed with variation in incident power. Finally, for fixed power and gas flows, a sharp response was observed with variations in sampling depth (distance from the load coil to the interface).

From all these observations concerning effect of voltages, applied radiofrequency (RF) power and gas flows, it is not surprising that if one is determining just a few elements of low or of high mass, sensitivity may be improved by adjusting these parameters to yield maximum response for a surrogate analyte similar in mass and ionization potential to the actual analytes. (Use of a surrogate is recommended because use of an actual analyte in high concentration could lead to memory effects if the concentration of the analyte were near the detection limit.)

Another concern is that the settings for maximum sensitivity may also be the ones yielding the most complications arising from polyatomic, oxide, or doubly charged interferences. In many cases it is desirable to optimize the ratio of analyte to analyte oxide or analyte to doubly charged species. Although absolute sensitivity may be reduced by about 50%, precision and accuracy will be improved.

Linear Range

A channel electron multiplier operating in the pulse-counting mode typically yields a linear range of at least a six orders of magnitude. For elements with multiple isotopes of varying relative abundance (less than 1% to more than 90%) and free of isobaric interference, two or more additional orders of magnitude may be readily attained. Most commercial instruments include hardware and software for analog detection as well as pulse counting, and this capability extends linear range by about three orders of magnitude. Software on most instruments analyzes data from a brief scan of all analytes, performs pulse counting for trace elements, and then switches to analog mode for more concentrated analytes, all without operator intervention. Using combined pulse-counting and analog detection, a linear range of eight to nine orders of magnitude is available for any given isotope.

Complementary or Related Techniques

- ICP-AES is a more rapid technique for determining trace and minor elemental components. Detection limits are typically a few parts per billion and many elements exhibit a linear calibration curve to approximately 1000 ppm. ICP-AES is also less prone to matrix effects but exhibits higher background spectral intensity and more numerous and challenging spectral interferences. ICP-AES instruments are less expensive and easier to maintain and, in the direct-reading design, afford greater sample throughput.

- Thermal ionization mass spectrometry achieves superior precision and accuracy for isotope ratio analysis, but the sample throughput rate is about one sample per hour.
- Spark source mass spectrometry may be used for direct analysis of solids, but lacks precision of ICP-MS.
- Glow-discharge mass spectrometry is a powerful alternative for direct analysis of conducting solids; nonconductors and semiconductors may be analyzed by mixing the sample with high-purity copper or a similar conducting material.
- X-ray fluorescence generally permits direct analysis of solids with greater precision and accuracy, although sensitivity is typically at or slightly below the part-per-million level.
- Secondary ion mass spectrometry (SIMS) provides superior capability for characterizing surface concentrations.
- Electron microscopy with energy-dispersive X-ray fluorescence (EDXRF) allows determination of elemental composition of extremely small regions of a solid.
- Neutron activation analysis allows precise and accurate determination of elemental composition with almost no sample preparation or matrix effects. However, this technique requires access to a source of neutrons at high flux, typically a nuclear reactor, and may require several weeks to obtain accurate data on some analytes.
- Ion chromatography and selective ion electrodes allow determination of halide and other anions in solution.
- Combustion techniques are used to determine C, H, N, O, and S.

Introduction

Just as ICP-AES was developed simultaneously and independently by researchers at Iowa State University in the United States and another research group in the United Kingdom, ICP-MS was pioneered at Iowa State University simultaneously with its appearance in England. Unlike ICP-AES, however, the pioneering investigators collaborated on a single publication (13).

ICP-MS afforded similar improvements in detection limits for plasma source elemental analysis as did the graphite furnace for atomic absorption spectrophotometry. Indeed, a major impetus to sales of ICP mass spectrometers has been their ability to perform simultaneous multielement analysis with detection limits similar or superior to those of single-element graphite furnace AAS.

An example of the power of ICP-MS is a request to the author for determination of trace elements in high-purity alumina. A detection limit of 1 ppm was required for all analytes. Initial efforts using ICP-AES were not successful. Detection limits, even in simple aqueous solutions, were still too high and spectral overlaps and background emission degraded detection limits by almost an order of magnitude for arsenic and selenium. Using ICP-MS, however, detection limits of no worse than a few tenths of a part per million were achieved for all analytes. The effect of the alumina matrix was negligible. The accuracy of the method was demonstrated in that spike recoveries for each of the eight analytes were between 90 and 120%. Precision for three replicates, including the sample preparation procedure, was better than 10% relative. A further advantage was that two or more isotopes were measured for several of the analytes, giving yet another cross-check of accuracy.

Although quadrupole-based mass spectrometers are adequate for resolving all monatomic ions, their resolution is no better than a few tenths of a dalton. This resolving power is not adequate to separate monatomic species from polyatomic species of similar mass-to-charge ratio. An example of the frustration this situation can cause was another request to the author for a complete ICP-MS scan of a ceramic material. Elements specified by the client as critical were silicon, phosphorus, potassium, calcium, and iron. None of these elements possess an isotope free of isobaric interferences. Although these analytes were all present at concentrations amenable to ICP-AES, the client was dissatisfied about having to pay for both ICP-AES and ICP-MS to obtain a complete analysis of the sample. Because these elements are of critical importance, especially in the semiconductor industry, high-resolution instruments (14) based on magnetic sector mass filters are commercially available. The cost of such an instrument in 1995 was two to three times that of a quadrupole-based mass spectrometer. Furthermore, sample throughput is typically a factor of more than two less than that obtained using a quadrupole mass filter. On the other hand, background levels are typically below 1 count per second, so detection limits are usually superior to those attained using a quadrupole mass filter.

Another drawback of most commercial instruments that are equipped with only one detector, especially for isotope-ratio measurements, is that precision and accuracy are degraded because only one isotope can be measured at a time. Consequently, fluctuations in the plasma are not corrected directly by an internal standard and precision of isotope-ratio measurements is degraded. Recently an instrument was developed that uses twin detectors (15). This approach greatly reduced the impact of flicker and other random variations on isotope measurement or even determination of elemental composition. Another approach to improving precision and accuracy is time-of-flight (TOF) mass spectrometry (16). In this experiment, simultaneous sampling of all analytes is performed and the sampled ions are measured sequentially. Although still not a true simultaneous technique because ions reach the detector sequentially, detector response is much more stable than are plasma excitation conditions. Thus, time-of-flight mass spectrometry may find a significant role in future commercial instrumentation.

Another improvement in recent ICP mass spectrometers has been reduced size and weight. Whereas the original instruments usually consisted of a large spectrometer and a separate and standalone generator, more recent instruments are smaller and use solid-state generators that can be located within the cabinetry of the spectrometer.

In 1995 approximately 1000 commercial ICP-MS instruments were in use throughout the world. Because extremely small amounts of impurities can have a highly deleterious impact on semiconductor performance, ICP-MS finds wide application in the semiconductor industry and by producers of raw materials for semiconductor applications. Geochemists are attracted by the capabilities of ICP-MS to perform virtually total elemental analysis as well as isotope ratio studies, although precision of conventional instruments for isotopic abundance is not adequate for demanding research. Geochemists especially value its sensitivity and relative freedom from interferences for rare-earth determinations.

As described in Chapter 21, the combination of ICP-MS with ICP-AES affords 11 orders of magnitude linear range and affords much greater confidence in accuracy of results. The major sources of interference differ between the two techniques, so just as one can use an atom and an ion line in ICP-AES to verify absence of matrix effects, discrepancies in values for the two techniques alert the operator to possible analytical errors.

To achieve the theoretical detection limits available from commercial ICP-MS instruments, great attention must be paid to purity of reagents and cleanliness of the laboratory. Torches, nebulizers, and spray chambers should be soaked in a 1:10 solution of nitric acid and rinsed several times with deionized water having a resistance greater than 18 $M\Omega$ before use. Sampler and skimmer cones should be cleaned daily to avoid cross-contamination. A class-100 or class-10 cleanroom is desirable for ultimate cleanliness. Solvents must be of the highest available purity; some laboratories even prepare their own acids and solvents by sub-boiling distillation to reduce background contamination. To reduce the probability of carryover from sample to sample, one typically washes with the most concentrated acid matrix expected to be encountered. This procedure reduces the probability that material deposited on the sampling cone will be dissolved by acid vapors from a more concentrated acid matrix, thereby giving rise to positive interference.

How It Works

Physical and Chemical Principles

ICP mass spectrometry is based on the fact that the ionization potential of argon, 15.8 eV, is great enough to ionize most other elements; only fluorine, helium, and neon have higher ionization potentials.

The ICP source is extremely hot. Flame temperatures in atomic absorption are typically less than 3000 °K. Arc and spark sources achieve temperatures of about 5000 °K. The ICP, however, is estimated to produce a maximum temperature of at least 6500 °K. This extremely high temperature is sufficient to break almost all chemical bonds in a sample, giving rise to monatomic ions that are virtually independent of one another. As a result, the technique exhibits high sensitivity, four or more orders of magnitude linear range, and much lower spectral interferences than emission techniques.

Samples are most commonly introduced into the plasma as aerosols. A wide variety of devices are available for sample introduction. Pneumatic nebulizers are the least expensive and most commonly used in commercial devices. The aerosol produced by the nebulizer is generally passed through a spray chamber to remove large droplets and produce a more homogeneous aerosol. While passing through the plasma, the aerosol is vaporized, atomized, and ionized. After leaving the extremely hot plasma, the ions are extracted from the plasma through an orifice (typically 1 mm in diameter) into a quadrupole mass spectrometer (see p. 656). A system of electronic lenses is used to focus the ion beam before entrance into the mass filter. As noted above, a quadrupole is the typical mass analyzer in ICP-MS, although some work has been done with time-of-flight (TOF) mass spectrometers (16) and magnetic sector mass analyzers have been used for higher resolution (13).

If information is desired about the various chemical forms of an element, one may pass the sample through a chromatograph before analysis and record the mass spectrum as a function of time (17, 18). Qualitative analysis is performed by measuring retention time and concentration by measuring either peak intensity or peak area.

Figure 22.1 is a block diagram of a typical ICP optical mass spectrometer system. The most obvious difference between ICP-MS instruments and conventional radial-view ICP-AES instruments is that the torch is mounted horizontally, rather than vertically. The reason is that mounting a mass spectrometer along the vertical axis of the plasma torch is a more difficult engineering task than mounting the spectrometer horizontally. (Indeed, Hieftje (19) has assembled a very viable ICP-MS system incorporating a vertically mounted torch.) Power for the excitation source is provided by the radiofrequency generator. Typical frequencies are 27.5 and 40 MHz; these frequencies are used in diathermy devices, so they do not interfere with commercial communications.

Figure 22.1 Block diagram of an ICP mass spectrometer.

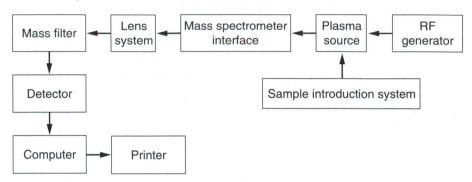

Incident power is typically 1.3 kW for aqueous solutions, although some special torches operate with lower power. For organic solutions the power is increased to 1.6 to 2.0 kW because more energy is absorbed by most organic solvents than by water. Furthermore, oxygen must be added to the argon to avoid deposition of elemental carbon on the sample orifice by converting the carbon to carbon monoxide and carbon dioxide.

Figure 22.2 shows a typical plasma torch, although other designs may be used. Three argon streams are supplied to the torch. The largest flow, normally called the plasma or cooling gas, is directed through the upper arm of the torch at about 15 L/min. A second flow, called the auxiliary gas, may be introduced through the lower arm. One usually operates with no auxiliary flow, but a small flow (about 1 L/min) may be useful for some matrices. The third argon stream, called the sample gas, passes through the nebulizer at about 0.7 to 1.0 L/min (depending on manufacturer and instrument geometry) and carries the sample aerosol into the plasma. The sample gas flowrate exerts considerable influence over both absolute and relative sensitivities of various analytes. This flowrate must be very precisely controlled, so most instruments use a mass flow controller to maintain a very constant sample gas flowrate.

In operation, the plasma torch is secured within the electromagnetic field arising from a conducting helical coil (made from copper or a similar conducting metal) that transfers power from the radiofrequency generator. Because the coil acts as an inductor, the source is called an inductively coupled plasma.

Sample is introduced into the plasma with the aid of a nebulizer. The concentric glass Meinhard nebulizer (Fig. 22.3) is very commonly used. It consists of a fine glass capillary concentric with a larger-diameter glass tube drawn to an internal diameter slightly larger than the outer diameter of the inner capillary. The sample is drawn through the central capillary and converted to an aerosol by an argon stream passing through the glass tubing surrounding the sample introduction capillary. Although the Venturi effect allows the device to draw sample unaided, typically a peristaltic pump is used to improve precision. To decrease the amount of water vapor (which gives rise to oxide and hydride formation) reaching the plasma, the spray chamber is typically thermostated near 0 °C with the aid of a recirculating constant-temperature bath.

When aerosol enters the plasma, the solvent is evaporated to yield a dry aerosol. The particles produced by evaporation are vaporized, atomized, and, for most elements, at least partially ionized. As the samples leave the plasma, the excited species are cooled. It is thought that very few polyatomic atoms survive within the plasma, but that recombination occurs as ions cool after leaving the plasma and that the polyatomic ions form before the samples reach the detector.

It is possible to reduce oxide interferences by using electrothermal atomization to vaporize the sample. The advantage of this technique is that the quantity of solvent is greatly reduced and vir-

Figure 22.2 Typical plasma torch. *(Reprinted with permission from the Annual Book of ASTM Standards, copyright American Society for Testing and Materials, 1916 Race Street, Philadelphia, PA 19103.)*

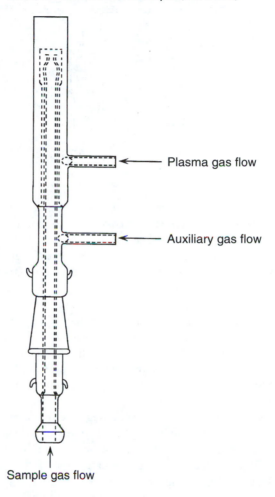

← Plasma gas flow

← Auxiliary gas flow

↑ Sample gas flow

tually all analyte species reach the plasma. Solvent reduction removes the major source of oxide and hydride interferences; increased sample transport efficiency lowers detection limits. The disadvantages are that a homogeneous solution is required, analyte vapors may condense before reaching the plasma, and some analytes form refractory carbides that severely reduce sensitivity. Furthermore, because the ICP-MS typically requires 100 ms for a full mass range scan, usually one may only determine a relatively few (less than 10) isotopes per determination.

Samples pass from the plasma into the mass spectrometer typically through two concentric orifices. The first, called the sampling orifice, is usually about 1 mm in diameter; the orifice is cooled by flowing water or a similar fluid to conduct unwanted heat from the plasma away from the interface. A second orifice, called the skimmer, is of a more conical shape and slightly smaller diameter. It is located within the Mach disk of the sampling orifice.

Figure 22.3 Meinhard concentric glass nebulizer. *(Reprinted with permission from the* Annual Book of ASTM Standards, *copyright American Society for Testing and Materials, 1916 Race Street, Philadelphia, PA 19103.)*

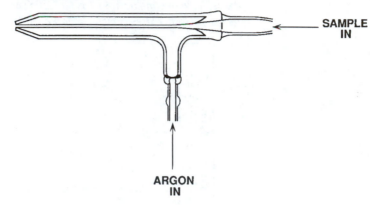

The vacuum system consists of two or more regions. The first region of reduced pressure is generally a moderate vacuum obtained with the aid of a mechanical pump. Early instruments used either a cryogenic pump or diffusion pumps to attain pressures below 10^{-5} torr in the regions in which the lens system and detector were located. Cryogenic pumps require regular shutdown for thawing and diffusion pumps require several hours to reach operating pressure after shutdown. Consequently, cryopumps and diffusion pumps are being replaced by turbomolecular pumps, which can operate continuously and reach operating pressure very quickly.

Because photons striking the detector can give rise to spurious signals, the ion path must be designed to minimize the number of photons emitted by the plasma that reach the detector. This reduction is often accomplished by means of a photon stop, a small plate located on the axis of the ion beam. Ions can pass around the stop, but photons strike the stop and do not enter the spectrometer.

When sample enters the orifice, a negative potential is applied to attract cations into an ion lens system that focuses the ion beam before passage through a mass filter. The typical mass filter is a quadrupole with resolution of about one dalton. This device is rugged, relatively inexpensive to fabricate and maintain, and has adequate resolution to provide acceptable abundance mass sensitivity (effect of a high concentration of M^+ on $(M + 1)^+$).

After being resolved by the mass filter, cations are passed to the detector. The typical detector is a channel electron multiplier, although other detectors may be used. The lifetime of the detector is shortened by exposure to intense ion beams, so most commercial instruments include software that turns off the detector if an excessive count rate occurs. Furthermore, data are typically not collected for masses less than 7 (H, He, and He-H), 12 to 22 (carbon through neon), 28 to 41 (nitrogen through argon hydride), and 80 (argon dimer). Data may be accumulated in either continuous scanning or peak-hopping mode. The continuous scanning mode requires excellent resolution to avoid interferences from adjacent peaks but is more forgiving of fluctuations in precise correlation between detected mass and quadrupole performance. Peak-hopping, on the other hand, requires extremely good reproducibility of mass position, but requires only moderate resolution. In theory, continuous scanning is slightly more accurate because response shifts and fluctuations in plasma parameters are less likely to affect total integrated signal, whereas peak-hopping is theoretically more sensitive because measurements are taken near the center of mass of the response peak. In practice, however, either technique provides acceptable sensitivity and accuracy.

What It Does

Limits of Detection

The ICP instrument detection limit (IDL) is typically determined by calibrating the instrument, then aspirating a solution containing an analyte concentration of three to five times the anticipated detection limit. Five to ten successive determinations of the apparent concentration are performed and the standard deviation of the measurements is calculated. The instrument detection limit is obtained by multiplying that standard deviation by three. The IDL varies from element to element, depending on the efficiency with which the plasma ionizes the element, the presence of polyatomic interferences, and the efficiency with which the ion is transmitted through the instrument and counted by the detector.

Most elements in simple aqueous solution exhibit a detection limit of less than one part per billion for a conventional quadrupole-based mass spectrometer using pneumatic nebulization and a 60-sec signal integration time for the entire mass range. For elements with low ionization potential, moderate mass, and no polyatomic interferences, such as cerium, detection limits below 1 part per trillion may be achieved. If only a few isotopes of similar mass are to be determined, detection limits may be lowered by optimizing instrumental parameters (lens voltage, sample gas flow rate, and instant power) and accumulating data only for the isotopes of interest.

High-resolution instruments routinely achieve sub–part-per-trillion detection limits, principally because there is virtually no response in the absence of analyte. Ultrasonic nebulization generally lowers detection limits by at least a factor of 10; in addition to greater sample transport efficiency, the ultrasonic nebulizer also reduces the amount of water vapor reaching the plasma, thereby decreasing oxide formation.

Linear Range

For a single isotope four orders of magnitude of linear range may be easily attained in pulse-counting mode and even greater linear range may be obtained in favorable cases. Some instruments afford the option of automatically switching between pulse counting and analog mode; this feature increases linear range to eight or more orders of magnitude.

Many elements exhibit isotopes of widely differing abundance. For some elements, as many as 12 orders of magnitude may be achieved if no interferences impede determination of trace isotopes.

Isotope-Ratio Measurements

In principle, isotope ratio determinations are very straightforward. One might anticipate that blank-subtracted ion counts could be determined for each isotope of interest, the sum of counts for all isotopes of a target element calculated, and simple ratios be performed. Unfortunately, this approach fails in most commercial mass spectrometers because the efficiency with which ions are counted varies from mass to mass. Ketterer (7) developed an internal standard method for correcting for this mass bias. He uses an element with a well-characterized isotope ratio as a surrogate analyte. By comparing the apparent isotope ratio obtained by ICP-MS to the theoretical value, a factor may be obtained to correct analyte isotope ratios for the mass discrimination effect. For example, the ratio of Tl-203 to Tl-205 may be used to correct the isotope ratios for the four isotopes of lead at 204, 205, 206, and 208. Analysis time depends on the number of isotopes to be determined, their relative isotopic abundance, the total concentration of the analyte, and the precision required. Typically precision and accuracy of better than 1% relative may be attained with an acquisition time of 75 sec per element.

Analytical Information

Qualitative

The ICP argon source is capable of at least partially ionizing all elements with an ionization potential less than that of argon (15.8 eV). The major impediment to qualitative analysis is interference from atmospheric gases, species arising from solvents and polyatomic ions. Unlike ICP-AES, in which analytes may emit literally thousands of lines, giving rise to complex spectra and significant problems of spectral overlap, each element exhibits at least one isotope free from isobaric interference from monatomic species. Unfortunately, polyatomic and doubly charged monatomic species are detected, giving rise to additional interferences. Furthermore, the intensities of polyatomic and doubly charged ion signals are matrix-dependent. Thus, some care is required in interpreting mass spectra. A useful diagnostic tool for elements with more than one usable isotope is the isotope ratio. The probability of interferences occurring in exactly the same ratio as the isotopic abundance is vanishingly small. Consequently, if the isotopic fingerprint of a suspected analyte in a complex sample matches that arising from a single-component solution of that suspected analyte, there is a high probability that the analyte is indeed present in the sample.

Quantitative

Quantitative analyses may be performed in either of two ways. The more commonly used method involves collecting data for all isotopes amenable to the technique. The advantage of this method is that if additional analyte species are identified after data are accumulated, one can retrieve intensities for elements for which no calibration was performed. One can then obtain a semiquantitative concentration by comparing signals from additional analytes to the known response for elements of similar mass and ionization potential (after correction for relative mass abundance). The second method is to collect data only for specified isotopes. The advantages of this method are greater sensitivity and better precision, because a much larger fraction of data collection time is devoted to target analytes than in full scanning mode. The disadvantage is that no data are collected for additional isotopes so the analysis must be repeated if additional isotopes are added to the target analyte list.

Three modes of multielement quantitation are typically afforded by commercial ICP-MS software: single internal standard, full quantitation, and isotope dilution.

Single Internal Standard

The instrument is calibrated by preparing a solution containing several analytes, mostly monoisotopic, that cover the entire mass range; a typical set would be ^{9}Be, ^{27}Al, ^{45}Sc, ^{59}Co, ^{89}Y, ^{115}In, ^{139}La, ^{159}Tb, and ^{232}Th. If all samples contain a similar matrix, the calibration set is prepared in that matrix. A response curve is generated, using Saha factors to correct ion intensities for differences in ionization potential. Response factors for all other elements are calculated by the software, based on their relative isotopic abundances and ionization potentials. A mid-mass element, typically indium, is added to all solutions in known concentration from a high-purity source to avoid contamination. Fifty ppb is often chosen as the concentration increment to give adequate response with minimal contamination. (Of course, the analyst must ensure that the element to be used as an internal standard is not already present in any of the samples.) Even for situations in which the sample matrix differs markedly from sample to sample, so that matrix-matching is not possible, results within a factor of two of the actual value are readily achieved. For samples with a matrix similar to the calibration standard solution, accuracy of 20% relative is easily attained. For favorable situations in which matrix matching is possible and care is taken to optimize Saha factors for all analytes, accuracies better than 10% relative are often achieved.

Full Quantitation

Solutions are prepared containing all analytes in known concentration. Multiple internal standard elements (as many as nine in some cases) are used to optimize match in ionization potential and mass between analyte and internal standard. Typically a blank and three solutions of varying concentration (5 times detection limit, 0.2 to 1 times the anticipated maximum concentration and 2 times the anticipated maximum concentration) are used. Calibration curves are generated for all analyte species. Accuracy better than 2% relative can be readily achieved with full quantitation. Obviously, preparing all the necessary standard solutions can be very tedious when large numbers of elements are to be determined.

Isotope Dilution

For elements with more than one isotope, isotope dilution is a powerful technique for optimal accuracy (10). Solutions are prepared from a source of the analyte element that has been enriched in a minor isotope. For example, ^{204}Pb is normally abundant at about 1%; lead compounds in which the ^{204}Pb content is more than 90% may be readily attained. The cost per gram of enriched isotopes may seem very large; however, when one considers that the amount of material required is exceedingly small (typically 1 µg per 10 mL of solution), the cost for enriched isotopic materials is not significant. Factors affecting accuracy include the accuracy with which the isotopic abundances of each isotope are known, the accuracy with which spiking solutions of the enriched isotope are prepared, and the degree to which the chemical form of the enriched isotope is atomized, ionized, and transmitted to the detector in a like manner to the corresponding analyte isotope. Isotope dilution techniques easily achieve accuracy and precision better than 1% relative but they do require considerably more time and slightly more cost than conventional analyses.

Accuracy and Precision

For simple solutions with known gross composition, accuracy better than 2% for analytes with concentrations greater than 50 or more times the detection limit may be achieved. Except for solutions of unknown composition for which matrix matching is not possible, accuracy better than 5% relative may be routinely achieved for analytes present at 50 or more times the detection limit. Isotope dilution techniques, in which known amounts of an enriched stable minor or trace isotope of each analyte of interest are added to all samples, enhance accuracy; isotope dilution, however, requires significantly more time than conventional analysis.

To obtain quantitative results for direct analysis of solids by laser ablation (20), an excellent match between bulk composition of standards and samples is necessary. In addition to possible matrix effects on response of analytes in the plasma and mass spectrometer, an additional complication is the efficiency with which the laser energy couples to the sample.

Applications

As noted above, ICP mass spectrometry may be used to determine the concentrations of more than 70 elements from low parts-per-trillion to parts-per-million levels in a single analysis requiring less than 3 min after calibration. A wide array of techniques have been developed to dissolve solid samples for analysis with virtually no contamination or loss. Solids may be analyzed directly by laser ablation techniques.

ICP-MS is heavily used in semiconductor and electronics industries. The requirements for high-purity materials and the problems arising from infinitesimal quantities of impurities require the sensitivity of ICP-MS to meet quality control standards (21).

ICP-MS is a major tool for geochemistry (22). Its speed, excellent sensitivity, accuracy, and long linear range allow rapid, precise analysis of a wide variety of minerals for minor and trace components. Furthermore, the ability of ICP-MS to determine isotope ratios at moderate precision allows use as a screening tool; novel instruments with multiple detectors afford precision and accuracy similar to thermal ionization mass spectrometry. ICP-MS affords superior resolution for determination of rare-earth elements (REEs). These elements typically emit hundreds or thousands of atomic emission lines in the range of 160 to 900 nm typically used for ICP-AES, so spectral overlap is a major problem. ICP-MS spectra are much simpler; furthermore, most REEs have sufficiently low second ionization potentials to allow determinations using the corresponding doubly charged ion as another alternative to minimizing spectral interference. Another useful attribute is superior sensitivity and detection limits for the platinum group elements.

Because ICP-MS is capable of determining virtually the entire periodic table, it is a powerful technique for qualitative and semiquantitative analyses. Laser ablation accessories allow rapid analysis of solids with no sample preparation. Laser ablation also allows comparison of different regions of a sample.

ICP-MS is also used to good advantage in biological and medical research (23). Many techniques have been developed for rapid dissolution or digestion of tissues or fluids. The speed, accuracy, and large number of elements that can be determined in a single analysis make ICP-MS a very cost-effective technique. In addition, enriched stable isotopes may be used to study metabolism of elements for which more than one isotope is available.

ICP-MS is also used extensively in the determination of radioactive and transuranic elements (24). Using bismuth as an internal standard, ultrasonic nebulization, and instrument parameters optimized for determining lead, the author has determined thorium and uranium at 1 part per trillion with an early version (Plasmaquad 2+) mass spectrometer. More recent instruments afford even lower detection limits. For highly radioactive samples, the instrument may be isolated and operated by robotics and a computer located outside the shielded room containing the instrument.

The USEPA has approved methods for most priority pollutant metals in a variety of matrices (25).

Nuts and Bolts

Relative Costs

A quadrupole ICP mass spectrometer in 1995 costs about $200,000, which is approximately twice the price of a direct-reading ICP emission spectrometer. High resolution ICP-MS instruments cost about $400,000, as do instruments equipped with multiple detectors for isotope ratio determination.

Vendors for Instruments and Accessories

Finnigan MAT
355 River Oaks Pkwy.
San Jose, CA 95134
phone: 800-538-7067
fax: 408-433-4823
email: williams@finnigan.com
Internet: http://www.finnigan.com

Hewlett-Packard
3495 Deer Creek Rd.
Palo Alto, CA 94025
phone: 800-229-9770
Internet: http://www.hp.com/go/chem

Perkin-Elmer Corporation
761 Main Ave.
Norwalk, CT 06859
phone: 800-762-4000
fax: 203-762-4228
email: info@perkin-elmer.com
Internet: http://www.perkin-elmer.com

Spectro Analytical Instruments
160 Authority Dr.
Fitchburg, MA 01420
phone: 508-342-3400
fax: 508-342-8695

Thermo Jarrell Ash/Baird Corp.
27 Forge Pkwy.
Franklin, MA 02038
phone: 508-520-1880
fax: 508-520-1732

Varian Instruments
P.O. Box 9000
San Fernando, CA 91340
phone: 800-926-3000

VG Elemental/Fisons Instruments
55 Cherry Hill Dr.
Beverly, MA 01915
phone: 800-999-5011
fax: 508-524-1100

Required Level of Training

A bachelor's degree in chemistry is sufficient to operate an ICP mass spectrometer. For routine analyses of many similar samples, autosamplers may be effectively used to free the analyst for other activities. Manufacturers typically present a brief orientation at the time of installation and encourage the new user to operate the instrument for several weeks. The user then attends a training course, typically 5 days, at the vendor's application facility to review fundamentals of the techniques and specific operating procedures for the exact instrument in question.

Method development and interpretation of data from unknown samples require a level of expertise comparable to that required for ICP-AES. Special care is required in developing protocols and maintaining facilities and equipment if optimal detection limits are to be achieved because detection limits are often influenced more by contamination of reagents, introduction of extraneous materials, or loss of analyte during sample preparation than by limitations of the instrument. Furthermore, matrix effects are more commonly encountered in ICP-MS than in conventional radial view ICP-AES.

Service and Maintenance

Routine ICP-MS maintenance is more time-consuming and expensive than ICP-AES maintenance. Vacuum systems require significant effort to maintain adequately low operating pressures. The plasma torch, spray chamber, and sample interface must be cleaned often to maintain optimal detection limits. Ion lenses must be cleaned about twice a year. The detector is usually replaced annually at a cost of about $1000 plus installation. Depending on amount and severity of use, the main power tube in the RF generator may have to be replaced about every 2 yr at a cost of $1000 to $2000.

The instrument typically consumes about 40 ft^3 of argon per hour; if one is using the instrument regularly, it is desirable to use liquid argon rather than gaseous argon from cylinder. A typical liquid argon cylinder will operate the instrument for 2 weeks if it is operated about 40 hr per week. Even purchased as a liquid, the argon required to operate the plasma costs less than $5 per hour.

Typical power consumption is 1 to 2 kW, so cost for electrical power is generally far less than $1 per hour. Similarly, water for cooling the load coil requires only a few liters per minute; if water is expensive or not readily available, one can use a recirculating water bath to minimize water consumption.

If the instrument is located in an environment from which acid fumes are excluded, and temperatures and humidity are well-controlled, the optical and electrical systems should require little maintenance.

ICP-MS instruments are typically more complex than ICP-AES instruments. In practice, most ICP mass spectrometrists come from a background in optical spectroscopy rather than mass spectrometry. As a result, they are not familiar with maintaining the mass spectrometer portion of the instrument.

As noted in the Introduction, regular cleaning of glassware and interface components is essential for controlling contamination. It is desirable to obtain a mass spectrum of a solution containing Be, Mg, Co, In, Ce, Pb, and U at least weekly. Comparison of responses for Be, In, and U provides information concerning relative high- and low-mass sensitivity. Examination of the response for the three isotopes of magnesium yields an assessment of low-mass resolution and examination of the four lead isotopes allows the analyst to assess high-mass resolution. Comparison of the responses at mass 70 (Ce^{2+}), 140 (Ce^+), and 156 (CeO^+) provides a measure of doubly charged and oxide interferences. Measurement at mass 150, for which no analyte or polyatomic species should be present, provides a measure of random detector noise. In case a service problem arises, a collection of this information over time can assist in diagnosing the problem.

Another aid is a daily record of the optimal gas flows, incident power, vacuum system pressures, and lens settings, as well as the response for ^{115}In. Such a record is also very helpful in understanding and correcting performance problems.

Normally, changes in lens voltages produce very large differences in response. If a plot of response versus applied voltage is not sharp, it is likely that the lens system is dirty. The lens system should be removed, carefully disassembled, cleaned with a solvent such as acetone, dried thoroughly in air, reassembled, and reinstalled. A channel electron multiplier has a typically useful life of about 2000 hours if not exposed to high count rates. Thus, the detector is typically changed annually.

Vacuum pump oil should be checked regularly to ensure that adequate levels are maintained. Most commercial instruments now use turbomolecular pumps that require little maintenance. Older instruments, however, may be equipped with oil diffusion pumps; the oil should be changed about every 6 months. Other older instruments use cryogenic pumps; these systems must be thawed regularly.

Most quadrupoles in modern instruments are very stable. As a precaution, however, the analyst should examine mass spectra daily to ensure that the centroids of mass peaks occur within 0.1 dalton of the true value; if not, mass calibration should be performed. Older styles of quadrupoles are not as stable as current versions, so analysts working with such instruments may have to perform mass calibration on a daily basis. Fortunately, the procedure is simple and rapid and may be performed while the instrument is warming up.

Vacuum integrity is critical to performance. The area requiring most maintenance is the seal between the sampling cone and the mass spectrometer. Other connections, especially O-rings, may fail from time to time. The log of operating pressure provides guidance concerning the need to examine and replace faulty O-rings.

Sampler and skimmer cones have a strong influence on sensitivity. Although such parts made from platinum are very expensive initially, they are much less prone to destruction by acid and other sample properties, so their extra lifetime usually makes them cost-effective. Aluminum samplers and skimmers usually exhibit acceptable lifetimes in many applications and are much less costly than platinum. Nickel is also a good choice for samplers and skimmers.

As with any instrument, the more knowledge one possesses and the more care one takes with preventive maintenance, the smaller the amount of unscheduled downtime and expensive repairs.

Suggested Readings

Books

HOLLAND, G., AND A. N. EATON, EDS., *Applications of Plasma Source Mass Spectrometry*. Cambridge, U.K.: Royal Society of Chemistry, 1991.

JARVIS, K. E., A. L. GRAY, AND R. S. HOUK, EDS., *Handbook of Inductively Coupled Plasma Mass Spectrometry*. Glasgow: Blackie, 1992.

Jarvis, K. E., and others, eds., *Plasma Source Mass Spectrometry*. Cambridge, U.K.: Royal Society of Chemistry, 1990.

Montaser, A., and D. W. Golightly, eds., *Inductively Coupled Plasmas in Analytical Atomic Spectrometry*, 2nd ed. New York: VCH, 1992.

Review Articles

Recent progress in ICP-MS was surveyed biannually from 1988 to 1992 in the section on Atomic Mass Spectrometry in the *Analytical Chemistry* Fundamental Reviews issue. The most recent such review is by D. Koppenaal, *Analytical Chemistry*, 64 (1992), 302R. The November 1994 issue of *Applied Spectroscopy* included a review of atomic mass spectrometry, including ICP-MS, by M. W. Blades. Advances in ICP-MS are also reviewed regularly in the *Journal of Applied Analytical Spectrometry*. The most recent such review is by J. G. Williams and others, *Journal of Analytical Atomic Spectroscopy*, 10 (1995), 253R.

Periodicals

Applied Spectroscopy

Atomic Spectroscopy

ICP Information Newsletter

Journal of Analytical Atomic Spectroscopy

Spectrochimica Acta, Part B

References

1. R. C. Hutton and others, *Journal of Analytical Atomic Spectroscopy*, 5 (1990), 463.

2. M. A. Vaughan and G. Horlick, *Applied Spectroscopy*, 40 (1986), 434.

3. S. H. Tan and G. Horlick, *Applied Spectroscopy*, 40 (1986), 445.

4. J. W. H. Lam and G. Horlick, *Spectrochimica Acta* 45B (1990), 1313.

5. J. W. H. Lam and J. W. McLaren, *Journal of Analytical Atomic Spectroscopy*, 5 (1990), 419.

6. F. G. Smith, D. R. Wiederin, and R. S. Houk, *Analytical Chemistry*, 63 (1991), 1458.

7. E. H. Evans and L. Ebdon, *Journal of Analytical Atomic Spectroscopy*, 4 (1989), 299.

8. J. J. Thompson and R. S. Houk, *Applied Spectroscopy*, 41 (1987), 801.

9. M. E. Ketterer, M. J. Peters, and P. J. Tisdale, *Journal of Analytical Atomic Spectroscopy*, 6 (1991), 439.

10. K. G. Heumann, *Chem. Anal. (N.Y.)*, 95 (1988), 301.

11. P. W. J. M. Boumans, *Theory of Spectral Excitation* (London: Hilger and Watts, 1966).

12. M. A. Vaughan, G. Horlick, and S. H. Tan, *Journal of Analytical Atomic Spectroscopy*, 2 (1987), 765.

13. R. S. Houk and others, *Analytical Chemistry*, 52 (1980), 2283.

14. N. Bradshaw, E. F. H. Hall, and N. E. Sanderson, *Journal of Analytical Atomic Spectroscopy*, 4 (1989), 801.

15. A. R. Warren and others, *Applied Spectroscopy*, 48 (1994), 1360.

16. D. P. Myers and others, *J. Am. Soc. Mass Spectrom.*, 6 (1995), 920.

17. H. Suyani and others, *J. Chromatogr. Sci.*, 27 (1989), 139.

18. N. S. Chong and R. S. Houk, *Applied Spectroscopy*, 41 (1987), 66.

19. D. A. Wilson and others, *Spectrochimica Acta*, 42B (1987), 29.

20. E. R. Denoyer, K. J. Fredeen, and J. W. Hager, *Analytical Chemistry*, 63 (1991), 445A.

21. H. Baumann and J. Pavel, *Mikrochim Acta*, 3 (1989), 413.

22. I. B. Brenner and H. E. Taylor, *Crit. Rev. Anal. Chem.*, 23 (1992), 355.

23. S. K. Aggarwal and others, *Crit. Rev. Clin. Lab. Sci.*, 31 (1994), 35.

24. G. L. Beck and O. T. Farmer, *Journal of Analytical Atomic Spectroscopy*, 3 (1988), 771.

25. U.S. EPA Methods 6020 (1995) and 200.8 (1994).

Chapter 23

Atomic Fluorescence Spectrometry

David J. Butcher

Western Carolina University
Department of Chemistry and Physics

Summary

General Uses

- Atomic fluorescence spectrometry (AFS) is used to determine the concentration levels of elements in samples (elemental analysis).
- For favorable elements (cadmium, lead, thallium), detection limits in the attogram range have been obtained; linear calibration curves extend 4 to 7 decades.
- Combined with chromatography, atomic fluorescence can provide qualitative and quantitative information on the chemical form of elements (metal speciation) in a sample.

Common Applications

- Determination of mercury at low part-per-trillion levels (1 ng/L) in environmental samples with commercial instrumentation
- Continuous monitoring of mercury in air to levels as low as 10 pg with commercial instrumentation
- Determination of arsenic, selenium, antimony, and tellurium in environmental samples, with detection limits between 10 and 50 ng/L, using commercial instrumentation
- Determination of femtogram (10^{-15} g) quantities of elements in samples by graphite furnace laser-excited atomic fluorescence spectrometry

Samples

State

Solids, liquids, and gases can be analyzed by atomic fluorescence, although most samples are converted to liquids before analysis.

Amount

Generally 0.1 to 1 g of a solid sample is dissolved to a volume of 50 to 250 mL. With the commercial AFS instrumentation, approximately 10 to 50 mL of solution is used in a continuous flow mode, 100 to 500 μL in a flow injection mode. In the commercial air analyzer for mercury samples, between 1 and 5 L of gas at a flow rate of 1 L/min is used.

Preparation

Solids are generally converted to liquids by a dissolution procedure involving acid digestion or fusion with an ionic solid (such as lithium metaborate). Limited work has been done for the direct analysis of solids by procedures that avoid a dissolution method. Liquids can be analyzed directly. Analysis of gases requires concentration of the analyte onto a solid material (for example, the adsorption of mercury to gold) from which it is released (as by the application of heat).

Analysis Time

The time for sample preparation varies considerably, depending on the sample and method, from several minutes to 24 hr or more. Analysis time depends on the instrumentation and method of calibration used; typically 15 min to 1 hr is required for calibration and sample determination.

Limitations

- AFS requires combination with chromatography to provide information regarding the chemical form of the analyte.
- Sample preparation, particularly for solids, is often the most time-consuming step.
- Sample analysis may involve chemical reactions between the analyte and other components of the sample (matrix) that may reduce the number of free atoms produced (chemical interferences).
- For the analysis of samples, it may be necessary to correct for spurious (non–analyte-induced) light (spectral interferences).
- In general, the technique is limited to the determination of metals and metalloids.

Complementary or Related Techniques

- Atomic emission and atomic absorption spectrometry are other techniques that involve electronic (valence electron) transitions of gaseous atoms.
- Inductively coupled plasma mass spectrometry is a sensitive, multielement technique that uses a plasma to produce ions that are analyzed by a mass spectrometer.
- X-ray fluorescence involves inner-shell (nonvalence electron) transitions of atoms; often it can be performed on solid materials with little sample preparation.
- Neutron activation analysis involves activation of nuclei with neutrons, followed by the emission of gamma rays whose wavelength (or frequency) is characteristic of the nucleus.

Introduction

Atomic fluorescence spectrometry (AFS) is the most recently developed of the basic analytical atomic spectrometric techniques for the determination of concentration levels of elements in samples. Like its complementary techniques of atomic absorption spectrometry (AAS) and atomic emission spectrometry (AES), AFS involves the conversion of samples to gaseous atoms. The analyte is determined by monitoring the interaction of light with its valence electrons at a wavelength that is unique to that element (at least in theory). The resulting changes in the energy levels of the electrons are called electronic transitions. AFS can be distinguished from AAS and AES because AFS involves both radiative excitation (absorption) and de-excitation (fluorescence). For analytical applications of AFS, radiative excitation and de-excitation processes are used in the ultraviolet/visible (UV/Vis) region of the electromagnetic spectrum between 180 and 800 nm.

AFS instrumentation is often characterized by the type of light source used for excitation. Conventional, or nonlaser, AFS instruments generally provide sensitivity comparable to that of AAS and AES for most elements, although AFS sensitivity may be slightly better or worse for selected elements. Consequently, conventional AFS has not demonstrated sufficient advantages compared to well-established AAS and AES techniques to stimulate widespread commercial development of instrumentation. Elements that can be determined very sensitively by AFS compared to AAS and AES include mercury, antimony, arsenic, selenium, and tellurium. Recently, commercial instrumentation has been developed for these elements, with detection limits between 1 and 50 ng/L.

Laser-excited atomic fluorescence spectrometry (LEAFS) has been shown to provide extremely high sensitivity, particularly when combined with a graphite furnace atom cell (GFLEAFS). Detection limits for most elements have been in the attogram (10^{-18} g) to femtogram (10^{-15} g) range, which is comparable to or better than that of other analytical techniques. Despite its high sensitivity, commercial GFLEAFS instrumentation has not thus far been developed because of the complexity and poor reliability of the laser systems used to date. In addition, LEAFS is truly a single-element technique, which is a significant disadvantage compared to multielemental techniques such as inductively coupled plasma mass spectrometry, which has comparable sensitivity for many elements.

This chapter provides an overview of the theory of AFS. Commercially available AFS instruments are described, as well as instrumentation for GFLEAFS. Several quantitative applications of AFS are also discussed.

How It Works

Physical and Chemical Principles

The basic principles of AFS are reviewed extensively elsewhere (1–3). Here a qualitative description of the AFS theory is presented, including discussions of AFS transitions and the effect of source irradiance on the fluorescence signal.

AFS Transitions

Commonly used spectroscopic transitions for analytical applications of atomic fluorescence are illustrated in Fig. 23.1 (3). In each case, photons provided by a light source are absorbed by the analyte to excite the atoms from a lower energy level (usually the ground state) to a higher energy level, followed by fluorescence of a photon to a lower energy level. In addition, nonradiative excitation and deactivation processes are included in some AFS transitions.

For most elements, the most intense AFS transitions are resonance transitions, in which the wavelength of the absorbed radiation is equal to the wavelength of the fluorescence radiation (Fig. 23.1(a)). Resonance transitions have been widely used for AFS with conventional sources because of their intensity. A disadvantage of these transitions, which is most significant for the analysis of complex matrices (such as blood, urine, and soil), is that light scattered off spectrometer compo-

Figure 23.1 Transitions for AFS. (a) Resonance, (b) nonresonance, and (c) two-step excitation (double resonance). The ground state is represented by 0; states 1, 2, and 3 represent excited states. Radiative transitions are represented by solid arrows ($\uparrow$ = absorption; $\downarrow$ = fluorescence); nonradiative transitions are represented by dashed arrows.

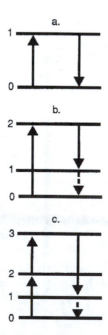

nents, or the sample matrix, cannot be distinguished from the fluorescence signal by the detection system because both are at the same wavelength.

Nonresonance transitions, such as Stokes direct-line fluorescence, which is illustrated in Fig. 21.1(b), involve absorption and fluorescent photons of different energies. Nonradiative processes are also involved in the activation or deactivation process; in Stokes direct-line fluorescence, light is absorbed to excite the atoms from the ground state (0) to an excited state (2), followed by fluorescence to an intermediate excited state (1) and subsequent nonradiative decay. Although the total amount of fluorescence produced by nonresonance transitions is generally lower than that produced by resonance transitions, scattered light can be separated from fluorescence light in the former by use of a wavelength selection device, and provides a better signal-to-noise ratio. Nonresonance transitions are widely used for GFLEAFS because they provide a convenient way to remove high levels of scattered laser radiation.

Two-step excitation, or double resonance fluorescence (Fig. 23.1(c)), involves the simultaneous use of two dye lasers to promote the analyte from the ground state (0) to an excited state (2), followed by immediate activation to an even higher energy level (3). Fluorescence is subsequently produced by de-excitation from the higher energy level to a third, lower-energy excited state (1). Two-step excitation is extremely sensitive, particularly for elements that do not have sensitive nonresonance transitions (such as cadmium), but its disadvantages include the expense of a second dye laser and the difficulty of aligning two laser beams.

Influence of Light Source Intensity on the AFS Signal

A logarithm/logarithm graph of the fluorescence signal versus source irradiance for a fixed analyte concentration is called a saturation curve (Fig. 23.2) (2). At relatively low source intensities, as would be obtained with a conventional (nonlaser) light source, the fluorescence signal is directly proportional to the source irradiance. This is called the linear region of the saturation curve. Consequently, a significant amount of research has been done to develop high-intensity conventional light sources for AFS to maximize the technique's sensitivity.

Lasers have sufficient energy to cause the rates of excitation (absorption) and de-excitation (fluorescence) to be equal, which is a condition called optical saturation (Fig. 23.2) (1), which is

Figure 23.2 Saturation curve for AFS. Note that above a certain source intensity, the analyte fluorescence signal becomes independent of source intensity.

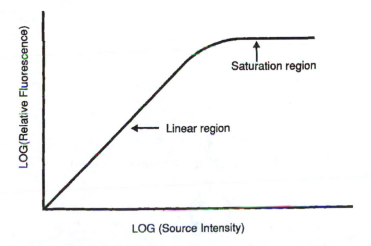

LOG(Relative Fluorescence)

Saturation region

Linear region

LOG (Source Intensity)

advantageous for AFS. Under these conditions, the maximum fluorescence signal is obtained for a given analyte concentration. In addition, saturation of an atom cell eliminates reduction of the fluorescence signal by collisional deactivation of excited atoms (quenching). It is generally advantageous to provide sufficient laser energy to just saturate an atomic transition to minimize the amount of scattered laser radiation produced.

Instrumentation

Instrumentation for atomic fluorescence consists of a light source, atom cell, wavelength selection device, and detection system (Fig. 23.3). The traditional arrangement for fluorescence collection involves orientation of the detection system at 90° to the direction of light source. This collection scheme, which is used in commercial AFS instrumentation, allows relatively easy instrumental alignment.

Light Sources

A wide variety of conventional (nonlaser) light sources have been used for AFS, including mercury discharge lamps, hollow cathode lamps, electrodeless discharge lamps, and xenon arc lamps. As discussed above, in order to maximize the sensitivity of AFS, it is desirable to use a high-intensity light source. Commercial AFS mercury analyzers use mercury discharge lamps because of their high spectral irradiance, reliability, and commercial availability. Conventional hollow cathode lamps (HCLs) are not sufficiently intense to achieve low detection limits by AFS, but the recent commercial production of boosted discharge hollow cathode lamps (BDHCLs), which are 10 to 100 times more intense than conventional HCLs, has provided a good conventional source for AFS. BDHCLs are used in a commercial AFS instrument for the determination of antimony, arsenic, selenium, and tellurium.

Modern laser systems for AFS are composed of three major components: a pump laser, a dye laser, and a frequency-doubling system (2). The pump laser emits high-energy light pulses of a single wavelength that are focused on a dye in the dye laser to produce tunable radiation between 380

Figure 23.3 Instrumentation for AFS.

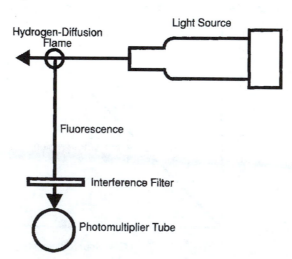

and 800 nm. In order to produce ultraviolet light between 200 and 380 nm, the wavelength of the dye laser light is halved with a frequency-doubling unit. Despite the advantages of laser excitation in terms of sensitivity, lasers have not been widely used for AFS because of their high cost and difficulty of operation.

Atom Cells

A variety of atom cells have been used for AFS, including cold vapor cells (for mercury), flames, furnaces, plasmas, and glow discharge devices (2–7). Here discussion is limited to atom cells used in commercial AFS instrumentation and GFLEAFS.

Cold Vapor Cells for Mercury Determination Cold vapor determination of mercury involves the conversion of dissolved mercury to elemental mercury by reaction with tin(II) chloride reagent (5, 6). The evolved mercury is then transported by a flow of gas to a quartz cell, which serves as the atom cell for an atomic fluorescence spectrometer. Advantages of this procedure include its simple design, low cost, and high sensitivity.

Hydride Generation Hydride generation (HG) (7–9) has been commonly used for the determination of the hydride-forming elements, which include antimony, arsenic, selenium, and tellurium. The sample or standard is treated with sodium borohydride and hydrochloric acid to produce a volatile hydride that is carried into an atom cell with a stream of inert gas. The atom cell is either an argon–hydrogen diffusion flame or a heated quartz cell, which decomposes the hydride into free atoms. The former approach is used in the commercial atomic fluorescence spectrometer (5, 8, 9).

Hydride generation is widely used for atomic absorption, and its advantages and disadvantages are well characterized (7). Hydride generation provides better sensitivity for these elements than conventional flame-techniques (AAS, AES, and AFS) because the efficiency of analyte transport by HG approaches 100%, but conventional pneumatic nebulization has an efficiency of approximately 10%. In addition, determination of arsenic and selenium is insensitive in a conventional air–acetylene flame because these elements' analytical wavelengths are located near the vacuum UV region (193.7 and 196.0 nm), where significant absorption of source light by air and flame gases occurs. However, this technique is susceptible to interferences, which are discussed in detail below.

Graphite Furnace LEAFS Graphite furnaces are widely used as atom cells for atomic absorption because of their higher sensitivity compared to flame atomization. Modern commercial atomic absorption graphite tube furnaces have been shown to reduce vapor phase interferences, which were common in open atomizers for involatile elements and the analysis of complex sample matrices. Early GFLEAFS work included open furnaces (such as cups, rods, or filaments), probably because it is relatively easy to collect fluorescence at a right angle (2). However, these atom cells have been shown to be impractical for most analyses because of vapor phase interferences, and modern work uses atomic absorption tube furnaces.

In order to collect fluorescence from an unmodified graphite tube, the preferred collection geometry involves the use of front surface illumination (2), in which fluorescence is collected at 180° to the direction of the laser beam (Fig. 23.4). (Fluorescence collection at 90° requires additional holes in the furnace to allow passage of the laser beam.) A mirror, which contains a hole for passage of the laser beam, is placed between the light source and the graphite furnace and oriented at 45° to the direction of the beam. The laser light passes through the hole in the mirror and through the graphite tube for excitation. Fluorescence is collected by the mirror and directed to a detection system. Front surface illumination allows the high sensitivity of LEAFS to be combined with modern AAS furnace technology that has been shown to minimize interferences.

Figure 23.4 Front surface illumination geometry for GFLEAFS.

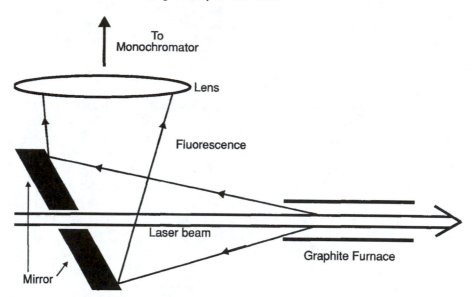

Detection Systems

Most laboratory-constructed detection systems for AFS include a monochromator as a wavelength selector (dispersive detection system). Recent work with the commercial hydride generation system (5, 6, 9) and graphite furnace LEAFS (2) demonstrates that optimum sensitivity is obtained with non-dispersive detection, in which a narrow-band filter is used to reject stray light. Nondispersive detection has the disadvantage that each element requires its own filter to obtain the maximum sensitivity.

Most AFS instrumentation uses a conventional photomultiplier tube (PMT) as the detector because of its sensitivity over a wide range of wavelengths. The commercial hydride generation atomic fluorescence spectrometry (HGAFS) instrument uses a solar-blind PMT (5, 9), which is extremely sensitive to wavelengths shorter than 280 nm and insensitive to wavelengths longer than 320 nm. This is desirable for the determination of As, Sb, Se, and Te, which have analytical wavelengths between 190 and 220 nm.

Preconcentration Methods for Mercury

Preconcentration methods are required in order to measure sub–part-per-trillion concentration levels of mercury (10). One procedure involves the use of gold-impregnated sand as a collection trap for mercury, which is adsorbed onto the surface. In order to determine the adsorbed mercury, the trap is heated to 500 °C to release the analyte.

Commercial AFS Instrumentation

Automated commercial AFS instrumentation for determination of mercury and the hydride-forming elements has been developed (5, 6, 9). Although different spectrometers are used, these systems use the same autosampler and continuous-flow sample introduction system (Fig. 23.5). From an autosampler, samples, standards, and reagents (tin(II) chloride for mercury, sodium borohydride for hydride elements) are delivered by a peristaltic pump into a mixing chamber. An

Figure 23.5 Schematic diagram of the commercially available continuous-flow vapor/hydride generation system for AFS. The dotted line represents the flow of the blank solution.

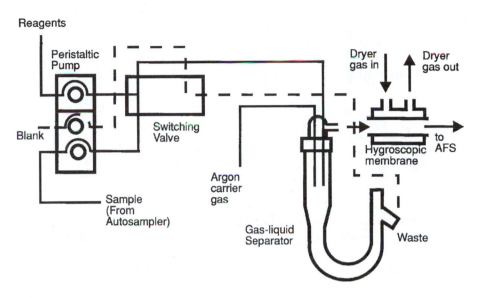

electronic switching valve allows alternation between the sample and blank. Elemental mercury or the volatile hydride is carried to a gas/liquid separator, where a stream of argon carries the analyte through a membrane dryer tube to remove moisture (11) and into the atom cell (cold vapor cell for mercury, argon–hydrogen diffusion flame for hydride elements). Instrument and data control are performed by software using an IBM-compatible computer.

Corns and colleagues (12) described the use of a flow-injection module as a substitute for the continuous-flow system for the determination of mercury. Sample volumes of 75, 100, or 200 µL were injected into a reagent stream. After the sample and reagents were mixed, the mercury was isolated with the gas/liquid separator, dried, and carried into the atom cell. Flow injection gave higher concentration detection limits than the continuous-flow system, although it was more sensitive on an absolute (mass) basis. In addition, flow injection uses smaller reagent volumes, and consequently lower disposal costs, than the continuous flow system. More work must be done with flow injection with commercial AFS instrumentation to characterize its advantages over the continuous-flow approach, particularly for the analysis of small sample volumes.

What It Does

Accuracy

The accuracy of atomic fluorescence analyses, like those of its complementary techniques of atomic absorption and atomic emission, depends primarily on the ability of the instrumentation to minimize interferences produced by the sample matrix (that is, components of the sample that are not the analyte). For AFS, the most significant types include chemical interferences, which involve a reduction in the percentage of gaseous analyte atoms produced in the atom cell to below 100% (that is, formation of molecular species containing the analyte) and spectral interferences, which involve the collection of light by the detection system that is not analyte atomic fluorescence (such as scattered light). Chemical interferences depend on the method of sample introduction and the atom cell used, not on the spectroscopic method used. A large volume of literature is available regarding chemical interferences present in atomic absorption, compared to the relatively few applications of AFS, and hence generalizations will be made based on AAS.

Chemical Interferences

For mercury analysis, a potential chemical interference involves the presence of species that inhibit formation of elemental mercury in the vapor generation procedure (6, 13), such as gold, platinum, selenium, and tellurium. For most samples, the relatively low concentration levels of these elements mean that few chemical interferences are encountered for cold vapor mercury generation. Chemical interferences in the cold vapor cell are probably negligible because the unique volatility of mercury minimizes interferences. For atomic fluorescence analysis, the use of argon as a carrier gas is preferable to nitrogen or air because it does not induce quenching (12).

Atomic absorption work with hydride generation indicates that chemical interferences are most significant in the vapor generation step, in which a number of transition metal ions have been shown to suppress formation of the volatile hydrides (7, 13). In addition, Bye (14) reports that accurate analysis by hydride generation requires quantitative conversion of the analyte to a particular oxidation state. For example, sodium borohydride converts selenium (IV) to its hydride, but selenium(VI) is unreactive with this reagent. In addition, Smith (15) reports the presence of chemical interferences in hydrogen diffusion flames caused by other volatile hydrides. The same interferences reported for atomic absorption would be expected to be present for atomic fluorescence analysis. In general, the use of hydride generation is relatively easy for simple matrices (such as water), but for complex matrices (such as blood, urine, and geological materials), dissolution and hydride generation protocols should be verified by the use of standard reference materials to verify their accuracy (7, 16, 17).

The use of modern graphite furnaces has greatly reduced chemical interferences in AAS, and consequently LEAFS with a modern graphite furnace would be expected to have relatively few interferences. The limited number of analyses done by GFLEAFS seems to confirm this generalization.

Spectral Interferences

Types of spectral interferences for AFS include scattered light, emission of light by the atom cell, nonanalyte atomic fluorescence, and molecular fluorescence (2). In general, scatter and atom cell emission are the most significant for most AFS work. Interferences caused by nonanalyte atomic fluorescence are relatively rare because the emission from line sources is sufficiently narrow to prevent absorption by other atoms. Consequently, low-resolution wavelength selectors can be used for AFS. This is in contrast to atomic emission, where a high-resolution monochromator is required to minimize spectral interferences. Molecular fluorescence is also relatively uncommon, particularly with nonresonance or two-step fluorescence, because a molecule would have to both absorb and fluoresce at the same wavelengths as the analyte in order to interfere, and such a coincidence is highly unlikely.

Spectral interferences are expected to be negligible for cold vapor determination of mercury. At ambient temperature, there is no atom cell emission, and only elemental mercury (and carrier gas) would reach the atom cell, eliminating scatter, nonanalyte atomic fluorescence, and molecular fluorescence (6, 13).

For the determination of the hydride-forming elements, the relatively cool hydrogen-diffusion flame has relatively low background emission levels, but may be unable to break down relatively involatile particles and small molecules, causing scatter and molecular fluorescence. In general, spectral interferences from hydride-generation are expected to be rare compared to the chemical interferences described above (7, 13).

The major spectral interferences for GFLEAFS are scatter and atom cell emission, with occasional reports of nonanalyte and molecular fluorescence, but, in general, the background levels are much lower than those for GFAAS (1). In order to perform accurate analyses by GFAAS, a method of background correction is necessary to account for relatively large background levels. Zeeman background correction, which uses a magnetic field to induce shifts in atomic energy levels and distinguish between signal and background, is generally accepted as the most accurate method for graphite furnace AAS (18). A modified atomic absorption Zeeman graphite furnace was used for graphite furnace LEAFS for the determination of lead and cobalt (19). Based on its success for atomic absorption, Zeeman background correction is expected to provide accurate analyses by GFLEAFS. However, background signals are relatively small for GFLEAFS, and a number of accurate analyses have been performed without background correction (1). More research must be done to determine whether background correction is required for GFLEAFS.

Precision

The precision of AFS depends on the method of sample introduction and the atom cell used. Precision reported for cold vapor determination of mercury by AFS was 2% for 90 measurements of a 1-μg/L standard (11), which is comparable to values obtained by AAS.

The precision of HGAFS is expected to be similar to that of HGAAS (7, 13), with typical values between 1 and 5%. In the presence of chemical interferences, the precision may be degraded to 10% in their presence. As discussed above, the accuracy of sample preparation procedures in AFS analyses must be verified to prevent interferences and obtain optimal precision.

In early GFLEAFS work, relatively poor precision was obtained because of the low laser repetition rates (less than 50 Hz) and poor laser pulse-to-pulse reproducibility (10 to 20%) (1). However, with modern laser systems and graphite furnaces, its precision is comparable to that of graphite furnace AAS, with values between 5 and 10% for dissolved samples.

Detection Limits

A comparison of detection limits for cold vapor determination of mercury and for hydride generation determination of antimony, arsenic, selenium, and tellurium by atomic fluorescence and atomic absorption is shown in Table 23.1. For mercury determination, the AFS detection limit is 20 times lower than that of atomic absorption using the same instrument. Consequently, atomic fluorescence is the method of choice for its determination at the low part-per-trillion level.

A direct comparison of the sensitivities of atomic absorption and atomic fluorescence for hydride generation is more difficult, primarily because different instruments and sample introduction procedures have been used. The hydride generation AAS (HGAAS) instrument uses a batch procedure in which a discrete volume of sample is reacted with discrete volumes of reagents (20). The atomic fluorescence spectrometer is used in conjunction with the continuous-flow system described above (8, 9). The HGAFS detection limits are 2 to 15 times lower than the HGAAS detection limits, although concentration detection limits can be improved for either technique by the use of larger sample volumes.

Table 23.1 Comparison of AAS and AFS for the determination of mercury, antimony, arsenic, selenium, and tellurium with commercial instrumentation.

| Element | Atomic Absorption | | Atomic Fluorescence | |
	Detection Limit (ng/L)	Sample Introduction System	Detection Limit (ng/L)	Sample Introduction System
Hg	20*	Continuous flow	1*	Continuous flow
As	50[†]	Batch	20[‡]	Continuous flow
Sb	250[†]	Batch	50[‡]	Continuous flow
Se	150[†]	Batch	10[‡]	Continuous flow
Te	100[†]	Batch	50[‡]	Continuous flow

*Measured on a commercial mercury analyzer equipped for absorption or fluorescence detection (8).
[†]Measured on a commercial AAS equipped with a hydride generation system using a 10-mL sample volume (20).
[‡]Measured on a commercial AFS equipped with a hydride generation system using a 10-mL sample volume (8).

The high sensitivity of GFLEAFS is illustrated in Table 23.2, in which its best detection limits are compared to those of the two most sensitive atomic spectrometric techniques, graphite furnace atomic absorption spectrometry (GFAAS) and inductively coupled plasma mass spectrometry (ICP-MS). In general, GFLEAFS is more sensitive than GFAAS by two to four orders of magnitude. For most elements, GFLEAFS detection limits are one to two orders of magnitude lower than those of ICP-MS. However, in spite of the higher sensitivity of GFLEAFS, GFAAS and ICP-MS are commercially available and much simpler to operate and maintain. ICP-MS has the added advantage of being a multielemental technique. Consequently, at the present time, GFLEAFS has been limited to a handful of research laboratories.

Analytical Information

Atomic fluorescence has been almost exclusively employed for quantitative analysis, so this section focuses on this application.

Quantitative

In order to use AFS for quantitative analysis, an instrument is generally standardized by use of a calibration curve, which is a logarithm/logarithm graph of the atomic fluorescence signal versus analyte concentration (Fig. 23.6) (3). Least-squares analysis is performed on these data to obtain the most representative linear relationship between fluorescence and concentration. The fluorescence intensity for a sample is then measured, and the analyte concentration in this sample is calculated from the linear relationship.

The slope of an AFS calibration curve across the linear range ideally has a value of 1. Compared to AAS, AFS has longer linear regions of its calibration curves (10^3 to 10^5, or 3 to 5 orders of

Table 23.2 Comparison of GFLEAFS, GFAAS, and ICP-MS detection limits.

Element	GFLEAFS		GFAAS (fg)	ICP-MS (fg)
	Absolute (fg)	**Concentration (ng/L)**		
Ag	10	0.5	500	5
Al	100	5	4,000	15
Au	10	0.2	10,000	5
Cd	0.5	0.01	300	12
Co	4	0.08	2,000	5
Fe	70	3.5	2,000	580
Ga	1	0.05	40,000	4
Hg (21)	90	9	25,000	—
In	2	0.01	9,000	2
Mn	100	5	1,000	6
P (22)	8,000	400	5,500,000	20,000
Pb	0.2	0.01	5,000	50
Sb	10	0.5	20,000	12
Sn	30	1.5	20,000	10
Te	20	1	10,000	10
Tl	0.1	0.005	10,000	3
Yb	220	11	4,000	—

Detection limits were taken from Reference 1 except as noted.

Figure 23.6 Typical calibration curves for AFS. (a) Continuum source and (b) line source.

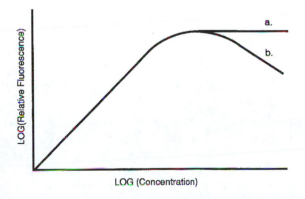

magnitude, compared to 10^2 to 10^3). Graphite furnace LEAFS has especially long linear ranges that may extend across seven orders of magnitude. At concentrations above the linear region of the calibration curve, the fluorescence signal will either reach a maximum level for a continuum source (such as a xenon arc lamp) or decrease for a line source (such as a hollow cathode lamp) (Fig. 23.6). The nonlinearity of calibration curves at high concentrations is caused by analyte-induced attenuation of the light source in front of the volume from which fluorescence is collected, or the absorption of fluorescence by analyte atoms. A more quantitative discussion of the nonlinearity of AFS calibration curves is given elsewhere (3).

Applications

Butcher et al. (1–3) have tabulated applications of AFS for real sample analysis; here some representative examples are discussed.

1. Speciation of Mercury in Environmental Samples.

In natural waters, microbial processes cause the methylation of mercury(II) to methylmercury (the predominant organic form) and dimethylmercury. The determination of individual species is significant because the toxicity of mercury depends on its chemical form. Consequently, a significant amount of research has been done to develop high-sensitivity methods of mercury speciation. As described above, AFS is well-suited for these analyses because of its sensitivity, although a preconcentration step is required to determine mercury at levels below 1 ng/L (1 part per trillion).

Saouter and Blattman (10) recently described a method for the determination of total and organic mercury with a commercial AFS spectrometer. The procedure for total mercury determination involved oxidation of all mercury species to mercury(II), subsequent reduction to elemental mercury, and concentration on gold-coated sand. The sand was heated rapidly to 550 °C to release the mercury, which was transferred to a cold vapor cell for AFS detection. Mercury speciation was performed by derivatization with sodium tetraethylborate and preconcentration on graphitized carbon. The carbon was heated to 350 °C to release the mercury derivatives into a gas chromatography (GC) column. An atomic fluorescence spectrometer served as the GC detector. Detection limits of 0.07 ng/L for total mercury and 0.05 ng/L for methylmercury were obtained. Analysis times for total and organic mercury were 5 and 10 min, respectively. Methylmercury was determined in a lobster-tissue standard reference material (SRM), and good agreement with the certified value was reported.

2. Determination of Mercury in Air.

Stockwell and colleagues (23) described a commercially available AFS instrument to monitor mercury in air or natural gas. A pump was used to introduce gas samples onto a gold-coated sand trap to concentrate the analyte. The trap was heated to 500 °C to release the mercury, and a flow of argon was used to carry the analyte to a cold vapor cell from which fluorescence was detected. Typically, 5 L of air were sampled at a flow rate of 1 L/min, and an absolute detection limit of 10 pg was reported.

3. Determination of Arsenic and Selenium by Hydride Generation AFS.

The commercial hydride generation AFS instrument was used for the determination of arsenic and selenium in water standard reference materials (9). In order to ensure complete volatilization of the analytes, arsenic(V) was reduced to arsenic(III) with iodide, and selenium(VI) was converted to selenium(IV) with hydrochloric acid. Good accuracy was reported for these analyses.

4. Determination of Cadmium by Vapor Generation AFS.

Although cadmium does not form a volatile hydride, reaction with sodium tetraethylborate produces a volatile cadmium compound, presumed to be diethylcadmium, which decomposes in an argon–hydrogen diffusion flame to elemental cadmium. Ebdon and colleagues (24) used this vapor generation procedure to determine cadmium with a commercially available continuous-flow system and atomic fluorescence spectrometer. A cadmium metal vapor lamp was used as the source, and a 228-nm bandpass filter was used for wavelength isolation. A detection limit of 23 ng/L was obtained, with a linear dynamic range of 3.5 orders of magnitude. Interferences were reported from transition metal ions (such as nickel(II) and copper(II)), but they were alleviated by the introduction of citric acid to mask their effects. Cadmium was determined in water and sewage sludge SRMs, and good accuracy was obtained.

5. Determination of Elements in Nickel-based Alloys by GFLEAFS.

Nickel-based alloys, used in aircraft engines, have been analyzed for a number of elements by GFLEAFS. Irwin and colleagues (25) accurately determined lead and thallium in nickel-based alloy SRMs by use of direct solid sampling, in which the sample is placed in the furnace in the form of a chip and analyte atomization occurs directly from the solid sample. Advantages of direct solid sampling include increased sensitivity (because no dilution is required in a dissolution procedure), reduced risk of contamination (because no dissolution reagents are required), and decreased sample preparation time. The GFLEAFS analyses were compared to previous solid-sampling analyses done by GFAAS. Lead is present at relatively high levels in these samples, and in order to determine lead by GFAAS, it was necessary to use an internal gas flow through the furnace to reduce the absorption signal sufficiently to be on the relatively small linear portion of the calibration curve. An internal gas flow is disadvantageous for graphite furnace analysis because it may increase chemical interferences (25). Conversely, the linear dynamic range of GFLEAFS (10^5 to 10^7) was sufficiently long to allow lead determination in these samples without an internal gas flow. A second advantage of GFLEAFS over GFAAS is the virtual absence of spectral background produced by the sample matrix in the former technique, whereas GFAAS backgrounds were approximately the same size as the analytical signal. Zeeman background correction was used for GFAAS, but no background correction was used for the GFLEAFS work.

Liang and colleagues (22) determined phosphorous in nickel-based alloy SRMs and samples by GFLEAFS. They were unable to accurately determine phosphorous in the SRMs with direct solid sampling because of the analyte's involatility in this matrix, but good agreement with the certified values was obtained after dissolution of the samples. The dissolution procedure was carried out in a microwave oven with nitric acid (2 mL), hydrofluoric acid (0.5 mL), and water (2 mL), followed by dilution to volume. The analytical precision of analyses was between 5 and 18%. A char step of 900 °C was required to remove a spectral interference caused by nitrogen monoxide, which was produced by the decomposition of nitric acid. The amount of phosphorous introduced into the graphite furnace for these analyses was approximately 2 ng, which is below the detection limit of GFAAS (Table 23.2). Consequently, a preconcentration step would be required to do these analyses by GFAAS.

Liang and colleagues (26) investigated the determination of tellurium and antimony in nickel-based alloy SRMs by GFLEAFS with solid and dissolution sampling. For tellurium, good agreement with the certified values was obtained with both methods of sample introduction. The relative standard deviations of the measurements were 13% and 9% for the solid sampling and dissolution methods, respectively, which were comparable to the precision obtained by GFAAS. Low recovery (50 to 60%) was obtained for the determination of antimony by solid sampling GFAAS, which was attributed to an involatile chemical form of antimony in the alloys. However, good agreement with certified values was obtained for the determination of antimony using the dissolution procedure described above, with an analytical precision of 7%. As was observed for phosphorous, the introduction of nitric acid produced a nitrogen monoxide background. A char step at 800 °C for 30 sec was used to remove this spectral interference.

Nuts and Bolts

Relative Costs

Commercial atomic fluorescence spectrometer (without vapor or hydride generation system)	$6–$12K
Commercial automated AFS mercury analyzer (includes autosampler, vapor generator, software, and computer)	$25K
Commercial automated AFS mercury analyzer for air samples	$39K
Commercial automated AFS hydride-element analyzer (includes autosampler, vapor generator, software, and computer)	$32K
GFLEAFS instrument assembled from components (laser system, graphite furnace, detection system, and computer)	$100–$250K

Commercial atomic fluorescence spectrometers, without a vapor or hydride generation system, are available at costs between $6,000 and $12,000. Fully automated commercial instrumentation for mercury or hydride elements can be purchased for $25,000 to $40,000. Components to construct GFLEAFS instrumentation are very expensive, with typical costs over $100,000. The high cost is primarily due to the complexity of a tunable laser system.

Vendors for Instruments and Accessories

Aurora Instruments
191 W. 6th Ave.
Vancouver, B.C. Canada V5Y 1K3
phone: 604-874-0227
fax: 604-874-0167
email: aurora@portal.ca

Brooks Rand, Ltd.
3950 Sixth Ave. NW
Seattle, WA 98107
phone: 206-632-6206
fax: 206-632-6017

PS Analytical
Arthur House
Sevenoaks UK TN15 6QY
phone: 732 763 416
U.S. supplier:
Questron
4044 Quakerbridge Rd.
Mercerville, NJ 08619
phone: 609-587-6898

Required Level of Training

Operation of a commercial AFS instrument for routine analysis can be performed by people with an associate college degree. Method development and troubleshooting require a bachelor's degree. Construction and operation of a GFLEAFS instrument requires at least a bachelor's degree and experience with operating a tunable laser system.

Service and Maintenance

Maintenance contracts for commercial AFS instrumentation may be purchased for approximately 10% of its price per year. The laser systems for GFLEAFS require considerable maintenance, which can often be done by the operator in consultation with the laser manufacturer, but annual maintenance costs may exceed 10% of the purchase price even with user-performed maintenance.

Suggested Readings

Articles

BUTCHER, D. J., AND OTHERS, "Conventional Source Excited Atomic Fluorescence Spectrometry," *Progress in Analytical Atomic Spectroscopy*, 10 (1987), 359–506.

BUTCHER, D. J., AND OTHERS, "Laser Excited Atomic Fluorescence Spectrometry in Flames, Plasmas, and Electrothermal Atomizers: A Review," *Journal of Analytical Atomic Spectrometry*, 3 (1988), 1059–78.

BUTCHER, D. J., "Atomic Fluorescence Spectrometry," *Spectroscopy*, 8, no. 2 (1993), 14–19.

JACKSON, K. W., AND G. CHEN, "Atomic Absorption, Atomic Emission, and Flame Emission Spectrometry," *Analytical Chemistry*, 68 (1996), 231R.

STOCKWELL, P. B., AND A. C. GRILLO, "Applications of a Mercury-Vapor Atomic Fluorescence Detector," *Spectroscopy*, 6, no. 7 (1991), 39–41.

References

1. D. J. Butcher, "Laser Excited Atomic and Molecular Fluorescence in a Graphite Furnace," in J. Sneddon, ed., *Advances in Atomic Spectroscopy*, Vol. 2 (Greenwich, CT: JAI Press, 1994).

2. D. J. Butcher and others, *Journal of Analytical Atomic Spectrometry*, 3 (1988), 1059–78.

3. D. J. Butcher and others, *Progress in Analytical Atomic Spectroscopy*, 10 (1987), 359–506.

4. A. D. Campbell, *Pure and Applied Chemistry*, 64 (1992), 227–44.

5. P. B. Stockwell and W .T. Corns, *Journal of Automatic Chemistry*, 15 (1993), 79–84.

6. R. G. Michel, "Atomic Fluorescence Spectrometry," in J. F. Riordan and B. L. Vallee, eds., *Metallobiochemistry*, Part A, Methods in Enzymology, vol. 158 (San Diego: Academic Press, 1988), pp. 222–43.

7. T. Nakahara, *Progress in Analytical Atomic Spectroscopy*, 6 (1983), 163–223.

8. P.S. Analytical, Commercial Literature, Kent, United Kingdom.

9. W. T. Corns and others, *Journal of Analytical Atomic Spectrometry*, 8 (1993), 71–7.

10. E. Saouter and B. Blattman, *Analytical Chemistry*, 66 (1994), 2031–37.

11. W. T. Corns and others, *Analyst*, 117 (1992), 717–20.

12. W. T. Corns and others, *Journal of Automatic Chemistry*, 13 (1991), 267–71.

13. P. J. Potts, *A Handbook of Silicate Rock Analysis* (New York: Chapman and Hall, 1987).

14. R. Bye, *Talanta*, 37 (1990), 1029–30.

15. A. E. Smith, *Analyst*, 100 (1975), 300–6.

16. B. Welz and P. Stauss, *Spectrochimica Acta, Part B*, 48B (1993), 951–76.

17. B. Welz, M. S. Wolynetz, and M. Verlinden, *Pure and Applied Chemistry*, 59 (1987), 927–36.

18. W. Slavin and G. R. Carnrick, *CRC Critical Reviews in Analytical Chemistry*, 19 (1988), 95–134.

19. R. L. Irwin and others, *Spectrochimica Acta, Part B*, 47B (1992), 1497–1515.

20. B. Welz and M. Schubert-Jacobs, *Atomic Spectroscopy*, 12 (1991), 91–104.

21. W. Resto and others, *Spectrochimica Acta, Part B*, 48B (1993), 627–32.

22. Z. Liang and others, *Journal of Analytical Atomic Spectrometry*, 7 (1992), 1019–28.

23. P. B. Stockwell, P. Rabl, and M. Paffrath, *Process Control and Quality*, 1 (1991), 293–8.

24. L. Ebdon and others, *Journal of Analytical Atomic Spectrometry*, 8 (1993), 723–9.

25. R. L. Irwin and others, *Journal of Analytical Atomic Spectrometry*, 5 (1990), 603–10.

26. Z. Liang, R. F. Lonardo, and R. G. Michel, *Spectrochimica Acta, Part B*, 48B (1993), 7–23.

Chapter 24

X-Ray Fluorescence Spectrometry

George J. Havrilla
Los Alamos National Laboratory

Summary

General Uses

- Qualitative identification and quantitative determination of elemental composition of a variety of samples for solids and liquids. Minimal sample preparation, wide dynamic range, and nondestructive methodology make X-ray fluorescence (XRF) the method of choice for many industrial analyses.
- Determination of sulfur in diesel fuel to meet environmental regulations.
- Control of lube oil additive concentrations with both off-line and on-line applications.
- Quality control and customer support for catalyst manufacture and plant use.
- Process control of steel and cement production.
- Determination of surface contamination in semiconductor production.
- Support of mineralogical and geological exploration and waste site field evaluation.
- Sorting of metal alloys.
- Forensic applications in evaluating evidence.
- Coating thickness and composition process monitoring for paper and metals industries.

Samples

State

Any solid or liquid sample can be analyzed. Hazardous materials require special sample cells to prevent instrument contamination.

Amount

Sample sizes range from micrograms to gram quantities depending on the methodology used.

Preparation

- Preparation is determined by analytical needs such as the level of accuracy and precision required.
- For simple analyses, no preparation is necessary; analyze as received, with rapid turnaround for qualitative and semiquantitative results.
- For minimal pretreatment, grind, homogenize, and press into pellets for semiquantitative and quantitative results.
- For maximum pretreatment, dry, fuse, and cast into glass disks; this provides the highest accuracy and precision for quantitative results.

Analysis Time

A qualitative spectrum on an energy-dispersive XRF (EDXRF) instrument can be obtained within 5 min to provide gross elemental composition. Quantitative programs for high accuracy and precision on a sequential wavelength-dispersive XRF (WDXRF) instrument can take up to 30 min for 15 elements, whereas a simultaneous instrument can do the same measurement within 5 minutes. This is just instrument time and does not take into account the level of sample preparation that might be required.

Limitations

General

- Elemental range is limited to boron and up. Detection of light elements ($Z < 11$) limited to solids and essentially surface composition.
- Matrix interferences can prevent or limit detection of some elements.
- Standards for quantitative analysis do not always match unknown matrix.

Accuracy

Depends on sample preparation and how well standards match the unknown matrix. In many situations, accuracy of less than 1% error can be achieved with proper care and attention to details.

Sensitivity and Detection Limits

This varies with element and sample matrix. In general, parts-per-million detection limits can be achieved for nominal matrices such as catalysts, steels, soils, petroleum products, and geological materials for transition row elements and above. The lighter elements (Z < 19) have detection limits that are typically higher, ranging from high tens of ppm to high hundreds of ppm for Mg and Na. Synchrotron radiation, total reflection XRF, and new methodologies for microsampling are pushing detection limits well below the parts-per-million level.

Complementary or Related Techniques

- Inductively coupled plasma (ICP) emission and mass spectrometry techniques provide greater routine sensitivity (parts-per-billion) but require all samples be dissolved (extensive sample preparation).

- Glow discharge mass spectrometry has similar sensitivity to ICP methods but can handle solid samples. It requires matrix matched standards for quantitative work.

- Atomic absorption is a single-element flame or furnace-based method that is simpler and cheaper than XRF yet labor-intensive for more than one or two elements.

- Atomic fluorescence compares with furnace AAS for sensitivity, but commercial instrumentation is limited and it requires similar sample preparation to ICP and AAS.

- Activation analysis has high sensitivity but requires nuclear reactor for activation. It is good for specialized applications.

Introduction

X-ray fluorescence (XRF) spectrometry is an atomic spectrometric method based on the detection of emitted X-ray radiation from excited atoms (Fig. 24.1). This technique is a two-step process that begins with the removal of an inner shell electron of an atom. The resulting vacancy is filled by an outer shell electron. The second step is the transition from the outer shell electron orbital to an inner shell electron orbital. The transition is accompanied by an emission of an X-ray photon. The fluorescent photon is characteristic of the element and is equal to the difference in energy between the two electron energy levels. Because the energy difference is always the same for given energy levels, the element can be identified by measuring the energy of the emitted photon. In turn, the intensity of the emitted photons determines the concentration of the element. Therefore the measure of the photon energy provides the identification of the element and the intensity of the photon emission provides a measure of the amount of the element.

This emission process is similar to other fluorescent measurement techniques, but it is restricted to the X-ray region of the electromagnetic spectrum that ranges from 0.1 to over 120 keV, or 11 to 0.1 nm. The typical X-ray analytical region is less than 50 keV. The photon energies detected are designated as K, L, or M X-rays, depending on the energy level being filled; for example, a K shell vacancy filled by an L level electron results in the emission of a K_α X-ray, as shown in Fig. 24.1. There are as many possible X-ray lines as there are inner shell electrons. However, the most analytically useful and most intense X-ray lines are the K shell electrons for elements from boron

Figure 24.1 X-ray fluorescence process and energy level diagram. An incident X-ray photon removes an inner shell electron. The vacancy is filled by an outer shell electron, which gives up an X-ray photon. The emitted X-ray photon is characteristic of the element. The energy or wavelength of the photon is determined by the energy gap between the two energy levels of the electron filling the inner shell vacancy.

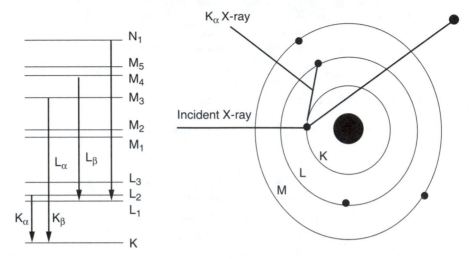

through cerium, whereas the L and some M lines are used for the remainder of the periodic table. Although the multitude of emitted X-ray lines could make for complex spectra, the relative low intensities of the lines below the L level allow for clear spectra with a minimum of interferences.

Overall, the analytical capability of X-ray fluorescence has been used since 1922 (1) to cover a wide dynamic range from trace levels of ng/mL to major composition as high as 100 wt%. The accuracy and precision of this method are unrivaled when standards are matrix matched to unknowns. This is evident in numerous industries where X-ray fluorescence is used to monitor product composition and ensure tight specification tolerances. The popularity of XRF is also a result of the limited sample preparation that is usually required. This affords the analyst fast turnaround for composition analyses of large volumes of valuable product.

X-ray fluorescence affords the analyst a wide variety of opportunities to answer the perennial question, "What is it?" The classic scenario is to determine the elemental composition of a complete unknown. This can be handled while the client waits, using energy-dispersive X-ray fluorescence (EDXRF). All elements from sodium through uranium can be detected simultaneously. Depending on the matrix and the element of interest, detection limits range from tens to hundreds of parts per million. The next step in this scenario is usually when the client asks, "About how much is present?" This again can be handled with EDXRF or, if the analyte is known, with wavelength-dispersive X-ray fluorescence (WDXRF). In this case, an assortment of equations and algorithms have been developed to determine elemental composition without the use of standards. This is a rapid way to obtain semiquantitative information on the sample composition. The final question in this process is "Exactly how much is present?" This is where WDXRF is most useful. Highly accurate and precise analyses are routinely achieved when calibration standards match the unknowns. This is why the metal, cement, oil, and petrochemical industries rely so heavily on XRF for composition analyses of product batches.

Although XRF is regarded as a mature technique, new developments in instrumental capabilities are pushing the limits of conventional XRF technology to lighter elements and lower limits of detection. In addition, recent developments, including synchrotron XRF, total reflection XRF, and X-ray microfluorescence, are changing the capabilities of XRF within analytical chemistry. These new developments are just part of XRF's potential in solving today's analytical problems.

How It Works

The basic principle of XRF is emission and detection of X rays. The energy or wavelength of the emitted X ray determines the element and the intensity of the X-ray emission defines the concentration of that element. There are two different ways in which the X rays are detected: energy-dispersive X-ray spectrometers and wavelength-dispersive X-ray spectrometers. Each technique has fundamental differences that provide some intrinsic advantages and disadvantages. Figure 24.2 illustrates the schematics of each approach.

In both cases, the process begins with excitation and a number of sources are used. The source of excitation also differentiates the various X-ray techniques. The most conventional and primary excitation sources are X-ray tubes. This distinguishes XRF from X-ray microanalysis in scanning electron microscopy (SEM), where electrons are used for excitation. The basic difference between the two methods is the depth of penetration of the excitation source and subsequent emission. The tube excitation source typically penetrates microns to millimeters within the sample, whereas electron excitation is limited to less than a few microns of specimen depth. X-ray tube excitation results in essentially a bulk determination with XRF, versus surface or spatially restricted composition determinations in SEM. Particle-induced X-ray emission (PIXE) uses protons or other heavy particles to induce X-ray emission. Radioisotopes are also used as excitation sources in XRF but their intensity is several orders of magnitude weaker than tube excitation. These sources are convenient for portable instrumentation and have limited sensitivity because of flux limitations. The last source type is synchrotron radiation (SR). The attractive features of SR include intensity, polarization, and collimation. The major drawback of SR is the limited access to the facility.

Figure 24.2 Schematic diagrams of wavelength- and energy-dispersive X-ray fluorescence instruments. (a) The wavelength-dispersive instrument relies on diffraction crystals to separate the X rays from the sample. (b) The energy-dispersive system has a solid-state SiLi detector that converts X-ray photons into pulses that are processed electronically.

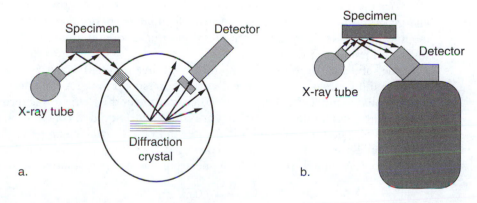

The dispersion of the emitted X rays allows the different X-ray energies to be measured and the intensities of each X-ray photon determined. This is where the two different measurement systems arise: wavelength-dispersive and energy-dispersive. In WDXRF, the emitted X rays are dispersed based on their wavelength using diffraction, as shown in Fig. 24.2. The planes of a crystal are used to disperse the emitted X-ray photons from the specimen based on Bragg's law:

$$n\lambda = 2d \sin\theta \tag{24.1}$$

where n is an integer, λ is the wavelength of the photon, d is the lattice spacing, and θ is the angle of incidence of the radiation. The d values of the typical crystals used in WDXRF range from 0.14 to 8 nm. The smaller d values are for natural crystals such as lithium fluoride and germanium. The larger d values are obtained with synthetic crystals. These are typically used for light element analysis; larger d values are used for longer wavelengths. The WDXRF system is a sequential measurement, where the instrument must step through the 2θ values of the goniometer (Fig. 24.2). This provides for higher spectral resolution and better sensitivity than with the EDXRF system. A variation of the sequential WDXRF is the simultaneous WDXRF instrument. This instrument has detectors at fixed 2θ values. This allows for rapid, sensitive analysis and is most often found in critical process control systems such as steel and cement plants.

In contrast to the sequential nature of WDXRF, the energy-dispersive system collects all the X-ray photons simultaneously onto the detector. Each photon generates an electrical pulse with an amplitude that is proportional to the energy of the X-ray photon. Further electronic processing involves amplification and analysis by a multichannel analyzer. The EDXRF system has the advantage of detecting all of the elements simultaneously, which means that analyses are rapid and unexpected elements are not missed. The major drawback, however, is that the overall resolution of the EDXRF system is not as good as the WDXRF. Consequently, there are tradeoffs in selecting which dispersive system is best for particular applications.

Three basic detectors are used in XRF instrumentation: gas ionization, scintillation, and solid-state semiconductors. The first two detectors are found in WDXRF systems. Gas ionization detectors consist of two electrodes, a wire anode in the center of a metal cylinder cathode, and a filler gas that is an argon (90%) and methane (10%) mixture (P10 gas). In this type of detector the X-ray photon enters through a window in the cylinder and ionizes the gas. The resulting ions and electrons are collected and the current is proportional to the intensity of the X-ray photon. These detectors are used for longer X-ray wavelengths of the lighter elements ($Z < 27$). There are both sealed and flow-proportional counters, with the flow-proportional counters used in most commercial instrumentation.

The scintillation detector consists of a thallium-doped sodium iodide crystal on the front of a photomultiplier tube. The X rays from the specimen strike the sodium iodide crystal and generate photons that illuminate the photocathode of the photomultiplier. The photons generate photoelectrons that are amplified and detected. The number of photons produced is proportional to the energy of the X rays. This detector is used for higher-energy X rays (short wavelength) for elements with $Z > 25$. Typically, both the proportional counter and scintillation detectors are used for the midrange elements $25 < Z > 35$.

The solid state detector is a lithium-drifted silicon (Si(Li)) wafer. The X ray strikes the Si(Li) detector and generates a series of pulses that correspond to the X-ray energy. The pulse height is proportional to the X-ray energy. The concentration of the element is determined by counting the pulses. The detector and preamplifier are cooled with liquid nitrogen to minimize current noise. The issue of spectral resolution is illustrated in Fig. 24.3, which compares the spectrum of a 316-grade stainless steel specimen obtained with WDXRF (Fig. 24.3(a)) and EDXRF (Fig. 24.3(b)) instruments. Although the WDXRF spectrum has better resolution, it took almost 30 min to

Figure 24.3 Analysis of 316-grade stainless steel. (a) Wavelength-dispersive spectrum shows the major and minor components. Note the identification of the manganese peak at around 53° 2θ. (b) Energy-dispersive spectrum. The same elements are shown in this spectrum, but the manganese peak is buried beneath the chromium and iron peaks. The difference in intensity is due to the overall longer integration time for the EDXRF spectrum; the WDXRF spectrum spent only several seconds at each point of the spectrum.

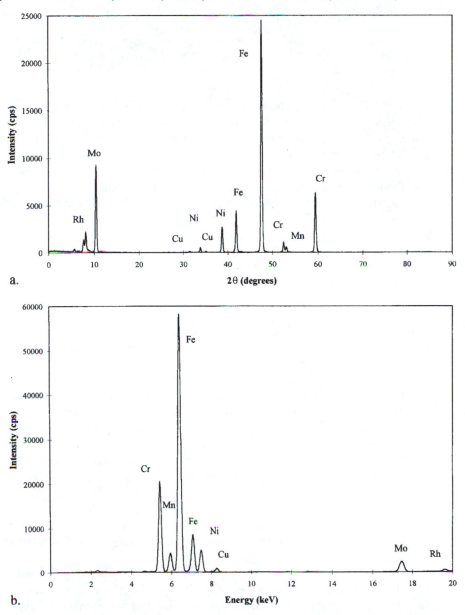

acquire, whereas the EDXRF spectrum took only 5 min. In addition, the manganese peaks are overlapped by the chromium and iron peaks in the EDXRF spectrum, yet resolved in the WDXRF spectrum. This illustrates the need to determine the best approach, arriving at a compromise among resolution, sensitivity, speed, and cost.

What It Does

Elemental X rays can be obtained from any material, both solids and liquids. XRF is a bulk method that collects X-ray emission from tens of microns to centimeters within a sample. The way a sample is treated and the information that can be obtained from the specimen dictate how the X-ray information from the specimen is interpreted. Depending on the analytical requirements, different sample preparation approaches result in different levels of analytical information. The analytes and the matrices determine the difficulty of sample preparation.

Liquids are straightforward in that they can usually be analyzed directly without much pretreatment. As long as the liquid is compatible with the sample cup and sealing film, there is not much effort in sample preparation other than pouring the sample into the cup and placing it into the instrument. The challenge in XRF is striking a balance between the perceived analytical requirements of the client and the constraints of time, money, and instrument capabilities. Liquids are usually easier to handle, particularly in terms of calibration standards, because these are homogeneous and easier to prepare.

Solids present unique opportunities where the art of XRF sample preparation is developed. Within the context of solids sample preparation, heterogeneity and interferences become important issues that must be resolved. Again, depending on the analytes and matrix involved, the issues of analyte sensitivity and interferences can be troublesome. A number of options are available in dealing with solids. One of the advantages of XRF is that solid samples can be analyzed directly and nondestructively. This can provide the client with critical information on the sample in a relatively short time. In this case the tradeoff is in accuracy and precision. However, the no-standard software that is available with commercial instrumentation usually provides reasonable numbers for unique sample unknowns.

The next level in solid sample preparation involves grinding the sample into a powder that is either analyzed as a loose powder or pressed into a pellet. The difference between the two approaches again involves the analytical needs of the client. The loose powder provides an increased level of homogeneity over the as-received solid. Pressing the powder into a pellet raises the homogeneity level and increases the uniformity of the unknown specimen with the calibration standards. In each case, as the level of sample preparation effort increases the accuracy and precision of the analysis increases as well.

The last level of solid sample preparation is fusion, where the ground sample is melted into a glass bead. This approach provides a high degree of uniformity and aids in reducing matrix interferences. This typically provides the highest level of accuracy and precision. It also is the most time-consuming and demanding of all the sample preparation approaches in XRF.

In addition to solids and liquids, thin films can be determined by XRF, providing both composition and thickness information. Instrumentation has been developed to determine elemental concentrations in surfaces based on total reflection of the X rays off the surface of a specimen (2). This approach involves a collimated beam from an X-ray tube that is directed onto the sample below the critical angle. This results in total reflection of the X rays off the surface and excitation of only a few nanometers of the specimen surface. An energy-dispersive detector is placed at 90° to detect the X-ray emission from the surface. The primary requirement for total reflection X-ray fluorescence (TXRF) is that the surface of the sample must be optically flat. This limits the application of the method to solids and dried solutions on special sample supports. Total reflection X-ray fluorescence has been embraced by the semiconductor industry for surface analysis. Specialized instruments have been developed for dedicated use in silicon wafer analysis by several vendors. Thin film analysis using XRF can be done using conventional WDXRF and EDXRF instrumentation. In many cases, films as thin as several angstroms can be measured by XRF, with appropriate standards providing both elemental and thickness measurements.

Spatially resolved X-ray fluorescence historically has been accomplished using scanning electron microscopy. Following demonstrations with capillary and apertured X-ray tubes, advances in small-spot X-ray tubes brought the routine use of X-ray microfluorescence (XRMF) into the analytical laboratory (3–6). The analytical potential of XRMF still must be developed. However, XRMF fills the gap between high-resolution scanning electron microprobe analysis and bulk elemental XRF. The primary advantages are rapid analysis, nondestructive determinations, and little or no sample preparation. X-ray spot size for commercial instrumentation is typically 50 to 100 μm, with capillary systems achieving 10 μm and smaller.

Synchrotron radiation XRF is a powerful method that affords trace element detection that rivals ICP analysis (7–8). This is due to the intensity of the synchrotron source. The radiation is generated by charged particles moving in a particle accelerator. In addition to being high-intensity sources, synchrotrons also provide tunable radiation, which enhances sensitivity for selected elements. Further improvements in the signal-to-noise ratio can be achieved by using the polarization of the X-rays to further reduce background signals. Although synchrotron XRF has sensitivity advantages over conventional XRF, difficulty in achieving access to a synchrotron limits its wide application except in certain circumstances.

Particle induced X-ray emission (PIXE) uses a beam of heavy charged particles to irradiate the sample (9). The signal-generation process is essentially the same as in conventional X-ray fluorescence except for the particles that induce the electron shell vacancies. Although there are some differences that govern the signal-generation process, PIXE has better sensitivity in the transition and heavy-element section of the periodic table than conventional XRF because of its lower background. However, ready access to a particle accelerator is limited, so PIXE is not widely used except in specialized cases.

Analytical Information

Qualitative

One of the most important uses of XRF is its capability to provide rapid, real-time qualitative elemental analysis on virtually any type of solid or liquid. The qualitative capabilities are fully realized using EDXRF spectrometers, which can provide a complete spectrum of 0 to 40 keV in a matter of minutes. The spectrum can provide identification of both expected and, most importantly, unexpected elements. Elemental identification can be confirmed by the presence of both K_α and K_β lines and, for the heavy elements, the L_α and L_β lines. When XRF is used in conjunction with X-ray diffraction and molecular spectroscopy, such as infrared and Raman, phase and compound identification are also possible.

Many issues can be resolved on the basis of qualitative information alone, such as in comparative analysis of a "good" and "bad" sample and identification of the ubiquitous unknown from a process line or reaction kettle bottom. The simple knowledge of the presence or absence of an element can be important to a process chemist in solving a process problem.

Quantitative

Determination of the amount of an element in a sample by XRF is usually based on a linear relationship between the emitted X-ray intensity and the concentration of the element. This is expressed by the simple equation

$$I = mC + b \qquad (24.2)$$

where I is the intensity of the emitted X ray, m is the slope of the line, b is the background, and C is the concentration. This simple relationship forms the basis of XRF quantitative elemental analysis. This is the ideal an analyst strives to attain in the face of complications such as matrix effects (multiplicative interferences) that can either enhance or suppress the analyte X-ray intensity.

In cases where the matrix is known and the standards closely match the unknowns, accuracy error of less than 1% can be routinely achieved. Interelement effects can be accounted for by determining appropriate calibration coefficients. Numerous approaches are available to accommodate any type of interelement effect and are common in the vendor-supplied software that accompanies modern X-ray spectrometers. Some of the more common matrix correction models involve intensity (Lucas–Tooth), concentration (LaChance–Traill), and secondary fluorescence (Rasberry–Heinrich) corrections. In each case, Eq. (24.2) is expanded to

$$C_i^x = (A_i + B_i I_i)(1 + \Sigma_j \alpha_{ij} \Delta C_j) \qquad (24.3)$$

where A = the reciprocal of the slope; B = the background; C_i^x = concentration of the unknown; A_i = reciprocal of the slope (constant for element i); B_i = background intensity (constant for element i); I_i = measured intensity of element i; α_{ij} = alpha coefficient (absorption coefficient of element j on element i); ΔC_j = average composition of sample. The second term of Eq. 24.3 defines the matrix effects and the coefficients of the ΔC_j term vary with the model used to define them.

Several factors underlie the capability of XRF to achieve such a high degree of accuracy. First of all, the sample preparation is critical in that both standards and unknowns must be handled in the same manner. In many cases, the sample preparation may be used to mitigate some of the matrix effects by attenuating the interfering element or by compensating by using internal standards, spiking, or double dilution. Next, the stability of the instrumentation is important because the emission intensity of the analyte is the basis of quantitative analysis. Commercial instrumentation typically has excellent long- and short-term stability to allow users to achieve analytical accuracy and precision of less than 1%. The ultimate accuracy and precision of an XRF analysis depends on the analytical needs of the client.

Applications

1. Quantitative Determination of 14 Elements in a Fluid Cracking Catalyst (FCC).

High accuracy and precision in quantitative analysis are demonstrated in the determination of the elemental composition of FCCs. These catalysts are used throughout the petroleum industry to refine crude oil into a variety of petroleum-based products. The advantage of this method is that major, minor, and trace elements can be determined simultaneously without the need for multiple analyses of the same sample or repeat analysis of dilutions. The key to the high level of precision is consistent sample preparation of both standards and unknowns. The procedure is as follows:

1. Heat 3 g of the FCC sample to 800 °C for 1 hour and allow to cool.
2. Weigh 1 g sample and mix with 4.2 g lithium tetraborate and 1.8 g lithium metaborate.
3. Add 200 μL saturated lithium bromide (6 g/25 mL).
4. Fuse the entire mixture and cast in platinum crucibles.

The instrument is calibrated by preparing high-purity metal oxide standards in the manner described above. Because there are 14 elements in the quantitative program, at least 32 standards should be prepared. It is important that the oxides be blended without any correlation among the elements. The results of this sample preparation provide precision on the order of less than 1%. The accuracy is demonstrated in Table 24.1. This method can provide routine results with 1-day turnaround and charge time of around 1 hr.

Table 24.1 Comparison of two laboratories analyzing the same FCC catalyst using different fusion procedures.

	wt%						
	Si	Al	La	Ce	Pr	Nd	Sm
Lab A	32.2	13.1	0.267	0.171	0.030	0.131	0.003
Lab B	31.9	14.3	0.175	0.044	0.121	—	<0.01
Precision of Lab A	±0.5	±0.4	±0.008	±0.003	±0.003	±0.004	±0.002

	wt%						
	Na	Mg	Ca	Ti	Fe	V	Ni
Lab A	0.168	0.011	0.040	0.622	0.355	0.007	0.002
Lab B	0.24	—	0.027	0.632	0.361	<0.01	<0.01
Precision of Lab A	±0.018	±0.008	±0.002	±0.011	±0.007	±0.001	±0.003

Table 24.2 Comparison of no-standard calculation of tubing composition with known elemental composition of specified and suspected tubing.

	wt%				
	Cr	**Fe**	**W**	**Mo**	**Ni**
Failed tubing	19	71	—	0.2	9
Hastelloy C	16	5	4	16	60
304-grade stainless steel	19	70	—	—	9

2. Identification of Metal Alloys Using EDXRF No-Standard Software.

The use of no-standard software provides rapid semiquantitative and, in some cases, reasonable quantitative analyses of unknown samples. This particular example illustrates the capability of no-standard software. Some new metal tubing installed in a pilot plant operation was failing. The tubing specified for the installation was Hastelloy C, which has a high nickel and molybdenum content. The failed tubing was analyzed as received, without any sample preparation. The resulting spectrum was background-subtracted and the net intensities were used by the no-standard software to calculate the composition. The calculated values for the major elements are given in Table 24.2.

The objective of the analysis is to provide the client with rapid identification of the metal tubing. The no-standard result identifies the tubing as 304-grade stainless steel, not the specified Hastelloy C. Another approach would be to base the comparison just on the qualitative results and net intensities if a sample of the specified tubing is available. In this case the client is informed that the wrong tubing was installed within 20 min of receipt of the sample.

3. Quantitative Analysis of Lube Oil Additive Elements.

Similar to the quantitative analysis of the FCC catalysts, liquids can also be analyzed with high precision and accuracy. In this case, the sample preparation is relatively simple and the application of routine analysis allows many samples to be analyzed per hour. Calibration of the instrument consists of blending known organometallic compounds in a pure oil base. Table 24.3 provides a comparison of the analysis of the additives with inductively coupled plasma atomic emission

Table 24.3 Comparison of ICP-AES and XRF values for four different additive samples.

	wt%					
	Ca, ICP	**Ca, XRF**	**Zn, ICP**	**Zn, XRF**	**S, ICP**	**S, XRF**
A	0.186	0.185	0.147	0.159	0.557	0.556
B	0.190	0.187	0.151	0.159	0.547	0.551
C	0.208	0.209	0.157	0.169	0.661	0.654
D	0.202	0.200	0.154	0.163	0.156	0.146

Table 24.4 Precision of XRF determination for given elements.

	wt%			
	Zn	**Ca**	**S**	**P**
Average	0.1492	0.1468	0.5293	0.1235
Precision	±0.0007	±0.0005	±0.003	±0.0009

spectrometry (ICP-AES) and Table 24.4 gives the precision of the XRF analysis over nine measurements (three triplicates on three different days).

4. X-Ray Microfluorescence.

There are several ways in which XRMF can be applied in the analytical laboratory: point analysis, line scan, and elemental map. Each one provides unique information about the sample and can be used in a variety of ways to solve problems. The following three examples demonstrate each of the features identified above.

Point Analysis

Localized analysis with a 100-μm aperture is used to identify the components in the material. Figure 24.4 shows the major elements identified in the spectrum, which include Al, Mn, Fe, and Zr. Peaks were also identified for the following impurities: V, Cr, Ni, Cu, Sn, and Ag. The spectrum

Figure 24.4 EDXRF point spectrum of metal alloy using 100-μm aperture. Both major and low-level impurities were detected.

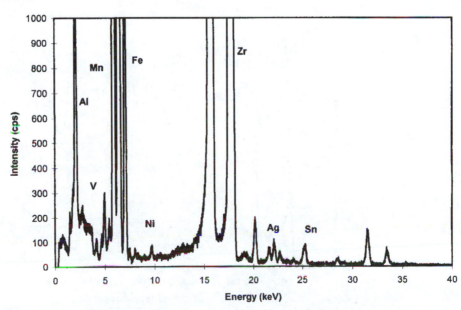

has been enhanced to show the lower-level impurity peaks. These impurities were surprising because these elements were unexpected and indicate that there is a problem with the process. The spectrum was acquired within 15 min of receipt of the sample.

Linescan

The linescan is done by starting at one edge of the specimen and stepping the X-ray beam across the width of the specimen. The data shown in Fig. 24.5 illustrate just three of the seven simultaneous linescans that can be acquired with a commercial instrument. The iron exhibits some apparent inhomogeneity even though it is one of the major constituents. The vanadium clearly is an impurity due to the spikes throughout the linescan. The vanadium spikes appear to correlate with peaks in the zirconium linescan and valleys in the iron linescan. This apparent correlation can help identify the source of the contamination. This information is confirmed by the elemental maps. The linescan involves several hours of data acquisition and data processing.

Elemental Maps

The elemental maps are merely multiple linescans averaged over several frames to increase sensitivity of detection. Figure 24.6 illustrates four of the major elements and contaminants. In this gray-scale image, the lighter the color, the higher the concentration of the element. Therefore, the white spots on the vanadium, nickel, and zirconium maps indicate higher concentrations of those elements, or hot spots. The iron map has dark areas that correspond to the hot spots in the other elemental maps. The important aspect of the elemental maps is that they were acquired over an area of 7 mm^2. This allows macroscale mapping of microscale impurities. The combination of all three sets of information provides a more precise picture of the sample. In addition to

Figure 24.5 Linescan obtained by moving the X-ray beam across the surface of the specimen in Fig. 24.4. This illustrates the heterogeneous nature of the alloy and indicates an apparent correlation between the zirconium and vanadium.

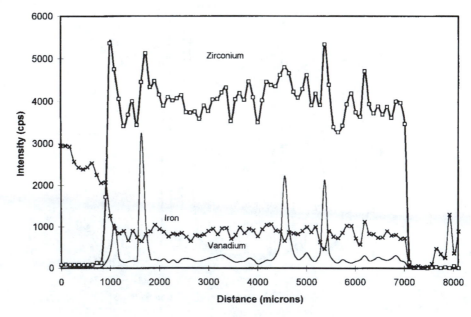

Figure 24.6 Elemental maps of (a) iron, (b) vanadium, (c) nickel, and (d) zirconium showing the extent of the impurities in the sample. The gray scale uses white as higher concentration and black as low concentration. The white areas in the maps are spots of high concentration of the particular element.

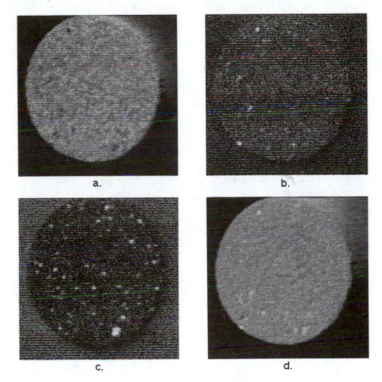

the detection of the impurities, an apparent correlation of some of the impurities with a major component has been discovered. This could not have been determined based only on the point analysis.

Numerous X-ray fluorescence applications are discussed in the literature. The few applications discussed here demonstrate some of the basic capabilities of X-ray fluorescence in solving analytical problems, not just providing numbers.

Nuts and Bolts

Relative Costs

The costs of X-ray spectrometers vary widely, depending on what type of analysis is required. Several factors affect the cost of the instrument:

- Simultaneous or sequential detection of analytes
- Detection limits

- Accuracy and precision
- Software

The most expensive commercial units are simultaneous WDXRF instruments designed for rapid and sensitive quantitative analysis for process control. These instruments are typically found in steel and cement plants, where tight control of the product specifications requires rapid and accurate information. The basic instrument is typically around $150K to $200K and each additional channel (element) is another $10K to $20K. When analyses require 10 to 20 elements as well as automated sample-handling systems and control interfaces, costs on the order of $400K to $500K are not unreasonable. Basic sequential WDXRF systems range from $150K to over $250K without many bells and whistles. Sample changers, helium flush for liquids, specialized specimen cups, and other accessories increase the purchase price accordingly.

Energy-dispersive systems are considerably less expensive, but the basic tradeoff is less sensitivity with EDXRF. Computerized benchtop EDXRF systems can be purchased for under $100K, but they are limited in performance. There are even radioisotope models that start at $30K for specific analytes. These systems are usually packaged for quick quantitative analyses in plant lab environments for single-element determinations. The more flexible and higher-performance EDXRF instruments start around $80K. When accessories and other essentials are added the usual cost is between $150K and $200K.

Vendors for Instruments and Accessories

Angstrom, Inc. (standards, presses, and fusion apparatus)
12890 Haggerty Rd.
Belleville, MI 48112
phone: 313-697-8058
fax: 313-697-3544

Applied Research Laboratories (wavelength-dispersive XRF)
55 Cherry Hill Dr.
Beverly, MA 07915
phone: 508-524-1000, 800-999-5011
fax: 508-524-1100

ASOMA Instruments Inc. (portable, on-line, benchtop equipment, and X-ray spectrometers)
11675 Jollyville Rd.
Austin, TX 78759
phone: 512-258-6608, 800-580-6608
fax: 512-331-9123
email: asoma@industry.net
Internet: http://www.industry.net/asoma

Baird Analytical (energy-dispersive XRF)
A Division of Imo Industries, Inc.
125 Middlesex Turnpike
Bedford, MA 01730-1468
phone: 617-276-6163
fax: 617-276-6510

Chemplex Industries, Inc. (standards, accessories, and equipment for XRF analysis)
160 Marbledale Rd.
Tuckahoe, NY 10707
phone: 914-337-4200, 800-424-3675
fax: 914-337-0160

Cianflone Scientific Instruments Corporation (portable, on-line XRF spectrometers)
228 RIDC Park West Dr.
Pittsburgh, PA 15275
phone: 412-787-3600, 800-569-9400
fax: 412-787-5022

Diano Corporation (wavelength-dispersive XRF)
30 Commerce Way
Woburn, MA 01801
phone: 617-935-4310
fax: 617-935-7953

EDAX International (energy-dispersive XRF)
91 McKee Dr.
Mahwah, NJ 07430
phone: 201-529-4880, 800-535-3329
fax: 201-529-3156

HNU Systems (portable and benchtop XRF)
160 Charlemont St.
Newton Highlands, MA 02161
phone: 617-964-6690, 800-724-5600
fax: 617-558-0056

Horiba Instruments Inc. (benchtop XRF)
17671 Armstrong Ave.
Irvine, CA 92714
phone: 714-250-4811, 800-446-7422
fax: 714-250-0924
email: stauffer@hii.horiba.com

Kevex Instruments (energy-dispersive XRF)
24911 Stanford Ave.
Valencia, CA 91355
phone: 805-295-0019, 800-865-3839
fax: 805-295-0419

Metorex, Inc. (portable XRF)
1900 N.E. Division St.
Suite 204
Bend, OR 97701
phone: 503-385-6748, 800-229-9209
fax: 503-385-6750
email: metorex@bendnet.com
Internet: http://www.metorex.fi

Outokumpu Electronics, Inc. (on-line, energy-dispersive XRF)
860 Town Center Dr.
Langhorne, PA 19047
phone: 800-229-9209

Oxford Instruments, Inc. (wavelength- and energy-dispersive XRF)
130A Baker Ave. Ext.
Concord, MA 01742
phone: 508-371-9009, 800-447-4717
fax: 508-371-0204

Philips Electronic Instruments Company (wavelength-dispersive XRF, TXRF)
85 McKee Dr.
Mahwah, NJ 07430
phone: 201-529-3800
fax: 201-529-2252

Rigaku/USA, Inc. (wavelength-dispersive XRF, TXRF)
Northwoods Business Park
199 Rosewood Dr.
Danvers, MA 01923
phone: 508-777-2446
fax: 508-777-3594
email: rigaku@aol.com

Siemens Industrial Automation, Inc. (wavelength-dispersive XRF)
Analytical X-Ray Instruments
6300 Enterprise Ln.
Madison, WI 53719-1173
phone: 608-276-3000, 800-276-9729
fax: 608-276-3006

SOMAR International Inc. (standards, accessories, and equipment for XRF)
115 Main St.
P.O. Box 395
Tuckahoe, NY 10707
phone: 914-961-1400
fax: 914-779-0153

Spectrace Instruments, Inc. (on-line, portable, and lab energy-dispersive XRF)
1275 Hammerwood Ave.
Sunnyvale, CA 94089
phone: 408-744-1414
fax: 408-744-1313

Spectro Analytical Instruments, Inc. (portable XRF)
160 Authority Dr.
Fitchburg, MA 01420
phone: 508-342-3400, 800-548-5809
fax: 508-342-8695

SPEX Industries, Inc. (standards, chemicals, accessories, and equipment for XRF)
3880 Park Ave.
Edison, NJ 08820
phone: 908-549-7144
fax: 908-603-9647

Veeco Instruments, Inc. (benchtop XRF)
Seiko Instruments Distributor
UPA Technology Division
One Terminal Dr.
Plainview, NY 11803
phone: 516-349-8300, 800-645-7566
fax: 516-349-8180

Required Levels of Training

Operation of Instrument

Routine analyses can be performed by people with a high school education or associate college degree. "Cookbook" routines are usually set up by vendors, who provide troubleshooting and technical support through application laboratories. Instruments can be programmed to lock out access to calibration routines and other sophisticated aspects of the instruments so that sample analysis is reduced to pushbutton operation. Some minimal knowledge of chemistry is useful, especially if the operator will be required to perform sample preparation.

XRF Analysis

This aspect of the operation includes laboratories where both routine and unique samples are analyzed. There is no substitute for a rapid elemental analysis using XRF, but a minimum of a bachelor's degree in chemistry is necessary to help in reconciling the elemental information from the instrument with the knowledge of the sample to determine whether the results are reasonable and how to interpret them to solve the analytical problem. Additional laboratory experience or advanced education is necessary where XRF is applied to a new sample or method development is involved.

Sample Preparation

This is one aspect of XRF that can come only with experience and is more art than science. Each new sample type presents new opportunities in sample preparation challenges for the XRF analyst, which are exceeded only by challenges in standard preparation and matrix matching.

Service and Maintenance

In most cases, the reliability of the instrumentation should be the primary concern after performance considerations, particularly in critical plant and laboratory installations. XRF instruments, with care and regular maintenance, can operate reliably for over 10 yr and up to 20 yr. Many commercial vendors have service contracts that ensure rapid response and turnaround. Typical contracts call for one or two maintenance visits, where instrumental performance is checked and documented and preventive maintenance is done. This service is usually 5 to 10% of the purchase price of the instrument. In addition, the operator is required to maintain the correct purity of required gases for flow detectors, fill liquid nitrogen dewars, and maintain proper cooling-water pressure and purity. The primary system that requires most maintenance on WDXRF systems is the flow detector. This involves replacing the detector wire and windows to ensure maximum sensitivity. Some vendors offer remote diagnostic maintenance services that allow the vendor service department to check instrument performance monthly and document possible trouble before it occurs. This is a big benefit where such systems provide crucial support for plant operations. The X-ray tube is the most expensive replacement part; it usually lasts 1 to 5 yr depending on the type of usage it experiences. Tube replacement cannot be predicted but costs range anywhere from $8K to over $20K depending on the make and model of the instrument and the X-ray tube. Tube life can be extended by maintaining a minimum voltage and current on the tube when it is not in use instead of cycling the tube on and off. These standby conditions vary by vendor, but usually mean around 20 kV and low current, depending on the power supply.

Suggested Readings

Books

Advances in X-ray Analysis: Proceedings of the Annual Conference on Applications of X-Ray Analysis (37 volumes to date). New York: Plenum Press.

BERTIN, EUGENE, *Principles and Practice of X-Ray Spectrometric Analysis*, 2nd ed. New York: Plenum Press, 1975.

JENKINS, RON, *An Introduction to X-Ray Spectrometry*. London: Heyden & Sons, 1974.

JENKINS, RON, R. W. GOULD, AND DALE GEDCKE, *Quantitative X-Ray Spectrometry*, 2nd ed. New York: Marcel Dekker, 1995.

LACHANCE, GERALD R., *Introduction to Alpha Coefficients*. Sainte-Foy, Quebec: Corporation Scientific Claisse Inc.

TERTIAN, R., AND F. CLAISSE, *Principles of Quantitative X-Ray Fluorescence*. London: Heyden, 1982.

VANDERCASTEELE, C., AND C. B. BLOCK, *Modern Methods for Trace Element Determination*. Chichester, UK: Wiley, 1993.

VAN GRIEKEN, RENE E., AND ANDRZEJ A. MARKOWICZ, *Handbook of X-ray Spectrometry: Methods and Techniques*. New York: Marcel Dekker, 1993.

Articles

ANALYTICAL METHODS COMMITTEE, "Evaluation of Analytical Instrumentation. Part VI Wavelength-Dispersive X-Ray Spectrometers," *Analytical Proceedings*, 27 (1990), 324–33.

BACON, JEFFERY R., AND OTHERS, "Atomic Spectrometry Update: Atomic Mass Spectrometry and X-Ray Fluorescence Spectormetry," *Journal of Analytical Atomic Spectrometry*, 19 (1994), 267R–306R.

DEGROOT, P. B., "SPC: What It Is and Why Should You Use It in Your X-Ray Analytical Laboratory?" *Advances in X-Ray Analysis*, 33 (1990), 537–42.

DEGROOT, P. B., "SPC Analysis of Optimal Strategies for Restandardization of X-Ray Fluorescence Analyses," *Advances in X-Ray Analysis*, 33 (1990), 543–8.

FORTE, MICHEL, "Fabrication and Use of Permanent Monitors and Standards," *X-Ray Spectrometry*, 12 (1983), 115–7.

GILFRICH, JOHN V., "Modern X-Ray Fluorescence Analysis," *Progress in Analytical Spectroscopy*, 12 (1989), 1–20.

GILFRICH, JOHN V., AND L. S. BIRKS, "Estimation of Detection Limits in X-Ray Fluorescence Spectrometry," *Analytical Chemistry*, 56 (1984), 77–9.

HARDING, ANTHONY, "On-Stream Elemental Analysis by EDXRF," *American Laboratory*, December (1987).

JANSSENS, KOEN H., AND FRED C. ADAMS, "New Trends in Elemental Analysis Using X-Ray Fluorescence Spectormetry," *Journal of Analytical Atomic Spectroscopy*, 4 (1989), 123–35.

LEYDEN, D. E., AND N. L. GILFRICH, "Developement and Comparison of Fundamental Parameters Software for X-Ray Spectormetry," *Trends in Analytical Chemistry*, 7 (1988), 321–7.

MANTLER, M., "Recent Methods and Applications of X-Ray Fluorescence," *Progress in Crystal Growth and Characterization*, 14 (1987), 213–63.

TOROK, S. B., AND OTHERS, "Application Reviews and Fundamental Reivews: BiAnnual," *Analytical Chemistry*, 68 (1996), 467R.

WEST, NORMAN, "X-Ray Fluorescence Spectrometry Applied to the Analysis of Enviornmental Samples," *Trends in Analytical Chemistry*, 3 (1984), 199–204.

Training Aids

American Chemical Society Short Course, X-Ray Spectrometry, offered at ACS meetings and PITTCON.

American Chemical Society Audio Course, X-Ray Spectrometry, available from the ACS.

Fundamentals of X-ray Fluorescence and Advanced Methods in X-ray Fluorescence (1-week courses)
International Centre for Diffraction Data
12 Campus Blvd.
Newton Square, PA 19703-3273
phone: 610-325-9814
fax: 610-325-9823

Workshops at the Annual Conference on Applications of X-Ray Analysis (the Denver X-Ray Conference), usually held the first week in August, in Colorado.
Sponsored by the Department of Engineering, University of Denver and International Centre for Diffraction Data.

References

1. A. Hadding, *Z. Anorg. Chem.*, 122 (1922), 195–200.

2. R. Klockenkamper and A. von Bohlen, *J. Anal. At. Spectr.*, 7 (1992), 273–9.

3. D. R. Boehme, *Adv. X-ray Anal.*, 30 (1987), 39–44.

4. M. C. Nichols and others, *Adv. X-ray Anal.*, 30 (1987), 45–51.

5. B. J. Cross and J. E. Augenstine, *Adv. X-ray Anal.*, 34 (1991), 57–70.

6. T. A. Anderson and others, *Spectroscopy*, July/Aug. (1991), 28–34.

7. A. Knochel, W. Petersen, and G. Tolkiehn, *Analytica Chim. Acta*, 173 (1985), 105–116.

8. H. Saisho, *Trends Anal. Chem.*, 8 (1989), 209–14.

9. R. P. H. Garten, *Trends Anal. Chem.*, 3 (1984), 152–7.

Chapter 25

Ultraviolet and Visible Molecular Absorption Spectrometry

James A. Howell

Western Michigan University
Department of Chemistry

Summary

General Uses

- Identification of many types of organic and inorganic molecules and ions
- Qualitative determination of some functional groups in organic molecules
- Quantitative determination of many biological, organic, and inorganic species
- Quantitative determination of mixtures of analytes
- Monitoring and identification of chromatographic effluents
- Determination of equilibrium constants
- Monitoring of environmental and industrial processes
- Automated trace-level analysis of clinical diagnostic samples
- Determination of stoichiometry of chemical reactions

Common Applications

- Identification of compounds by spectrum matching with reference spectra
- Trace analysis of inorganic, organic, and biological species

- Enzyme assays
- Monitoring of reaction rates (stopped-flow and temperature-jump chemical kinetics)
- Detector for chromatography (particularly high-performance liquid chromatography) and electrophoresis
- Environmental remote sensing (hydrogeologic, aquatic, and atmospheric)
- Field testing (pH, metals, nonmetals, and organics)
- Study of equilibrium systems (complexation, pH, and redox)
- Trace analysis via reaction rate determination (catalytic and enzyme)
- Pharmaceutical analysis
- Surface analysis by reflectance measurements (total, specular, and diffuse)

Samples

State

- Any transparent solid, liquid, or gaseous sample can be used.
- Opaque solids and liquids can be measured with reflectance or photoacoustic techniques.
- Turbid samples can sometimes be accommodated with derivative techniques.

Amount

- Typically samples of a few milliliters or a few milligrams may be used.
- It is not unusual for microliter or microgram sample sizes to be used.

Preparation

- An appropriate high-purity transparent solvent must be chosen.
- Aqueous systems often require the use of an appropriate buffer system.
- The use of clean glassware is imperative.
- Sample cleanup (isolation of analyte from interfering species) may be required.
- Some trace analysis techniques require preconcentration.
- Turbid samples may require filtration to reduce light scattering.

Analysis Time

- Typical analysis times range from 2 to 30 min per sample.
- Automated analysis schemes (flow-injection techniques) can reduce time of analysis per sample to a few seconds.
- Samples involving extensive cleanup or derivatization steps can take up to several hours per sample.

Limitations

General

- Generally, sample concentrations must be low (absorbances less than 2).
- Photosensitive compounds may be difficult to analyze.

Accuracy

- Light scattering can limit the precision of measurements.
- Overlapping absorption bands in complex mixtures can reduce the precision of measurements.

Sensitivity and Detection Limits

Limits of detection can be considerably higher than in fluorescence or chemiluminescence methods.

Complementary or Related Techniques

- Flow injection analysis
- Reflectance spectroscopy
- Chromatography
- Photoacoustic spectrometry
- Tristimulus colorimetry and color specification
- Photometric titrations

Introduction

Historical

The act of identifying materials based on their color was probably one of the earliest examples of qualitative molecular absorption spectrophotometry. Also, the first recognition that color intensity can be an indicator of concentration was probably the earliest application of employing molecular absorption spectroscopy for quantitative estimates. The first measurements were made by using the human eye as a detector and undispersed sunlight or artificial light as the light source. Later it was found that accuracy and precision could be improved by isolating specific frequencies of light using optical filters. Further improvement of measurements came with the use of prism and grating monochromators for wavelength isolation. Photoelectric detectors were soon developed, but were quickly replaced with phototubes and photomultiplier tubes. The development of solid-state microelectronics has now made available a wide range of detector types and highly sophisticated readout electronic systems.

Although many new sophisticated spectrophotometers feature refinements over their earlier counterparts, many rudimentary instruments are still widely used as portable field test devices. Swimming pool pH and chlorine levels are commonly monitored by the addition of a color-producing indicator or reagent to the sample, followed by visual color comparison with reference standards using undispersed light. Many other field test devices use the above techniques and some incorporate filters as wavelength isolation devices for improved accuracy and precision. Several portable, battery-operated spectrophotometers using an internal light source, grating monochromator, and phototransistor detector are currently available for quantitative spectrophotometric analysis.

Modern spectrophotometers designed for laboratory applications cover a wide range of sophistication depending on their intended application. Many still use filters for wavelength isolation, although grating instruments are becoming more common because of the availability of high-quality, low-cost holographic gratings. A wide range of spectrophotomers are now available featuring ultraviolet and visible (UV/Vis) measurements, low stray light, small effective bandwidths, solid-state detectors, digital data acquisition and processing, and double- or single-beam operation. Because of the high stability of the electronic systems and light sources, coupled with the increased speed of data acquisition, double-beam operation often does not seem to offer sufficient advantages over the single-beam systems to offset the cost. Most modern instruments still use mechanical wavelength scanning and phototransistor detectors, but many newer spectrophotometers feature rapid digital scanning using either diode array or charge coupled device (CCD) detectors.

Current Use

Field testing has become an extremely important application of ultraviolet and visible spectrophotometry, particularly in the area of environmental testing and monitoring. Here much interest is focused on monitoring pH, metal and nonmetal ion concentrations, and concentrations of specific organic contaminants. Remote sensing using fiberoptic systems has been found useful for in-situ studies of ground water systems and for highly contaminated sites, where the safety of workers might be of concern.

Industrial applications of spectrophotometry are wide ranging; process monitoring and control are important applications. Remote sensing with fiberoptic devices has become very important for process monitoring. In the area of pharmaceutical production process monitoring and product assay, UV/Vis spectrophotometric measurements remain an important tool. Perhaps equally important in this area is the application of high-performance liquid chromatography (HPLC) with ultraviolet or visible detection for the analysis of pharmaceutical preparations.

Medical diagnostics commonly include a wide range of spectrophotometric techniques in carrying out a battery of automated analyses for biologically important analytes. Many of these systems use enzyme-specific color reactions for specificity and flow-injection techniques for high sample throughput. These automated systems commonly incorporate separation and sample preparation tasks before the actual analysis. It is not uncommon for the systems to perform the analysis of a dozen or more assays and provide the physician with a printout of the patient's name, assay results, and norms.

In the research environment, spectrophotometric measurements are still an important tool used to study and elucidate molecular structure and electronic states. The availability of rapid data acquisition with spectrophotometers has made them one of the more popular analytical tools of kineticists interested in fast reactions. The techniques of stopped-flow and temperature-jump are readily adaptable to spectrophotometric measurements.

In all of the above areas, highly sophisticated computer analyses of the data are used to achieve the highest degree of precision and accuracy. Signal-to-noise (S/N) enhancement commonly involves the Golay–Savitzky data-smoothing algorithm. A variety of chemometric techniques, including the K-factor method, principal component analysis, partial least squares, Kalman filtering, and full-spectrum quantitation (FSQ) are available for greater refinement of spectrophotometric data.

Analytical techniques are often combined, as in gas chromatography mass spectrometry (GC-MS). UV/Vis spectrophotometry was one of the earliest techniques to be combined with other techniques when chromatographic columns incorporated UV/Vis detection. Although it is common to use fixed wavelength detection, rapid digital scanning can permit essentially continuous spectral scanning of chromatographic effluents. The newly emerging technique of capillary electrophoresis (CE, see Chap. 10) commonly uses fixed-wavelength ultraviolet absorption detectors.

How It Works

Physical and Chemical Principles

The ultraviolet region of the spectrum is generally considered to range from 200 to 400 nm and the visible region from 400 to 800 nm. The corresponding energies for these regions are about 150 to 72 and 72 to 36 kcal mole^{-1}, respectively. Energy of this magnitude often corresponds to the energy difference between electronic states of many molecules. For example, single-bond systems tend to exhibit transitions from sigma bonding orbitals to sigma antibonding orbitals ($\sigma \rightarrow \sigma^*$) and occasionally transitions of nonbonding electrons to sigma antibonding orbitals ($n \rightarrow \sigma^*$), when nonbonding electrons can become involved. Thus, these molecular electronic transitions can occur when light with energy of this magnitude is absorbed by the molecule. Typically, $\sigma \rightarrow \sigma^*$ and $n \rightarrow \sigma^*$ transitions can be observed by absorption bands below 200 nm. Nevertheless, a few of these may occasionally be seen at wavelengths greater than 200 nm. Absorption bands resulting from electronic transitions involving double-bond or pi systems and nonbonding electrons are much more common in the UV/Vis region. These are the transitions from pi bonding orbitals to pi antibonding orbitals ($\pi \rightarrow \pi^*$) and transitions of nonbonding electrons to pi antibonding orbitals ($n \rightarrow \pi^*$).

Transition metals commonly exhibit absorption bands in the UV/Vis region of the spectrum. These often result from the energy differences in the various d-electron states arising from electron interactions of coordinated donor atoms. Analogous behavior can be seen, although to a lesser degree, with many metals from the lanthanide series, resulting from differences in their f-electron states.

Absorption bands can also arise from certain molecules where the energy required to remove an electron from one atom and place it on another falls within the UV/Vis region. This process is known as a charge transfer excitation. Although charge transfer bands are not at all uncommon, it is not always an easy matter to predict them based on molecular structural properties.

Molecules with the ability to exhibit the above types of electronic transitions are said to possess chromophores (from the Greek words *chroma*, meaning color, and *phoros*, meaning producer). Chromophores are often associated with certain molecular groups. In order to observe a $\pi \rightarrow \pi^*$ transition, there must be a molecular group containing a double bond (such as ethylenes, acetylenes, carbonyls, and azo compounds); to observe an $n \rightarrow \pi^*$, nonbonding electrons in addition to a double bond must be present (such as carbonyls, nitro groups, and azo groups). Molecules exhibiting $n \rightarrow \sigma^*$ transitions necessarily have a single bond and nonbonding electrons (such as alcohols, amides, and water.) Other groups are called auxochromes (from the Greek words *auxein*, meaning to increase, and *chroma*), which may not exhibit absorption themselves but can affect the wavelength or intensity of the absorption band of a chromophore. It is common for two chromophores to exhibit auxochromic effects on each another. In these instances it is best to consider the two interactive chromophores a new single chromophoric group. Wavelength shifts are classified as either bathochromic (red shift or shift to longer wavelengths) or hypsochromic (blue shift or shift to shorter wavelengths). Enhanced absorption is called a hyperchromic effect and a decrease in absorption intensity is a hypochromic effect. Table 25.1 lists some common chromophoric groups.

Transition metals having unfilled *d*-orbitals usually exhibit absorption bands in the UV/Vis region. Samples that absorb significantly in the visible region are always colored because color results when a band of frequencies is absorbed from white light. The actual wavelength of *d–d* transitions depends on the metal involved (the number of *d*-electrons initially present), the number of coordinating groups, the strength (basicity) of the donor atoms, and the geometry of the coordinating groups.

The classic process of making physical measurements of molecular absorption of ultraviolet or visible light involves passing light through the sample. The radiant power of the incident beam of light can be designated P_o; the power of the transmitted beam is P. Using these two parameters, it is possible to define transmittance, T, as follows:

$$T = \frac{P}{P_o} \qquad (25.1)$$

Unfortunately, the absorption phenomenon of the molecular species is logarithmically related to the transmittance. Consequently, it is helpful to define the absorbance, A, as follows:

$$A = -\log T \qquad (25.2)$$

Table 25.1 Typical chromophoric groups in the ultraviolet and visible region.

Group	Structure	Transitions
Acetylenic	$>C \equiv C<$	$\pi \rightarrow \pi^*$
Amide	$-CONH_2$	$\pi \rightarrow \pi^*$
		$n \rightarrow \pi^*$
Carbonyl	$>C = O$	$\pi \rightarrow \pi^*$
		$n \rightarrow \pi^*$
Carboxylate	$-COO^-$	$\pi \rightarrow \pi^*$
		$n \rightarrow \pi^*$
Ester	$-COOR$	$\pi \rightarrow \pi^*$
		$n \rightarrow \pi^*$
Ethylenic	$>C = C<$	$\pi \rightarrow \pi^*$
Nitro	$-NO_2$	$\pi \rightarrow \pi^*$
		$n \rightarrow \pi^*$
Oxime	$>C = N-$	$\pi \rightarrow \pi^*$
		$n \rightarrow \pi$

Bouguer in 1729 (1) and Lambert in 1760 (2) stated that the absorbance was directly proportional to the depth or thickness of sample being measured if concentration was constant. This has become known as the Lambert–Bouguer law and may be stated as

$$A = kb \qquad (25.3)$$

where b is the sample thickness and k is a proportionality constant. In 1852, Beer (3) determined that if the sample thickness was held constant, the absorbance was directly proportional to the concentration of the absorbing species if certain limitations were observed. Thus, Beer's law can be written as

$$A = k'c \qquad (25.4)$$

where c is the concentration and k' is a proportionality constant. Although the Lambert–Bouguer law and Beer's law were independently developed primarily from an empirical treatment of absorption data, one can derive the combined Beer–Lambert–Bouguer law by first imposing certain limitations on the system:

- The incident radiation, P_o, is monochromatic (only one wavelength).
- All rays of the incident radiation, P_o, travel equidistant parallel paths through the absorbing sample (highly collimated beam with no internal reflections).
- The incident radiant power, P_o, is not sufficient to significantly alter the ground-state population of the absorbing molecules (avoid nonlinear optics such as high-power laser light sources).
- The absorbing sample must be homogeneous and not scatter or reflect the incident radiation.
- The absorbing species must behave as independent moieties (no molecular interactions with other like or unlike molecules, only at high dilution, for example).

Strong (4), applying the above limitations, presented the derivation of the combined Beer–Lambert–Bouguer law (often simply called Beer's Law) from beginning principles. This expression is written as

$$A = abc \qquad (25.5)$$

where a is a proportionality constant known as the absorptivity, and b and c are the sample thickness and concentration, respectively. The absorptivity is a constant for a given molecular species but is dependent on wavelength. Its units depend on those chosen for the sample thickness and for the concentration. When the sample thickness is measured in centimeters and the concentration in moles per liter, the absorptivity is known as the molar absorptivity, and is designated as ε with units of $L\ mol^{-1}\ cm^{-1}$.

In applying the Beer–Lambert–Bouguer law, one must always be aware of the five limitations described above. Deviations from the law due to the failure to strictly adhere to the limitations can be classified into three categories: physical processes, instrumental processes, and chemical processes. These are discussed in detail later in this chapter.

Instrumentation

The terminology for describing instruments for measuring UV/Vis absorption is somewhat confused because of improper usage over the years. Much of this stems from the fact that many of the same terms are used in both UV/Vis absorption measurements and color specification or tristimulus colorimetry. The former discipline concerns itself with determining the spectral properties or concentration of molecular species. The latter area of study is interested in quantitatively defining visible colors in terms of three primary colors. Because the earliest absorption measurements dealt only with visible measurements, it is not surprising that it became known as colorimetry and the instruments used to make the measurements were called colorimeters. In the strictest sense, the term *colorimeter* should apply to an instrument for color specification measurements.

The earliest devices for measuring molecular absorption of radiation were visual comparisons of samples with standards of known concentration observed in ambient light. These devices were known as color comparators, and often consisted of a rack of test tubes holding a number of tubes containing various concentrations of a reference material. Later some of these devices included filters to achieve wavelength specificity. These devices continue to be widely used for field work, particularly where portability and ruggedness are required and a modest degree of accuracy and precision will suffice.

A photometer is a device using filters for wavelength selection and some type of photoelectric detection. Later versions of these instruments featured the ability to measure the ratio of two beams of light separated in time (single-beam) or space (double-beam), thus allowing the direct measurement of transmittance. Most photometers are designed to be operated in a laboratory environment; however, the advent of modern microelectronics has given rise to rugged portable versions of these instruments for field use.

The absorptiometer or spectrophotometer is an instrument capable of providing transmittance or absorbance data as a function of wavelength. These instruments use either grating or prism monochromators for wavelength isolation. A few of these are limited to the visible region, but most are constructed with ultraviolet transparent optics (quartz) to allow measurements in the UV/Vis range. As was the case with the photometer, spectrophotometers often have double-beam capability.

Instrument Components

Figure 25.1 depicts a block diagram of the essential components of a typical ultraviolet or visible instrument. Some instruments position the sample/reference compartment before the wavelength isolation device. This arrangement permits significantly greater intensity of incident light falling on the sample and reference but may result in photodegradation of some thermally sensitive samples, possibly creating baseline problems with highly fluorescent samples.

Light Source The ideal light source produces a stable continuum of uniformly high-intensity radiant energy over the spectral region of interest. It should be rugged, have a long life, and be relatively inexpensive to acquire and operate. A well-regulated (electronic or battery-operated) power supply is an essential component of the light source. Some systems may also incorporate electronic modulation or mechanical chopping of the light source to reduce certain types of noise.

The incandescent tungsten filament lamp with a glass envelope operated at 2000 to 3000 °K produces a continuum in the wavelength range of 320 to 2500 nm, with a spectral radiance of about $10 \text{ mW cm}^{-2} \text{ nm}^{-1} \text{ sr}^{-1}$ at 500 nm. This lamp approximates the emittance of a black-body radiator, with its maximum intensity at about 1000 nm and falling off rapidly to either side. The glass envelope absorbs essentially all radiant energy below 320 nm, thus limiting the lamp's use to the visible and near infrared regions of the spectrum. Higher operating temperatures can produce greater emittance; however, the envelope quickly becomes darkened by the deposition of volatilized tungsten and ultimately failure of the filament.

The quartz–iodine lamp is a variation of the incandescent tungsten filament lamp where a quartz envelope is used and the lamp operated at temperatures up to 3600 °K. This lamp has a wave-

Figure 25.1 Block diagram of a typical instrument for making UV/Vis absorption measurements.

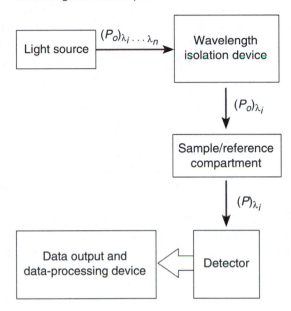

length range of 200 to 3000 nm, with a spectral radiance of about 50 mW cm^{-2} nm^{-1} sr^{-1} at 400 nm, nearly five times greater than the simple tungsten lamp. The spectral output of this lamp is similar to that of the incandescent tungsten filament lamp except that it is not limited in the ultraviolet region by the glass envelope. The introduction of iodine in the lamp permits operation at higher temperatures because it can combine with volatilized tungsten, forming tungsten iodide, which diffuses back to the hot filament, where it is decomposed back to iodine and tungsten. This lamp is superior to the simple incandescent tungsten filament lamp in the visible region. It is also useful in the ultraviolet region, although other sources of greater intensity in this region are available.

The hydrogen or deuterium arc discharge lamp operating with 2 to 3 torr pressure with a quartz window exhibits an intense output in the region of 180 to 370 nm, with a spectral radiance of about 5 mW cm^{-2} nm^{-1} sr^{-1} at 250 nm. The deuterium lamp is generally preferred over the hydrogen lamp because its spectral radiance is nearly three times greater. Deuterium lamps are probably the most widely used sources for spectrophotometers in the ultraviolet region.

Wavelength Isolation Device The purpose of the wavelength isolation device is to separate the many wavelengths of light coming from the continuum produced by the light source and isolate the particular wavelength of interest. The effectiveness of a device to perform this function is measured in terms of its throughput, resolution, and bandwidth. Throughput is a measure of how much radiant energy of the desired wavelength is transmitted by the device and is quantitatively defined by the lens formula for calculating the f-number. Resolution is a measure of how well the device separates adjacent wavelengths from each other and is related to the reciprocal dispersion, D^{-1}, which is defined as

$$D^{-1} = \frac{d\lambda}{dy} \tag{25.6}$$

where $d\lambda$ is the differential change in wavelength and dy is the corresponding differential change in y, the distance separating wavelengths on the dispersion axis. The bandwidth of transmitted radiant energy is measured in terms of the band's full width at half maximum, or fwhm. The significance of the fwhm is that it represents the width of the wavelength range of 75% of the radiant energy emerging from the device. This is known as the spectral bandwidth, SBW, and depends on the slit width, reciprocal dispersion, slit curvature, optical aberrations, and Rayleigh diffraction. Instrument technology has made the latter three factors relatively insignificant, so the SBW is given by the following approximate relationship:

$$SBW = wD^{-1} \tag{25.7}$$

where w is the slit width of the dispersive device and D^{-1} is its reciprocal dispersion.

The earliest device to perform this function was the filter. A filter is a nondispersive device that, when illuminated with a polychromatic band of light, absorbs or destructively destroys certain frequencies of light and transmits only a narrow band. In general they have high energy throughput but are somewhat limited by the fact that they transmit wide bandwidths. The criteria of interest in selecting a filter are its wavelength of maximum transmittance, λ_{max}; the peak transmittance, T_{max}; the bandwidth of the transmitted band, fwhm; short and long wavelength blocking, if present; and the order of bands transmitted, where applicable. Interference filters, though more costly, usually exhibit significantly narrower bandwidths than absorption filters.

Spectrophotometer monochromators use either diffraction gratings or prisms as dispersive devices. Diffraction gratings are classified as either classically ruled or holographic gratings. The classically ruled grating is manufactured by mechanically scribing the lines, whereas the holographic grating is made by means of producing an interference image of lines with a laser on an optically flat photosensitive coating, which is subsequently etched to produce the grooves. Holographic gratings are currently the preferred choice due to their significantly lower cost, fewer aberrations, and lower stray light (greater wavelength purity). Many different geometric arrangements for positioning the grating and slit have evolved but are beyond the scope of this discussion. The reciprocal dispersion for either type of grating can be expressed as

$$D^{-1} = \frac{d}{nF} \tag{25.8}$$

where d is the distance between adjacent grooves, n is the order of diffraction, and F is the focal length of the monochromator system.

Prism monochromators are another popular choice as a wavelength isolation device for spectrophotometers. There are many geometric designs for prisms, but the Cornu (60°–60°–60°) and the Littrow (30°–60°–90°) prisms are the most widely used and may be arranged in a variety of configurations. In a prism monochromator, light from a polychromatically illuminated slit is focused on the face of the prism at some angle such that refraction of the incident light occurs. Because different wavelengths of light experience differing degrees of refraction, their separation and isolation are possible. The reciprocal dispersion for a prism is more complicated than that of a grating:

$$D^{-1} = \frac{d\eta}{d\lambda} \cdot \frac{dr}{d\eta} \tag{25.9}$$

where η is the refractive index of the prism material (a wavelength-dependent parameter), λ is the wavelength, and r is the angle of refraction. The first term, $d\eta/d\lambda$, is called the optical dispersion term; the second term, $dr/d\eta$, is the geometric dispersion term. It is worth noting that the reciprocal dispersion of a grating is essentially constant with respect to wavelength (Eq. (25.8)); however, it is obvious that the reciprocal dispersion of a prism varies with wavelength (Eq. (25.9)). The refractive index of prism materials increases nearly exponentially with wavelength. This results in a prism having greater resolving power at short wavelengths.

Nondispersive wavelength isolation devices such as the optical interferometer coupled with fast Fourier transform (FFT) for data processing are extremely popular for infrared spectrophotometers. However, they have not enjoyed the same degree of popularity for UV/Vis spectrophotometers. This is probably due to the higher tolerances required of the interferometer working in the UV/Vis region. As technology evolves, these instruments may become important developments in UV/Vis spectrophotometers of tomorrow.

Sample and Reference Compartments Only double-beam instruments have a reference compartment. The double-beam instrument is usually configured so that monochromatic light from the wavelength isolation system is divided by means of an optical beam splitter and passed through both the sample and reference compartments continuously or rapidly alternating between the compartments. This allows compensation for small variations in the light source and permits continuous blank corrections. The high stability of modern light sources coupled with rapid measurement systems allows many high-quality instruments to incorporate a single-beam optical system without significantly sacrificing measurement stability and without the additional expense of the double-beam optics system.

The sample and reference compartments must be lightproof and provide a kinematically engineered cell-holding device to ensure reproducible placement of the sample and reference cells in the light beams. A well-designed cell compartment has a highly collimated light beam traversing the cell holder. This is important to reduce reflective light losses and multiple reflection errors. In this respect it is also important that the cell holder position the cell such that both faces are at right angles to the incident and exiting light beam. If long cells are to be used, it is imperative that the light beams be highly collimated to avoid reflections off the walls of the cells. It is desirable for the cell compartments to have provisions for purging atmospheric gases. Thermostatic control of the sample and reference compartments can permit operation at subambient and elevated temperatures, which is a desirable feature found in some spectrophotometers. The sample/reference compartments often feature provisions for installation of autosampling devices.

Detector The earliest photoelectric detectors were either photovoltaic or photoconductive cells that functioned as a low-impedance voltage source or a variable resistance source whose output was a function of the radiant power incident on the cells' active surfaces. These devices are still widely used for many field instruments and low-cost filter photometers. The sensitivities of these devices generally limit their use to the visible and near infrared regions of the spectrum.

The vacuum phototube with its photoemissive cathode provides greater sensitivity than the photovoltaic and photoconductive cells. These detectors have been largely replaced by the photomultiplier and certain solid-state detectors. The photomultiplier has many photoemissive surfaces, called dynodes, each at a successively higher positive voltage. Only the first dynode is illuminated, resulting in the production of photoelectrons that are accelerated and focused onto the next dynode, producing more electrons to be focused onto subsequent dynodes and finally collected at the anode, producing a measurable photocurrent. This cascading effect makes the photomultiplier one of the most sensitive photodetectors available. Phototubes and photomultipliers are available with a wide range of spectral sensitivities ranging from approximately 185 to 1000 nm. A more extensive discussion of phototubes and photomultipier detectors can be found in the literature (5).

Solid-state detectors such as the photodiode and phototransistor generally exhibit satisfactory sensitivity, are rugged, and are relatively inexpensive. These devices operate on the principle that irradiation of a *p–n* junction will produce charge carriers known as electron-hole pairs. Once the number of photoinduced carriers exceeds those induced by thermal processes (dark current), the photocurrent is proportional to the incident radiation. Their spectral response ranges from the ultraviolet to the near infrared regions, but their greater response is generally from about 850 to

1000 nm. The responsivity of these devices is significantly less than that of the photomultiplier. Their responsivity is adequate for today's light sources and their simplicity, ruggedness, and low cost readily offset this reduced sensitivity.

Solid-state microelectronics has produced a variety of detectors, of which the diode array is currently one of the more popular ones found in modern spectrophotometers. Instead of having a single photodiode, these devices are designed with a linear array of as many as 1024 or 2048 photodiode elements. The physical spacing between these elements is typically 25.4 to 50 μm. Dispersed radiation from the monochromator is focused on the array, with each element receiving a bandwidth of radiation corresponding to a given wavelength. Each diode is operated in parallel with a charged capacitor. When light falls on the diode element, photoinduced charge carriers allow the capacitor to partially discharge. A solid-state switch disconnects the diode element from the capacitor and the signal is read by the amount of current necessary to recharge the capacitor to its initial state. The array is electronically scanned in a matter of milliseconds and digital data processing produces a spectrum.

Two solid-state devices that offer much promise as potential detectors are the charge-coupled devices (CCDs) and charge-injection devices (CIDs). These detectors are commonly produced as two-dimensional arrays and thus can function as imaging devices. Their sensitivities under the proper circumstances can approach those of many photomultiplier systems. These devices are described in greater detail in the literature (6).

Data Output and Data-Processing Device　Signal output devices can be as simple as an analog absorbance or transmittance meter where the data are read, recorded, and processed by the operator. Some systems use logic circuitry to provide digital readouts in transmittance, absorbance, or concentration. Most modern spectrophotometers now incorporate microprocessors for control and monitoring of instrument operation. These systems commonly provide an interface to a computer system along with software to control instrument operations, data collection, and data processing.

Instrument Operation

The basic function of a UV/Vis absorption colorimeter, photometer, or spectrophotometer is to measure transmittance or absorbance. Very few instruments measure absorbance directly; generally, they measure transmittance and subsequently convert the measurements to absorbance units. The measurement of transmittance involves adjusting the instrument to establish the upper and lower limits of transmittance (0% T and 100% T). Setting the 0% T is actually the process of electrically nulling the detector dark current. This is done by blocking the optical beam with an opaque shutter and adjusting the instrument output to read 0% T. The 100% T limit is set by placing a blank (a cell containing solvent and any other substances found in the sample excluding the chromophoric group) into the optical beam and adjusting the instrument output to read 100%. With the upper and lower limits established, the sample is placed in the sample beam and its transmittance is read.

The manner in which the dark current adjustment is made is consistent among all instruments and simply consists of applying a variable opposing current against the darkened detector output until a null condition is obtained. Establishing the 100% T limit can be carried out in essentially three different ways. The first way is simply to increase or decrease the amplifier gain with the blank in the optical beam so that a 100% T reading is obtained. This procedure is common with fixed-slit instruments. Despite its simplicity, when scanning wavelength or reading at different wavelengths, the gain must be changed to compensate for differences in source output and detector response. This results in readings taken with different signal-to-noise ratios at different wavelengths. A second procedure avoids the problem of having spectral scans with varying S/N by adjusting the intensity of the reference beam when a blank is placed in the sample beam. This is done by increasing or decreasing the slit width of the instrument. Unfortunately, while avoiding spectra with varying S/N, this procedure gives spectra with varying resolution. These two modes of operation are called autoslit (variable resolution) or autogain (variable S/N).

The preceding discussion covered double-beam operation. Single-beam operation is accomplished in the same manner except that the same beam is used to establish the upper and lower limits; the adjustments are separated in time. Finally, another mode of making transmittance measurements is quite common with modern instruments using single-beam operation but digital data-handling capability. Here, the dark current is established by closing a shutter and measuring the dark current, i_d, and storing it in memory. The shutter is opened, a blank is placed in the optical beam, and the detector output, i_{100}, is read and stored. The sample is then placed in the beam and the detector output, i_{samp}, read again. The transmittance is then computed as $(i_{samp} - i_d)/(i_{100} - i_d)$. Scanning instruments scan wavelength during both the blank and sample readings.

A few instruments use a measurement technique commonly used in many double-beam infrared instruments that adjust the 100% T limit by adjusting the position of a comb (beam attenuator) in the reference beam when a blank is placed in the sample beam. With the sample placed in the sample beam, the comb in the reference beam is adjusted until the intensities of the two light beams are equal. The adjustment position of the comb is calibrated in terms of transmittance or absorbance. This technique is known as an optical null balance.

What It Does

Data Format

Most color comparators provide readout in terms of the quantity being measured (such as pH or concentration). Simple colorimeters and photometers usually output data in terms of transmittance, absorbance, and sometimes concentration. Many data outputs are available in modern spectrophotometers depending on the operator's application.

Accuracy and Precision

The accuracy and precision of spectrophotometric methods depend on three major factors: instrumental limitations, chemical variables, and operator skill. Instrumental limitations are often determined by the quality of the instrument's optical, mechanical, and electronic systems. These may vary widely depending on the cost of the instrument. Chemical variables are determined by purity of standards, reagent and chromophore stability, reaction rates, reaction stoichiometry, pH, and temperature control. These factors are usually determined by the methodology chosen for the analysis. Finally, operator skill can obviously contribute greatly to the degree of accuracy and precision. Because there is little one can do to improve instrumental limitations, one attempts to use methods and operator skills that are good enough to make the instrumental factors the limiting factor. Under ideal conditions, it is possible to achieve relative standard deviations in concentration as low as about 0.5%. A number of publications deal in depth with the factors affecting the precision of spectrophotometric measurements (7–13).

Detection Limits

Spectrophotometric detection limits depend on molar absorptivities of the chromophoric groups being measured. The molar absorptivities, ε, of strong absorption bands generally range from 10^4 to 10^5; values of 10^3 or less are classified as weak bands, often resulting from forbidden transitions. Thus, with an instrument capable of accurately measuring absorbances as high as 3.0000 and using 10.00-cm cells, one could conceivably analyze as little as 3.00 μM/L. Detection limits can be reduced somewhat by solvent selection because molar absorptivities depend on the solvent system. Another technique used to increase the detection limit is to use indirect determinations, where a stoichiometric gain in the number of chromophores may result or the newly formed chromophore may have a higher molar absorptivity. Reaction rate methods can sometimes have lower detection limits than do conventional spectrophotometric measurements.

Limitations

The limitations of UV/Vis spectrophotometric methods are most easily discussed in terms of their effect on the accuracy and precision of the measurements, which were described in terms of instrumental limitations, chemical variables, and operator skill.

Probably one of the most commonly encountered but unrecognized problems in measuring absorbance is stray light error. All wavelength isolation devices tend to produce some low-intensity radiation at wavelengths other than the desired one. This is usually due to optical imperfections, or simply from scattered light due to dust particles on optical surfaces. Because one has usually selected a wavelength at which the compound of interest absorbs most strongly, it is not surprising that the stray light falling on the sample is of wavelengths at which the compound does not absorb strongly. Thus, one might expect that stray light errors will result in a negative bias for absorbance readings. This is seen in Eq. (25.10):

$$T_{\text{obs}} = \frac{T_{\text{true}} + \rho}{1 + \rho} \tag{25.10}$$

where ρ is the fraction of all the light coming from the wavelength isolation device, which is stray light, and T_{obs} and T_{true} are the observed and true transmittances, respectively. Table 25.2 illustrates the effect of stray light on absorbance readings. Normally the absolute amount of stray light tends to be relatively constant with respect to wavelength. Unfortunately, the fraction of stray light is highly wavelength-dependent because the amount of energy of the selected wavelength depends on the source intensity at that wavelength. Thus, stray light errors are most predominant at long and short wavelengths and when high absorbances are being measured.

A commonly encountered stray light artifact seen in spectra is the stray light peak. This peak, occasionally reported as real, is usually encountered as the wavelength scan approaches the lower limit of the spectrometer's range, where the source intensity is rapidly decreasing and the sample is exhibiting end absorption (the beginning of a strong absorbance band whose maximum lies beyond the wavelength range of the instrument). As the scan proceeds to lower wavelengths, the fraction of stray light is continually increasing due to the reduced source intensity. Because stray light is not being absorbed to any great degree, the rate of increase of absorbance decreases until finally the stray light becomes the predominant light reaching the detector. At this point the absorbance scan reaches a maximum and then declines because stray light, which is not being absorbed, becomes dominant. One should point out that during this process, stray light, being of longer wavelengths, probably has greater detector response than the short wavelength being scanned. Obviously, under these circumstances negative deviation from Beer's law will result be-

Table 25.2 The effect of stray light on absorbance readings.

True Absorbance	Stray Light (%)	Observed Absorbance	Error (%)
1.000	1.0	0.9629	3.8
	0.1	0.9961	0.4
	0.01	0.9996	0.04
	0.001	1.0000	0.004
2.000	1.0	1.7033	14.8
	0.1	1.9590	2.1
	0.01	1.9957	0.2
	0.001	1.9996	0.02
3.000	1.0	1.9629	34.6
	0.1	2.6994	10.0
	0.01	2.9587	1.4
	0.001	2.9957	0.1

cause the incident light is not monochromatic. Instrument manufacturers usually state stray light characteristics of their instruments in their specifications. A number of investigators have described methods for determining stray light (14–16).

Another error commonly encountered when making spectrophotometric measurements is called the finite slit width effect. The exit slit of the monochromator subtends a portion of the dispersed continuum from the grating or prism. If any light is to pass through the slit it must have a finite width. However due to its width, more than one wavelength of light, called the bandwidth, emerges. Although most of the energy emerging from the slit is of the selected or nominal wavelength, a small percentage is of adjacent wavelengths, called spectral bandwidth (SBW). SBW is simply the wavelength span, centered on the nominal wavelength, containing 75% of the radiant energy emerging from the slit. Thus, the narrower the SBW, the better conformity to Beer's law. If the SBW is too wide, negative deviation from Beer's law is likely, resulting in a false absorbance reading. Thus it is desirable to work with as narrow an SBW as possible. However, reducing the SBW usually requires decreasing the slit width (Eq. 25.7). Of course, this results in less light reaching the detector and consequently causes a reduced signal-to-noise (S/N) ratio. The certainty of determining an accurate absorbance is reduced by too wide a slit caused by Beer's law violation; when the slit is too narrow, errors due to increased noise can result. Obviously, this is a situation where the SBW should be optimized.

Another difficulty associated with a too-wide SBW is the loss of resolution. Wide slits lead to the loss of the fine structure of sharp bands. When an absorption band has been completely resolved, the observed peak height is equal to the true peak height and the observed band width (OBW) (fwhm) will be equal to the natural band width (NBW). Figure 25.2 illustrates the relationship between the ratio of the observed peak height to the true peak height versus the ratio SBW/NBW. When the SBW is one-tenth of the NBW, the deviation between the observed peak height and the true peak height is less than 0.5% (13). The simplest way to optimize the SBW is simply to obtain a succession of spectral scans, each with a narrower SBW, until the peak height

Figure 25.2 Relationship between the ratio of spectral bandwidth to natural bandwidth versus the ratio of observed peak height to true peak height.

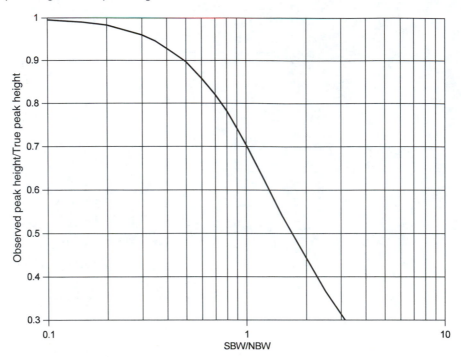

or the OBW no longer changes. Further reduction of the SBW will not improve the characteristics of the spectrum, but simply decreases the S/N.

A common practice that can sometimes lead to erroneous absorbance measurements occurs when a distilled water blank is used instead of a true blank for establishing the 100% transmittance or baseline reading. One may think that this corrects for absorbing species in the solvent and re-agents and because there are no known absorbing species in reagents or solvent, distilled water can be used instead. Unfortunately, the difference in refractive indices between the sample solution and reference solution must be kept reasonably close or reflective losses at the cell windows may not be the same. Even when the incident light is highly collimated and falls on the cell window at normal incidence, a small fraction of light is reflected back at each interface where there is a re-fractive index difference, at the two air–window interfaces, and the two window–solution interfac-es. Because the sample and reference cells are of the same composition, reflections from the air–window interfaces are compensated for. However, reflections from the solution–window in-terfaces may be different if the refractive indices of the sample and blank are not nearly the same. The relationship between the fraction of light reflected at an interface varies with wavelength and is given by Eq. (25.11):

$$\rho(\lambda) = \frac{(\eta_2 - \eta_1)^2}{(\eta_2 + \eta_1)^2} \tag{25.11}$$

where ρ is the fraction of the incident light reflected, η is the refractive index, and the subscripts refer to the solution and window. Failure to properly match the refractive indices of the sample and blank usually results in shifts of the baseline. The use of distilled water blanks is often responsible for the presence of negative baselines (T > 1.000). Reflections can also produce another type of error called multiple reflection path errors. This can occur when the light reflected from the second solution–window interface traverses across the cell, striking and being reflected from the first window–solution interface back to and through the second window. This ray has traversed the sample three times and thus may exhibit a positive bias in the absorbance read. Fortunately, this error is significant only where a great deal of reflection is present and with highly absorbing solutions (17).

Improper handling of cells can lead to a variety of measurement errors in spectrophotometry. It is important that cells be routinely cleaned in accordance with the manufacturer's recommendations in order to prevent plaqueing (the accumulation of absorbing substances on the surfaces of the cell windows). Positioning of both the sample and reference cells is extremely important if one is to achieve the highest degree of precision in the absorbance readings. This is one of the principal reasons for requiring the instrument to have a highly collimated light beam and to provide a sound, kinematically designed cell holder. Even then, the operator must make some effort to place the cells into the instrument cell compartment in a reproducible manner. One study has shown that it is not uncommon to encounter positioning errors as large as 0.4% (18). In order to avoid this kind of problem, some manufacturers have developed a sipper cell, which is not removed from the cell compartment to change samples. These cells simply draw sample into them from the sample container. Unfortunately, the use of sipper cells requires repeated rinsing between sample changes and they are somewhat troublesome to clean.

The chemical system producing the chromophoric group of interest may itself impose a number of limitations. Buffers should be considered to prevent acid–base equilibria from adversely affecting Beer's law. Sometimes it may be necessary to use mixed solvents (aqueous–organic) to reduce dissociation of the complex ion or molecule. Stability of the chromophore with respect to temperature, atmospheric oxidation, light sensitivity, and hydrolysis should always be kept in mind when selecting a spectrophotometric method, as these are all potential sources of error.

The operator should have proper training and conscientious attitude if operator errors are to be kept to a minimum. Although these two conditions may be met, there is never any assurance that operator errors will not occur. Therefore, it is important for any laboratory to perform replicate analyses and to install an effective quality assurance program with active recordkeeping of calibration tests routinely carried out, check analyses with standards, and periodic testing of reagents and standards.

Analytical Information

Qualitative

UV/Vis absorption spectra are often very useful in qualitative identification of molecular species. This often is accomplished by comparing the spectrum of an unknown species with a spectrum of a suspected substance from a library (19, 20) of spectra. These spectra are usually in a format of transmittance or absorbance versus wavelength (nm). Transmittance spectra, although never exceeding the 0% T and 100% T limits for any concentration, do not have intensities proportional to concentration and tend to lose spectral sharpness at low transmittance values. On the other hand, absorbance spectra exhibit bands that are proportional to concentration but are often limited to absorbances of less than 2. Also, absorbance spectra of the same compound but at different concentrations may not resemble one another too closely. This problem can be overcome somewhat by comparing $\log A$ with λ displays of the unknown and reference spectra. The spectral characteristics of these plots tend to be independent of concentration because $\log A$ is proportional to the absorptivity, a. These plots are displaced from one another by a factor proportional to the $\log bc$, the log of the product of the cell path length and the concentration.

One way of obtaining greater spectral detail is to display the first ($dA/d\lambda$ versus λ) or second ($d^2A/d\lambda^2$ versus λ) derivative of the absorbance spectrum. Derivatives produce maximums or minimums where the original spectrum exhibited inflection points and zero crossings where the original exhibited a maximum. Thus derivatives provide more easily recognizable features than are seen with the single maximum of the original spectrum. A disadvantage of these comparisons is that the shapes of derivative plots are sensitive to concentration. This problem can be alleviated by normalizing the absorbance spectra before taking the derivatives.

Molecular bonding and structural information can be correlated to the ultraviolet and visible absorption spectra. Theoretical treatment of the UV/Vis data can be found in a number of classic references (21–25). A more applied approach to the interpretation of UV/Vis absorption spectra can be found in other sources (26–29).

UV/Vis absorption spectral data are widely used for determining stoichiometry because one can measure one or more of the active species without perturbing the equilibrium under consideration. Many procedures have been developed for determining the stoichiometry of complex ions and sometimes the determination of the equilibrium constant. These methods incorporate different assumptions regarding the nature of the complex ion being studied and thus may be more applicable to certain types of complexes than the others. Some of these are method of continuous variations (30, 31), mole ratio method (32), slope ratio method (33, 34), and Bjerrum's method (35).

Quantitative

The greatest use of UV/Vis absorption spectroscopy lies in its application to quantitative measurements. The reasons for this stem from the ease with which most spectrophotometric measurements can be made, their sensitivity and precision, and the relatively low cost of instrument purchase and operation. A variety of techniques have been developed for different types of samples. Direct determinations are made when the analyte molecule contains a chromophore, thus allowing the direct measurement of its absorbance. Standards must be used to determine the absorptivity so that concentration can be calculated by Eq. (25.5) or by establishing a calibration plot from which the concentration can be determined by graphic interpretation or by regression analysis. Indirect determinations are commonly used when the analyte molecule does not contain a suitable chromophore. In these instances the analyte is made to quantitatively react with a molecule containing a chromophore and correlating the diminution of absorbance with the concentration of the analyte or by reacting with a reagent, which produces a chromophoric group.

Mixtures of absorbing analytes can be determined by simultaneous multicomponent analysis. Two wavelengths are chosen that produce the greatest differences in absorptivities of the two analytes. This is often at the wavelength maximum of each component. Because both analytes are assumed to absorb at both wavelengths, the measured absorbance is the sum of the absorbances by each analyte. This can be expressed as follows:

$$A_{\lambda 1} = {}^1A_{\lambda 1} + {}^2A_{\lambda 1} = b({}^1a_{\lambda 1}{}^1C + {}^2a_{\lambda 1}{}^2C) \qquad (25.12)$$

$$A_{\lambda 2} = {}^1A_{\lambda 2} + {}^2A_{\lambda 2} = b({}^1a_{\lambda 2}{}^1C + {}^2a_{\lambda 2}{}^2C) \qquad (25.13)$$

where subscripts $\lambda 1$ and $\lambda 2$ denote the wavelengths at which analytes 1 and 2 absorb most strongly, and superscripts 1 and 2 denote analytes 1 and 2. The absorptivities of both analytes are measured at both wavelengths using standard solutions of the analyte. The sample absorbance is measured at both wavelengths, $A_{\lambda 1}$ and $A_{\lambda 2}$. With the cell path b known, Eqs. (25.12) and (25.13) can be solved for the analyte concentrations ^1C and ^2C by substitution or by using determinants.

Theoretically, there is no limit to the number of analytes that can be determined in this manner, provided the proper number of wavelengths can be found. However, a loss of precision should be expected to occur with larger and larger numbers of analytes.

Reaction rate methods have gained measurably in popularity during the past 20 years. This stems from the fact that in general they are rapid procedures, they often do not require blank measurements, and because they are based on relative measurements, instrumental factors affecting the absorbance tend to be insignificant. However, they do require that the rates be measurably slower than the mixing rate but fast enough to provide a reasonable analysis time and not be significantly affected by instrument drift. In reaction rate methods it is not uncommon to find that the pH and temperature effects are much more critical than in conventional spectrophotometric methods. Several techniques have been developed for reaction rate methods. In all these procedures it is imperative that measurements be made during the interval established just after mixing and before the effects of equilibrium become significant. The slope method (7) involves computing the derivative (slope) of the response curve. The initial concentration of the rate determining species is proportional to the slope. The fixed time method (8) operates by measuring the absorbance change over a predetermined time interval. Under these conditions the change in absorbance should be proportional to the initial concentration. Another technique known as the fixed-signal method (8) measures the time required for a certain predetermined change in absorbance to occur.

Applications

Spectrophotometric analysis continues to be one of the most widely used analytical techniques available. Many methods are available for a variety of analytes (such as colored, colorless, natural, synthetic, inorganic, and organic analytes) and sample types ranging from in-situ biological assays to the determination of trace elements in steels. UV/Vis detection of chromatographic and electrophoretic eluents remains one of the most popular choices among chromatographers. Many medical diagnostic test kits use photometric measurements. Diabetics commonly use blood-glucose analysis kits based on the glucose oxidase enzyme reaction that secondarily produces a colored product. In the food industry, winemakers have long recognized the effect of iron levels on the taste of wines and consequently are one the largest users of 1,10-phenanthroline for determining iron photometrically. A common field test for chlorine in swimming pools and drinking water is based on the color produced by the action of chlorine on o-tolidine.

Many compilations of methodology for a variety of analytes and sample types that are regularly updated are available (36–38). Other general sources for spectrophotometric analysis are commonly consulted and found helpful (39–41). Methods specific for metals (42–44), and nonmetals (45) should be consulted when dealing with these analytes. Standard methods specific to certain industries and areas of study are very useful sources when specific sample types are being considered, such as water and wastewater (46) and pharmaceuticals (47).

Nuts and Bolts

Relative Costs

Color comparator systems for specific analytes can be purchased for as little as $25 to as much as several hundred dollars. The cost of filter photometers ranges from several hundred dollars to about $1,000. The cost of a spectrophotometer can range from about $5,000 to $60,000 depending on the quality of the optical and electronic system and the number and types of accessories purchased.

Spectro-grade solvents are perhaps the most costly consumable required of photometric and spectrophotometric analysis. A good-quality microanalytical balance can be a great cost-saving device, allowing the weighing of very small amounts of standards and thus avoiding multiple dilutions with costly spectro-grade solvents. It also reduces the amounts of costly reference standards. Both solvents and standards can be purchased from most chemical suppliers. Cells or cuvettes are also available from these suppliers and many instrument manufacturers.

Vendors of Instrumentation and Accessories

A number of factors must be considered when buying a UV/Vis spectrophotometer, among which is a good understanding of the terminology and implications of instrument specifications (48). One must then consider what type of UV/Vis absorption measurements are to be made (such as high or low absorbance, narrow or wide band absorptions, wavelength range, and precision and accuracy required) and the nature of the samples themselves (such as gas, liquid, or solid, sample stability, and sample throughput). Below are listed manufacturers that provide a variety of photometric instruments and accessories. This list is not all-encompassing. The article *Ultraviolet and Light Absorption Spectrometry*, featured in the Fundamental Reviews (on even-numbered years) in *Analytical Chemistry* (38) discusses new instrumentation introduced in the past 2 years. A number of trade publications issue yearly buyer's guides, which might be helpful in determining which firms manufacture photometric instrumentation.

Acton Research Corp.
525 Main St.
Acton, MA 01720
phone: 508-263-3584
fax: 508-263-5086
email: mc@acton-research.com
Internet: http://www.acton-research.com

Beckman Instruments Inc.
2500 Harbor Blvd.
Fullerton, CA 92634
phone: 800-742-2345
fax: 800-643-4366

Ciba Corning Diagnostics—Gilford Systems
132 Artino St.
Oberlin, OH 44074
phone: 800-445-3673

Fisher Scientific Co.
711 Forbes Ave.
Pittsburgh, PA 15219
phone: 800-766-7000
fax: 800-926-1166
email: info@fisher1.com
Internet: http://www.fisher1.com

GBC Scientific Equipment Inc.
3930 Ventura Dr.
Arlington Heights, IL 60004
phone: 800-445-1902
fax: 708-506-1901

Groton Technology Inc.
45 Winthrop St.
Concord, MA 01742
phone: 508-371-1900
fax: 508-371-2045

Hach Corporation
5000 Lindbergh Dr.
Loveland, CO 80538
phone: 800-227-4224
fax: 970-669-2932

Hewlett-Packard
3495 Deer Creek Rd.
Palo Alto, CA 94025
phone: 800-227-9770
Internet: http://www.hp.com/go/analytical

Hitachi Instruments Inc.
3100 N. 1st St.
San Jose, CA 95134
phone: 800-455-4440
fax: 408-432-0704
email: info@hii.hitachi.com
Internet: http://www.hii.hitachi.com

Instruments S.A., Inc.
3880 Park Ave.
Edison, NJ 08820
phone: 800-438-7739
fax: 908-549-5125
email: john@isainc.com

JASCO Inc.
8649 Commerce Dr.
Easton, MD 21601
phone: 800-333-5272
fax: 410-822-7526
email: jascoinc.com

On-Line Instrument Systems Inc.
130 Conway Dr.
Bogart, GA 30622
phone: 800-852-3504
fax: 706-353-1972
email: olis@mindspring.com
Internet: http://www.olisweb.com

Perkin-Elmer Corporation
761 Main Ave.
Norwalk, CT 06859
phone: 800-762-4000
fax: 203-762-4228
email: info@perkin-elmer.com
Internet: http://www.perkin-elmer.com

Shimadzu
7102 Riverwood Dr.
Columbia, MD 21046
phone: 800-477-1227
fax: 410-381-1222
Internet: http://www.shimadzu.com

Spectronic Instruments
820 Linden Ave.
Rochester, NY 14625
phone: 800-654-9955
fax: 716-248-4014
email: info@spectronic.com

Varian Instruments
P.O. Box 9000
San Fernando, CA 91340
phone: 800-926-3000

Carl Zeiss Inc.
1 Zeiss Dr.
Thornwood, NY 10594
phone: 800-233-2343
fax: 914-681-7446

Required Level of Training

One of the more attractive features of spectrophotometric analysis is the speed and ease with which one can learn to operate the instrumentation. People with no experience in operating any kind of spectrophotometer can operate even some of the more sophisticated models after only a few hours training. Also, the instrument manuals provided by most manufacturers provide excellent tutorials for operating their instruments. With a minimum amount of experience and some consultation with certain literature sources (49–51), the operator can perform high-quality spectrophotometric measurements in a short time.

Unfortunately, detailed interpretation of UV/Vis spectral data can require a great deal of experience and training. However, a number of literature sources are available to assist the analyst in qualitative identification of compounds by the interpretation of spectral data (26, 29, 52, 53). The interpretation of quantitative spectrophotometric data is generally straightforward and permits the analyst to apply the data with a minimum amount of training.

Service and Maintenance

Except for the simplest types of instrument repair, service is best left to local service personnel. It is usually wise to call the service department of your local instrument vendor and discuss the problem with them. Often, problems can be solved by instructions taken over the phone. Even when this is not possible, it is advisable to provide the service person with as much information as possible before his or her arrival. That way, if special parts or tools are required for the service call, he or she will be prepared.

Routine maintenance is imperative if one is to get the best performance from the instrument and extend its useful lifetime. Most manufacturers describe routine maintenance regimens for their instruments. Good maintenance is responsible for the fact that many spectrophotometers are still functioning within original specifications after more than 50 years of service. To ensure that the instrument is operating within its specifications, it is advisable to do routine tests with standards of known spectral characteristics and concentration.

A variety of standard reference materials (SRMs) and procedures for calibration of wavelength scales, transmittance and absorbance scales, and stray light characteristics can be found in the *Standard Reference Materials Catalog* of the National Institute of Standards and Technology (NIST) (54). Table 25.3 describes and summarizes the available SRMs for calibrating spectrophotometers. NIST has published a number of documents discussing various aspects of spectrophotometry, ranging from general aspects of accuracy (55), the use of acidic potassium dichromate solutions as a standard (56), didymium glass filters as a wavelength standard (57), and holmium oxide solution as a wavelength standard (58). A set of standard materials and instructions on their use for calibrating spectrophotometers is available from the Milton Roy Company (59).

Table 25.3 NIST standard reference materials for molecular absorption.

SRM	Type	λ Range (nm)	Description
930e	Transmittance	440–635	3 glass filters
931e	Absorbance	302–678	12 ampules, liquid filters
935a	UV absorbance	235–350	15 g potassium dichromate
1930	Transmittance	440–635	3 glass filters
2030a	Transmittance	465.0	1 glass filter
2032	Stray light	240–280	25 g potassium iodide
2034	Wavelength	240–650	1 sealed cuvette, Ho_2O_3 solution

References

1. P. Bouguer, *Essai d'optique sur la gradation de la luminère* (Paris: Claude Tombert, 1729).

2. J. H. Lambert, *Photometria sine de mensura et gradibus luminis colorum et umbrae* (Augsburg: Eberhard Klett, 1760).

3. A. Beer, *Ann. Phys. Chem.*, 86 (1852), 78.

4. F. C. Strong, *Analytical Chemistry*, 24 (1952), 338.

5. F. E. Lytle, *Analytical Chemistry*, 46 (1974), 545A.

6. J. D. Ingle and S. R. Crouch, *Spectrochemical Analysis* (Englewood Cliffs, N.J.: Prentice-Hall, 1988), pp. 116–17.

7. H. L. Pardue, "Applications of Kinetics to Automated Quantitative Analysis," in C. N. Reilley and F. W. McLafferty, eds., *Advances in Analytical Chemistry and Instrumentation*, vol. 3 (New York: Wiley-Interscience, 1968).

8. W. J. Blaedel and G. P. Hicks, "Analytical Applications of Enzyme-Catalyzed Reactions," in C. N. Reilley and F. W. McLafferty, eds., *Advances in Analytical Chemistry and Instrumentation*, vol. 3 (New York: Wiley-Interscience, 1968).

9. L. D. Rothman, S. R. Crouch, and J. D. Ingle, Jr., *Analytical Chemistry*, 47 (1975), 1226.

10. J. D. Ingle, Jr., and S. R. Crouch, *Analytical Chemistry*, 44 (1972), 1375.

11. H. L. Pardue, T. E. Hewitt, and M. J. Milano, *Clin. Chem.*, 20 (1974), 1028.

12. J. O. Erickson and T. Surles, *Amer. Lab.*, 8, no. 6 (1976), 41.

13. *Optimum Parameters for Spectrophotometry* (Palo Alto, CA: Varian Instruments Division, 1977).

14. W. Slavin, *Analytical Chemistry*, 35 (1963), 561.

15. D. D. Tunnicliff, *J. Opt. Soc. Am.*, 45 (1955), 963.

16. R. E. Poulson, *Appl. Opt.*, 3 (1964), 99.

17. R. W. Burnett, *Analytical Chemistry*, 45 (1973), 383.

18. T. Surles and J. Erickson, *Varian Instrum. Appl.*, 8 (1974), 5.

19. Sadtler Research Laboratories, *Ultraviolet Reference Spectra* (Philadelphia: Sadtler Research Laboratories, updates at regular intervals).

20. L. Lang, ed., *Absorption Spectra in the Ultraviolet and Visible Regions* (New York: Academic Press, continuing series since 1961).

21. H. H. Jaffé and M. Orchin, *Theory and Applications of Ultraviolet Spectroscopy* (New York: Wiley, 1962).

22. C. N. R. Rao, *Ultra-Violet and Visible Spectroscopy: Chemical Applications*, 3rd ed. (London: Butterworths, 1975).

23. R. P. Bauman, *Absorption Spectroscopy* (New York: Wiley, 1962).

24. L. E. Orgel, *An Introduction to Transition Metal Chemistry, Ligand-Field Theory* (New York: Wiley, 1960).

25. F. A. Cotton, A. Wilkinson, and P. L. Gauss, *Basic Inorganic Chemistry*, 2nd ed. (New York: Wiley, 1987).

26. R. M. Silverstein, G. C. Bassler, and T. C. Morrill, *Spectrometric Identification of Organic Compounds*, 5th ed. (New York: Wiley, 1991), pp. 299–305.

27. L. M. Fieser and M. Fieser, *Steroids* (New York: Reinhold, 1959), pp. 15–24.

28. R. B. Woodward, *J. Am. Chem. Soc.*, 63 (1941), 1123; 64 (1942), 72, 76.

29. A. I. Scott, *Interpretation of the Ultraviolet Spectra of Natural Products* (New York: Pergamon Press, 1964).

30. P. Job, *Ann. Chim.*, 9 (1928), 113.

31. J. H. Yoe and A. L. Jones, *Ind. Eng. Chem., Anal. Ed.*, 61 (1944), 111.

32. W. C. Vosburgh and G. R. Cooper, *J. Am. Chem. Soc.*, 63 (1941), 437.

33. S. D. Christian, *J. Chem. Educ.*, 45 (1968), 713.

34. F. J. C. Rossotti and H. Rossotti, *The Determination of Stability Constants* (New York: McGraw-Hill, 1961).

35. D. T. Sawyer, W. R. Heineman, and J. M. Beebe, *Chemistry Experiments for Instrumental Methods* (New York: Wiley, 1984), pp. 198–205.

36. ASTM, *Annual Book of ASTM Standards* (66 volumes in 16 sections) (Philadelphia: American Society for Testing and Materials, 1987).

37. S. Williams, ed., *Official Methods of Analysis of the Association of Official Analytical Chemists*, 16th ed. (Arlington, VA: Association of Official Analytical Chemists, 1995).

38. J. A. Howell and L. G. Hargis, *Anal. Chem.*, 68 (1996), 169R.

39. D. Eckroth, ed., *Encyclopedia of Chemical Technology*, 3rd ed. (New York: Wiley, 1984).

40. L. C. Thomas and G. J. Chamberlin, *Colorimetric Analytical Methods*, 9th ed. (Salisbury, England: Tintometer Press, 1980).

41. Z. Marczenko, *Spectrophotometric Determination of Elements* (New York: Halsted, 1975).

42. F. D. Snell, *Photometric and Fluorometric Methods of Analysis*, parts 1 and 2 (New York: Wiley, 1978).

43. E. B. Sandell and H. Onishi, *Photometric Determination of Traces of Metals: General Aspects*, 4th ed. of part I of *Colorimetric Determination of Traces of Metals* (New York: Wiley, 1978).

44. H. Onishi, *Photometric Determination of Traces of Metals*, parts IIA and IIB, 4th ed. (New York: Wiley, 1989).

45. D. F. Boltz and J. A. Howell, *Colorimetric Determination of Nonmetals*, 2nd ed. (New York: Wiley, 1978).

46. American Public Health Association, *Standard Methods for the Examination of Water and Wastewater*, 17th ed. (Washington, D.C.: American Public Health Association, 1992).

47. United States Pharmacopeial Convention, *United States Pharmacopoeia*, 23rd rev. (New York: U.S. Pharmacopoeia Convention, 1995).

48. L. G. Hargis and J. A. Howell, *Visible and Ultraviolet Spectrophotometry*, in B. W. Rossiter and R. C. Baetzold, eds., *Physical Methods of Chemistry* (New York: Wiley, 1993).

49. D. A. Skoog and J. J. Leary, *Principles of Instrumental Analysis*, 4th ed. (Saunders College Publishing, 1992), Chs. 5–8.

50. R. D. Braun, *Introduction to Instrumental Analysis* (New York: McGraw-Hill, 1987), Ch. 9.

51. G. D. Christian and J. E. O'Reilly, *Instrumental Analysis*, 2nd ed. (Boston: Allyn & Bacon, 1986), Ch. 7.

52. V. M. Parikh, *Absorption Spectroscopy of Organic Molecules* (Reading, MA: Addison-Wesley, 1974).

53. C. J. Creswell, O. Runquist, and M. M. Campbell, *Spectral Analysis of Organic Compounds*, 2nd ed. (Minneapolis: Burgess Publishing Company, 1972).

54. National Institute of Standards and Technology, *Standard Reference Materials Catalog 1995–1996*, NIST Special Publication 260 (Gaithersburg, MD: National Institute of Standards and Technology, 1995).

55. National Bureau of Standards, *Accuracy in Analytical Spectrophotometry*, NBS Special Publication 260-81, Washington, D.C., 1983.

56. National Bureau of Standards, *Certification and Use of Acidic Potassium Dichromate Solutions as an Ultraviolet Absorbance Standard: SRM 935*, NBS Special Publication 260-54, Washington, D.C., 1977.

57. National Bureau of Standards, *Didymium Glass Filters for Calibrating the Wavelength Scale of Spectrophotometers: SRM 2009, 2010, 2013, and 2014*, NBS Special Publication 260-66, Washington, D.C., 1979.

58. National Bureau of Standards, *Holmium Oxide Solution Wavelength Standard from 240 to 640 nm: SRM 2034*, NBS Special Publication 260-102, Washington, D.C., 1986.

59. Milton Roy Company, *Spectronic Standards: User's Manual* (Rochester, N.Y.: Milton Roy Company, Analytical Products Division, 1990).

Chapter 26

Molecular Fluorescence and Phosphorescence Spectrometry

Earl L. Wehry

University of Tennessee
Department of Chemistry

Summary

General Uses

- Quantification of aromatic, or highly unsaturated, organic molecules present at trace concentrations, especially in biological and environmental samples
- Can be extended to a wide variety of organic and inorganic compounds via chemical labeling and derivatization procedures

Common Applications

- Determination of trace constituents in biological and environmental samples
- Detection in chromatography (especially high-performance liquid chromatography) and electrophoresis
- Immunoassay procedures for detection of specific constituents in biological systems
- Environmental remote sensing (hydrologic, aquatic, and atmospheric)

- In-situ analyses in biological systems (such as single cells) and cell sorting (flow cytometry)
- Studies of the molecular microenvironment of fluorescent molecules (fluorescent probe techniques)
- DNA sequencing
- Studies of ligand binding in biological systems
- Studies of macromolecular motions via polarized fluorescence measurement
- Fundamental studies of ultrafast chemical phenomena

Samples

State

Almost any solid, liquid, or gaseous sample can be analyzed, although solid samples may require a special sample compartment. Highly turbid liquid samples may cause difficulty.

Amount

Samples can be extremely small (for example, specialized techniques for examining nanoliter sample volumes have been developed).

Preparation

- It may be necessary to convert analyte to fluorescent derivative or tag analyte with a fluorescent compound.
- Use of high-purity solvents and clean glassware is essential.
- Complex mixtures may need extensive cleanup and separation before analysis.
- Turbid samples may require filtering or more extensive cleanup to minimize scatter background.
- Solvents that absorb strongly in the ultraviolet (UV) (such as toluene) must usually be avoided.

Analysis Time

Normally 1 to 10 minutes is needed to obtain a spectrum. Analysis time is determined primarily by time required for preliminary sample cleanup (which may be extensive and lengthy) rather than time required to obtain spectral data.

Limitations

General

- Analysis is limited to aromatic and highly unsaturated molecules unless derivatization or tagging procedure is used.

- Mixtures may need extensive cleanup before measurement.
- The possibility of quenching in mixtures means that care must be exercised in calibration (such as the use of standard additions).

Accuracy

Accuracy is highly dependent on the complexity of the sample and care must be used in calibration. Accuracies of 1% relative or better are possible if sufficient care is exercised.

Sensitivity and Detection Limits

Sensitivity is highly dependent on fluorescence quantum yield of analyte and the extent to which blank signals (such as impurity fluorescence and Rayleigh and Raman scatter) are minimized. For intensely fluorescent compounds (such as polycyclic aromatic hydrocarbons), detection limits of 10^{-11} to 10^{-12} M can readily be achieved using commercial instrumentation. Using specialized apparatus and great care, highly fluorescent nonphotosensitive analytes can be detected at or near single-molecule quantities.

Complementary or Related Techniques

- UV/visible (UV/Vis) absorption spectroscopy is much less sensitive but more nearly universal and often more accurate.
- Chemiluminescence is less widely applicable, but uses simpler apparatus than fluorescence, is more sensitive for some analytes, and often exhibits much lower blank because no light source is used.

Introduction

Photoluminescence is a type of optical spectroscopy in which a molecule is promoted to an electronically excited state by absorption of ultraviolet, visible, or near infrared radiation. The excited molecule then decays back to the ground state, or to a lower-lying excited electronic state, by emission of light. The emitted light is detected. Photoluminescence processes are subdivided into fluorescence and phosphorescence (1). For simplicity, we use the term *fluorescence* to mean both fluorescence and phosphorescence.

The key characteristic of fluorescence spectrometry is its high sensitivity. Fluorometry may achieve limits of detection several orders of magnitude lower than those of most other techniques. This is known as the fluorescence advantage. Limits of detection of 10^{-10} M or lower are possible for intensely fluorescent molecules; in favorable cases under stringently controlled conditions, the ultimate limit of detection (a single molecule) may be reached. Because of the low detection limits, fluorescence is widely used for quantification of trace constituents of biological and environmental samples. Fluorometry is also used as a detection method in separation techniques, especially liquid chromatography and electrophoresis. The use of fluorescent tags to detect nonfluorescent molecules is widespread and has numerous applications (such as DNA sequencing).

Because photons can travel through transparent media over large distances, fluorescence is applicable to remote analyses. Atmospheric remote sensing is an example of this type of application. Remote sensing often requires the use of specialized apparatus (such as fiberoptics or laser sources).

The spectral range for most molecular fluorescence measurements is 200 to 1000 nm (10,000 to 50,000 cm^{-1}). Hence, optical materials used in UV/Vis absorption spectrometry are suitable for molecular fluorescence. Instrumentation for molecular fluorescence spectrometry is available from numerous manufacturers. Although commercial fluorescence spectrometers are useful in many situations, there are several important specialized applications (such as atmospheric remote sensing) for which truly suitable commercial instrumentation is not readily available.

Excitation of a molecule does not automatically produce fluorescence; many molecules exhibit very weak fluorescence. Most intensely fluorescent organic molecules contain large conjugated π-electron systems (1, 2). For example, most polycyclic aromatic hydrocarbons are intensely fluorescent. Very few saturated organic molecules, and relatively few inorganic molecules, exhibit intense fluorescence. To extend the applicability of fluorometry to the many compounds that do not exhibit intense native fluorescence, chemical reactions can be used to convert (derivatize) nonfluorescent molecules to fluorescent derivatives, or a nonfluorescent molecule may have chemically attached to it a fluorescent tag or label (3–6).

For a sample needing no preliminary cleanup, a fluorometric analysis can be carried out in 10 min or less. However, analyses of complex materials often require considerable preliminary cleanup. Fluorescence measurements may be carried out on liquid, gaseous, and solid samples. Solvents do not interfere unless they absorb at the wavelength used to excite the analyte or act to decrease the efficiency with which the excited analyte molecule fluoresces. However, fluorescent contaminants in solvents and laboratory glassware can be a major nuisance.

Fluorometry is more selective than UV/Vis absorption spectrometry for two reasons. First, many molecules absorb strongly in the UV or visible range but do not exhibit detectable fluorescence. Second, two wavelengths (excitation and emission) are available in fluorometry, but only one wavelength is available in absorptiometry. If two sample constituents with similar absorption spectra fluoresce at different wavelengths, they may be distinguished from one another by appropriate choice of emission wavelength. Similarly, two compounds that have similar fluorescence spectra but absorb strongly at different wavelengths may be distinguished by proper choice of excitation wavelength (selective excitation).

The selectivity of fluorometry is limited by the broad, featureless nature of the absorption and fluorescence spectra of most molecules. In UV/Vis absorption and fluorescence spectra, bandwidths of 25 nm or more are common, especially for polar or hydrogen-bonding molecules in polar solvents. The positions of spectral bands are not sensitive to the finer details of molecular structure, such as the presence of specific functional groups, and often cannot be predicted a priori with confidence. Hence, fluorometry is not generally useful for molecular identification.

The absorption and fluorescence spectra of a mixture of fluorescent compounds may be a jumble of overlapping bands. Such samples must be subjected to preliminary cleanup and separation (which may be quite time-consuming) or specialized sample-preparation and measurement techniques may be used; these may be instrumentally complex or time-consuming.

Because fluorescence measurements are rapid and use relatively inexpensive and rugged instrumentation, fluorescence can be used to screen samples, to generate preliminary data that allow an analyst to decide whether a sample requires detailed characterization, perhaps by a more expensive and complex technique such as gas chromatography/mass spectrometry (7). This is especially appropriate for environmental samples, which usually are very complex. Small, portable fluorometric instruments are suitable for performance of such screening operations in the field (8).

The fluorescence spectrum and intensity of a molecule often depend strongly on that molecule's environment. For example, changes in the "polarity" or hydrogen-bonding ability of a solvent may cause dramatic changes in the fluorescence behavior of a solute (6, 9). Thus, the fluorescence characteristics of probe molecules may be used to make inferences about their immediate microenvironments. Fluorescent probe studies are a very important application of fluorometry, especially in biological and polymer science.

Molecular fluorescence spectrometry is usually blank-limited. Limits of detection are often governed not by one's ability to induce or detect the fluorescence of the analyte, but rather by the ability to distinguish the analyte fluorescence from Rayleigh and Raman scatter radiation generated by the sample, as well as from fluorescence of other sample constituents, fluorescent contaminants in the solvent, and fluorescent species adsorbed on the walls of the sample container. Whenever fluorescence spectrometry is used for trace analyses, scrupulous care must be devoted to maximizing the fluorescence signal produced by the analyte while minimizing blank fluorescence and scattering signals. Use of solvents (10) and sample containers that are as free as realistically possible of fluorescent contaminants is essential.

How It Works

Physical and Chemical Principles

The initial step in a fluorescence measurement is electronic excitation of an analyte molecule via absorption of a photon. Once formed, an excited molecule has available a variety of decay processes by which it can rid itself of the energy imparted to it by absorption. In addition to fluorescence (the desired decay route), there are nonradiative decay processes, leading to release of energy in the form of heat rather than light. Other sample constituents may interact with an excited analyte molecule in such a way as to prevent it from fluorescing; such processes are called quenching. Also, an electronically excited molecule may undergo chemical reaction (photodecomposition).

The fraction of electronically excited molecules that decay to the ground state by fluorescence is called the fluorescence quantum yield. The maximum possible value for the fluorescence quantum yield is 1.00. The number of molecules known to exhibit fluorescence quantum yields of unity can be counted on the fingers of one hand. Most intensely fluorescent molecules (that is, those with fluorescence efficiencies of 0.05 or greater) have extended π-electron systems (such as polycyclic aromatic hydrocarbons) (1, 2).

In fluorescence, the spin multiplicities of the ground and emissive excited states are the same. In most organic molecules, the ground state is a singlet state (all spins paired). Fluorescence occurs when a molecule has been promoted to an excited singlet state by absorption, and then decays back to the ground singlet state by emission. Fluorescence generally occurs only from the first excited singlet state (that is, the excited singlet state of lowest energy), irrespective of which excited singlet state was initially produced by absorption.

Phosphorescence is a light emission process in which the excited and ground states have different spin multiplicities. In an organic molecule whose ground state is a singlet, there are several energetically accessible triplet excited states (two unpaired spins). Following excitation into the manifold of singlet excited states by absorption, a molecule may undergo nonradiative decay (intersystem crossing) to the manifold of triplet states. The triplet state may emit a photon as the molecule decays back to the ground singlet state (phosphorescence).

Information Available from Fluorescence Measurements

Two basic types of spectra can be produced by a fluorescence spectrometer. In a fluorescence spectrum, or emission spectrum, the wavelength of the exciting radiation is held constant (at a wavelength at which the analyte absorbs) and the spectral distribution of the emitted radiation is measured. In an excitation spectrum, the fluorescence signal, at a fixed emission wavelength, is measured as the wavelength of the exciting radiation is varied. Because an analyte can fluoresce only after it has absorbed radiation, an excitation spectrum identifies the wavelengths of light that the analyte is able to absorb. Thus, subject to certain constraints, the excitation spectrum of a molecule should be the same as its UV/Vis absorption spectrum.

The excitation spectrum for a compound should not change as the emission wavelength is varied. Whenever the excitation spectrum varies with choice of emission wavelength, there is good reason to believe that two or more different substances are responsible for the observed fluorescence.

The maximum in the fluorescence spectrum of a compound occurs at longer wavelength than the maximum in the absorption spectrum. The wavelength difference between the absorption and fluorescence maxima is called the Stokes shift. Often, the Stokes shift is large (20 to 50 nm), especially for polar solutes in polar solvents. There is often some overlap, but not a great deal, between the absorption and fluorescence spectra of a compound. Both spectra may exhibit wavelength shifts whenever the solvent is changed; again, these effects are largest for polar solutes dissolved in polar, hydrogen-bonding solvents (6, 9).

In many fluorescence spectrometers, one can simultaneously vary the wavelengths of the exciting and emitted radiation. Such measurements, commonly called synchronous scanning, are useful in the analysis of mixtures (11).

Fluorometry is a multidimensional technique (12). Several types of information (in addition to spectra and signal intensities) can be obtained. The fluorescence of a molecule may be partially or fully polarized. Measurements of fluorescence polarization can provide important information, particularly for macromolecules; the use of polarized fluorescence measurements is widespread in biological (13) and polymer (14) science.

Also, the singlet excited states responsible for fluorescence of organic molecules have finite lifetimes, usually nanoseconds. (The triplet excited states responsible for phosphorescence of organic molecules have much longer lifetimes, often milliseconds or longer.) The fluorescence or phosphorescence rate of a molecule can be measured, and changes in fluorescence spectra as a function of time (time-resolved spectra) can be obtained. Measurements of time-resolved spectra or decay times can aid in analytical applications of fluorometry, and can also provide unique fundamental information in the study of very fast chemical and physical phenomena.

General-Purpose Instrumentation for Molecular Fluorescence Measurements

A block diagram of a fluorometer is shown in Fig. 26.1. In addition to the optical components shown, most fluorometers have dedicated computers, which control instrumental operating parameters (excitation and emission wavelengths, wavelength scan rates, monochromator slit widths, detector parameters) and the acquisition of spectral data, and also may be used for postprocessing of the data.

Figure 26.1 Block diagram of fluorescence spectrometer using conventional right-angle fluorescence collection.

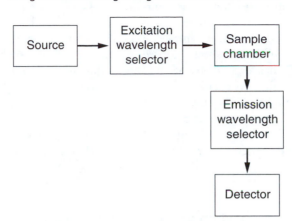

Sources

The signal produced by an analyte is proportional to the number of excited analyte molecules formed per unit time. Thus, the source must produce high optical power (that is, a large number of photons per unit time). Because molecular absorption spectra usually are broad, a highly monochromatic source is not generally needed; an intense continuum source that emits throughout the UV, visible, and near infrared regions is adequate. The source used in most commercial fluorometers is the xenon arc lamp, the characteristics of which are described elsewhere (15).

For certain applications, it is preferable to use a laser excitation source (16). Few fluorescence spectrometers using laser sources are commercially available; most such instruments are intended for highly specific applications such as analyses of uranium in the nuclear industry (17). A Raman spectrometer (which uses a laser source) often can serve as an excellent, albeit expensive, instrument for laser-induced fluorescence spectrometry.

Wavelength Selectors

Portable, inexpensive fluorescence spectrometers use filters as wavelength selectors. Such instruments (filter fluorometers) are used when it is sufficient to measure fluorescence intensity at a single excitation and emission wavelength. These instruments are used in environmental field screening (8), hospital or clinical settings (18), and other applications in which low cost and small size are crucial. Moreover, filters can transmit a very large number of photons from source to sample and from sample to detector. Thus, filter instruments may be used in ultratrace analysis, wherein it is crucial to maximize the fluorescence signal that impinges on the detector, at the cost of decreased selectivity.

Most fluorometers in laboratory environments use grating monochromators as excitation and emission wavelength selectors. Usually, only moderate spectral resolution (1 to 2 nm) is needed. Parameters governing the choice of a monochromator for use in a fluorescence spectrometer are discussed in detail elsewhere (19).

Sample Illumination

The most common arrangement is the right-angle geometry in Fig. 26.1, wherein fluorescence is viewed at a 90° angle relative to the direction of the exciting light beam. This geometry is suitable for weakly absorbing solution samples. For solutions that absorb strongly at the excitation wavelength, and for solids (or samples adsorbed on solid surfaces, such as thin-layer chromatography plates), a front-surface geometry often is preferable; here, fluorescence is viewed from the face of the sample on which the exciting radiation impinges.

For solution samples, rectangular 1-cm glass or fused silica cuvettes with four optical windows are usually used. For specialized applications, when very low limits of detection are required or it is necessary to illuminate a very small volume of solution, various flow (20, 21) or windowless (22) cells have been designed. Fiberoptics also are widely used in fluorometry.

Detectors

The fluorescence signal for an analyte present at low concentration is small; thus, a key requirement for a detector is its ability to detect weak optical signals. A photomultiplier tube (PMT) is used as the detector in most fluorescence spectrometers. PMTs used in fluorometry are chosen for low noise and high sensitivity, and are sometimes operated at subambient temperatures to improve their signal-to-noise ratios.

The main shortcoming of a PMT is that it is a single-channel detector. To obtain a spectrum, one must mechanically scan the appropriate monochromator across the wavelength range encompassed by the spectrum, which may be 50 nm or more. Thus, it is difficult to obtain spectra of transient species or analytes that remain in the observation region for a short time (such as eluents from chromatographic columns). It has long been recognized that a multichannel instrument using an array of detectors would be preferable for such applications because the entire spectrum could be viewed at once. UV/Vis absorption spectrometers with array detectors are commercially available and widely used.

Until recently, no electronic array detector has been competitive with a PMT in the detection of weak optical signals. That situation is changing as new classes of electronic array detectors are developed and improved. At present, the most promising electronic array detector for fluorometry is the charge-coupled device (CCD) (23). Fluorescence instruments using CCDs or other high-performance array detectors are not numerous, but will become more common in the future.

Corrected Spectra

Most fluorometers are single-beam instruments (see Fig. 26.1). Excitation and fluorescence spectra obtained using such an instrument are distorted, due to variation of source power or detector sensitivity with wavelength. Spectra of the same sample obtained using two different fluorometers may therefore be quite dissimilar (1); even changing the source or detector in a fluorometer may alter the apparent fluorescence or excitation spectrum of a compound. It is possible instrumentally to eliminate these artifacts, and several manufacturers offer instruments that can generate corrected spectra. Because most published fluorescence spectra are uncorrected, they cannot readily be reproduced by other investigators. Hence, there are few extensive and broadly useful data bases of fluorescence spectra.

That a fluorescence spectrometer is a single-beam instrument also means that fluctuations in the power output of the excitation source produce noise. This problem may be solved by splitting off a portion of the source output and viewing it with a second detector, and electronically ratioing the observed fluorescence signal to the output of the detector that is used to monitor the source power. High-performance commercial fluorometers have this capability.

More detailed discussions of instrumentation for measurement of fluorescence are available elsewhere (1, 6, 13, 24).

Specialized Types of Fluorescence Measurements

Several types of useful fluorescence measurements may not be possible using the simple fluorometer illustrated in Fig. 26.1. Some examples of more specialized fluorometric techniques are considered below.

Synchronous Fluorescence and Excitation–Emission Matrices

As noted above, it is possible to scan the excitation and emission monochromators simultaneously (synchronous fluorometry) (11). Often, synchronous fluorometry is carried out by scanning the excitation and emission monochromators at the same rate while keeping the wavelength difference (or offset) between them constant.

The main purpose of synchronous scanning is to generate spectra having decreased bandwidths. For example, Fig. 26.2(a) shows the excitation and fluorescence spectra of perylene, an aromatic hydrocarbon, in ethanol solution. The Stokes shift for perylene is relatively small (3 nm). If one carries out a synchronous scan with the wavelength offset between the excitation and emission monochromators held at 3 nm, one obtains the spectrum shown in Fig. 26.2(b) (25). The width of the synchronous spectrum is much smaller than that of either the excitation or fluorescence spectrum. When dealing with a mixture of fluorescent components, the beneficial effect of synchronous scanning is to greatly simplify the spectrum and decrease the extent of spectral overlaps (11, 25).

In some instances, it may be preferable to use a constant wavenumber (rather than wavelength) offset between the monochromators in synchronous fluorometry (26).

Because the synchronous spectrum for a compound depends on both the excitation and fluorescence spectra, it is strongly dependent on the wavelength (or wavenumber) offset. If one were to run many synchronous spectra for a particular compound, each at a different wavelength offset, one could acquire all information present in the absorption and fluorescence spectra of the compound. This information could then be set up in the form of a two-dimensional matrix called an excitation–emission matrix (27, 28). Acquiring an excitation–emission matrix by running many synchronous scans is slow; it can be generated much more rapidly by using an array-detector fluorometer designed expressly for this purpose (27).

The data in an excitation–emission matrix are often visually presented in the form of a fluorescence contour plot, wherein points of equal detector signal are connected to produce what looks like a topographic map. Examples of such contour plots for several complex samples are shown in Fig. 26.3 (29). These plots can be quite useful for distinguishing complex samples from one another. Each one serves as a spectral fingerprint for a particular complex material (such as petroleum, coal-derived liquids, or biological fluids) (29, 30), although it does not directly identify the fluorescent constituents of the sample in any obvious way.

Fiberoptic Sensors

An optical fiber is a light pipe that may be used to transmit light beams over long distances. A fiber may be used to transmit exciting light to a fluorescent analyte or transmit the emitted fluorescence to a detector. Thus, fiberoptics may be used to deal with extremely small objects (such

Figure 26.2 Comparison of conventional excitation and Fluorescence spectra. (a) Perylene in ethanolic solution; (b) synchronous fluorescence spectrum using a 3-nm offset between excitation and emission monochromators. *(Reprinted with permission from T. Vo-Dinh et al., "Synchronous Spectroscopy for Analysis of Polynuclear Aromatic Compounds,"*Environmental Science & Technology, *12, pp. 1297–1302. Copyright 1978 American Chemical Society.)*

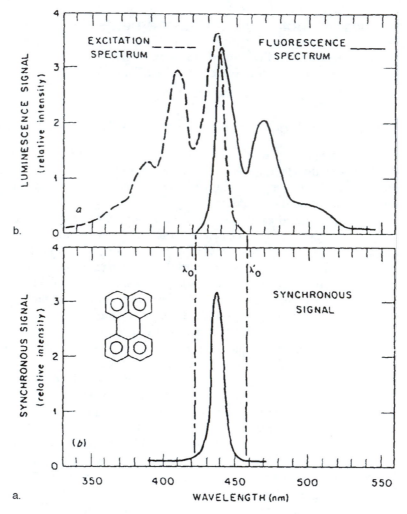

as electrophoresis capillaries (31)), inaccessible samples (such as groundwater (32)), radioactive or otherwise hazardous materials (33), or materials that may be difficult to sample, store, and transmit to a laboratory (such as sea water (34)).

Numerous types of fiberoptic sensors based on fluorescence have been designed (35–37). These devices are sometimes called optrodes because they often are used for the same general purposes as ion-selective electrodes. It may be necessary to use in such a device a reagent that can incorporate chemical selectivity into the fluorometric analysis; because of their high selectivities, enzymes and antibodies may be used for this purpose (38).

Difficulties can be experienced in the use of fiberoptic sensors. For example, when a reagent is incorporated into a sensor, the sensing tip may not be as long-lived as one would wish, the reagent may not be sufficiently selective for the intended analytical use, and the chemical reactions may be slower than desirable.

Figure 26.3 Fluorescence contour maps for four coal-derived liquids in ethanolic solution. The vertical axis is excitation wavelength; the horizontal axis is emission wavelength. *(Reprinted with permission from P. M. R. Hertz and L. B. McGown, "Organized Media for Fluorescence Analysis of Complex Samples: Comparison of Bile Salt and Conventional Detergent Micelles in Coal Liquids," Analytical Chemistry, 64, pp. 2920–2928. Copyright 1992 American Chemical Society.)*

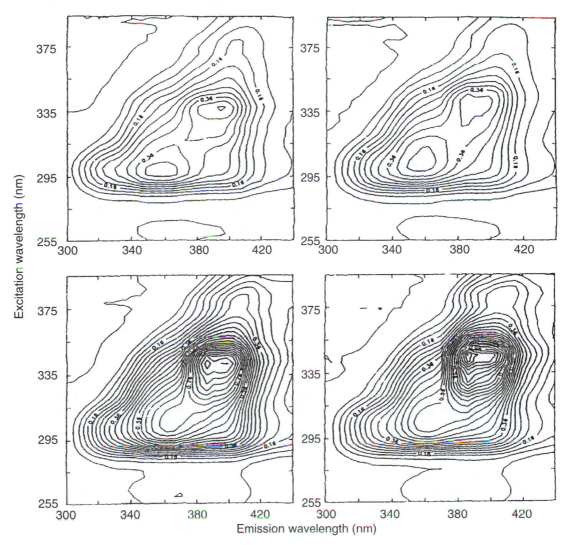

Polarized Fluorescence Measurements

When a molecule is excited by polarized light, its fluorescence may be partially or fully polarized (13). A fluorometer can be used to measure fluorescence polarization by placing polarizing prisms or sheets in the excitation and emission beams. High-quality instrumentation for polarized fluorescence measurements is commercially available. Fluorescence polarization measurements are widely used for studying rotational motions of electronically excited molecules and to detect the binding of relatively small molecules to macromolecules. Polarized fluorescence measurements are often used in fluoroimmunoassay procedures that are widely used in the life sciences (39).

Laser-Induced Fluorescence

For measurements that require very high excitation source power, monochromatic exciting radiation, the ability to illuminate a very small sample volume, excitation with short light pulses, or propagation of the exciting light over large distances, it may be necessary to use a laser source (16, 40). For solution samples, laser-induced fluorescence (LIF) tends to exhibit somewhat better limits of detection than lamp-excited fluorescence (41). However, the detection-limit advantage of LIF tends to be much less dramatic than one might expect (42) because of the blank-limited nature of fluorometry.

Because the wavelength of the exciting light must correspond to an absorption band of the analyte, fixed-wavelength lasers are not generally suitable for fluorescence spectrometry. A laser source for fluorometry should exhibit wavelength tunability over a fairly wide range in the UV or visible ranges. Thus, many applications of LIF use a dye laser (43) as the excitation source.

Most dye lasers are expensive (the pump source, usually another laser, may cost $20,000 to $65,000) and require some expertise to use effectively. The pump laser may consume a great deal of electrical power. Eventually, dye lasers may largely be replaced in fluorometry by compact solid-state lasers with low electricity requirements. Small solid-state lasers have been used in fluorometric analyses of compounds that absorb in the visible or near infrared (44). However, these devices cannot presently produce tunable UV (250 to 380 nm) output—needed for many applications of fluorometry—at high enough power to exploit the fluorescence advantage.

Laser sources (usually argon ion lasers) are used in fluorescence flow cytometers, which are commercial instruments used to count and sort biological cells and other particles (12, 45).

Fluorescence Decay Times and Time-Resolved Spectra

A fluorescence decay time is a measurement, at fixed wavelength, of fluorescence signal as a function of time. A time-resolved fluorescence spectrum is a spectrum measured within a narrow time window during the decay of the fluorescence of interest.

Two different approaches are used in such measurements. In pulse fluorometry, an excitation source is used that produces light pulses whose durations are short in comparison with the excited-state lifetime of the fluorescent molecule. The decay of the fluorescence is then measured directly, using a fast detector (1, 40). Because fluorescence decay times are often 1 ns or less, short pulses are needed. Lasers that can produce ultrashort pulses (10^{-9} to 10^{-12} sec or less) are commercially available, but tend to be expensive and touchy to operate. Also, fast detection systems are needed, and considerable care is needed in the choice of cables, connectors, and other ancillary components (46, 47).

An alternative technique, phase-modulation fluorometry (often called frequency-domain fluorometry) uses a source (lamp or laser) that is amplitude-modulated at one or more frequencies. Measurement of the phase or demodulation of the fluorescence signal can be used to generate fluorescence decay times and time-resolved fluorescence spectra (1, 48). Such instruments are generally simpler than those used for pulse fluorometry, although high modulation frequencies (1 GHz or greater) are needed to measure very fast fluorescence decays. Commercial instrumentation is available for these types of measurements.

Time-resolved measurements are instrumentally sophisticated, but they can improve both the sensitivity and selectivity of fluorometry. Measurements of fluorescence decay times or time-resolved spectra are used for several purposes, such as distinguishing sample constituents whose fluorescence spectra overlap one another (40, 49, 50) and distinguishing the fluorescence of an analyte from background scattering or luminescence of other sample constituents (44, 51); this approach to background suppression is widely used in fluorescence immunoassays (52, 53).

Low-Temperature Fluorometry

The absorption and fluorescence spectra of a molecule may undergo dramatic narrowing if the molecule is inserted in a solid matrix at low temperature (77 °K or lower) (16, 54) or expanded, in the gas phase, to form a supersonic free jet (55). These techniques exhibit much greater selectivity than conventional fluorescence measurements in liquid solution. Often, use of a laser source is required to exploit fully the opportunities for increased selectivity offered by the low-temperature techniques. Use of the methods also entails an investment in cryogenic apparatus (for the solid-state techniques) or vacuum hardware (supersonic jet spectroscopy), and the time required to obtain a spectrum can be substantial. However, these methods can effect major savings in sample pretreatment before measurement of fluorescence.

What It Does

Analytical Information: The Fluorescence Advantage

The main analytical application of molecular fluorescence spectrometry is detection and quantification of species present at concentrations so low that most other techniques are not useful. The origin of the fluorescence advantage in detection limit capabilities can be understood in the following way. Consider Beer's law, the fundamental relationship in quantitative absorption spectrometry:

$$A = \varepsilon bc = \log (P_0/P) = \log P_0 - \log P \tag{26.1}$$

where ε is the absorptivity of the analyte, b is the optical pathlength of the sample, c is the concentration of the analyte, P_0 is the excitation power (photons sec^{-1}) incident on the sample, and P is the power transmitted by the sample. P is the quantity that varies with c: As c decreases, P increases. When c is small, P is slightly smaller than P_0; thus, one must measure a small difference between two large numbers.

In contrast, the relationship between a measured fluorescence signal (F, in photons sec^{-1}) to the analyte concentration is

$$F = k\phi_F P_0(1 - 10^{-\varepsilon bc}) \tag{26.2}$$

where ϕ_F is the fluorescence quantum yield, k is the fraction of the photons emitted by excited analyte molecules that actually are detected (often 0.10 or less), and the other symbols have the same meanings as in Eq. (26.1). If the product $\varepsilon bc \leq 0.02$, as is often the case in analytical applications of fluorometry, Eq. (26.2) simplifies to

$$F = k\phi_F P_0 \varepsilon bc \tag{26.3}$$

According to Eq. (26.3), if the analyte concentration is 0, the measured fluorescence signal is 0. If the analyte concentration is small, F is a small number. Hence, when the analyte concentration is low, the measurement situation in fluorometry—distinguishing between a small signal and zero signal—is more favorable than that encountered in absorption spectroscopy (measuring a small difference between two large numbers).

Eq. (26.3) predicts the fluorescence signal to be linear in the analyte concentration and the excitation power, P_0. The product $\varepsilon\phi_F P_0$ determines the sensitivity of fluorometry for the analyte.

In the most favorable case, the analyte absorbs strongly at the excitation wavelength, the source generates a large number of photons per unit time at that wavelength, and the excited analyte molecules so produced exhibit a high probability of decaying via fluorescence. If all three of these conditions are satisfied and the detector exhibits high sensitivity at the wavelength at which the analyte emits, then it is possible to achieve extremely low limits of detection for the analyte, much lower than can be achieved by absorption spectroscopy. This is the fluorescence advantage.

In real life, when the analyte concentration is zero, the observed signal is greater than zero, due to background signals from fluorescence of other constituents of the sample or contaminants in the solvent or sample cell. Rayleigh or Raman scattering of source radiation onto the detector also generates background. Thus, fluorescence spectrometry is ordinarily blank-limited (42). Therefore, achieving low limits of detection in fluorometry is possible only if great care is taken to minimize the background arising from unwanted fluorescence or scattering, and to distinguish the analyte fluorescence from the background (10, 51, 56).

The sensitivity of fluorometry also depends on the efficiency with which light emitted by the sample is collected and caused to impinge on the detector. This, in part, determines the value of k in Eq. (26.2). The collection efficiency often is low (less than 10%). Sample compartments and cuvettes for commercial fluorometers can be modified to increase the fraction of emitted light that is actually collected (by placing mirrors behind the cell, for example). Because such alterations also increase the amount of scatter and spurious fluorescence that reach the detector, in practice they may be of little real benefit.

The low sensitivity implied by a small value of k can be improved by exploiting the fact that an analyte molecule may be excited more than once (thus, one may obtain many emitted photons from one analyte molecule (57); this is sometimes called the photon-burst effect). This opportunity works out strongly to the analyst's advantage only if the analyte exhibits a very low quantum yield for photodecomposition (56, 57); photosensitive molecules will be destroyed before they have the chance to emit many photons.

Interferences

It is useful to consider interferences in two classes: additive and multiplicative (58). In additive interference, background fluorescence is emitted by concomitants in the sample or contaminants in solvents or glassware, causing a finite blank. In multiplicative interference, the interferent does not itself fluoresce, but causes the fluorescence signal for the analyte to be smaller or (occasionally) larger than that observed in the absence of the interferent. One important phenomenon—quenching—can simultaneously cause additive and multiplicative interference. It is generally easier to correct for multiplicative interference (by standard additions (59), for example) than additive interference.

Additive Interference: Background Fluorescence and Scattering

Molecular absorption and fluorescence bands tend to be broad (25 nm or more) and featureless, especially in solution. In multicomponent samples, therefore, the absorption and fluorescence spectra of the various sample constituents may overlap. Background fluorescence (additive interference) from other sample constituents, or contaminants in the solvent or sample cell, can be a major problem, especially in biological and environmental samples. Several steps can be taken to deal with the problem:

- Using solvents and sample cells that are free of fluorescent contaminants is essential. Care must be exercised to avoid contamination of laboratory glassware by fluorescent substances (such as detergents used to clean glassware).

- Whenever possible, analyte fluorescence should be excited and measured at wavelengths at which other sample constituents do not absorb or fluoresce. Most organic molecules absorb and fluoresce in the UV or visible range (200 to 550 nm). Hence, fluorescence background often can be decreased by exciting and measuring analyte fluorescence at longer wavelengths in the near infrared (600 nm or more) (60, 61). Numerous fluorescent tags and derivatives have been developed that absorb and emit in the near infrared (60, 62).

- Synchronous scanning often is helpful in dealing with overlapping absorption and fluorescence spectra of mixture constituents.

- The background fluorescence may have a different decay time from that of the analyte. Time resolution can be used to distinguish the analyte fluorescence from the background. This is most successful if the analyte fluorescence is longer-lived than the background. In such a situation, one simply waits until the interfering fluorescence and scattering have decayed away to negligible levels, and then measures the remaining analyte fluorescence (44, 49, 52). Fortunately, the background fluorescence observed in many samples is relatively short-lived (fluorescence from trace contaminants in many solvents has a decay time of 2 ns or less (10) and that in many biological samples has a decay time of 20 ns or less (44)). Moreover, Rayleigh and Raman scatter background is instantaneous.

- There may be an excitation wavelength at which only the analyte absorbs appreciably. If so, its fluorescence can be selectively excited. Use of low-temperature techniques greatly increases the possibility of achieving selective excitation of analyte fluorescence in complex mixtures (16, 54, 55).

- Fluorescent sample constituents may be eliminated by separating the analyte from the interferents before measurement of fluorescence. Accordingly, fluorometric analyses of complex samples often are preceded by, or coupled directly to, chromatographic or electrophoretic separations.

- Mathematical (chemometric) techniques may in some cases be able to decompose spectra that consist of overlapping bands, produced by several different compounds, into contributions from the individual sample constituents (63).

Combinations of the above techniques may be especially useful. For example, spectral overlaps may be dealt with more effectively by combined use of synchronous scanning and time resolution than by either method used individually (50).

Multiplicative Interference: Inner-Filter Effects and Quenching

The fluorescence signal generated by an analyte may be altered—perhaps even totally suppressed—by other sample constituents. One form of multiplicative interference is the inner-filter effect, in which an interferent that absorbs in the same wavelength range as the analyte decreases the radiant power available to excite the analyte. Another form of inner-filter effect occurs when an interferent absorbs at the wavelength at which the analyte fluoresces, thus causing the number of emitted photons that escape the sample and reach the detector to diminish. Instrumental correction for inner-filter effects is possible (64); however, such procedures may be difficult to implement in some commercial fluorometers.

Quenching is another type of multiplicative interference. Quenching is any process in which a sample constituent decreases the fluorescence quantum yield for the analyte. Among the most common fluorescence quenchers is O_2; removal of oxygen from a sample before fluorometric analysis (65) is often advisable.

One way that fluorescence quenching can occur is intermolecular electronic energy transfer:

$$M^* + Q \rightarrow M + Q^* \tag{26.4}$$

Here an excited analyte molecule (M^*) transfers excitation energy to a quencher molecule Q, causing de-excitation of M and forming an excited quencher molecule, Q^*. If Q^* is a fluorescent species, its fluorescence (called sensitized fluorescence) may then be observed. This phenomenon can allow one to observe fluorescence from a molecule (Q) that may be difficult to excite directly. More often, however, these processes are a nuisance. Not only do they cause a decrease in the fluorescence signal observed for a given concentration of analyte (M) in the sample, but they may produce unwanted background fluorescence signals; that is, Q may act both as a multiplicative and an additive interference.

Quenching often follows the Stern–Volmer equation:

$$\frac{\phi_F^0}{\phi_F} = 1 + K_{SV}(Q) \tag{26.5}$$

where ϕ_F^0 and ϕ_F are the fluorescence quantum yields for the analyte in the absence of quencher and presence of quencher at concentration (Q), respectively, and K is the quenching constant (a measure of the efficiency with which Q quenches analyte fluorescence). Because the analyte fluorescence signal depends on the quencher concentration, one can determine Q, via its quenching action, in a sample that contains a fluorescent compound. Numerous procedures that use fluorescence quenching to determine species (most notably O_2) that are not themselves fluorescent but can act as efficient fluorescence quenchers have been devised (66, 67).

Equation (26.5) shows that the effect of a quencher decreases as the sample is diluted. Thus, quenching can be circumvented simply by diluting the sample, provided that the diluent is not itself a quencher and assuming that one does not thereby decrease the analyte concentration below its limit of detection.

Figure 26.4 is a dramatic example of the influence of inner-filter effects and intermolecular energy transfer on fluorescence spectra in mixtures (68). The sample, obtained in an industrial setting, is a complex mixture of polycyclic aromatic hydrocarbons. At the higher concentration in Fig. 26.4, the spectrum is badly perturbed by inner-filter effects and intermolecular energy transfer, causing it to be depleted of contributions from compounds that emit at high energy (short wavelength). The effect of energy transfer is to quench fluorescence from sample constituents that fluoresce at shorter wavelengths and sensitize fluorescence from the compounds that fluoresce at longer wavelengths. When the sample is diluted sufficiently, the various quenching phenomena cease to occur and the appearance of the spectrum changes dramatically. Checking for the occurrence of such phenomena by measuring fluorescence spectra of complex mixtures before and after dilution is a useful precaution.

The fluorescence quantum yield for a given compound can vary dramatically from one sample to another, much more so than the molar absorptivity. Accordingly, the accuracy of fluorometric analysis is much more susceptible to errors caused by multiplicative interferences (and, thus, by improper or inadequate calibration) than is UV/Vis absorption spectrometry. Most complex samples contain one or more components that can quench the fluorescence of the analyte. Thus, it often is necessary to subject complex samples to be analyzed by fluorometry to extensive prior cleanup to remove potential quenchers.

Alternatively, one may try to provide a uniform microenvironment for the analyte (and thus a reproducible fluorescence yield) from sample to sample by any of several ploys, such as adding a micelle-forming surfactant to each sample (29). This is based on the fact that fluorescent molecules in solution may be partially or fully hidden from quenchers by incorporating them into organized media such as surfactant micelles (29) or cyclodextrin cavities (69).

Figure 26.4 Fluorescence spectrum of solvent extracts of a sample of particulate matter from an industrial environment. The samples are identical except that one was diluted 1:100 (dashed line) and the other 1:1000 (solid line) with ethanol. Extensive intermolecular energy transfer or inner-filter effects are occurring in the more concentrated sample. *(Reprinted with permission from T. Vo-Dinh, R. B. Gammage, and P. R. Martinez, "Analysis of a Workplace Air Particulate Sample by Synchronous Luminescence and Room-Temperature Phosphorescence,* Analytical Chemistry, *53, pp. 253–258. Copyright 1981 American Chemical Society.)*

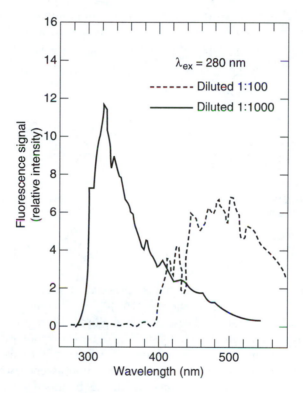

Triplet states of organic molecules have much longer lifetimes than singlet excited states, and thus are more susceptible to quenching. Hence, phosphorescence is much more likely to be quenched than is fluorescence, so the experimental conditions needed to observe phosphorescence are more stringent than those required to detect fluorescence. Observation of phosphorescence of useful intensity from solutes in liquid-solution samples is rare.

Historically, phosphorescence measurements were made in low-temperature glasses formed by freezing liquid solutions. Phosphorescence received a major boost as an analytical technique when it was discovered that molecules in triplet states can be protected from quenching agents by adsorbing them on filter paper or other solid supports (70) or (as noted above for fluorescence) sequestering them in surfactant micelles or cyclodextrin cavities in liquid media. Room-temperature phosphorescence using these approaches has become a popular technique (70, 71), whereas the classic low-temperature procedure is now rarely used.

Accuracy and Precision

The accuracy and precision of analyses performed by molecular fluorescence can range from extremely high to abysmally poor, depending on the following factors:

- The care taken to prevent introduction of fluorescent contaminants into the sample (by use of high-quality solvents and clean glassware, for example).

- The complexity of the sample, number and types of interfering species present, and effectiveness of preliminary cleanup procedures. Loss of analyte, or contamination of the sample, during preliminary cleanup can cause large errors.

- The care exercised in calibration of the fluorescence measurement and use of appropriate mathematical techniques for evaluating the experimental data.

- The quality of the instrumentation used. For example, an instrument that monitors the power output of the source (and thus is able to correct measured fluorescence signals for fluctuations in P_0) will produce much more precise data than an instrument that does not do so.

Applications

General Considerations

Major classes of applications of fluorometry include the following:

- Detection and quantification of trace-level species, especially in biological-clinical (72–74) and environmental (75) samples.

- Detection in separation techniques, especially liquid (16, 76, 77) and thin-layer chromatography (77) and electrophoresis (16, 44, 78). Coupling of laser-induced fluorescence to electrophoresis for rapid base sequencing of DNA fragments (79, 80) may have considerable significance in biotechnology. Use of derivatization or labeling reactions to convert nonfluorescent compounds to fluorescent entities (3–6) may be necessary. Another way to deal with nonfluorescent species is indirect fluorescence, wherein a nonfluorescent analyte displaces a fluorescent molecule (which may be added to the mobile phase); the change in concentration (and hence in fluorescence intensity) of the fluorescent species is measured and related to the concentration of the nonfluorescent constituent (81).

- On-line analyses and remote sensing, using fiberoptic sensors or laser-induced fluorescence. Monographs (35) and review articles (36, 37, 82) on this subject are available.

- Detection and quantification in immunoassay. Fluoroimmunoassay procedures are discussed in detail elsewhere (39, 83, 84).

- Identifying, sorting, and counting particles (most notably biological cells) via fluorescence flow cytometry. Monographs (45) and review articles (16) on flow cytometry are available. Flow cytometers are also used to determine fluorescent analytes bound to particles or adsorbed on particulate surfaces (85).

- In-situ imaging, mapping, and quantification of species in biological systems, such as tissues and single cells. Detailed reviews of these types of applications are available (6, 86).

- Preliminary screening of complex mixtures (especially environmental samples), using the fluorometric data to decide whether it is appropriate to subject a sample to more detailed characterization (7, 8).

- Studies of the microenvironments of fluorescent probe molecules. Such properties of materials as viscosity, pH, "polarity," and temperature may be inferred from measurements of fluorescence spectra, decay times, and polarization of suitable probe molecules. Fluorescent probes are widely used in biology and materials science to obtain information regarding the nature and accessibility of binding sites in biological macromolecules, dynamics of motions of polymer chains, homogeneities of polymer samples, the properties of micelles, the nature of domains on solid surfaces, the concentrations of specific ions in biological cells, and the spatial distributions of specific molecular species in biological membranes, to list a few of the possibilities. Reviews discussing the principles and practice of fluorescence probe studies, and precautions that must be exercised in the interpretation of these experiments, are available (6, 14, 87).

Selected Example: Determination of Polycyclic Aromatic Hydrocarbons

Polycyclic aromatic hydrocarbons (PAHs), some of which can undergo metabolic conversion to carcinogens, are widely dispersed in the environment, often in small quantities. There are many different PAHs and PAH derivatives; many are intensely fluorescent. Thus, fluorometry is often used for the determination of PAHs and PAH derivatives in environmental and biological samples. Virtually every approach used in the fluorometric analysis of mixtures has been applied to PAHs. Hence, we can use the determination of PAHs to exemplify the strategies available for quantification of fluorescent analytes present in complex samples. Specific examples are considered below.

Fluorometry Preceded by Separation

An excellent example of this approach is described by May and Wise (88), who wished to determine PAHs sorbed on urban airborne particulate matter. The results were to serve as part of the data set to certify the sample as a standard reference material for 13 PAHs. Many other PAHs and PAH derivatives were present.

It was first necessary to remove the PAHs from the particulate samples via Soxhlet extraction. (Some sort of extraction is unavoidable; the extraction could be done with a supercritical fluid rather than a liquid solvent.) Then, each extract was concentrated and passed over a small silica gel column to remove the most polar compounds. Next, each extract was subjected to normal phase liquid chromatography (LC). This LC procedure separated an extract into several fractions, according to number of aromatic rings—three-ring, four-ring, and five-ring fractions were obtained—but isomeric PAHs were not separated from one another, nor were alkylated PAHs (such as the methylchrysenes) separated from their parents.

Each of the fractions obtained via the normal phase LC separation was then subjected to a second, reversed phase LC separation. Using this column, it was possible to separate isomeric PAHs from one another and alkylated PAHs from their parents. In the latter separation, a molecular fluorescence detector (using numerous excitation and emission wavelengths appropriate for the specific PAHs of interest) was used. The analyses were calibrated via internal standard techniques.

This procedure is typical in that interferences are minimized by carrying out a series of separations before the fluorescence measurement. The time needed to make the fluorescence measurement is trivial compared with that required for the various cleanup steps needed to get the sample to the point at which useful measurements can be made.

Synchronous Fluorometry

Synchronous scanning may decrease the extent of overlap in the spectra of mixtures of fluorescent compounds. This does not automatically eliminate the need for sample cleanup, but may enable the analyst to use fewer or less extensive cleanup steps than in conventional fluorometry.

A good example is described by Vo-Dinh and colleagues (68), who studied the PAH content of industrial airborne particulate matter. The sample was first subjected to solvent extraction to remove the PAHs from the particles, and the extract was then fractionated by LC. Seven LC fractions were produced (one fraction contained aliphatics, three contained primarily PAHs, and three contained primarily polar aromatics). One of the PAH fractions was examined by synchronous fluorometry, using a wavelength offset of 3 nm; the conventional and synchronous fluorescence spectra of this fraction are compared in Fig. 26.5. Nine resolved features are observed in the synchronous spectrum; it is possible to identify and quantify (via standard additions) all nine PAHs responsible for these features.

In comparison with the separation approach of May and Wise, we note that use of synchronous fluorometry allowed Vo-Dinh and colleagues to omit one LC separation used by May and Wise; the LC step in question required at least 1 hr. Of course, several synchronous scans, at different wavelength offsets, may be needed to locate all sample constituents of interest. However, numerous synchronous scans can be run in the time needed for one LC separation in samples of this complexity. Thus, whenever one wishes to quantify a small number of fluorescent constituents of a complex sample, use of synchronous fluorometry may save time by decreasing the amount of sample cleanup needed to obtain useful fluorescence data.

Temporal Resolution

If two spectrally similar PAHs have different fluorescence decay times, they can be distinguished from one another via time resolution, provided that the decay time difference is sufficiently large. Structurally similar PAHs that are difficult to distinguish spectrally may also have similar fluorescence decay times. For example, the isomers benzo[a]pyrene and benzo[e]pyrene have fluorescence decay times in acetonitrile solution of 14.9 ns and 16.9 ns, respectively (89), making it quite difficult to distinguish between them solely on the basis of decay-time measurements. More encouraging examples can also be cited. For example, in acetonitrile solution the isomers dibenzo[a,h]pyrene, dibenzo[a,e]pyrene, and dibenzo[a,i]pyrene have decay times of 5.5, 18.5, and 26.5 ns, respectively (90) and the isomers benzo[k]fluoranthene and benzo[b]fluoranthene have decay times of 7.8 and 27.3 ns, respectively (89). Thus, even if time resolution by itself cannot realistically be expected to unravel a complex mixture of spectrally similar PAHs, it may add valuable selectivity to fluorometric procedures. For instance, decay-time data may be capable of distinguishing between PAHs that are not completely separated chromatographically, and may be used to ascertain if what appears to be a single peak in a chromatogram is actually due to two or more coeluting compounds (89, 91). Also, the fluorescence decay time may be used, in conjunction with chromatographic retention-time data, to identify fluorescent compounds as they elute; it may be easier experimentally to use decay times than emission wavelengths for this purpose (90). Of course, time resolution is also used in conjunction with fluorescence detection in liquid chromatography to reduce scatter or luminescence background interference.

Fiberoptic Sensors

Because PAHs are environmental contaminants, techniques have been developed for determining these compounds in natural waters, biological fluids, and industrial process streams and effluents.

Figure 26.5 Fluorescence spectra of extract of a particulate matter sample obtained in an industrial environment. (a) Conventional; (b) synchronous (wavelength offset: 3 nm). All peaks can be assigned to a single PAH constituent of the sample except peak 2, which is attributed to two PAHs. *(Reprinted with permission from T. Vo-Dinh, R. B. Gammage, and P. R. Martinez, "Analysis of a Workplace Air Particulate Sample by Synchronous Luminescence and Room-Temperature Phosphorescence, Analytical Chemistry, 53, pp. 253–258. Copyright 1981 American Chemical Society.)*

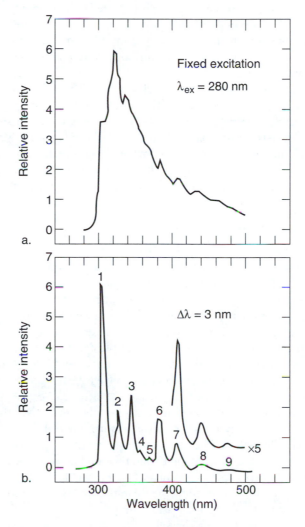

Fiberoptic sensors are a potentially attractive way of achieving rapid, sensitive determinations of PAHs in such samples. The obvious problem is how to achieve selectivity, given that most PAH-containing samples contain numerous fluorescent compounds.

Chemical selectivity can be achieved by using, at the sensor tip, a reagent that selectively binds the analyte, thereby increasing its concentration (and decreasing the concentrations of interferents) at the probe tip. Immunochemical systems can provide exceptionally high selectivities. For example, Tromberg and colleagues (38) designed a fiberoptic sensor for a fluorescent carcinogenic metabolite of the PAH benzo[a]pyrene (BaP), in which the probe tip contains a monoclonal antibody for which the BaP metabolite is a hapten. The large equilibrium constant for formation of the antibody–hapten complex results in low limits of detection for the analyte, due to the preconcentration achieved at the fiber tip. Selectivity is enhanced by separating the reagent layer from the bulk sample by a membrane through which the analyte must diffuse; some potential interferents either may not penetrate the membrane or may not be soluble in the solvent inside the reagent compartment.

A shortcoming of such sensors is the need to produce an appropriate antibody for each analyte of interest (not necessarily an easy or inexpensive task), and of course the method is inapplicable to any analyte for which no antibody exists. One must also worry about the stability of the antibody and the rates of formation of the antibody–antigen complex and diffusion of analyte through the membrane.

Temporal selectivity can be achieved in fiberoptic sensors for PAHs by using, as excitation source, a pulsed laser and exploiting differences in fluorescence lifetimes for discriminating between PAHs, as in natural waters (92). A relatively inexpensive pulsed laser may be used, and no reagent layer is needed, simplifying the design and of the probe tip and eliminating the kinetic problems that may be encountered when designs including reagents at the fiber tip are used. This approach ultimately is limited by the fact, noted previously, that many PAHs have similar fluorescence decay times; it is most useful for PAHs (such as pyrene) with relatively long decay times.

Low-Temperature Fluorometry

In the best of all possible worlds, one could subject a complex sample to no (or minimal) cleanup and selectively excite fluorescence of the analyte (using an excitation wavelength at which only the analyte absorbs appreciably). In the absence of intermolecular energy transfer, one would obtain a fluorescence spectrum of the analyte that could be used directly for quantitative purposes.

Because of the broad, featureless absorption spectra of most molecules in liquid and gas phases, such a procedure is seldom possible for complex liquid or gaseous samples. However, the absorption and fluorescence spectra of sample constituents may undergo dramatic sharpening if the sample is incorporated into a low-temperature solid (16, 54) or expanded in the gas phase in a supersonic free jet (58). Such procedures have been applied extensively to the determination of PAHs in complex samples.

When PAHs are dissolved in n-alkane solvents that are then frozen at 77 °K or lower temperature, extraordinarily highly resolved absorption and fluorescence spectra may be obtained (the Shpol'skii effect (93)). The absorption spectra of sample constituents may be sharpened to the extent that an excitation wavelength can be found at which only the analyte absorbs appreciably. Also, in solid matrices, fluorescence quenching is less efficient than in solution. Thus, very high selectivity, freedom from additive and multiplicative interferences, and low limits of detection may be achieved with minimal sample cleanup.

For example, D'Silva, Fassel, and colleagues were able to identify and quantify individual PAHs in very complex materials, including crude petroleum and shale oils, without preliminary cleanup (93, 94). The samples were simply dissolved in a Shpol'skii solvent (such as n-octane), residual insoluble matter was filtered off, and the solutions were frozen rapidly to 77 °K. The spectral resolution is so high that a PAH often can be distinguished from its deuterated analog (see Fig. 26.6). Besides demonstrating the high selectivity of the technique, this fact shows that the deuterated analog can be used as an internal standard for quantitative purposes.

Even with this level of selectivity, preliminary sample cleanup may be needed in very demanding cases. For example, Garrigues and colleagues wished to determine each of the 12 possible iso-

Figure 26.6 Fluorescence spectra of benzo[a]pyrene (BaP), in a shale oil sample with 10 ppb perdeuterobenzo[a]pyrene (BaP-d_{12}) added as internal standard. The spectra were obtained in *n*-octane frozen solution (a Shpol'skii matrix) at 15 °K. *(Reprinted with permission from Y. Yang, A. P. D'Silva, and V. A. Fassel, "Deuterated Analogues as Internal Reference Compounds for the Direct Determination of Benzo[a]pyrene and Perylene in Liquid Fuels by Laser-Excited Shpol'skii Spectrometry," Analytical Chemistry, 53, pp. 2107–2109. Copyright 1983 American Chemical Society.)*

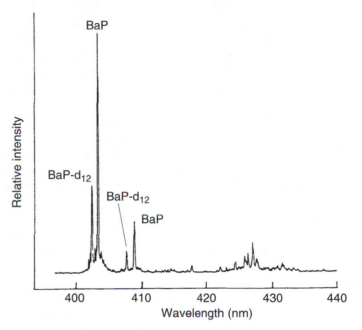

meric methylbenzo[a]pyrenes in a coal tar extract (95). The effect of changing the position of a methyl group on the absorption and fluorescence spectra of these compounds is very small; thus, not only is very high spectral selectivity needed, but some degree of separation of the various isomers is required before measurement of fluorescence. Thus, the coal tar extract was subjected to a liquid chromatographic (LC) fractionation that produced several fractions, according to the number of aromatic rings. Then, the five-ring fraction was subjected to a second stage of LC, which separated the methylbenzo[a]pyrenes from the other five-ring PAHs and (to some extent) from each other. Chromatographic fractions were then diluted with a Shpol'skii solvent (n-octane) and cooled to 15 °K. As a result of this combination of chromatographic and spectroscopic selectivity, it was possible to quantify 8 of the 12 possible isomers (3 were not present in the sample, and 1 could not be distinguished from the parent PAH).

Many fluorescent molecules do not exhibit the Shpol'skii effect. For such compounds, chemical conversion to derivatives that exhibit highly resolved spectra in Shpol'skii matrices may be possible (96). Another approach using low-temperature solid matrices is fluorescence line narrowing (sometimes known as site selection), in which the choice of solvent is less critical than in Shpol'skii fluorometry. Very highly resolved fluorescence spectra may be obtained for molecules that do not exhibit the Shpol'skii effect. The principles of fluorescence line narrowing are described in detail elsewhere (16, 54, 97).

An impressive demonstration of the analytical capabilities of fluorescence line narrowing is a series of studies of adducts of carcinogenic PAH metabolites with DNA (97, 98). To detect, identify, and quantify these species in real samples requires extraordinary selectivity and sensitivity;

numerous spectrally similar adducts of a particular metabolite may be present, the quantity of each adduct may be extremely small, and the scatter and fluorescence background interferences in biological materials are always a source of concern.

Shpol'skii fluorometry and fluorescence line narrowing use frozen liquid solution samples. An alternative approach is matrix isolation, wherein the sample is sublimed under vacuum and mixed with a gaseous diluent (the matrix). The resulting gas-phase mixture is deposited on a cold surface for spectroscopic examination as a solid (54, 99). The main advantage of matrix isolation is that analytes are dissolved in the solvent in the gas phase. Hence, solubility problems do not arise, and the solvent thus can be chosen for spectroscopic rather than chemical reasons. Shpol'skii matrices can be used in matrix isolation, and fluorescence line narrowing experiments also can be carried out.

Matrix isolation is difficult to apply to involatile analytes, but can be applied to extremely difficult samples. For example, Perry and colleagues detected and quantified individual PAHs in intractable solid solvent-refined coal samples by matrix-isolation Shpol'skii fluorometry without any preliminary sample cleanup (100). The extremely high selectivity that is possible is shown in Fig. 26.7, which compares the fluorescence spectrum of one sample constituent (7,10-dimethylbenzo[a]pyrene) with that of the pure compound. It is possible to combine low-temperature fluorometric measurements with time resolution to achieve even higher selectivity in mixture analysis (49).

Selective and sensitive as these methods are, there are several caveats. Specialized, rather expensive apparatus may be needed to implement them properly. For example, it is difficult to exploit fully the selectivity of these techniques unless a laser is used as excitation source. The techniques may be time-consuming (although not necessarily more so than the separations often required to clean up a complex sample in conventional fluorometry) and require expertise and experience on the part of the analyst. Finally, authentic samples of the analytes should be available, so that the conditions for determining them (such as optimal excitation wavelengths) can be identified. Considerable trial and error may be required to identify the best conditions. Whether this process is more time-consuming than the separation steps that would be needed to eliminate interferences before carrying out a conventional fluorometric analysis varies from sample to sample.

Figure 26.7 Fluorescence spectra of pure 7,10-dimethylbenzo[a]pyrene (right) and the same PAH in a solid coal-derived material (left). Both spectra were obtained by matrix isolation in *n*-octane (a Shpol'skii matrix) at 15 °K. *(Reprinted with permission from M. B. Perry, E. L. Wehry, and G. Mamantov, "Determination of Polycyclic Aromatic Hydrocarbons in Unfractionated Solid Solvent-Refined Coal by Matrix Isolation Fluorescence Spectrometry," Analytical Chemistry, 55, pp. 1893–1896. Copyright 1983 American Chemical Society.)*

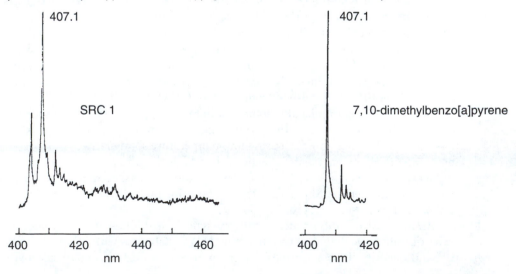

Nuts and Bolts

Relative Costs

The simplest instruments for molecular fluorescence spectrometry (filter fluorometers or low-resolution scanning monochromator systems with no bells and whistles) cost less than $20,000. Instruments with reasonably high spectral resolution capable of synchronous scanning, generating corrected spectra, and computer postprocessing of spectra are usually in the $20,000–$30,000 price range. More sophisticated instruments offering more advanced components or capabilities (such as electronic array detector for rapid acquisition of spectra, high spectral resolution, accurate fluorescence polarization measurements, measurement of phosphorescence decay times, and special sample chambers for nonroutine types of samples) can range from $30,000 to $75,000 or more. Instrumentation for fluorescence decay time and time-resolved fluorescence measurements (both pulsed and phase-modulation methods) is available; such instruments generally cost at least $50,000 and perhaps much more, depending on their capabilities. For certain specialized applications (such as laser-induced remote sensing or low-temperature fluorometry), assembly of an instrument from components (such as a laser, monochromator, and detector) may be necessary; in such cases it is easy to spend $100,000 or more.

A Raman spectrometer can double as a high-resolution, low stray-light fluorescence spectrometer; however, few (if any) laboratories would purchase a Raman spectrometer solely for fluorescence measurements!

Vendors for Instrumentation and Accessories

Many of these companies manufacture special-purpose fluorescence instrumentation, such as that used in polarized fluorescence measurements or determination of fluorescence lifetimes or time-resolved spectra.

Spectrometers
Buck Scientific Inc.
58 Fort Point St.
East Norwalk, CT 06855
phone: 800-562-5566
fax: 203-853-0569
email: 102456.1243@compuserve.com
Internet: http://ourworld.compuserve.com/homepages/Buck_Scientific

Hamamatsu Corp.
360 Foothill Rd.
Bridgewater, NJ 08807
phone: 800-524-0504
fax: 908-231-1218
email: hamacorp@interramp.com

Hitachi Instruments Inc.
3100 N. 1st St.
San Jose, CA 95134

phone: 800-455-4440
fax: 408-432-0704
email: info@hii.hitachi.com
Internet: http://www.hii.hitachi.com

Instruments S.A., Inc.
3880 Park Ave.
Edison, NJ 08820
phone: 800-438-7739
fax: 908-549-5125
email: john@isainc.com

ISS Inc.
2604 N. Mattis Ave.
Champaign, IL 61821
phone: 217-359-8681
fax: 217-359-7879

McPherson Inc.
530 Main St.
Acton, MA 01720
phone: 800-255-1055
fax: 508-263-1458
email: 72234.2257@compuserve.com

On-Line Instrument Systems Inc.
130 Conway Dr.
Bogart, GA 30622
phone: 800-852-3504
fax: 706-353-1972
email: olis@mindspring.com
Internet: http://www.olisweb.com

Perkin-Elmer Corporation
761 Main Ave.
Norwalk, CT 06859
phone: 800-762-4000
fax: 203-762-4228
email: info@perkin-elmer.com
Internet: http://www.perkin-elmer.com

Photon Technology International
1 Deerpark Dr.
South Brunswick, NJ 08852
phone: 908-329-0910
fax: 908-329-9069

Shimadzu
7102 Riverwood Dr.
Columbia, MD 21046
phone: 800-477-1227
fax: 410-381-1222
Internet: http://www.shimadzu.com

Spectronic Instruments
820 Linden Ave.
Rochester, NY 14625
phone: 800-654-9955
fax: 716-248-4014
email: info@spectronic.com

Turner Designs Inc.
845 W. Maude Ave.
Sunnyvale, CA 94086
phone: 408-749-0994
fax: 408-749-0998

Varian Instruments
P.O. Box 9000
San Fernando, CA 91340
phone: 800-926-3000

Cells

Buck Scientific (see listing above)

Hellma Cells Inc.
P.O. Box 544
Borough Hall Sta.
Jamaica, NY 11424
phone: 718-544-9534
fax: 718-263-6910

Wilmad Glass
Route 40 and Oak Rd.
Buena, NJ 08310
phone: 609-697-3000
fax: 609-697-0536
email: cs@wilmad.com
Internet: www.wilmad.com

Optical Parts

Esco Products Inc.
171 Oak Ridge Rd.
Oak Ridge, NJ 07438
phone: 201-697-3700
fax: 201-697-3011

Melles Griot
1770 Kettering St.
Irvine, CA 92714
phone: 800-835-2626
fax: 714-261-7589

Oriel Corp.
250 Long Beach Blvd.
Stratford, CT 06497

phone: 203-377-8262
fax: 203-378-2457
email: 73163.1321@compuserve.com

Detectors

Burle Industries Inc.
1000 New Holland Ave.
Lancaster, PA 17601
phone: 800-326-3270
fax: 717-295-6097

EG&G Reticon
345 Potrero Ave.
Sunnyvale, CA 94086
phone: 408-738-4266
fax: 408-738-6979

Hamamatsu Corp. (see listing above)

Princeton Instruments Inc.
3660 Quakerbridge Rd.
Trenton, NJ 08619
phone: 609-587-9797
fax: 609-587-1970
email: postmaster@prinst.com

For vendors of fluorescence detection systems for liquid chromatography and electrophoresis, see the chapters on those topics.

Consumables

Users of fluorometers often need to purchase sample cells and various optical parts (such as lamps, mirrors, gratings, detectors, polarizers, and fiberoptics). Although these can often be purchased from the manufacturer of the fluorometer, it may be advantageous (and cheaper) to procure these items from specialty vendors.

Solvents and chemicals may be purchased from any of the major chemical manufacturers and supply houses. As noted earlier, solvent purity is a key issue. Do not scrimp here. At the very least, spectrophotometric-grade solvents should be used; in some instances, HPLC-grade or equivalent solvents may be needed. The latter are expensive but their use may save much time and money in the long run.

Required Level of Training

The steepness of the learning curve for operating a fluorescence spectrometer is strongly dependent on the complexity of the instrument. Virtually all manufacturers will install the instrument at your site, check to ensure that the instrument meets specifications, and provide a rudimentary overview of instrument operation to prospective operators. Often, this is all that is needed; most instruments use menu-driven computer software, so that any operator familiar with the terminology of fluorometry can readily access the instrument's capabilities. Anyone who can operate a

UV/Vis absorption spectrometer should be able to learn to operate a basic type of fluorescence spectrometer without significant difficulty. More complex instruments (especially those using laser sources) are correspondingly more difficult to operate.

Operation of basic instrumentation for routine, well-established fluorometric methods can be performed by a person with an associate degree. Adaptation of existing procedures for nonstandard samples, or use of more complex instrumentation, requires a bachelor's degree background in analytical chemistry (and perhaps biochemistry or organic chemistry). Development of new instruments or new types of applications, or operation of highly sophisticated instruments (as in laser-excited fluorescence for atmospheric remote sensing) usually requires some graduate-level education.

Service and Maintenance

The key to long life and trouble-free operation of a fluorescence spectrometer is the same as that for any optical spectrometric instrument: House it in a climate-controlled environment free from dust, chemical fumes, and excessive heat, humidity, and vibration. If this is done, the only maintenance that should normally be required is periodic replacement of the source. Lamps (even those obtained from the same manufacturer) tend to be highly variable; some last for years, but others last only several months. Lamps age and their power output tends to decrease slowly with time. Periodic checks of instrumental sensitivity using samples (such as quinine or anthracene solutions) recommended in the manufacturer's instruction manual should be carried out; when the sensitivity has degraded to an unacceptably low level, the lamp probably must be replaced. High-pressure xenon lamps must be handled carefully to avoid breakage and possible injury. The manufacturer's instruction manual should explicitly describe the precautions to be taken in changing the lamp. Follow those precautions rigorously.

Although a manufacturer may offer a yearly service contract, these are probably unnecessary unless the instrument is to be used under very harsh conditions or the user's installation has virtually no capabilities for even minor instrument repairs.

Most manufacturers maintain service departments that can provide assistance over the telephone (instruction manuals tend to be lamentably inadequate in the area of troubleshooting) or, in case of severe difficulty, will dispatch a service engineer for on-site repairs. For a properly housed and properly used fluorometer, such service visits should be infrequent.

Suggested Readings

Books

GUILBAULT, G. G., *Practical Fluorescence*, 2nd ed. New York: Marcel Dekker, 1990. Multiauthor treatise accurately described by its title. The best available compendium of information on analytical applications of fluorometry.

PARKER, C. A., *Photoluminescence of Solutions*. Amsterdam: Elsevier, 1968. Despite its age, this is still an excellent introduction to the principles and practice of fluorometry. The discussion of applications, though useful, is very dated.

RENDELL, D., *Fluorescence and Phosphorescence Spectroscopy*. New York: Wiley, 1987. Open-learning introductory text; very readable and contains many useful examples and helpful hints.

SCHULMAN, S. G., *Molecular Luminescence Spectroscopy: Methods and Applications*. New York: Wiley, 1985, 1988, 1993. Three-volume multiauthor compendium of recent advances; many excellent chapters on specific topics.

SLAVIK, J., *Fluorescent Probes in Cellular and Molecular Biology.* Boca Raton, FL: CRC Press, 1994. Much less limited in scope than the title implies; contains much useful introductory material, especially on instrumentation.

Book Chapters and Review Articles

Analytical Chemistry, in even-numbered years, publishes a Fundamental Review Article on Molecular Fluorescence, Phosphorescence, and Chemiluminescence; the most recent such article is SOPER, S. A., L. B. McGOWN, AND I. M. WARNER, *Analytical Chemistry*, 68 (1996), 73R. Recent developments in instrumentation and techniques are surveyed authoritatively in these articles. In odd-numbered years, some reviews survey applications of fluorescence; for example, see DIAMANDIS, E. P., *Analytical Chemistry*, 65 (1993), 454R.

HARRIS, T. D., AND F. E. LYTLE, "Analytical Applications of Laser Absorption and Emission Spectroscopy," in D. S. Kliger, ed., *Ultrasensitive Laser Spectroscopy.* New York: Academic Press, 1983. Authoritative (albeit somewhat dated) and readable survey of the virtues and shortcomings of laser-induced fluorescence techniques.

HURTUBISE, R. J., "Trace Analysis by Luminescence Spectroscopy," in G. D. Christian and J. B. Callis, eds., *Trace Analysis: Spectroscopic Methods for Molecules.* New York: Wiley, 1986. Particularly strong on instrumentation and applications.

WEHRY, E. L., "Molecular Fluorescence and Phosphorescence Spectroscopy," in B. W. Rossiter and R. C. Baetzold, eds., *Determination of Electronic and Optical Properties.* 2nd ed., vol. 8. Physical Methods of Chemistry. New York: Wiley, 1993. Detailed coverage of instrumentation and techniques.

References

1. E. L. Wehry, in B. W. Rossiter and R. C. Baetzold, eds., *Physical Methods of Chemistry*, vol. 8 (New York: Wiley, 1993), p. 109.

2. E. L. Wehry, in G. G. Guilbault, ed., *Practical Fluorescence*, 2nd ed. (New York: Marcel Dekker, 1990), p. 75.

3. W. R. Seitz, *Crit. Rev. Anal. Chem.*, 8 (1980), 367.

4. D. W. Fink, *Trends Anal. Chem.*, 1 (1982), 254.

5. H. Lingemann and others, *J. Liq. Chromatogr.*, 8 (1985), 789.

6. J. Slavik, *Fluorescent Proves in Cellular and Molecular Biology* (Boca Raton, FL: CRC Press, 1994).

7. M. M. Krahn and others, *Environ. Sci. Technol.*, 27 (1993), 699.

8. J. P. Alarie and others, *Rev. Sci. Instrum.*, 64 (1993), 2541.

9. E. L. Wehry, in G. G. Guilbault, ed., *Practical Fluorescence*, 2nd ed. (New York: Marcel Dekker, 1990), p. 127.

10. T. G. Matthews and F. E. Lytle, *Analytical Chemistry*, 51 (1979), 583.

11. T. Vo-Dinh, in E. L. Wehry, ed., *Modern Fluorescence Spectroscopy*, vol. 4 (New York: Plenum, 1981), p. 167.

12. T. Ndou and I. M. Warner, *Chemical Reviews*, 91 (1991), 493.

13. J. R. Lakowicz, *Principles of Fluorescence Spectroscopy* (New York: Plenum, 1983).

14. M. A. Winnik, *Photophysical and Photochemical Tools in Polymer Science* (Dordrecht, The Netherlands: Riedel, 1986).

15. R. Phillips, *Sources and Applications of Ultraviolet Radiation* (New York: Academic Press, 1983).

16. J. W. Hofstraat, C. Gooijer, and N. H. Velthorst, in S. G. Schulman, ed., *Molecular Luminescence Spectroscopy: Methods and Applications*, vol. 3 (New York: Wiley, 1993), p. 323.

17. R. Brina and A. G. Miller, *Analytical Chemistry*, 64 (1992), 1413.

18. J. F. Brennan and others, *Appl. Spectrosc.*, 47 (1993), 2081.

19. C. A. Parker, *Photoluminescence of Solutions* (Amsterdam: Elsevier, 1968), p. 131.

20. J. M. Harris, *Analytical Chemistry*, 54 (1982), 2337.

21. L. W. Hershberger, J. B. Callis, and G. D. Christian, *Analytical Chemistry*, 51 (1979), 1444.

22. G. J. Diebold and R. N. Zare, *Science*, 196 (1977), 1439.

23. P. M. Epperson, R. D. Jalkian, and M. B. Denton, *Analytical Chemistry*, 61 (1989), 282.

24. C. A. Parker, *Photoluminescence of Solutions* (Amsterdam: Elsevier, 1968), p. 128.

25. T. Vo-Dinh and others, *Environ. Sci. Technol.*, 12 (1978), 1297.

26. E. L. Inman, Jr., and J. D. Winefordner, *Analytical Chemistry*, 54 (1982), 2018.

27. G. D. Christian, J. B. Callis, and E. R. Davidson, in E. L. Wehry, ed., *Modern Fluorescence Spectroscopy*, vol. 4 (New York: Plenum, 1981), p. 111.

28. T. Vo-Dinh, *Appl. Spectrosc.*, 36 (1982), 576.

29. P. M. R. Hertz and L. B. McGown, *Analytical Chemistry*, 64 (1992), 2920.

30. D. Eastwood, in E. L. Wehry, ed., *Modern Fluorescence Spectroscopy*, vol. 4 (New York: Plenum, 1981), 251.

31. J. A. Taylor and E. S. Yeung, *Analytical Chemistry*, 64 (1992), 1741.

32. S. Barnard and D. R. Walt, *Environ. Sci. Technol.*, 25 (1991), 1301.

33. C. Moulin and others, *Appl. Spectrosc.*, 47 (1993), 2007.

34. R. B. Thompson and E. R. Jones, *Analytical Chemistry*, 65 (1993), 730.

35. O. S. Wolfbeis, *Fiber Optic Chemical Sensors and Biosensors* (Boca Raton, FL: CRC Press, 1991).

36. W. R. Seitz, *Crit. Rev. Anal. Chem.*, 19 (1988), 135.

37. M. J. Sepaniak, B. J. Tromberg, and T. Vo-Dinh, *Progr. Anal. Spectrosc.*, 11 (1988), 481.

38. B. J. Tromberg and others, *Analytical Chemistry*, 60 (1988), 1901.

39. M. C. Gutierrez, A. Gomez-Hens, and D. Perez-Bendito, *Talanta*, 36 (1989), 1187.

40. T. D. Harris and F. E. Lytle, in D. S. Kliger, ed., *Ultrasensitive Laser Spectroscopy* (New York: Academic Press, 1983), p. 369.

41. R. J. van de Nesse and others, *Anal. Chim. Acta*, 281 (1993), 373.

42. F. E. Lytle, *J. Chem. Educ.*, 59 (1982), 915.

43. J. Hecht, *The Laser Guidebook* (New York: McGraw-Hill, 1992), p. 263.

44. K. J. Miller and F. E. Lytle, *J. Chromatogr.*, 648 (1993), p. 245.

45. H. M. Shapiro, *Practical Flow Cytometry* (New York: Liss, 1988).

46. J. N. Demas, *Excited State Lifetime Measurements* (New York: Academic Press, 1983).

47. F. E. Lytle, *Analytical Chemistry*, 46 (1974), 545A, 817A.

48. L. B. McGown and K. Nithipatikom, in T. Vo-Dinh, ed., *Chemical Analysis of Polycyclic Aromatic Compounds* (New York: Wiley, 1989), p. 201.

49. R. B. Dickinson, Jr., and E. L. Wehry, *Analytical Chemistry*, 51 (1979), 778.

50. K. Nithipatikom and L. B. McGown, *Appl. Spectrosc.*, 41 (1987), 395.

51. J. N. Demas and R. A. Keller, *Analytical Chemistry*, 57 (1985), 538.

52. E. P. Diamandis and T. K. Christopoulos, *Analytical Chemistry*, 62 (1990), 1149A.

53. N. T. Azumi and others, *Appl. Spectrosc.*, 46 (1992), 994.

54. E. L. Wehry, in E. H. Piepmeier, ed., *Analytical Applications of Lasers* (New York: Wiley, 1986), p. 211.

55. J. M. Hayes and G. J. Small, *Analytical Chemistry*, 55 (1983), 565A.

56. S. A. Soper and others, *Analytical Chemistry*, 63 (1991), 432.

57. M. D. Barnes and others, *Analytical Chemistry*, 65 (1993), 2360.

58. T. C. O'Haver, in J. D. Winefordner, ed., *Trace Analysis: Spectroscopic Methods for Elements* (New York: Wiley, 1976), p. 15.

59. G. L. Campi and J. D. Ingle, Jr., *Anal. Chim. Acta*, 224 (1989), 225.

60. G. Patonay and M. D. Antoine, *Analytical Chemistry*, 63 (1991), 321A.

61. S. A. Soper, Q. L. Mattingly, and P. Vegunta, *Analytical Chemistry*, 65 (1993), 740.

62. A. J. G. Mank and others, *Analytical Chemistry*, 65 (1993), 2197.

63. J. Zhang and others, *Anal. Chim. Acta*, 279 (1993), 281.

64. D. R. Christman, S. R. Crouch, and A. Timnick, *Analytical Chemistry*, 53 (1981), 2040.

65. M. E. Rollie, G. Patonay, and I. M. Warner, *Analytical Chemistry*, 59 (1987), 180.

66. W. Xu and others, *Analytical Chemistry*, 66 (1994), 4133.

67. R. L. Plant and D. H. Burns, *Appl. Spectrosc.*, 47 (1993), 1594.

68. T. Vo-Dinh, R. B. Gammage, and P. R. Martinez, *Analytical Chemistry*, 53 (1981), 253.

69. R. P. Frankewich, K. N. Thimmiah, and W. L. Hinze, *Analytical Chemistry*, 63 (1991), 2924.

70. R. J. Hurtubise, in G. G. Guilbault, ed., *Practical Fluorescence*, 2nd ed. (New York: Marcel Dekker, 1990), p. 431.

71. T. Vo-Dinh, *Room-Temperature Phosphorimetry for Chemical Analysis* (New York: Wiley, 1984).

72. W. R. G. Baeyens, in S. G. Schulman, ed., *Molecular Luminescence Spectroscopy: Methods and Applications* (New York: Wiley, 1985), p. 29.

73. G. G. Guilbault, *Practical Fluorescence*, 2nd ed. (New York: Marcel Dekker, 1990).

74. E. P. Diamandis, *Analytical Chemistry*, 65 (1993), 454R.

75. E. L. Wehry, in G. G. Guilbault, ed., *Practical Fluorescence*, 2nd ed. (New York: Marcel Dekker, 1990), p. 367.

76. P. Froehlich and E. L. Wehry, in E. L. Wehry, ed., *Modern Fluorescence Spectroscopy*, vol. 3 (New York: Plenum, 1981), p. 35.

77. A. Hulshoff and H. Lingeman, in S. G. Schulman, ed., *Molecular Luminescence Spectroscopy: Methods and Applications*, vol. 1 (New York: Wiley, 1985), p. 621.

78. E. Arriaga and others, *J. Chromatogr.*, 652 (1993), 347.

79. X. C. Huang, M. A. Quesada, and R. A. Mathies, *Analytical Chemistry*, 64 (1992), 2149.

80. K. Ueno and E. S. Yeung, *Analytical Chemistry*, 66 (1994), 1424.

81. S. I. Mho and E. S. Yeung, *Analytical Chemistry*, 57 (1985), 2253.

82. D. R. Walt, *Proc. IEEE*, 80 (1992), 903.

83. I. A. Hemmila, *Applications of Fluorescence in Immunoassay* (New York: Wiley, 1991).

84. H. T. Karnes, J. S. O'Neal, and S. G. Schulman, in S. G. Schulman, ed., *Molecular Luminescence Spectroscopy: Methods and Applications*, vol. 1 (New York: Wiley, 1985), p. 717.

85. J. Frengen and others, *Clin. Chem.*, 39 (1993), 2174.

86. R. Y. Tsien, *ACS Symp. Ser.*, 538 (1993), 130.

87. B. Valeur, in S. G. Schulman, ed., *Molecular Luminescence Spectroscopy: Methods and Applications*, vol. 3 (New York: Wiley, 1993), p. 25.

88. W. E. May and S. A. Wise, *Analytical Chemistry*, 56 (1984), 225.

89. W. T. Cobb and L. B. McGown, *Analytical Chemistry*, 62 (1990), 186.

90. D. J. Desilets, P. T. Kissinger, and F. E. Lytle, *Analytical Chemistry*, 59 (1987), 1830.

91. L. B. McGown, *Analytical Chemistry*, 61 (1989), 839A.

92. R. Niessner, U. Panne, and H. Schroder, *Anal. Chem. Acta.*, 255 (1991), 231.

93. A. P. D'Silva and V. A. Fassel, *Analytical Chemistry*, 56 (1984), 985A.

94. Y. Yang, A. P. D'Silva, and V. A. Fassel, *Analytical Chemistry*, 53 (1981), 2107.

95. P. Garrigues, J. Bellocq, and S. A. Wise, *Fresenius' J. Anal. Chem.*, 336 (1990), 106.

96. S. Weeks and others, *Analytical Chemistry*, 62 (1990), 1472.

97. R. Jankowiak and G. J. Small, *Analytical Chemistry*, 61 (1989), 1023A.

98. P. D. Devanesan and others, *Chem. Res. Toxicol.*, 5 (1992), 302.

99. E. L. Wehry and G. Mamantov, *Analytical Chemistry*, 51 (1979), 643A.

100. M. B. Perry, E. L. Wehry, and G. Mamantov, *Analytical Chemistry*, 55 (1983), 1893.

Chapter 27

Chemiluminescence

Timothy A. Nieman

University of Illinois
Department of Chemistry

Summary

General Uses

- Quantitation of trace concentrations of various inorganic and organic species
- Usually coupled with a selective physical or chemical step such as chromatography, electrophoresis, enzyme reactions, or antigen–antibody reactions

Common Applications

- Detection of sulfur compounds and nitrogen compounds in gas chromatography
- Detection of fluorophores, amines, amino acids, and labeled species in liquid chromatography
- Measurement of nitrosamines
- Measurement of atmospheric ozone and oxides of nitrogen
- Immunoassay
- DNA probe assays, polymerase chain reaction (PCR) methods
- Characterization of polymer stress and aging

Samples

State

It is possible to perform chemiluminescence measurements in the gas, liquid, or solid state. However, measurements in the liquid or gas phase are most common. With the exception of polymer characterization applications, solid samples are usually dissolved before analysis. Use of water or a mixed solvent containing water is most common for liquid-phase measurements.

Preparation

The preparation required depends on the specific application, but is generally dictated by the methodology preceding the chemiluminescence reaction (such as chromatography or antigen–antibody binding). For determination of certain gas-phase species (such as NO, SO, or O_3) with sufficiently selective reactions, there is no sample preparation required at all.

Analysis Time

This is highly dependent on the sample preparation requirements and specifics of the analytical method and the chemiluminescence reaction. However, use of chemiluminescence detection is generally not the time-limiting factor. Rather, it is the methodology preceding chemiluminescence detection (such as chromatography or antigen–antibody binding) that is time-limiting.

Limitations

General

- In most cases, the reaction is destructive of the analyte and the chemiluminescence reagent.
- Although chemiluminescence is generally considered a spectroscopic method, the necessity to carry out a chemical reaction with the analyte precludes the possibility of remote sensing, as is done with fluorescence and Raman.
- Chemiluminescence generally cannot be used for multicomponent determinations on mixtures of analytes without a separation step.

Accuracy

Chemiluminescence intensities are sensitive to environmental factors, so there can be accuracy problems if factors such as temperature, viscosity, pH, and the presence of quenching species between running standards and unknowns are not controlled.

Sensitivity and Detection Limits

Sensitivity is a strength of chemiluminescence. Typical detection limits are 10 pmol NO using ozone, 0.1 pmol sulfur compounds using a hydrogen flame followed by ozone, 1 fmol fluorophores using peroxyoxalates, and 0.1 fmol peroxidase using luminol.

Complementary or Related Techniques

Fluorescence is the most similar technique because it is also a luminescence method. Many of the measurement concerns are similar. However, fluorescence measurements allow versatility not possible with chemiluminescence: Selection of emission and excitation wavelengths allows selectivity control, adjustment of source intensity allows sensitivity control, and adjustment of excitation beam size and location allows control of the sample region probed.

Other mass flow sensitive techniques such as amperometry have the same sensitivity to measurement issues of mass transport, reaction time, and observation time as does chemiluminescence.

Introduction

Chemiluminescence (CL) reactions yield light as one of the products. Some of the most familiar CL systems are fireflies, chemical lightsticks for emergency lighting, and novelties at amusement parks. Because the emission intensities are a function of the concentrations of chemical species involved in the CL reaction, measurement of emission intensities can be used for analytical purposes to quantitate those chemical concentrations. Although CL has been observed in living systems (such as fireflies, bacteria, and marine organisms) since antiquity, and in synthetic CL compounds since the late 1800s, investigation of CL for analytical use began in the 1970s. Routine application of CL as an analytical tool dates from the 1970s for gas-phase and from the 1980s for liquid-phase reactions. In more recent years, analytical interest in and application of CL has grown at a rapid pace, and CL finds application in a variety of areas, including environmental and biomedical analysis, toxicology, detection for chromatography and electrophoresis, immunoassay and nucleic acid assays, polymer characterization, and biosensors.

CL involves both a chemical reaction and a luminescence process. In a CL reaction between species A and B, some fraction of the product species, P, are formed in electronically excited states, P^*, which can subsequently relax to the ground state with emission of a photon:

$$A + B \rightarrow P^* \rightarrow P + \text{Light} \tag{27.1}$$

The number of reactions that result in significant CL emission is relatively small. One important requirement for CL is that there be sufficient energy available for formation of an electronically excited state. This energy comes from the chemical reaction. For emission in the visible region, a reaction liberating 40 to 70 kcal/mol is required. Reactions this energetic are generally redox reactions. The liberation of this much energy usually comes from bond cleavage or electron transfer. In systems involving bond cleavage (as with luminol or peroxyoxalates), the CL molecule can be used only once. Some electron transfer reactions (such as with aromatic radical ions from rubrene or p-benzoquinone or from metal complexes such as tris(2,2′-bipyridyl) ruthenium(II)) result in CL emission without bond cleavage or rearrangement, so those systems can be recycled, as is common with some types of electrogenerated chemiluminescence.

A second requirement is that the excited state be capable of losing this energy by emission of a visible photon. As a result, the products of CL reactions generally include one species that is highly fluorescent, and the CL emission is from that species. Alternatively, the excited state formed in the CL reaction can transfer its energy to a fluorescent species added to the reaction mixture.

The intensity of light emission depends on the rate of the chemical reaction, the efficiency of production of excited states, and the efficiency of achieving light emission from those excited states:

$$I_{CL} = \Phi_{CL}\frac{dP}{dt} = \Phi_{EX}\Phi_{EM}\frac{dP}{dt} \qquad (27.2)$$

where I_{CL} is the CL emission intensity (photons emitted per second), dP/dt is the rate of the chemical reaction (molecules reacting per second), Φ_{CL} is the chemiluminescence quantum yield (photons emitted per molecule reacted), Φ_{EX} is the excitation quantum yield (excited states produced per molecule reacted), and Φ_{EM} is the emission quantum yield (photons emitted per excited state). CL systems used in analysis generally have Φ_{CL} in the range of 0.01 to 0.20 (1 to 20%).

How It Works

The basic measurement requirements for CL are simple. One needs only to bring together the reactants and then measure the resulting CL emission. Because the instrumentation does not involve a light source, CL measurements do not suffer from source fluctuation noise or from source scatter; as a result, CL sometimes is capable of lower detection limits than fluorescence. Although there is no source noise or background in CL, there is chemical noise and background that limits detection; reagents and solvents must be of the highest purity. To appreciate analytical applications of CL, it is necessary to understand the instrumental approaches to the measurement and the chemistry of reactions that have proven utility for analysis.

Chemical Reactions

A large number of CL reactions are known. Some of the most important ones are explained here and applications based on these reactions are described later.

In the gas phase, the most important reactions involve ozone. Several of these are as follows:

$$O_3 + \text{Olefins} \rightarrow \text{Numerous excited species} \rightarrow \text{Light (300 to 600 nm)} \qquad (27.3)$$

$$O_3 + NO \rightarrow O_2 + NO_2^* \rightarrow \text{Light (590 nm to 3\mu m)} \qquad (27.4)$$

$$O_3 + SO \rightarrow O_2 + SO_2^* \rightarrow \text{Light (260 to 480 nm)} \qquad (27.5)$$

In all three of these examples the emission intensity is linearly dependent on the concentration of either reactant over a 4- to 6-decade range.

In solution-phase CL, some of the most useful reagents are luminol, acridinium esters, aryl oxalates, dioxetanes, and tris(2,2′-bipyridyl)ruthenium(III) or Ru(bpy)$_3^{3+}$. The structures of these reagents are shown in Fig. 27.1. With luminol, emission is in the blue, centered about 425 nm. Oxidants such as permanganate, hypochlorite, or iodine can be used, but the most useful oxidant is hydrogen peroxide. With peroxide a catalyst is required, and typical catalysts are transition metal ions (such as Co^{2+}, Cu^{2+}, and Fe^{3+}), ferricyanide, or certain metallocomplexes (such as hemin, hemoglobin, and peroxidases).

$$\text{Luminol} \xrightarrow{\text{OH}^-, \text{ Oxidant, Catalyst}} (\text{3-aminophthalate})^* \rightarrow \text{Light)} \qquad (27.6)$$

Figure 27.1 Structures of analytically useful CL reagents for solution-phase assays.

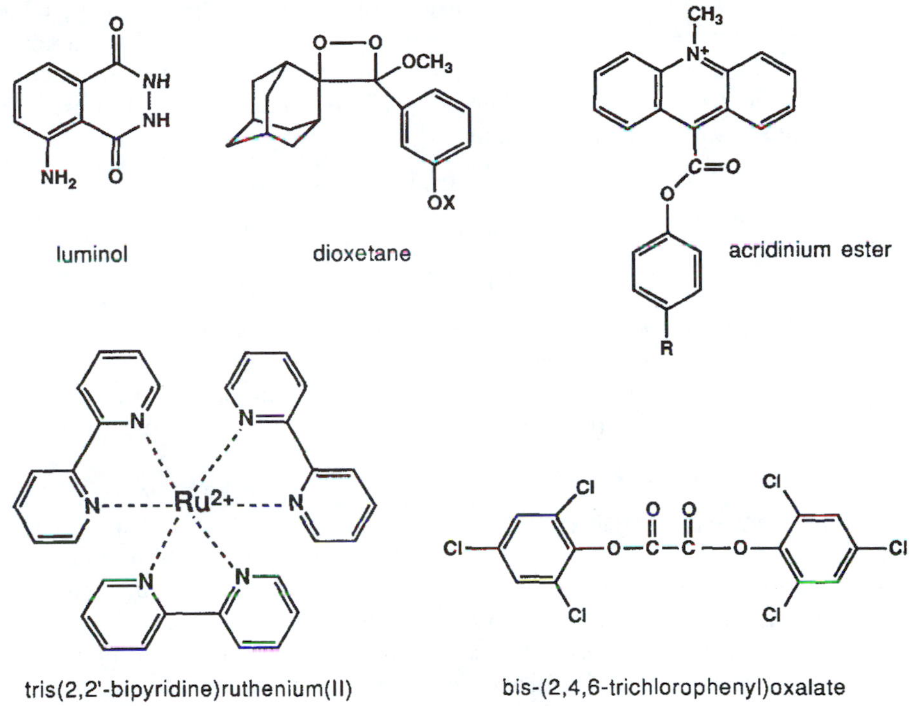

luminol   dioxetane   acridinium ester

tris(2,2'-bipyridine)ruthenium(II)   bis-(2,4,6-trichlorophenyl)oxalate

The acridinium ester reaction is similar, involving an oxidation with alkaline hydrogen peroxide; no catalyst is needed. Emission is centered about 440 nm.

$$\text{Acridinium ester} \xrightarrow{\text{OH}^-, \text{H}_2\text{O}_2} (\text{N-methylridone})^* \rightarrow \text{Light} \qquad (27.7)$$

Peroxyoxalate CL reactions are the hydrogen peroxide oxidations of aryl oxalate esters. These reactions are among the most efficient nonbiological CL reactions, with quantum efficiencies as high as 20 to 30%. A high-energy intermediate is formed by the reaction of the aryl oxalate and hydrogen peroxide. If an efficient fluorophore is present, energy transfer to that fluorophore raises it to the excited state. Emission is then from the fluorophore. Unlike most other CL reactions, here the CL emission spectrum is determined not by any species involved in the chemical reaction, but by this additional fluorophore. TCPO or bis-(2,4,6-trichlorophenyl)oxalate is a commonly used oxalate.

$$\text{TCPO} \xrightarrow{\text{H}_2\text{O}_2, \text{Fluorophore}} 2 \text{ trichlorophenol} + 2\,\text{CO}_2 + \text{Fluorophore}^* \rightarrow \text{Light} \qquad (27.8)$$

Dioxetanes are four-membered cyclic peroxides. Many are unstable, and have been proposed as intermediates in several CL reactions. Substituted 1,2-dioxetanes have been prepared that are stable at room temperature but can be chemically triggered to produce CL. Such dioxetanes have an adamantyl group on one side of the ring and a substituted aryl group on the other side. CL is generated by removal of a protecting group X from the aryl-OX group to yield an unstable aryloxide that spontaneously decomposes. A common protecting group is phosphate, which is cleaved from the dioxetane by the enzyme alkaline phosphatase. Emission comes either from the resulting excited aryl ester or from an added fluorescent enhancer via energy transfer.

$$\text{Stable dioxetane} \xrightarrow{\text{Trigger Reagent}} \text{Unstable dioxetane} \rightarrow \text{Fluorophore}^* \rightarrow \text{Light} \qquad (27.9)$$

In the $Ru(bpy)_3^{2+}$ system, orange emission centered at 610 nm results from excited state $Ru(bpy)_3^{2+*}$. This excited species can be obtained in several ways: by reaction of $Ru(bpy)_3^{3+}$ and $Ru(bpy)_3^{+}$, by reaction of $Ru(bpy)_3^{+}$ with certain oxidants, and by reaction of $Ru(bpy)_3^{3+}$ with certain reductants. Electron transfer, not bond cleavage and rearrangement of the species, leads to the excited state, and the $Ru(bpy)_3^{2+}$ species is regenerated. A commonly used reductant is either oxalate anion or an aliphatic amine.

$$Ru(bpy)_3^{2+} \rightarrow Ru(bpy)_3^{3+} + e^- \tag{27.10}$$

$$Ru(bpy)_3^{3+} + C_2O_4^{2-} \rightarrow [Ru(bpy)_3^{2+}]^* + 2CO_2 \tag{27.11}$$

$$[Ru(bpy)_3^{2+}]^* \rightarrow Ru(bpy)_3^{2+} + Light \tag{27.12}$$

For all of these reactions, the emission intensity can be linearly proportional to the concentration of any one reagent, given suitable adjustment of the other reagent concentrations.

CL reactions involving solids are much less common. One significant example is the weak CL that accompanies peroxide formation, chain linking, and chain cleavage in many polymers.

Sample and Reagent Introduction

Upon mixing of the necessary reactants, the CL reaction begins and the emission intensity rises rapidly. As the CL reaction proceeds, reactants are consumed. The reduction in reagent concentrations causes a decrease in the rate of reaction and a corresponding reduction in the CL intensity. This means that the CL emission intensity is inherently transient, as illustrated in Fig. 27.2, which depicts the variation in CL emission intensity with time after mixing aliquots of reactants. The time scale for emission depends on the specific reaction and sometimes the reactant concentrations involved, but can range from a short flash lasting less than a second to a continuous glow lasting minutes to hours.

There are two basic measurement approaches with CL: static samples and flowing streams. With the static sample approach, discrete portions of the CL reagents and the analyte are mixed in front of a detector and the entire CL emission intensity-versus-reaction time profile is observed. Generally, either the maximum intensity ($(I_{CL})_{max}$) or intensity integrated over the whole peak (proportional to total moles reacted) is measured and correlated with analyte concentration, although occasionally other values, such as peak width or rate of decay from maximum, are used.

Figure 27.2 CL emission versus time for reactions with emission of short, medium, and long duration.

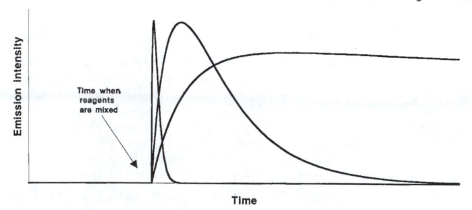

Figure 27.3 Simplified CL instrumentation for static sample measurements. To the right is the resulting emission intensity-versus-time plot showing the parameters typically quantitated.

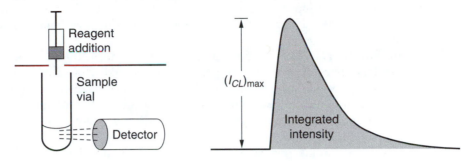

Such monitoring of emission versus time with a static sample is common in solution-phase CL for immunoassay or nucleic acid assays. The instrument for measurements on static solutions is generally called a luminometer and several are commercially available. Provision is made for addition of the appropriate trigger solution directly in the sample compartment (Fig. 27.3).

Flowing stream methods are almost always used with gas-phase CL and are very common with solution-phase CL. CL reagent and analyte streams continuously flow and are mixed together, with the mixture continuing on in the flowing stream (Fig. 27.4). Observation occurs at a fixed position after mixing. The volume between the point of mixing and the point of observation, plus the flowrate in this region, determines the time in the CL emission intensity-versus-reaction-time profile at which observation occurs, and the observed emission intensity remains constant at this value. For maximum sensitivity, it is desirable to adjust the flowrate, observation cell volume, and transfer line volumes so that the point of observation comes at the peak in the CL intensity-versus-time profile.

Figure 27.4 Simplified CL instrumentation for flow stream measurements. To the right are signal traces for a continuously flowing sample (top) and for injections of a sample into a carrier stream (bottom).

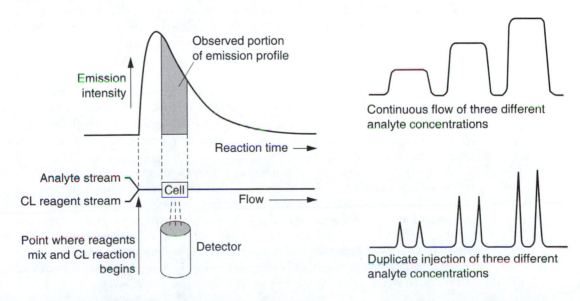

For applications in chromatographic detection (either gas or liquid) or for flow injection, the analyte carrier stream continues to flow, but the analyte concentration is not constant with time because the analyte appears as injected bands transported by the carrier stream into the reaction and detection zones. The resulting signal is a narrow peak as each analyte aliquot passes through the cell.

For flow stream CL with a continuous flow of analyte, a large-volume cell results in observation of a large portion of the CL emission profile and therefore a high signal. However, in flow stream CL with small injected aliquots of analyte, it is generally necessary to use a much smaller cell in order to prevent excessive dilution of the small analyte volume and to maintain sufficient temporal resolution between signal peaks for adjacent analyte injections or chromatographic bands.

Gas-phase flow systems can use a single vacuum pump at the outlet. With solution-phase flow systems, it is most common to have separate positive-pressure pumping for each of the analyte and reagent channels. Instrumentation for flow injection or high-performance liquid chromatography (HPLC) applications of CL generally is custom built or adapted from other instrumentation. Suitable mixing of sample and reagent streams must be incorporated; this can be a concern with multiple solvent systems, as in peroxyoxalate CL.

With both static and flow methods, the magnitude of the emission intensity seen by the detector is proportional to the volume of the cell. A common geometry for flow streams (especially for solution-phase measurements) is a cell that is long (about 1 cm) in the direction of flow and thin (50 to 100 µm) perpendicular to the direction of flow. This thin dimension minimizes band broadening and also provides a thin optical path to minimize inner filtering effects on the emission intensity (1). The cell requires a suitable window; glass and quartz are most versatile, but the visible emission wavelengths involved allow acrylic or other plastics to be used if they are compatible with the reagents and solvents.

Detection of Emitted Light

It is most common to direct the emitted radiation directly to a photodetector without an intervening emission monochromator for wavelength selection. The reason for this is because with most CL reactions, the wavelength range emitted is a function of the CL reagent, not the analyte. For a given CL reaction, the emission spectrum remains constant regardless of the analyte. Because the CL is generated chemically, emission by impurities in solvents or interfering sample constituents is less common than in fluorescence spectroscopy, in which the emission is excited by a light source. This factor also allows one to avoid using an emission monochromator in CL measurements.

The intensity of CL emission is low when minute analyte concentrations are involved. As a result, it is necessary to use very sensitive photodetectors, and the most commonly used in CL is a photomultiplier tube. For measurement of very low emission intensities, it is desirable to use photon counting instrumentation. The operation of photomultiplier tubes and photon counting instrumentation is described elsewhere (2).

Imaging detection can be advantageous with static sample measurements. Some simple instruments for immunoassay have used photographic detection, especially with multiple samples in microtitre places; by inclusion of several standards at different concentrations, it is possible to achieve semiquantitation by comparison of the different degrees of exposure of sample and standard spots in the same film. When CL is used for detection with slab gel electrophoresis of proteins or nucleic acids, photographic detection is also done. CL imaging applications can be performed with sensitive charge-coupled device (CCD) array detectors (3). Use of an array detector facilitates quantitation of the signal as well as storage and manipulation of the image.

Electrogenerated Chemiluminescence

Some CL reactions can be triggered electrochemically either by direct electrochemical initiation of the CL reaction or by electrochemical generation of a necessary component of the reaction. This is called electrogenerated chemiluminescence (ECL). Electrochemical initiation of a CL reaction can provide a way to control when or where the reaction (and emission) occurs. By controlling whether the electrode potential is in an appropriate potential range to initiate the reaction, one can effectively turn the CL reaction on or off. If the reaction is sufficiently fast, then the emission is confined to the electrode surface, and only the region of sample that is very close to the electrode is probed. It is important to remember that one no longer is dealing with homogeneous solution reactions when doing ECL. The working electrode is a reactive surface, and material transport to and from that surface probably limits the overall response.

Instrumentation for electrogenerated ECL requires suitable electrodes and a potentiostat to control the potential difference between them. Gold, platinum, or carbon working electrodes are placed in the observation cell and the counter and reference are placed downstream. Choice of the particular working electrode depends on the potential range involved. The reference electrode is typically Ag/AgCl. A common flow cell design is similar to the thin-layer amperometric cell for electrochemical detection in HPLC, with the inclusion of a window to allow viewing the region by the electrodes.

Of the CL reactions discussed above, luminol has been used to some extent in ECL mode, and $Ru(bpy)_3^{2+}$ is often used via ECL. It is possible to electrochemically initiate the CL reaction of H_2O_2 and luminol at an electrode that is held at about +0.5 V (compared to Ag/AgCl). The electrode takes the place of the conventional dissolved catalyst. The CL reaction is fast enough to confine emission close to the electrode surface. For determination of species labeled with luminol, two electrodes can be used, one at –1.0 V to generate the necessary H_2O_2 and one at about +0.5 V to initiate the CL reaction. Detection limits are the same with electrogenerated CL as with a conventional dissolved catalyst. The $Ru(bpy)_3^{2+}$ CL system easily lends itself to electrochemical control due to facile electrochemical interconversion of the key oxidation states. $Ru(bpy)_3^{2+}$ is the stable species in solution and the other two can be generated from it: $Ru(bpy)_3^{+}$ by reduction at about –1.3 V (compared to Ag/AgCl) and $Ru(bpy)_3^{3+}$ by oxidation at about +1.3 V. Additionally, because the $Ru(bpy)_3^{2+}$ species is regenerated during the CL reaction, it can be advantageous to immobilize $Ru(bpy)_3^{2+}$ on the surface of an electrode (4).

What It Does

Limitations

Like fluorescence, chemiluminescence emission intensities are decreased in the presence of species that quench the excited state. Oxygen is occasionally a concern in this regard; use of $Ru(bpy)_3^{2+}$ at low reactant concentrations is noticeably affected by oxygen quenching. CL emission intensity is also affected by factors that alter the reaction rate or the quantum efficiency of chemi-excitation. Thus, CL measurements can be sensitive to environmental factors (such as pH, ionic strength, solution composition, viscosity, and temperature).

Like other mass flow detection approaches (amperometry in liquid chromatography or flame ionization detection in GC) fluctuation in the flowrate causes fluctuation in the rate of delivery of reactants, and thus results in fluctuation in signal intensity.

Analytical Information

Qualitative

There is inherently very little qualitative information available from CL. There are often several species that can produce CL emission from the same CL reagent. Because the emitting species generally comes from the CL reagent rather than the analyte, all analytes yield the same emission spectrum with a given CL reagent. Luminol, for example, yields CL emission with many oxidants (such as hydrogen peroxide, organic peroxides, and hypochlorite), transition metal ions (such as Fe(II), Fe(III), Cr(III), and Co(II)), and organometallic species (such as horseradish peroxidase, hemoglobin, and reduced vitamin B_{12}). For all these species the emitter is the oxidation product of luminol (3-aminophthalate) rather than any species related to the oxidant or metal-ion catalyst, so there is no spectroscopic way to identify which oxidant or metal-ion species was responsible for the emission. Some reactions are essentially compound-specific. Although many metal ions cause light emission from luminol in the presence of H_2O_2, Fe^{2+} is the only species to cause significant emission in the absence of H_2O_2. Other examples are tetrakis(dimethylamino)ethylene, which gives CL emission only with O_2, and aqueorin, which gives CL emission only with Ca^{2+}.

Qualitative information usually comes from the physical or chemical step that is coupled with CL detection. CL detection could be used to allow detection of the qualitative information available from a chromatography, electrophoresis, or immunoassay method. Usually, the only qualitative information obtained directly from CL is whether you obtained significant emission from a sample. There are instances where that information can be valuable; one example is the forensic use of luminol CL to determine whether small specks of blood-colored material really are blood, based on the intense CL emission from luminol/H_2O_2 in the presence of hemoglobin.

Quantitative

Recall that the basic relationship governing CL intensities is

$$I_{CL} = \Phi_{CL}\frac{dP}{dt} = \Phi_{EX}\Phi_{EM}\frac{dP}{dt}$$

It is because the reaction rate is a function of chemical concentrations that CL is useful for quantitation of chemical concentrations. High sensitivity is a major attribute of CL. Because the technique is simultaneously a kinetic method and a luminescence method, it provides very low detection limits and wide dynamic ranges with emission intensity linearly proportional to concentration over several decades in concentration. Typical detection limits are 10 pmol NO using ozone, 0.1 pmol sulfur compounds using a hydrogen flame followed by ozone, 1 fmol fluorophores using peroxyoxalates, and 0.1 fmol peroxidase using luminol. Linear ranges typically cover three to four decades of concentration for solution reactions and six to seven decades for gas-phase reactions.

Applications

1. Monitoring Atmospheric Pollutants.

A few reactions are selective enough that they are used directly to monitor atmospheric species without need for a separation. The reaction between ozone and olefins is used for monitoring atmospheric O_3 (5). A constant flow of ethylene is supplied as a reagent. Quantitation is possible over a 4-decade range with a detection limit of 3 ppb by volume air. The CL reaction of ozone with NO is used for monitoring of atmospheric NO by using a constant flow of ozone as the reagent. Response is linear for over 6 decades, with a detection limit of about 5 ng NO or 1 ppt by volume of air. Emission is conveniently to the red of emission from ozone with other potential interferences (such as olefins).

2. Detection in Gas Chromatography.

The reaction between ozone and olefins can be used with O_3 as the controlled reagent for a GC detector for hydrocarbons; at low temperatures, only alkenes are detected, but above 150 °C aromatics react, and above 200 °C saturated hydrocarbons also react. As a GC detector, detection limits are in the nanogram range for most hydrocarbons.

The reaction of O_3 with NO can be applied to a variety of different analytes by converting the analyte to an equivalent amount of NO. Carcinogenic N-nitrosamines are sensitively detected at picogram to subnanogram levels by using pyrolytic decomposition at 300 to 350 °C to yield NO (6). A chromatographic detector using this approach is commercially available and is called a thermal energy analyzer. Catalytic oxidation by O_2 at 800 to 1000 °C yields NO from all oxidizable nitrogen-containing compounds; this method allows use of the O_3/NO reaction in place of a Kjeldahl determination for total nitrogen (7).

The reaction between ozone and SO forms the basis for sulfur CL detectors, which are commercially available for use in GC and have been adapted for use in SFC and HPLC (8). The chromatographic effluent is passed through a heated chamber or a H_2 flame, where SO is created from S-containing analytes. Then in a reaction cell, this SO reacts with O_3. Detection limits are approximately 10 to 15 pg of a sulfur-containing compound, or about 20 times better than the flame photometric sulfur detector, and about the same sensitivity for sulfur as the flame ionization detector has for carbon. Response is linear for over 5 decades, as opposed to the nonlinear response of the flame photometric sulfur detector.

3. Detection in HPLC and Capillary Electrophoresis.

An important application of peroxyoxalate CL is in HPLC (usually reverse phase) as a postcolumn reaction for determination of fluorophores (9). The aryl oxalate and H_2O_2 are added postcolumn to initiate CL. Detection limits of 1 fmol or lower are achieved for fluorescent analytes (polycyclic aromatic hydrocarbons and polycyclic aromatic amines such as anthracene, perylene, benzopyrenes, aminoanthracenes, and aminopyrenes) and for fluorescent derivatives. A number of derivatizing agents have been described that yield 1 to 50 fmol detection limits for a variety of analytes: dansyl chloride for amino acids, amines, catecholamines, steroids, and amphetamines; o-phthalaldehyde and naphthalene dialdehyde for amines; fluorescamine for catecholamines; and N-[4-(6-dimethylamino-2-benzofuranyl)phenyl]maleimide for thiols. Because peroxyoxalate fluorescence is most efficient with easily oxidized fluorophores, peroxyoxalate CL detection results in different relative sensitivities and generally simpler chromatograms than are obtained with conventional fluorescence detection.

Ru(bpy)$_3^{2+}$ has been used for detection of oxalate, amines, amino acids, and some antibiotics following separation on reverse phase and ion-exchange columns (10). Amino acids can be detected without derivatization, but with a wide variation in detection limits. Lower and more uniform detection limits are obtained for amino acids that have been derivatized (as with dansyl chloride) to contain a 3° amine. However, the 2° and 3° amine-containing species that can be sensitively detected without derivatization are most attractive for Ru(bpy)$_3^{2+}$ ECL detection because many of them are difficult to detect by more conventional means. The herbicide glyphosate and the antibiotic erythromycin are examples of two such compounds. Detection limits are generally 1 pmol or below. Ru(bpy)$_3^{3+}$ ECL can be used over a very wide pH range of about 2 to 11. Background emission increases somewhat in strongly alkaline solutions because of reaction between Ru(bpy)$_3^{3+}$ and OH$^-$. Amino acids and most other amines give considerably greater emission intensity at pH 9 to 11 than at pH 7 or below, presumably due to deprotonation of the amine nitrogen.

Luminol has been used for determination of transition metal ions separated by ion exchange, for determination of ligands (catecholamines, amino acids, and proteins) that complex these transition metal catalysts, and for determination of species that have been precolumn derivatized with a luminol-like reagent (such as amines, carboxylic acids, or amino acids). Luminol CL requires a strongly alkaline pH, so generally a postcolumn pH change is required.

Very limited work has appeared to date with CL detection in capillary electrophoresis (11). The systems briefly demonstrated are luminol, acridinium esters, peroxyoxalate, and Ru(bpy)$_3^{2+}$. There are some technological difficulties with postseparation addition and mixing of reagents in capillaries, and the very small volumes involved mean that emission intensities are very low. But the high power of capillary separations will stimulate development of CL detection systems designed for the capillary environment.

4. Detection of Enzyme Reaction Products: Flow Streams and Biosensors.

To increase the selectivity of CL reactions and to extend CL to analytes not directly involved in a CL reaction, it is common practice to precede a CL reaction by an enzyme reaction for which the desired analyte is the substrate and one of the products is sensitively detected by CL. This is most commonly done in flowing systems with columns of immobilized enzyme. Recently, interest has been directed toward biosensor designs using optical fibers.

Oxidase enzymes that generate H$_2$O$_2$ are commonly used. Not only can H$_2$O$_2$ be quantitated with several CL systems, but the necessary oxidant (O$_2$) is already present in most samples. Assuming quantitative conversion by the enzyme, substrates can be determined down to 10 to 100 nM, just as can H$_2$O$_2$. Substrates detected this way include glucose, cholesterol, choline, uric acid, amino acids, aldehydes, and lactate. For example:

$$\text{Uric acid} + \text{O}_2 \xrightarrow{\text{Uricase}} \text{Allantoin} + \text{H}_2\text{O}_2 \qquad (27.13)$$

The approach can be extended by using sequential enzyme steps to ultimately convert analyte to CL reactant. In this way, other sugars, glucosides, cholesterol esters, creatinine, and acetylcholine have been measured (12). For example:

$$\text{Sucrose} + \text{H}_2\text{O} \xrightarrow{\text{Invertase}} \alpha\text{-D-glucose} + \text{Fructose} \qquad (27.14)$$

$$\alpha\text{-D-glucose} \xleftarrow{\quad\text{Mutarotase}\quad} \beta\text{-D-glucose} \tag{27.15}$$

$$\beta\text{-D-glucose} + O_2 \xrightarrow{\quad\text{Glucose oxidase}\quad} \text{Gluconic acid} + H_2O_2 \tag{27.16}$$

Luminol plus a peroxidase catalyst is the best system for H_2O_2 determination. Peak CL intensity is reached in about 100 msec; the solvent is completely aqueous, but compatible with some organic components; the detection limit is around 0.1 μM, with linearity for 3 to 4 decades.

Another useful category of redox enzymes is dehydrogenases, which generate NADH (from NAD^+) in amounts proportional to the analyte concentration. Because NADH contains an aliphatic 3° amine (and NAD^+ does not), it is possible to sensitively detect the produced NADH with $Ru(bpy)_3^{2+}$ ECL. This approach has been used with glucose dehydrogenase to measure glucose (13). NADH generation can also be coupled with detection using bacterial bioluminescence.

In several instances, very useful detection capabilities have been created by joining in sequence HPLC, an immobilized enzyme reactor, and a CL reaction. Now one must simultaneously consider the solution composition requirements of the separation, the enzyme reactions, and the CL reaction. Relatively large amounts of acetonitrile are acceptable with many immobilized enzymes; however, methanol usually causes loss of enzyme activity. β-D-glucosides can be separated by reverse phase with an acetonitrile/aqueous buffer mobile phase. This mobile phase is compatible with the enzyme reactor containing β-glucosidase and glucose oxidase that follow. The generated H_2O_2 is quantitated by merging the effluent with a stream containing luminol and horseradish peroxidase (14).

5. Immunoassay and Nucleic Acid Assays.

CL detection is attractive for the detection of proteins and oligonucleotides in immunoassay and nucleic acid assays (DNA fingerprinting, DNA sequencing, and detection of DNA following electrophoresis and blotting) due to advantages of sensitivity (equal to or better than radioactive labels such as ^{125}I), dynamic range, simple instrumentation, long shelf life, and low cost (15). In chronological order, the CL systems that have found application here are luminol, acridinium esters, dioxetanes, and $Ru(bpy)_3^{2+}$. The assay can involve either direct labeling with the CL species or indirect labeling with a species that catalyzes a CL indicator reaction. After incubation to achieve binding and then separation (if necessary) of the bound and unbound material, suitable reagents are added to initiate the CL reaction. All the necessary labeled species and other reagents are available from several companies for a variety of analytes.

Direct labels include luminol, acridinium esters, and $Ru(bpy)_3^{2+}$. With direct labeling, CL emission is a short flash lasting 1 to 5 sec because the CL reagent is consumed. This short emission time scale requires instrumentation that allows reaction initiation directly in front of the detector. Although one can monitor peak intensity, it is also common to integrate the entire light output. With direct CL labels, each label molecule can react only once and produce at most only one photon (actually less because the quantum efficiency is less than unity). $Ru(bpy)_3^{2+}$ is the exception here, as it can be continuously recycled and re-excited.

Catalysts for CL reactions can be used as indirect labels. Each such label can catalyze the reaction of many CL substrate molecules and lead to emission of many photons. For enzyme immunoassay, the most common labels are horseradish peroxidase (HRP), alkaline phosphatase, and β-galactosidase, and for DNA detection the most common nonisotopic label is alkaline phosphatase. CL systems are available for all these: enhanced luminol for HRP, and dioxetanes for alkaline phosphatase and β-galactosidase. If a sufficiently large concentration of the CL reagent is added that its concentration remains essentially constant during measurement, the rate of the CL reaction is limited by the enzyme catalyst concentration. Emission is then in the form of "continuous" glow lasting many minutes to hours, so that it is no longer necessary for instrumentation or procedures to provide reaction initiation directly in front of detector and it is possible to scan microtiter plates or electrophoresis gels or make multiple recordings of emission intensity. With luminol for detection of HRP, maximum intensity is reached in a few minutes, with steadily decreasing intensity; measurement is usually made between 2 and 30 minutes after reaction initiation. Dioxetanes require many minutes to an hour to reach maximum intensity, but then can remain constant for hours. Quantitation with enzyme labels is generally via monitoring of peak light intensity or integration over a fixed time rather than over the entire emission-versus-time profile; as the catalyst concentration changes, the rate of reaction changes but the equilibrium position remains constant.

6. Materials Characterization.

Polyolefins, polyamides, rubber, epoxies, lubricating oils, and edible oils are examples of some materials characterized by CL. CL measurements on solids involve holding the solid at a controlled temperature (25 to 250 °C) in front of the detector and monitoring the intensity of CL emission versus time. The shapes of CL intensity–time curves are characteristic of particular materials and can help determine whether given materials are identical in composition or whether identical materials have identical histories. If materials are heated in the presence of oxygen, CL emission is proportional to the rate of oxidation. If materials are heated in the absence of oxygen, CL emission is due to decomposition of previously formed peroxide groups. Because of the sensitivity of CL, it is possible to detect the difference in oxidative durability of stabilized and unstabilized polymers in less than a minute. By using a CCD camera to record the image of CL emission from a polymer surface, it is possible to see patterns of stress and to follow nonuniform aging across the surface (16). A related application is to use imaging of luminol or $Ru(bpy)_3^{2+}$ ECL to visualize the difference in electron-transfer properties across the surface of carbon fibers (17).

Nuts and Bolts

Relative Costs

Chemiluminescence instrumentation can be fairly inexpensive because of the simplicity of the instrumental demands. A sensitive detector is required because of the low-level signals involved. The basic instrument is merely a detector coupled with a sample compartment or flow cell and is approximately $10,000. Addition of sample and reagent pumping capability raises the price to about $15,000. For a commercial immunoassay unit with automated sample handling and data manipulation, the price can be $30,000 or more.

Vendors for Instruments and Accessories

Instrumentation for chemiluminescence measurements is acquired from various sources depending on the specific application. Much of the instrumentation is assembled by individual users. Commercially available luminometers and reagents for chemiluminescence have been reviewed and periodically updated over the last several years (18).

Suppliers of instrumentation for gas-phase chemiluminescence measurements (oxides of nitrogen, ozone, sulfur chemiluminescence detectors for chromatography) include the following:

Antek Instruments, Inc.
300 Bammel Westfield Rd.
Houston, TX 77090
phone: 713-580-0339, 800-365-2143
fax: 713-580-0719

Sievers Instruments
6185 Arapahoe St.
Boulder, CO 80303
phone: 303-444-2009
fax: 303-444-9543
email: sievers@url.com
Internet: http://www.sieversinst.com

Thermo Electron Corp.
81 Wyman St.
Waltham, MA 02254
phone: 617-622-1000
fax: 617-622-1207
Internet: http://www.thermo.com

For general-purpose solution-based instrumentation it is possible to use fluorescence instruments with the source either turned off or blocked. Additional sensitivity can be gained by setting the emission monochromator to the zero-order position to pass all radiation without wavelength selection.

There are several suppliers of instruments designed for immunoassay or nucleic acid hybridization:

Analytical Luminescence Laboratory
11760 Sorrento Valley Rd.
San Diego, CA 92121
phone: 800-854-7050
fax: 619-455-9204

Ciba Corning Diagnostics
132 Artino St.
Oberlin, OH 44074
phone: 800-445-3673

EG&G Berthold
4 Tech Circle
Natick, MA 03063
phone: 800-343-6345
fax: 508-651-3250

Integrated BioSolutions Inc.
4270 U.S. Route 1
Monmouth Junction, NJ 08852
phone: 908-274-1778
fax: 908-274-1733

Perkin-Elmer Corporation
Applied Biosystems Div.
850 Lincoln Centre Dr.
Foster City, CA 94404
phone: 800-345-5224
fax: 415-572-2743

Turner Designs, Inc.
845 W. Maude Ave.
Sunnyvale, CA 94086
phone: 408-749-0994
fax: 408-749-0998
email: tdesigns@ix.netcom.com
Internet: http://www.turnerdesigns.com

Instrumentation for chemiluminescence detection in flow injection or HPLC is often assembled by the user. However, there are a few suppliers of commercially available CL detection units for HPLC and flow injection:

Analytical Luminescence Laboratory
(see listing above)

Coulter Corp.
P.O. Box 169015
Miami, FL 33116
phone: 800-523-3713
fax: 305-883-6877

JASCO Inc.
8649 Commerce Dr.
Easton, MD 21601
phone: 800-333-5272
fax: 410-822-7526
email: jascoinc.com

JM Science Inc.
5820 Main St.
Buffalo, NY 14221
phone: 800-387-7187
fax: 716-634-1970

On-Line Instrument Systems
130 Conway Dr.
Bogart, GA 30622
phone: 800-852-3504
fax: 706-353-1972
email: olis@mindspring.com
Internet: http://www.olisweb.com

Perkin-Elmer Corp.
(see listing above)

Reagents for solution-phase chemiluminescence are available from general chemical suppliers:

Sigma-Aldrich Corp.
P.O. Box 14508
St. Louis, MO 63178
phone: 800-325-3010
fax: 800-325-5052
email: custserv@sial.com
Internet: http://www.sigma.sial.com

For the specialized reagents and kits used in chemiluminescence immunoassay or nucleic acid assays, one should consult the following vendors:

Amersham Life Science Inc.
26111 Miles Rd.
Cleveland, OH 44128
phone: 800-321-9322
fax: 216-464-5075
email: alsclev@ix.netcom.com

Analytical Luminescence Laboratory
(see listing above)

Boehringer Mannheim Corp.
9115 Hague Rd.
Indianapolis, IN 46250
phone: 800-262-1640
fax: 800-845-7355

Ciba Corning Diagnostics
(see listing above)

IGEN Inc.
1530 E. Jefferson St.
Rockville, MD 20852
phone: 800-336-4436
fax: 301-230-0158

Sigma-Aldrich Inc.
(see listing above)

Tropix Inc.
47 Wiggins Ave.
Bedford, MA 01730
phone: 800-542-2369
fax: 617-272-8581

Required Level of Training

Operation of instrumentation for routine, well-established assays can be performed by people with an associate college degree. A knowledge of the chemistry of the sample is useful. Adaptation of existing procedures for nonstandard samples requires a bachelor's degree in organic chemistry and analytical chemistry. Familiarity with the basics of gas and liquid chromatography, immunoassay, or enzyme kinetics can be useful given the specific application. Development of new instrumentation, new reagent systems, or new types of applications requires graduate-level training.

Service and Maintenance

The types of problems that can occur with the instrumentation are generally simple and can be handled by trained laboratory personnel. The most common problems are plumbing problems in the sample handling components. The detector may have to be replaced if it is exposed to high levels of ambient light while powered.

Suggested Readings

Books

BIRKS, JOHN W., ED., *Chemiluminescence and Photochemical Reaction Detection in Chromatography*. New York: VCH, 1989.

BURR, JOHN G., ED., *Chemi- and Bioluminescence*. New York: Marcel Dekker, 1985.

CAMPBELL, A. K., *Chemiluminescence. Principles and Applications in Biology and Medicine*. Chichester, England: Ellis Horwood, 1988.

WEEKS, IAN, *Chemiluminescence Immunoassay*. Amsterdam: Elsevier, 1992.

Articles and Chapters

COULET, P. R., AND L. J. BLUM, "Bioluminescence/Chemiluminescence Based Sensors," *Trends in Analytical Chemistry*, 11 (1992), 57–61.

HAGE, D. S., "Chemiluminescent Detection in High-Performance Liquid Chromatography," in G. Patonay, ed., *HPLC Detection, Newer Methods*. New York: VCH, 1992, pp. 57–75.

KANKARE, J., "Chemiluminescence: Electrogenerated Chemiluminescence," in R. Macrae, ed., *Encyclopedia of Analytical Science*. New York: Academic Press, 1995.

KNIGHT, A. W., AND G. M. GREENWAY, "Occurrence, Mechanisms, and Analytical Applications of Electrogenerated Chemiluminescence," *Analyst*, 119 (1994), 879–90.

LANCASTER, J. S., "Chemiluminescence: Gas Phase Chemiluminescence," in R. Macrae, ed., *Encyclopedia of Analytical Science*. New York: Academic Press, 1995, pp. 621–6.

MENDENHALL, G. D., "Chemiluminescence Techniques for the Characterization of Materials," *Angewandte Chemie (International Edition)*, 29 (1990), 362–73.

NAKASHIMA, K., AND K. IMAI, "Chemiluminescence," in S. G. Schulman, ed., *Molecular Luminescence Spectroscopy*, part 3. New York: Wiley, 1993, pp. 1–23.

NIEMAN, T. A., "Chemiluminescence: Overview of Techniques," in R. Macrae, ed., *Encyclopedia of Analytical Science*. New York: Academic Press, 1995, 608–12.

NIEMAN, T. A., "Chemiluminescence: Liquid Phase Chemiluminescence," in R. Macrae, ed., *Encyclopedia of Analytical Science*. New York: Academic Press, 1995, pp. 613–21.

PRINGLE, M. J., "Analytical Applications of Chemiluminescence," *Advances in Clinical Chemistry*, 30 (1993), 89–183.

ROBARDS, K., AND P. J. WORSFOLD, "Analytical Applications of Liquid-Phase Chemiluminescence," *Analytica Chimica Acta*, 266 (1992), 147–73.

References

1. E. H. Ratzlaff and S. R. Crouch, *Analytical Chemistry*, 55 (1983), 348–52.

2. H. A. Strobel and W. R. Heineman, *Chemical Instrumentation: A Systematic Approach* (New York: Wiley, 1989), pp. 278–93.

3. C. S. Martin and I. Bronstein, *J. Biolum. Chemilum.*, 9 (1994), 145–53.

4. T. M. Downey and T. A. Nieman, *Analytical Chemistry*, 64 (1992), 261–68.

5. J. A. Hodgeson and others, *Analytical Chemistry*, 42 (1970), 1795–1802.

6. T. A. Gough, K. S. Webb, and R. F. Eaton, *J. Chromatogr.*, 137 (1977), 293–303.

7. M. W. N. Ward, C. W. I. Owens, and M. J. Rennie, *Clin. Chem.*, 26 (1980), 1336–39.

8. A. L. Howard and L. T. Taylor, *HRCC: J. High Resol. Chromatogr.*, 14 (1991), 785–94.

9. P. J. M. Kwakman and U. A. Th. Brinkmann, *Anal. Chim. Acta*, 266 (1992), 175–92.

10. W. Y. Lee and T. A. Nieman, *J. Chromatogr.*, 659 (1994), 111–18.

11. W. R. G. Baeyens and others, *J. Microcol. Sep.*, 6 (1994), 195–206.

12. C. A. K. Swindlehurst and T. A. Nieman, *Anal. Chim. Acta*, 205 (1988), 195–205.

13. A. F. Martin and T. A. Nieman, *Anal. Chim. Acta*, 281 (1993), 475–81.

14. P. J. Koerner and T. A. Nieman, *J. Chromatogr.*, 449 (1988), 217–28.

15. H. A. H. Rongen and others, *J. Pharm. Biomed. Anal.*, 12 (1994), 433–62.

16. B. Mattson and others, *Polym. Test.*, 11 (1992), 357–72.

17. P. Hopper and W. G. Kuhr, *Analytical Chemistry*, 66 (1994), 1996–2004.

18. P. E. Stanley, *J. Biolum. Chemilum.*, 9 (1994), 123–25.

Mass Spectrometry

Chapter 28

Introduction

Charles L. Wilkins

University of California
Department of Chemistry

Analytical mass spectrometry (MS) is a technique that dates back to the early twentieth century, when J. J. Thomson first described the fundamentals of the technique (1). From the outset, mass spectrometry was established as a unique analytical tool. Its first major triumph was Thomson's use of mass spectrometry to demonstrate the existence of two isotopes of neon. For many years, mass spectrometry was confined to analysis of gaseous samples. It was not until the 1940s that organic mass spectrometry began to be widely practiced. At that time, the first commercial mass spectrometers appeared and they were immediately used in the war effort, where the search for natural rubber substitutes and improved petroleum products was vital. Not surprisingly, the early applications of organic mass spectrometry were also confined to volatile low-molecular-weight species. As will become clear in Chapters 29–34, mass spectrometry has developed well beyond its modest beginnings and is now one of the most sensitive, specific, and versatile of modern analytical methods.

The first two chapters of this section are devoted to the fundamentals of mass spectrometry. In Chapter 29, J. Throck Watson discusses the mass spectometry of volatile analytes. Here the basic elements of electron ionization (EI) and chemical ionization (CI) mass spectrometry are presented. This chapter also discusses the basic mechanisms of organic mass spectral fragmentations and contrasts the types of data obtained from EI and CI measurements. Examples of both types of spectra are presented, together with brief interpretations of the data. Chapter 30, by Vidavsky and Gross, focuses on high-resolving-power mass spectrometry. This chapter introduces the ideas associated with high-resolution mass spectrometry using sector or Fourier transform mass spectrometry (FTMS). The fundamentals of both techniques are described and the rationale for needing the type of information these instruments provide is discussed. Examples of high-resolving-power mass spectrometry are discussed in light of a specific application: analysis of environmental samples suspected to contain 2,3,7,8-tetrachlorodibenzodioxin (TCDD).

Chapters 31–34 are devoted to discussions of various means of introducing samples into a mass spectrometer, illustrated with typical examples of usual applications. In Chapter 31, Ronald Hites describes gas chromatographic mass spectrometry (GC-MS), which is one of the most common mass spectrometric techniques. He discusses the common GC-MS interfaces and the requirements for data processing and mass spectral information which they impose. Then, the discussion turns to both qualitative and quantitative applications of GC-MS. Illustrative examples of typical analytical applications are included. Chapter 32, by Robert Minard, is devoted to fast atom bombardment (FAB) and liquid secondary ion mass spectrometry (LSIMS). These techniques, which were developed in the 1980s, are designed for use with low-volatility or nonvolatile samples and are the forerunners to electrospray ionization (ESI) and matrix-assisted laser desorption/ionization (MALDI) methods. This chapter contains a discussion of both the limitations and the particular analytical advantages of the FAB and LSIMS methods. In Chapter 33, Chhabil Dass discusses the current state of the art of high-performance liquid chromatography electrospray ionization mass spectrometry (HPLC-ESI-MS). He mentions the requirements for an ideal LC-ESI-MS system and discusses the advantages and limitations of the most commonly used combinations.

Chapter 34 describes analytical laser mass spectrometry. It is limited to laser desorption methods (direct and matrix assisted) because those are the techniques that have proved most analytically useful. Direct laser desorption MS, using either time-of-flight mass spectrometry (TOFMS) or FTMS, is discussed. Recent major advances in laser desorption mass spectrometry made possible by use of MALDI are described. Use of both TOF (exceptional mass range) and Fourier tranform mass spectrometers (exceptional mass resolution) for MALDI are discussed. Several examples illustrate the practical implications for biochemical and polymer analysis. (See also Chapter 22, which addresses inductively coupled plasma mass spectrometry, a popular technique for inorganic analysis.)

It should be clear, from the information in these chapters, that the capabilities of analytical mass spectrometry have evolved significantly in recent years. As a consequence of developments in both source design and mass analyzers, MS is now a truly general analytical tool. Both organic and inorganic chemical MS analysis are now possible, with outstanding sensitivity and accuracy and almost unlimited mass range. Much current research is aimed at adapting these exciting research advances to routine use in analytical laboratories. Increased use of modern computers and electronics technology is resulting in compact and highly efficient mass spectrometers. There is no doubt that mass spectrometry has become one of the most versatile, specific, and sensitive chemical analysis tools.

Vendors of Mass Spectrometry Instruments and Accessories

Bruker Instruments, Inc.
19 Fortune Dr.
Billerica, MA 01821
phone: 508-667-9580
Internet: www.bruker.com

Finnigan FT/MS
6416 Schroeder Rd.
Madison, WI 53711-2424
phone: 608-273-8262
Internet: www.ftms.com

Finnigan MAT
355 River Oaks Parkway
San Jose, CA 95134
phone: 408-433-4800, 800-538-7067
Internet: www.finnigan.com

Hewlett-Packard
1601 California Ave.
Palo Alto, CA 94304

phone: 415-857-6028, 800-227-9770
Internet: www.hp.com

Hitachi Instruments
3100 N. First St.
San Jose, CA 95134
phone: 800-548-9001
Internet: www.hii.hitachi.com

Ionspec Corporation
18009 Skypark Circle, Suite F
Irvine, CA 92714-6516
phone: 714-261-7743
Internet: www.ionspec.com

JEOL USA, Inc.
11 Dearborn Rd.
Peabody, MA 01960
phone: 508-535-5900
Internet: www.jeol.com

Kratos Analytical, Inc.
Wharfside
Trafford Wharf Rd.
Manchester M17 1GP
United Kingdom
phone: 44-0161-888-4400, 800-935-0213
Internet: www.kratos.com

Micromass UK, Ltd.
Floats Rd.
Wythenshawe
Manchester M23 9LZ
United Kingdom
phone: 44-0-161-946-0565
Internet: www.micromass.co.uk

Perkin-Elmer
Applied Biosystems Division
850 Lincoln Center Dr.
Foster City, CA 94404-1128
phone: 800-345-5224
Internet: www.perkin-elmer.com:80

PerSeptive Biosystems
500 Old Connecticut Path
Framingham, MA 01701
phone: 501-383-7700, 800-899-5858
Internet: www.pbio.com

Reference

1. J. J. Thomson, *Rays of Positive Electricity and Their Application to Chemical Analysis* (London: Longmans, Green & Co., 1913).

Chapter 29

Mass Spectrometry of Volatile Analytes

J. Throck Watson

Michigan State University
Department of Chemistry

Summary

General Uses

- Identification of organic compounds by virtue of fragmentation pattern
- Determination of molecular weight
- Quantitation of organic compounds at low levels
- Elucidation of structural features in an unknown molecule
- Determination of isotope incorporation of stable isotopes
- Estimation of elemental composition based on appearance and multiplicity of isotope peaks

Common Applications

- Identification of common environmental contaminants
- Identification of drugs and drug metabolites
- Verification or confirmation of product or products of organic synthesis
- Determination of the extent of incorporation of stable isotope–labeled tracer compound into biological pool of hormones or other chemical transmitters

- Determination of the kinetics of biosynthesis or organic chemical turnover as a precursor or as a metabolite in a biological system

Samples

State

Gases as well as liquids and solids can be analyzed as long as some of the compound can be converted to the vapor state thermally without degradation. Any compound that is amenable to gas chromatography is also amenable to electron or chemical ionization mass spectrometry.

Amount

Between 1 and 10 pmol of most organic compounds will provide an interpretable electron or chemical ionization mass spectrum having a signal-to-background ratio of at least 10.

Preparation

A principal concern is achieving sample volatilization without degradation. It may be necessary to transform the analyte chemically to meet these criteria. For example, a given carboxylic acid may not be sufficiently volatile or stable to produce a mass spectrum, whereas preparation of the methyl ester of the same carboxylic acid should produce a suitable spectrum. The second major concern in sample preparation is sample purity. Mass spectrometry of impure samples is rarely useful unless one knows the identity of the impurities. The use of gas chromatography mass spectrometry is the best way for analyzing impure samples.

Analysis Time

The analysis time is typically on the order of minutes, and the greatest fraction of time is consumed by ensuring that the instrument is in proper calibration and that no interfering background signals interfere with acquiring the mass spectrum of the analyte. Once the analyte has been introduced into the ion source of the mass spectrometer, the mass spectrum can be acquired in a few seconds. Another significant fraction of the analysis time is in selecting a scan file that contains the most interpretable mass spectrum for identification purposes.

Limitations

Electron and chemical ionization mass spectrometry are applicable only to samples in which the analyte can transform some fraction of its quantity into the vapor state for interaction with energetic electrons or proton-rich reagent gas ions for purposes of ionization.

Complementary or Related Techniques

Combined gas chromatography mass spectrometry is the most desirable electron impact or chemical ionization technique because the specialized inlet system of the gas chromatograph ensures that the analyte is in the vapor state and that most of the components of a mixture are introduced independently into the mass spectrometer in purified form. Other complementary spectroscopic techniques include nuclear magnetic resonance spectrometry and infrared and ultraviolet (UV) absorption spectrometry for verification of specific functional groups.

Introduction

Mass spectrometers were originally designed by physicists at the turn of the century to determine the mass-to-charge ratio (m/z) of charged particles by virtue of their behavior in magnetic and electric fields. The early interest of the physicists was in proving the existence of isotopes of various elements. Organic mass spectrometry was cultivated by scientists in the petroleum industry starting in the 1940s. Mass spectrometry expanded into biomedical applications in the 1970s.

In general, mass spectrometry is a microanalytical technique that requires some energetic process for converting a significant number of molecules of the analyte to a charged species so that the m/z ratio of the charged form of the analyte may be determined. A given compound is converted to a charged species by one of a variety of available ionization processes. Conditions appropriate for a particular ionization process are established in the ion source of a given mass spectrometer. The operating conditions for particular ionization processes are often mutually exclusive or incompatible, and not easily converted from one form of ionization to another. Some analytes are amenable to more than one type of ionization; others are amenable to only one type of ionization.

This chapter is limited to descriptions of ionization by electron (often called electron impact) ionization and chemical ionization. Both electron ionization (EI) and chemical ionization (CI) require the analyte molecule to be in the vapor state for effective interaction with 70-eV electrons (in EI) or with proton-rich reagent ions (in CI). The volatile compounds discussed in this chapter are not restricted to those that are fixed gases at ambient conditions. Compounds such as acetone and diethyl ether readily meet the description of a volatile compound, but it is also possible to obtain the mass spectrum of acetylsalicylic acid from an aspirin tablet, for example, because the partial pressure of the acid in equilibrium with the solid provides a sufficient number of vaporous molecules to the ionization chamber.

How It Works

Electron Ionization (EI)

Interaction of molecules in the vapor state with energetic electrons can be accomplished with a hardware configuration, as represented in Fig. 29.1. A very low pressure is necessary during EI ionization to extend the lifetime of the energetic electrons and to provide the nascent ions with a collision-free path through the ion source and into the mass analyzer. Furthermore, the low pressure is important during EI so that molecular ions retain enough internal energy to drive the fragmentation process, rather than losing the excess internal energy through collisional deactivation that might occur at elevated pressures. The pressure in the ion source housing, as generally shown in the large volume in Fig. 29.1, is 10^{-6} torr (mm Hg). The pressure in the center of the ionization chamber, buried within the ion source block, is on the order of 10^{-4} to 10^{-5} torr.

The key to ionization of molecules in the vapor state is the availability of a well-defined beam of energetic electrons. Such a beam can be formed by simply heating a wire filament with direct current so that it emits electrons. As indicated in Fig. 29.1, the electrons are directed across the ionization chamber at right angles to the path taken by the newly formed ions. The energy of electrons is controlled by the difference in potential between the emitting filament and the metallic source block of the ionization chamber; this difference in potential is typically 70 V to produce 70-eV electrons. Sample vapor is admitted to the ionization chamber, such that a beam of molecules is directed across the electron beam in a direction orthogonal with the electron beam and the resulting ion beam, as illustrated in Fig. 29.1.

Chemical Ionization (CI)

Chemical ionization uses all components of EI as illustrated in Fig. 29.1, but at a relatively high pressure, so additional hardware is needed, as illustrated in Fig. 29.2. Because CI requires a pressure of approximately 1 torr in the ionization chamber, provisions must be available for removing the high gas flow through the ion source housing so that the pressure in the analyzer is not adversely affected. The use of differential pumping, as illustrated schematically in Fig. 29.2, allows the relatively high pressure in the ion source to be used effectively without significantly increasing the pressure in the mass analyzer.

Differential pumping of two separate volumes is accomplished as illustrated in Fig. 29.2 by providing an independent pumping system for each volume while allowing the volumes to com-

Figure 29.1 Schematic diagram of an EI ion source. *(Figure reproduced from* Introduction to Mass Spectrometry, *2nd edition by J. T. Watson, 1985, with permission from Raven Press.)*

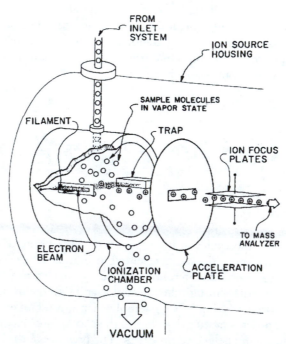

Figure 29.2 Schematic diagram illustrating the concept of differential pumping.

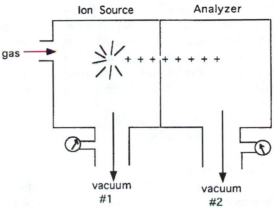

municate only through a small aperture. In this way, ions that are formed in the high-pressure ion source can be transmitted through the small aperture to the chamber housing the mass analyzer. The relatively high pressure in the ion source does not adversely affect the pressure in the analyzer when using differential pumping, because the gas load tends to go directly to the vacuum pump on the ion source rather than travel through the small aperture into the mass analyzer and its vacuum pump.

Mechanism of Ionization

Electron Ionization

When a molecule in the vapor state interacts with a 70-eV electron, some of the kinetic energy of the electron is absorbed by the electrons of the molecule. In a successful interaction between the 70-eV electron and the molecule, as represented in Scheme 29.1, on average only about 20 eV are absorbed by the molecule. However, the 20 eV of excess energy is more than sufficient to expel an electron from the molecule because the ionization energy of the molecule is typically only about 10 eV. Thus, the excited molecule, with 20 eV of excess internal energy, expels one of its electrons to form a molecular ion, an odd-electron species, with an excess of internal energy amounting to 10 eV in this example. In reality, fewer than 0.1 percent of the molecules become ionized as a result of absorbing a distribution of internal energies ranging from 10 to 30 eV from bombarding electrons. Molecular ions having the larger excess of internal energy are the most likely to undergo fragmentation processes. It is interesting to note that although the EI process is relatively inefficient, remarkably low detection limits (into the femtomole range) can be achieved by EI mass spectrometry.

Scheme 29.1

$$M + e^-_{70\ eV} \longrightarrow M^{+\bullet} + e^-_{50\ eV} + e^-_{thermal}$$

Chemical Ionization

In chemical ionization (CI), a reagent gas is converted into reagent ions, which react with the analyte to achieve its ionization, usually by protonation. The ion source of a CI mass spectrometer is remarkably similar to that described above for EI, except that the ionization chamber is tighter so that the local pressure can be elevated to approximately 1 torr. Most ion sources are designed to produce a continuous flow of ions, and it is necessary to have a high pressure, namely 1 torr, of the reagent gas in the ionization chamber so that the electron beam selectively ionizes the reagent gas. At a partial pressure of 1 torr, the reagent gas molecules outnumber the analyte molecules in a typical sample by a factor of 10^4 or more and, thus, the vast majority of ions that result from the EI process are those of the reagent gas for statistical reasons.

As illustrated in Scheme 29.2, once some of the reagent ions are converted to molecular ions of the reagent gas, many decompose into fragment ions. The residual molecular ions and the fragment ions of the reagent gas react via ion–molecule reactions with neutral molecules of the reagent gas to form proton-rich ions such as CH_5^+, $C_2H_5^+$, and $C_3H_5^+$. The concentration of these proton-rich reagent ions from the reagent gas is higher than that of the analyte, so the analyte molecule is more likely to interact with one of the proton-rich ions of the reagent gas than it is to interact with some of the electrons. If the analyte molecule has a higher proton affinity than the reagent gas, a proton is transferred from the proton-rich reagent ion to the analyte, which then becomes ionized as a result of protonation.

Scheme 29.2

$$CH_4 + e^-_{70\ eV} \longrightarrow CH_4^{+\bullet} + e^-_{thermal} + e^-_{50\ eV} \qquad \text{Ionization}$$

$$CH_4^{+\bullet} \longrightarrow CH_3^+ + H^\bullet \qquad \text{Fragmentation}$$

$$CH_4^{+\bullet} \longrightarrow CH_2^{+\bullet} + H_2 \qquad \text{Fragmentation}$$

$$CH_4^{+\bullet} + CH_4 \longrightarrow \underline{CH_3^+} + CH_3^\bullet \qquad \text{Ion/molecule reaction}$$

$$CH_3^+ + CH_4 \longrightarrow \underline{C_2H_5^+} + H_2 \qquad \text{Ion/molecule reaction}$$

$$CH_2^{+\bullet} + CH_4 \longrightarrow C_2H_3^{+\bullet} + H_2 + H \qquad \text{Ion/molecule reaction}$$

$$C_2H_3^+ + CH_4 \longrightarrow \underline{C_3H_5^+} + H_2 \qquad \text{Ion/molecule reaction}$$

Note: The underlined species are not reactive with molecular CH_4, and thus they tend to accumulate in the ionization chamber.

As a result of ionization (protonation) by proton-rich ions in a chemical ionization source, the analyte is distinguished in the mass spectrum by an intense peak representing the intact molecule at an m/z value one mass unit (u) higher than the value of its molecular weight. The reason for the abundance of the protonated molecule is due in large measure to the fact that it is an even-electron species, and most of the excess energy transferred to the protonated molecule during the ionization process is lost due to collisional deactivation in the high-pressure environment. It is proper to refer to the protonated form of the analyte as a protonated molecule, not as a protonated molecular ion. The latter name suggests some combination of a proton (positively charged) and a molecular ion (also positively charged, as described in the section on EI), which surely does not happen because there are substantial repulsive forces between M^+ and H^+, and furthermore, such a species would have two positive charges associated with it. Because of a high abundance of the protonated molecule (MH^+) and its tendency not to fragment, the process of chemical ionization is called soft ionization.

It is important to realize that the high pressure in a chemical ionization source is localized in the relatively small volume of the ionization chamber. That is to say, in the internal volume of approximately half a cubic centimeter, the pressure of the reagent gas is approximately 1 torr. The gas from the ionization chamber rapidly leaks out into other volumes in the ion source housing, where the pressure drops as the gas expands into the relatively large volume, so that the average pressure in the ion source housing is approximately 10^{-4} torr. With proper differential pumping in the mass spectrometer, most of the gas that causes the elevated pressure in the ion source housing is denied access to the mass analyzer, where the pressure can easily be maintained at 10^{-6} torr.

Negative Ion Mass Spectrometry

Negative ions as formed by EI have been studied for many years. However, the practice of negative ion EI mass spectrometry is not very popular because of the relative inefficiency of generating molecular anions with conventional EI sources. Because of the very low flux of thermal electrons available from a typical EI filament, micromolar samples must be used, and thus there is no analytical advantage according to detection limits. The vast majority of electrons available from a typical EI filament preclude formation of a stable molecular anion and cause its immediate dissociation, according to Scheme 29.3.

Scheme 29.3

$ABC + e^- \ (0.03-0.1 \ eV) \longrightarrow ABC^-$		Resonance capture
$ABC + e^- \ (0.1-10 \ eV) \longrightarrow AC + B^-$		Dissociative capture
$ABC + e^- \ (10-15 \ eV) \longrightarrow AB^+ + C^-$		Ion-pair formation

There are significant analytical advantages in producing negative ions of certain classes of compounds provided that a reasonable flux of thermal electrons is available. Noting that the EI filament is a relatively poor source of thermal electrons, the realization that a very high flux of thermal electrons is available in a typical chemical ionization source drew much attention to the use of the CI source in negative ion mass spectrometry. The high flux of thermal electrons in the chemical ionization source derives from the first equation in Scheme 29.2. Ionization of methane (or any gas molecule) at a partial pressure of 1 torr yields a high flux of reagent ions and an equally high flux of thermal electrons that are expelled from the reagent gas molecules during the ionization process. Because of the high flux of thermal electrons in a chemical ionization source, introduction of an electrophilic analyte causes the generation of a large flux of molecular anions of that particular analyte via electron capture, leading to an obvious analytical advantage in detectability.

Concerning nomenclature, it is important to recognize the processes involved in the classic chemical ionization source. Positive ions, most often protonated molecules, are generated via an ion–molecule reaction between a proton-rich ion of the reagent gas and the analyte molecule to give the protonated molecule of the analyte. This is taken as the prototype process for positive chemical ionization, namely an ion–molecule reaction between a positive ion and the analyte molecule. By analogy, negative chemical ionization connotes an ion–molecule reaction between a reagent anion and an analyte molecule to produce an anion that represents the analyte. Unfortunately, *negative chemical ionization* (NCI) is often misused in analytical terminology. The analytically attractive process of an electrophilic analyte capturing an electron to produce a high abundance of molecular anions is *not* a process of negative chemical ionization; it is a process of electron capture. In an effort to use correct nomenclature for the actual physical process of anion formation, many investigators use the term *electron capture negative ionization* (ECNI); others use the term *negative ion mass spectrometry* (NIMS).

What It Does

Analytical Information

Considerable information about the structure of the analyte can be gathered from a cursory inspection of the mass spectrum. The bulk of this segment of the chapter is devoted to a description of EI mass spectra; CI mass spectra can be explained in part by complementary processes because the molecular ionic species in CI is an even-electron species, whereas the molecular ion in EI is an odd-electron species.

One of the most important steps in interpreting a mass spectrum is identifying the peak that corresponds to the intact analyte. Of course, such a peak for the molecular species would occur at the high-mass end of the mass spectrum, and it should occur at an m/z value that is consistent with the nitrogen rule as defined below. The candidate for the molecular ion should then be tested for logical losses during fragmentation processes, as indicated by the mass difference between the candidate for the molecular ion and the various fragment ions. An indication of the elemental composition of the analyte can be obtained from the multiplicity of isotope peaks representing the various ions.

Recognition of Isotope Peaks

An appreciation for the natural abundance of stable isotopes of the various elements is essential in interpreting a mass spectrum. For the elements commonly encountered in organic mass spectrometry, it is a fortunate fact that the lightest isotope of the element is the most abundant. This fortunate circumstance helps the analyst to recognize the nominal mass of the ion as represented by usually distinct clusters of peaks.

By definition, the nominal mass of a given ion is represented by the peak at the low-mass edge of a cluster of peaks representing the ion in question. The nominal mass of any ionic species is taken to be the mass that is equal to the nonisotopic mass (actually the lightest isotope) of the element multiplied by the number of atoms of that element present. Thus, even though the atomic weight of bromine is commonly reported as 80, the nominal mass of atomic bromine is 79 in mass spectrometry. In nature, bromine occurs as two isotopes. The major isotope (50.5% abundance) occurs at mass 79, and the other major isotope (49.5% natural abundance) occurs at mass 81. Obviously, the average mass of bromine is 80, but ions for the two isotopic species of atomic bromine would be detected at individual m/z values in the mass spectrum. Thus, it is important to agree that the isotope of lowest mass will be called the nominal mass of the element.

In the case of bromoethane, the nominal molecular weight is 108. This results from the sum of the nominal atomic masses of three elements: 79 for bromine, 24 for two carbons, and 5 for the five hydrogens. The cluster of peaks representing the molecular ion for bromoethane consists of a major peak at m/z 108 in the mass spectrum (see Fig. 29.3) accompanied by another major peak at m/z 110; the latter peak would be almost as intense as the peak at m/z 108, as these two peaks are indicative primarily of the natural abundance of bromine itself. In addition, there would be mi-

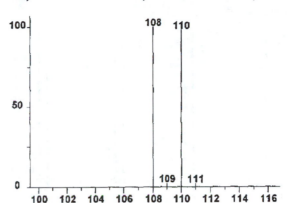

Figure 29.3 Abbreviated bargraph mass spectrum of ethyl bromide in the vicinity of the molecular ion peaks.

nor but discernible peaks at m/z 109 and m/z 111 of equal intensity, both being approximately 2.2% of the intensities at m/z 108 and 110. These minor peaks are approximately 2.2% of those at m/z 108 and 110 because they represent the probability of the molecules containing one of the carbon atoms as ^{13}C, which occurs at a natural abundance of approximately 1.1% in nature.

The multiplicity of isotope peaks for bromine or chlorine or combinations of the two are the easiest of the isotope peak clusters to recognize because the isotope peaks are quite intense and are spaced at 2-mass-unit intervals. For purposes of reference, the graphical appearance of various combinations of chlorine and bromine are illustrated in Fig. 29.4. Note that the major peaks in each cluster occur at 2-mass-unit intervals. Also note in Fig. 29.4 that the mass axis for each of the clusters of peaks is not designated numerically, but rather represented as the letter X to correspond to the nominal mass for a particular ion, and that adjacent isotope peaks occur at multiples of 2 mass units higher than X. In this way, the reference patterns are applicable to chlorine- and bromine-containing ions at any location along the mass scale in a given mass spectrum.

Consider the abbreviated mass spectra for three different molecules, each containing two chlorine atoms, in Fig. 29.5. Each of these abbreviated mass spectra shows a region in the vicinity of the cluster of peaks representing the molecular ion of the molecule. The nominal molecular weight of a molecule of chlorine is 70 mass units; that is, $2 \times 35 = 70$. It can be seen that this abbreviated mass spectrum of a molecule of chlorine is identical to the reference pattern represented as the second entry in the top row of Fig. 29.4. The second abbreviated mass spectrum in Fig. 29.5 represents the molecular ion region for dichloromethane, with a nominal molecular weight of 84; this is consistent with a peak representing the nominal molecular ion at m/z 84. Again, note that even though the cluster of peaks representing the molecular ion of dichloromethane starts at m/z 84 in Fig. 29.5, the multiplicity is identical to the major isotopic reference peak pattern in Fig. 29.4 for the second entry in the first row. Similarly, in Fig. 29.5 the multiplicity of major peaks representing the molecular ions starting at m/z 146 for dichlorobenzene has the same pattern as that for the other species even though the cluster of peaks occurs in a higher-mass region in the mass spectrum. Thus, this example indicates that the predominant contribution of chlorine to the multiplicity of isotope peaks is dominant regardless of whether the chlorines are by themselves as in a molecule of chlorine or attached to one or six carbon atoms, as illustrated by the other examples in Fig. 29.5. It is this simple observation of the striking multiplicity of isotope peaks for chlorine or bromine that has been used to advantage in analyses of environmental samples to recognize the presence of halogenated organocompounds.

Figure 29.4 Isotope peak intensity patterns for ions containing the indicated number of chlorine or bromine atoms. (*Figure reproduced from* Introduction to Mass Spectrometry, *2nd edition by J. T. Watson, 1985, with permission from Raven Press.*)

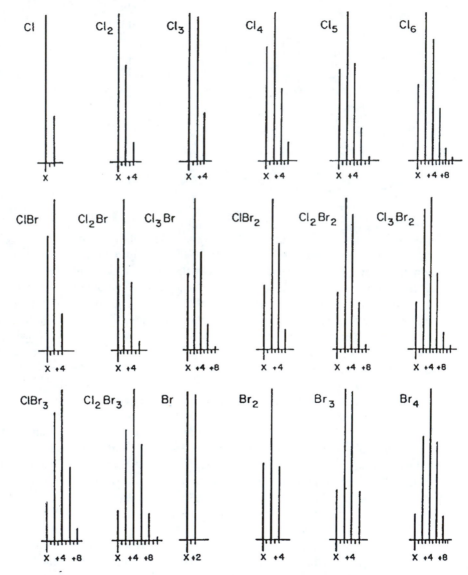

In addition to the appearance of the dominant peaks representing two chlorines in each of the three mass spectra in Fig. 29.5, there are minor peaks that give an indication of the number of carbon atoms in each of the molecules. Note that in the case of chlorine, there are no peaks at X + 1 and X + 3. Note, however, that in the mass spectrum of dichloromethane there is a small but discernible peak at X + 1, namely at m/z 85; there is also a minor peak at X + 3 (m/z 87), although it is not readily discernible. Furthermore, note that in the mass spectrum of dichlorobenzene, there is a more significant peak at X + 1, namely m/z 147, as well as at X + 3, namely m/z 149; there is also a minor but not readily distinguishable peak at m/z 151. These minor peaks at X + 1, X + 3, and so on are dependable indicators of the maximum number of carbon atoms present in the ion.

Figure 29.5 Abbreviated mass spectra in the molecular ion region for the indicated molecules. *(Figure reproduced from* Introduction to Mass Spectrometry, *2nd edition by J. T. Watson, 1985, with permission from Raven Press.)*

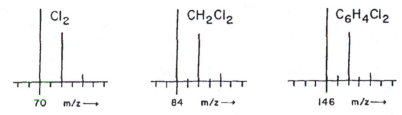

Because the natural abundance of ^{13}C in ^{12}C is approximately 1.1% and because there is only one carbon atom possible in dichloromethane, the peak at m/z 85 representing the species containing one ^{13}C atom is only 1.1% as intense as that at m/z 84, the nominal mass in which carbon is present as ^{12}C. In the mass spectrum of dichlorobenzene, the peak at X + 1 is much more obvious and, in fact, is six times as intense as that at X + 1 in the mass spectrum of dichloromethane because there are six times as many carbons in the dichlorobenzene molecule. Because there are six carbons in dichlorobenzene, the probability of one ^{13}C atom being present in the molecule is equal to six times the 1.1% abundance for a single carbon atom. That is, for dichlorobenzene containing six carbons, the probability of the molecule containing at least one ^{13}C is 6.6% of the probability of it containing all carbons as ^{12}C; therefore, the peak at X + 1 is 6.6% of that at X in the mass spectrum of dichlorobenzene. The peak at X + 3 (m/z 149) is 6.6% of the intensity at X + 2 (m/z 148) for the same reason; that is, the peak at m/z 148 represents the ionic species containing all six carbons as ^{12}C and one of the chlorines present as ^{37}Cl, with the other present as ^{35}Cl. The ion containing one heavy isotope of chlorine also has a 6.6% chance of having one of its carbons being present as ^{13}C, and this species is represented by the peak at m/z 149.

In the mass spectrum of an unknown compound, the relative intensities of the isotope peaks place a limit on the number of atoms of a given element that should be considered constituents in the elemental composition of the ion. For example, in Fig. 29.5 the peak at m/z 147 is only 6.6% of that at m/z 146, and if this were an unknown spectrum, one would not expect the analyte to contain more than six carbon atoms as an upper limit. More on this strategy is illustrated in the following examples.

Concerning nomenclature, the abbreviation for mass unit is *u*. The popular and meaningful abbreviation *amu* is no longer officially correct because this abbreviation refers to the definition of mass unit before 1961, when the atomic mass of all elements was measured relative to atomic oxygen as 16.00000. By convention, the International Union of Pure and Applied Chemistry declared in 1961 that all atomic masses would be reported relative to ^{12}C being 12.00000, and that mass unit would be represented by *u*.

The Fragmentation Pattern

The fragmentation process offers ways in which the molecular ion can dispose of some of its excess internal energy left over from the ionization process following electron impact. Although kinetic factors may limit the abundance of fragment ions formed via a given pathway, the stabilities of the fragment ion and the neutral radical species are also important determinants of one pathway over another that is available for decomposition of the molecular ion. Thus, an intense fragment ion peak in the mass spectrum corresponds to an ionic species that is more stable than its precursor (most often the molecular ion). An intense fragment peak also indicates that formation of the ion proceeds via a kinetically favorable pathway.

There are four distinct classes of fragmentation: sigma bond cleavage, radical-site–initiated cleavage, charge-site–driven cleavage, and intramolecular rearrangement. Each of the four types of fragmentation pathways is available to the molecular ion formed during electron impact ionization. A given molecular ion will generally undergo fragmentation by only one of the pathways. The mass spectrum obtained from a mass spectrometer is the result of billions of molecular ions undergoing discrete pathways of fragmentation. That is, the mass spectrum is a frequency distribution diagram of the various pathways available for decomposition of the molecular ion.

Sigma Bond Cleavage

In sigma bond cleavage, the molecular ion exists as an odd-electron species by virtue of one of its sigma bonds consisting of only one electron. Because this one-electron sigma bond is the weakest in the molecular ion, it is the site of fragmentation. Two different arrangements of a one-electron bond are possible during the formation of a molecular ion by losing an electron from a given sigma bond; consider the loss of one of the two electrons between carbons 3 and 4 in tetradecane, as illustrated in Scheme 29.4. Depending on which atom retains the odd electron, the charge resides on the opposite member of the two carbon atoms that were bonded by the pair of electrons in the original sigma bond. Thus, as shown in Scheme 29.4, the two different forms of the single-electron sigma bond lead to potentially different fragment ions having complementary mass values, but not necessarily equal abundance, as indicated by the peaks at m/z 43 and m/z 155 (M-43) in the mass spectrum of tetradecane (Fig. 29.6).

Scheme 29.4

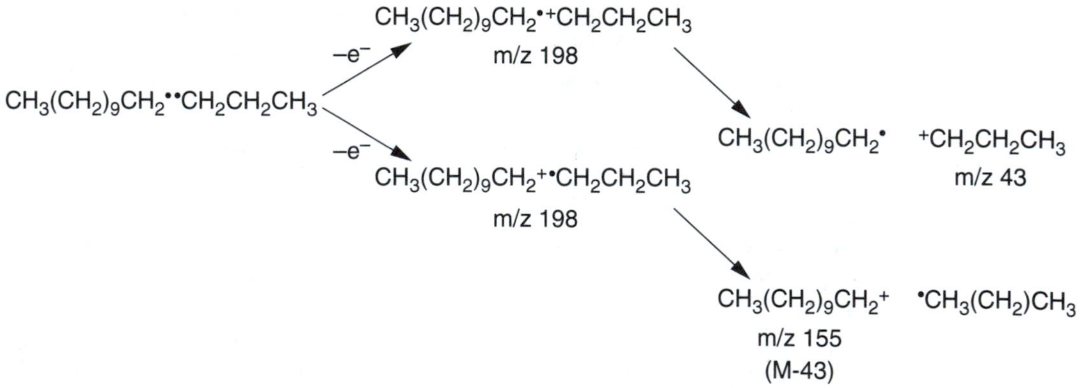

Because of the improved relative stability of larger radical species, the fragmentation process involving loss of the larger alkyl radical is favored. For example, consider the example illustrated in Scheme 29.5, which shows the possibility of losing either an ethyl radical or a methyl radical to produce a secondary carbonium ion. The loss of the methyl radical would generate a secondary carbonium ion of mass 57, whereas that involving loss of an ethyl radical would form an ion of m/z 43; in such a case, the peak intensity at m/z 43 may be as much as twice that at m/z 57, indicating the more favorable loss of the larger species and thereby leading to a more energetically favorable result.

Because of the large number of sigma bonds available in a given molecule, the sigma bond cleavage pathway usually does not lead to structurally useful fragment ions. An exception is the case of a branch point in a carbon chain where sigma bond cleavage leads to the formation of secondary carbonium ions. Because secondary carbonium ions are more stable than primary carbon-

Figure 29.6 EI mass spectrum of tetradecane. *(Data acquired by Naxing Xu on a Finnigan 4600 quadrupole mass spectrometer in the MSU Mass Spectrometry Facility.)*

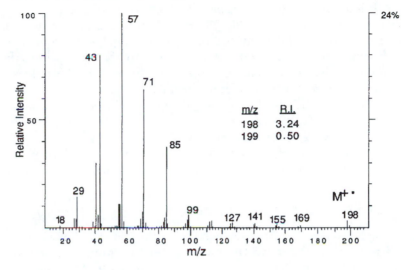

m/z	R.I.
198	3.24
199	0.50

ium ions that would result from sigma bond cleavage along a normal chain, for example, peaks in the mass spectrum that correspond to cleavage at a branch point are much more intense than those resulting from sigma bond cleavage at other sites. This phenomenon can be used to advantage in recognizing the sites of branching in long-chain branched hydrocarbons, for example. The smooth envelope of monotonous peaks in the mass spectrum of a normal chain hydrocarbon (Fig. 29.6) represent primary carbonium ions formed through sigma bond cleavage. A branch point in an aliphatic hydrocarbon would have one or two intense peaks in its mass spectrum that would disrupt the smooth pattern seen in Fig. 29.6; these intense peaks can be correlated with the location of the branch point in the main carbon chain.

Scheme 29.5

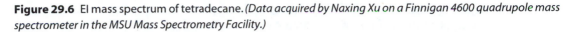

The mass spectrum of anthracene, an aromatic hydrocarbon, is shown in Fig. 29.7. Comparison of Figs. 29.6 and 29.7 shows the marked contrast between the mass spectra of aliphatic and aromatic hydrocarbons. The molecular ion of the aromatic hydrocarbon is quite stable because of its cyclic structure and delocalized electron system of π-orbitals; thus, it resists fragmentation. As a result, the molecular ion of anthracene in Fig. 29.7 accounts for approximately 47% of all ionization.

It is also instructive to examine the multiplicity of isotope peaks in the mass spectra in Figs. 29.6 and 29.7. Note that in Fig. 29.6, the relative intensity for the peak representing the nominal molecular ion is 3.24% and that for the first isotope peak at m/z 199 is 0.50%. For purposes of indicating elemental composition, it is the ratio of the M + 1 peak to the M peak that is important; in this case, that corresponds to $0.50 \div 3.24 \times 100 = 15.4$. This ratio is consistent with the number of carbons present, namely 14, for which the expected ratio of the M + 1 peak intensity to the M peak intensity is 15.4%. In Fig. 29.7, the ratio of the observed isotope peak intensity for the molecular ion at m/z 179 to that at m/z 178 is 14.6%. This compares reasonably well with the expected value of 15.4% for an ion containing 14 carbon atoms. The disparity between the observed and expected ratio of isotope peaks in Fig. 29.7 is a little more than expected with GC-MS data, where the precision on peak intensities may be ±5% relative standard deviation or worse.

Radical-Site–Driven Fragmentation

Homolytic cleavage of a given bond is usually stimulated by its proximity to a radical site. The driving force for homolytic fission is the movement of a single electron from a covalent bond to pair up with the odd electron at the original radical site. The movement of single electrons during homolytic bond fission is represented by singly barbed arrows or fishhooks as indicated in Scheme 29.6. In homolytic fission, the charge remains at the original site in the fragment ion also as shown in Scheme 29.6.

Scheme 29.6

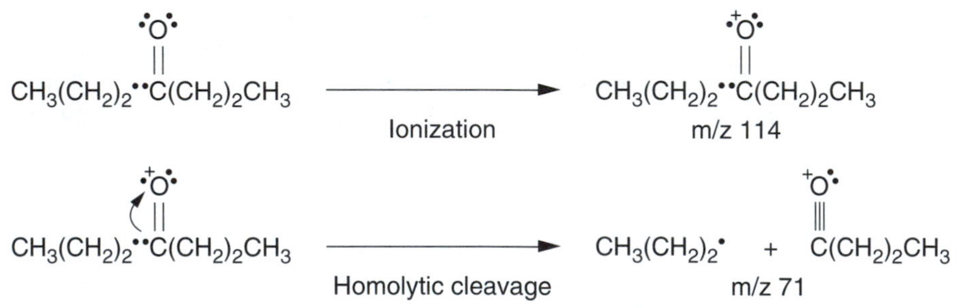

Radical-site cleavage, also called homolytic cleavage, often occurs at one bond removed from a heteroatom. The fragmentation near a heteroatom is due to the localization of the radical site in a nonbonding orbital of the heteroatom. Because the electron in a nonbonding orbital is easier to remove from a molecule than an electron from a sigma bond, for example, the predominant form of the molecular ion containing a heteroatom is that with the radical site located on the heteroatom, as illustrated in Scheme 29.6. The process of homolytic fission represented in Scheme 29.6 is often called alpha cleavage because a new bond alpha (adjacent) to the heteroatom holding the radical site is formed during the process. F. W. McLafferty defines alpha cleavage as cleavage of a bond on an atom adjacent to the atom bearing the odd electron (but not the bond to the latter atom). The major peak at m/z 71 in Fig. 29.8 represents an ion formed by alpha cleavage in Scheme 29.6.

Charge-Site–Driven Fragmentation

As shown in the ionization step in Scheme 29.6, the charge site as well as the radical site is localized on a heteroatom in the analyte. In heterolytic fission, both electrons from a nearby bond are drawn

Figure 29.7 EI mass spectrum of anthracene. Intensity axis on the right ordinate indicates that 47% of total ionization is represented by the molecular ion peak. *(Data acquired by Naxing Xu on a Finnigan 4600 quadrupole mass spectrometer in the MSU Mass Spectrometry Facility.)*

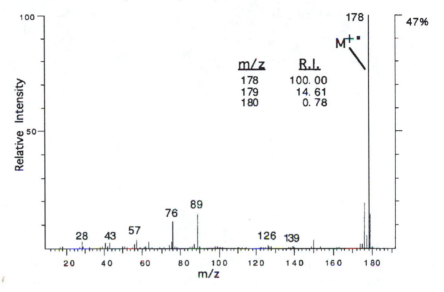

to the charge site, as illustrated in Scheme 29.7. This movement of both electrons to the charge site effectively neutralizes the charge on the heteroatom and causes the charge to move to the atom from which both electrons of the covalent bond were removed. In heterolytic fission, the movement of two electrons is indicated with a double-barbed arrow, and the charge moves from its original site in the molecular ion. The peak at m/z 43 in Fig. 29.8 represents an alkyl ion as formed by heterolytic fission or the charge-site drive mechanism illustrated in Scheme 29.7.

Scheme 29.7

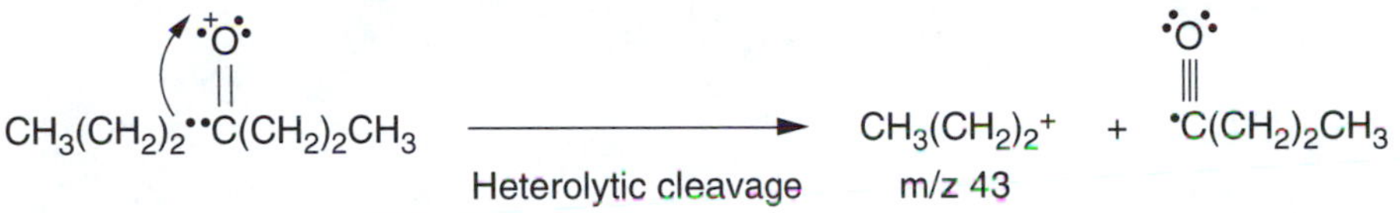

$$CH_3(CH_2)_2 \cdot \cdot C(CH_2)_2CH_3 \xrightarrow[\text{Heterolytic cleavage}]{} CH_3(CH_2)_2{}^+ + \cdot C(CH_2)_2CH_3 \\ \text{m/z 43}$$

Intramolecular Rearrangement

Most fragmentation involving rearrangements occurs in a folding of the molecule to accommodate shifts of single electrons and a hydrogen radical through a four- or, more commonly, a six-membered ring. As illustrated in Scheme 29.8, rearrangement of a gamma hydrogen combined with beta cleavage is often observed in the mass spectra of molecules that have a double bond and a gamma hydrogen relative to the location of the double bond. From the standpoint of counting electrons, note that the molecular ion is an odd-electron species and that an intramolecular rearrangement as illustrated in Scheme 29.8 produces a fragment ion that also contains an odd number of electrons. Note also that the mass of the odd-electron ion has an even mass number because it contains no nitrogens.

Scheme 29.8

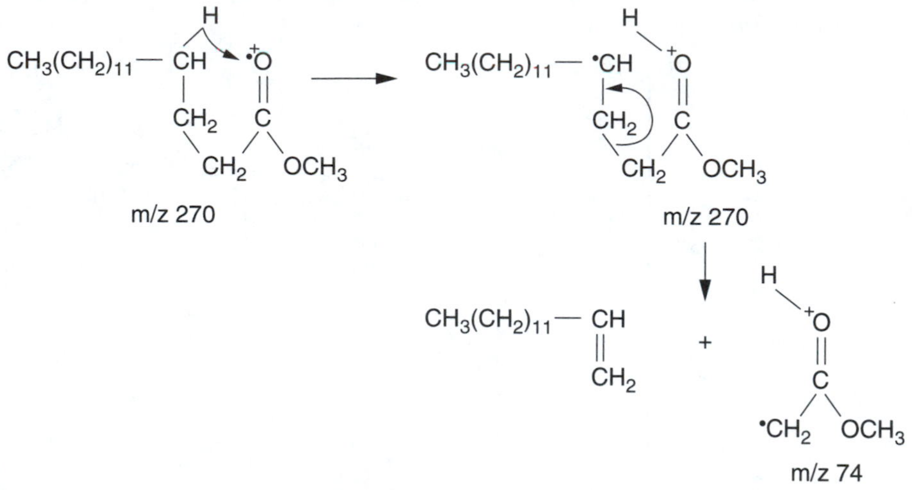

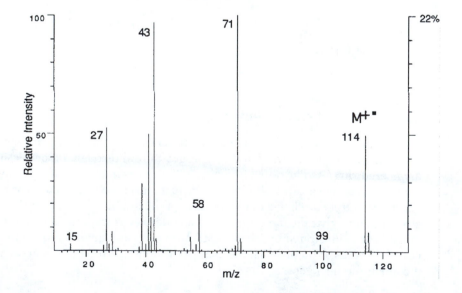

This feature of the even-mass versus odd-mass of certain ions makes it relatively easy to recognize that a rearrangement mechanism is involved in the fragmentation process. For example, note that in the mass spectrum in Fig. 29.9, the molecular ion peak is at an even m/z value, and all of the fragment ion peaks occur at odd values of m/z except for that at m/z 74. Thus, the peak at m/z 74 must represent a special circumstance in that it represents either a rearranged fragment ion or the molecular ion of a contaminating compound. In this case, the peak at m/z 74 represents the rearranged atoms at the carbomethoxy end of methyl palmitate, as illustrated in Scheme 29.8.

Even though beta cleavage associated with migration of a gamma hydrogen was first described by A. J. C. Nicholson, it is commonly known as the McLafferty rearrangement because it has been studied so extensively by F. W. McLafferty during the last four decades. A McLafferty rearrangement is also possible in the mass spectrometry of 4-heptanone, which has a double bond between

Figure 29.8 EI mass spectrum of 4-heptanone. *(Data acquired by Naxing Xu on a Finnigan 4600 quadrupole mass spectrometer in the MSU Mass Spectrometry Facility.)*

Figure 29.9 EI mass spectrum of methylpalmitate. *(Data acquired by Naxing Xu on a Finnigan 4600 quadrupole mass spectrometer in the MSU Mass Spectrometry Facility.)*

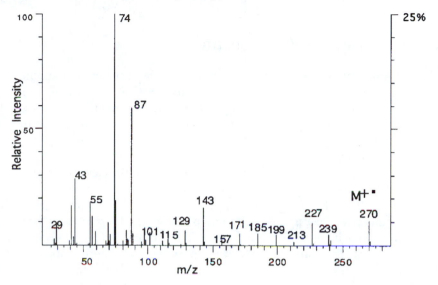

carbon-4 and the oxygen, and has six gamma hydrogens. Thus, in principle, there could be a rearrangement of a gamma hydrogen from either end of the side chains of the ketone to cause an elimination of ethylene with cleavage between carbon-2 and carbon-3. As illustrated in Scheme 29.9, the McLafferty rearrangement should produce a species of mass 86; however, no peak at m/z 86 is observed in Fig. 29.8. Because there are two alkyl chains each having gamma hydrogens, a second McLafferty rearrangement can occur that leads to a species of mass 58. A peak is present in the mass spectrum in Fig. 29.8 at m/z 58. Observing a peak at an even value of m/z, such as 58, provides evidence that a rearrangement has occurred.

Scheme 29.9

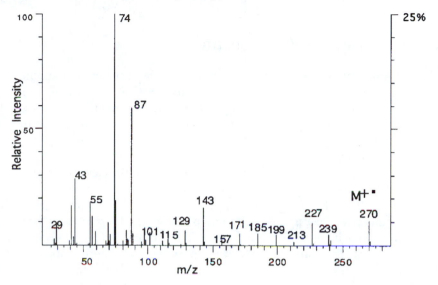

Applications

Comparison of EI and CI Mass Spectra

The main attribute of CI mass spectra is that they usually show an intense peak that represents the intact molecule or a minimally fragmented component of the main molecule. In this regard, the CI mass spectrum often gives a good indication of the molecular weight of the analyte, whereas the EI mass spectrum shows considerable fragmentation and, in some cases, may not show a very intense peak for the molecular ion. Thus, the spectra from the two different ionization processes complement one another.

The mass spectrum in Fig. 29.10 results from the electron impact ionization of stearyl amine, which contains 18 carbons in a normal aliphatic chain with a primary amino group. As can be seen in Fig. 29.10, the principal ion current is carried by the fragment ion of m/z 30. This ion is formed by homolytic cleavage, as illustrated in Scheme 29.10. Formation of the methyl immonium ion as shown in Scheme 29.10 is facile and, thus, is represented by a very intense peak in the mass spectrum of stearyl amine. Note that the peak representing the molecular ion is either very small or not discernible in the EI spectrum. On the other hand, the CI spectrum (Fig. 29.11) of stearyl amine shows an intense peak at m/z 270, which corresponds to the protonated molecule (MH$^+$). One often obtains a simplified mass spectrum upon chemical ionization of a molecule that has a higher proton affinity than that of the reagent gas, in which case the analyte abstracts a proton from a proton-rich reagent ion to protonate itself. Clearly, in the comparison between Figs. 29.10 and 29.11, the CI mass spectrum confirms the molecular weight of the analyte in Fig. 29.10 as being 269 Da.

Figure 29.10 EI mass spectrum of stearyl amine (molecular weight = 269). *(Data acquired by Naxing Xu on a Finnigan 4600 quadrupole mass spectrometer in the MSU Mass Spectrometry Facility.)*

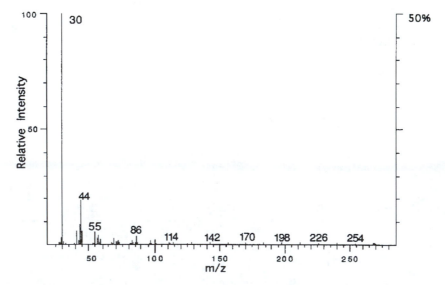

Scheme 29.10

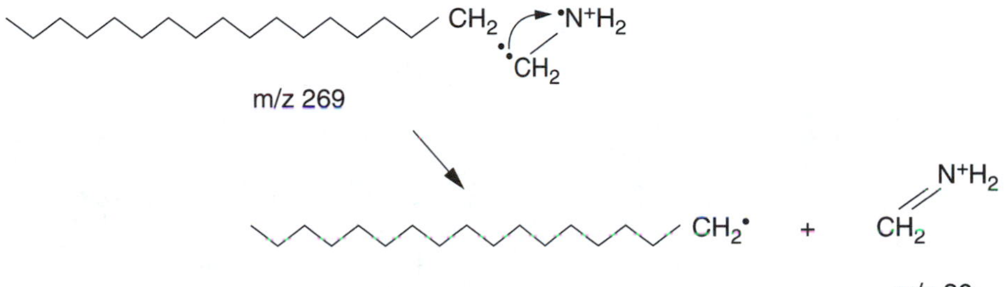

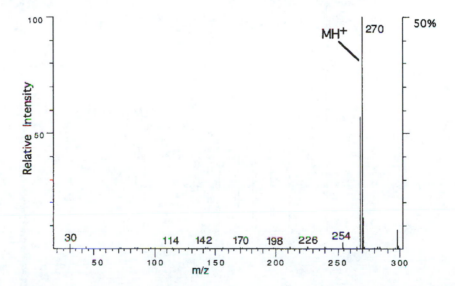

The EI mass spectrum of dodecanol is shown in Fig. 29.12. Typically, the EI mass spectra of long-chain alcohols do not show a discernible peak for the molecular ion. This is the case for dodecanol, which does show a minor peak at m/z 168, which in this case represents the molecular ion of the dehydrated species. The main ion current in primary alcohols is carried by a series of hydrocarbon-like alkyl ions of m/z 29, 43, 57, 71, etc., and a series of alkynl ions represented by peaks at m/z 27, 41, 55, 69, etc. In addition, there is a unique ion formed by homolytic fission as illustrated in Scheme 29.11; this oxonium ion is very stable, and it is represented by a significant peak at m/z 31 in Fig. 29.12. The methane CI mass spectrum of dodecanol is shown in Fig. 29.13, which shows a significant peak in the mass spectrum at m/z 185, which corresponds to $(M-H)^+$. Hydride extraction by the reagent ion is often seen in the CI spectra of alcohols, as with any other type of molecule where the hydride ion affinity (HIA) of the reagent ion is greater than the HIA of the $(M-H)^+$ ion of the analyte.

Figure 29.11 CI mass spectrum of stearyl amine using methane as reagent gas.*(Data acquired by Naxing Xu on a Finnigan 4600 quadrupole mass spectrometer in the MSU Mass Spectrometry Facility.)*

Figure 29.12 EI mass spectrum of dodecanol (molecular weight = 186). *(Data acquired by Naxing Xu on a Finnigan 4600 quadrupole mass spectrometer in the MSU Mass Spectrometry Facility.)*

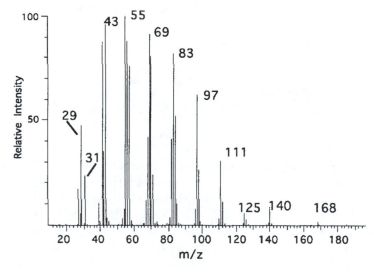

Figure 29.13 CI mass spectrum (using methane as reagent gas) of dodecanol (molecular weight = 186). *(Data acquired by Naxing Xu on a Finnigan 4600 quadrupole mass spectrometer in the MSU Mass Spectrometry Facility.)*

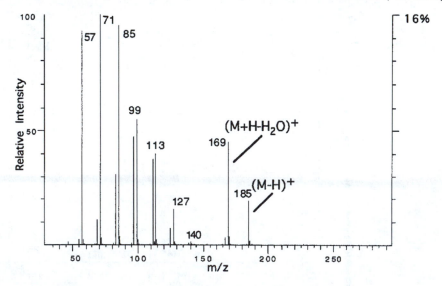

Scheme 29.11

m/z 186

m/z 31

Summary

In both EI and CI, neutral molecules of the analyte in the vapor phase are converted to charged species, which can be mass analyzed. Considerable fragmentation occurs in EI that can be related to structural features in the original molecule. In some cases, fragmentation in EI is so extensive that there is no discernible peak for the molecular ion and thus no indication of the molecular weight. In such cases CI, a soft ionization technique, may be used to generate ions of the intact molecule, thereby complementing the data available through EI.

Nuts and Bolts

See Chap. 30, pages 605–607.

Suggested Readings

CHAPMAN, J. R., *Practical Organic Mass Spectrometry*, 2nd ed. New York: Wiley, 1993.

HARRISON, A. G., *Chemical Ionization Mass Spectrometry*, 2nd ed. Boca Raton, FL: CRC Press, 1992.

HITES, R. A., *Handbook of Mass Spectra Environmental Contaminants*, 2nd ed. Lewis Publishers, Boca Raton, FL: 1992.

McLAFFERTY, F. W., AND F. TURECEK, *Interpretation of Mass Spectra*, 4th ed. University Science Books, Mill Valley, CA: 1993.

WATSON, J. T., *Introduction to Mass Spectrometry*, 2nd ed. Raven Press, New York: 1985.

Chapter 30

High-Resolution Mass Spectrometry of Volatiles and Nonvolatiles

Ilan Vidavsky and Michael L. Gross

Washington University
Department of Chemistry

Summary

General Uses

- Determination of exact molecular weights and elemental compositions of molecular substances
- Resolution of isotopic envelopes for high-molecular-weight materials (by obtaining at least sufficient resolving power to separate adjoining mass peaks)
- Determination of one material in the presence of isobaric interfering materials in complex mixtures such as petroleum
- Detection and quantification of targeted substances at trace (part-per-trillion and lower) levels with high certainty

Common Applications

- Determination of molecular weight and elemental composition of synthetic materials and natural products in lieu of a combustion or other elemental analysis

- Determination of the elemental composition of fragment ions and, if the elemental composition of the molecular ion is known, the accompanying neutrals formed in the fragmentation of unknown compounds
- Determination of fragmentation pathways in mechanistic studies of known compounds
- Determination of the charge of multicharged ions produced in electrospray ionization (ESI)
- Determination of the presence of target compounds in complex mixture, as are often encountered in biological and environmental samples

Samples

State

All pure or nearly pure volatile materials including gases, liquids, and solids can be handled by direct introduction into the mass spectrometer ion source. Components of mixtures can be introduced after separation by chromatographic methods coupled on-line to the mass spectrometer, as in gas chromatography mass spectrometry (GC-MS).

Nonvolatile samples are introduced into the mass spectrometer by desorption/ionization methods such as fast atom bombardment (FAB), laser desorption (LD), and matrix-assisted laser desorption ionization (MALDI). Other soft ionization methods such as thermospray (TSP) and ESI are used by injecting directly a solution containing the sample or by first separating via high-performance liquid chromatography (HPLC), for example, and introducing solutions containing mixture components.

Amount

The amount of sample used for exact mass determinations of substances having molecular weights in the 100 to 800 range is usually 1 to 10 ng, but in most cases more is needed to permit easy handling during introduction to the mass spectrometer. When analysis is done in conjunction with a separation method such as GC-MS or HPLC-MS, lower amounts can be accommodated. For high-resolution multiple ion monitoring of GC eluents, subpicograms of targeted compounds can be identified and quantified.

Preparation

None or very little preparation is needed for volatile samples. Usually only loading onto a sample introduction probe or injecting the sample into a heated reservoir is required. A calibration compound is cointroduced with the sample into the mass spectrometer source. Nonvolatile samples must be mixed with a suitable liquid matrix in the case of FAB, or cocrystalized with a solid matrix when MALDI is used. For samples that require ionization by TSP and ESI, a solvent system must be found so that ionization is facilitated. Often with FAB or MALDI, a calibration compound must be cointroduced carefully because the calibrant may suppress the desorption of the analyte or vice versa.

Analysis Time

Analysis time is usually very short, only a few minutes, once the instrument is set up, tuned, and calibrated. When coupled to a separation method, such as GC or HPLC, the analysis usually requires 10 to 30 min or as much as 60 min, depending on the elution times of the analytes.

Limitations

When using sector instruments, larger sample quantities and more analysis time are required for a high-resolution determination than when a low-resolution spectrum is acquired. The precision and accuracy of the mass measurements are poorer when sample quantities are smaller. The precision and accuracy can be lower than 5 ppm (relative) for sample size introduction of 1 μg. High-resolution electron ionization (EI) mass spectra are most easily obtained in a scanning mode, but for experiments requiring higher accuracy and precision (approximately 1 ppm) or involving nonvolatile samples and desorption ionization, a one-ion-at-a-time peak matching experiment should be considered. Peak matching is a much more demanding and much slower experiment.

Complementary or Related Techniques

Elemental composition of pure molecular substances can be determined by methods that reveal the elemental composition. The percentages of C and H, for example, can be determined by combustion analysis. For organometallic and coordination compounds, atomic absorption or titration analyses can be used to obtain percentage composition of metals. Some preliminary indication of elemental composition for organic and organometallic compounds can be obtained by examining the isotopic peaks in a mass spectrum. For example, the presence and number of chlorines and bromines are revealed by isotopic patterns of the molecular and other ions.

Introduction

Sector Mass Spectrometers

Mass spectrometry has come a long way since it was first introduced by Thomson (1). In the intervening years, many designs of mass spectrometers have been built and evaluated. In this section, we examine the principles of double-focusing mass spectrometers.

Ions are formed in a suitable ion souce of the mass spectrometer and are accelerated by an electric voltage V and leave the ion source via a slit with a kinetic energy eV, where e is the charge. When they pass through a region with a magnetic field, B, they follow a trajectory where the radius is given by

$$r = \left(\frac{2mV}{eB^2}\right)^{\frac{1}{2}} \tag{30.1}$$

Equation 30.1 shows that large-mass ions are deflected less than small-mass ions (provided their charges are equal). It can also be shown that if the ions enter the magnetic sector at right angles and the object and image slits are positioned symmetrically, a directional focusing is achieved (Fig. 30.1).

However, the trajectory radius for all ions of the same mass-to-charge (m/z) ratio is not a constant. Ions leave the source with a directional and velocity spread. For a symmetrical magnetic analyzer (that is, one in which the object and image slits are located symmetrically with respect to the magnet position), the width of the image at the collector slit is equal to that of the source slit plus any aberration that might be present in the magnetic field region. Velocity spread and aberrations broaden the image and the resulting peak.

Figure 30.1 Directional focusing of an ion beam that is at right angles to a magnetic sector.

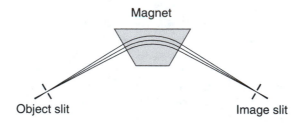

The magnetic analyzer provides no velocity focusing. There are three reasons why ions of the same m/z ratio have different velocities even though the accelerating voltage is constant. First, ions are formed from neutral molecules that have a Boltzmann distribution of velocities (energies). Second, there are small electric field gradients in the ion source, causing ions that are formed in different regions to experience different accelerations. Third, fragment ions are formed with isotropic conversion of internal energy into translational energy (called kinetic-energy release). The resulting spread of velocities or energies causes a spreading of the radii of ion trajectories, leading to a blurring or broadening of the signal (image) at the detector.

To compensate for these problems, an electrostatic analyzer is placed in series with the magnetic analyzer. If an ion beam enters a radial electric field E at right angles to the field as shown in Fig. 30.2, a circular trajectory results, and its radius is given by Eq. (30.2).

$$r = \frac{2V}{E} \qquad (30.2)$$

Ions that have different velocities can be viewed as experiencing different accelerations or different V values; thus, they have different radii through the electric sector. We may describe the situation by saying that the electric sector produces a velocity dispersion, and that dispersion is opposite to the velocity dispersion that occurs in the magnetic analyzer. The electric sector compensates for the velocity dispersion in the magnetic field, and the ion beam is now sharply focused because the velocity aberration is almost completely eliminated. The directional focusing at the same time is preserved. A mass spectrometer using this combination is said to be double focusing.

One design is the Mattauch–Herzog (2) double-focusing mass spectrometer, as shown in Fig. 30.3. This arrangement has the advantage of being double-focusing simultaneously for all m/z ra-

Figure 30.2 Directional focusing of an ion beam that is at right angles to an electric sector.

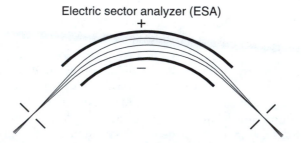

Figure 30.3 Double-focusing focal-plane mass spectrometer of the Mattauch–Herzog type.

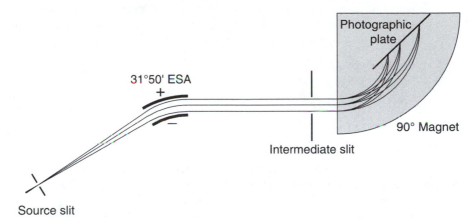

tios. As a result, the design has attracted interest recently because the inconvenient photoplate can be replaced with an electronic version called an array detector.

Second-order directional and first-order velocity focusing with the proper arrangement of the electric and magnetic analyzers was shown by Nier and Johnson (3). In this arrangement, the double focusing is achieved for ions of one m/z ion at a time, as shown in Fig. 30.4. Because only one ion is collected at a time, an electron multiplier detector is used. These detectors have fast response and high sensitivity. Thus, these instruments are more commonly used than those with Mattauch–Herzog design even though the Nier–Johnson spectrometer does not accommodate an array detector as readily as does the Mattauch–Herzog spectrometer.

Modern mass spectrometers use several methods to compensate for dispersion, such as shaped magnetic pole faces, z-directional focusing to increase sensitivity, and additional electrostatic devices to correct for the effect of fringe fields and other aberrations. Examples of these latter devices are hexapoles, shaped lenses, and quadrupole doublets.

The resolving power (defined as M divided by ΔM) that can be achieved on high-end, commercial mass spectrometers exceeds 100,000, as defined by peak width at 10% height. Instruments with even higher resolving power have been built for research purposes. For example, it was shown by Matsuo and Matsuda (4) that super-high resolving power can be achieved on a four-sector mass spectrometer.

Figure 30.4 Double-focusing mass spectrometer of the Nier–Johnson type.

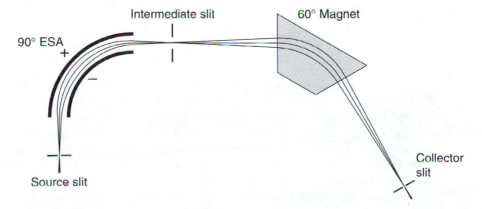

Because the atomic weight of the elements is not unity and each element has a different mass defect (a different deviation from unit mass), it is possible to calculate an elemental composition of a given ion from its exact mass. For example, the compositions of the isobaric ions listed below can determined by measuring their masses with high accuracy.

$C_6H_{14} = 86.1095$

$C_5H_{10}O = 86.0731$

$C_4H_6O_2 = 86.0367$

$C_4H_{10}N_2 = 86.0843$

$C_4H_6S = 86.0190$

The use of high-resolution mass spectrometry and exact mass measurement to determine the elemental composition was first promoted by Beynon (5). The method used involved recording the ratio of accelerating potentials needed to focus identically an unknown mass ion beam and a reference or known mass ion beam onto the collector slit, while keeping the magnetic field constant. This method is usually called peak matching. It is also possible to obtain exact mass measurements in a scanning experiment in which all the ions in a mass spectrum are assigned. These two approaches are discussed later.

Exact mass measurements are used for many purposes: determination of the elemental composition of synthetic materials and natural products, determination of elemental composition of product ions formed by fragmentation to assist with interpretation of mass spectra of unknown compounds, elucidation of fragmentation pathways of known compounds, and detection at high certainty of trace (part-per-trillion) components in biological and environmental samples.

Fourier Transform Mass Spectrometers

Although high-resolution or exact mass measurements are done mainly on sector mass spectrometers, other types of mass spectrometers can be used for this type of measurement. One instrument that has very high resolving power is the Fourier transform ion cyclotron resonance mass spectrometer, first introduced by Comisarow and Marshall (6). In Fourier transform (FTMS), ions are transferred to or formed inside a trap or cell that is located in a high static magnetic field. Any ion in the field will move in a circular trajectory whose radius is given by Eq. (30.3).

$$r = \frac{mv}{eB} \qquad (30.3)$$

The frequency or number of complete circular paths executed per second by the cyclotroning ion is given by Eq. (30.4).

$$f = \frac{eB}{2\pi m} \qquad (30.4)$$

Because the frequency depends only on mass and magnetic field strength, B, in the absence of electric fields, and can be measured very accurately, precise and accurate measurement of its mass is possible. If ions are irradiated by alternating electric fields of frequency equal to their natural cyclotron frequency, they gain energy and can be detected by the image currents they impose on the cell plates. The experiment consists of ion production or introduction to the cell, excitation by radiofrequency irradiation, and detection. Fourier analysis of the decaying ion signal, which is a time domain signal, transforms it into frequency domain signals; because frequency is inversely related to mass, the frequency domain information can be readily converted to a mass spectrum. A consequence of the frequency/mass relationship is that resolving power is inversely proportional to mass.

The advantages of this method are its nondestructiveness (the ions at the end of the experiment are still in the trap and can be used for another measurement) and its multichannel advantage (the whole mass range is recorded at once instead of by scanning one ion mass after another). The mass resolution attainable by this method has been shown to be several million.

One disadvantage is that the FT method, because it is a pulsed method, is incompatible with continuous ion sources, which are used in most other mass spectrometers. Of course, this disadvantage becomes an advantage when one wishes to use a pulsed ionization method such as matrix-assisted laser desorption/ionization (MALDI). Another disadvantage is that the pressure in the cell region must be very low (approximately 10^{-8} torr) to ensure good performance. Thus, high-pressure ion sources (such as CI and ESI) must be located outside the cell. Transfer of ions from an external source, or even from another adjacent cell, can be complicated. Furthermore, the calibration of the instrument is not trivial because the cyclotron frequency also depends to a small extent on the electric fields that are overtly introduced to trap the ions and also generated by the ions themselves.

Another kind of instrument that shows potential for high resolving power is the radiofrequency ion trap, where high resolution can be achieved by very slow scanning of electric fields. There is yet no indication that high accuracy and precision can be achieved routinely with this instrument.

Accuracy and Precision of Mass Measurement

The accuracy and precision of exact mass measurement are two different issues. High precision depends on instrumental factors such as stability of power supplies, rate of data acquisition, and overall instrument design. A method for statistical evaluation of accurate mass measurement quality was outlined by Sack, Lapp, Gross, and Kimble (7). Their method was applied to a high-performance mass spectrometer and the standard deviation of all mass measurement errors was determined to be 2.5 ppm (an error of 1 ppm corresponds to a deviation of 0.00032 for a measurement of 319.6938). The precision does not necessarily depend on high resolution, but because most exact mass measurements involve an internal calibrant, high resolution is needed to separate ions from reference compounds that have the same nominal mass as those of the unknowns (consider, for example, C_2F_5, a common reference ion of exact mass 118.99202, and C_8H_7O, ion of exact mass 119.04969). The accuracy of the measurement depends on a stable instrument and good calibration. Usually the more calibration points, the higher the accuracy, hence the popularity of reference compounds such as perfluorokerosene (PFK) that have many abundant peaks in their mass spectra.

How It Works

High-resolution mass spectra are obtained by measuring both the abundance and exact masses of ions that make up a mass spectrum. In the case of a sector instrument, ions that are formed in the source by a variety of techniques are accelerated by an electric field, which, depending on the instrument, is generated by applying to the ion source a voltage in the range of 4 to 10 kV. The ions are brought to focus at the source slit (object slit of the instrument) by electrostatic lenses. The slit width for high resolution is usually very small, on the order of microns. The width depends on the particular instrument and the required resolving power. The ion beam then passes through electric and magnetic fields, where the ions are separated according to their mass and focused on a collector or detector slit. This slit has a width similar to that of the source, but it is usually slightly wider to permit reasonable sensitivity. The spectrum is obtained by scanning of the magnetic field such that only one different m/z ratio ion at a time is focused onto the collector slit.

Before data acquisition can take place, the instrument must be tuned. Tuning is done by varying voltages to several electrostatic lenses in the source region or after it, the purpose of which is to focus the ion beam onto the source slit and to compensate for aberrations so that maximum resolving power and sensitivity are obtained. Usually the source slit is closed first and the instrument is tuned to get maximum sensitivity and a symmetrically shaped, flat-topped peak. The shape of the peak is very important because asymmetry or tailing causes excessive signal loss as higher resolving power is sought. After that, the collector slit is closed until the peak just becomes Gaussian in shape and further closing only decreases its intensity. The procedure is repeated until the desired static resolving power is obtained. Dynamic resolving power (the one observed during scanning of the magnet) in most cases would be lower than static, so one would tune the mass spectrometer to a static resolving power that is higher than is sought in the dynamic or scanning mode. It is also possible to monitor the resolving power during a magnet scan by using the data system and making adjustments in an iterative way. Tuning is normally done by using one of the intense peaks from the reference material because the reference usually gives a steady signal for a long time, whereas the amount of sample to be analyzed is often limited.

The shape and height of the peak can be observed on an oscilloscope or a computer screen that simulates a scope screen. A small sawtooth waveform is imposed on the accelerating voltage and the electrostatic analyzer voltage to modulate them while the magnetic field is held constant. This permits rapid scanning over a small mass range, so that the peak shape is readily observed and can be adjusted.

Detection

Most of the original high-resolution mass spectrometric detection was done with photoplates. The advantages are the integrating nature of the photoplate and the sensitivity in comparison to the electronic detection methods of the time. The photoplate also has a multichannel advantage, meaning that many ions of different m/z ratios are recorded at the same time. The disadvantages are that the spectra cannot be seen during acquisition and considerable time is consumed to develop the plate and quantify the results with a densitometer. The development of electron multiplier detectors and high-speed electronics in tandem with the development of computers has replaced photoplate detection. In recent years, however, there is renewed interest in the electronic equivalent of the photoplate (that is, the array detector) because it has a multichannel advantage but none of the disadvantages of the photoplate.

Scanning

There are two modes for doing exact mass measurements. The first is a full-scan mode, where the magnetic field is scanned to cover a large mass range that should encompass the mass spectrum of the sample. In this mode, a reference compound is cointroduced with the sample to correct for scan-to-scan variations.

The spectrum is produced by exponential scanning over the mass range. The resulting analog signal from the detector is digitized at equal time intervals. The intensity values are saved in computer memory together with the peak-center times that are registered from the beginning of the scan. The center of gravity or the centroid and the area of each peak are usually recorded (the centroid is calculated as the intercept of two lines; one line is calculated by averaging the points on the rising edge of the peak and the other on the falling edge of the peak). Most instruments use fast preprocessing for computing the centroid values to lighten the burden on the data system.

Ideally, the mass of any peak is an exponential function of the elapsed time from start of the scan, where M is the mass, M_{max} is the highest mass in the scan, t is the time elapsed from beginning of scan, and τ is the time constant of the scan, as shown in Eq. (30.5).

$$M = M_{max}\, \exp^{-t/\tau} \tag{30.5}$$

It can be seen that the mass can be established accurately if M_{max}, t, and τ are known accurately. The use of a reference compound with many peaks across the mass spectrum allows the mass scan law of Eq. (30.5) to be established in the region of most unknown peaks. The scan rate, τ_{10}, is usually measured in seconds/decade of mass (one decade of mass corresponds to a mass range of m/z 10 to 100 or m/z 50 to 500, for example) and can be calculated as shown in Eq. (30.6).

$$\frac{M}{10} = M\, \exp^{-t/\tau} \qquad t = \tau\, \ln_{10} = \tau_{10} \tag{30.6}$$

Resolving power is usually defined as $R = M/\Delta M$, where ΔM is the exact mass difference measured across the peak at some specified peak height (such as 10%); see Fig. 30.5.

The time to scan across a peak can be calculated from the scan law, Eq. (30.5).

$$dM = -\frac{M_{max}}{\tau}\, \exp^{-t/\tau} dt \tag{30.7}$$

For small ΔM and Δt, this can be approximated as

$$\Delta M = -\frac{M_{max}}{\tau}\, \exp^{-t/\tau} \Delta t \tag{30.8}$$

Figure 30.5 Definition of resolving power.

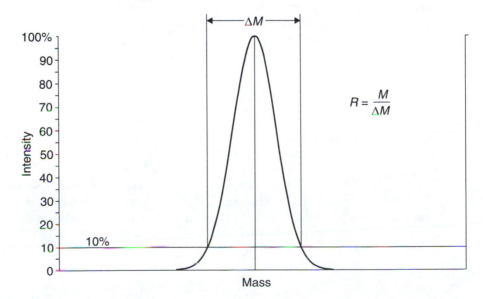

Division of Eq. (30.8) by 5 gives

$$\frac{\Delta M}{M} = -\frac{\Delta t}{\tau} \quad \text{or} \quad \Delta t = -\frac{\tau}{R}$$

and

$$t = -\frac{\tau_{10}}{R \ln_{10}} \tag{30.9}$$

From Eq. (30.9), we see that the time to scan across a peak is mass independent; thus, data acquisition at constant time intervals (at constant frequency) is justified. This is one reason for choosing an exponential scan; another is they are easily generated as a capacitance discharge in the magnet's reference power supply.

The time to scan across a peak at resolution of 10,000 and scan rate of 10 sec/dec, for example, is

$$f = \frac{10}{10,000 \times 2.3026} = 434 \ \mu\text{sec} \tag{30.10}$$

Because 20 points across a peak are needed to calculate accurately a peak centroid, an acquisition rate of one point per 20 μsec (50 kHz) is sufficient. This rate can be easily accommodated by modern data systems. For fast scanning rates such as those needed for GC/high-resolution MS (such as 1 sec/dec) and a mass resolving power of 10,000, the acquisition rate must be 500 kHz, which is above or close to the limit of those curently used with commercial instruments. It is also more difficult to achieve high sensitivity at high acquisition rates.

As mentioned above, the spectra are calibrated by introducing a reference compound along with the sample. The spectra are recorded in the form of a data table containing peak intensities and times for the peak centroids. After acquisition, the data system software uses the reference compound peaks to calibrate the spectrum (that is, to establish accurately the scan law). Usually the procedure involves the following steps. A few lock masses are identified by their relatively high abundances over a wide mass window. The lock masses are starting points for the calibration procedure. An early version of the scan law is calculated on the basis of the times and lock masses. The remainder of the reference peaks are then identified over a much narrower prediction window. Adjustment of the prediction window parameters is sometimes needed to identify correctly all the reference peaks. One parameter that can help to judge a calibration is the deviation from the predicted masses of reference peaks. Some software packages allow the operator to skip certain reference peaks or correct the computer identification of certain peaks to improve the calibration.

Perfluorokerosene (PFK) has several desirable properties that make it a good reference compound. All the ions in its mass spectrum have a negative mass defect, so they are easily distinguished from molecular and fragment ions of organic compounds made up of C, H, N, and O; these latter compounds have a positive mass defect (for example, the mass of $C_2F_5^+$ from PFK is 118.9920, whereas the mass of $C_9H_{11}^+$ is 119.0860 and that of $C_8H_7O^+$ is 119.0496; the mass of $C_3F_7^+$ from PFK is 168.9888, whereas the mass of $C_{10}H_{17}O_2^+$ is 169.1227 and that of $C_{11}H_{21}O^+$ is 169.1591). The many reference peaks ensure that the interpolation is short and accurate. PFK ions can be used in EI, CI, and sometimes also in negative CI ionization methods. A single reference compound for use in desorption ionization method is much more difficult to find because little fragmetation takes place with FAB or electrospray. Furthermore, FAB suffers from suppression effects, and considerable care must be taken when selecting a reference compound and adding it to the FAB mixture.

Peak Matching

The most precise and accurate method for determining the mass of a few different mass ions in a mass spectrum is peak matching. In this method the magnet field is fixed, but the accelerating and ESA voltage are switched between two values that are adjusted until the reference compound peak and the sample peak are focused identically at the detector slit. The peaks are displayed on a storage oscilloscope screen, and the ratio between the two voltages can be adjusted until the two peaks are nearly perfectly overlaid. The ratio of the voltages is then read with high accuracy, taking advantage of a six- to eight-figure voltage divider (or its equivalent in a data system). Preferably this ratio should be 1.00 ±0.10 to avoid ion beam defocusing in the source. Alternatively, the accuracy of the method can be improved by using two reference peaks, one higher and one lower.

In cases where using an internal standard is difficult (such as FAB), it is possible to record the peak shape of the references and then, without changing the magnetic field and focusing parameters, remove the reference compound and introduce the sample. Some probes are designed to permit loading of the reference compound and sample on two different sides or stages and then switching between them without removing the probe. Peak matching can be done more accurately with the aid of software for averaging weak signals, smoothing and centroiding the peaks.

Reference compounds for FAB include a mixture of CsI or NaI in glycerol or a mixture of polypropylene glycols and KCl. In cases where the nature of sample is known (such as peptides), a set of materials that are closely related to the sample can be used as reference compounds so that FAB discrimination is avoided. For a selection guide for reference compounds, see Reference 8.

Gas Chromatography High-Resolution Mass Spectrometry (GC-HRMS)

In GC-MS, especially when using capillary GC columns, high scan speed is very important. The width of each chromatographic peak is several seconds or less, and a scan speed that would give a scan cycle time equal to or less than half of the chromatographic peak width time is essential. It is preferred to obtain 5 to 10 scans for each chromatographic peak. Although this is possible for GC low-resolution MS, the required fast scanning for GC-HRMS produces inadequate sesitivity. As shown before, full-scan GC-HRMS is very demanding on the data system. If narrower scans and less accuracy in the mass determination (fewer points across a peak) can be accommodated, full-scan GC-HRMS is usually possible but always under compromised conditions (9).

On the other hand, selected ion monitoring (SIM) can benefit considerably from high-resolution measurement. For example, in the analysis of samples for trace (parts-per-trillion) levels of chlorinated dioxins and related substances, high-resolution mass spectrometry can minimize interferences from isobaric materials and increase the confidence level for identification of a targeted compound.

The common approach is to use high-resolution SIM and select a few characteristic analyte ions (including one from an isotopically labeled internal standard, which serves as both a quantification and mass standard). The mass spectrometer is tuned to a resolving power of typically 10,000 and then adjusted to dwell on the peak tops of the analyte and internal standard ions, one after another. This procedure is called peak-top monitoring and is relatively easy to set up. Two disadvantages apply. First, it is not known whether the peak top includes contributions from interferences that are difficult to resolve. For example, when monitoring the molecular ion of a tetrachlorodioxin, $C_{12}H_4O_2{}^{35}Cl_3{}^{37}Cl$ at 321.8936, interference by comparable ion intensities of dichlordiphenylethylene (a metabolite of DDT that is still found in many biological samples) at 321.9292 is readily removed, but an interference by a polychlorinated biphenyl fragment ion at

321.8677 ($C_{12}H_3{}^{35}Cl_5$) is more difficult to eliminate by using resolving power of 10,000. Second, peak-top monitoring does not allow the analyst to establish the exact molecular mass of the detected compound because one is never certain that the center of the peak is being detected.

Both of these problems can be solved and the certainty of the assignment of the detected material can be increased by using peak-profile monitoring. To accomplish this, the mass spectrometer is set up as for peak-top monitoring, but instead of dwelling only on the peak top, the mass spectrometer makes a narrow scan (usually over a fraction of one m/z value) and a peak profile is obtained and stored in the computer. Because the analyst can view the entire peak, the exact mass and correct intensity can be established from the centroid, and the presence of overlapping interferences can be assessed. Furthermore, if unexpected materials are present in the sample, their identification can be expedited because their exact masses can be calculated. A demonstration of the advantages of peak-profile monitoring was recently published (10) and is discussed later.

Fourier Transform Mass Spectrometry

Fourier transform ion cyclotron resonance mass spectrometry (FTMS) can yield the highest resolving power and the most accurate mass determination. Indeed, the masses of isotopes and of organic and biomolecules can now be measured with higher accuracy by using FTMS than with magnet sector instruments. Thus far, however, the improvement in accuracy is not so significant that analysts are turning to what is still a quite expensive alternative.

The heart of the FT mass spectrometer is relatively simple, as shown in Fig. 30.6. It is a cubic cell consisting of six isolated plates grouped in three opposing pairs. Trapping plates, along the axis of the magnetic field, are held at a small positive potential, forming a saddle electric field inside the cell and preventing the ions from escaping along the magnetic axis. The magnetic field constrains the ions to move in circular paths (cyclotron orbits) and prevents them from escaping along the other axes. A radiofrequency voltage is applied to the excitation plates to excite the ions in the trap to higher cyclotron orbits. Excitation can take place by frequency sweeps or by applying tailored waveforms to excite only ions of certain m/z values. Ion signals are recorded with the detection plates, where image currents due to moving packets of excited ions are induced. Very low pressures, on the order of 10^{-8} torr, are needed for the highest-resolution operation of the instrument.

Figure 30.6 FTMS cubic cell.

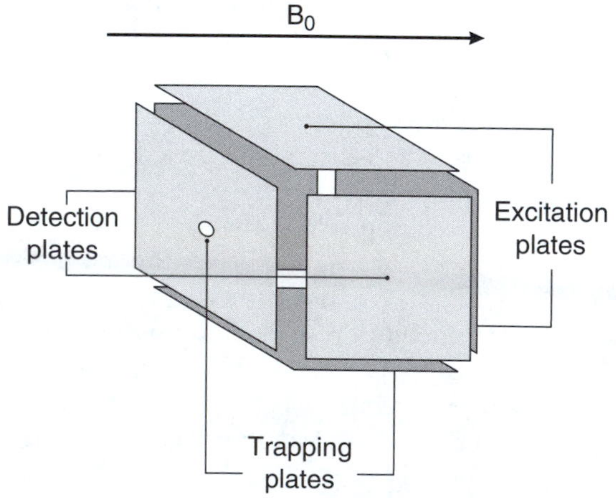

B$_0$

Detection plates

Excitation plates

Trapping plates

Ions are formed inside the cell by an electron or laser beam passing through holes in the trapping plates. Alternatively, ions can be formed outside the magnetic field by using higher-pressure techniques such as FAB, CI, and ESI and then transferred to and trapped in the cell. Transfer and trapping are not trivial because the ions must be transferred along the magnetic axis to overcome the magnetic mirror effect, which will reflect any ion not perfectly on the center, and then stopped in the cell. A gas pulse can be used to aid the trapping, but the efficiency of the trapping is usually low. To achieve high resolution, the pressure in the cell must be as low as possible to prevent collisions between ions and background molecules; these collisions cause dephasing of the ion packet and lead to a short transient signal. Furthermore, the trapping voltage must be lowered as much as possible to decrease the electric fields in the cell and minimize magnetron motion (that is, the ion motion caused by crossed electric and magnetic fields). There is also a limit on the number of ions in the cell; there should be less than a few hundred thousand to minimize coulombic effects.

The transient signal is acquired by a fast analog-to-digital converter. To acquire a full spectrum, a large number of data points are required because the frequency band (mass range) is usally large and one must sample at a rate determined by the lowest mass (highest frequency) that one wishes to observe. The requirement for fast acquisition can be alleviated by heterodyning if a target mass is sought. In this method, a reference frequency is mixed with the signal and only the difference frequency is digitized, which, of course, is much smaller and requires a lower sampling rate and fewer data points.

The fundamental frequency of each ion in the trap, in the absence of an electric field, depends on mass and magnetic field alone (see Eq. (30.4)) and, in principle, accurate mass determination is simple. The electric trapping field and the number of ions in the trap, however, cause shifts in the ion frequencies. This problem can be overcome in part by introducing reference materials together with the sample and by developing mass calibration laws that include these electric field effects (11).

Most high-resolution experiments are performed on sector instruments, but from time to time the need for super-high resolving power arises. In the analysis of petroleum products, the distinction of $C_{40}H_{82}S$ and $C_{43}H_{78}$, for example, requires a resolving power on the order of 200,000 (see Fig. 30.7). Distinctions of this type are needed to understand the origin of sulfur in oil and other energy materials (sulfur is a cause of air pollution and acid rain).

Another need for super-high resolving power is in the determination of high-molecular-weight materials, where resolving powers on the order 50,000 to 100,000 are needed to separate the isotopic envelopes of moderately sized proteins. Even in electrospray, where the m/z ratios of ions are usually on the order of 1000, this resolving power is needed to separate the isotopic envelope and elucidate the charge state of each ion. Because mass spectrometry measures m/z and ESI produces multiply charged ions, the charge state of various m/z ions must be established before the mass (molecular weight) can be computed.

What It Does

Analytical Information

Calibration of high-resolution mass spectra results in a list of exact masses and abundances of the ions in the spectrum. This table can be output as a list or in the conventional mass spectrometric graphic format (abundance versus m/z). Exact masses in the spectrum are compared with candidate masses to determine the composition of the ions. The computer program considers possible com-

Figure 30.7 Theoretical peak shapes at mass resolving power 200,000, which is needed to resolve $C_{40}H_{82}S$ from $C_{43}H_{78}$.

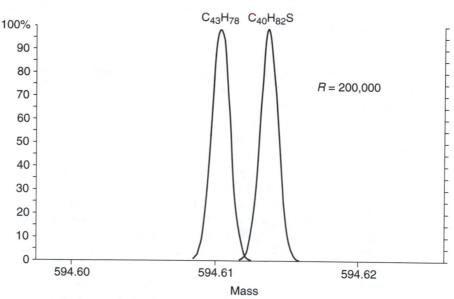

binations of the elements that the analyst expects to make up the analyte molecule; some knowledge of the sample simplifies the process. Apart from very low masses, the number of possible combinations is formidable, and knowledge of the sample is one way to eliminate some. For example, in the determination of a synthetic compound, information on the expected composition comes from the chemical reactions used for the synthesis. In this case, the list of elements to be considered can be very short. When the compound to be determined contains halogens or certain metals that produce a distinct isotopic pattern for the molecular ion, the number and combination of these atoms can be fixed, eliminating many unlikely combinations. If the calculation is limited to the lowest major isotope of each element (such as ^{12}C and not ^{13}C, ^{14}N and not ^{15}N, or ^{16}O and not ^{18}O), a large number of combinations is also eliminated, but this can be applied only to the low mass peak of each isotopic cluster. Chemical reasoning can also be used; in EI-MS, the molecular ion must be an odd electron (OE) species so only (OE) combinations must be taken into account (for example, C_7H_7 should not be considered if the molecular ion has a mass of 91). In CI and FAB, the molecular ion is typically an even electron (EE) species, and the fragments are also EE species. It is also possible to restrict the number of double bonds and rings if there is some knowledge on the nature of the compound (for example, $C_{16}H_{34}$ should not be considered for a molecular ion of nominal mass 178 if the sample is thought to be a polycyclic aromatic hydrocarbon). The deviation of the calculated mass from the measured one can also be used to rule out combinations; the value should be set to one representing the instrument precision (7). This value is usually between 5 and 20 ppm.

The number of compositions to be considered changes as function of the mass. At lower mass (12), where the accuracy in absolute mass units is high (at m/z 100, an error of 5 ppm corresponds to 0.5 mDa), the number of candidate combinations at a given precision level is small. But as the masses become higher, the absolute accuracy goes down, and the number of combinations to be considered within a certain precision window goes up very fast. An example from the American Society for Mass Spectrometry policy on "The Use of High-Resolution Mass Spectral Data for Structure Confirmation" shows that no two candidate formulae at nominal m/z 118 are 34 ppm apart for element limits of C_{0-100}, H_{3-74}, O_{0-4}, N_{0-4}. At nominal m/z 500, there are five compositions with a molecular weight that is within 5 ppm, using the limits C_{0-100}, H_{25-100}, O_{0-15}, and N_{0-15}. At m/z 750.4, there are 626 candidate formulae that have molecular weight within 5 ppm. Thus, for measurement at m/z 118, an error of less than 34 ppm is sufficient to distinguish two candidates, but at m/z 750 an error of less than 0.018 ppm would be needed. Clearly more information to limit the number of candidates is needed at high m/z than at low m/z.

A sample of an elemental composition report is shown in Fig. 30.8. In this sample, a synthetic compound molecular ion with nominal m/z of 391 and expected composition of $C_{23}H_{22}NO_3P$ was determined. The sample was investigated by using FAB ionization. At the top of the form are the elemental limits of the calculation and the error window (in millidaltons) for which compositions are considered. For each peak, several candidates are considered, and the difference between their mass and the experimental mass is reported. Because the instrument error in this experiment is expected to be ±5 ppm (within one standard deviation) only the composition, $C_{23}H_{23}NO_3P$, which is the expected one, must be considered for the $[MH]^+$ at m/z 392.

Detection Limits

The absolute intensity of the ion beam in high-resolution mode is lower than in a comparable experiment done in the low-resolution mode because the ion beam must be attenuated to achieve high resolution. Another limiting factor is that the sampling number goes down (number of averaged measurements for each data point) because the acquisition rate under high-resolution conditions is high. This problem can be overcome by scanning slowly or by using only the peaks of interest, such as in a peak matching experiment.

Sensitivity is measured as a ratio of charge per quantity of sample (coulombs/μg) or as a signal-to-noise (S/N) ratio, whereas detection limits are expressed as a quantity of sample (in picograms or picomols) that can be detected at a specified S/N (such as 3:1). In some ionization modes where the background is relatively low (such as EI), the detection limits under high-resolution conditions would be 10 to 100 times poorer than those in the low-resolution experiment. For other types of ionization such as FAB, the sensitivity is limited because the matrix background is high and, in

Figure 30.8 A typical elemental composition report.

Heteroatom max:	20	Ion:	Both even and odd							
Limits:										
389.384	5.0				−0.5	15	10	0	0	1
395.139	100.0	10.0			20.0	30	40	2	5	1

Mass	%RA	mDa	ppm	Calc. Mass	DBE	C	H	N	O	P
393.144445	24.3	−3.6	−9.2	393.140829	17.5	27	22		1	1
		4.9	12.6	393.149382	13.0	23	24	1	3	1
		7.6	19.4	393.152062	17.5	26	22	2		1
		−7.6	−19.4	393.136806	13.5	22	22	2	3	1
392.141371	100.0	0.2	0.5	392.141557	13.5	23	23	1	3	1
		2.9	7.3	392.144237	18.0	26	21	2		1
		−8.4	−21.3	392.133004	18.0	27	21		1	1
		8.7	22.3	392.150111	9.0	19	25	2	5	1
391.133771	72.5	0.0	−0.1	391.133732	14.0	23	22	1	3	1
		2.6	6.8	391.136412	18.5	26	20	2		1
		8.5	21.8	391.142286	9.5	19	24	2	5	1
		−8.6	−22.0	391.125179	18.5	27	20		1	1

DBE = Double bond equivalents

some cases, the detection limit can be actually improved by turning to a high-resolution experiment.

In FTMS, the difference between experiments at high resolution and low resolution is the greater time required to acquire high-resolution data, assuming the pressure is the same. An increase in the resolving power does not require a slit to be closed; the S/N for high resolution can be greater than that for low resolution. The conditions of the experiment, however, must be controlled carefully. Besides keeping the pressure as low as possible, the number of ions in the trap must be limited to decrease coulombic effects, and the acquisition time must be extended.

Applications

This section describes one application taken from the author's work in environmental chemistry. The application illustrates the use of exact mass measurements to identify an unknown compound discovered in a GC-HRMS experiment in which chlorinated dioxins and dibenzofurans were targeted.

The objective of the research was to determine whether chlorinated dioxins and dibenzofurans were accumulating in the tissue of crabs in the Newark Bay system of New York harbor. Crab tissue was chosen because crabs are bottom feeders, and it is known that dioxins, if present, will be found in the sediments of the harbor. Furthermore, crab tissue is consumed by recreational fishermen in the area of Newark Bay, and human health may be jeopardized if the levels are excessive (greater than a few tens of parts per trillion). Because the levels are expected to be below 1 ppb, the analytical method must have high sensitivity and high certainty. To achieve the latter, high-resolution mass spectometry was chosen as the detection method.

In all the samples of crab tissue that were analyzed, measurable levels of 2,3,7,8-tetrachlordibenzodioxin (TCDD) were detected by using GC-HRMS in the peak-profile monitoring mode. The levels are in the range of 40 to over 1000 parts per trillion.

In addition to 2,3,7,8-TCDD, another material was detected in all the samples. The material was first thought to be another isomer of TCDD because its retention time nearly matched that of 1,2,8,9-TCDD. It was clear that the material had an exact mass similar to that of TCDD because the detection was made in the high-resolution mode. Multiple ion monitoring at high resolving power (see Fig. 30.9(a)) shows that the level of the material is somewhat greater than that of 2,3,7,8-TCDD (its chromatographic profile is indicated by a question mark). When one looks at the mass profile of the ion peaks, however, it is immediately clear that the unknown material is not a TCDD (see Fig. 30.9(b)). The mass profiles of the unknown do not overlap with those of a TCDD, and from the centroids of the unknown peaks, the exact mass of the unknown material was determined. On the basis of the exact mass, the formula was determined to be $C_{12}H_4Cl_4S$, a compound in which the two oxygens of a dioxin are replaced by a single sulfur.

A material of this elemental composition (2,4,6,8-TCDD) had been previously synthesized for the Environmental Protection Agency and we were able to obtain a small sample. Its retention time, full mass spectrum, and exact masses are identical to those of the unknown. Its levels in various samples correlated with those of 2,3,7,8-TCDD, but the levels of the new material are 5 to 10 times of those of 2,3,7,8-TCDD. The origin of this material is not yet extablished, but its presence would have been extremely difficult to prove were it not for high-resolution mass spectrometry. A complete description of the experiment and findings was published (13).

Figure 30.9 GC-HRMS peak profile from a sample of crab tissue from Newark Bay. (a) Mass profile of TCDD; (b) mass profile of the unknown compound 2,4,6,8-TCDT, in comparison to that of 2,3,7,8-TCDD.

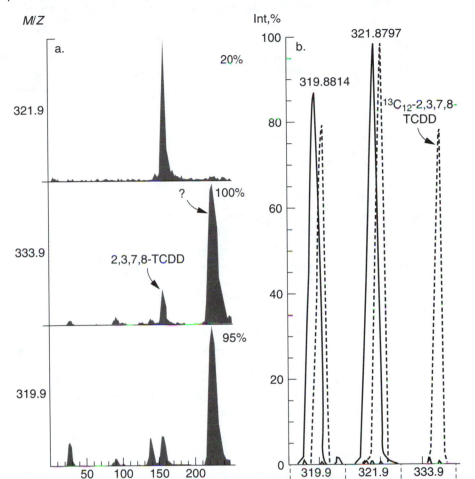

Nuts and Bolts

Relative Costs

A high-resolution sector mass spectrometer costs approximately $200,000 to over $500,000 (1995 U.S. dollars). At the time of this writing, little is known about the lowest-cost instrument because the first model was only recently announced by a Japanese company. It is likely that the instrument will be useful for EI and CI and will have a rather low upper mass limit and mass resolving power.

The second tier of instruments is priced around $300,000. These spectrometers have upper mass limits on the order of 600 and ultimate resolving powers of 10,000, are capable of EI and CI, and are interfaced to gas chromatographs. They are capable of very fast scanning (greater than 10 spectra per second) for GC-MS and are designed to support synthetic efforts where low-molecular-weight samples, which can be introduced by a direct solids probe, need confirmation by exact mass measurements.

The third tier of instruments is priced in the range of $400,000 to $450,000. These spectrometers have upper mass limits of approximately 3,000 and ultimate resolving powers of 60,000. Their scanning rates (approximately five spectra per second) are not quite as rapid as those of the second-tier instruments (the magnets are larger), but they are still adequate for GC-MS. This tier of instruments can be equipped with FAB ionization sources, interfaces to HPLC, and array detectors, all at an extra cost. Two instruments can also be coupled by the manufacturer to produce a tandem mass spectrometer, which is most useful in structural studies of molecules that are introduced by FAB.

The uppermost tier of instrument is priced at $500,000 to $550,000. The upper mass limit and the ultimate resolving power are now increased to 10,000 and 100,000, respectively. The features and upgrades of the instrument are very similar to those of the third-tier instruments. Performance specifications such as mass measurement accuracy and precision are improved.

Vendors for Instruments and Accessories

Four manufacturers worldwide offer double-focusing sector mass spectrometers: Finnigan-MAT of San Jose, CA (the sector mass spectrometers are manufactured in Germany); VG Analytical of Manchester, UK; JEOL of Tokyo, Japan; and Hitachi of Japan. Mass spectrometers from the latter company are no longer sold outside Japan.

A directory of companies that supply mass spectrometers, components, and services was recently published (14) and should be consulted for details on purchasing parts and arranging service. See also the list in Chap. 28 (p. 564). It is noted here that less expensive parts can often be found with vendors other than the original manufacturer.

Required Level of Training

Operation of Instrument

Operation of the instrument can be done by people with a high school education and significant training in electronics or chemistry. Preparing difficult samples and using introduction techniques such as GC or LC require a trained person with an undergraduate degree in science and extensive experience in mass spectrometry.

Processing Qualitative and Quantitative Data

Qualitative and quantitative interpretation of the results can be done by an experienced person with at least an undergraduate degree in chemistry. More advanced interpretation is best done by a person with an advanced degree in chemistry, preferably with emphasis in mass spectrometry.

An excellent means of continuing education is via national societies for mass spectrometry. In North America, for example, the American Society for Mass Spectrometry holds an annual meeting in late spring where hundreds of papers are presented and short courses are offered. Furthermore, members of the society receive the journal of the society as a benefit of membership.

Service and Maintenance

Some instrument servicing, including source replacement and cleaning, filament replacement, and pump oil replacement, can be done by the operator. Other problems can be solved by an operator with a background in electronics and vacuum systems, preferably in an engineering area. More serious problems require a service engineer. Regular source cleaning and pump maintenance are essential for good operation of the instrument. From time to time, other parts must be cleaned or baked out. These include lenses, flight tubes, and collision cells, if any are used.

Suggested Readings

ASAMOTO, B., ED., *FT-ICR/MS: Analytical Applications of Fourier Transform Ion Cyclotron Resonance Mass Spectrometry*. New York: VCH Publishers, 1991.

CHAPMAN, J. R., *Practical Organic Mass Spectrometry*. New York: Wiley, 1985.

GASKEL, S., ED., *Mass Spectrometry in Biomedical Research*. New York: Wiley, 1986.

GROSS, M. L., ED., *High Performance Mass Spectrometry: Chemical Applications*. Washington, D.C.: American Chemical Society, 1978.

GROSS, M. L., ED., *Mass Spectrometry in Biological Sciences: A Tutorial*. Boston, MA: Kluwer Academic Publishers, 1992.

HARRISON, A. G., *Chemical Ionization Mass Spectrometry*, 2nd ed. Boca Raton, FL: CRC Press, 1992.

LAMBERT, J. B., AND OTHERS, *Introduction to Organic Spectroscopy*. New York: Macmillan, 1987.

MATSUO, T., AND OTHERS, EDS. *Biological Mass Spectrometry Present and Future*. New York: Wiley, 1994.

MCCLOSKEY, J. A., ED., *Methods in Enzymology*, vol. 193. New York: Academic Press, 1990.

MCDOWELL, C. A., ED., *Mass Spectrometry*. New York: McGraw-Hill, 1963.

MCLAFFERTY, F. W., AND F. TURECEK, *Interpretation of Mass Spectra*, 4th ed. Mill Valley, CA: University Science Books, 1993.

WATSON, J. T., *Introduction to Mass Spectrometry*, 2nd ed. New York: Raven Press, 1985.

WHITE, F. A., AND G. M. WOOD, *Mass Spectrometry Application in Science and Engineering*. New York: Wiley, 1986.

References

1. J. J. Thomson, in *Rays of Positive Electricity and Their Application to Chemical Analysis* (London: Longmans, Green & Co., 1913).

2. J. Mattauch and R. Herzog, *Z. Physic*, 80 (1934), 786.

3. E. G. Johnson and A. O. Nier, *Phys. Rev.*, 91 (1953), 10.

4. J. Matsuo and H. Matsuda, *Int. J. Mass Spectrom. Ion Proc.*, 91 (1989), 27.

5. J. H. Beynon, in R. M. Elliott, ed., *Advances in Mass Spectrometry*, vol. 2 (London: Pergamon Press, 1962).

6. M. B. Comisarow and A. G. Marshall, *Chem. Phys. Lett.* 25 (1974), 282.

7. T. M. Sack and others, *Int. J. Mass Spectrom. Ion Proc.* 61 (1984), 191.

8. R. M. Milberg, in J. A. McCloskey, ed., *Methods in Enzymology*, vol. 193 (New York: Academic Press, 1990), p. 305, Appendix 2.

9. B. J. Kimble, in M. L. Gross, ed., *High Performance Mass Spectrometry: Chemical Application* (Washington D.C.: American Chemical Society, 1978), 120.

10. H. Y. Tong and others, *Analytical Chemistry*, 63 (1991), 1772.

11. E. B. Ledford, Jr., D. L. Rempel, and M. L. Gross, *Analytical Chemistry*, 56 (1984), 2744.

12. M. L. Gross, *J. Am. Soc Mass Spectrom.*, 5 (1994), 57; A. B. Giordany and P. C. Price, Proceedings of the 42nd ASMS Conference on Mass Spectrometry and Allied Topics, May 29–June 3,1994, Chicago; K. Biemann, *Methods in Enzymology*, vol. 193 (New York: Academic Press, 1990), 295.

13. Z. Cai and others, *Environ. Sci. and Technol.*, 28 (1994), 1535.

14. S. A. Lammert and R. G. Cooks, *Rapid Communications in Mass Spectrometry*, 6 (1992), 75.

Chapter 31

Gas Chromatography Mass Spectrometry

Ronald A. Hites

Indiana University
School of Public and Environmental Affairs
and Department of Chemistry

Summary

General Uses

- Identification and quantitation of volatile and semivolatile organic compounds in complex mixtures
- Determination of molecular weights and (sometimes) elemental compositions of unknown organic compounds in complex mixtures
- Structural determination of unknown organic compounds in complex mixtures both by matching their spectra with reference spectra and by a priori spectral interpretation

Common Applications

- Quantitation of pollutants in drinking and wastewater using official U.S. Environmental Protection Agency (EPA) methods
- Quantitation of drugs and their metabolites in blood and urine for both pharmacological and forensic applications

- Identification of unknown organic compounds in hazardous waste dumps
- Identification of reaction products by synthetic organic chemists
- Analysis of industrial products for quality control

Samples

State

Organic compounds must be in solution for injection into the gas chromatograph. The solvent must be volatile and organic (for example, hexane or dichloromethane).

Amount

Depending on the ionization method, analytical sensitivities of 1 to 100 pg per component are routine.

Preparation

Sample preparation can range from simply dissolving some of the sample in a suitable solvent to extensive cleanup procedures using various forms of liquid chromatography.

Analysis Time

In addition to sample preparation time, the instrumental analysis time usually is fixed by the duration of the gas chromatographic run, typically between 20 and 100 min. Data analysis can take another 1 to 20 hr (or more) depending on the level of detail necessary.

Limitations

General

Only compounds with vapor pressures exceeding about 10^{-10} torr can be analyzed by gas chromatography mass spectrometry (GC-MS). Many compounds with lower pressures can be analyzed if they are chemically derivatized (for example, as trimethylsilyl ethers). Determining positional substitution on aromatic rings is often difficult. Certain isomeric compounds cannot be distinguished by mass spectrometry (for example, naphthalene versus azulene), but they can often be separated chromatographically.

Accuracy

Qualitative accuracy is restricted by the general limitations cited above. Quantitative accuracy is controlled by the overall analytical method calibration. Using isotopic internal standards, accuracy of ±20% relative standard deviation is typical.

Sensitivity and Detection Limits

Depending on the dilution factor and ionization method, an extract with 0.1 to 100 ng of each component may be needed in order to inject a sufficient amount.

Complementary or Related Techniques

- Infrared (IR) spectrometry can provide information on aromatic positional isomers that is not available with GC-MS; however, IR is usually 2 to 4 orders of magnitude less sensitive.
- Nuclear magnetic resonance (NMR) spectrometry can provide detailed information on the exact molecular conformation; however, NMR is usually 2 to 4 orders of magnitude less sensitive.

Introduction

Like a good marriage, both gas chromatography (GC; see Chapter 8) and mass spectrometry (MS; see Chapter 30) bring something to their union. GC can separate volatile and semivolatile compounds with great resolution, but it cannot identify them. MS can provide detailed structural information on most compounds such that they can be exactly identified, but it cannot readily separate them. Therefore, it was not surprising that the combination of the two techniques was suggested shortly after the development of GC in the mid-1950s.

Gas chromatography and mass spectrometry are, in many ways, highly compatible techniques. In both techniques, the sample is in the vapor phase, and both techniques deal with about the same amount of sample (typically less than 1 ng). Unfortunately, there is a major incompatibility between the two techniques: The compound exiting the gas chromatograph is a trace component in the GC's carrier gas at a pressure of about 760 torr, but the mass spectrometer operates at a vacuum of about 10^{-6} to 10^{-5} torr. This is a difference in pressure of 8 to 9 orders of magnitude, a considerable problem.

How It Works

The Interface

The pressure incompatibility problem between GC and MS was solved in several ways. The earliest approach, dating from the late 1950s, simply split a small fraction of the gas chromatographic effluent into the mass spectrometer (1). Depending on the pumping speed of the mass spectrometer, about 1 to 5% of the GC effluent was split off into the mass spectrometer, venting the remaining 95 to 99% of the analytes into the atmosphere. It was soon recognized that this was not the best way to maintain the high sensitivity of the two techniques, and improved GC-MS interfaces were designed (2). These interfaces reduced the pressure of the GC effluent from about 760 torr to 10^{-6} to 10^{-5} torr, but at the same time, they passed all (or most) of the analyte molecules from the GC into the mass spectrometer. These interfaces were no longer just GC carrier gas splitters, but carrier gas separators; that is, they separated the carrier gas from the organic analytes and actually increased the concentration of the organic compounds in the carrier gas stream.

The most important commercial GC carrier gas separator is called the jet separator; see Fig. 31.1 (3). This device takes advantage of the differences in diffusibility between the carrier gas and the organic compound. The carrier gas is almost always a small molecule such as helium or hydrogen with a high diffusion coefficient, whereas the organic molecules have much lower diffusion coefficients. In operation, the GC effluent (the carrier gas with the organic analytes) is sprayed through a small nozzle, indicated as d_1 in Fig. 31.1, into a partially evacuated chamber (about 10^{-2} torr). Because of its high diffusion coefficient, the helium is sprayed over a wide solid angle, whereas the heavier organic molecules are sprayed over a much narrower angle and tend to go straight across the vacuum region. By collecting the middle section of this solid angle with a skimmer (marked d_3 in Fig. 31.1) and passing it to the mass spectrometer, the higher-molecular-weight organic compounds are separated from the carrier gas, which is removed by the vacuum pump. Most jet separators are made from glass by drawing down a glass capillary, sealing it into a vacuum envelope, and cutting out the middle spacing (marked d_2 in Fig. 31.1). It is important that the spray orifice and the skimmer be perfectly aligned.

These jet separators work well at the higher carrier gas flow rates used for packed GC columns (10 to 40 mL/min); however, there are certain disadvantages. Packed GC columns are an almost infinite source of small particles upstream of the jet separator. If one of those particles escapes from the column, it can become lodged in the spray orifice and stop (or at least severely reduce) the gas flow out of the GC column and into the mass spectrometer. Part of this problem can be eliminated with a filter between the GC column and the jet separator, but eventually a particle will plug up the orifice. In fact, sometimes it is not a particle at all, but rather tar (mostly pyrolyzed GC stationary phase) that has accumulated in the spray orifice over time. Clearly, these devices require maintenance.

Figure 31.1 The jet separator, a device for interfacing a packed column GC with an MS. The three distances are typically d_1, 100 µm; d_2, 300 µm; and d_3, 240 µm.

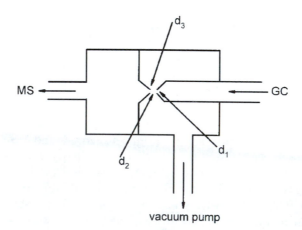

Currently, the most common strategy, which is ideally suited for capillary GC columns, is to pass all of the carrier gas flow into the ion source of the mass spectrometer (4). This works only if the GC gas flow is sufficiently small and the pumping speed of the mass spectrometer's vacuum system is sufficiently high to handle the gas flow. For most capillary GC columns, the gas flow is 1 to 2 mL/min, and for most modern mass spectrometers, the pumping speed is at least 300 L/sec. The development of flexible, fused silica capillary columns has made this approach routine. In fact, the only time a jet separator is now used is for a few applications that require packed or thick stationary phase GC columns (for example, for permanent gas analysis).

In practice, most GC-MS interfacing is now done by simply inserting the capillary column directly into the ion source. Fig. 31.2 is a diagram of one such system. The fused silica column runs through a 1/16-in.-diameter tube directly into the ion source. Other gases, such as methane for chemical ionization, are brought into the ion source by a T joint around the capillary column. One of the other two lines into the ion source is used for a thermocouple vacuum gauge tube so that the pressure in the ion source can be roughly measured. The remaining line into the ion source is for the delivery of the mass spectrometer calibration standard, perfluorotributylamine. Most joints are welded together to avoid leaks when this inlet system is thermally cycled or vented. The only removable (Swagelok) fitting is at the junction of the GC column and the far end of the inlet tube (marked with an asterisk in Fig. 31.2). This fitting uses Vespel ferrules. Once the ferrules are on the GC column and it is in the ion source, it is desirable to cut off a few centimeters of the column, if possible. This eliminates the possibility of fine particles partially occluding the end of the column.

Figure 31.2 A typical GC-MS interface for fused silica capillary GC columns. The end of the GC column enters the ion source of the mass spectrometer.

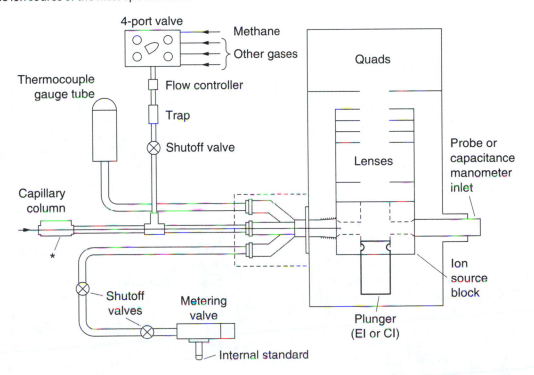

If the end of the column cannot be placed directly in the ion source, the material in the GC-MS interface becomes important. The interface is held at 250 to 280 °C; thus, it should not include a reactive metal (such as copper). In some interfaces, glass-lined stainless steel tubing has been used, even though this tubing is difficult to bend properly.

In summary, for capillary GC-MS, the best interface is no interface at all; run the flexible, fused silica GC column directly into the ion source. Using a column that is 25 to 30 m long by 220 to 250 µm inner diameter gives an ion source pressure of 10^{-6} to 10^{-5} torr, a more than acceptable pressure at which to obtain electron impact spectra. This gives a helium or hydrogen GC carrier gas velocity of 25 to 35 cm/sec or a flow of about 1 to 2 mL/min. The GC columns most widely used for GC-MS are those in which the stationary phase has been chemically bonded to the fused silica; DB-5 is a common trade name. Occasionally, there have been problems with the plastic cladding on the outside of the GC column. This cladding is usually hot (typically 250 °C) and under vacuum. Thus, it may decompose, giving background ions in the mass spectrum or weakening the fused silica itself.

The Data System

The amount of data that can be produced during one GC-MS experiment is overwhelming. In a typical GC-MS experiment, the mass spectrometer might be scanned every 2 sec during a 90-min GC run, whether GC peaks are entering the mass spectrometer or not. Assuming that each mass spectrum has an average of 100 mass/intensity measurements, one such GC-MS experiment will give 270,000 mass/intensity pairs. Because these data have several significant figures and because other ancillary data are also obtained, the data output from a typical GC-MS experiment is about 1 megabyte. To manage this high data flow, computers are required; thus, it is virtually impossible to purchase a GC-MS system without a powerful (but small) computer acting as a data system.

How do data systems work? Two things are going on at the same time (5). There are two different rates within the system. There is a slow rate that times the start and stop of the mass spectrometer scan. This is usually set such that 10 to 15 mass spectra are obtained across a typical GC peak. Because these peaks are usually on the order of 20 to 30 sec wide, the mass spectrometer scan speed is usually set at 2 to 3 sec per spectrum. While this scan is going on, the computer must read the output of the electron multiplier at a rate fast enough to define the mass peak profile. In most commercial GC-MS data systems, the voltage output from the preamplifier on the electron multiplier is converted from an analog signal to a digital value (using an analog-to-digital converter) at a rate of 10,000 to 100,000 times per sec. This process generates large amounts of data: If the analog-to-digital converter worked at 10,000 conversions/second, each minute of the GC-MS experiment would generate 600,000 numbers. This would quickly fill most bulk storage devices; thus, to avoid saving all of these data, most data systems find the mass peaks in real time and convert them into mass intensity pairs, which are then stored on the computer's hard disk. Once the most recent mass spectral scan is stored, this cycle is repeated until the end of the gas chromatogram is reached. Each of the spectra stored on the hard disk has a retention time associated with it, which can be related directly to the gas chromatogram itself. The latter is usually reconstructed by the GC-MS data system by integrating the mass spectrometer output. All modern GC-MS data systems are capable of displaying the mass spectrum on the computer screen as a bar plot of normalized ion abundance versus mass-to-change (m/z) ratio (often called mass). Like the other parts of the GC-MS instrument, the data system must be calibrated. Typically this is done by running a standard compound, such as perfluoro-tributylamine.

What It Does

Gas chromatographic mass spectrometry is the single most important tool for the identification and quantitation of volatile and semivolatile organic compounds in complex mixtures. As such, it is very useful for the determination of molecular weights and (sometimes) the elemental compositions of unknown organic compounds in complex mixtures. Among other applications, GC-MS is widely used for the quantitation of pollutants in drinking and wastewater. It is the basis of official EPA methods. It is also used for the quantitation of drugs and their metabolites in blood and urine. Both pharmacological and forensic applications are significant. GC-MS can be used for the identification of unknown organic compounds both by matching spectra with reference spectra and by a priori spectral interpretation. The identification of reaction products by synthetic organic chemists is another routine application, as is the analysis of industrial products for control of their quality.

To use GC-MS, the organic compounds must be in solution for injection into the gas chromatograph. The solvent must be volatile and organic (for example, hexane or dichloromethane). Depending on the ionization method, analytical sensitivities of 1 to 100 pg per component are routine. Sample preparation can range from simply dissolving some of the sample in a suitable solvent to extensive cleanup procedures using various forms of liquid chromatography. In addition to the sample preparation time, the instrumental analysis time is usually fixed by the duration of the gas chromatographic run, typically between 20 and 100 min. Data analysis can take another 1 to 20 hr (or more) depending on the level of detail necessary.

GC-MS has a few limitations. Only compounds with vapor pressures exceeding about 10^{-10} torr can be analyzed by GC-MS. Many compounds that have lower pressures can be analyzed if they are chemically derivatized (for example, as trimethylsilyl ethers). Determining positional substitution on aromatic rings is often difficult. Certain isomeric compounds cannot be distinguished by mass spectrometry (for example, naphthalene versus azulene), but they can often be separated chromatographically. Quantitative accuracy is controlled by the overall analytical method calibration. Using isotopic internal standards, accuracy of ±20% relative standard deviation is typical.

Mass Spectrometer Components

Electron ionization (Chapter 30) is most commonly used to produce ions from the compounds separated by the GC. Chemical ionization may also be used. Quadrupole (p. 656), ion trap (p. 656), and time-of-flight analyzers may be used to separate ions in the MS. These analyzers have rapid response times and relatively low costs.

Analytical Information

GC-MS is used both for the qualitative identification and for the quantitative measurement of individual components in complex mixtures. There are different data analysis strategies for these two applications.

Qualitative

There are three ways of examining GC-MS data. First, the analyst can go through the gas chromatogram (as reproduced by the mass spectrometer) and look at the mass spectra scanned at each GC peak maximum. This has the advantage of being relatively quick but the disadvantage of missing components of the mixture that are not fully resolved by the GC column. The second approach is to look at each mass spectrum in turn, in essence stacking up the mass spectra one behind the other and examining them individually. This has the advantage of completeness but the disadvantage of tedium. The third approach is to look at the intensity of one particular mass as a function of time.

This third approach makes use of the three-dimensional nature of GC-MS data. Two of these dimensions are the mass versus intensity of the normal mass spectrum; the third dimension is the GC retention time over which the mass spectral data are acquired. This idea is illustrated in Fig. 31.3. The x-axis represents GC retention time, the y-axis represents intensity, and the z-axis represents mass (or more properly, m/z ratios). As shown in Fig. 31.3, a mass spectrum is displayed in the y–z plane. Because a mass spectrum is scanned every 1 to 3 sec, it is also possible to examine the data in the x–y plane. This is a plot of the intensity of one selected mass as a function of time. This plot is called a mass chromatogram (6).

Figure 31.3 A diagram demonstrating the three-dimensional nature of GC-MS data. The abbreviated mass spectrum extending onto the foreground is that of methyl stearate; the mass chromatogram of m/z 298 (the molecular ion of methyl stearate) is also shown.

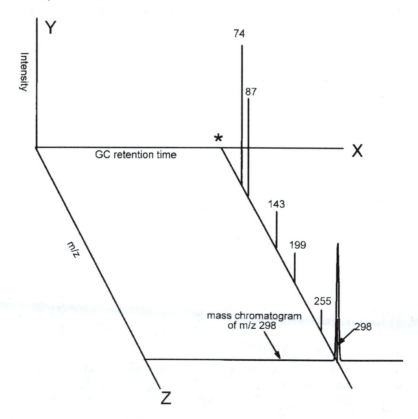

An example may make this concept clear. At the retention time marked with an asterisk in Fig. 31.3, the abbreviated mass spectrum extending into the foreground (the y–z plane) was observed. This happens to be the mass spectrum of methyl stearate; note the relatively abundant ion at m/z 298, which is this compound's molecular weight. The mass chromatogram of m/z 298 is shown in the x–y plane in Fig. 31.3. Note that this mass chromatogram shows one peak, which corresponds to the retention time of methyl stearate. In other words, of all the hundreds of mass spectra taken during this GC-MS experiment, m/z 298 is present in only a very few spectra. Only compounds with m/z 298 in their mass spectra will show up in the mass chromatogram of that mass.

Mass chromatograms can be thought of as a very selective gas chromatographic detector, in this case, one that responds only to methyl stearate. Other compounds can be selectively detected by picking other masses. For example, m/z 320 would be a good mass to use for the selective detection of a tetrachlorodibenzo-p-dioxin because this is its molecular weight. If a mass that is present in the mass spectra of a class of compounds is selected, that compound class can be selectively detected. For example, m/z 149 is present in the mass spectra of alkyl phthalates for alkyl chain lengths greater than two carbon atoms. Thus, the mass chromatogram of m/z 149 would selectively show all the phthalates in a sample.

Mass chromatograms are also useful for determining whether a given mass belongs in a given mass spectrum. For example, if the liquid phase from a GC column is beginning to thermally decompose, all of the mass spectra taken during a GC-MS experiment with that column might show a moderately abundant ion at m/z 207. However, the mass chromatogram of m/z 207 will not show peaks because the source of this ion is bleeding continuously from the column and is not a discrete compound. In fact, the mass chromatogram of 207 will probably track the temperature program of the GC column; see trace (a) in Fig. 31.4. Using this approach, it is easy to distinguish between the ions that really belong in a given mass spectrum and those from background. By looking at sets of mass chromatograms, it is possible to determine whether various ions come from the same compound even if the compounds are not completely resolved by the GC column. If ions belong to-

Figure 31.4 Hypothetical mass chromatograms of four masses. (a) A background ion from the GC column bleed (m/z 207, for example); (b) an ion from a later-eluting compound; (c, d) two ions from an earlier-eluting compound. Note the offset between (b) and (c) or (d).

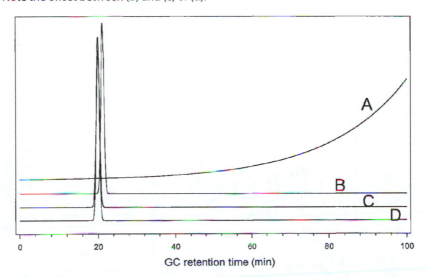

gether (that is, they come from the same GC peak), the mass chromatograms for all these ions will be superimposable in time; see traces (c) and (d) in Fig. 31.4. In fact, these mass chromatograms should all have the same peak shape because they all came from the same GC peak. If the mass chromatograms are not superimposable in time or in shape, the corresponding ions are probably from different compounds, which may have come out of the GC column at slightly different retention times; compare trace (b) to traces (c) or (d) in Fig. 31.4.

Because GC-MS experiments are somewhat complicated, there is always a possibility for something to go wrong. To prevent this, stringent quality assurance procedures are necessary. The following is a nonexhaustive list of some of these problems and what can be done about them.

First, because the analyst is often working at ultra trace levels (a few nanograms, for example), it is possible for a compound that was not originally in the sample to sneak in during the analytical procedure. In particular, sample contamination can come from solvents and glassware. The former problem can be prevented by using high-quality (and expensive) solvents, the latter by heating the glassware to 450 °C after solvent and acid washing. The most common contaminant is probably di(2-ethylhexyl)phthalate. Its mass spectrum is shown in Fig. 31.5. Note the important ions at m/z 149, 167, and 279. It pays to remember this spectrum; it was once published as that of a natural product (7). Phthalates are extremely common as plasticizers. They are particularly abundant (5 to 20%) in polyvinylchloride-based plastic products such as Tygon tubing.

Second, if components in the sample decompose before or after workup, the analyst will not obtain accurate results. Under these conditions, it is possible to identify (and quantitate) a compound that was not originally in the sample, or the analyte of interest could have vanished from the sample. Thus, both false positives and false negatives can result from sample decomposition. This can happen while the sample is waiting to be analyzed or during the analysis itself. For example, a GC injection port held at 250 to 300 °C can cause thermal decomposition of some compounds. One useful procedure is to add (or spike) the analyte into a sample at a known

Figure 31.5 Mass spectrum of di(2-ethylhexyl)phthalate, a very common experimental contaminant. The structures of some ions are shown.

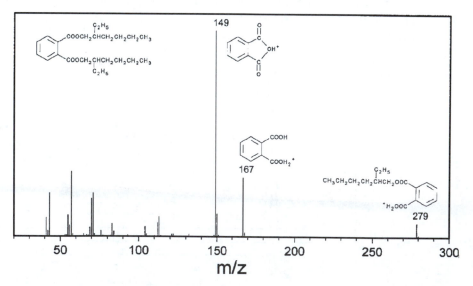

concentration. If there is a substantial loss of this compound or the suspicious formation of another compound, sample decomposition may be a problem.

Third, if the GC column or GC-MS interface is not working properly, the whole GC-MS experiment is in jeopardy. Cold spots are a common problem, as are catalytic surfaces that selectively remove some compounds from the GC gas stream. These problems can be identified using a mixture of standard compounds of varying polarities and acidities.

Fourth, either the mass spectrometer itself or the data system may not be working properly. In this case, incorrect isotope ratios, mass discrimination (ions at higher masses appear less abundant than they should), or mass assignment errors could occur. The key to identifying these problems is to run an overall mass spectrometer performance standard. The one recommended (mandated in many cases) by the EPA is decafluorotriphenylphosphine, the mass spectrum of which is shown in Fig. 31.6. This compound is a good standard. It is easy to run by GC-MS, and it has ions up to about m/z 450. The abundances for the various ions, as required by the EPA, are published in the Federal Register (8) and in various other official EPA methods (9). These requirements change from time to time, but they are available from the EPA.

To ensure qualitative identification of an organic compound using GC-MS, several criteria should be met: First, the mass spectra of the unknown compound and of the authentic compound must agree over the entire mass range of the spectra. It is particularly important to compare the patterns within narrow mass ranges (for example, from m/z 50 to 60 in Fig. 31.5); these patterns should agree almost exactly. In this case, the spectrum of the authentic compound might come from a library of reference spectra or from the actual compound itself. In the latter case, the compound could be purchased or synthesized. Second, the GC retention times of the unknown compound and of the authentic compound must agree within about ±1 to 2 sec. It is often convenient to do this experiment by coinjecting the unknown mixture and the authentic compound. The GC peak in question should increase in size by the correct factor. Third, a compound cannot be considered fully identified in a mixture unless two other questions are addressed: Is the identification

Figure 31.6 Mass spectrum of decafluorotriphenylphosphine, an EPA-mandated standard. The structures of some ions are shown.

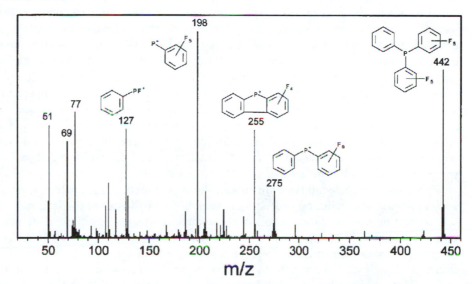

plausible? Why is it present in a given sample? If an identification is implausible or if there is no reason for a compound to be present in a sample, the identification could be wrong or the compound could be a contaminant.

Quantitative

GC-MS can also be used to measure the concentration of one or more analytes in a complex mixture. Quantitation can be based on peak areas from mass chromatograms or from selected ion monitoring. The latter requires more explanation (10).

With the selected ion monitoring technique, the mass spectrometer is not scanned over all masses; instead, the instrument jumps from one selected mass to another. The advantage of this approach is that the mass spectrometer spends much more time at a given mass, the signal-to-noise ratio at that mass improves, and the overall sensitivity of the experiment increases by a factor of 100 to 1000. For example, the mass spectrometer might integrate for 500 msec at mass m_1, jump to mass m_2 in 10 msec, integrate for 500 msec at that mass, and jump back to m_1 in 10 msec, repeating this cycle for the duration of the GC run. In essence, the intensities of the two masses are recorded as a function of GC retention time, with measurements made every 1.02 sec. In practice, rather than only two masses, 5 to 10 masses are usually monitored simultaneously for 100 to 200 msec each. In this manner, the GC-MS response of 5 to 10 compounds, depending on the specificity of the selected masses, can be measured.

The difference between mass chromatograms and selected ion monitoring is significant. With the latter technique, the responses from only a few preselected masses are recorded. With mass chromatograms, all of the masses are scanned; thus, no preselection is required. This is the necessary tradeoff for higher sensitivity. Clearly, the mass spectrum of the analyte must be known so that the masses that uniquely characterize it can be selected. This information can be obtained from the literature (or from a library of reference spectra) or from the laboratory. Each selected set of masses can be monitored for the duration of the complete GC run or for only selected GC retention times (often called time windows). Selected ion monitoring is almost fully software driven; thus, it is very flexible. Different sets of masses, different time windows, and different integration times can be easily set up.

To convert the peak areas to mass of analyte, whether from mass chromatograms or from selected ion monitoring, the peak areas must be calibrated. The two main strategies are based on external and internal standards. With external standards, the area of one or more mass chromatogram is calibrated with a known amount of the analyte injected into the GC-MS in a different experiment. Detection limits of a few nanograms can be achieved with this technique. However, the strategy that gives the most accurate quantitative results is the use of internal standards, which are known amounts of compounds added to the sample before isolation of the analytes begins. After sample extraction and cleanup, only the ratio of response between the analyte and the internal standard must be measured. This ratio multiplied by the amount of the internal standard gives the amount of the analyte injected into the GC-MS system. This can be converted to concentration using the correct dilution factors.

The best internal standards are chemically very similar to the analyte; thus, any losses of the analyte during the analytical procedure are duplicated by losses of the internal standard, so it is a self-correcting system. Homologues of the analyte can be used as internal standards, but the very best are isotopically labeled versions of the analyte. Using isotopically labeled standards and selected ion monitoring, it is possible to get sensitivities of less than 1 pg. Depending on the relationship of the internal standard to the analyte, the precision and accuracy of most analyses are improved by at least a factor of 2 to 3 over external calibration. The tradeoff is complexity and cost.

Ideally, an internal standard for each analyte in a mixture should be used, and isotopic standards can sometimes cost several hundred dollars for a few milligrams.

Applications

The analysis of octachlorodibenzo-p-dioxin (OCDD) in sediment from Lake Ontario is a useful example of a quantitative measurement made with GC-MS. This example also demonstrates the use of isotopically labeled internal standards. These experiments were done in the author's laboratory using electron capture negative ionization, but the principle is the same regardless of the ionization technique selected.

OCDD has a molecular weight of 456 (usually called M). The mass spectrum of the unlabeled (native) compound is dominated by an ion cluster corresponding to M-Cl, the most intense peak of which is m/z 423, which is the first isotope peak (the one containing one ^{37}Cl) in this cluster. The isotopic standard used for this experiment was per-^{37}Cl-labeled OCDD, which has a molecular weight of 472. Its M-Cl ion is at m/z 435. Because there is no ^{35}Cl in this molecule, there are no isotope peaks. Selected ion monitoring of m/z 423 and 435 was used for the measurements.

The first step in this procedure is to calibrate the internal standard against a known amount of native OCDD. This calibration results in a response factor relating the response of the native compound to the labeled compound. Standard solutions of each were prepared such that 1 μL had 85 pg of native and 40 pg of labeled OCDD. Injection of 1 μL of this standard gave the selected ion monitoring results shown in Fig. 31.7(a); the areas of the two peaks are given as values of A in the figure. The response factor is

$$\text{Response factor} = \frac{\text{Area}_{423}/\text{pg Native OCDD}}{\text{Area}_{435}/\text{pg Labeled OCDD}} \tag{31.1}$$

In this case, the response factor is (16,581/85 pg) divided by (24,073/40 pg), which is 0.32. This is what it should be, given equimolar responses of the two compounds and given their difference in isotopic composition.

The selected ion monitoring data for an unknown sample are shown in Fig. 31.7(b). In this case, 1000 pg of the labeled OCDD was added to 1.8 g (dry weight) of Lake Ontario sediment before extraction. After extensive cleanup on silica and alumina, two clean GC-MS peaks were obtained with the areas shown in the figure. The calculation of the final concentration of OCDD in the sediment proceeds as follows: First, the amount of the internal standard in the sediment is divided by the weight of dry sediment; in this case, 1000 pg/1.8 g = 560 pg/g. Second, a corrected area of the native OCDD is calculated by dividing the area of this peak by the response factor; in this case, this corrected area is 37,011/0.32 = 115,660. (Note that these areas have consistent but arbitrary units.) Third, the ratio of the native and labeled areas is 115,660/12,569 = 9.2. This is the factor by which the concentration of native OCDD exceeds the labeled compound. Fourth, the concentration of the labeled material is multiplied by this factor to obtain a final concentration of native OCDD in the sediment. This value is 9.2 × 560 pg/g = 5100 pg/g. This measurement was part of a larger study that indicated that the major source of polychlorinated dibenzo-p-dioxins and dibenzofurans to the atmosphere was the combustion of municipal or chemical waste rather than coal (11).

A qualitative application has been selected from the work of a colleague in the author's department who has been carrying out studies of the electrochemical reduction of phthalide (12); see Fig.

Figure 31.7 Selected ion monitoring records of m/z 423 (from native octachlorodibenzo-p-dioxin) and m/z 435 (from $^{37}Cl_8$-octachlorodibenzo-p-dioxin) showing quantitation by the isotopic internal standard method. Only the gas chromatographic retention time range between 62 and 69 min is shown. The values of A are the areas of the peaks, in arbitrary units.

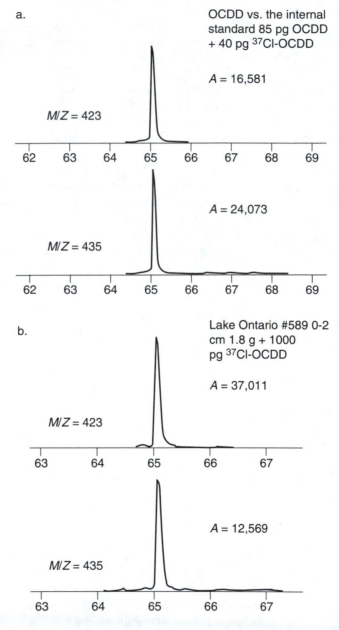

31.8 for all structures and data. Dimethylformamide was the solvent and tetra-n-butylammonium perchlorate was the supporting electrolyte. Products derived from the electrolysis were analyzed by GC-MS, and the gas chromatogram and mass spectra shown in Fig. 31.8 were obtained. The peak at 18.16 min was due to phthalide, and the peak at 18.34 min was due to n-tetradecane, which had been added in a known amount as an internal standard for purposes of quantitating the products. The identities of the other two peaks were not immediately clear.

Figure 31.8 Gas chromatogram of a reaction mixture from the electrochemical reduction of phthalide. The solvent was dimethylformamide and the supporting electrolyte was tetra-n-butylammonium perchlorate. The two compounds at 15.21 and 19.40 min were unknown; their mass spectra are shown.

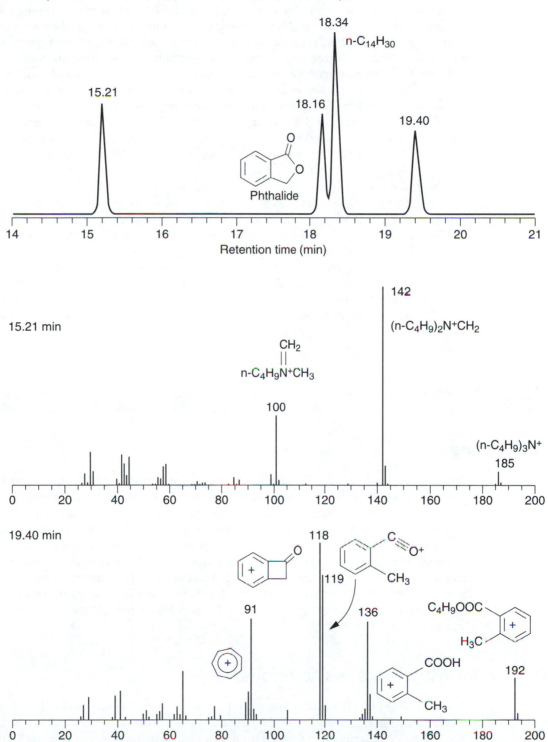

The mass spectrum of the peak at 15.21 min (see Fig. 31.8, middle) indicates that this compound has a molecular weight of 185 Da; because this is an odd number, this compound probably has at least one nitrogen atom. There is a substantial loss of 43 Da to give the ion at m/z 142, and there is also a major loss of 42 Da to give the ion at m/z 100. These two losses suggest the presence of at least two C_3H_7 moieties. Remembering that the supporting electrolyte was tetra-n-butylammonium perchlorate, it was reasonable to suggest that this peak was tri-n-butylamine. This assignment was verified by comparison of this mass spectrum and gas chromatographic retention time with those of authentic material. The structures of the ions at m/z 100, 142, and 185 are suggested on the mass spectrum; however, it is important to remember that these ion structures are a result of the interpretation of the spectrum and are not produced by the data system.

The mass spectrum of the peak at 19.40 min (see Fig. 31.8, bottom) indicates that this compound has a molecular weight of 192 Da. There was a substantial loss of 56 Da to give the ion at m/z 136. The ion at m/z 91 was probably a tropylium ion ($C_7H_7^+$), and this suggested that this molecule was derived from phthalide. The meaning of the ions at m/z 118 and 119 was not obvious, but the difference between 91 and 119 suggested the presence of a carbonyl group (28 Da) in this molecule. Putting these ideas together and noting the components of the reaction mixture, it was hypothesized that this compound was butyl 2-methylbenzoate. Again, the structures of the major ions are suggested on the mass spectrum. The validity of the identification of this GC peak was verified by comparison of the mass spectrum and gas chromatographic retention time with those of the authentic compound. Apparently, this compound results from a reaction between an electrogenerated intermediate (radical anion of phthalide) and the tetra-n-butylammonium cation (of the supporting electrolyte). Moreover, there is reason to believe that the process leading to butyl 2-methylbenzoate actually occurs in the heated injection port of the gas chromatograph.

Nuts and Bolts

Relative Costs

Although in principle GC-MS experiments can be performed on magnetic sector instruments, in practice almost all GC-MS today is done on quadrupole or ion trap instruments. These instruments are relatively inexpensive and are simple to control by a computer. The major factor influencing the cost of a quadrupole- or ion-trap–based GC-MS system is the ionization methods available on the instrument and the mass range of the mass spectrometer. Simple quadrupole or ion trap instruments that use only electron impact ionization and have a mass range of 20 to 700 cost about $50,000. Those capable of both positive and negative chemical ionization and with mass ranges of 20 to 2000 cost about $200,000. Operating costs include instrument maintenance, GC carrier gases and columns, and spare parts. In most laboratories, these costs are about 5% of the instrument cost per year.

Vendors for Instruments and Accessories

The following list is not exhaustive; several smaller companies enter the field each year and several leave. The following are some of the larger companies that deal with complete GC-MS systems. See Chapter 28 (p. 564) for a detailed list of vendors, including addresses for the following vendors.

The Finnigan Corp. (San Jose, CA) sells several instruments, some based on traditional quadrupole technology, some based on ion trap technology, some based on triple quadrupole technology, and a few based on magnetic sector technology. Various ionization methods are available.

Micromass UK (Manchester, U.K.) also has a wide range of instruments available with a wide range of ionization methods. Some are quadrupole based; some are magnetic sector based.

The Hewlett-Packard Corp. (Palo Alto, CA) markets several quadrupole-based instruments ranging from small benchtop instruments designed for the chromatographer to versatile, stand-alone instruments that can accommodate both gas and liquid chromatographic inlet systems.

Varian Associates, Inc. (Walnut Creek, CA) sells ion-trap–based GC-MS systems. These are typically small instruments with a good price-to-performance ratio.

With all of these companies, the recent trend has been to produce smaller and smaller instruments. Thirty years ago a GC-MS system with its data system occupied a whole room (or even two). Now these systems fit on the top of a small bench or table. Not only does this trend save space, but it saves manufacturing costs, some of which are passed on to the instrument purchaser. These smaller instruments also have fewer parts, making them less costly to maintain. There is also a trend toward increasing automation. Modern instruments are often equipped with an automatic injection system, and once filled with samples, the data system can control all functions of the instrument including sample introduction. Thus, once a trained person has developed the methodology and set up the data system, the instrument will almost run itself.

Required Level of Training

The required level of training and expertise varies as a function of the level of data interpretation. At the simplest level, because of the computer interface, most GC-MS instruments can be operated by people with no formal training in mass spectrometry or chemistry. A high school education is often sufficient. Maintenance of the instrument requires some mechanical and electronic skills, but again no formal training in mass spectrometry or chemistry is needed. For interpretation of the data, some chemistry training is needed, particularly organic chemistry. Many graduates of high-quality undergraduate programs in chemistry and most graduates of graduate programs in analytical or organic chemistry acquire these skills through their course work. Given a normal undergraduate course sequence in organic chemistry, most technically trained people can acquire specific training in mass spectrometry through 1- to 2-week courses offered through professional societies (such as the American Chemical Society or the American Society for Mass Spectrometry).

Service and Maintenance

Unlike most other spectrometers, in which radiation is passed through the sample, with mass spectrometry, the sample is inserted directly into the instrument. Thus, these instruments require more care than most others. The analyst should expect to clean the ion source every 2 to 4 mo and change the GC column every 3 to 6 mo. The instrument is electronically complex but highly modularized; thus, most electronic failures are corrected by replacement of a printed circuit board. The data systems are relatively hardy and are furnished with diagnostic software for both the computer and the mass spectrometer. Downtime for most modern instruments should be less than 5 to 10%, and maintenance costs should be less than 5% of the instrument cost per year.

Suggested Readings

CHAPMAN, J. R., *Practical Organic Mass Spectrometry*, 2nd ed. New York: Wiley, 1993.

HITES, RONALD A., *Handbook of Mass Spectra Environmental Contaminants*, 2nd ed. Boca Raton, FL: Lewis Publishers, 1992.

KARASEK, FRANCIS W., AND RAY E. CLEMENT, *Basic Gas Chromatography–Mass Spectrometry: Principles & Techniques*. Amsterdam: Elsevier, 1988.

MCLAFFERTY, FRED W., *Registry of Mass Spectral Data*, 5th ed. New York: Wiley, 1989.

MCLAFFERTY, FRED W., *Registry of Mass Spectral Data with Structures* (CD-ROM), 5th ed. New York: Wiley, 1989.

MCLAFFERTY, FRED W., AND FRANTISEK TURECEK, *Interpretation of Mass Spectra*, 4th ed. Mill Valley, CA: University Science Books, 1993.

NATIONAL INSTITUTE OF STANDARDS AND TECHNOLOGY, *NIST/EPA/NIH Mass Spectral Library for Windows*™ (61.4 megabytes). Gaithersburg, MD: NIST Standard Reference Data, 1995.

WATSON, J. THROCK, *Introduction to Mass Spectrometry*, 2nd ed. New York: Raven Press, 1985.

References

1. R. S. Gohlke, *Analytical Chemistry*, 31 (1959), 535–41.

2. J. T. Watson and K. Biemann, *Analytical Chemistry*, 37 (1965), 844–51.

3. R. Ryhage, *Analytical Chemistry*, 36 (1964), 759–64.

4. T. E. Jensen and others, *Analytical Chemistry*, 54 (1982), 2388–90.

5. R. A. Hites and K. Biemann, *Analytical Chemistry*, 40 (1968), 1217–21.

6. R. A. Hites and K. Biemann, *Analytical Chemistry*, 42 (1970), 855–60.

7. P. Kintz, A. Tracqui, and P. Mangin, *Fresenius Journal of Analytical Chemistry*, 339 (1991), 62–3.

8. *Fed. Regist.*, 49 (1984), 43234–439.

9. *EPA Method 525.1, Rev 2.2*, May 1991; NTIS order numbers PB-89-220461 and PB-91-108266.

10. C. C. Sweely and others, *Analytical Chemistry*, 38 (1966), 1549–53.

11. J. M. Czuczwa and R. A. Hites, *Environmental Science Technology*, 20 (1986), 195–200.

12. M. L. Vincent and D. G. Peters, *Journal of Electroanalyical Chemistry Interfacial Electrochemistry*, 327 (1992), 121–35.

Chapter 32

Fast Atom Bombardment and Liquid Secondary Ion Mass Spectrometry

Robert D. Minard

*Pennsylvania State University
Department of Chemistry*

Summary

General Uses

- Molecular weights can be determined for single components or multiple components of a mixture for all types of organic, organometallic, and bioorganic compounds, especially peptides or small proteins.
- Molecular weights of very polar, nonvolatile compounds can be determined.
- An upper mass limit of about 3000 Da is normal for routine analysis, but molecular weights over 10,000 Da have been determined.
- Microgram to picogram quantities of material can be detected.
- Isotope patterns and accurate mass determinations can yield elemental composition information.
- Quantitative analysis is possible using isotopically labeled internal standards.
- Effluents from directly coupled high-performance liquid chromatography (HPLC) can be identified.
- Structural information, such as peptide sequences, can be derived from fragment ion masses, although collisionally activated fragmentation is often required.

Common Applications

- Molecular weight verification and structural analysis of polar or ionic, thermally labile compounds.
- Structural determination of biopolymers (peptides and proteins, oligosaccharides, and oligonucleotides), glycoconjugates, and phospholipids, organometallic and coordination compounds, metabolites, natural products, antibiotics, drug and steroid conjugates, bile acids, dyestuffs, and quaternary ammonium salts.

Samples

State

Almost any unknown solid or high-boiling liquid compound with molecular weight less than 3000 Da can be analyzed. Most solids or high-boiling liquid compounds of known structure with molecular weights up to 6000 Da can be analyzed. Certain impurities can be tolerated. Mixtures can be analyzed.

Amount

Microgram to picogram quantities can be detected depending on the type of sample analyzed.

Preparation

Dissolution in a low-volatility liquid matrix. Type of matrix and analyte concentration are critical.

Analysis Time

Analysis takes from 15 min to 2 hr depending on the sample.

Limitations

General

- Sample should have at least one polar functional group.
- Salt contamination can totally suppress organic analyte signal.
- High background makes it difficult to intepret spectra.

Accuracy

If isotopically labeled internal standards are used, precision is ±1%. Reproducibility is comparable to electron impact mass spectra.

Complementary or Related Techniques

- Nuclear magnetic resonance provides additional information on detailed molecular structure.
- Infrared spectroscopy provides evidence of functional groups.
- Chemical or biochemical sequencing provides confirmatory or complementary evidence for biopolymer sequence.

Introduction

Although mass spectrometry (MS) has always provided more bang for the buck (structural information per microgram) than any other analytical technique, it was the advent of fast atom bombardment (FAB) ionization that allowed mass spectrometry to greatly expand its versatility. All at once, hundreds of mass spectrometers could be modified to allow facile analysis of whole new classes of thermally labile, polar or ionic, higher-molecular-weight compounds without the need for extensive purification or derivatization. To be sure, field desorption and plasma desorption ionization had previously allowed mass analysis of these types of compounds, but neither of these methods were as readily adaptable or as easily applied as FAB. The rapid acquisition of FAB ionization in mass spectrometry facilities throughout the world quickly led to its successful application to a variety of sample types: organometallics, synthetic polymers, proteins, peptides, glycoconjugates, oligosaccharides, oligonucleotides, antibiotics, metabolites, steroids, and phospholipids. The majority of samples were from the biochemistry arena and there is little doubt that FAB played a major role in ushering in the modern era of bioanalytical mass spectrometry (1).

The intent of this chapter is to emphasize the practical aspects of FABMS as practiced routinely in a service mass spectrometry facility whose major goal is providing verification to chemists and biochemists that they have indeed isolated what they hoped to have isolated. Each mass spectral analysis is really an experiment, and the better we understand the principles by which the experiment works, the more likely the experiment is to produce the desired result. Unfortunately, there are a number of variables involved in FABMS analysis for which the underlying principles are poorly understood, so that even when the sample composition is known, it is impossible to predict the result from the interaction of these variables. Therefore, much experimental design in FABMS analyses is based on empirical rules and on an iterative trial-and-error approach guided by experience. Extensive reviews have been written on the underlying principles in FABMS (2, 3) and the interested reader should refer to these references.

How It Works

Traditional ionization methods such as electron ionization or chemical ionization (see Chap. 30) require that the sample be vaporized so that the gas state molecules can be ionized by electrons or reagent gas ions. This means a compound must have a finite vapor pressure (about 10^{-6} torr) at reasonable temperatures, about 300 °C depending on the thermal lability of the compound. At normal heating rates, many molecules (essentially all polar molecules weighing more than 1000 Da) thermally decompose before vaporizing sufficiently. Because the rate constant for vaporization is often higher than the rate constant for decomposition, the more rapidly a high temperature can be attained in the vicinity of the molecules, the more vaporization would be favored over decomposition (4). This is the basis of desorption chemical ionization and surface bombardment or irradiation techniques.

Analysis of the secondary ions emitted when a surface is bombarded with an energetic primary ion beam is an established technique for examining the surface layers of inorganic materials such as metals, alloys, and semiconductors (5). Bombarding a surface with high-energy ions (such as 5 kV Ar^+ ions) produces energy spikes (instantaneous high-temperature regions) and the surface material is ejected or sputtered and ionized to yield secondary ions. When secondary ion mass spectrometry (SIMS) was attempted for analysis of organic molecules coated on a surface, only transient ion signals were observed and this was attributed to extensive sample destruction by the primary ion beam. In the late seventies, Benninghoven and colleagues found that using low-nanoampere primary ion beam fluxes greatly reduced sample damage, but ion signals were quite weak (6).

In 1981, this situation changed dramatically when M. Barber and colleagues discovered that protonated molecular ions, $(M+H)^+$, could be generated abundantly by bombarding a solution of the sample in a liquid matrix such as glycerol with keV energy atoms. It was possible to determine the molecular weights of previously intractable molecules such as peptides up to a few thousand daltons with an accuracy of better than 1 Da, even with rather complex mixtures (7). In 1982, it was reported that the neutral atom beam of FAB could be replaced with an ion beam with a measurable improvement in sensitivity; this related technique is known as liquid SIMS (LSIMS) (8). Experience has shown that for all practical purposes the results obtained using FAB and LSIMS are very similar, so they will be considered as interchangeable techniques. Because the term *FAB* is more firmly entrenched in the literature, it will be used most often in this chapter.

As a consequence of this ability to produce ions of high mass, the mass range of commercially magnetic deflection mass spectrometers was soon extended to reach 10,000 to 15,000 Da at full accelerating voltage (8 to 10 keV). The molecular ion region observed by the FABMS analysis of 1 mg of human insulin ($C_{257}H_{383}N_{65}O_{77}S_6$) soon became the specification for high mass resolution and sensitivity. This is shown in Fig. 32.1 and illustrates one of the many initially disorienting effects of high mass analysis; that is, for molecules containing over 100 carbon atoms, the abundance of the $(M+H)^+$ ion containing only ^{12}C becomes minor compared to species containing two, three, or more ^{13}C atoms. As the ion signal becomes distributed over such a large number of isotopic peaks, it becomes difficult to identify the all-^{12}C peak and the correct molecular weight of an unknown. In practice, FAB analyses are often run at a lower resolution that merges all of the isotopic species into one unresolved envelope whose centroid represents the average molecular weight, equivalent to the chemical molecular weight.

Both the practice and principles involved in the FAB method and the closely related LSIMS method are described concisely below and illustrated in Fig. 32.2.

Sample Preparation and Introduction

Sample preparation is critical in FAB analysis. The sample is introduced as a few microliters of a dilute solution (10^{-3} to 10^{-5} *M*) in a polar low-volatility liquid called a matrix. Because liquid matrix selection and sample preparation are so important to the success of FABMS analysis, much of the FAB literature is devoted to this topic (2, 3, 9–12); guidelines are elaborated in the next section.

The sample/matrix solution is placed on a clean metal probe tip, normally stainless steel, as a small droplet 2 to 5 mm in diameter. The stainless steel probe tip is normally cleaned with dilute

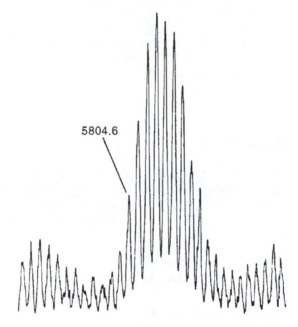

Figure 32.1 FAB mass spectrum of the molecular ion region of human insulin. *(Reprinted from B. S. Larsen, Mass Spectrometry of Biological Materials, C. N. McEwen, B. S. Larson (eds.), 1990 pp. 197–214, Marcel Dekker Inc., N.Y., by courtesy of Marcel Dekker Inc.)*

5804.6

Figure 32.2 The FAB or LSIMS sample insertion probe, inserted into the ion source of a magnetic mass spectrometer. The sample is bombarded by either Xe^0 atoms or Cs^+ ions.

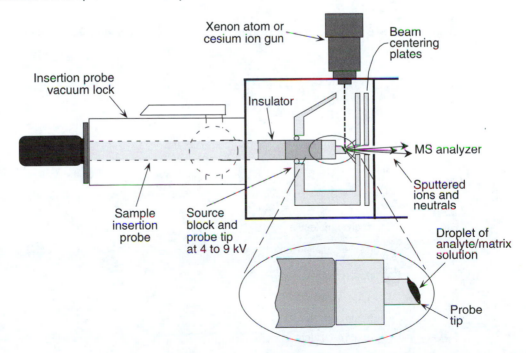

acid or abrasives to ensure that the sample droplet wets and adheres to the surface. The probe tip is attached to a sample introduction probe assembly and this is inserted into the ion source of the mass spectrometer through a vacuum lock, as shown in Fig. 32.2.

Particle Beam Production

A beam of energetic particles, normally xenon atoms (Xe^0) for FAB and cesium ions (Cs^+) for LSIMS, are directed to strike the sample at an acute angle of between 20° and 40° to the plane of the droplet surface. Typical particle energies and fluxes are 5 to 10 keV and 30 to 100 µA for FAB and 4 to 35 keV and 1 to 3 µA for LSIMS.

Particle Impact and Sputtering

The impact of these energetic particles produces an intense thermal spike leading to the ejection or sputtering of the sample/matrix solution in all directions. A view of this process at the molecular level is depicted in Fig. 32.3. Although the exact nature of the processes taking place has been subject to considerable research and is still not fully understood, phenomenologically we observe or can deduce the following:

- If the sample consists of preformed positive ions (for example, quaternary ammonium ions) or negative ions (for example, carboxylate ions), these will be ejected and readily detected, as one would predict.
- More critical to the general utility of FAB and LSIMS, however, is the fact that neutral sample molecules are converted into both positive and negative ions. In most cases, posi-

Figure 32.3 FAB bombardment and sputtering processes viewed at the molecular level.

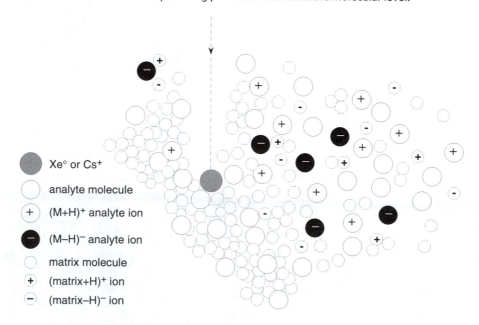

tive ions arise from protonation of the sample molecule (as in chemical ionization) to yield $(M+H)^+$ ions (also symbolized as MH^+). In most cases, negative ion formation involves loss of a proton, yielding $(M–H)^-$ ions. These are ejected along with a preponderance of neutral species. The relative amounts of positive versus negative ions depends on the structure of the sample molecules and the nature of the matrix (for example, the matrix pH). The more massive the particles of the primary beam, the higher the secondary ion yield.

- The molecules and ions are quickly cooled by the rapid expansion of the sample/matrix from the condensed phase to the gas phase. Thus, thermally labile molecular ions can survive the initial thermal spike and are usually much more abundant in the ion beam than fragment ions.

- As expected, large amounts of positive or negative matrix ions are also produced. In addition, charged matrix clusters (dimer, trimer, tetramer, pentamer, and so on) are observed, although their relative intensity falls off rapidly with increasing size.

- If inorganic ions are present, such as Na^+, K^+, or Cl^-, these will also attach to sample molecules and be detected as the cationized species MNa^+ or MK^+ in positive ion MS analysis or MCl^- in negative ion analysis, although the latter is less common.

- Occasionally, matrix or sample molecules will attach to protonated or cationized molecular ions to form matrix adducts ($M \cdot matrix+H)^+$ or protonated dimers $(M_2+H)^+$.

Mass Analysis

Ions sputtered with the correct trajectories (only a very small proportion) pass through the source slit or exit aperture into the mass analyzer and their mass-to-charge (m/z) ratio is measured. The mass analyzer is usually a double-focusing magnetic mass spectrometer, although quadrupole, time-of-flight, Fourier transform ion cyclotron resonance (FTICR), or ion trap instruments can be used. One problem with mass analysis at m/z over 1000 is that the detection efficiency of an electron multiplier falls off rapidly. The solution to this problem is to increase the velocity of the ion at the detector with the use of a postacceleration detector. A concise explanation of the theory and operation of these mass analyzers and detectors can be found in references by Das (13) and Chapman (4).

The utility of the FABMS method also stems from the fact that it produces high ion currents, comparable to electron impact, and that these currents are sustainable for many minutes, permitting signal averaging, high-resolution precise mass determination, and tandem analysis.

Matrix Selection and Sample Preparation

Sample preparation and introduction, particularly matrix selection, are critical to successful FABMS or LSIMS analysis. For example, the trimethylsilyl derivative shown in Fig. 32.4 shows absolutely no MH^+ when run in glycerol, but a strong MH^+ when analyzed in 15-crown-5. Considerable effort has been devoted to the problem of selecting the most appropriate sample matrix, adjusting the matrix pH or composition with additives, and setting the proper instrument operating conditions such as positive or negative ion detection and source temperature (2, 3, 9–12).

The ideal sample matrix should have low volatility (that is, a high boiling point), dissolve the sample analyte, produce very few interfering peaks (particularly in the mass region of interest for the analyte), be fluid enough (not too viscous) to allow diffusion of the analyte to and mixing at the droplet surface, not react with the sample (or, if desired, react in a predictable manner to provide

Figure 32.4 The effect of the matrix on the production of an MH⁺ ion for a trimethylsilyl derivative of a multi-functionalized molecule. (a) FAB-MS spectrum of the sample analyzed in a glycerol matrix. No MH⁺ is detected. (b) Spectrum of the sample analyzed in 15-crown-5 shows a strong MH⁺ as well as some fragmentation.

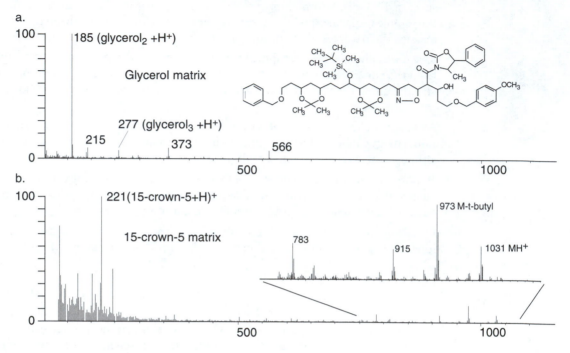

enhanced ion yields or structurally useful product ions), and be a medium in which the analyte can act as a surfactant and form a stable monolayer at the droplet surface.

More than 90% of the reported FABMS or LSIMS analyses are performed with one of the matrices shown in Table 32.1, with glycerol being a common choice. Certain matrices are best used for certain classes of compounds as shown in the table. Most matrices contain hydroxyl (–OH) or thiol (–SH) groups having weakly acidic active hydrogens. These would be incompatible with highly reactive organometallic samples, which must be analyzed using more inert matrices such as 2-nitrophenyl octyl ether, 15-crown-5, or other nonhydroxylic liquids (11, 12).

The matrix selection process can be straightforward if you are working with a particular class of compounds, such as proteins, phospholipids, or saccharides whose solubility, charge state, and approximate molecular weight characteristics are known. Unfortunately, when working with unknown or poorly characterized samples, there is little assurance that using a particular liquid matrix with a particular compound class will guarantee success. Often, more than one matrix must be tried before useful spectra are produced. Ideally, enough sample (tens of micrograms) is available to allow experimentation.

Our experience has been that when confronted with a new compound type, a predicted matrix will produce usable data only about 50% of the time and other matrices will need to be tried. For this reason, the matrix is never added directly to the whole sample because it would be very difficult to remove the matrix if it did not work. Either matrix is added to a small portion of the sample or, more routinely for the small amounts of material often available, the sample is dissolved in a common volatile solvent such as methanol, dichloromethane, or water, one that dissolves the sample and is itself soluble in the matrix. Dimethylsulfoxide can be used for difficult-to-dissolve samples. Addition to the matrix of trace amounts of solubilizing agents such as Triton X-100 may help (3, 10).

Table 32.1 Common liquid matrices used for FABMS analysis.

Matrix	Structure	Molecular Weight	bp/mp	Used for	Comment
Glycerol		92	182 °C at 20 mm	Peptides and polar or polyhydroxylic compounds	Good first choice when compound behavior is hard to predict; reduction of analyte possible.
Thioglycerol		108	118 °C at 5 mm	Proteins, peptides	More acidic than glycerol; can be used as admixture with glycerol; reduction of disulfide bonds is a possible side reaction.
3-Nitrobenzyl alcohol		153	175–180 °C at 3 mm	Small proteins; organometallics; less polar compounds	Can oxidize sample.
Diethanolamine		105	217 °C at 150 mm	Saccharides, sulfonates, and fatty acids	Usually used in negative ion detection mode.
3-Amino-1,2-propanediol		91	264 °C at 739 mm	Saccharides, sulfonates, and fatty acids	Usually used in negative ion detection mode; reduces suppression by impurities.
Triethanolamine		149	190–193 °C at 5 mm	Saccharides, sulfonates, and fatty acids	Usually used in negative ion detection mode.
Dithioerythritol: dithiothreitol 1:5 ("magic bullet")	erythro and threo	154	Erythro: mp = 83 °C Threo: mp = 43 °C Mixture is a liquid	Organometallics, proteins, and peptides	Can react with organometallic; can reduce –S–S– linkages in peptides/proteins.
2-Nitrophenyl octyl ether		251	197–198 °C at 11 mm	Organometallics and less polar materials	Nonhydroxylic, so it is less reactive to water-sensitive compounds; can oxidize sample, but will not reduce it.
15-Crown-5		220	100–135 °C at 0.2 mm	Reactive organometallics	Nonhydroxylic, so it is less reactive to water-sensitive compounds; may sequester Na^+.

All matrices are available from Aldrich Chemical (Milwaukee, WI) except for 2-nitrophenyl octyl ether, which can be obtained from Fluka Chemical Corp.(Ronkonkoma, NY).

The concentration of the analyte is another important variable, one that is confounded by the fact that it depends on the matrix used and that it actually changes continuously in the mass spectrometer as the matrix evaporates or is sputtered away. Certain compounds need to be fairly high in concentration, approximately 10^{-3} M, to produce results. Other compounds will not produce useful spectra at high concentrations and a good indication that the concentration is too high is the production of a burst of initial ion current, which rapidly drops to zero within a few seconds. When diluted by a factor of 10 to 100, however, these compounds can yield good spectra. Given the small quantities (less than 1 mg) often available, it is impractical to prepare these solutions quantitatively and because the correct concentration cannot be determined a priori for new compounds, a simple trial-and-error approach is often used.

- For visible amounts of solids, one drop of a volatile solvent is added to a small crystal or lump of solid (about .5 mm) and the solid dissolved.

- For barely visible microgram quantities of sample, 5 to 10 μL of volatile solvent is added to the whole sample and it is dissolved.

Using a glass or plastic micropipet, a microliter or two of this volatile solvent solution is added to a droplet of matrix (about 5 μL), which has been dabbed onto the probe tip with a small glass capillary. Much of the volatile solvent evaporates in the vacuum lock before final insertion into the FAB ion source.

Modern mass spectrometry data systems allow the display of a mass region of interest as the instrument is making repetitive mass scans, thus providing feedback on whether important peaks are present. Spectra are acquired for about 5 min, even if nothing is seen initially, because it is quite common to see the molecular ion signal appear or increase with time. After each analysis, the mass spectra are examined and if the expected result has not been obtained, the spectra can provide clues as to what may be the problem. Because FAB spectra are quite complex compared to those obtained by other ionization methods, a discussion of their interpretation is in order.

What It Does

The Appearance and Interpretation of FAB Mass Spectra

Confirmation of molecular weight by determination of MH⁺ (positive ion detection) or (M–H)⁻ (negative ion detection) molecular ion species is the primary objective of FABMS analysis. Formation of fragment ions is usually minor, although this can change with the type of matrix and whether positive or negative ion spectra are obtained. Figure 32.5 shows the electron impact, chemical ionization, field desorption, and FAB ionization spectra of phenobarbital, molecular weight 232, in glycerol. The FAB spectrum is much more complex than the other three, showing ions at m/z values beyond 600. The MH⁺ at m/z 233 is actually a minor peak. This is due to the high background arising from the matrix and the formation of adduct ions of analyte, matrix, impurity, or additives. New users of FABMS can be put off by the messy appearance of spectra. Experience with reading these spectra builds confidence that useful information can be obtained with care. The following features of FAB mass spectra are important.

Figure 32.5 Mass spectra of phenobarbital. Ionized via (a) electron impact, (b) chemical ionization (CI isobutane), (c) field desorption (FD), and (d) FAB. Note that this is an artificial illustration because phenobarbitol would normally be analyzed by the simpler electron impact or CI, not FD or FAB.

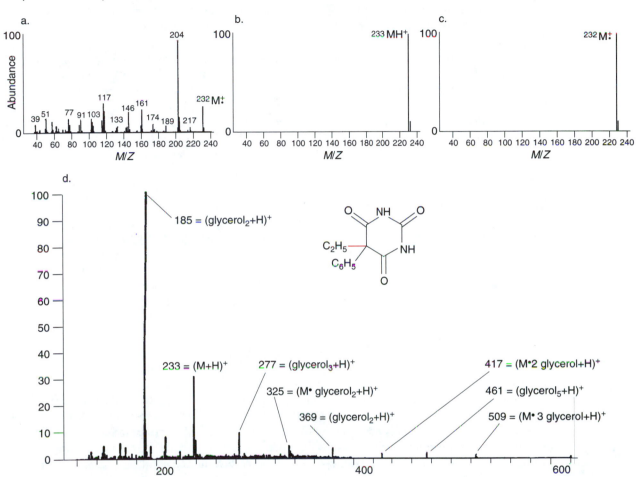

Matrix Peaks

Not only is a strong MH^+ or $(M-H)^-$ derived from the matrix observed, but cluster ions of composition $(matrix_n+H)^+$ or $(matrix_n-H)^-$ can also be very prevalent. Because scanning lower than m/z 100 is rarely done, the MH^+ at m/z 93 is not seen in Fig. 32.5(d), but peaks at m/z 185 (glycerol$_2$H$^+$), 277 (glycerol$_3$H$^+$), and 369 (glycerol$_4$H$^+$) are prominent, the intensity dropping rapidly with increasing cluster size. Glycerol forms cluster ions more than other matrices and these can sometimes be detected above m/z 1000. Most of the other matrices in Table 32.1 form only protonated dimers or occasionally trimers.

Adduct Ions

The presence of sodium, potassium, or ammonium ions, even in trace quantities, can lead to the reduction in the MH^+ peak and the appearance of peaks at $(M+Na)^+$, $(M+K)^+$, or $(M+NH_4)^+$. This has the disadvantage of dividing the molecular ion signal over several species, lowering the signal-to-noise ratio. It has the advantage, however, of providing confirming determinations of the molecular weight.

These same inorganic cations can form adducts with the matrix. For example, the presence of an ion at m/z 207 (glycerol$_2$Na$^+$) in spectra run in glycerol is good evidence of the presence of sodium.

Those submitting samples for mass spectral analysis should be advised to remove all contaminating salts from samples before FABMS analysis by using reversed phase chromatography or solid-phase extraction, dialysis, or acid ion exchange columns. Too much salt can totally suppress the organic sample ion current. Phosphate and other inorganic buffers should be replaced with volatile buffers such as ammonium acetate or formate.

Adduct ions can occasionally arise from protonated clusters of sample and matrix molecules. Fig 32.5d shows three ions of this type. Oligosaccharides also tend to show this behavior. Negative adduct ions with anions such as Cl$^-$ are observed only rarely.

Chemical Noise

Another common feature of FAB spectra is the appearance of a peak at every m/z starting at masses above the molecular ion and often increasing steeply with decreasing m/z. This chemical noise, or incoherent fragmentation, is thought to arise from extensive molecular damage by the impacting primary beam. High-resolution analysis shows that each peak consists of a number of isobaric ions of different composition.

Chemical Reactions

FAB spectra reflect analyte acid–base equilibria in the matrix solution. This has been demonstrated by lowering the pH of a basic analyte-containing matrix and observing increased intensities for the MH$^+$ and, for larger molecules with multiple protonatable centers, MH$_2^{2+}$ and MH$_3^{3+}$ ions. This effect is often taken advantage of by adding HCl, trifluoroacetic acid, or p-toluenesulfonic acid to glycerol or by using a matrix containing the slightly acidic thiol (–SH) grouping such as thioglycerol or dithioerythritol/dithiothreitol. As one would predict, raising the pH of the matrix enhances the production of (M–H)$^-$ ions from acidic species such as carboxylic acids, sulfonic acids, and organophosphates. The basic amine matrices shown in Table 32.1 are used for this purpose.

Certain types of compounds, such as carbohydrates, appear to form adduct ions with Na$^+$ or NH$_4^+$ ions more readily than protons, so controlled amounts of these cations have been used to yield improved FAB detection.

One or two electron reductions can be affected by the primary particle beam (14). For example, nucleosides show [M+2H]$^+$ and [M+3H]$^+$. Reductions of quinones, dyestuffs, and peptides are also possible as well as reductive dehalogenation of organohalides (15).

Fragmentation

Generally, FAB spectra show very little fragmentation and fragment ions are often difficult to see within the high matrix background common in FAB. There are numerous exceptions to this rule, however. For example, no (M–H)$^-$ ion is observed in the negative ion spectra of glycerophosphocholine lipids. Instead, intense fragment ion peaks at M–15 [–CH$_3$], M–60 [–NH(CH$_3$)$_3$] and M–86 [–CH$_2$=CHN(CH$_3$)$_3$] are observed and can be used dependably to define the molecular weight. Weak coordination bonds as in metal carbonyl complexes are readily broken, showing a complete series of carbonyl losses as shown in Fig. 32.6.

Figure 32.6 FABMS of an organometallic complex showing extensive fragmentation caused by loss of carbonyl (C≡O) groups. The matrix was 18-crown-6 containing 10% tetraglyme. Matrix cluster peaks arising from the mix are labeled 18-cr-6. *(Ref. 16.)*

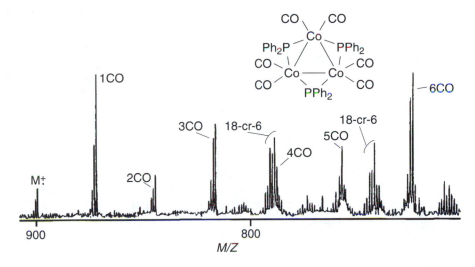

Minimal fragmentation is a major advantage of FABMS for analyzing mixtures and this is discussed in the next section.

In summary, interpretation of FAB spectra requires consideration of a large number of possibilities for ion formation. Many peaks attributable to matrix clustering, adduct formation, or analyte reaction or fragmentation may need to be accounted for before it can be determined whether useful data have been acquired. If not, the sample is reanalyzed after another matrix is chosen or the analyte concentration is modified. Unless the sample is badly contaminated with salts (this can be determined by examining the spectra for matrix adduct ions, such as glycerol$_2$Na$^+$), in which case it must be cleaned up, satisfactory results are normally obtained after two or three attempts. Changing the matrix from a hydroxylic one such as glycerol to a nonhydroxylic one such as 15-crown-5 or vice versa leads to dramatic changes in analyte behavior and often yields positive results (see Fig. 32.4).

If the acquired FAB mass data are to be reported in the literature, it is essential to report the matrix. All too often investigators fail to do so. An example of the concise format used in the *Journal of Organic Chemistry* experimental section for negative FABMS data would be FABMS (glycerol) m/z (relative intensity) 481 (100, M–H), 269 (25, M-ribose phosphate).

Mixture Analysis

Because of the extensive fragmentation induced by electron impact ionization, EI-MS analysis of mixtures produces spectra in which it is difficult to determine whether a given peak is a molecular ion of one mixture component or a fragment ion of another. In contrast, fragmentation by FAB ionization is usually minor and each component should produce a single peak (such as MH$^+$, MNa$^+$, or (M–H)$^-$) characteristic of the molecular weight of each component. If the components are very similar in nature, then the relative intensities of these peaks are representative of the relative amounts of each in the mixture.

Mixture analysis is complicated by the fact that, if the mixture components are dissimilar in nature, they often display dramatic differences in desorption efficiencies. For example, the signal from a hydrophilic peptide is strongly suppressed in the presence of hydrophobic peptides, because the hydrophilic peptide is unable to compete effectively for the surface layer of the matrix. This problem can be overcome in certain cases by derivatization or addition of additives. Also, selectivity can be used as an advantage by suppressing signals from components that are not of interest.

FAB Tandem Mass Spectrometry

FAB spectra often exhibit little fragment ion intensity and therefore lack the structurally elucidating information that chemists and biochemists often desire. Fortunately, at the same time that FABMS was being developed, methods for fragmenting ions and recording spectra of these fragment ions were being introduced as tandem mass spectrometry (MS/MS). A full discussion of MS/MS is beyond the scope of this chapter, but can be found in the literature (4, 13). Tomer published a comprehensive review on FAB combined with MS/MS for the determination of biomolecules (16). An example of a peptide sequencing using this method is shown in Fig. 32.7.

There are two important points to be made regarding FAB MS/MS. First, it is an extremely powerful structural elucidation tool that has and will undoubtedly continue to provide answers to questions of profound significance in areas of biochemistry such as immunology, neuroscience, and cancer research. Second, from a practical standpoint, it involves complex, expensive instrumentation that must be operated by a knowledgeable staff experienced in both mass spectrometry and biomaterial handling. For this reason, it is not practiced in very many labs worldwide.

Flow FAB

Ito and colleagues and Caprioli and colleagues modified the normal FAB probe by inserting a 50- to 75-μM fused silica capillary through the shaft, thus allowing the analyte to be carried in a continuous flow of water, water/methanol, or water/acetonitrile solution containing 3 to 10% glycerol to the FAB probe tip at a rate of about 5 μL/min (19). At the tip, the solution flows out as a thin

Figure 32.7 An example of oligopeptide sequencing by FAB MS/MS. The daughter ion spectrum of the MH⁺ (m/z 2165.0) for a derivatized octadecapeptide. The sequencing information is marked; cm = carbamidomethyl. (*K. Biemann, "Mass Spectrometric Methods for Protein Sequencing," Analytical Chemistry, 58(13), 1986, 1288A–1300A. Reprinted by permission from Humana Press.*)

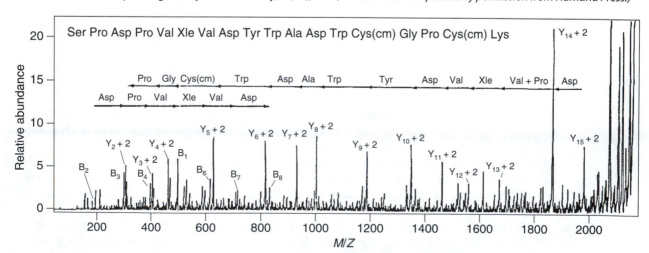

film and undergoes FAB ionization before evaporating or flowing off the tip. Flow FAB has a number of advantages:

- Convenient sample introduction by injection into the flow stream. Samples can be injected as often as once every 2 min.
- Greatly reduced background because the matrix is present at much lower levels.
- Temporal definition of the analyte signal. Flow injection produces an ion current profile attributable to the analyte rising above a constant baseline of background signal. This allows accurate background subtraction.
- Improved sensitivity stemming largely from the reduced background.
- Elimination of suppression effects when mixtures are analyzed. This is probably caused by constant mixing in the continuous flow at the probe tip and the greater surfactancy of many samples in water.
- Improved quantitation, because the areas of well-defined ion profiles from flow injection are much more accurately measured.
- The ability to monitor reactions dynamically on-line.
- Direct coupling of HPLC, capillary electrophoresis (CE), or any other flowing system. For example, on-line pharmokinetic drug studies have been carried out in a live rat by monitoring the effluent of an implanted flow-through microdialysis probe connected to flow FAB.

Source temperature is very critical for stable performance in flow FAB. If it is too low, the probe ices up and if it is too high, bump boiling occurs. In either case, the signal is very erratic. Also, working at such low flowrates requires special equipment and extra care in removing dead volume from all connections. Caprioli has written extensively on the details of the method and has edited a volume titled *Continuous-Flow Fast Atom Bombardment Mass Spectrometry* that describes many applications of flow FAB (19).

Other Considerations

Calibration

This is a relatively straightforward process in FAB ionization because alkalai metal iodides, in particular cesium iodide, give a regular pattern of cluster ions of composition $(CsI)_nCs^+$ for positive FAB or $(CsI)_nI^-$ for negative FAB with masses extending to 10,000 Da and higher, depending on the mass range of the instrument and the sensitivity of the detection system. A droplet of a 10% aqueous solution of CsI is applied to the FAB probe tip, the water is pumped off in the vaccum insertion lock, and the dry sample provides strong ion currents for long periods of time, allowing the instrument to be tuned and calibrated.

Analytical Reproducability

An interlaboratory comparison was carried out by Murphy and colleagues (20) that showed that the coefficients of variation of the relative intensities in the FAB data were comparable to those derived from electron impact data. Figure 32.8 shows the variation in the negative FAB mass spectrum of a mixture of glycerophosphocholines obtained in the author's laboratory on a KRATOS MS-50RF instrument (Fig. 32.8a) and the FAB spectrum of the same sample obtained on a JEOL-HX110/HX110 instrument in a different laboratory (Fig. 32.8(b)) (21).

Figure 32.8 Analytical reproducibility of negative FAB mass spectra of a sample containing a mixture of phosphoglycerocholines obtained on two different instruments. Triethanolamine was used as the matrix in both cases.

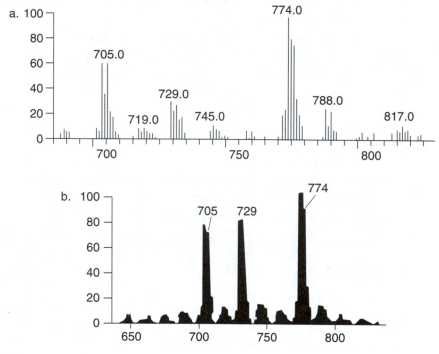

Quantitative Analysis

Numerous quantitative investigations using electron impact, chemical ionization, or field desorption have proven that internal standardization using isotopically labeled (usually ^{2}H or ^{13}C) internal standards is the method of choice for this purpose. Studies by Lehmann and colleagues (22) show that accurate isotope abundance measurements can also be carried out using FAB. For example, for a dansylated phenylalanine sample containing a dansylated penta- and heptadeutero-Phe internal standard, quantitative results are within ±1% of those obtained by other methods. Desiderio has carried out quantitative studies of neuropeptides using deuterated peptide standards (23).

Exact Mass Measurement

High-resolution mass measurements allow the elemental composition possibilities to be calculated for a molecular ion species and are often required by journals when reporting the synthesis or isolation of a new compound. The rationality of this requirement at masses above 1000 is questionable because at higher masses, the number of compositional possibilities within accepted error tolerances becomes enormous. As with other ionization methods, the best way to carry out precise high-resolution mass measurements on FAB-derived ions is to have defined reference masses in the vicinity of the mass being measured during the data acquisition. Sometimes the matrix cluster ions are used or a small amount of polyethylene glycol is added to the matrix to produce reference peaks without suppressing the sample signal. Split probes with the sample on one half and a reference such as CsI on the other have been used, although the tuning can be different from one side to the other and this can lead to a systematic error.

Related Techniques

Since the appearance of FAB in 1981, two other ionization techniques have been developed and commercialized and are now widely used for analyzing thermally labile, ionic, or high-molecular-weight compounds. Brief descriptions are given below for comparison purposes. Greater detail can be found in the work of Das (13) and Chapman (4) and references cited therein.

Matrix-Assisted Laser Desorption Ionization (MALDI)

The sample is imbedded in a matrix of a solid organic compound, such as sinapinic acid. Sample and matrix are desorbed or sputtered by rapid energy deposition in the matrix by pulses from an ultraviolet laser in a manner similar to FAB. Mass analysis is typically by time-of-flight MS (TOFMS), although FTICRMS is also used.

By this means, ionized proteins with molecular masses greater than 200 kDa have been observed with reasonable sensitivity. The instruments are compact and sample introduction can be automated. However, the precision of mass analysis by TOFMS is not as good as FABMS on a magnetic instrument in the molecular weight range accessible to both systems. Also, MS/MS methods are not currently available on TOF instruments.

Electrospray Ionization (ESI)

This ionization method was brought to fruition by Fenn and colleagues in 1989 (24) and many commercial instruments are now available. It is not a desorption method like FAB and MALDI, but rather a spray ionization method. Liquids, normally methanol/water mixtures, are sprayed from a capillary into electrical field of a few kilovolts. The charged droplets are desolvated by flowing gases, by differential pumping, and by heating, and the residual ions, often multiply charged, enter the mass analyzer through a small orifice.

The major advantage of ESI is that the ions are multiply charged, the number of charges increasing from 1 to 3 for a molecule weighing approximately 1000 Da to 30 or more above 50,000 Da. This yields an m/z ratio that is almost always below 2000, a mass range within the means of all mass spectrometers, including quadrupole instruments. A classic ESI sample is albumin (about 66,000 Da). Thus, ESIMS extends the mass range considerably higher than that attainable by FAB and it does this with a significantly reduced background with correspondingly lower detection limits. However, although sensitivity is good for polar molecules able to accept multiple charges, it does not not work well for nonpolar molecules. If the ESI source can also be run in the atmospheric pressure ionization (API) mode, this problem can be partially overcome.

Applications of FABMS and LSIMS

The principal applications of FAB and LSIMS have been structural determinations of biopolymers, with most of the work involving the analysis of peptides and proteins, particularly those that were not amenable to traditional protein-sequencing methods because they were blocked at the N-terminus or post-translationally modified in a way (such as glycosylation or sulfonation) that could not be characterized by any other method (1, 18). Most proteins and peptides have better FABMS sensitivity than almost any other class of compounds. Natural and recombinant proteins and monoclonal antibody light chains with molecular weights up to 25,000 Da have been analyzed (25), although peptides weighing less than 3000 Da represent most of the work in this area. The use of FAB or LSIMS combined with MS/MS for amino acid sequencing of peptides has undoubtedly been the most impressive and highly publicized application of these methods (17, 18). The structure elucidation and sequencing of oligosaccharides (1, 26) and oligonucleotides (27) has proved more difficult, but tremendous progress has been made.

Many of the samples reported in the literature as analyzed by FABMS involve highly polar or ionic species of more modest molecular weights. Examples are glycoconjugates (28) and phospholipids (20). An application area that has proved FABMS to be a workhorse in the author's laboratory and yet does not seem to get much attention is the analysis of organometallic compounds and complexes (11, 12). Chapman (4) provides references for application of FABMS to antibiotics, drug and steroid conjugates, bile acids, dyestuffs, and quaternary ammonium salts. It is probably in many of these areas that FABMS will continue to serve the needs of chemists, whereas the higher-mass analysis of biopolymers will be handled by ESIMS, ESIMS/MS, or MALDI.

Nuts and Bolts

Relative Costs

Commercial FABMS instruments are all high-performance double-focusing magnetic instruments with extended mass range magnets, with the exception of the Finnigan FTMS system. Typical costs are in the $300–$500K range with data systems. Tandem mass spectrometers for FABMS/MS cost roughly twice this amount.

Vendors for Instruments and Accessories

See Chap. 28 (p. 564) for a detailed list of vendors.

Required Level of Training

The operator should have a minimum of a bachelor's degree in chemistry and supervision by an experienced chemist/mass spectrometrist is normally required. Training requires at least 1 wk, and a year's experience is very valuable in facilities that must carry out FABMS on a range of sample types. Familiarity with other mass spectrometric methods and derivatization techniques is also important in order to ensure that correct answers can be achieved for a variety of samples. Data system operation, backup, and updating require a knowledge of computers.

Service and Maintenence

FABMS instruments and their associated data acquisition and analysis systems can be quite demanding in terms of service and maintenance, as evidenced by the expensive service contracts associated with them ($25,000 or more annually). Facilities that do their own service will need access to an electronics repair shop, a machine shop, and a glass-blowing shop. Routine maintenence involves maintaining a high vacuum system (turbo or diffusion pumps, mechanical pumps, and ion gauges) and cleaning of the ion source and FAB gun at least semiannually, more often if the sample load is high or flow FAB is used.

Suggested Readings

CHAPMAN, J. R., *Practical Organic Mass Spectrometry, a Guide for Chemical and Biochemical Analysis*. New York: Wiley, 1993.

DESIDERIO, D. M., ED., *Mass Spectrometry, Clinical and Biomedical Applications*, vol. 2. New York: Plenum Press, 1994.

GROSS, M. L., ED., *Mass Spectrometry in the Biological Sciences, a Tutorial*, Dordrecht: Kluwer Academic, 1992.

MCCLOSKEY, J. A., ED., *Methods in Enzymology*, vol. 193 *Mass Spectrometry*. San Diego: Academic Press, 1990.

References

1. S. A. Carr and others, *Analytical Chemistry*, 63, no. 24 (1991), 2802–24.

2. C. Fenselau and R. J. Cotter, *Chemical Reviews*, 87 (1987), 501–12.

3. E. DePauw, A. Agnello, and F. Derwa, *Mass Spectrometry Reviews*, 10, no. 4 (1991), 283–301.

4. J. R. Chapman, *Practical Organic Mass Spectrometry, a Guide for Chemical and Biochemical Analysis* (New York: Wiley, 1993), 132–81.

5. J. A. McHugh, *Methods of Surface Analysis* (Amsterdam: Elsevier, 1975), Ch. 6.

6. R. J. Day, S. E. Unger, and R. G. Cooks, *Analytical Chemistry*, 52 (1980), 557A–72A.

7. M. Barber and others, *Analytical Chemistry*, 54, no. 4 (1982), 645A–57A.

8. W. Aberth, K. M. Straub, and A. L. Burlingame, *Analytical Chemistry*, 54, no. 12 (1982), 2029–34.

9. B. S. Larsen, in C. N. McEwen and B. S. Larsen, eds., *Mass Spectrometry of Biological Materials* (New York: Marcel Dekker, 1990), 197–214.

10. K. L. Bush, in D. M. Desiderio, ed., *Mass Spectrometry of Peptides* (Boca Raton, FL: CRC Press, 1991), 174–97.

11. M. I. Bruce and M. J. Lidell, *Applied Organometallic Chemistry*, 1 (1987), 191–226.

12. J. M. Miller, *Mass Spectrometry Reviews*, 9, no. 3 (1990), 319–47.

13. C. Das, in D. M. Desiderio, ed., *Mass Spectrometry, Clinical and Biomedical Applications*, vol. 2 (New York: Plenum Press, 1994), pp. 1–52.

14. R. L. Cerny and M. L. Gross, *Analytical Chemistry*, 57 (1985), 1160–63.

15. K. L. Busch, *Chemtracts: Anaytical Physical Inorganic Chemistry*, 2, no. 5 (1990), 449–53.

16. R. D. Minard and G. L. Geoffroy, *Proceedings of the 30th Annual Conference on Mass Spectrometry and Allied Topics*, June 6–11, 1982, Honolulu, Hawaii.

17. K. B. Tomer, *Mass Spectrometry Reviews*, 8, no. 6 (1989), 445–82.

18. K. Biemann, *Analytical Chemistry*, 58, no. 13 (1986), 1288–1300A; K. Biemann, in K. A. Walsh, ed., *Methods in Protein Sequence Analysis* (Clifton, N.J.: Humana Press, 1987).

19. R. M. Caprioli, ed., *Continuous-flow Fast Atom Bombardment Mass Spectrometry* (New York: Wiley, 1990); R. M. Caprioli and M. J. F. Suter, *International Journal of Mass Spectrometry and Ion Processes*, 118–19 (1992), 449–76.

20. R. C. Murphy and K. A. Harrison, *Mass Spectrometry Reviews*, 13, no. 1 (1994), 57–75.

21. The author thanks C. Fenselau and D. Fabris (University of Maryland, Baltimore County) for help in obtaining these spectra.

22. W. D. Lehmann, M. Kessler, and W. A. Konig, *Biomedical Mass Spectrometry*, 11 (1984), 217–22.

23. D. Desiderio, in D. M. Desiderio, ed., *Mass Spectrometry, Clinical and Biomedical Applications*, vol. 2 (New York: Plenum Press, 1992–1994), pp. 133–65.

24. J. B. Fenn and others, *Science*, 246 (1989), 64–71.

25. M. M. Siegel and others, *Analytical Chemistry*, 62, no. 14 (1990), 1536–42.

26. A. Dell, in J. A. McCloskey, ed., *Methods in Enzymology*, vol. 193. *Mass Spectrometry* (San Diego: Academic Press, 1990), pp. 647–60; H. Sasaki and others, *Journal of Biological Chemistry*, 262 (1987), 12059–76.

27. R. A. Laine, in J. A. McCloskey, ed., *Methods in Enzymology*, vol. 193. *Mass Spectrometry* (San Diego: Academic Press, 1990), pp. 539–53.

28. Jasna Peter-Katalinic, *Mass Spectrometry Reviews* 13, no. 1 (1994), 77–98.

Chapter 33

High-Performance Liquid Chromatography Electrospray Ionization Mass Spectrometry

Chhabil Dass

*The University of Memphis
Department of Chemistry*

Summary

General Uses

- Separation and identification of complex mixtures of nonvolatile and thermally labile biochemical, inorganic, and organic compounds
- Determination of the molecular weight (MW) of macromolecules (peptides, proteins, oligonucleotides, carbohydrates, and organic polymers)
- Determination of the primary structure of compounds (such as peptides and proteins); probing the higher-order structures of proteins
- Quantitative analysis of compounds
- Kinetic analysis of chemical and enzymatic reactions

Common Applications

- Peptide and carbohydrate mapping
- Determination of the purity and selective detection of a variety of compounds

- Identification of metabolic products
- Identification of posttranslational modifications in proteins and peptides
- Analysis of drugs, toxins, and endogenous compounds from biological specimens
- Confirmation of peptide synthesis and identification of the side products

Samples

State

Solids and liquids can be analyzed.

Amount

Attomole to picomole amounts are sufficient.

Preparation

Almost no sample preparation is required; the sample is dissolved in an appropriate aqueous/organic solvent.

Analysis Time

Sample preparation requires minimal time; an electrospray ionization (ESI) mass spectrum can be obtained in 2 to 10 min; overall analysis time of the high-performance liquid chromatography (HPLC) ESI mass spectrometry (MS) analysis depends on efficiency of the HPLC separation and may vary between 5 and 40 min.

Limitations

General

- The sample must be soluble in the electrospray solvent.
- The MS analysis time is increased by combining HPLC with MS.
- Liquids with high conductivity and surface tension are difficult to electrospray.
- Coeluting isomers are not easily discriminated by ESIMS.
- Demands on the MS data system increase.

Accuracy

The MW of a protein often can be determined with an accuracy of 0.01% or better.

Sensitivity and Detection Limits

Attomole levels are possible under favorable conditions; femtomole to picomole levels are routinely obtained.

Complementary or Related Techniques

- Capillary zone electrophoresis (CZE) ESIMS can be used alternatively.
- Liquid chromatography (LC) continuous-flow (CF) fast atom bombardment (FAB) MS can be substituted for most of the applications except for the MW determination of macromolecules.
- Matrix-assisted laser desorption/ionization (MALDI) time-of-flight (TOF) MS is a useful technique for MW determination of macromolecules. LC aerosol MALDI-TOFMS, although at the development stage, holds potential as a substitute for LC-ESI-MS.

Introduction

The technique of LC-ESI-MS is one of the most exciting developments of recent times in analytical methodology. The combination of HPLC with MS is a marriage between the two most powerful standalone analytical techniques. The analytical applications of HPLC and MS are well established. HPLC is a method of resolving a mixture of compounds into its individual components (Chapter 9). It is applicable to several classes of polar and thermally labile compounds (small as well as very large) not amenable to the complementary technique of gas chromatography (GC). However, on its own, HPLC does not provide unambiguous confirmation of the identity of the analyte.

On the other hand, MS is an excellent technique for identification of compounds (Chapter 30). MS offers unique advantages of high molecular specificity and detection sensitivity. MS provides a fingerprint mass spectrum, which contains the MW and structure-specific fragment ion information. The major drawback of MS is its limitation in handling mixtures of compounds. Tandem mass spectrometry (MS/MS) can alleviate this problem only to a certain extent. Interfacing of HPLC with MS thus benefits mutually from the high-resolution separation capability of HPLC and highly sensitive and structure-specific detection capability of MS. MS comes very close to being a universal detector for LC. The primary advantages of LC/MS are as follows:

- Capabilities of both HPLC and MS are synergistically enhanced. On one hand, the need for high-resolution separation by HPLC is reduced. On the other hand, a low-cost low-resolution mass spectrometer can be used as an efficient mass detector.
- Molecular specificity of the analysis is further improved because the analyte is identified both by chromatographic retention time and structure-specific MS data.
- Mixtures of varying complexities are easily analyzed.
- Sensitivity of analysis is improved because the sample enters the mass spectrometer in the form of a narrow-focused HPLC band.
- A number of laborious off-line experimental steps are avoided, resulting in minimal sample loss and reduction in overall analysis time.
- Less sample is required than in off-line LC-MS analysis.
- Because of sample cleanup by HPLC, mutual signal suppression in the MS analysis is minimized and the quality of mass spectral data is improved.

Although attempts were made in the 1970s to combine LC with MS, an efficient and dependable combination has been difficult to achieve because of fundamental differences in their operating conditions. Conventional wide-bore analytical columns function best at a liquid flowrate of 0.5 to 1.5 mL/min, whereas mass spectrometers operate under high vacuum (10^{-5} to 10^{-8} torr). Unlike GC-MS, this liquid flow is not compatible with the MS vacuum system. In addition, the LC-MS interface should also be able to handle the aqueous content, organic and ionic modifiers, and buffers of the mobile phase. Since 1970, several different LC-MS interfaces have been developed, but only a few have found wide acceptance. Some commercially marketed interfaces are the direct liquid introduction (DLI) probe (1), moving belt (2), thermospray (3), particle beam interface (4), CF-FAB (5), and atmospheric pressure ionization (API) interface (6). API with the ESI source is unique in that it has a great potential for the analysis of macromolecules, provides high sensitivity, and is easy to couple with LC.

How It Works

An LC-MS instrument consists of three major components: an LC (to resolve a complex mixture of compounds), an interface (to transport the analyte into the ion source of a mass spectrometer), and a mass spectrometer (to ionize and mass analyze the individually resolved components).

High-Performance Liquid Chromatography

In LC, a sample is applied to the top of a separating column filled with a solid stationary phase and eluted by a stream of a liquid mobile phase. Components of the mixture are separated and emerge from the column one after another. HPLC is a variant of LC in which, to increase the efficiency of separation, the mobile phase is forced to flow through the column at a very high pressure. Identification of a compound is based on its retention time (that is, the time spent by each component in the column). The basis of separation in HPLC involves partitioning of the analyte molecules between the liquid mobile phase and the solid stationary phase. Essential components of a complete HPLC system are solvent reservoirs, a solvent delivery and gradient forming system, fixed-volume loop injector, packed column, detector, data system, and recorder (Fig. 33.1).

When used in the normal phase, the separation of analytes takes place between polar silica microparticles and a nonpolar organic mobile phase. However, this mode is not suitable for the electrospray process.

Reversed phase (RP) HPLC is a widely preferred mode of chromatography and is a major contributing factor to advances made in several areas of biomedical research. In RPHPLC, the stationary phase consists of a nonpolar matrix and elution is carried out with a polar mobile phase (water mixed with a polar organic modifier such as methanol, isopropanol, or acetonitrile).

The basic principle of RPHPLC separation is the hydrophobic interaction between the nonpolar hydrocarbonanaceous matrix of the column material and the hydrophobic groups of the analyte. Two different mechanisms, adsorption and partition, are responsible for retention of solutes in the stationary phase. In the adsorption model, a solute is adsorbed on the hydrophobic surface of the solid support and remains adsorbed until the attractive forces are weakened by a sufficiently high concentration of the organic modifier in the mobile phase. At this critical concentration, the adsorbed solute molecules are replaced by the molecules of the organic modifier and eluted from the column, with little further interaction with the stationary phase. The process can be regarded as endothermic and entropically driven.

Figure 33.1 Basic configuration of an HPLC instrument, where A is a solvent reservoir, B a pumping system, C a fixed-volume sample injector, D a guard column, E an anlaytical column, F a detector, G a data system, and H a printer.

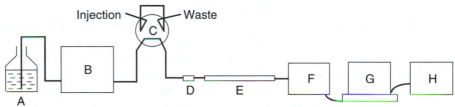

In the partition model, the solid surface is considered a hydrophobic bulk phase and equilibrium is achieved when a solute partitions between this solid phase and the mobile phase. With the downward flow of the mobile phase, the solute moves in and out of the stationary phase. Solutes with higher equilibrium constants are retained longer in the column. The equilibrium can be shifted toward the liquid phase by increasing the concentration of the organic modifier.

Retention of small molecules such as alkylphenones is governed largely by partition, whereas molecules such as peptides interact with the solid support via an adsorption type of mechanism. Intermediate-size molecules such as peptides may follow both partition and adsorption mechanisms. Whatever the mechanism, compounds with sufficient hydrophobicity are retained longer by the stationary phase. Therefore, in practice, relatively hydrophilic compounds elute earlier with the aqueous mobile phase. Strongly retained hydrophobic compounds are eluted by increasing the strength of the organic modifier.

When composition of the mobile phase is held constant, the practice is called isocratic elution. In contrast, in gradient elution, which is preferred for separation of complex mixtures, composition of the mobile phase is changed stepwise or linearly by mixing two or more solvents. The use of binary, ternary, and quaternary solvent gradients is very common.

The mobile phase composition is very critical in achieving selectivity in RPHPLC separation. An ion-pairing reagent is added to adjust the pH of the mobile phase and to alter the hydrophobic character of a solute (7, 8). For example, the positively charged (amino) and negatively charged (carboxylic) functional groups impart hydrophilic character to a peptide. When auxiliary ions are added to the mobile phase, the resulting ion-pairing complex between these groups and the hydrophobic counterions increases hydrophobicity of the peptide. As a consequence, the peptide displays enhanced affinity with the hydrophobic stationary surface. Although a large number of buffer systems have been used in conventional RPHPLC, only the volatile ion-pairing reagents can be used in LC-ESI-MS analysis.

A wide variety of RPHPLC columns are available. Most columns are silica based. Silica offers good mechanical stability. A typical stationary phase is formed by chemically bonding a long-chain hydrocarbon group to porous silica. Typical ligands are n-octadecyl (C_{18}), n-octyl (C_8), n-butyl (C_4), diphenyl, and cyanopropyl. In general, the shorter-chain phases perform well for more hydrophobic solutes and the longer-chain phases for hydrophilic compounds. For example, with respect to the analysis of peptides and proteins, the C_4 phase works better for proteins and large peptides such as those generated by the CNBr treatment of a protein. The C_8 phase is better suited for small hydrophilic proteins, peptide maps, and synthetic peptides. The tryptic fragments of a polypeptide, small hydrophilic peptides, and synthetic peptides are better resolved with C_{18} columns. Silica-based columns are not stable at basic pHs. Because of this problem, columns with cross-linked poly(styrene-divinylbenzene) packing are gaining popularity. Those columns are stable in the pH range of 1 to 14.

With regard to the particle size of packing material, smaller particles offer better resolution. For analytical separations, most columns are packed with 5 µm particles. The average pore diameter of the packing should be roughly 10 times larger than the molecular diameter of analytes. A 300-Å pore size packing is used for separation of proteins and large peptides, whereas a pore size of 100 Å or less is adequate for very small hydrophilic peptides.

HPLC columns are available in different lengths and diameters. Sensitivity, efficiency, sample-loading capacity, and sample size are important criteria in the selection of column dimensions. The longer columns offer little advantage in the separation of large peptides and proteins. On the contrary, sample losses are greater with longer columns. A 20- to 50-mm-long column is preferred for separation of proteins and a 150- to 250-mm-long column is preferred for small peptides. Columns are available ranging in internal diameter from 0.1 to 4.6 mm. Small-diameter columns provide increased sensitivity and are used when the amount of sample is limited; sensitivity can be increased fourfold by decreasing the column diameter by one-half. The diameter of the column also determines flowrate of the mobile phase. Although flowrate does not affect the resolution of proteins, it does matter in the separation of small peptides because their retention in the column is controlled by both adsorption and partition mechanisms. Table 33.1 lists a few of the characteristics of various HPLC columns.

Electrospray Ionization

The development of ESI has revolutionized research in the biochemical field. Now, the on-line combination of ESIMS with HPLC is able to solve with relative ease the real-world problems dealing with high-mass peptides, proteins, and other biopolymers. ESI is an atmospheric pressure ionization (API) technique in which production of ions occurs by electrospraying the solution of an analyte into a chamber maintained at nearly atmospheric pressure (9, 10). ESI source serves as both an LC interface and a source of generating ions. A simple schematic of a typical ESI source is shown in Fig. 33.2. The heart of the ESI source is a stainless steel capillary tube through which a solvent (typically an LC effluent or a mixture of water and methanol containing acetic acid or other suitable additive) flows continuously at the rate of 2 to 5 μL/min. The flow is assisted by an infusion syringe or an LC pump. A high voltage is applied to the tip of the capillary to produce an electrostatic field sufficiently strong to disperse the emerging solution into a fine mist of charged droplets.

Table 33.1 Characteristics of typical RPHPLC columns

Column	Dimensions (length × internal diameter, mm)	Flowrate (μL/min)	Injection volume (μL)	ESI Setup	
Analytical (wide-bore)	150 × 4.6	500–2000	10–500	C	Postcolumn split
				D	No split
Narrow-bore	150 × 2.1	50–500	5–50	C	Postcolumn split
				D	No split
Microbore	150 × 1.0	10–100	1–20	C	Postcolumn split
				B	No split
Capillary LC	300 × 0.3	1–10	0.05–2	A	Precolumn split
Nanoscale LC	500 × 0.05	0.05–0.5	0.01–1		Pure ESI with make-up sheath liquid or micro-ESI

A, B, C, and D refer to Fig. 33.4.

Figure 33.2 Schematic of a typical ESI source.

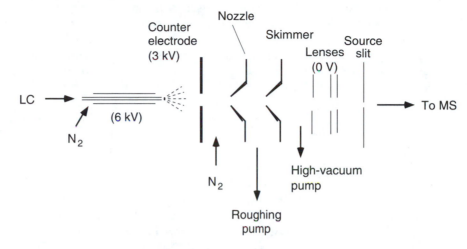

Figure 33.3 depicts the field-assisted desorption of ions from the initially formed charged droplets. As the droplets drift down field through the atmospheric gas, they shrink in size. When the charge density on their surface reaches the Rayleigh instability limit, the repulsive coulombic forces cause the droplets to explode into smaller charged droplets. This fission process is repeated several times until the droplet is reduced to a size when the electric field at its surface is strong enough to assist direct ion evaporation.

The ions are sampled through a conical nozzle and a skimmer into regions maintained at progressively reduced pressures, where most of the neutrals are removed by auxiliary pumping. The free ions then enter the mass analyzer for subsequent mass analysis. The formation of positive or negative ions depends on the capillary bias.

ESIMS is capable of analyzing small to very large compounds. The basis of high-mass analysis is the formation of a series of multiply charged ions of a molecule. The benefit of multiple-charging is to reduce significantly the m/z ratio of the intact analyte, bringing very-high-mass analytes within the usable mass range of a mass spectrometer.

Several parameters must be controlled to obtain a stable spray of the emerging solvent. Flowrate, conductivity, and surface tension of the solvent profoundly influence operation of the electrospray process. A flowrate of 1 to 10 μL/min is optimum for maintaining a stable spray. At higher flowrates, larger droplets are formed, leading to electrical breakdown.

For stable operation, it is important that the current due to ion migration to the electrospray tip be appropriate for the electrolyte concentration. Liquids with conductivity, σ, in the range $10^{-13}/\Omega/$ cm to $10^{-5}/\Omega/$cm are easily electrosprayed. Many of the polar solvents commonly used in HPLC such as methanol, ethanol, isopropanol, and acetonitrile are suitable for electrospray operation. In

Figure 33.3 Schematic representation of field-assisted ion formation in ESI.

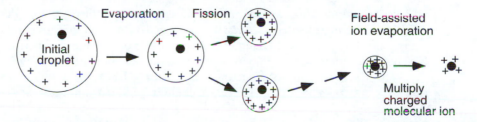

contrast, nonpolar solvents are not dispersed. As a consequence, HPLC in the normal phase mode cannot be used with ESI interfaces unless a polar solvent is admixed with nonpolar mobile phase. Fluids with high surface tension (such as water) are also difficult to electrospray.

Electrolytes generally influence the analyte ion current. Analyte detection sensitivity decreases as the total electrolyte concentration increases in the electrospray solvent. High concentrations of salts, ion-pairing reagents, plasticizers, or detergents, often encountered in protein chemistry, are harmful to the ESI process. The detection sensitivity for many analytes (such as organic bases) is also pH dependent.

Nevertheless, high detection sensitivities (attomole to femtomole) have been reported for many classes of compounds. There is a potential for further improvement in detection sensitivities. Ionization efficiency (that is, the fraction of the neutral analyte molecules converted to ions) of the ESI process is very high but the transport efficiency (that is, the number of ions transported to the mass analyzer) of the interface is still limited and has a room for improvement. The detection sensitivity is also affected by the presence of solvated ions and other cluster ions. These ions reduce the molecular ion signal and increase the chemical background. Both factors have a detrimental effect on detection sensitivity. ESI also exhibits a wide dynamic range.

To overcome some of the limitations of the conventional ESI source and make it more desirable for LC experiments, several different designs have emerged. Some designs have incorporated an auxiliary gas flow (such as N_2), heating through the sampling region, or low-energy collisions to assist evaporation of droplets and reduce cluster formation. In pneumatically assisted electrospray, also called ion spray, a sheath of a nebulizing gas is used to disperse larger droplets, enabling higher liquid flows (100 to 200 μL/min) to be electrosprayed. This type of source is very useful for micro–LC-MS operation (11). To accept the mL/min flowrates of the wide-bore columns directly, a grounded liquid shield is introduced after the ESI needle and the ion sampling capillary is heated (12). In ultrasonically assisted ESI source, the ESI needle is held in an ultrasonic nebulizer (13). This source can produce stable spray from any mobile phase composition. An ESI source suitable for nL/min liquid flows is also available (14).

LC-ESI-MS Setup

Because ion formation in ESI takes place by electrospraying a liquid solution, the technique has led to one of the most successful interfaces between LC and MS. The actual experimental setup is dictated by the sample size, column dimensions, and the available ion source (see also Table 33.1). A few on-line LC-MS configurations are illustrated in Fig. 33.4.

In Fig. 33.4(a), a packed capillary column (about 0.3 mm internal diameter) is directly connected to a pure ESI source. The rate of liquid flow (1 to 5 μL/min) of this column is compatible with the flow requirements of pure electrospray process. Capillary columns offer reduced sample load and high sensitivity. However, they have limitations with respect to reproducing flowrate, gradient, and uniform packing and are limited to nanoliter injection volumes. Therefore, optimum efficiency may not be achieved with these columns. Precolumn flow splitting is recommended for obtaining reliable gradients (15). The solvents (S_1 and S_2) are delivered by micro-LC pumps at higher flow rates (200 to 500 μL/min) and mixed in a static mixer, and the required flow is diverted to the capillary column via a flow splitter. An accurate microflow processor (such as the Accurate Microflow Processor from LC Packings, San Francisco, CA) can be substituted for the solvent mixer and flow splitter for reproducible flowrates. An optional ultraviolet (UV) detector is connected between the column and the ESI interface. A low-volume injector loop (60 or 500 nL, available from Valco, Houston, TX) is used.

With the availability of a pneumatically assisted ESI source, which can accept higher liquid flowrates, it is possible to use microbore columns (0.8 to 1 mm internal diameter) for RPHPLC separation (16). As shown in Fig. 33.4(b), no precolumn or postcolumn splitting is required. A 5-µL sample loop injector is used with this column.

For analysis requiring high efficiency and large sample loads, standard analytical columns (2.1 to 4.6 mm internal diameter) are preferred. As shown in Fig. 33.4(c), the high liquid flow used with wide-bore columns mandates postcolumn splitting with the conventional or pneumatically assisted ESI sources (17). Although flow splitting may lead to peak broadening, there is no serious loss in sensitivity because a mass spectrometer behaves as a concentration-sensitive detector. However, flow splitting has the advantage that a large fraction of the peak material can be recovered for other uses. Alternatively, as shown in Fig. 33.4(d), wide-bore columns can be used with megaflow ESI source without flow splitting (12).

Mass Spectrometer

A mass spectrometer produces a beam of ions, separates them according to their m/z ratios, and records their relative abundances. A mass spectrum is the plot of m/z ratios of the ions versus their relative abundances. Essential components (Fig. 33.5) of a mass spectrometer are sample inlet, ion source, mass analyzer, detector, and data system (18). After leaving the ion source, the ions enter

Figure 33.4 Various configurations used in HPLC-ESI-MS analysis.

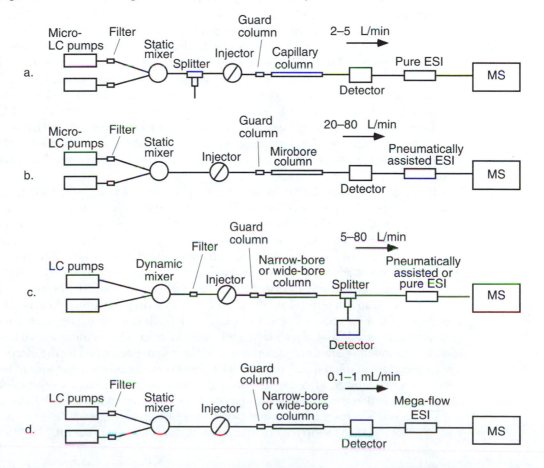

Figure 33.5 Schematic of a mass spectrometer. *(From C. Dass, in D. M. Desiderio, ed., Mass Spectrometry, Clinical and Biomedical Applications, vol. 2, New York: Plenum Press, 1994, pp. 1–52. Reprinted with permission.)*

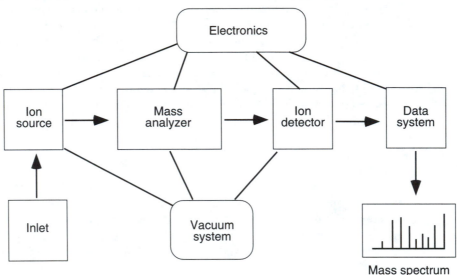

Mass spectrum

the mass analyzer, where they are resolved in terms of their m/z ratios. To obtain a mass spectrum, the mass analyzer is scanned to allow each ion to register its response at the detector.

To achieve full benefit of an LC-MS combination, the mass spectrometer should be of low cost and have high sensitivity, high scan speed, adequate mass range, and reasonable mass resolution. Although several types of mass analyzers are available, the quadrupole mass filter offers most of these desirable features. Because the quadrupole ion source operates at low accelerating potentials, the ESI ion source is easy to outfit with the quadrupole mass filter. High scan speed is a very useful feature of the GC or LC detector. The sample concentration in a GC or LC peak is always not the same; it is zero in the beginning and at the end and maximum in the middle. Because of this changing sample profile, a high scan speed in data collection is required so that a reasonable number of scans can be accumulated and averaged to provide a representative spectrum of the analyte.

In quadrupole mass spectrometers, mass separation is accomplished solely by using electric fields. The mass filter consists of four cylindrical rods. Opposite rods are electrically connected. A direct current (DC) potential and radiofrequency (RF) voltage are applied to the rods. With proper selection of these voltages, a path of stability is created in which only ions of a certain m/z ratio pass through. A mass spectrum is obtained by scanning these voltages. The mass resolution of a quadrupole typically is limited to unit mass.

Although they are still at an early experimental stage, LC-ESI-MS experiments have been performed with an ion trap (IT) (19). The IT, a cousin of the quadrupole mass filter, operates by trapping and storing a batch of ions by means of RF fields in a volume between a doughnut-shaped central ring electrode and two endcap electrodes. At certain RF amplitudes, all ions above a specified m/z ratio follow stable trajectories within the volume described by the electrodes. To obtain a mass specrtum of the trapped ions, the RF voltage is increased such that ions of higher m/z ratios become sequentially unstable and exit through perforations in the endcap electrodes. For detection of those ions, a detector is placed just outside one of the electrodes. Like the quadrupole mass filter, ITMS offers certain unique advantages, such as high scan speed, high ion transmission, low cost, compactness, and simplicity of operation. It also holds potential for MS/MS experiments.

Time-of-flight mass spectrometers (TOFMS) also have a number of useful features, such as unlimited mass range, high detection sensitivity (due to their high ion transmission), very high spectrum acquisition rate, multiplex detection capability, and low cost, that are attractive for LC-MS combination. In this instrument, a pulse of ion is allowed to drift in a field-free region, where ions are separated on the basis of their velocities. Because all ions exit the source with the same kinetic energy, the lighter ions travel faster and reach the detector earlier than the heavier ions. The coupling of ESI with TOFMS is complicated because, unlike scanning instruments, TOFMS detects ions in pulses and ESI produces ions in continuous mode. The coupling of ESI with TOFMS is achieved by using the orthogonal ion extraction approach. ESI-produced ions are stored between each duty cycle and injected into drift tube of the TOF instrument in the pulse mode (20–22).

Often, a need exists for high-resolution mass measurements. To take advantage of the high-resolution capability of the double-focusing magnetic sector instruments (an electric sector followed by a magnetic sector or vice versa), ESI sources have been coupled to these instruments. However, the source design is somewhat complicated because of the need for high voltages in ion acceleration. A mass spectrum of the ions ejected into the mass analyzer is acquired by scanning the magnetic field.

Fourier transform MS (FTMS) based on the ion cyclotron resonance (ICR) principle also holds immense potential for the analysis of ESI-produced ions (23).

What It Does

On-line LC-ESI-MS is useful in determining the MW and structure of the individual components of the mixture resolved by the LC column. The data output is in the form of ion chromatogram and mass spectrum. Collision activation in the source or in the intermediate region of a tandem mass spectrometer provides structural information.

Ion Chromatograms

In LC-ESI-MS analysis, a sample dissolved in a suitable solvent (such as water, methanol, or acetonitrile) is applied to the top of the column by injecting through a fixed-volume loop injector. While substances elute from the column, MS is simultaneously scanned within a certain mass range to furnish mass spectra of the LC effluents. Data flow in an LC-MS run is very large; hundreds and even thousands of mass spectra are acquired during one complete HPLC run. Therefore, a computer-based data system is an absolute necessity for LC-MS operation. A computer stores the sum of ion signals from all the ions in the spectrum as a total ion current (TIC). The TIC of each mass scan is plotted against the scan number to give a TIC chromatogram. This TIC profile is similar to a normal chromatogram obtained using a conventional UV detector. Although a TIC chromatogram is nonspecific in nature, it serves a unique purpose in directing the analyst to choose any individual or group of scans for further examination.

Often, in a mixture, a series of homologous compounds is present, ionization of which produces a common ion. When the ion current of that characteristic ion versus scan number is plotted, a more selective chromatogram, known as a mass chromatogram or ion extraction chromatogram, is obtained, from which the compounds belonging to that homologous series can be identified easily. The mass chromatogram is also useful in distinguishing the coeluting components.

Rather than acquiring a full-scan mass spectrum, it is also possible to record the signal arising from one or more compound-specific ions only. This technique, known as selected ion monitoring (SIM), is very useful in selective detection of a homologous series of compounds and quantification of a target compound. During a full scan, ion current from all ions except the one arriving at the detector is wasted. In SIM, the detector is dedicated to monitoring the specified type of ion, resulting in significant enhancement in the detection sensitivity over a full scan.

Determination of the Molecular Mass of Macromolecules

One major benefit of ESIMS is its applicability to determination of the MW of macromolecules. This facility is the result of the multiple-charging phenomenon. For example, in positive-ion ESI analysis, protonation of basic amino acid residues and the terminal amino group in proteins leads to generation of multiprotonated ions of the general composition $(M + nH)^{n+}$. Similarly, abstraction of protons from acidic amino acids of a protein results in the formation of multiply charged negative ions, such as $(M - nH)^{n-}$. The molecular mass M of a macromolecule may be determined from the m/z values and charge state of any two multiply charged ions in a series. Here, m_1 is the m/z of the $(M + nH)^{n+}$ ion (here, $H = 1.007829$ is the mass of the proton) and m_2 $(m_1 > m_2)$ of the adjacent ion such that

$$m_1 = \frac{M + nH}{n} \tag{33.1}$$

and

$$m_2 = \frac{M + (n + 1)H}{n + 1} \tag{33.2}$$

These two simultaneous equations can be solved for n and M to give

$$n = \frac{m_2 - H}{m_1 - m_2} \tag{33.3}$$

and

$$M = n(m_1 - H) \tag{33.4}$$

Several values of M are calculated from the successive pairs of adjacent ions and the best value of M is obtained by taking average of all of the individually calculated mass values. Thus, when more types of ions are included in the calculation, better precision is achieved. Computer programs for these calculations are common with commercial instruments.

Collision-Induced Dissociation (CID)

Whereas the molecular ions formed in the ESI process are useful for obtaining the MW of an analyte, structural information is not available, as fragment ions are usually absent. However, fragmentation can be induced by imparting excess energy to the ions. Before mass analysis, collision activation is accomplished in the source by increasing the nozzle-skimmer voltage. However, this technique is limited to pure eluting components.

Tandem Mass Spectrometry (MS/MS)

To obtain structural information for substances eluting in the real-time format, an on-line combination of LC with a tandem mass spectrometer is used. In contrast to the nozzle-skimmer voltage CID, LC-MS/MS is useful even when an LC peak contains more than one component. This instrument consists of two stages (MS_1 and MS_2) of mass analysis and an intermediate region equipped with a device to induce fragmentation. From the pool of ions present in the ion source, a specified ion is mass selected by the MS_1 into the intermediate region, where collisions with an inert gas (such as He, Ar, or N_2) facilitate fragmentation. MS_2 is scanned to provide mass analysis of the product ions formed by dissociation of the mass-selected precursor ion. Although state-of-the-art four-sector magnetic tandem instruments are superior for high-quality MS/MS data (high-energy CID and higher mass resolution for precursor ion mass selection and product ion mass analysis), a triple-stage quadrupole (Q_1qQ_2) tandem mass spectrometer is more than adequate for LC-MS/MS methodology. Here, Q_1 acts as an MS_1, the RF-only quadrupole q as the intermediate region, and Q_2 as MS_2.

Analytical Information

Qualitative

LC-ESI-MS has been primarily used for identification of several classes of thermally labile compounds. The basis of identification is the retention time (obtained from ion chromatograms) and the MW and structure-specific fragment ions information (obtained from the ESI mass spectrum).

Quantitative

Because of its inherent high detection sensitivity, ESI can be easily adapted to quantitative determination (24). The area beneath a peak in an ion chromatogram is proportional to the sample ion concentration. A TIC chromatogram, a mass chromatogram, or an SIM chromatogram can provide quantitative information. However, SIM is invariably used because it provides lower detection limits. A calibration curve in the range of analysis should be obtained. The use of an internal standard is recommended. A stable-isotope–labeled analog has the advantage over the homologous and structurally analogous internal standards in that it can accurately account for any sample losses in various sample-handling and chromatographic steps and fluctuations in the ionization and detection processes. Although the analyte and its stable-isotope–labeled analog both coelute from the HPLC column, they exhibit MS response at two distinct m/z values.

In situations where another compound has the same MW as that of the analyte or its internal standard coelutes, one must use the MS/MS technique known as selected reaction monitoring (SRM). The molecular ions of the analyte and internal standard are alternately mass selected and allowed to fragment in the intermediate region, and one or more product ions exclusively formed from these molecular ions are monitored.

Applications

LC-ESI-MS has found applications in several areas of research such as sequencing of proteins, identification of mixtures of compounds, tryptic maps, and posttranslational modifications in proteins, structure elucidation of metabolic products, analysis of drugs, pesticides, and toxins, and quantification of a variety of compounds.

1. Peptide Mapping.

This is a generalized approach used for determining a complete or partial sequence of a protein. The protein is cleaved into smaller, manageable fragments by enzymatic or chemical treatment, the fragments are separated, and their structure is determined. The on-line combination of LC with ESIMS is a very viable approach for this application. As an example, the trypsin digestion of β-lactoglobulin A (with N-tosyl-L-phenylalanine chloromethyl ketone–treated trypsin for 16 to 20 hr at 37 °C with a substrate-to-enzyme ratio of 50:1 in 50 mM ammonium bicarbonate buffer at pH 8.5) produces several small fragments (16). The tryptic digests are separated on a C_8 microbore (1 mm internal diameter) RPHPLC column with gradient elution using acetonitrile in water (both containing trifluoroacetic acid, TFA) as the mobile phase. The MW of each tryptic

Figure 33.6 MS/MS spectrum of a singly charged tryptic peptide derived from β-lactoglobulin A. *(Reprinted by permission of Elsevier Science Inc. from "LC/MS and LC/MS/MS Determination of Protein Tryptic Digests" by E. C. Huang and J. D. Henion, Journal of the American Society for Mass Spectrometry, Vol. 1, pp. 158–165. Copyright © 1990 by the American Society for Mass Spectrometry.)*

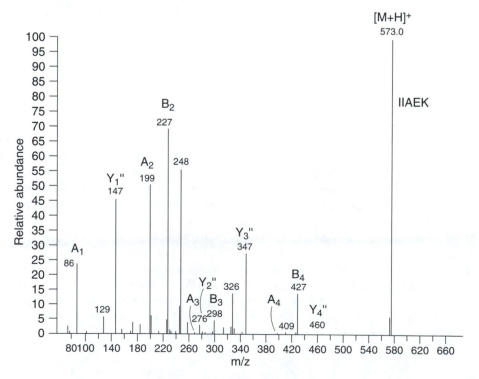

fragment eluting from the LC column is determined by ESI. After the molecular ion of each fragment is identified, a second HPLC run with the MS operating in the MS/MS mode provides the amino acid sequence of those tryptic peptides. Figure 33.6 shows the MS/MS spectrum of one of the tryptic fragments IIAEK (H$_2$N–Ile–Ile–Ala–Glu–Lys–OH). In that figure, the peaks labeled as A$_1$, A$_2$, etc. and B$_2$, B$_3$, etc. are the N-terminal fragments formed by cleavage of the –CHR–CO– and –CO–NH– bonds, respectively, in the peptide's backbone (25). The corresponding C-terminal ions formed by cleavage of the –CO–NH– bond are annotated as Y$_1''$, Y$_2''$, etc.

2. Selective Detection of Compounds in a Complex Mixture.

This objective is illustrated in Fig. 33.7, which shows a negative ion TIC chromatogram of a mixture of 16 peptides (3 phosphorylated and 13 nonphosphorylated) separated on a C$_{18}$ packed capillary column (0.32 mm internal diameter) (26). For selective detection of phosphorylated peptides, the orifice voltage of the pneumatically assisted ESI source is increased and the phosphate marker ions, PO$_2^-$ and PO$_3^-$, are monitored in the SIM mode. The SIM chromatogram shows only three distinct peaks at the retention times of phosphorylated peptides.

Similarly, the selective detection of glycopeptides is accomplished by monitoring the carbohydrate-specific ions such as the N-acetylhexosamine oxonium ion at m/z 204 (27).

Figure 33.7 Negative-ion LC-ESI-MS analysis of a 16-component peptide mixture. The TIC chromatogram is shown by the shaded area. The combined ion extracted chromatogram (dark line) for m/z 63 (PO$_2^-$) and 79 (PO$_3^-$) demonstrates the presence of three phosphopeptides. "RIC" stands for reconstructed ion chromatogram. *(Reprinted by permission of Elsevier Science Inc. from "Selective Determination of Phosphopeptides in Complex Mixtures by Electrospray Liquid Chromatography/Mass Spectrometry" by M. J. Huddleston et al., Journal of the American Society for Mass Spectrometry, Vol. 14, pp. 710–717. Copyright © 1993 by the American Society for Mass Spectrometry.)*

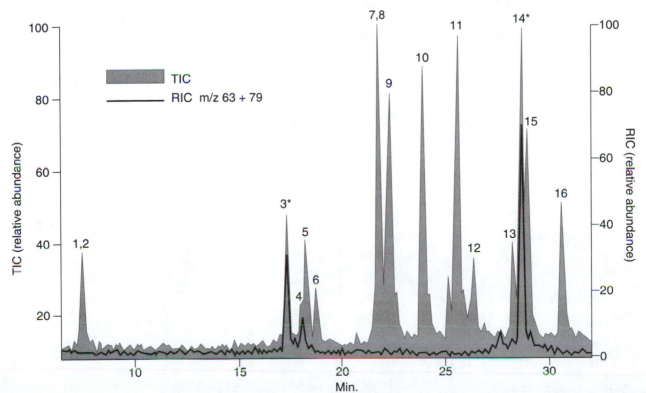

3. Solving Real-World Biomedical Problems.

LC-ESI-MS has been used in identification of peptides associated with viral infection of cells. This study has great implications in immunotherapy. LC-ESI-MS was able to determine the primary structure of a peptide associated with class I major histocompatibility complex (MHC) HLA-A2.1 recognized by melanoma-specific human cytotoxic T cell lines (28).

4. Identification of Metabolites.

Many hormones, neuropeptides, drugs, and toxins are metabolically modified when administered in the body. Figure 33.8 illustrates the use of the negative-ion LC-ESI-MS for the structure elucidation of an unknown metabolic product of a herbicide bromacil (29). A polystyrene-divinylbenzene–based polymer capillary column (C_{18}, 0.32 mm internal diameter) was used for RPHPLC separation.

5. Quantification of Compounds in Biological Matrix.

Quantification of a peptide drug (Ac–Arg–Pro–Asp–Pro–Phe–NH_2) in human and rabbit plasma was accomplished using LC-ESI-MS (30). After addition of the deuterated internal standard, the sample was desalted, proteins were precipitated, and the drug was extracted by solid-phase extraction. HPLC was performed in the isocratic mode with a mobile phase consisting of 1% methanol and 20% acetonitrile in 10 mM ammonium formate (pH 5.2) using an analytical column (2.0×250 mm). A flow splitter was used to allow only a one-twentieth flow to enter the ESI source.

Figure 33.8 ESI negative-ion mass spectrum of bromacil goat urine metabolite of bromacil. *(R. W. Reiser and A. J. Fogiel, Rapid Communications in Mass Spectrometry, 8, pp. 252–257. Copyright © 1994 by R. W. Reiser and A. J. Fogiel. Reprinted by permission of John Wiley & Sons, Ltd.)*

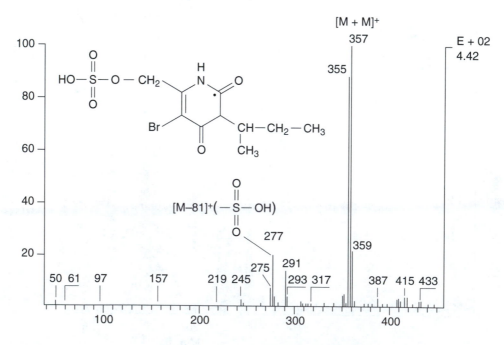

Nuts and Bolts

Relative Costs

LC-ESI single quadrupole instrument	$170–$200K
LC-ESI triple-stage quadrupole instrument	$350–$400K
LC-ESI double-focusing magnetic sector instrument	$400–$450K
LC-ESI tandem magnetic sector instrument	>$600K

A major factor in the cost of LC-MS instrument is its MS/MS capability. In general, magnetic sector instruments are more expensive than quadrupole instruments.

Vendors for Instruments and Accessories

See Chap. 28 (p. 564) for a detailed list of vendors.

Required Level of Training

Operation of Instrument

Routine analysis can be performed by person with a college degree in chemistry. However, an experienced operator is required for obtaining optimum results.

Processing Qualitative and Quantitative Data

For qualitative interpretation of the data, a person with a graduate-level course in MS is helpful. Quantitative determination requires a college degree in chemistry with analytical chemistry background.

Service and Maintenance

Maintenance of a mass spectrometer requires trained personnel. Cleaning the ion source and changing the pump oil are the major service items. Servicing the electronic components requires advanced electronics training, although diagnostic software may help in troubleshooting. Often, services of the vendor's engineer are needed. A service contract is useful.

Suggested Readings

ALLEN, M. H., AND B. I. SHUSHAN, "Liquid Chromatography/Mass Spectrometry," *LC-GC*, 11 (1993), 112–26.

MANT, C. T., AND R. S. HODGES, *High-Performance Liquid Chromatography of Peptides and Proteins*. Boca Raton, FL: CRC Press, 1991.

MATUSO, T., AND OTHERS, *Biological Mass Spectrometry, Present and Future*. New York: Wiley, 1994.

McCloskey, J. A., *Methods in Enzymology*, vol. 193. San Diego, CA: Academic Press, 1990.

Yergey, A. L., and others, *Liquid Chromatography/Mass Spectrometry: Techniques and Applications.* New York: Plenum Press, 1990.

References

1. A. P. Bruins and B. F. H. Drenth, *J. Chromatogr.*, 271 (1983), 71–82.

2. W. H. McFadden, H. L. Schwartz, and D. C. Bradford, *J. Chromatogr.*, 122 (1976), 389–96.

3. M. L. Vestal, *Analytical Chemistry*, 56 (1984), 2590–92.

4. R. C. Willoughby and R. F. Browner, *Analytical Chemistry*, 56 (1984), 2625–31.

5. R. M. Caprioli, T. Fan, and J. S. Cottrell, *Analytical Chemistry*, 58 (1986), 2949–54.

6. E. C. Huang and others, *Analytical Chemistry*, 62 (1990), 713A–25A.

7. D. Guo, C. T. Mant, and R. S. Hodges, *J. Chromatogr.*, 386 (1987), 205–22.

8. C. Dass, P. Mahalakshmi, and D. Grandberry, *J. Chromatogr. A*, 678 (1994), 249–57.

9. J. B. Fenn and others, *Science*, 246 (1989), 64–71.

10. R. D. Smith and others, *Mass Spectrom. Rev.*, 10 (1991), 359–452.

11. A. P. Bruins, T. R. Covey, and J. D. Henion, *Analytical Chemistry*, 59 (1987), 2642–46.

12. G. Hopfgartner and others, *Analytical Chemistry*, 65 (1993), 439–46.

13. J. F. Banks, Jr., and others, *Analytical Chemistry*, 66 (1994), 406–14.

14. M. R. Emmett and R. M. Caprioli, *J. Am. Soc. Mass Spectrom.*, 5 (1994), 605–13.

15. E. C. Huang and J. D. Henion, *Analytical Chemistry*, 63 (1991), 732–39.

16. E. C. Huang and J. D. Henion, *J. Am. Soc. Mass Spectrom.*, 1 (1990), 158–65.

17. T. R. Covey, E. C. Huang, and J. D. Henion, *Analytical Chemistry*, 63 (1991), 1193–1200.

18. C. Dass, in D. M. Desiderio, ed., *Mass Spectrometry, Clinical and Biomedical Applications*, vol. 2 (New York: Plenum Press, 1994), pp. 1–52.

19. S. A. McLuckey and others, *Analytical Chemistry*, 63 (1991), 375–83.

20. J. G. Boyle and C. M. Whitehouse, *Analytical Chemistry*, 64 (1992), 2084–89.

21. O. A. Mirgorodskaya and others, *Analytical Chemistry*, 66 (1994), 99–107.

22. A. I. Verentchikov, W. Ens, and K. G. Standing, *Analytical Chemistry*, 66 (1994), 126–33.

23. C. C. Stacy and others, *Rapid Commun. Mass Spectrom.*, 8 (1994), 513–16.

24. C. Dass and others, *J. Am. Soc. Mass Spectrom.*, 2 (1990), 149–56.

25. P. Roepstorff and J. Fohlman, *Biomed. Mass Spectrom.*, 11 (1984), 601.

26. M. J. Huddleston and others, *J. Am. Soc. Mass Spectrom.*, 4 (1993), 710–17.

27. M. J. Huddleston, M. F. Bean, and S. A. Carr, *Analytical Chemistry*, 65 (1993), 877–84.

28. A. L. Cox and others, *Science*, 264 (1994), 716–19.

29. R. W. Reiser and A. J. Fogiel, *Rapid Commun. Mass Spectrom.*, 8 (1994), 252–57.

30. J. Crowther and others, *Analytical Chemistry*, 66 (1994), 2356–61.

Chapter 34

Laser Mass Spectrometry

Kathleen L. Walker and Charles L. Wilkins

University of California, Riverside
Department of Chemistry

Summary

General Uses

- Mass information may be obtained for molecules ranging from 200 to 1,000,000 Da.
- Structural information often may be obtained.
- Biological, organic, inorganic, and polymer samples can be analyzed.
- It is generally not quantitative.
- Low sample quantities are required.
- Mass accuracy is as good as a few parts per million.

Common Applications

- Mass information for nonvolatile samples
- Identification of posttranslational modifications
- Identification of proteins that are blocked to Edman degradation
- Sequence and structure information through collisional dissociation
- Polymer mass and sequence information
- Ablation of materials for composition information
- Analysis of samples isolated by chromatographic and electrophoretic methods

Samples

State

Solid or liquid samples that will not volatilize within 5 min under vacuum (10^{-4} to 10^{-8} torr) can be analyzed.

Amount

Required quantities vary with different analytes and analysis conditions.

Solids Nanomoles to subpicomole samples are sufficient.

Liquids Millimolar to micromolar solutions are sufficient.

Preparation

Laser Desorption

- Solids are diluted in appropriate solvents and deposited directly onto a sample probe. If required, a chemical ionization reagent is added at this time.
- Liquids are diluted (if necessary) and similarly deposited directly onto the sample probe.

Laser Ablation

- Solids are affixed directly to the laser probe.
- In matrix-assisted laser desorption, solids are diluted in appropriate solvents, mixed with the appropriate matrix, and deposited directly onto a sample probe. If required, a chemical ionization reagent is added at this time.
- Liquids are diluted (if necessary) and similarly mixed and deposited directly onto the laser probe.

Analysis Time

Total sample preparation and analysis time for routine samples can be as little as 10 min. Highly automated systems allow analysis of multiple samples on the same sample probe; in this case, the average analysis time decreases.

Limitations

General

- It is not routinely quantitative.
- Matrix-assisted laser desorption/ionization (MALDI) samples must usually be soluble in solvents that are compatible with an appropriate matrix.
- Samples must contain an easily ionized atom (such as O or N). If not, either suitable chemical ionization reagents or special matrices must be used.

- Laser desorption samples must absorb at the laser wavelength used.
- Direct laser desorption samples are typically limited in mass range to samples below 10 kDa.
- MALDI samples should have little or no absorption at the laser wavelength.
- The mass spectrum of mixtures is often unrepresentative of both composition and quantity.

Accuracy

Accuracy varies according to the mass analyzer. Time of flight is the most common mass analyzer, and with external calibration, 0.1 to 0.03% mass accuracy is routine. With internal standards, accuracy can be improved to a few tenths of a percent. Fourier transform mass spectrometers can be expected to give much higher mass accuracy, with errors often as low as a few parts per million.

Sensitivity and Detection Limits

Detection limits are routinely between 1 and 10 pmol deposited on a probe of area 3.14 mm^2.

Complementary or Related Techniques

- X-ray crystallography
- Fluorescence spectroscopy
- Gel electrophoresis
- Polymer methods such as gel permeation chromatography
- Nuclear magnetic resonance (multidimensional analyses can provide structural information for macromolecules)
- Infrared and Raman spectroscopy (structural information for molecular identification)
- Chromatographic and electrophoretic methods (molecular identification by retention or electrophoretic migration time)

Introduction

Laser desorption ionization techniques were developed for the analysis of nonvolatile samples, generally macromolecules. Fast atom bombardment (FAB) techniques are useful for nonvolatile samples ranging in molecular weight up to 2000 Da, but are not generally useful for higher-molecular-weight samples.

In all mass spectrometric techniques, mass analysis is based on formation of ions, which are detected at their mass-to-charge (m/z) ratios. The formation of almost exclusively intact, singly charged, molecular ions is the primary attractive feature of laser desorption (LD) techniques and is an advantage over FAB (even for analytes of mass less than 2000 Da). Direct laser desorption often involves some fragmentation of the analyte, and is generally limited to analytes of mass less than 10 kDa. To address the problem of molecular fragmentation and extend the available mass range of nonvolatile applications, matrix-assisted laser desorption/ionization (MALDI) was introduced independently by the groups of Tanaka and Hillenkamp in 1989 (1, 2). MALDI is the method of choice when no fragmentation is desired, or for analytes of mass greater than 10 kDa.

When structural information is desired from fragmentation, secondary experiments designed specifically for that purpose can be carried out. These include mass assignments based on postdesorption formation of product ions. These ions can be formed in several ways. If the desorption process imparts energy to the ions that exceeds their dissociation limit, they may undergo intramolecular rearrangements and dissociate to product ions. This process is called metastable decay. Dissociation of analytes may also be induced by more finely controlled experiments such as collision-induced dissociation (CID) and surface-induced dissociation (SID) (3, 4). CID is commonly used with Fourier transform mass spectrometers, where the ions are trapped in the presence of an inert bath gas. When the ions are excited in the presence of the gas, collisional transfer of energy can occur. If this process is efficient, ions will again have sufficient internal energy to dissociate and form product ions. SID occurs by applying a voltage to one of the Fourier transform mass spectrometer trapping plates, which forces the ions to collide with the other uncharged trap cell walls. If the ions collide with sufficient force, they will fragment and form product ions.

Laser desorption (LD) occurs when the analyte absorbs the incident laser radiation and is subsequently volatilized into the gas phase and ionized. Infrared lasers are often used for this technique because all molecules absorb in the infrared. The exact mechanism of LD ionization has not been defined.

MALDI relies on mixing the analyte with a matrix that has a strong UV absorption at the wavelength of the laser used. The matrix–analyte mixture is deposited on the laser probe and allowed to dry. During the drying process the analyte and matrix usually cocrystallize, with the analyte forming an inclusion complex in the matrix crystal. When the laser radiation impinges on the crystals, it is absorbed by them and the matrix and analyte desorb into the gas phase. Abundant intact analyte molecular ions are formed during this process. Generally, MALDI is used with ultraviolet lasers operating at 337 or 355 nm. Again, the mechanism of MALDI is not well defined, although many investigators have published explanations of their findings as they relate to both the sample preparation and ionization process. Unfortunately, these explanations are usually based on indirect observations, not on direct empirical evidence.

Mass spectral databases such as those available for organic compounds have not been widely established for macromolecules. Because the technique is soft and does not induce fragmentation, it is not reliant on spectral interpretation for mass assignments, so databases are usually not required. However, databases providing information on protein sequences as examined by mass spectrometry do exist, and can be useful for structure elucidation applications.

How It Works

Physical and Chemical Principles

Mass spectra of macromolecules are measured as a function of an ion's m/z ratio. Commercially available laser mass spectrometers consist of a laser ion source, a mass filter, and an ion detector, which is compatible with the mass filter. With Fourier transform mass spectrometry, data collection and storage are carried out with computer workstations. In the case of time-of-flight mass spectrometry, data acquisitions are very fast (microseconds) and, in most instruments, high-speed digital oscilloscopes (2 to 4 nsec time resolution) acquire the data. Data are then transferred to computers for storage and processing.

For the sake of brevity, all instruments are described here for use as MALDI instruments. They work similarly for LD. Thus, their fundamental principles need not be reiterated for laser desorption.

MALDI combines a chemical ionization process with LD. A laser is fired at a surface on which the matrix–analyte mixture has cocrystallized. The mixture desorbs into the gas phase and analyte ions of the general compositions $(M+H)^+$, $(2M+H)^+$, $(M+2H)^+$, and their negative ion $(M–H)^-$ analogs are most commonly formed during MALDI. The exact mechanism of ion formation in MALDI has not been determined at the time of this writing.

Matrices

Several matrices have been popularized by the original developers of MALDI and are very effective for many uses. These are 2,5-dihydroxybenzoic acid (common name gentisic acid), sinnapinic acid, and alpha-cyano-4-hydroxy cinnamic acid. These are most commonly used for peptides, proteins, and heteroatom-containing polymers such as polyethylene glycol. Picolinic acids are most popular for applications to oligonucleotide polymers. Other matrices have been developed to facilitate the acquisition of improved-quality spectra in specific applications and the reader is advised to consult the recent literature to obtain the most up-to-date information on matrices for MALDI.

Lasers

The lasers used for LD and MALDI are capable of generating nanosecond pulses with typical output energies of up to several millijoules. Focused laser powers are on the order of 10^6 to 10^7 W/cm^2. The most popular ultraviolet (UV) MALDI laser is the nitrogen laser, which operates at 337 nm. Low-power nitrogen lasers are extremely simple, inexpensive, and very compact. Unfortunately, in many cases the laser pulse energy varies by as much as 20% from pulse to pulse. Because mass spectra are usually the average of several laser shots, this energy variance can degrade spectral resolution. When better specifications are required, higher-powered nitrogen or other types of UV lasers (which are more expensive and complex) can be used. Neodymium/YAG lasers offer an excellent higher-end option for MALDI mass spectrometers. The output of neodymium/YAG lasers is at 1064 nm, but for MALDI this is usually frequency-tripled to shift the output to 355 nm. Carbon dioxide lasers are commonly used when work in the infrared region is desired. These lasers are intermediate in complexity between nitrogen and neodymium/YAG lasers.

MALDI Time-of-Flight

MALDI time-of-flight mass spectrometry (MALDI-TOFMS) is the most commonly purchased MALDI instrument, and was the original type of instrument used for development of MALDI. There are several reasons for the popularity of time-of-flight (TOF) instrumentation. Among them are economy, ease of use and maintenance, sensitivity, and high upper mass limit (up to 1 million Da). The major shortcoming of MALDI-TOFMS is its low mass resolution (routinely 100 to 250 when defined as $M/\Delta M$, full width at half maximum). For analyses of analytes of ions with m/z ratios above 50 kDa resolution rarely exceeds 100. This limiting factor has not prevented TOF from being basis of the most popular of MALDI instruments and providing useful mass data for many scientists who have no other way to characterize the mass of their nonvolatile macromolecules.

TOF instruments are very simple mass analyzers. The laser desorption source is maintained at a high voltage (5 to 30 kV), and after ion formation occurs, the ions are accelerated by this potential out of the source region and down a flight tube of length 1 to 2 m. At high potential, the ion kinetic energies are approximately the same, and velocities vary inversely with the square root of mass. By traversing a long flight path, the ions become sufficiently spatially resolved to be detected separately in time by the detector at the end of the flight tube. In the most common commercial instrumental configurations, ion flight times are on the order of tens to hundreds of microseconds. Short analysis times are an important advantage of TOF because many of the macromolecular ions formed in MALDI have sufficient energy to undergo rearrangements to product ions (called metastable decay) if sufficient time is allowed (usually milliseconds). Another advantage is the ability to monitor an entire spectrum from a single laser pulse. Both positive and negative ions are usually formed in MALDI. Both types of ions can be analyzed. In order to accomplish this the source polarity is maintained at the same bias as the desired ions (e.g. positive for positive ions).

Mass detection usually uses some type of ion multiplier. Microchannel plates or hybrid detectors with dynode-backed microchannel plates are most common and offer good sensitivity. These detectors are usually biased between 3 and 5 kV. Unfortunately, they suffer from dynamic range problems. If the relative abundance of an ion with one m/z ratio is high relative to others, the signal of the lower-abundance ion may be suppressed if its m/z ratio is close to that of the ions generating the large signals. This is one source of difficulty in mixture analyses by MALDI. Many commercial instruments eliminate the dynamic range problems introduced by the matrix by providing a variable ion gate. This gate is biased at the same potential as the ions and pulses quickly after the laser pulse to deflect matrix ions away from the flight path to the detector.

Because the mass measurements occur on the microsecond time scale, data from the detector must be fed directly to a transient recorder, usually a fast oscilloscope (50 MHz or greater). After the data are collected by the transient recorder they are transferred to a computer with a data processing software package, where the data are stored and manipulated. Data are collected as flight times and must be converted to mass data by correlating the flight times with particular masses. This is done by measuring the spectrum of a standard of known mass and using a TOF equation to calibrate the instrument. Most software packages contain a routine that performs these calculations. Due to instrumental fluctuations, calibration is generally carried out on a daily basis. When extremely high mass accuracy is required (0.03% or more), an internal standard of known mass can be mixed with the analyte and used to calibrate the mass scale. Higher mass accuracies are obtained in this manner because inconsistencies that occur from spectrum to spectrum are lessened when the ions from the mass standard and analyte are in the same spectrum.

When higher resolution of analytes is required, the resolution can be improved by use of an ion reflector. Ion reflectors (commonly called reflectrons) are available in the costlier instruments and can significantly enhance the resolution of MALDI ions of lower m/z ratios (typically those of 8 kDa or less). Ion reflectors improve resolution by focusing the kinetic energy differences of an ion packet of a single m/z ratio. This is done by designing the reflectron to be a repulsive potential well, with the highest potential at the bottom of the well. The ion packet enters the well and the ions penetrate the well according to their kinetic energy. The ions with the highest velocity can go further into the well before the repulsive potential overcomes their kinetic energy and forces them to turn around and exit the well. In ideal cases, the well's potentials are closely matched to the kinetic energy spread of the ions and the ion packet is extremely well focused upon exiting the well. In this case, the resolution for ions up to 3 kDa can be 3000 or more. The inherent difficulty with ion reflectors is that the potential well is tuned to a fixed set of parameters, and thus has an optimum m/z range, which is focused, and masses outside this range are not as well focused. Commercially available reflectrons are currently most effective for m/z ratios no higher than a few thousand daltons.

Fourier Transform

MALDI/Fourier transform mass spectrometry (FTMS) or Fourier transform ion cyclotron resonance (FTICR) is becoming increasingly popular as an analytical tool for determination of molecular masses with high accuracy and resolution. Since the introduction of MALDI-FTMS in 1991 (5), several manufacturers have been producing these instruments, and they are becoming an important part of the modern mass spectrometry facility.

The underlying phenomenom exploited by FTMS is the precession of a charged ion in a magnetic field. The frequency of this motion is called the cyclotron frequency (ω), and is a function of an ion's m/z ratio, $\omega = qB/M$, where q is the charge of the ion, B is the magnetic field strength, and M is its mass. The ions are trapped in an ion cyclotron resonance cell, which most commonly is of cubic geometry and is constructed from thin conductive metal plates. To trap the charged ions, the plates are held at a potential of a few volts and the ions precess at their cyclotron frequency within the plates. Subsequently, they are excited to cause them to move as a phase-coherent packet. To detect the ions, two cell plates are configured to detect the cyclotron frequency and each time the ions pass that plate, they are detected. This information is stored as frequency data and transformed by the data system into mass information. FTMS instruments need infrequent calibration, assuming a stable magnetic field. A calibration mixture is occasionally mass analyzed and the instrumental data system used to determine a calibration, which is stored. Daily corrections to the calibration are performed by using the data system to compare the most recent m/z data with the previously stored calibration equation.

When using MALDI-FTMS instruments, mass resolution well above that required to resolve carbon isotopes is routinely obtained. In the hands of a skilled operator, resolution of several hundred thousand can be obtained for molecules ranging in mass up to 1500 Da and resolving power of almost 100,000 has been demonstrated for proteins with masses greater than 5,000 Da. Extremely high mass accuracies (mass errors of 1 or 2 ppm) are also common for species in this molecular weight range. Detection limits are as low as 1 pmol. The major limiting factor in the utility of MALDI-FTMS is the available upper mass limit. Resolution degrades rapidly for analytes of molecular weight greater than 12 kDa. Much effort is being focused on instrument design and methods development, which will allow FTMS to be more effective for higher masses, and the upper mass limit should be improved in the future. Because FTMS traps the ions, it is a useful tool in sequencing and ion chemistry studies. The trapped ions can be dissociated by gas phase collision, SID, or photodissociation. The resulting mass information from fragment ions can be used for structure determination.

FTMS does require low pressures be maintained in the analyzer cell (10^{-8} torr or lower) if the resolving power and accuracy cited above are to be achieved.

The laser desorption and MALDI ion sources associated with these mass analyzers are much like those used on TOF. The main variation in MALDI-FTMS sources lies in their location (either just outside of the ICR cell or external to the magnetic field). Placing the source directly outside the cell possibly allows greater sensitivity and minimizes the time between ionization and detection (which can be a critical issue as ions formed by MALDI often contain energy in excess of their dissociation limit and may experience metastable decay after a few microseconds or milliseconds). External MALDI sources desorb the ions outside the magnet in a differentially pumped region and then guide them into the ICR cell. These sources allow lower pressures (10^{-9} torr) to be maintained in the analyzer cell region, potentially enhancing the resolution over the internal source approach, although this has not yet been demonstrated to be a routinely attainable result.

The main limitations of FTMS are the low limit of the upper mass range and the relatively high cost to purchase and maintain. It should also be noted that mass resolving power varies inversely with mass. However, when high mass accuracy is required, FTMS is the method of choice for analytes within its molecular weight range.

Other MALDI mass spectrometers are in the developmental stage and are not discussed here. Many possibilities exist for the interface of laser desorption and currently existing mass analyzers. These new instruments will undoubtedly be an important part of the future of laser mass spectrometry.

What It Does

Sample classes cover a broad range and include polymers, fullerenes, dendrimers, inorganics, proteins (including those that are glycosylated), and DNA. Samples range from highly pure, well-characterized materials to heterogeneous unknowns or mixtures of unknowns that may be contaminated.

Organic polymers, like most synthetic organic products, are often very chemically clean, and often highly homogeneous. Many are easily analyzed by LD or MALDI using routine matrices and sample preparation. Polymers containing easily ionized groups such as oxygen or nitrogen (called heteroatoms) fall into this category and are routinely analyzed using 2,5-dihydroxybenzoic acid as matrix. Some heteroatom-containing polymers respond better to a matrix recently introduced for this purpose, trans-indole acrylic acid (6). Hydrocarbon polymers do not contain heteroatoms and are usually not water-soluble, so they present a problem for analysis by MALDI. Currently available methods for hydrocarbons include direct laser desorption and direct laser desorption in the presence of metal ions such as silver (7). MALDI has not been widely applied to purely hydrocarbon macromolecules, although one matrix exists for this purpose (8).

Samples of biological origin often contain contaminants introduced during isolation of the sample, such as salts from separation buffers or solubilizing detergents such as sodium dodecyl sulfate. In addition, biological samples are often heterogeneous. Despite these complications, MALDI can be performed with many such samples. Often the analyses require modifications in the sample preparation procedure to facilitate carrying out the MALDI process in the presence of these substances.

As previously mentioned, MALDI is a soft ionization technique. However, both biological and synthetic macromolecules may contain weak bonds, which can dissociate during the MALDI process. Oligonucleotides are a specific example of a class of macromolecules that are subject to fragmentation in MALDI. Special sample preparation can improve analyses in these cases. Alternative methods of mass analysis, such as electrospray ionization, may be preferable in the case where molecular fragmentation is debilitating to the analysis. In the case of unknowns, preliminary size data (such as that provided by gel permeation chromatography or gel electrophoresis) are very helpful to the analyst.

Sequence information for polymers (biological and synthetic) can be obtained by two or more stages of mass spectral anaysis (MS/MS measurements). This is accomplished by forming product ions from the primary parent molecular ion. Product ions are formed either by making use of the excess energy imparted to the analyte during desorption or by imparting more energy to the molecular ion through collision-induced dissociation (CID) or surface-induced dissociation (SID). In MALDI-TOF, use is made of the excess desorption energy. As the molecular ions traverse the flight tube, some decompose to form fragment ions. If the masses of these fragment ions can be correlated to specific fragments, structural information about the molecule can be obtained. Reflectors are often used in these applications because they lengthen the ion's flight path and allow more time for the dissociations to occur. New MALDI-TOF instruments are being marketed that contain dual reflectrons or collision cells for use in obtaining structural information. The ICR cell of the FTMS instrument is ideal for use as a collision cell or for carrying out SID. FTMS has the added advantage of allowing the operator to impart energy to the ions before CID or SID by applying an excitation frequency to the ions.

Analytical Information

Data Format

Data are recorded as ion flight times (TOF) or frequency (FTMS). After manipulation with the data systems, data are reported as m/z ratios. m/z data are usually reported both as plots of relative signal intensity versus mass and as peak lists indicating the same information in the form of list.

Qualitative

Determination of analyte mass and purity are the ultimate qualitative data provided by mass spectrometry. The analyses outlined in the Applications section cover the various options available for qualitative determinations.

Quantitative

LD and MALDI methods are not considered to routinely provide quantitative information. Whenever possible, more traditional methods of analyte quantification should be used. In instruments with very stable laser systems, it is possible to obtain crude estimates of analyte quantities by making use of internal standards and sensitivity correction factors. Relative quantities of analytes in a mixture cannot be directly determined from the relative ion signal intensities of that mixture because the different components may have different ionization efficiencies, and competitive ionization effects within the mixture may also affect the relative ion current. To address this issue, mixture components can be analyzed separately to determine their relative ion current and correction factors determined and applied to the data. Unfortunately, this tedious procedure cannot be expected to give quantitative data to better than ±10% accuracy.

Accuracy

Mass accuracy is usually limited by the quality of the calibration. In the case of FTMS, measured masses can differ from the theoretical by as little as a few parts per million. TOFMS is inherently less accurate and mass accuracies with external calibration range from 0.1 to 0.03%, and internal calibrations range from 0.03% to tens of parts per million.

Detection Limits

With TOF measurements, routine detection limits are considered to be 1 to 10 pmol. Exceptions are reported and hundreds of femtomoles are the current detection limits. The detection limit for FTMS can be equally low, but picomole detection limits are routine. These methods are rapidly improving, and attomole detection limits may soon be the norm for both types of instrumentation.

Limitations

Direct LD and MALDI are generally not quantitative. LD causes fragmentation of the analyte and usually does not give intact molecular ions for analytes ranging in mass above 10 kDa. MALDI has a higher upper mass limit, but the presence of easily ionizable groups in the molecule (such as O, N, and S) is required unless special MALD/chemical ionization techniques are used. Molecules with masses less than 500 Da are rarely analyzed by MALDI not only because of interference from matrix dimers and other matrix fragments and adducts that populate the spectra with chemical noise below 500 Da, but also because of the abundance of other mass specral methods available for that mass range.

Applications

Confirmation of Structure

No matter how carefully a synthesis is carried out, synthetic products require characterization and confirmation.

1. The Mass Spectrum of a Hydrocarbon Dendrimer.

Mass confirmation by mass spectrometry is shown in Fig. 34.1. The phenylacetylene dendrimer was synthesized by a new method and confirmation of the expected structure and synthetic purity were required. The dendrimer was analyzed by UV MALDI-TOF. Sample preparation was carried out by dissolving the dendrimer in methylene chloride and mixing it with the retinoic acid matrix at a ratio of 1:500. Before deposition of this mixture, a poly(ethylene glycol)-1000 polymer (PEG) was dissolved in methylene chloride, deposited on the sample probe and allowed to dry. An aliquot of the matrix/dendrimer mixture corresponding to 71 pmol of the dendrimer was deposited on top of the PEG layer. The spectrum shows the presence of the expected dendrimer $(M+H)^+$, with no impurities, and the PEG spectrum that was used as an internal mass standard. The measured mass was within 40 ppm of the expected mass, which was determined from the empirical formula obtained from combustion analysis. The resolving power for the molecular ion is 1100.

Figure 34.1 The mass spectrum of cascade polymer 48-cascade:benzene[3-1,3,5]:(5-ethynyl-1,3-phen-lylene):5-ethynyl-1,3-di(tert-butyl)benzene, containing 94 phenylacetylene monomer units (D-94) analyzed by UV MALDI-TOF. The measured mass accuracy is 40 ppm using the PEG as an internal mass standard. The spectral resolution of the molecular ion is 1100.

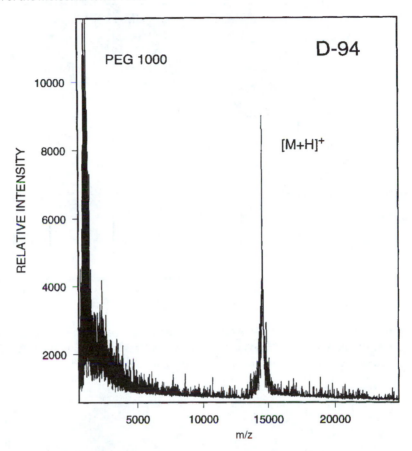

2. The Mass Spectrum of a Substituted Fullerene.

This spectrum is pictured in Fig. 34.2. In this case, high mass accuracy was required and UV MALDI-FTMS was used. The analyte was dissolved in toluene and mixed with the 2,5-dihydroxybenzoic acid matrix at a ratio of 1:5000. The sample was deposited by aerosol nebulization of approximately 46 nmol of the matrix–analyte mixture onto a 2-cm-diameter stainless steel sample probe. The M^- molecular ion of the fullerene is the prominent peak in the spectrum. The mass measurement accuracy was found to agree with that calculated for the proposed structure to within 3.7 ppm, externally calibrated, and the resolving power for the molecular ion is 10,000.

3. Sequence Information.

Sequence information is obtained for the peptide dermankephaline using a TOF instrument with a curved field reflectron designed for this purpose (Fig. 34.3) (9). The product ions result from

Figure 34.2 The mass spectrum of a substituted fullerene obtained with UV MALDI-FTMS. The mass measurement accuracy was found to agree with that of the proposed structure to within 3.7 ppm, externally calibrated, and the spectral resolution is 10,000.

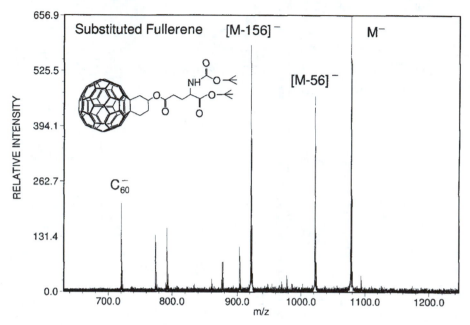

Figure 34.3 A mass spectrum of the peptide dermankephalin, [M+H]⁺, m/z 955, and its product ions. *(From T. J. Cornish and R. J. Cotter, Rapid Communications in Mass Spectrometry, 8, p. 781. Copyright © 1994 by T. J. Cornish and R. J. Cotter. Reprinted by permission of John Wiley & Sons, Ltd.)*

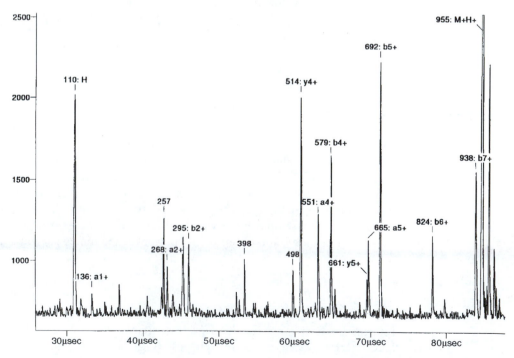

normal metastable decay processes occurring as a result of excess internal energy imparted to the parent ion during the MALDI event. These ions are collected and mass analyzed using the extended flight path and curved field of the reflectron. Commercially available tandem reflecting TOF instruments provide similar data derived from CID of parent ions.

Mass Determinations

Mass analysis of the intact molecular ion of proteins is a primary application of MALDI.

4. High-Resolution UV MALDI-FTMS.

A high-resolution UV MALDI-FTMS spectrum of bovine insulin in shown in Fig. 34.4. 2,5-Dihydroxybenzoic acid was used as the matrix, with a matrix-to-analyte ratio of 5000:1. To reduce molecular ion fragmentation, fructose was added as a comatrix, also in a ratio of 5000:1 to the analyte. The mixture was 3.8 mM in the analyte and 0.5 mL of this solution was aerosol nebulized onto a stainless steel probe tip. The average spectral resolving power is 60,000 and the average mass accuracy is 70 ppm, externally calibrated. The (M+H)$^+$ molecular ion is the predominant peak in the spectrum and the smaller peak corresponds to (M+H)$^+$ – 17 Da. The resolution is sufficient to resolve the different peaks resulting from the carbon isotope distribution, and thus the molecular ion appears as a distribution.

Figure 34.4 A high-resolution UV MALDI-FTMS spectrum of the protein standard bovine insulin. The lower-intensity distribution corresponds to the mass (M+H)$^+$ – 17 Da. The average resolution of the molecular ion peak is 60,000 and the average mass accuracy is 70 ppm, externally calibrated.

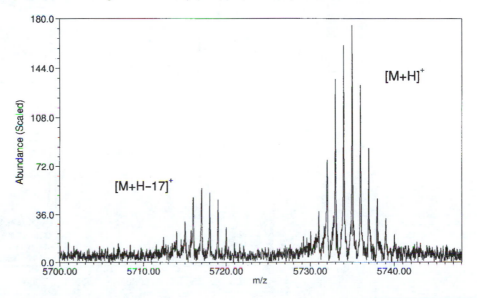

5. Lysosyme Fraction Spectrum.

A spectrum of a 500-fmol lysosyme fraction isolated from a protein mixture by capillary electrophoresis is shown in Fig. 34.5. The spectrum was analyzed using MALDI-TOF. The matrix used was α-cyano-4-hydroxy cinnamic acid, with a matrix-to-analyte ratio of approximately 50,000 to 1. The protein was isolated in the presence of 20 mM sodium acetate buffer. Mass spectral signal strength was improved in the presence of this buffer by using a matrix solvent that was 50/50 volume/volume acetonitrile and 10% formic acid. The mass accuracy of the (M+H)$^+$ molecular ion was found to be 0.01% with external calibration.

Polymer Distributions

The reported average masses of synthetic polymers (that is, the center of the polymer distribution) are often determined by gel permeation chromatography. Mass spectral measurements have shown that the oligomer distributions often vary from that predicted by the manufacturer. Poly-

Figure 34.5 A spectrum of a 500-fmol lysosyme fraction isolated from a protein mixture by capillary electrophoresis. The spectrum was analyzed using UV MALDI-TOF. The mass accuracy was found to be 0.01% with external calibration.

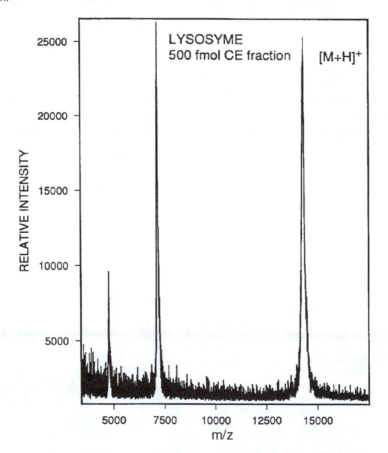

mer mass distributions can be determined more accurately by laser desorption mass spectrometry than by gel permeation chromatography.

6. Mass Distribution of PEG-6000.

Figure 34.6 depicts the mass distribution of the polymer poly(ethylene glycol)-6000. This PEG was found to have its center of the distribution at masses 6207 and 6251 Da. The spectrum was obtained using UV MALDI and FTMS. Sample preparation was carried out by dissolving the polymer to a concentration of 4.7 mM in 0.1% trifluoroacetic acid in methanol, mixing with the 2,5-dihydroxybenzoic acid matrix in a ratio of 1:500, and by aerosol deposition of approximately 50 nmol of the solution on a 2-cm-diameter stainless steel probe. The average mass measurement accuracy was found to agree with that calculated for the proposed structure to within 68 ppm, externally calibrated. The mass distribution of synthetic polymer unknowns can be similarly determined.

7. High-Resolution Mass Spectrum of PEG-4000.

A partial high-resolution mass spectrum of poly(propylene glycol)-4000 is shown in Fig. 34.7. The spectrum was obtained using infrared laser desorption (IRLD) and FTMS and shows the resolution of the carbon isotopes for several oligomers. Mass determinations were carried out by dissolving the polymer in methanol doped with potassium chloride salts followed by deposition on a 2-cm-diameter stainless steel sample probe. For the potassium-attached oligomers shown, the average mass accuracy is 8.59 ppm and resolving power is 60,000 (10).

Figure 34.6 The mass distribution of the sodium-attached oligomers of the polymer poly(ethylene glycol)-6000. This PEG was found to have its center of the distribution at masses 6207 and 6251 Da. The spectrum was obtained using UV MALDI and FTMS. The average mass measurement accuracy is 68 ppm with external calibration.

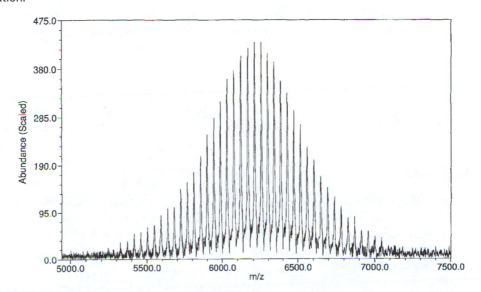

Figure 34.7 A portion of the high-resolution mass spectrum of poly(propylene glycol)-4000. It was acquired by IRLD-FTMS. The average mass accuracy of the potassium-attached oligomers displayed is approximately 8.59 ppm, externally calibrated, and the resolution is 60,000. *(Reprinted from C. F. Ijames and C. L. Wilkins, "High Resolution Mass Spectrum of poly(propylene glycol)-4000,"Journal of the American Chemical Society, 110, p. 2687. Copyright 1988 American Chemical Society.)*

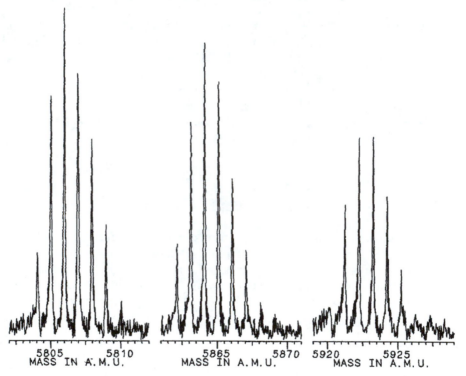

8. Identification of Impurities.

Synthetic products often contain impurities, as do those of natural origin. Mass spectra showing peaks other than those expected can give valuable insight into the efficiency of a synthetic methods or into the nature of the natural/biological product isolated.

Nuts and Bolts

Relative Costs

Relative costs of mass spectrometers are for complete systems with data system, workstation, and lasers included. Laser prices are also listed separately because they are a significant portion of the purchase price and additional lasers are often added to systems. Prices shown are the minimum; lasers can range to much higher prices for high-quality lasers that have very stable outputs and high peak powers.

Time-of-flight mass spectrometer	$100–$300K
Fourier transform mass spectrometer	$400–$1000K
Nitrogen laser	$7–$15K
YAG laser	$10–$30K
Maintenance (per year)	$5–$50K
Supplies (per year)	$2–$10K

Maintenance and supply costs vary greatly depending on the type of mass spectrometer used and the applications of the instrument. Fourier transform mass spectrometers have higher maintenance costs because their superconducting magnets must be kept filled with liquid nitrogen and liquid helium at all times.

Vendors for Instruments and Accessories

There are excellent comprehensive directories of vendors published (11) that should be consulted for a complete picture of available products. The list below gives a brief overview of the most popular vendors. See Chap. 28 for detailed contact information.

Time-of-Flight Mass Spectrometry

Bruker Instruments, Inc., Billerica, MA

Finnigan MAT, San Jose, CA

Kratos Analytical, Inc., Ramsey, NJ

PerSeptive Biosystems/Vestec, Cambridge, MA

Micromass UK, Ltd., Manchester, UK

Fourier Transform Mass Spectrometry

Bruker Instruments, Inc., Billerica, MA

Ionspec, Irvine, CA

Finnigan FT/MS, Madison, WI

Required Level of Training

Any literate, technically oriented person can be trained to run a mass spectrometer for routine analyses. For complex applications, a minimum of a bachelor's degree in the sciences is needed. For research and development applications, an advanced degree or several years of experience with mass spectrometry are useful.

Service and Maintenance

People experienced in mass spectrometry instrumentation can extend their knowledge to the instruments described in this chapter. Service contracts are available with most instruments and may be more convenient in applications where the use of the instrument is constant and necessary (such as in a service facility).

Suggested Readings

Books

COTTER, R. J., ED., *Time of Flight Mass Spectrometry*. ACS Symposium Series 549. Washington, D.C.: American Chemical Society, 1994.

LUBMAN, D. M., ED., *Lasers and Mass Spectrometry*. New York: Oxford University Press, 1990.

MATSUO, T., AND OTHERS, EDS., *Biological Mass Spectrometry Present and Future*. New York: Wiley, 1994.

RUSSELL, D. H., ED., *Experimental Mass Spectrometry*. New York: Plenum Press, 1994.

STANDING, K. G., AND W. ENS, EDS., *Methods and Mechanisms for Producing Ions from Large Molecules*. New York: Plenum Press, 1994.

Articles

BURLINGAME, A. L., R. K. BOYD, AND S. J. GASKELL, *Analytical Chemistry*, 66 (1994), 634R.

CARR, S. A. AND OTHERS, "Integration of Mass Spectrometry in Analytical Biotechnology," *Analytical Chemistry*, 63 (1991), 2802.

COTTER, R. J., "Lasers and Mass Spectrometry," *Analytical Chemistry*, 56 (1984), 485A.

COTTER, R. J., "Time-of-Flight Mass Spectrometry for the Structural Analysis of Biomolecules," *Analytical Chemistry*, 64 (1992), 1027A.

KÖSTER, C., AND OTHERS, "Fourier Transform Mass Spectrometry," *Mass Spectrometry Reviews*, 63 (1992), 495.

References

1. K. Tanaka and others, *Rapid Commun. in Mass Spectrom.*, 2 (1988), 151.

2. M. Karas and F. Hillenkamp, *Analytical Chemistry*, 60 (1988), 2299.

3. J. W. Gauthier, T. R. Trautaman, and D. B. Jacobson, *Anal. Chim. Acta*, 246 (1991), 211.

4. C. F. Ijames and C. L. Wilkins, *Analytical Chemistry*, 62 (1990), 1295.

5. R. L. Hettich and M. V. Buchanan, *J. Amer. Soc. for Mass Spectrom.*, 2 (1991), 22.

6. P. O. Danis and D. E. Karr, *Org. Mass Spectrom.*, 28 (1993), 923.

7. M. S. Kahr and C. L. Wilkins, *J. Amer. Soc. for Mass Spectrom.*, 6 (1993), 453.

8. K. L. Walker and others, *J. Amer. Soc. for Mass Spectrom.*, 5 (1994), 731.

9. T. J. Cornish and R. J. Cotter, *Rapid Commun. in Mass Spectrom.*, 8 (1994), 781.

10. C. F. Ijames and C. L. Wilkins, *J. Amer. Chem. Soc.*, 110 (1988), 2687.

11. S. A. Lammert and R. G. Cooks, *Rapid Communications in Mass Spectrometry*, 5 (1991), 425.

Electroanalytical Techniques

Chapter 35

Introduction

Janet Osteryoung

North Carolina State University
Department of Chemistry

Electroanalytical techniques include varied electrochemical approaches to obtaining information about the composition of a sample and the molecular nature of its constituents. All have in common the use of an electrochemical cell, good to excellent sensitivity, excellent accuracy, highly developed and accurate theory, and poor resolution and selectivity. The next four chapters describe a selection of the more widely used electroanalytical techniques. This chapter provides a general introduction and also gives brief sketches of techniques that deserve mention but are not generic enough in their application to deserve chapters of their own.

Electrochemical Measurements

Conductance

The conductance of a solution is a physical property that can give important quantitative information regarding the composition of the solution. Conductance is a measure of the ability of a solution to carry current and depends on the concentration, mobility, and charge of ions in the solution, and on temperature. The effect of temperature occurs mainly through the dependence of mobility on temperature. Mobility also depends on the nature of the solvent and the overall composition of the solution as well as on the identity of the ion. A wide variety of commercial instruments and cells are available for measurement of conductance. The details of the instrumentation depend on the purpose of the measurements. In some cases measurements of high accuracy are desired; in others convenience or stability with time for monitoring applications are the primary considerations. The great virtues of routine conductance measurements are ease, robustness, and nondestructiveness. The main weakness is that the conductance measurement by itself gives no molecular information and cannot identify the species that carry the current.

Conductance has been widely used in monitoring applications, especially for natural waters, for decades. Two particularly interesting applications apply to natural waters. Conductance data obtained at gauging stations in rivers and streams can be used together with temperature and flow data and occasional measurements of composition to estimate composition of the water. In seawater, conductance is used to determine the fundamental property of salinity.

The measurement of conductance can be used as a detector for the endpoint of a titration. This is a means of using the conductance measurement to provide quantitative information regarding the concentration of specific species. In addition, as described in detail elsewhere, conductance is used for detection in ion chromatography. Conductance is described in Chapter 39.

Potentiometry

All of the other electrochemical techniques involve some net change in the sample. In the case of potentiometry, modern instrumentation permits measurements in which the net flow of current through the cell is so small (<1 pA) that the sample is not perturbed at all. Thus potentiometry can be considered a completely nondestructive technique. The potentiometric electrode is often called a probe or a sensor for this reason, because it can be used to obtain information about the sample without affecting it. Potentiometry, in contrast with conductance, has some specificity for chemical species. The potentiometric principle is used in many clever ways, often using the techniques of biotechnology, to create new electrodes that expand the range of molecules that can be determined or to improve selectivity.

Potentiometry has the further interesting property that the signal depends not on the concentration of the analyte, but rather on its activity. Scientific applications of potentiometry have not been able to respond to this unique capability. Instead, potentiometric methods invariably involve methods of calibration designed to measure concentration rather than activity. The single, highly important exception is the measurement of pH. The direct use of activity measurements remains a challenge, which could yield major benefits to both research and routine practice in fields such as oceanography and medicine. Potentiometry is described in Chapter 38.

Voltammetry

In voltammetry, the potential of an electrode immersed in solution is controlled so as to force a redox reaction to occur at the electrode surface. The rate of this reaction is proportional to the current flowing through this electrode. A voltammogram is a plot of this current versus the potential of the electrode.

Usually the conditions are chosen so that the rate of this reaction is controlled by the rate of transport of the reacting species to the electrode surface. In that case the rate of the reaction, and hence the current flowing through the cell, is proportional to the concentration of the reacting species. This is the basis for analytical applications of voltammetry. In some cases, the current is controlled by the rate of electron transfer at the electrode surface or, in the case of more complex reactions, by the rate of a coupled reaction that occurs in solution. Proper control of conditions in these more complicated cases ensures that current is also proportional to concentration.

Although some analyte is converted to product by the redox reaction, usually the extent of reaction is quite small, with less than 1% of the material in the cell converted to product. Therefore, this technique is also essentially nondestructive. Like all rate measurements, voltammetry is highly sensitive, and the conditions must be selected and controlled to ensure that the mechanism controlling the current does not change.

Modern analog electronics under control by software create a bewildering array of potential modulation and current sampling schemes. However, the underlying principles are the same for all techniques. The present state of commercial development provides a reasonable suite of techniques with associated rules and algorithms for analysis of data. Therefore it is not necessary to be an expert to obtain reliable analytical information. (The situation is much like that of nuclear magnetic resonance.)

The potential dependence of the current response in voltammetry contains information about the identity of the analyte. Although voltammetry has very poor resolution, the position of the signal on the potential scale can depend strongly on the exact chemical form of the reactant. Thus it gives molecular information rather than elemental composition. Voltammetry is described in Chapter 37.

Amperometric Detection

The current for the reaction of analyte can be measured at constant potential in a flow cell. The flow of solution maintains the proportionality between current and concentration, and measurement at constant potential minimizes transient background currents. This scheme is an ideal way to measure concentrations in process streams, where the sample is well known, and in the effluent from chromatographic columns, where the separation step compensates for poor resolution. The application to chromatographic detection is described in Chapter 36.

Amperometry is also used for detection of the endpoint in titrations. This mode of operation avoids the most common limiting factor in amperometric measurements: difficulty with estimating the background current. Amperometric detection schemes are available for all of the common types of titration reaction: oxidation-reduction, precipitation, and complex formation.

Other Electrochemical Techniques

Coulometry

The most fundamental, accurate, and robust electroanalytical technique is coulometry. Coulometry is broadly applicable to analytical problems, but occupies niches rather than being widely used; therefore, it is not accorded a separate chapter in this book. Coulometry is based on completely converting the analyte to product through electrochemical reaction and measuring the total charge required. Thus, in contrast with the other techniques described here, coulometry completely consumes the analyte. The total amount of analyte in the sample is related to the charge required by

$$Q = nFVc = nFwM_w^{-1} \tag{35.1}$$

where Q is the quantity of charge (C), n is the number of equivalents of charge required per mole of reaction (equiv mol^{-1}), F is the value of the faraday (96,486 C equiv^{-1}), V the volume of solution (L), c the concentration of analyte (mol L^{-1}), w the weight of analyte, and M_w its molecular weight. This relation contains only fundamental quantities. Thus coulometric determinations are absolute. Unlike all other instrumental methods of analysis, they require no calibration. In other words, the local power company supplies the universal, pure, precalibrated reagent, the electron, at a readily measurable rate, the current. As a result of this important feature, coulometry is favored when good accuracy is required.

Coulometry can be carried out by exhaustive electrolysis at constant potential, which offers some discrimination among analytes through choice of potential. The resulting current decays exponentially, and the experiment is finished when the current drops to an acceptably low fraction of the initial value. Charge is obtained by electronic or numerical integration of the current.

In contrast, coulometry can be done with constant current, using a redox buffer to maintain the potential in the range required to ensure that all of the current goes, directly or indirectly, to the reaction of analyte. A significant fraction of the current converts the buffer to product, which then reacts quantitatively with the analyte, resulting in 100% current efficiency. Because of this feature, this mode of operation is often called constant current coulometric titration.

In the case of constant potential coulometry, the experiment is stopped when an acceptable fraction (such as 99.99%) of the analyte has been consumed. For constant current coulometry, an endpoint must be detected. The charge related to the amount of analyte is then given by the product of the (constant) current and the time required to reach the endpoint.

The notion of absolute accuracy is extended to titrations by using constant current coulometry to generate reagents. This is particularly attractive for unstable reagents, such as V(II). However, its accuracy makes this an attractive strategy even for stable reagents such as H^+. The equipment is simple and compact and generates small, accurately known amounts of reagent on demand. It can be operated remotely, unattended for long periods of time.

In the case of coulometric titrations, it is often convenient to use amperometric detection of the endpoint because the characteristics of the chemistry that make the titration reaction work reliably also enable accurate and precise amperometric detection of the endpoint.

A special case of coulometry is that in a thin-layer cell. These are cells in which a thin film of solution is trapped between two electrodes, which often are chosen to be transparent. With these cells, one can obtain the concentration of the analyte by coulometry, do voltammetry to characterize the potential dependence of the response, and examine both reactants and products using optical spectroscopy. The techniques for handling these types of cells are a bit fussy, but they can provide full characterization of small amounts of material, and thus have been used widely in studies of redox-active molecules of biological interest.

Impedance

Chapter 39 describes methods for determining the resistivity of a sample without regard to the capacitance involved in coupling the measurement system to the sample. A broader approach is to use electrodes in solution with an imposed alternating current (AC) voltage of very low amplitude (about 5 mV) and to detect the impedance of the system. The impedance, Z, is a complex quantity given by

$$Z = R - j/\omega C \qquad (35.2)$$

where R is the resistance, C the capacitance, ω the frequency of the AC perturbation, and $j = \sqrt{(-1)}$. By measuring the impedance as a function of frequency, one can characterize the sample, which may be a solid or a solution. In contrast with the conductance measurement, which does not work well if redox reactions occur at the electrode surface, the impedance method may be used with or without species present that will react at the electrode surface. Impedance techniques are somewhat specialized, but they have been highly developed, especially for the detection and characterization of corrosion of conducting materials.

Chronopotentiometry

The technique of determining the time dependence of the potential when a constant current is passed through an electrochemical cell is called chronopotentiometry. Quantitative information is obtained from chronopotentiometry by measuring the time required for the passage of current to occasion an abrupt change in potential. Chronopotentiometry has not proven to be a widely useful analytical tool because accurate measurement of this time is difficult. However, there are special situations, such as the determination of concentration of a reducible or oxidizable material in very concentrated solution, in which this technique works extremely well. The advantage of chronopotentiometry for this application, in comparison with voltammetry, is that the instrumentation is simpler and therefore more robust and better suited to process control or remote monitoring applications.

Applications

The electrochemical techniques described in this section are used widely in various combinations in commercial analyzers and in officially sanctioned methods. Here several examples are given to show the flexibility and practical nature of this approach to chemical analysis. The examples are illustrative rather than comprehensive and complement the specific examples given in the chapters devoted to specific techniques.

Determination of Residual Chlorine by Titration with Phenylarsine Oxide with Amperometric Endpoint Detection

In acetate buffer, pH 4.8, the various forms of residual chlorine (including the chloramines) react rapidly and quantitatively with excess phenylarsine oxide, a form of As(III) that is stable in dilute solution, to form chloride and As(IV). The endpoint of the titration is detected amperometrically by monitoring the current for the reduction of chlorine at a platinum electrode maintained at a potential of 0.00 V versus SCE (saturated calomel electrode, see below). The current is proportional to the concentration of chlorine, and the titration is continued to the point at which addition of further reagent does not cause the current to decrease (the dead-stop endpoint). The detection limit for this procedure is typically on the order of 1 ppb chlorine. Quantitation is established by standardization of the phenylarsine oxide solution against coulometrically generated iodine solution.

Similar methods are used for chlorine dioxide and for iodine. This method can be used for any analyte that reacts quantitatively with iodide to give iodine (or triiodide).

Determination of Dissolved Oxygen by Amperometric Sensing

Oxygen can be reduced readily in acid solution at a platinum electrode to form water. The amperometric sensor for oxygen employs this reaction in a cell that isolates the electrode from the sample solution by using a membrane permeable to oxygen but not to larger molecules, which might interfere in the determination. The conditions of stirring, thickness of the membrane, and temperature are chosen so that the rate of diffusion (flux) of oxygen through the membrane, and hence the amperometric current, is proportional to the concentration of oxygen in the solution. Quantitation is established by determination of standard solutions of oxygen.

The membrane electrode for determination of oxygen is especially effective for colored or turbid solutions, for field measurements, and for applications requiring continuous monitoring or measurements in situ. An important application is the determination of oxygen in blood. The amperometric electrode for determination of oxygen in blood is usually known as the Clark electrode, named for its developer, Leland Clark.

Coulometric Determination of Dissolved Organic Halogen

The first step of this method is to adsorb organic halogen on activated carbon. The activated carbon is then washed with an aqueous solution of nitrate to displace any adsorbed inorganic halogen. By these means the organic halogen is separated from inorganic halogen. The activated carbon, together with the adsorbed organic halogen, is then pyrolyzed to form CO_2 and HX. The HX is then swept into a carrier gas stream that takes it into a microcoulometric cell. The cell contains a silver electrode through which a current is passed to oxidize Ag^0 to Ag^+, which precipitates as AgX. The concentration of Ag^+ in the cell, as determined potentiometrically, is maintained at a constant value by automatic adjustment of the current so that the rate of production of Ag^+ equals the rate of transport of X^- to the cell. The amount of organic halogen in the sample is then obtained by integration of the current required for oxidation of the equivalent amount of silver.

Standard samples of chloroform are used to calibrate the entire process. Commercial analyzers are available. The method does not respond to fluorinated organic compounds and is subject to interference by inorganic halogens and pseudohalogens. This method provides an efficient means for screening samples in cases where those yielding results below a threshold value require no further examination.

General References

For many years *Analytical Chemistry* has published a Reviews issue each year. In even-numbered years this issue is devoted to fundamentals, and in odd years to applications. The most recent Fundamental Reviews (*Analytical Chemistry* 66, no. 12 [1994]) contains the articles "Chemical Sensors," which covers thoroughly the subject of potentiometry, and "Dynamic Electrochemistry: Methodology and Application," which covers voltammetry and combinations of electrochemistry and spectroscopy. The Applications Reviews (such as *Analytical Chemistry* 67, no. 12 [1995]) cover a wide range of topics, from air pollution to water analysis; many of the articles discuss applications of electrochemical techniques.

Chapter 36

Amperometric Techniques

Chester T. Duda and Craig S. Bruntlett

Bioanalytical Systems, Inc.

Summary

General Uses

- A general mode for determining the concentration of electrochemically active species in static or flowing solutions.
- A quantitative technique for the determination of oxygen and hydrogen peroxide that forms the basis of many biosensors.
- A mode for determining the endpoint in a titration.

Common Applications

- The largest number of applications of amperometric detection are when coupled to liquid chromatography in biomedical research. Table 36.1 is a list of the most common analytes detected.
- Also listed are analytes of industrial and environmental importance.
- The oxygen electrode is based on amperometry and used extensively as the basis of other sensors where oxygen is produced or consumed.
- Hydrogen peroxide is also easily detected amperometrically and its determination is the basis of many biosensors in research, under development, or commercially available. Analytes include glucose, lactate, ethanol, acetaminophen, choline, and acetylcholine.

Table 36.1 Candidates for electrochemical detection.

Biomedical		Environmental and Industrial	
Acetylcholine*	Neutral phenols	Analines	Herbicides
Amino acids*	Nitrosothiols	Antioxidants	Naphthols
Benzoic acids	Oxalate*	Aromatic amines	PCB metabolites
Cinnamic acids	Peptides*	Biphenyls	Peroxides
Coenzymes	Phenylpropionic acids	Chelating agents	Pesticides
DNA adducts	Phenylpyruvic acids	Ethylenethiourea	Phenols
Enzymes	Thiols and disulfides	Explosives	
Estrogenic hormones	Tryptophan metabolites		
Glucose*	Tyrosine metabolites	**Typical Functional Groups**	
Lactic acid*	Vitamins		
Mandelic acids			

Oxidizable	Reducible
Aromatic amines	Aliphatic nitro
Ascorbic acids	Aromatic nitro
Hydroquinones	Azo compounds
Indoles	Azomethine
Phenols	Nitrosamines
Phenothiazenes	N-oxides
Thiols	Organometallics
Vanillyl	Peroxides
Xanthines	Quinones
	Thioamides

Ions

Bromide	Nitrite
Cyanide	Sulfite
Iodide	

Pharmaceutical

Alkaloids	Disulfides
Analgesics	L-DOPA and related compounds
Antibiotics	Nitrogen heterocycles
Anticancer	Phenothiazines
Antimalarial	Thiols
β-mimetics and β-blockers	Tricyclic antidepressants

* Require chemical or enzymatic derivatization before detection.

Samples

State

- Analytes are most often materials that can be dissolved in polar solvents such as water.
- Solids and gaseous samples are analyzed in special cases.

Amount

Sample mass is usually determined by volume; this ranges from a few nanoliters injected onto a capillary electrophoresis (CE) system to a few milliliters needed for a direct measurement by an amperometric probe. The largest application is liquid chromatographic detection or some form of flow system, and in these cases the typical volume is tens of microliters.

Preparation

Often some form of analyte isolation from interferences is required before detection. This may be a liquid chromatograph or some other separation technique, a discriminating covering for the electrode, or a purging of the solution with an inert gas to remove oxygen.

Analysis Time

Depends on the number of interfering substances present in the sample and the detection limit required but most often from a few seconds to a few minutes per sample.

Limitations

General

- Relatively poor analyte resolution; that is, it is difficult to differentiate compounds of similar structures.
- Only electrochemically active materials are detected but this is often an advantage.
- Not only must the electron transfer reaction be considered when using this technique but also the movement of analyte (mass transport) to and from the electrode.
- Best results are obtained when the electron transfer rate is relatively fast, which generally means that the potential is as low as possible to obtain a maximum current response. (That is, a large overpotential is not required for the reaction to occur at a mass transfer limited rate.)
- No chemical structure information is available or obtained.
- Because electrochemistry often involves some surface chemistry, the electrode periodically will react with solvent or solution species. This degradation may occur with each sample, which necessitates a disposable sensor, or it may occur over months or years as in some applications of liquid chromatography–electrochemistry (LCEC).
- Redox measurements must be carried out in ionically conductive solutions.
- Hydrolysis of water limits usable potentials to ±1.0 volts in most cases.
- Oxygen is often an interference at reductive potentials and must be removed by purging all solutions with an inert gas.

Accuracy

Precision is typically less than ±2%.

Sensitivity and Detection Limits

- A virtue of amperometric detection is the high current response relative to the amount of material (the absolute sensitivity).
- Many variables determine detection limits but typical values are from a few femtomoles injected onto a liquid chromatography (LC) system (10^{-9} M) to attomolar (10^{-18} M) or lower when using some amperometric biosensors. A major advantage of amperometry over other

finite current techniques is the use of a fixed potential, so there is no background current contribution from charging the electrode or from relatively slow electrode reactions. The result is low baseline noise.

Complementary or Related Techniques

- All voltammetric techniques are complementary because they provide basic current-potential information needed to develop an amperometric method. The most commonly used technique is cyclic voltammetry, which automatically scans a potential range and monitors the current response. From this basic information the fixed potential required for the amperometric measurement is determined.

- Techniques based on potentiometry, an electrochemical equilibrium measurement, are the ones most familiar to analysts: pH and ion-selective electrodes (ISE).

- Conductivity measurements.

- Often an amperometric detection mode can be coupled to spectroscopy and photochemistry such as electrochemically generated luminescence, electrochemiluminescence, or photolytic cleavage followed by amperometric detection.

- Coupling electrochemical cleaning of the electrode surface, analyte adsorption (concentration), and then amperometric detection is the basis of what is commonly called pulsed amperometric detection.

Introduction

Amperometry is fundamentally very simple but has excellent analytical features. As its name implies the technique is simply the measurement of current. This current is the result of material coming into contact with an electrode surface that is at a fixed, nonvarying potential of sufficient electron energy for an electrochemical reaction to take place. The current response is most often measured as a function of time. It also can be measured as a function of the volume of a titrant, for example. The key feature of the technique is that the measured current is proportional to the concentration of the analyte as long as the movement of material to the electrode is constant.

The simplicity of the technique leads to many applications. For this chapter, the focus is on amperometric detection with flow cells, that is, precisely controlled movement of material to the electrode. The coupling of liquid chromatography with amperometric detection (LCEC) is the most widely used form of finite current electrochemistry for chemical analysis.

Another rapidly growing application area where amperometric principles are advantageous is in the development of biosensors. The sensor is usually meant to stand alone. The selectivity is achieved by the placement of a discriminating film on the electrode to remove unwanted interferences. The film may also contain enzymes or antibodies to enhance the response, but the final quantitation is by amperometry. Examples include a sensor for oxygen and biosensors for glucose, lactate, choline, acetylcholine, and drugs. Environmental analysis is also an area where specific sensors are being developed. The principles of such sensors is briefly discussed.

A third broad application area is using amperometry as an endpoint indicator for titrations. This is well established and is used extensively. Although it is important, this chapter does not discuss this area and refers the reader to many basic quantitative analysis texts on the subject.

There are different uses of amperometric detection but the basics are exactly the same.

How It Works

Principles of Amperometric Detection

Amperometric detection is the measurement of current at a fixed potential. Either an oxidation or reduction is forced to occur by judicious selection of the potential applied to an electrode by a controlling potentiostat. The electrode acts as an oxidizing or reducing agent of variable power. In order to use amperometric measurements effectively, it is important to recognize that electrochemical detection is a surface technique, which means molecules not adjacent to the electrode must be moved to the surface to react.

Most amperometric transducers are based on thin-layer hydrodynamic chronoamperometry, which is the measurement of current at controlled potential as a function of time in a stirred solution. The various electrodes are placed in a flowing stream (hydrodynamic) configured as a thin film (Fig. 36.1). The film thickness is variable, with typical values being 15 to 125 μm.

The potential applied to the electrochemical cell between the reference and working electrode serves as the driving force for the detection redox reaction to occur. In a solution with sufficient electrolyte concentration, nearly all of the potential is applied across a very thin, interfacial region (typically less than 50 Å) between the working electrode surface and the bulk solution. The electric field in this zone is therefore very large, of the order of 10^5 to 10^6 V/cm.

All amperometric determinations ultimately depend on Faraday's law:

$$Q = nFN \tag{36.1}$$

where Q is the number of coulombs used in converting N moles of material, n is the number of electron equivalents lost or gained in the transfer process per mole of material, and F is Faraday's constant (96,500 C equiv^{-1}). Differentiation (d) of Eq. (36.1) with respect to time (t) yields current (I), which is the measure of the rate at which material is converted:

$$\frac{dQ}{dt} = I = nFA\frac{dN}{dt} \tag{36.2}$$

When an amperometric detector is used, a sufficient potential is applied that a reduction (or oxidation) reaction is highly favored, which results in all electrochemically active material that comes into contact with the electrode being converted to product. Under these conditions, current depends on mass transport. This is determined by diffusion coefficient, D, and by the concentration gradient, the change in concentration with distance (x) evaluated at the electrode surface ($x = 0$):

$$\frac{dc_{x,t}}{dx_{x=0}} \tag{36.3}$$

The rate of mass transport (mol cm^{-2} s^{-1}) is given by

$$\frac{dN}{dt} = -D\left(\frac{dc_{x,t}}{dx}\right)_{x=0} \tag{36.4}$$

Figure 36.1 Amperometric detection in a thin-layer flow cell configuration (exploded view near the electrode under hydrodynamic conditions). Analyte$_{ox}$ is carried by the flow stream to the diffusion layer boundary near the electrode surface. Analyte$_{ox}$ diffuses to the electrode and undergoes the electrochemical reaction (reduction in this example), and the product, Analyte$_{red}$, diffuses back to the diffusion flow boundary to be swept away. The potential at the electrode is great enough that all analyte reaching the electrode surface undergoes the electrochemical reaction. The diffuse layer, δ, is defined by the concentration gradient at the electrode surface as shown by the solid line. The dashed line indicates more of what the actual concentration profile may be. *(Reprinted with permission of Bioanalytical Systems, Inc.)*

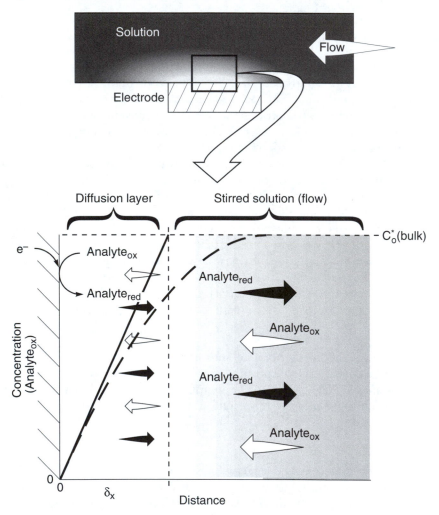

or in terms of the current response

$$I = -nFAD\left(\frac{dc}{dx}\right)_{x=0} \tag{36.5}$$

Under constant flow or controlled hydrodynamic conditions, the concentration gradient is constant because the diffusion layer, δ, (Fig. 36.1) is nonvarying:

$$I = nFAD\frac{c_O^*}{\delta} \tag{36.6}$$

where c_O is the unperturbed concentration of reactant. Equation (36.6) relates a measurable quantity, the current, to the concentration of reactant passing through the electrochemical transducer cell.

Amperometry has an advantage over most analytical detection techniques in that it involves a direct conversion of chemical information to an electrical signal without the use of optical or magnetic carriers. If a reduction takes place, electrons flow from the electrode to the molecule in a heterogeneous transfer; conversely, an oxidation is the transfer of electrons in the opposite direction. Under steady-state conditions, the current measured is contributed from three sources: the background electrolyte, the electrode material itself, and the analyte. The medium and electrode are chosen so that the contributions of the first two sources are as small as possible and the small residual current from these two sources is electronically removed before quantitation of the analyte.

The working potential is usually determined by producing a current–potential plot. This is accomplished by monitoring the current at various potentials. In this experiment, initially, at low positive (negative) potentials, the energy applied to the cell is insufficient to cause any reaction to occur. As the applied potential increases in the positive (negative) direction, the energy requirement for the reaction is now partially met, and a faradaic current ensues. As the potential is further increased, the faradaic current rises until a potential is reached, past which no further improvement in the current response is noted. This current versus voltage plot is called hydrodynamic voltammogram (HDV).

Analogous electrochemical information may be obtained by scanning potential in a linear sweep and measuring the current that arises as a function of the potential at any point along the sweep (Fig. 36.2). The peak potential may be used as an approximate indication of the voltage required for an amperometric detector. The preferred method of determining this parameter is HDV. Electroactive functional groups (Table 36.1) have zones of characteristic redox potentials.

Figure 36.2 Voltammograms of an electrochemically oxidizable compound (norepinephrine). Line A is a cyclic voltammogram produced by monitoring the current during a linear potential sweep (0 to +1.0 to −0.4 to 0 volts versus Ag/AgCl) in a stationary analyte/electrolyte solution; line B represents the current produced during a linear scan from 0 to +1.0 volts of the same solution in line A but it was stirred during the scan, generating a hydrodynamic voltammogram (HDV); line C is also an HDV but was generated in an LCEC system by changing the potential at the cell, and injecting the same mass of analyte at each different potential. *(Reprinted with permission of Bioanalytical Systems, Inc.)*

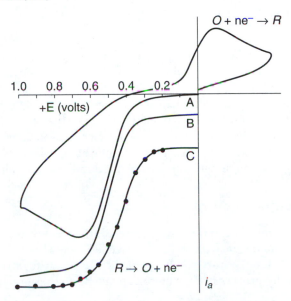

Instrumentation

Hydrodynamic amperometry has relatively poor molecular selectivity among the classes of compounds that are redox active. For this reason it is necessary to incorporate a separation step before the amperometric detector. Present-day reversed-phase or ion-exchange chromatography is ideally suited to this purpose. The basic components of such a system are depicted in Fig. 36.3.

Pump/Mobile Phase

The heart of a system is a pump that provides a constant flow of mobile phase, the electrolyte, to an injection valve, where the sample is introduced. The flow should be as pulseless as possible to minimize baseline noise. Amperometric detectors respond to pressure pulsation. Dual-piston pumps with a pulse dampener are most commonly used. Because amperometry involves a surface reaction between the electrode and the mobile phase and analyte, it is not surprising that the composition of the mobile phase plays an essential role in the successful experiment. The requirements are that it must have low electrochemical activity (that is, low background currents) and it must contain an electrolyte, usually 0.01 to 0.1 M in ionic strength, to minimize ionic resistance.

Degasser

Gases dissolved in the mobile phase can contribute to baseline noise and poor performance. It is desirable that the pump receive mobile phase with gases below saturation, minimizing the chance of air bubbles forming during the pump refill stroke or in the detector cell.

Injection

Sample injection may be manual or automated but should be of short duration. Because the detector responds to pressure changes, it is affected by the movement of the injector valve from the load to inject position. In addition, there is a small pressure drop as the volume within an injection loop is compressed as it enters the pressurized mobile phase flow stream.

Flow Injection Analysis Column

An injected sample may pass through a tube directly to the amperometric cell or into a column for separation of individual components. The former situation is called flow injection analysis (FIA) and produces a single signal representing all electroactive compounds in the sample. Generally, FIA is useful only if a single electroactive analyte is being determined.

   If the sample contains many electroactive components, then an appropriate liquid chromatographic separation is needed (Chapter 9). The type of analytical column used depends on the application but most often consists of reversed-phase material, 3- to 10-μm particles, packed into stainless steel tubing ranging in size from what is called microbore (< 1 mm) to a standard size (2 to 5 mm) and ranging in length from a few millimeters to hundreds of millimeters.

Temperature Control

Electrochemical detectors are adversely influenced by temperature changes. The redox kinetics of background processes and diffusion coefficients have high temperature dependence. Changes in ambient temperature result in serious baseline drift and redox response. In addition, the chromatography process itself is quite dependent on thermodynamics (retention) and kinetics (peak

Figure 36.3 Basic components of an LCEC system.
(Reprinted with permission of Bioanalytical Systems, Inc.)

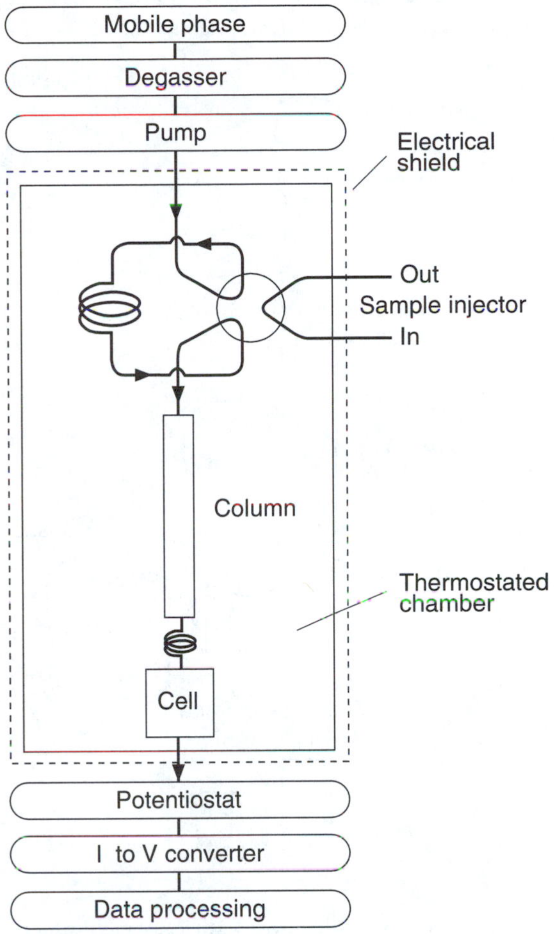

width). The best approach is to have the cell, column, and injector (heat sink) in the same temperature-controlled environment. Temperature gradients across these three components should also be prevented. The absolute temperature per se is not as critical as maintaining a constant temperature.

Transducer Cell

Figure 36.4 is an illustration of a typical amperometric transducer cell. The thin-layer channel is defined by a gasket held between a stainless-steel block and a polymeric block. The stainless-steel block is the auxiliary electrode and provides a compartment for the reference electrode. The polymeric block contains the working electrodes. This design and, in fact, nearly all cell designs incorporate three electrodes: working, auxiliary, and reference. The potential selected by the user is applied between the reference and working electrodes while the current is passed between the auxiliary and working electrodes. Detection occurs at the working electrode in the thin-layer region. Electrodes of the same or different materials may be interchanged by simply swapping the working electrode half of the thin-layer cell. Carbon paste, glassy carbon, mercury on gold, platinum, and

Figure 36.4 Thin-layer amperometric cell used in an LC system. The sandwich is assembled by clamping the working electrode block against the auxiliary electrode block. The dual working electrode may be configured in a parallel or series mode relative to the flowing stream. Gasket thickness controls the cell dead volume. *(Reprinted with permission of Bioanalytical Systems, Inc.)*

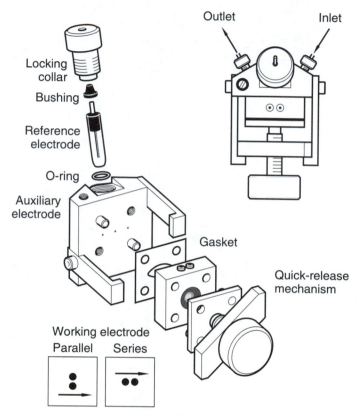

silver have all been used. This design allows easy collection of solute bands without appreciable dispersion. The cell volume can also be reduced for microbore LC to less than 300 nL by simple gasket changes.

The thin-layer design has consistently demonstrated some of the lowest detection limits reported because signal-to-noise increases as the electrode size decreases. The practical limitation is how small can a well-sealed electrode be made and how small a current can be measured without introducing substantial electronic noise. Electrodes having diameters of a few millimeters have proved to be reasonable compromises. Figure 36.4 also schematically depicts a thin-layer cell with two working electrodes, which can be arranged in a series or parallel configuration relative to a flow stream. A different potential can be applied to each electrode and the current of each monitored.

Capillary electrophoresis (CE) has become a useful alternative analytical method because of its high separation efficiency. Because extremely small sample volumes (nanoliters) are used in this technique, amperometric detection has been used to lower detection limits. However, the current of an electrophoresis system gives rise to noise at the electrochemical cell and must be electrically isolated. This has most often been achieved through a conducting joint in the capillary tubing before the electrochemical cell, providing an electrical path to ground for the electrophoresis current. Construction of this joint as well as cell design has been varied. Extreme sensitivity has been reported, such as determination of the concentration of catecholamines in single nerve cell.

Electronic Control

Because electrochemical experiments involve a direct conversion of chemical information to electricity, the instrumentation can be relatively simple. For example, there is no need for high-quality power supplies to drive a light source or operate a photomultiplier tube. On the other hand, because the process often measures nanoamperes (or less) of current, electrochemical detectors can be subject to noisy components and electrical interferences, and proper grounding can be crucial to successful experiments at high current-to-voltage gains. All electronics modules must provide certain control features. These include potential, the applied potential difference between the working and reference electrodes; range, the full-scale current sensitivity; offset, a cluster of calibrated controls dealing with nulling the background or residual current; and filter, which reduces noise in the output signal. The basic circuitry may be expanded by the addition of a waveform generator, allowing pulse experiments to be performed. A triple-pulse waveform has been used to detect carbohydrates at platinum and gold working electrodes.

Data Processing

The output on most amperometric detectors is an analog signal of 0 to 1.0, 0 to 0.1, or 0 to 0.01 volts. A strip chart recorder, integrator, or computer with an A/D interface and software package can be used to collect and process the output signal.

Amperometric Detection and Biosensors

The advantages of amperometric detection can be extended to the in situ determination of important molecules, including those of biological significance. The most widely studied application of biosensors is the determination of blood glucose. There are many other applications but the strategies for determining glucose apply to most of these.

The reasons for modifying an electrode are to improve selectivity, prevent electrode fouling, reduce background response and noise, or increase sensitivity. Glucose is difficult to electrochemically oxidize, and at potentials where it could be directly measured other blood components are active. Analyte selectivity is usually obtained by using biomolecules such as enzymes or antibodies. The logic is to immobilize these agents on the electrode in a thin film and measure the result of the analyte interacting with the enzyme or antibody–enzyme complex. It is usually not possible to measure directly the oxidized or reduced form of the enzyme because the electroactive group is not readily accessible to the electrode and is site specific. There are ways around this. These include determining a product or cofactor related to the enzyme activity (for example, determining the hydrogen peroxide produced as the result of oxygen reduction during a reaction catalyzed by glucose oxidase). (See Fig. 36.5a.)

Many applications using oxidase enzymes are based on this simple principle. The common problem is that hydrogen peroxide requires a relatively high potential at a platinum electrode, so easily oxidized materials interfere.

Another route to link the electrode with the enzyme is via a chemical-electron shuttle, an electron transfer mediator (cofactor). This is illustrated in Fig. 36.5b. In this case the enzyme oxidizes the analyte and in turn reduces the mediator. The reduced mediator contacts the electrode, where it is oxidized, ready to interact once again with reduced enzyme. Although in principle this is very simple, in practice these devices are difficult to make and meet commercial criteria of long-term stability, reproducibility, and speed of analysis, but the benefit is sufficiently great that there is much ongoing research and development from both academic and commercial sources.

Figure 36.5 Basis of two approaches for amperometric biosensors. (a) Monitoring the hydrogen peroxide produced by an oxidase enzyme specific for the analyte of interest; (b) measuring a reduced enzyme cofactor (an electron transfer mediator) that can be detected at a favorable potential. *(Reprinted with permission of Bioanalytical Systems, Inc.)*

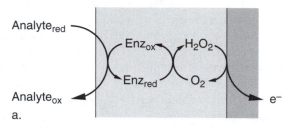

What It Does

Liquid chromatography and hydrodynamic electrochemistry (LCEC) are compatible technologies that, in combination, yield important advantages for a number of trace determinations, namely selectivity and low detection limits. Thus, LCEC results in the quantitation of a few specific analytes present in a complex mixture of hundreds of other compounds. An example is the determination of a neurotransmitter (adrenaline) or drug (acetaminophen) in plasma or urine. The chromatograph carries out sample preparation and separation of the analytes of interest, whereas the amperometic detector produces a signal necessary for their quantitation.

Analytical Information

Qualitative

Although most routine LCEC experiments use a single working electrode, it is a relatively simple matter to monitor simultaneously the current at several working electrodes. In practice, the use of more than three or four electrodes is rather awkward and expensive. Figure 36.4 depicts a thin-layer cell with two working electrodes. In the parallel mode, one can monitor the current at two different potentials. This is analogous to dual-wavelength UV absorbance detection. The ratio of the output of these two electrodes can indicate peak purity and, with experience, a general idea of the group undergoing the electrochemical oxidation (or reduction) reaction.

Quantitative

Accuracy and precision are similar to those of a chromatography system using absorbance or fluorescence detectors and, depending on total variance of system, including injection process, constant flow, and column, should be ±2%.

Detection limits vary depending on column configuration (dimensions) and on-column concentration of injected sample. The smaller-diameter columns result in less sample dilution; the bands entering the electrode (detector) are more concentrated, resulting in lower detection limits. Using a low-dispersion injector and considering only the column format, then for 150×0.32 mm, 3-µm particles, C_{18}, and an injection volume of 0.5 µL, a detection limit of 50 fg (0.3 fmoles) has been observed. Using larger columns 100×3.2 mm of the same packing material, 500 fg would be a reasonable limit to expect, and for a 250×4.6 mm, 5 µm, C_{18}, 5.0 pg would be reasonable. Detection limit is also influenced by the potential required to give maximum response (that is, to carry out the oxidation/reduction reaction): The higher the potential the higher the detection limit. Rate of electron transfer (slope of HDV), cell volume, flow rate, and temperature influence detection limits directly at the level of output signal or through their effect on baseline noise. Typically one should expect detection limits of 2 to 0.02 pmoles injected.

Applications

Listing every compound that is electrochemically active is not practical, but listing general groups of compounds is, and these can be found in Table 36.1. There is some overlap between the sections, but the biomedical list is confined to endogenous analytes and the pharmaceutical list to manufactured compounds. Vitamins and L-DOPA, for example, can fit in both categories. The category of typical functional groups, although not all-inclusive, indicates some functional groups and compounds that are known to be electrochemically active (oxidized or reduced.)

In some cases, analytes that are not electrochemically active within practical potential limits, acetylcholine and amino acids for example, must first be derivatized before detection. This can be accomplished by chemical reaction, precolumn in the case of amino acids and postcolumn via enzyme-catalyzed reactions for acetylcholine determination. The former method involves gradient separation, the latter isocratic separation. Postcolumn derivatization, but before the electrochemical cell, can also be accomplished by UV irradiation. This photolytic derivatization occurs on-line and forms electroactive products, in a continuous manner, from compounds that are non-electroactive or difficult to oxidize (reduce).

The nature of the working electrode, particularly its bulk composition and surface treatment, is critical to detector performance. Many organic compounds react at significantly different rates depending on the electrode used. It is normally desirable to carry out the electrode reactions at the greatest possible rate in order that the current be limited only by mass transport of molecules to the surface and not by their reaction rate at the surface. This situation affords the greatest sensitivity and stability without sacrificing selectivity.

Figure 36.6 Chromatograms of a standard solution containing cystein (CSH), glutathione (GSH), and their disulfides CSSC and GSSG. The four compounds were separated on a reversed-phase column and detected using a dual Hg/Au electrode cell in a series mode. The upstream (W1) electrode had an applied potential of –1.0V versus Ag/AgCl and the downstream (W2) electrode an applied potential of +0.15V. The disulfides were reduced to their respective thiols at the upstream electrode (On) and oxidized (detected) at the downstream electrode. The thiols (already in a reduced form) passed unchanged over the upstream electrode and were oxidized at the downstream electrode. The disulfides could be detected only after an initial electrochemical reduction reaction (Off). *(Reprinted with permission of Bioanalytical Systems, Inc.)*

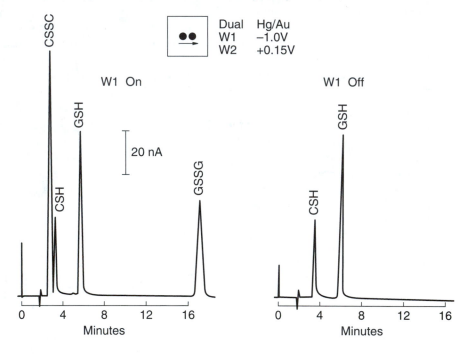

Many electrode materials are commonly used. Glassy (vitreous) carbon, mercury-amalgamated gold (Hg/Au), platinum (Pt), and silver (Ag) have all been used in electrochemical cells. The choice depends on range of available potential, chemical and physical compatibility with the mobile phase, and long-term stability. In the usual case, all three considerations are closely related. The amagalm electrode Hg/Au is rather specific for thiol groups (Fig. 36.6) but also reacts with chelating agents and halides, and has been used in reductive applications due to its high hydrogen overvoltage.

Ag is more limited in its use, being mainly used to determine iodide, bromide, and cyanide. In LCEC, Pt has been used to determine sulfite, but is more widely used to determine H_2O_2. The source of the H_2O_2 is mainly an enzyme- (oxidase) catalyzed reaction. An example is the determination of acetylcholine/choline, where both of these substrates undergo reactions ending in the production of stoichiometric amounts of H_2O_2. Glassy carbon is by far the most widely used electrode material in LCEC systems (Fig. 36.7). This material has excellent chemical, mechanical, and electrical properties, is relatively free of impurities, can take a high polish, and has low background current and noise at a wide range of potentials.

Figure 36.7 Chromatogram of a standard solution of nine analytes separated on a microbore column (150 × 1.0 mm) and detected amperometrically on a glassy carbon electrode at +0.75 volts versus Ag/AgCl. 1-norepinephrine, 2-epinephrine, 3-2,5-dihydroxybenzoic acid, 4-3,4-dihydroxyphenylacetic acid, 5-2,3-dihydroxybenzoic acid, 6-dopamine; 7-5-hydroxyindol-3-acetic acid, 8-homovanillic acid, 9-serotonin. *(Reprinted with permission of Bioanalytical Systems, Inc.)*

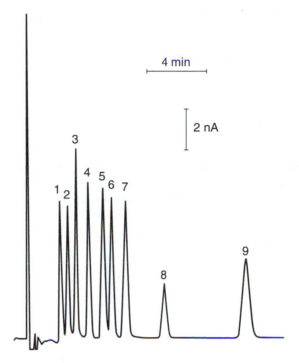

Nuts and Bolts

Relative Costs

Amperometric detectors (electronics and transducer cell) $2400–$15,000

Price varies with single- or dual-electrode control of potential, pulsed and scan capability, temperature control, injector, and remote control.

LCEC system (detector and chromatograph) $16,000–$65,000

The degrees of automation and options dramatically influence total cost. The lowest price starts with the basics, as in Fig. 36.3 with manual injection and a strip chart recorder, whereas the high range includes a gradient system with dual detectors and software-controlled data reduction.

Vendors for Instruments and Accessories

Bioanalytical Systems, Inc.
2701 Kent Ave.
West Lafayette, IN 47906
phone: 765-463-4527
fax: 765-497-1102
email: bas@bioanalytical.com
Internet: http://www.bioanalytical.com

Dionex Corp.
1228 Titan Way, P.O. Box 3603
Sunnyvale, CA 94088-3603
phone: 408-737-8522
fax: 408-730-9403
Internet: http://www.infoweavers.com/DIONEX

EG & G Princeton Applied Research
P.O. Box 2565
Princeton, NJ 08543
phone: 609-530-1000
fax: 609-883-7259

ESA
22 Alpha Rd.
Chelmsford, MA 01824
phone: 508-250-7000
fax: 508-250-7090
Internet: http://www.esainc.com

Gilson Company, Inc.
P.O. Box 677
Worthington, OH 43085-0677
phone: 614-548-7298
fax: 614-548-5314
email: gilson@bronze.coil.com
Internet: http://www.globalgilson.com

Hewlett-Packard
2850 Centerville Rd.
Wilmington, DE 19808-1610
phone: 302-633-8696, 800-227-9770
fax: 302-633-8901
Internet: http://www.hp.com/go/chem

Rainin Instrument Co.
Mack Rd.
Woburn, MA 01801-4026
phone: 617-935-3050, 800-472-4646
fax: 617-938-8157

Shimadzu Analytical & Laboratory Instruments
7100 Riverwood Dr.
Columbia, MD 21046
phone: 410-381-6996, 800-388-6996
fax: 410-290-9140
Internet: http://www.shimadzu.com

Waters Chromatography
24 Maple St.
Milford, MA 01757
phone: 800-252-4752
fax: 508-872-1990
email: info@waters.com
Internet: http://www.waters.com

Required Level of Training

Any persons trained in chromatography can use an amperometric detector. Assembly of the transducer cell and setting control parameters (such as potential, gain, and filtering) can easily be accomplished with a minimal degree of mechanical aptitude. It would be useful to have college-level chemistry courses when working with unique electroactive compounds or where a detection potential is not known (as in nonroutine determination). Quantitative determinations require a minimum of an introductory analytical course or previous experience using a chromatography system with another type of detector.

Service and Maintenance

In an LCEC system the bulk (98%) of the troubleshooting is associated with the chromatography (LC) portion, which generally adversely affects the amperometric detector (that is, the cell). Because EC detection requires ions (salts) in the mobile phase, precautions must be taken to prevent corrosion of the stainless steel auxiliary electrode and components of the chromatograph.

The electronics of the detector require no special care, but avoid having them come into contact with liquid (possible when flowing fluid is around). Cell maintenance has two requirements: proper handling of the reference electrode in order to ensure its role in maintaining a constant potential, and cleaning or polishing the working electrode when required, as indicated by a loss in response, by electrochemical, mechanical, or chemical procedures.

Suggested Readings

General Electrochemistry

Kissinger, P. T., and W. R. Heineman, eds., *Laboratory Techniques in Electroanalytical Chemistry*. New York: Marcel Dekker, 1984.

Liquid Chromatography/Electrochemistry

Duda, C. T., and P. T. Kissinger, "Determination of Biogenic Amines, Their Metabolites and Other Neurochemicals by Liquid Chromatography/Electrochemistry," in S. H. Parvez, M. Naoi, T. Nagatsu, and S. Parvez, eds., *Methods in Neurotransmitter and Neuropeptide Research, Part 1*. Amsterdam: Elsevier Science, 1993, 41–81.

KRULL, I. S., AND OTHERS, "Derivatization and Post-Column Reaction for Improved Detection in Liquid Chromatography/Electrochemistry," *Journal of Liquid Chromatography*, 8 (1985), 2845–70.

LAVRICH, C., AND P. T. KISSINGER, "Liquid Chromatography–Electrochemistry: Potential Utility for Therapeutic Drug Monitoring," in S.H.Y. Wong, ed., *Therapeutic Drug Monitoring and Toxicology by Liquid Chromatography*. New York: Marcel Dekker, 1985, 191–233.

SELAVKA, C. M., AND I. S. KRULL, "The Forensic Determination of Drugs of Abuse Using Liquid Chromatography with Electrochemical Detection: A Review," *Journal of Liquid Chromatography*, 10 (1987), 345–75.

SHOUP, R. E. "Liquid Chromatography/Electrochemistry," in C. S. Horvath, ed., *High-Performance Liquid Chromatgraphy*, vol. 4. New York, Academic Press, 1986, 91–194.

Amperometric Sensors and Biosensors

There are reviews of this technology but the R&D is causing rapid changes. The emphasis is in the materials and (bio)chemistry near or on the electrode rather than any basic electrochemistry per se. Almost any issue of leading analytical journals will contain new approaches for solving problems in many application areas. A couple of more focused journals in this area are as follows:

Biosensors and Bioelectronics, edited by A. P. F. Turner. Elsevier Science Publishing Co., Inc., New York (includes a monthly, worldwide update of publications in the area).

Electroanalysis, edited by J. Wang. VCH Publishers, Inc., New York (frequent reviews and original articles).

General Sensor

FREW, J. E., AND M. J. GREEN, "Biosensors for Clinical Analysis," *Analytical Proceedings*, 25 (1988), 276.

FREW, J. E., AND M. J. GREEN, "Amperometric Biosensors," *Analytical Proceedings*, 26 (1989), 334.

HART, J. P., AND S. A. WRING, "Screen-Printed Voltammetric and Amperometric Electrochemical Sensors for Decentralized Testing," *Electroanalysis*, 6 (1994), 617.

KAUFFMANN, J. M., AND G. G. GUILBAULT, "Enzyme Electrode Biosensors: Theory and Applications," in C. H. Suelter, ed., *Bioanalytical Applications of Enzymes*, vol. 36. New York: Wiley, 1992, 63.

MURRAY, R. W., A. G. EWING, AND R. A. DURST, "Chemically Modified Electrodes: Molecular Design for Electroanalysis," *Analytical Chemistry*, 59 (1987), 379A.

Sensors for Environmental Analysis

FLEET, B., AND H. GUNASINGHAM, "Electrochemical Sensors for Monitoring Environmental Pollutants," *Talanta*, 39 (1992), 1449.

Chapter 37

Voltammetric Techniques

Samuel P. Kounaves

Tufts University
Department of Chemistry

Summary

General Uses

- Quantitative determination of organic and inorganic compounds in aqueous and nonaqueous solutions
- Measurement of kinetic rates and constants
- Determination adsorption processes on surfaces
- Determination electron transfer and reaction mechanisms
- Determination of thermodynamic properties of solvated species
- Fundamental studies of oxidation and reduction processes in various media
- Determination of complexation and coordination values

Common Applications

- Quantitative determination of pharmaceutical compounds
- Determination of metal ion concentrations in water to sub–parts-per-billion levels
- Determination of redox potentials
- Detection of eluted analytes in high-performance liquid chromatography (HPLC) and flow injection analysis

- Determination of number of electrons in redox reactions
- Kinetic studies of reactions

Samples

State

Species of interest must be dissolved in an appropriate liquid solvent and capable of being reduced or oxidized within the potential range of the technique and electrode material.

Amount

The amounts needed to obtain appropriate concentrations vary greatly with the technique. For example, cyclic voltammetry generally requires analyte concentrations of 10^{-3} to 10^{-5} M, whereas anodic stripping voltammetry of metal ions gives good results with concentrations as low as 10^{-12} M. Volumes may also vary from about 20 mL to less than a microliter (with special microelectrode cells).

Preparation

The degree of preparation required depends on both the sample and the technique. For determination of Pb(II) and Cd(II) in seawater with a microelectrode and square-wave anodic stripping voltammetry (ASV), no preparation is required. In contrast, determination of epinepherine in blood plasma at a glassy carbon electrode with differential pulse voltammetry (DPV) requires that the sample first be pretreated with several reagents, buffered, and separated.

Analysis Time

Once the sample has been prepared, the time required to obtain a voltammogram varies from a few seconds using single-sweep square-wave voltammetry, to a couple of minutes for a cyclic voltammogram, to possibly 30 min (or more) for a very-low-concentration ASV determination.

Limitations

General

- Substance must be oxidizable or reducible in the range were the solvent and electrode are electrochemically inert.
- Provides very little or no information on species identity.
- Sample must be dissolved

Accuracy

Accuracy varies with technique from 1 to 10%.

Sensitivity and Detection Limits

Detection limit varies with technique from parts per thousand to parts per trillion.

Complementary or Related Techniques

- Other electroanalytical techniques may provide additional or preliminary information for electrochemical properties.
- Simultaneous use of spectroscopic methods can identify species undergoing reaction.
- Liquid chromatography is often used to separate individual analytes before analysis.

Introduction

Historically, the branch of electrochemistry we now call voltammetry developed from the discovery of polarography in 1922 by the Czech chemist Jaroslav Heyrovsky, for which he received the 1959 Nobel Prize in chemistry. The early voltammetric methods experienced a number of difficulties, making them less than ideal for routine analytical use. However, in the 1960s and 1970s significant advances were made in all areas of voltammetry (theory, methodology, and instrumentation), which enhanced the sensitivity and expanded the repertoire of analytical methods. The coincidence of these advances with the advent of low-cost operational amplifiers also facilitated the rapid commercial development of relatively inexpensive instrumentation.

The common characteristic of all voltammetric techniques is that they involve the application of a potential (E) to an electrode and the monitoring of the resulting current (i) flowing through the electrochemical cell. In many cases the applied potential is varied or the current is monitored over a period of time (t). Thus, all voltammetric techniques can be described as some function of E, i, and t. They are considered active techniques (as opposed to passive techniques such as potentiometry) because the applied potential forces a change in the concentration of an electroactive species at the electrode surface by electrochemically reducing or oxidizing it.

The analytical advantages of the various voltammetric techniques include excellent sensitivity with a very large useful linear concentration range for both inorganic and organic species (10^{-12} to 10^{-1} M), a large number of useful solvents and electrolytes, a wide range of temperatures, rapid analysis times (seconds), simultaneous determination of several analytes, the ability to determine kinetic and mechanistic parameters, a well-developed theory and thus the ability to reasonably estimate the values of unknown parameters, and the ease with which different potential waveforms can be generated and small currents measured.

Analytical chemists routinely use voltammetric techniques for the quantitative determination of a variety of dissolved inorganic and organic substances. Inorganic, physical, and biological chemists widely use voltammetric techniques for a variety of purposes, including fundamental studies of oxidation and reduction processes in various media, adsorption processes on surfaces, electron transfer and reaction mechanisms, kinetics of electron transfer processes, and transport, speciation, and thermodynamic properties of solvated species. Voltammetric methods are also applied to the determination of compounds of pharmaceutical interest and, when coupled with HPLC, they are effective tools for the analysis of complex mixtures.

How It Works

The electrochemical cell, where the voltammetric experiment is carried out, consists of a working (indicator) electrode, a reference electrode, and usually a counter (auxiliary) electrode. In general, an electrode provides the interface across which a charge can be transferred or its effects felt. Because the working electrode is where the reaction or transfer of interest is taking place, whenever we refer to the electrode, we always mean the working electrode. The reduction or oxidation of a substance at the surface of a working electrode, at the appropriate applied potential, results in the mass transport of new material to the electrode surface and the generation of a current. Even though the various types of voltammetric techniques may appear to be very different at first glance, their fundamental principles and applications derive from the same electrochemical theory. Here we summarize some of the electrochemical theory or laws common to all of the voltammetric techniques. Where necessary, more specific details are given later under the discussion of each technique.

General Theory

In voltammetry, the effects of the applied potential and the behavior of the redox current are described by several well-known laws. The applied potential controls the concentrations of the redox species at the electrode surface ($C_O{}^0$ and $C_R{}^0$) and the rate of the reaction (k^0), as described by the Nernst or Butler–Volmer equations, respectively. In the cases where diffusion plays a controlling part, the current resulting from the redox process (known as the faradaic current) is related to the material flux at the electrode–solution interface and is described by Fick's law. The interplay between these processes is responsible for the characteristic features observed in the voltammograms of the various techniques.

For a reversible electrochemical reaction (that is, a reaction so fast that equilibrium is always reestablished as changes are made), which can be described by $\mathbf{O} + ne^- \Leftrightarrow \mathbf{R}$, the application of a potential E forces the respective concentrations of $\mathbf{O}$ and $\mathbf{R}$ at the surface of the electrode (that is, $c_O{}^0$ and $c_R{}^0$) to a ratio in compliance with the Nernst equation:

$$E = E^0 - \frac{RT}{nF} \ln \frac{c_R^0}{c_O^0} \tag{37.1}$$

where R is the molar gas constant ($8.3144\ J\ \mathrm{mol^{-1}K^{-1}}$), T is the absolute temperature (K), n is the number of electrons transferred, F = Faraday constant ($96{,}485$ C/equiv), and E^0 is the standard reduction potential for the redox couple. If the potential applied to the electrode is changed, the ratio $c_R{}^0/c_O{}^0$ at the surface will also change so as to satisfy Eq. (37.1). If the potential is made more negative the ratio becomes larger (that is, $\mathbf{O}$ is reduced) and, conversely, if the potential is made more positive the ratio becomes smaller (that is, $\mathbf{R}$ is oxidized).

For some techniques it is useful to use the relationship that links the variables for current, potential, and concentration, known as the Butler–Volmer equation:

$$\frac{i}{nFA} = k^0 \{ c_O^0 \exp[-\alpha\theta] - c_R^0 \exp[(1-\alpha)\theta] \} \tag{37.2}$$

where $\theta = nF(E - E^0)/RT$, k^0 is the heterogeneous rate constant, α is known as the transfer coefficient, and A is the area of the electrode. This relationship allows us to obtain the values of the two analytically important parameters, i and k^0.

Finally, in most cases the current flow also depends directly on the flux of material to the electrode surface. When new **O** or **R** is created at the surface, the increased concentration provides the force for its diffusion toward the bulk of the solution. Likewise, when **O** or **R** is destroyed, the decreased concentration promotes the diffusion of new material from the bulk solution. The resulting concentration gradient and mass transport is described by Fick's law, which states that the flux of matter (Φ) is directly proportional to the concentration gradient:

$$\Phi = -AD_O(\partial c_O / \partial x) \tag{37.3}$$

where D_O is the diffusion coefficient of **O** and x is the distance from the electrode surface. An analogous equation can be written for **R**. The flux of **O** or **R** at the electrode surface controls the rate of reaction, and thus the faradaic current flowing in the cell. In the bulk solution, concentration gradients are generally small and ionic migration carries most of the current. The current is a quantitative measure of how fast a species is being reduced or oxidized at the electrode surface. The actual value of this current is affected by many additional factors, most importantly the concentration of the redox species, the size, shape, and material of the electrode, the solution resistance, the cell volume, and the number of electrons transferred.

In addition to diffusion, mass transport can also occur by migration or convection. Migration is the movement of a charged ion in the presence of an electric field. In voltammetry, the use of a supporting electrolyte at concentrations 100 times that of the species being determined eliminates the effect of migration. Convection is the movement of the electroactive species by thermal currents, by density gradients present in the solution, or by stirring the solution or rotating the electrode. Convection must be eliminated or controlled accurately to provide controlled transport of the analyte to the electrode.

Many voltammetric techniques have their own unique laws and theoretical relationships that describe and predict in greater detail the various aspects of the i–E behavior (such as curve shape, peak height, width, and position). When appropriate, these are discussed in more detail.

Instrumentation

The basic components of a modern electroanalytical system for voltammetry are a potentiostat, computer, and the electrochemical cell (Fig. 37.1). In some cases the potentiostat and computer are bundled into one package, whereas in other systems the computer and the A/D and D/A converters and microcontroller are separate, and the potentiostat can operate independently.

The Potentiostat

The task of applying a known potential and monitoring the current falls to the potentiostat. The most widely used potentiostats today are assembled from discrete integrated-circuit operational amplifiers and other digital modules. In many cases, especially in the larger instruments, the potentiostat package also includes electrometer circuits, A/D and D/A converters, and dedicated microprocessors with memory.

A simple potentiostat circuit for a three-electrode cell with three operational amplifiers (OA) is shown in Fig. 37.2. The output of OA-1 is connected to the counter electrode with feedback to its own inverting input through the reference electrode. This feedback decreases the difference between the inverting and noninverting inputs of OA-1 and causes the reference electrode to assume the same potential as E_{in} of OA-1. Because the potential difference between the working electrode and the reference electrode is zero the working electrode is set to the same potential as applied to the OA-1 input. With the reference electrode connected to E_{in} through the high

Figure 37.1 Block diagram of the major components of an electroanalytical system for performing voltammetric analysis.

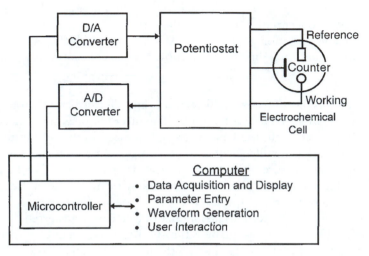

impedance of OA-3, the current must flow through the counter electrode. Current flow through the reference not only is undesirable because of its higher resistance but also would eventually cause its potential to become unreliable. A three-electrode system is normally used in voltammetry for currents in the range of microamperes to milliamperes. With the use of micron-sized electrodes, currents are in the pico- to nanoampere range, and thus two electrodes are often used (that is, the counter and reference are tied together). An OA acting as a current-to-voltage converter (OA-2) provides the output signal for the A/D converter.

Most voltammetric techniques are dynamic (that is, they require a potential modulated according to some predefined waveform). Accurate and flexible control of the applied potential is a critical function of the potentiostat. In early analog instruments, a linear scan meant just that, a continuous linear change in potential from one preset value to another. Since the advent of digital

Figure 37.2 The basic potentiostat circuit composed of operational amplifiers.

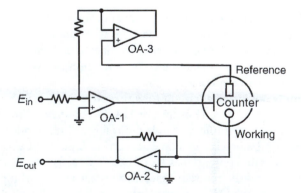

electronics almost all potentiostats operate in a digital (incremental) fashion. Thus, the application of a linear scan is actually the application of a "staircase" modulated potential with small enough steps to be equivalent to the analog case. Not surprisingly, digital fabrication of the applied potential has opened up a whole new area of pulsed voltammetry, which gives fast experiments and increased sensitivity. In the simpler standalone potentiostats the excitation signal used to modulate the applied potential is usually provided by an externally adjustable waveform generator. In the computer-controlled instruments, the properties of the modulation and the waveform are under software control and can be specified by the operator. The most commonly used waveforms are linear scan, differential pulse, and triangular and square wave.

The use of micro- and nanometer-size electrodes has made it necessary to build potentiostats with very low current capabilities. Microelectrodes routinely give current responses in the pico- to nanoampere range. High-speed scanning techniques such as square-wave voltammetry require very fast response times from the electronics. These diverse and exacting demands have pushed potentiostat manufacturers into providing a wide spectrum of potentiostats tailored to specific applications.

The Electrodes and Cell

A typical electrochemical cell consists of the sample dissolved in a solvent, an ionic electrolyte, and three (or sometimes two) electrodes. Cells (that is, sample holders) come in a variety of sizes, shapes, and materials. The type used depends on the amount and type of sample, the technique, and the analytical data to be obtained. The material of the cell (glass, Teflon, polyethylene) is selected to minimize reaction with the sample. In most cases the reference electrode should be as close as possible to the working electrode; in some cases, to avoid contamination, it may be necessary to place the reference electrode in a separate compartment. The unique requirements for each of the voltammetric techniques are described under the individual techniques.

Reference Electrodes The reference electrode should provide a reversible half-reaction with Nernstian behavior, be constant over time, and be easy to assemble and maintain. The most commonly used reference electrodes for aqueous solutions are the calomel electrode, with potential determined by the reaction $Hg_2Cl_2(s) + 2e^- = 2Hg(l) + 2Cl^-$ and the silver/silver chloride electrode (Ag/AgCl), with potential determined by the reaction $AgCl(s) + e^- = Ag(s) + Cl^-$. Table 37.1 shows the potentials of the commonly used calomel electrodes, along with those of some other reference electrodes. These electrodes are commercially available in a variety of sizes and shapes.

Counter Electrodes In most voltammetric techniques the analytical reactions at the electrode surfaces occur over very short time periods and rarely produce any appreciable changes in bulk concentrations of **R** or **O**. Thus, isolation of the counter electrode from the sample is not normally necessary. Most often the counter electrode consists of a thin Pt wire, although Au and sometimes graphite have also been used.

Working Electrodes The working electrodes are of various geometries and materials, ranging from small Hg drops to flat Pt disks. Mercury is useful because it displays a wide negative potential range (because it is difficult to reduce hydrogen ion or water at the mercury surface), its surface is readily regenerated by producing a new drop or film, and many metal ions can be reversibly reduced into it. Other commonly used electrode materials are gold, platinum, and glassy carbon.

Table 37.1 Reference electrodes of the type || KCI/MCI(satd.)/M.

MCI/M	KCI	$E^{\circ\prime}$ at 25 °C 25	$\dfrac{d(E^{\circ\prime})}{dt}$ (mV deg^{-1} at 25 °C)
AgCl/Ag	3.5 M (at 25 °C)	0.205	−0.73
	Satd.	0.199	−1.01
Hg2Cl2/Hg	0.1 M (at 25 °C)	0.336	−0.08
	1.0 M (at 25 °C)	0.283	−0.29
	3.5 M (at 25 °C)	0.250	−0.39
	Satd.	0.244	−0.67

What It Does

This section of the chapter discusses in more detail some of the more common forms of voltammetry currently in use for a variety of analytical purposes. The uniqueness of each rests on subtle differences in the manner and timing in which the potential is applied and the current measured. These differences can also provide very diverse chemical, electrochemical, and physical information, such as highly quantitative analyses, rate constants for chemical reactions, electrons involved on redox reactions, and diffusion constants.

Polarography

Even though polarography could be considered just another variation of technique within voltammetry, it differs from other voltammetric methods both because of its unique place in the history of electrochemistry and in respect to its unique working electrode, the dropping mercury electrode (DME). The DME consists of a glass capillary through which mercury flows under gravity to form a succession of mercury drops. Each new drop provides a clean surface at which the redox process takes place, giving rise to a current increase with increasing area as the drop grows, and then falling when the drop falls. Figure 37.3 shows a polarogram for a 1 M solution of HCl that is 5 mM in Cd^{2+}. The effect of drop growth and dislodging can be clearly seen. The potential when the current attains half the value of the plateau current is called the half-wave potential and is specific to the analyte's matrix. The plateau current is proportional to the concentration of analyte. For example, Fig. 37.4 shows a differential pulse polarogram for the acetyl derivative of chlordiazepoxide. In this case the peak height is proportional to the analyte concentration.

The current for the polarographic plateau can be predicted by the Ilkovic equation:

$$i_d = 708n\mathrm{D}^{1/2}m^{2/3}t^{1/6}c^0 \tag{37.4}$$

where m is the rate of flow of the Hg through the capillary, t is the drop time, and c^0 is the bulk analyte concentration.

Even though polarography with the DME is the best technique for some analytical determinations, it has several limitations. Mercury is oxidized at potentials more positive than +0.2 V ver-

Figure 37.3 Classic polarogram taken at a DME showing background taken in 1 *M* HCl (line A) and 1 *M* HCl + 0.5 mM Cd(II) (line B). *(From D. T. Sawyer and J. L. Roberts,* Experimental Electrochemistry for Chemists, *copyright © 1974 John Wiley & Sons, Inc. Reprinted by permission of John Wiley & Sons, Inc.)*

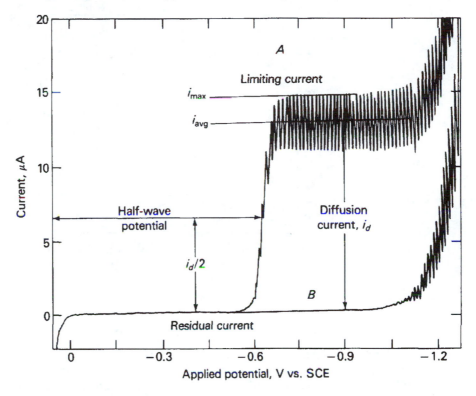

sus SCE, which makes it impossible to analyze for any analytes in the positive region of potential. Another limitation is the residual current that results from charging of the large capacitance of the electrode surface.

By manipulating the potential and synchronizing potential pulses with current sampling, the same basic experiment can be made to yield a more useful result.

Cyclic Voltammetry

Cyclic voltammetry (CV) has become an important and widely used electroanalytical technique in many areas of chemistry. It is rarely used for quantitative determinations, but it is widely used for the study of redox processes, for understanding reaction intermediates, and for obtaining stability of reaction products.

This technique is based on varying the applied potential at a working electrode in both forward and reverse directions (at some scan rate) while monitoring the current. For example, the initial scan could be in the negative direction to the switching potential. At that point the scan would be reversed and run in the positive direction. Depending on the analysis, one full cycle, a partial cycle, or a series of cycles can be performed.

The response obtained from a CV can be very simple, as shown in Fig. 37.5 for the reversible redox system:

Figure 37.4 Differential pulse polarogram of the seven-acetyl analog of chlordiazepoxide. *(Reprinted from Anal. Chim. Acta, 74, M. A. Brooks, et al., p. 367, copyright 1975 with kind permission of Elsevier Science—NL, Sara Burgerhartstraat 25, 1055 KV Amsterdam, The Netherlands.)*

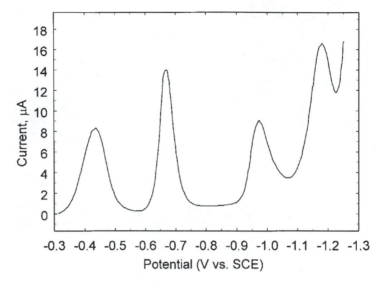

$$Fe(CN)_6^{-3} + e^- = Fe(CN)_6^{-4} \tag{37.5}$$

in which the complexed Fe(III) is reduced to Fe(II).

The important parameters in a cyclic voltammogram are the peak potentials (E_{pc}, E_{pa}) and peak currents (i_{pc}, i_{pa}) of the cathodic and anodic peaks, respectively. If the electron transfer process is fast compared with other processes (such as diffusion), the reaction is said to be electrochemically reversible, and the peak separation is

$$\Delta E_p = |E_{pa} - E_{pc}| = 2.303 \; RT/nF \tag{37.6}$$

Thus, for a reversible redox reaction at 25 °C with n electrons ΔE_p should be 0.0592/n V or about 60 mV for one electron. In practice this value is difficult to attain because of such factors as cell resistance. Irreversibility due to a slow electron transfer rate results in $\Delta E_p > 0.0592/n$ V, greater, say, than 70 mV for a one-electron reaction.

The formal reduction potential (E^o) for a reversible couple is given by

$$E^o = \frac{E_{pc} + E_{pa}}{2} \tag{37.7}$$

For a reversible reaction, the concentration is related to peak current by the Randles–Sevcik expression (at 25 °C):

$$i_p = 2.686 \times 10^5 n^{3/2} A c^0 D^{1/2} v^{1/2} \tag{37.8}$$

where i_p is the peak current in amps, A is the electrode area (cm^2), D is the diffusion coefficient (cm^2 s^{-1}), c_0 is the concentration in mol cm^{-3}, and v is the scan rate in V s^{-1}.

Cyclic voltammetry is carried out in quiescent solution to ensure diffusion control. A three-electrode arrangement is used. Mercury film electrodes are used because of their good negative potential range. Other working electrodes include glassy carbon, platinum, gold, graphite, and carbon paste.

Figure 37.5 Cyclic voltammograms of 5 mM Fe(CN)$_6^{-3}$ in 1 M KCl with v = 500 mV/s.

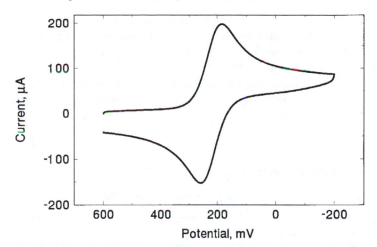

Pulse Methods

In order to increase speed and sensitivity, many forms of potential modulation (other than just a simple staircase ramp) have been tried over the years. Three of these pulse techniques, shown in Fig. 37.6, are widely used.

Figure 37.6 Potential waveforms and their respective current response for (a) differential pulse, (b) normal pulse, and (c) square-wave voltammetry.

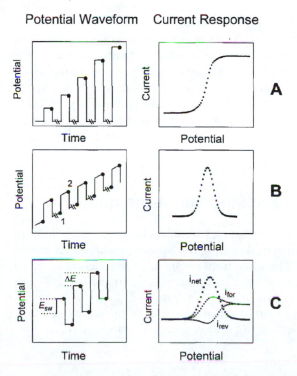

Normal Pulse Voltammetry (NPV)

This technique uses a series of potential pulses of increasing amplitude. The current measurement is made near the end of each pulse, which allows time for the charging current to decay. It is usually carried out in an unstirred solution at either DME (called normal pulse polarography) or solid electrodes.

The potential is pulsed from an initial potential E_i. The duration of the pulse, τ, is usually 1 to 100 msec and the interval between pulses typically 0.1 to 5 sec. The resulting voltammogram displays the sampled current on the vertical axis and the potential to which the pulse is stepped on the horizontal axis.

Differential Pulse Voltammetry (DPV)

This technique is comparable to normal pulse voltammetry in that the potential is also scanned with a series of pulses. However, it differs from NPV because each potential pulse is fixed, of small amplitude (10 to 100 mV), and is superimposed on a slowly changing base potential. Current is measured at two points for each pulse, the first point (1) just before the application of the pulse and the second (2) at the end of the pulse. These sampling points are selected to allow for the decay of the nonfaradaic (charging) current. The difference between current measurements at these points for each pulse is determined and plotted against the base potential.

Square-Wave Voltammetry (SWV)

The excitation signal in SWV consists of a symmetrical square-wave pulse of amplitude E_{sw} superimposed on a staircase waveform of step height ΔE, where the forward pulse of the square wave coincides with the staircase step. The net current, i_{net}, is obtained by taking the difference between the forward and reverse currents ($i_{for} - i_{rev}$) and is centered on the redox potential. The peak height is directly proportional to the concentration of the electroactive species and direct detection limits as low as $10^{-8}\,M$ are possible.

Square-wave voltammetry has several advantages. Among these are its excellent sensitivity and the rejection of background currents. Another is the speed (for example, its ability to scan the voltage range over one drop during polarography with the DME). This speed, coupled with computer control and signal averaging, allows for experiments to be performed repetitively and increases the signal-to-noise ratio. Applications of square-wave voltammetry include the study of electrode kinetics with regard to preceding, following, or catalytic homogeneous chemical reactions, determination of some species at trace levels, and its use with electrochemical detection in HPLC.

Preconcentration and Stripping Techniques

The preconcentration techniques have the lowest limits of detection of any of the commonly used electroanalytical techniques. Sample preparation is minimal and sensitivity and selectivity are excellent. The three most commonly used variations are anodic stripping voltammetry (ASV), cathodic stripping voltammetry (CSV), and adsorptive stripping voltammetry (AdSV).

Even though ASV, CSV, and AdSV each have their own unique features, all have two steps in common. First, the analyte species in the sample solution is concentrated onto or into a working electrode. It is this crucial preconcentration step that results in the exceptional sensitivity that can be achieved. During the second step, the preconcentrated analyte is measured or stripped from the electrode by the application of a potential scan. Any number of potential waveforms can be used for the stripping step (that is, differential pulse, square wave, linear sweep, or staircase). The most common are differential pulse and square wave due to the discrimination against charging current. However, square wave has the added advantages of faster scan rate and increased sensitivity relative to differential pulse.

The electrode of choice for stripping voltammetry is generally mercury. The species of interest can be either reduced into the mercury, forming amalgams as in anodic stripping voltammetry, or adsorbed to form an insoluble mercury salt layer, as in cathodic stripping voltammetry.

Stripping voltammetry is a very sensitive technique for trace analysis. As with any quantitative technique, care must be taken so that reproducible results are obtainable. Important conditions that should be held constant include the electrode surface, rate of stirring, and deposition time. Every effort should be made to minimize contamination.

Anodic Stripping Voltammetry

ASV is most widely used for trace metal determination and has a practical detection limit in the part-per-trillion range (Table 37.2). This low detection limit is coupled with the ability to determine simultaneously four to six trace metals using relatively inexpensive instrumentation.

Metal ions in the sample solution are concentrated into a mercury electrode during a given time period by application of a sufficient negative potential. These amalgamated metals are then stripped (oxidized) out of the mercury by scanning the applied potential in the positive direction. The resulting peak currents, i_p, are proportional to the concentration of each metal in the sample solution, with the position of the peak potential, E_p, specific to each metal. The use of mercury limits the working range for ASV to between approximately 0 and -1.2 V versus SCE. The use of thin Hg films or Hg microelectrodes along with pulse techniques such as square-wave voltammetry can substantially lower the limits of detection of ASV.

With more than one metal ion in the sample, the ASV signal may sometimes be complicated by formation of intermetallic compounds, such as ZnCu. This may shift or distort the stripping peaks for the metals of interest. These problems can often be avoided by adjusting the deposition time or by changing the deposition potential.

Table 37.2 Relative sensitivity of some electrochemical techniques for metals.

Technique	Limits of Detection for Pb(II)
Ion selective electrode	10^{-5} M
DC polarography at DME	10^{-6} M
Differential pulse polarography at SMDE	10^{-7} M
Differential pulse ASV at HMDE	10^{-10} M*
DC ASV at mercury film	10^{-11} M*
Square-wave ASV at mercury film	10^{-12} M*

*Deposition for 360 seconds; LOD varies with deposition time; S(H)MDE = static (hanging) mercury drop electrode

Cathodic Stripping Voltammetry

CSV can be used to determine substances that form insoluble salts with the mercurous ion. Application of a relatively positive potential to a mercury electrode in a solution containing such substances results in the formation of an insoluble film on the surface of the mercury electrode. A potential scan in the negative direction will then reduce (strip) the deposited film into solution. This method has been used to determine inorganic anions such as halides, selenide, and sulfide, and oxyanions such as MoO_4^{2-} and VO_3^{5-}. In addition, many organic compounds, such as nucleic acid bases, also form insoluble mercury salts and may be determined by CSV.

Adsorptive Stripping Voltammetry

AdSV is quite similar to anodic and cathodic stripping methods. The primary difference is that the preconcentration step of the analyte is accomplished by adsorption on the electrode surface or by specific reactions at chemically modified electrodes rather than accumulation by electrolysis. Many organic species (such as heme, chlorpromazine, codeine, and cocaine) have been determined at micromolar and nanomolar concentration levels using AdSV; inorganic species have also been determined. The adsorbed species is quantified by using a voltammetric technique such as DPV or SWV in either the negative or positive direction to give a peak-shaped voltammetric response with amplitude proportional to concentration.

Analytical Information

Qualitative

As shown in Figs. 37.3 through 37.5, voltammetric techniques give rise to current signals that appear at a characteristic position on the potential scale. The potential at which the signal appears gives qualitative information about the reactant. However, the ability of the potential of the signal to identify the reactant is not very large because the position of the signal depends on the reactant conditions and the resolution is poor. Thus, a characteristic potential excludes many possibilities for the identity of the reactant; in particular, the voltammetric response absolutely excludes all non-electroactive substances. If the response is the same as that of a known substance, obtained under exactly the same conditions, then the known substance is a good hypothesis for the identity. However, in general voltammetric techniques are not good tools for qualitative identification of analytes.

Quantitative

The main virtue of voltammetric techniques is their good accuracy, excellent precision (<1%), sensitivity, and wide dynamic range. In the special case of stripping voltammetry, detection limits routinely are lower than the amount of signal due to contamination of sample. An impression of the relative ability of many electrochemical techniques to measure small concentrations of analytes in solution is given in Table 37.2. This table applies to routine practice with standard equipment. The detection limits given should be attainable, for example, in an undergraduate instructional laboratory.

Nuts and Bolts

Relative Costs

The size, power, sophistication, and price of the potentiostats for voltammetry vary from large research-grade instruments (20 to 30 kg with a ±10-volt potential and 1 A to 100 nA current ranges, $15 to 20K) to simple battery-powered units (3 to 1 kg with a ±2.5-volt potential and 6 mA to 50 pA current ranges, $3 to 8 K). The choice of instrument depends on the type of voltammetric analysis to be performed, the information desired, and somewhat on the size of the electrodes. Cyclic voltammetry experiments using 5-mm-diameter disk electrodes with scan rates no larger than 1 Vs⁻¹ are easily performed with most potentiostats. To determine quantitatively trace amounts of an analyte in an organic solvent using a 1-μm-diameter microelectrode and high-frequency square-wave voltammetry requires the more expensive instrumentation. More detailed information is presented in Table 37.3.

Vendors for Instruments and Accessories

In the United States there are several companies that manufacture electroanalytical instrumentation capable of performing voltammetric analyses and several who are distributors for U.S. or non-U.S. manufacturers. Table 37.3 lists the major vendors and a sample of the available models.

Table 37.3 Manufacturers and distributors of voltammetric instrumentation.

Manufacturers/Distributors	Model	Techniques*	Price
BioAnalytical Systems, Inc.	CV27	Limited	$3,995
	CV50W	Sufficient	11,500
	100BW	All	19,500
Cypress Systems, Inc.	OMNI90	Limited	1,800
	CS1090	Sufficient	12,950
	CS2000	All	21,500
EG&G Princeton Applied Research	264	Limited	6,500
	263	All	11,440
	273A	All	18,695
Pine Instruments	AFRDE4	Limited	2,175
Brinkman Instruments (Metrohm)	693	Sufficient	18,670
Taccusell, France (ASI, Inc., U.S. distributor)	PJT	Sufficient	NA
Eco Chemie, Netherlands	PSTAT10	All	12,500

*Limited = performs only a few techniques such as linear scan voltammetry; Sufficient = performs most of the common techniques; All = performs all or almost all of the major techniques.

BioAnalytical Systems, Inc.
2701 Kent Ave.
West Lafayette, IN 47906
phone: 765-463-4527
fax: 765-497-1102
email: bas@bioanalytical.com
Internet: http://www.bioanalytical.com

Cypress Systems, Inc.
2500 West 31st St., Suite D
Lawrence, KS 66047
phone: 800-235-2436
fax: 913-832-0406

EG&G Princeton Applied Research
P.O. Box 2565
Princeton, NJ 08543
phone: 609-530-1000
fax: 609-883-7259

Pine Instruments
101 Industrial Dr.
Grove City, PA 16127
phone: 412-458-6391
fax: 412-458-4648
Internet: http://www.pineinst.com

Brinkman Instruments (Metrohm)
One Cantiague Rd.
P.O. Box 1019
Westbury, NY 11590-0207
phone: 800-645-3050
fax: 516-334-7506
email: info@brinkmann.com
Internet: http://www.brinkmann.com

Eco Chemie B.V.
P.O. Box 513
3508 AD Utrecht
The Netherlands
phone: +31 30 2893154
fax: +31 30 2880715
email: autolab@ecochemie.nl
Internet: http://www.ecochemie.nl

Required Level of Training

With modern commercial instrumentation, routine analytical voltammetry is made fairly straight-forward by the manufacturer, who typically supplies not simply the instrument but rather a complete analytical system, including cell, electrodes, and software for data analysis. In cases for which the analyte is known and the method specified (often provided by the vendor), general training in chemistry at the postsecondary level is adequate. In less well-defined cases that involve some aspect of method development, baccalaureate training and some specific experience with voltammetry are desirable. In the case of stripping methods, considerable experience with the specific techniques and problems of interest is often required, due not to increased complexity of the electrochemical technique but rather to general requirements for trace analysis involving sample handling, blank subtraction, and calibration.

Service and Maintenance

Trouble with voltammetric procedures almost always arises in a part of the system external to the instrument. Thus, the first recourse when a problem arises is not to an electronics or software expert, but to someone with electrochemical experience. Most equipment manufacturers provide telephone consulting as well. Because of the integrated nature of the commercial equipment, repair of instruments is almost always done by returning the instrument to the factory. Typically no routine maintenance is required other than installation of software upgrades provided by the manufacturer. An instrument that functions well when first set up is most likely to do so for many years.

Suggested Readings

BAARS, A., M. SLUYTERS-REHBACH, AND J. H. SLUYTERS, "Application of the Dropping Mercury Microelectrode in Electrode Kinetics," *Journal of Electroanalytical Chemistry*, 364 (1994), 189.

BARD, A. J. AND L. R. FAULKNER, *Electrochemical Methods*. New York: Wiley, 1980.

BERSIER, B. M., "Do Polarography and Voltammetry Deserve Wider Recognition in Official and Recommended Methods?," *Analytical Proceedings*, 24 (1987), 44.

BRETT, C. M. A., AND A. M. O. BRET, *Electrochemistry: Principles, Methods and Applications*. Oxford: Oxford University Press, 1993.

CHRISTENSEN, P. A., AND A. HAMNET, *Techniques and Mechanisms in Electrochemistry*. New York: Chapman & Hall, 1994.

GOSSER, D. K., *Cyclic Voltammetry: Simulation & Analysis of Reaction Mechanisms*. New York: VCH Publishers, 1993.

KISSINGER, P. T., AND W. R. HEINEMAN, "Cyclic Voltammetry," *J. Chem. Ed.*, 60 (1983), 702.

KISSINGER, P. T., AND W. R. HEINEMAN, *Laboratory Techniques in Electroanalytical Chemistry*. New York: Marcel Dekker, 1984.

KOUNAVES, S. P., AND OTHERS, "Square Wave Anodic Stripping Voltammetry at the Mercury Film Electrode: Theoretical Treatment," *Analytical Chemistry*, 59 (1987), 386.

O'DEA, J. J., J. OSTERYOUNG, AND R. A. OSTERYOUNG, "Theory of Square Wave Voltammetry for Kinetic Systems," *Analytical Chemistry*, 53 (1981), 695.

OSTERYOUNG, J., AND R. A. OSTERYOUNG, "Square Wave Voltammetry," *Analytical Chemistry*, 57 (1985), 101A.

RUDOLPH, M., D. P. REDDY, AND S. W. FELDBERG, "A Simulator for Cyclic Voltammetric Response," *Analytical Chemistry*, 66 (1994), 589A.

VAN DEN BERG, C. M. G., "Potentials and Potentialities of Cathodic Stripping Voltammetry of Trace Elements in Natural Waters," *Anal. Chim. Acta*, 250 (1991), 265.

WANG, J., *Stripping Analysis*. Deerfield Beach, FL: VCH Publishers, 1985.

Chapter 38

Potentiometric Techniques

Richard S. Hutchins and Leonidas G. Bachas

University of Kentucky
Department of Chemistry
and Center of Membrane Sciences

Summary

General Uses

- Selective, quantitative determination of many organic and inorganic ions in solution
- Determination of ions in a specific oxidation state within a sample
- Determination of stability constants of various complexes
- Determination of reaction rates and mechanisms
- Quantitative determination of acidic and basic gases
- Quantitative determination of enzymatic reaction products

Common Applications

- Industrial process analysis of ions in batch or flow-through configurations
- Determination of pollutant gases and continuous monitoring of air quality
- Determination of electrolytes in physiological fluids for clinical analysis
- Development of biosensors based on immobilized enzymes and electrodes
- Determination of ion constituents in agricultural, environmental, and pharmaceutical samples

- pH determination
- Endpoint determination in acid, base, and redox titrations

Samples

State

Most liquid and gaseous samples can be readily analyzed. Solid samples may be analyzed if they can be prepared in solution form.

Amount

The detection limits for conventional electrodes are approximately 10^{-5} to 10^{-6} M. For gas sensors detection limits range from 0.01 to 5.0 ppm.

Preparation

Little preparation is required for liquid and gaseous samples. In both cases the pH of the sample may need to be adjusted. This serves to free complexed ions, avoid precipitation of metal ions, or, in the case of potentiometric gas sensors, to convert ions to a gaseous form. Solid samples must be prepared in solution. Organic solids that cannot be simply dissolved (such as food, vegetation, and pharmaceuticals) may be first ashed, and then the ions extracted using an appropriate solvent. Many samples must be buffered to prevent OH^-/H^+ interference and adjusted to a constant ionic strength. Distilled, deionized water is preferable for use in preparing all sample solutions.

Analysis Time

The time required for analysis varies based on the electrode used, the analyte determined, and the analyte concentration. A rapidly responding electrode, such as the pH electrode, can be both calibrated and used to determine the pH of a sample in 1 min or less. Typical sample analysis times, not including calibration, for conventional ion-selective electrodes range from 5 to 60 sec, whereas gas and enzyme sensors require 1 to 5 min or longer for the determination of a single sample. Non-solid samples can be prepared for testing within 5 min.

Limitations

General

- There are many ions for which no selective electrode exists.
- Most electrodes require frequent calibration for use in accurate quantitative analysis.
- A buffered sample is often required to avoid OH^-/H^+ interference.
- Matrix effects (that is, ionic strength differences, electrolytes present in the sample and their influence on the junction potential, and the presence of species that may foul the active surface of the electrode) must be accounted for.

Sensitivity

- Generally an analyte concentration higher than 10^{-6} M is required for most potentiometric determinations.

Complementary or Related Techniques

- Ion chromatography, which can be used for both the quantitative determination of ions and the separation of ions
- Atomic absorption/emission spectroscopy, used to determine the total analyte present, as opposed to free ions only, and to determine concentration rather than activity
- Amperometry, a more sensitive technique available for the determination of ions that are electroactive
- Coulometry, used for determining concentration by monitoring the total charge required for its oxidation or reduction

Introduction

Potentiometry can be simply described as the measurement of a potential in an electrochemical cell. It is the only electrochemical technique that directly measures a thermodynamic equilibrium potential and in which essentially no net current flows. The roots of potentiometry as an analytical technique extend back to pioneers such as Luigi Galvani (1737–1798) and, more recently, Walther Nernst (1864–1941). Nernst was instrumental in deriving the thermodynamic equilibrium relationship between the galvanic cell potential and the activity of an ion in solution. One of the unique features of potentiometry is the ability to monitor the activity of an ion in a sample rather than the concentration. The direct determination of activity results from the thermodynamic equilibrium relationship between the activity of an ion and the potential of a cell. Potentiometry thus provides one of the few methods available for distinguishing between free (ionized) and bound (complexed) ions in a sample as well as between the activities of different oxidation states of a given ion that may be present. A calibration plot is constructed for the determination of an unknown activity by measuring the cell potential (E_{cell}) at various known activities and graphing the results (E_{cell} against the logarithm of the activity). Two electrodes are used in measuring cell potentials: a reference and an indicator (working) electrode (see Fig. 38.1).

The potential difference between these two electrodes, the indicator electrode potential changing with the activity of analyte present and the potential of the reference electrode remaining constant, is the E_{cell} observed experimentally. Under ideal thermodynamic equilibrium conditions and in the absence of interfering species, the Nernst equation (see Eq. 38.4), dictates the slope of the linear portion of the calibration plot.

One of the most common and earliest applications of potentiometry is pH determination. Before the late 1960s potentiometry as an analytical technique was limited to the pH and redox electrodes, useful in titrations for the determination of electroactive species. Since that time, a renaissance has occured in the development of electrodes that are selective for a given ion, or ion-selective electrodes (ISEs). Through the development of electrodes that selectively determine targeted ions, potentiometry in general (and more specifically ISEs) is replacing many older,

Figure 38.1 Typical instrument configuration for a potentiometric analysis.

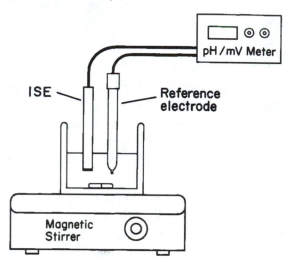

more expensive, and time-consuming techniques for analytically monitoring and measuring ion activities. Recent developments in the ISE branch of potentiometry have made the monitoring of inorganic, organic, gaseous, and biologically important ions possible. The flexibility in the available configurations allows these ions to be monitored either in a single sample solution (batch mode) or continuously in a flow-through apparatus (flow-injection analysis). One precaution about many commercially available ISEs is that they are often labeled as selective for a given ion despite responding to interferents. Only if the sample matrix in which such an electrode will be used is known and the interferent is not present at amounts sufficient to cause interference can such an electrode be used successfully for the determination of the desired ion.

How It Works

Physical and Chemical Principles

Potentiometric analysis involves measuring the potential difference, E_{cell}, between an indicator and a reference electrode, which depends on the membrane potential, E_m. In Eq. (38.1) the contributing sources to the observed E_{cell} are shown.

$$E_{cell} = (E_m + E_{ref1}) - (E_{ref2} + E_j) \tag{38.1}$$

The internal reference potential of the indicator electrode (E_{ref1}) and the potential of the reference electrode (E_{ref2}) are fixed. They do not change during the course of the experiment, although the E_{cell} does depend on the reference electrode used. The junction potential, E_j, results from the different mobilities of individual ions and the separation of charge that results across the junction, known as a salt bridge, of two solutions of varying compositions. Ions that have higher mobilities diffuse across the interface ahead of slower moving ions, resulting in this separation of charge. A steady state is soon established between the charge separation (E_j) and the ion movement across the interface. The liquid junction potential can be rather large (>50 mV), but there are steps that can be taken to minimize and stabilize E_j. Using an electrolyte in which the cation and anion are approximately equal in mobility (such as KCl rather than HCl) minimizes E_j. Using a high concentration of electrolyte in each solution tends to stabilize E_j, as only minimal concentration differences would then result when the analyte is added to the sample solution. With all of the contributing potentials mentioned above fixed, E_{cell} in Eq. (38.1) can be rewritten as follows:

$$E_{cell} = K + E_m \tag{38.2}$$

where K is a constant and E_m is the membrane potential,

$$E_m = E_{pb1} + E_{pb2} + E_d \tag{38.3}$$

where E_{pb1} is the phase boundary potential at the external surface of the membrane, which is created by the separation of charge resulting from the partitioning of a given ion into the membrane. The potential E_{pb2} arises in a similar fashion on the interior of the indicator electrode membrane. The potential E_d is the potential arising from the separation of charge due to the diffusion of ions through the membrane. Generally E_d is small on the time scale of potentiometric measurements, and it can therefore be ignored. The potential E_{pb2} can be held constant by fixing the concentration of the ions in contact with the inner membrane. Hence, the entire E_{cell} can be directly related to the ability of a membrane in contact with the sample solution to be permselective (allow only ions of one charge to partition into the membrane), resulting in a charge separation (E_{pb1}) that can be measured. The selectivity of the membrane for a targeted ion depends on how specific the interaction of the membrane is with the analyte.

External reference electrode	Aqueous sample	Ion-selective membrane	Internal filling solution	Internal reference solution
E_{ref2} $\quad$ E_j		E_{pb1} $\quad$ $\longleftarrow E_d \longrightarrow$	E_{pb2}	E_{ref1}

Using the Henderson approximation, which assumes a linear concentration gradient of each ion across the membrane, and assuming the membrane is only permeable to a single ion, one arrives, through thermodynamic derivation (1), at the Nernst equation:

$$E_{cell} = K + \frac{RT}{z_i F} \ln(a_i) \tag{38.4}$$

Here R is the gas constant, K is a constant of the measurement system, T is the temperature in K, z_i is the charge (including the sign) on the analyte ion i, F is Faraday's constant, and a_i is the activity of the analyte ion i. This relationship between the activity of an ion and the E_{cell} is the backbone for potentiometric analysis. At 25 °C and with a charge of +1, E_{cell} versus log a_i should be linear over the working range of the electrode with a slope of 59 mV/decade change in activity. Because membranes are never completely specific for a given ion (that is, they allow only one ion to partition into the membrane), a more general form of this relationship is described by the Nicolsky–Eisenman equation (2):

$$E_{cell} = K + \frac{RT}{z_i F} \ln(a_i + K_{ij} a_j^{z_i/z_j}) \tag{38.5}$$

where a_i is the activity of the analyte ion, a_j the activity of the interfering ion, z_i and z_j the charges of ions i and j, respectively, and K_{ij} the selectivity coefficient. This equation is a rough approximation that provides a conceptual and empirical framework for handling interferences. If more than one interfering ion is present, additional $K_{ij} a_j^{z_i/z_j}$ terms are needed to describe the behavior of the electrodes (Eq. 38.6) (2), as there are additional ions that partition into the membranes; the magnitude of interference in the measurement of a_i is determined by each K_{ij}.

$$E_{\text{cell}} = K + \frac{RT}{z_i F} \ln\left(a_i + \sum_j K_{ij} a_j^{z_i/z_j} \right) \qquad (38.6)$$

If $K_{ij} = 1$, then the membrane responds equally to i and j. The smaller K_{ij} is, the less interference is observed from ion j. The selectivity coefficients are determined experimentally, and for many commercially available electrodes selectivity coefficients have been determined and are provided by the vendor. When applying the Nernst and other equations, care must be taken in using the proper sign notation. Anions have a negative charge and hence result in a negative change in the cell potential with increasing activities, whereas positively charged analytes produce a positive change with increasing activities.

Another potential source that contributes to E_{cell} and has not been mentioned yet is the asymmetry potential, E_{asym}. This potential arises because of the unavoidable differences that exist in the internal and external surfaces of the indicator membrane. Because E_{asym} can change over time, frequent calibration of the electrode is required. In most applications E_{asym} is assumed to be constant (part of K in Eqs. 38.2, 38.5, and 38.6) because it contributes only a very small potential (typically about 1 to 5 mV) to the overall E_{cell}.

Instrumentation

The instrumentation required to perform potentiometric measurements includes a reference electrode, an indicator electrode, and a high-input-impedance mV (pH/pIon) meter (see Fig. 38.1). The two most popular types of reference electrodes are the saturated calomel and silver/silver chloride systems. Both types of reference electrodes exhibit many of the ideal characteristics sought in a reference electrode. These include maintaining a fixed potential over time and temperature, having long-term stability, and returning to the initial potential after being subject to small currents.

The saturated calomel electrode (SCE) is composed of metallic Hg, solid mercurous chloride (Hg_2Cl_2), and a saturated solution of KCl at equilibrium. Using a saturated KCl solution gives the added advantage of fixing the Cl^- concentration in this solution. Consequently, the potential of the SCE (+0.241 V versus the standard hydrogen electrode) remains constant, even if some of the liquid evaporates over time. The disadvantage of using saturated KCl is that changes in temperature require long periods for the SCE to reestablish the solubility equilibrium of KCl. Fixed, nonsaturating concentrations of KCl are more effective and efficient under changing temperature conditions. Both types of calomel electrodes can be purchased commercially. The SCE is more popular and, with a constant temperature bath, the hysteresis caused by fluctuating temperatures can be eliminated. The SCE can be used as a reference electrode in samples that do not exceed 80 °C.

The silver/silver chloride reference electrode includes a silver wire, coated on one end with the insoluble AgCl salt. When the electrode is immersed in a saturated KCl solution, its potential at 25 °C depends only on the Cl^- concentration (which is fixed in the saturated solution) and is +0.192 V versus the standard hydrogen electrode. The Ag/AgCl electrode should not be used in solutions that contain species that can precipitate or complex with silver (such as halides, proteins, and sulfide).

Reference electrodes should be prepared and maintained so that the level of the internal liquid is kept above the sample solution being tested, to avoid infusion of sample into the reference electrode. Failure to do this may result in plugging of the junction due to precipitates forming with either Ag^+ or Hg^+. A double junction electrode, shown in Fig. 38.2, is commonly used as a precaution to avoid any contamination of the sample by Cl^-, Ag^+, or Hg_2^{2+} ions.

Figure 38.2 An Orion Ag/AgCl double-junction reference electrode. *(Reprinted with permission from Orion Research, Inc.).*

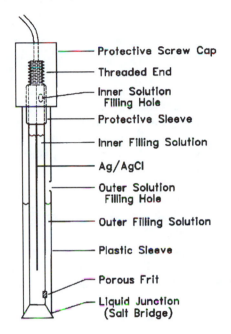

Reference electrodes are discussed rigorously in physical chemistry textbooks and have been highly elaborated by manufacturers and discussed in practical detail in their literature. For example, the Ag/AgCl reference electrode exists in a multitude of designs including those for high temperature applications (especially when sterilization is required), hydrogel designs for maintenance-free industrial applications, and pressurized electrodes for high pressure applications.

There are many types of indicator electrodes available, the selection of which depends on the applications desired. Most types can be classified as either metallic or membrane electrodes. The descriptions of specific indicator electrodes can be found in the following section. The change in potential of the indicator electrode with respect to the fixed potential of the reference electrode is the analytical parameter observed. This potential difference is monitored using a voltmeter. This meter must have a high input impedance to facilitate the measurement of small potential changes and the use of electrodes with high resistances, as well as to ensure that no current is drawn while measuring the potential difference. Most meters can be temperature programmed to accommodate measurements under the given experimental conditions. Several meters may be purchased with a temperature compensator probe accessory that, when inserted into the sample, corrects the potential value for changes in the temperature of the sample. Battery-operated, portable pH meters are also available, many of which include an internal logger and a communications port, for use in field work. Although the meter is typically the biggest expense in potentiometric analysis, a simple, inexpensive, functional meter may be adequate (for a possible design see Reference 3). In potentiometric experiments it is common to perform analyses in a jacketed beaker connected to a water bath circulator. This also eliminates problems that can arise due to temperature fluctuations in the sample.

If needed, strip-chart recorders can be used to record the cell potential. They are rugged and fairly inexpensive. A more sophisticated approach involves the use of a data acquisition package, run by computer, which can eliminate the need for strip-chart recorders. Several electrodes may be monitored simultaneously depending on the acquisition package. Internal boards or external hardware are usually required, and these are additional expenses.

What It Does

The most important consideration in using potentiometric techniques is the type of indicator electrode to be used. Depending on the application, a variety of electrodes are available, each with inherent advantages and disadvantages. In this section the most common indicator electrode configurations are described, along with a brief discussion of the strengths and weaknesses of each specific electrode type.

Metallic Electrodes

There are four types of metallic indicator electrodes: electrodes of the first, second, and third kinds, and redox electrodes. For redox electrodes and electrodes of the first kind, the exposed metal surface may be dipped in acid (nitric acid works well) to clean the surface before use.

An electrode of the first kind is simply a metal wire (mesh or solid plate) that responds to its own metal cation in solution. One example is a silver wire and Ag^+ in solution. These electrodes exhibit poor selectivity, usually responding preferentially to cations that are most easily reduced. Additionally, electrodes of the first kind do not respond to the analyte if the metal ion has been oxidized. Likewise, the surface of the electrode itself can be oxidized in the presence of air, making it impossible to obtain accurate measurements at low ion activities. The most widely used electrodes of the first kind are Ag/Ag^+ and Hg/Hg^+. If precautions are taken (such as removal of O_2 by deaerating the sample solution), other metals may also be used as electrodes of the first kind, such as Cu/Cu^{2+}, Cd/Cd^{2+}, Sn/Sn^{2+}, and Pb/Pb^{2+}.

An electrode of the second kind consists of a metal immersed in a saturated solution containing one of its sparingly soluble salts, or coated with such a salt. This indicator electrode is responsive to the anion of the metal salt. The electrode responds to the activity of the anion, despite the absence of direct electron transfer between the anion and the electrode. The activity of the free metal ion (to which the electrode does respond via a direct electron transfer) is controlled by complexation or precipitation with the anion. Thus, the observed potential can be related indirectly to the activity of the anion. Selectivity under these conditions may be a problem. Any other anion present that can form an insoluble salt or complex with the metal ion will interfere with the measurement of the targeted anion. Both the Ag/AgCl and SCE electrodes described earlier (reference electrodes) are examples of electrodes of the second kind. Halides and other anions that form slightly soluble salts with silver (or other metals) may be detected using an electrode of the second kind.

Like electrodes of the second kind, electrodes of the third kind depend on solution equilibria, but in this case two solution equilibria are involved. These electrodes respond to a cation other than that of the electrode metal. In order for this electrode to respond properly, the complex between the metal cation and the anion must be more stable than the complex between the targeted cation and the anion, and there must be an excess of uncomplexed cation in solution. One example of an indicator electrode of this type is a mercury electrode, with ethylenediaminetetraacetic acid (EDTA) as the common anion that responds to calcium. Although this dual equilibria indicator electrode is useful for pure samples, there are too many sources of interferents for electrodes of the third kind to function well in complex sample matrices.

Redox indicator electrodes can be used to detect redox species. An inherent advantage of potentiometry is the ability to measure the activity of an analyte (rather than simply measuring molecular concentration). Hence, the determination of an analyte in various oxidation states is possible, as the potential depends on the ratio of activities of both species in a redox couple. The best redox electrodes are those made of the most inert metals. Gold, palladium, and platinum all work well as redox electrodes. A carbon surface, on which redox reactions are typically fast, may also be used. The drawback of these electrodes is that electron transfer is not always reversible, resulting in nonreproducible potentials.

Membrane Electrodes

Several types of membranes have been used that allow for a variety of different analytically useful ion-specific electrodes. Although the pH electrode is the most popular potentiometric sensor in use today, many innovative membrane electrodes have been developed in recent years that make possible the monitoring of a plethora of additional ions in different sample matrices. A brief description of the general classes of membrane electrodes and their specific configurations follows.

Glass Electrodes

The pH electrode responds to hydrogen ions as a result of the ion-exchange sites on the surface of a hydrated glass membrane. The electrode consists of a thin layer of glass, typically about $50 \, \mu m$ thick. Charge is transported across the membrane by sodium or lithium ions within the glass. The internal filling solution is a chloride solution that is also saturated in AgCl. This solution keeps the H^+ activity within the electrode constant, and hence, fixes the internal potential. Ag/AgCl wire is immersed in this acidic medium. A combination pH electrode, illustrated in Fig. 38.3, contains both the indicator and reference electrodes in the same unit.

It is crucial that the pH electrode be sufficiently hydrated before being used, or erratic and inaccurate responses will result. When not in use, the electrode should be stored in an aqueous solution because once it is dehydrated, several hours are required to rehydrate it fully. The high resistance inherent to glass electrodes (10^7 to $10^9 \, \Omega$) mandates the use of a high input impedance mV meter, such as that described in the previous section of this chapter.

The pH electrode may be calibrated using commercially available standards. Specific calibration instructions accompany all pH meters, but a multipoint calibration is best, chosen so that the expected pH of the sample is bracketed by that of the standard buffers used. The accuracy of the measurement can be maximized by allowing sufficient time for equilibration of each sample, using highly accurate pH buffers in the calibration, and by calibrating the pH meter at the same temperature as the sample to be analyzed. In very basic pH solutions the apparent pH is usually less than the real pH (alkaline error) due to the response of the membrane to sodium ions. This error can be minimized by altering the glass composition; electrodes especially designed for measurements at high pH are available (4). For determinations of high pH values, it is important to check the reported accuracy of an electrode in the basic pH region before purchasing it. Over time, glass electrodes eventually change composition and the response worsens. It is possible to rejuvenate glass electrodes by washing in concentrated acid, followed by soaking in water. Additionally, glass electrodes selective for other cations have been developed using different glass compositions. Glass electrodes are commercially available for Li^+, Na^+, K^+, Rb^+, Cs^+, Ag^+, Fe^{3+}, Pb^{2+}, and Cu^{2+}. Some glass electrodes are fragile and the use of a protective plastic cage around the glass membrane, during both measurements and storage, helps to prevent breakage. However, manufacturers have been quite successful at constructing robust glass electrodes, especially for field applications. Autoclavable electrodes are also available.

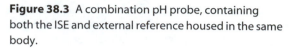

Figure 38.3 A combination pH probe, containing both the ISE and external reference housed in the same body.

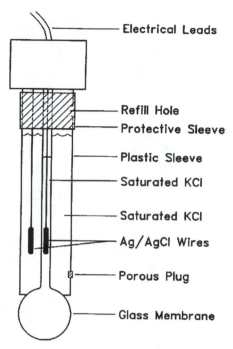

Solid-State Electrodes

Solid-state electrodes come in the form of crystals or pressed pellets containing the salts of the target ion. The LaF_3 crystal is the basis for the most well-known single crystal-based ISEs, selective for fluoride ion. Because low conductivity is a problem for most solid-state sensors, some sort of doping material is commonly used to improve electrical conductivity. The LaF_3 crystal is doped with Eu^{2+}, and the electrode responds to F^- through an equilibration reaction. The selective response to F^- occurs, in part, because of the size-exclusion phenomenon of the crystal lattice (that is, the channels of the lattice are complementary to the F^- ion in size). The LaF_3- based electrode is unresponsive to HF, and thus cannot be used in acidic solutions to determine total flouride content.

Whether a crystal or pressed-pellet membrane is used, the solid-state material must not be soluble in the solvent used for analysis (typically H_2O). For this reason, silver salts are used to make pressed-pellet electrodes. To increase both the conductivity and mechanical strength of the pellets, Ag_2S is added in an equal mole ratio to the Ag^+ salt (5). There is a similarity between the pressed-pellet electrode and the silver/silver chloride electrode of the second kind, both of which respond to the anion of the Ag^+ salt. Pressed-pellet electrodes suffer from interferences similar to those for electrodes of the second kind because many common anions complex with Ag^+.

Liquid Membrane ISEs

The membranes of liquid membrane ISEs usually contain three main components: a solid support, a lipophilic solvent, and an ionophore. The support in which the other two components are embedded is most commonly polyvinyl chloride, although other polymers are also used. By combining the three ingredients, dissolving them in a volatile solvent such as tetrahydrofuran, and allowing the tetrahydrofuran to evaporate, a flexible polymer membrane is obtained. Although the liquid solvent within the polymer (plasticizer) can exert a significant effect on the performance of the electrode, it is the ionophore that is responsible for the selective response of the ISE. Ionophores are prepared as either charged cation or anion exchangers or as neutral carriers, selective for a given ion. The mechanism of response is different for the various types of ionophores, but in all cases the generation of a potential (due to charge separation) across the membrane produces the signal observed. The selectivity of an electrode is determined by the degree to which an ionophore complexes selectively with the analyte ion. The development of selective lipophilic complexing agents for use as ionophores has produced a large number of novel selective sensors in recent years and has been the main thrust of research in the area of potentiometry. The liquid membrane-based ISEs are limited in lifetime due to leaching of both the plasticizer and ionophore over time from the membrane. Sensors with increased lifetimes have been developed by increasing the lipophilicities of both the plasticizer and ionophore, as well as by covalently attaching the ionophores directly to the polymers.

Another configuration of polymer electrodes uses membranes coated directly onto a metal wire. The main interest in the coated-wire electrodes (CWE) is in the area of miniaturization and in vivo ion analysis. Such electrodes contain no internal reference solution and, as a result, can be easily designed in miniature, with tip diameters as small as 0.1 μm.

Gas Sensing Probes

Gas sensing probes can be used both to monitor gases directly and to determine the concentration of ions whose conjugated acid or base is a gaseous species. The ISE and external reference electrode are placed behind a thin, gas-permeable membrane through which the gas diffuses. When the gas redissolves in the internal solution, between the ISE and gas-permeable membranes, two sensing schemes are possible. These are illustrated in Fig. 38.4, which shows an ammonia gas sensor.

An ammonium ISE may be used behind the gas-permeable membrane, provided that the internal solution is buffered at a pH at which the NH_3 gas is converted to NH_4^+. Alternatively, an unbuffered NH_4Cl solution can be used behind the gas-permeable membrane. With this arrangement the change in pH induced by the hydrolysis of gaseous NH_3, as it redissolves in the internal solution, can be sensed using a pH electrode.

Additional selectivity over conventional ISEs can be obtained in the determination of ions for which the conjugate acid or base is a gaseous species. The buffer used outside the gas-permeable membrane is one in which the analyte is converted to a gas. Unless interferant ions also exist as a gas under these conditions, they cannot penetrate the gas-permeable membrane. By using an ISE other than a pH electrode behind the gas-permeable membrane, it is possible to enhance selectivity further by discriminating against other gaseous species. In particular, pH electrodes respond to any gas that crosses the gas-permeable membrane and changes the pH of the internal filling solution. Conversely, an ISE that is selective for the conjugated acid or base of the gaseous species discriminates against other gases. Although gas sensors typically have long response times (1 to 5 min), the reduction in interference from nonvolatile ions makes them extremely useful, highly selective sensors. However, only a few ions may be detected using this configuration; commercial sensors for only NH_3, CO_2, and NO_x gases are available.

Biocatalytic Probes

The identification of a biocatalyst capable of responding selectively to an analyte of interest can often be used to develop a selective biocatalytic probe. Enzymes are a good example of a biocatalyst that exhibits high selectivity. By attaching a thin layer of a biocatalytic material to the surface of an ISE or gas-sensing electrode, a selective sensor may be achieved. The high cost of enzymes may

Figure 38.4 An Orion NH$_3$ series 95 gas-sensing probe. The species present behind the gas-permeable membrane as NH$_3$ gas penetrates the outer membrane are shown. *(Reprinted with permission from Orion Research, Inc.)*

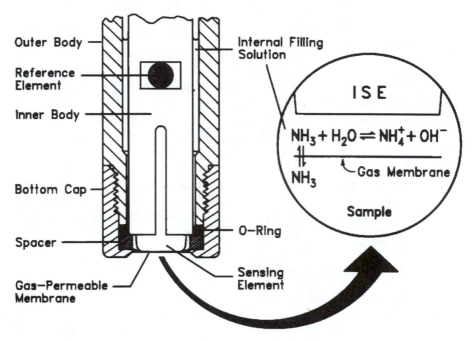

be avoided by incorporating a small amount of enzyme in an immobile layer. The biocatalytic layer can be attached in a variety of ways, including immobilization in a gel, covalent attachment to the polymer support of the ISE, and direct adsorption onto the surface of the electrode. The response times of biocatalytic sensors are typically longer than those of ISEs, but they may be reduced by keeping the biocatalytic layer thin. As a result of the high selectivity of enzymes, coupled with the selectivity of a membrane responsive to the product of the enzyme-catalyzed reaction, very few species interfere with biocatalytic probes. However, enzyme inhibitors, if present, can potentially interfere very strongly. In fact, biocatalytic ISEs can be used to determine the concentration of inhibitors indirectly through the effect of the inhibitor on the rate of the enzyme reaction. The ability to monitor selectively biologically important molecules under mild conditions (pH, temperature) with the simplicity of a potentiometric measurement make this configuration important for many analytical applications.

Analytical Information

Qualitative

Ion-selective electrodes provide only negative qualitative information. If an electrode for a specific ion (for example, a fluoride electrode) does not respond at all in the analyte solution, then the concentration for that ion is below the detection limit of the electrode. However, the most usual circumstance is that the analyte solution causes some response, often a mixed response to several ions present, and it is not possible to deduce from that response the identity of the ions responsible. Thus the most important applications of ion-selective electrodes are to samples for which the general composition of the sample is known and it is desired to analyze accurately and efficiently many samples.

Quantitative

Electrodes must be calibrated before use in potentiometric analysis. The E_j, which is unknown and changes with ionic strength, and the E_{asym}, which may vary over time, make it impossible to calculate K in the Nernst equation (Eq. 38.4). The assumption is made that this constant does not change during the analyte determination. However, this is never truly the case. The ionic strength (affecting E_j) of the calibration standards and the sample is seldom exactly the same, and the E_{asym} can change over time. Hence, frequent calibration is required. Using a high concentration of background electrolyte in the standards and sample, whenever possible, helps in fixing E_j.

The concentration of ions may also be determined using potentiometry. Concentration is directly related to activity (a) by a proportionality constant known as the activity coefficient (f):

$$a = c \times f \qquad (38.7)$$

The activity coefficient is typically calculated using the extended Debye–Hückel equation,

$$\log f = \frac{-0.51 z^2 \sqrt{\mu}}{1 + (\alpha \sqrt{\mu} / 305)} \qquad \text{at 25 °C} \qquad (38.8)$$

where z is the charge of the ion, a is the effective hydrated radius of the ion (in angstroms), and μ is the ionic strength of the solution. In samples with low total concentrations ($<10^{-3}\,M$) of ions the activity coefficient approaches unity and the concentration and activity of an ion are essentially equal. As the concentration and ionic strength increase, the activity coefficient decreases, resulting in activity values that diverge from the determined concentration. This divergence is even more pronounced for ions with multiple charges. Because ISEs respond strictly to activity, when a measurement of concentration is desired certain precautions must be taken, particularly when $\geq 10^{-3}\,M$ ions are present in the sample. To ensure that the accuracy and precision of the desired ion concentration measurement are maintained, the ionic strength of both the standard and sample solutions is fixed by using an ionic strength adjuster (a solution of known, high ionic strength). Thus, the activity coefficient remains essentially unchanged and activities are directly proportional to concentrations.

There are two general methods for determining an unknown concentration. The first involves making a calibration curve by plotting the E_{cell} against the logarithm of the concentration for standard analyte solutions. When the E_{cell} in a sample is determined, the calibration curve may then be used to determine the corresponding concentration of the analyte ion. The drawback of this method is that matrix effects may result in calculated concentrations that deviate from the actual concentration of the sample. An alternative method that may be used is the standard additions method (6). After determining the E_{cell} of the unknown, one or two small additions of standard solution are made to the sample and the new E_{cell} is recorded. The analyte concentration is again determined graphically. Measurements obtained in this fashion are less likely to suffer from matrix interferences. In order to achieve the highest accuracy, the standard additions should cause minimal changes in ionic strength and overall volume (both are assumed to be fixed).

Detection limits in potentiometric analysis are typically in the range 10^{-5} to $10^{-6}\,M$, although this varies with each sensor. There have been some reports of ISEs with detection limits as low as $10^{-9}\,M$ for well-buffered solutions. Often a linear logarithmic response is achievable over four or five decades of concentration. The selectivity depends on the specificity of the electrode for the targeted analyte. Selectivity coefficients for many common ions are provided by vendors and may be determined experimentally if the values have not been provided for a given species.

Potentiometric titrations are one of the most accurate quantitative analysis techniques available. The high accuracy and the added advantage of analyzing turbid or colored solutions make this method extremely useful for many applications. Although it is impossible to achieve the high accuracy of the titrations in direct potentiometric applications (due mainly to E_j and E_{asym}), nonetheless frequent calibration of the electrode can easily result in relative errors of less than $\pm 2\%$. If this is tolerable, then the speed of potentiometric analysis, low cost of equipment, and minimal training needed to operate such a system make this an attractive method of analysis.

Applications

1. Determination of L-ascorbic Acid by Direct Potentiometric Titration.

The objective of this analysis is to determine the concentration of L-ascorbic acid in pharmaceutical preparations using a potentiometric titration (7). A pharmaceutical preparation containing ascorbic acid (in liquid form) is placed quantitatively in a calibrated flask and diluted to a known volume with distilled water to give an ascorbic acid concentration of about 1 mg/mL. This is performed under inert atmosphere to avoid oxidation of the ascorbic acid in air. A volume of 4.0 mL of the sample is placed in a beaker containing a NH_4SCN solution buffered at pH 5.3. A copper(II)-selective electrode and a single-junction reference electrode are connected to a mV meter and placed in the stirred sample solution. The titration is performed by using a copper(II) sulfate solution as the oxidant under N_2 atmosphere. During the titration the Cu(I) formed precipitates as a white CuSCN salt. A sudden potential increase of 10 to 20 mV/0.05 mL of titrant indicates the endpoint.

Solid tablets are also analyzed as described above, after being dissolved in distilled water and diluted to a known volume. Replicate titrations produce results that are reproducible within $\pm 0.1\%$. The titration of standard ascorbic acid samples, using the described procedure, resulted in average recoveries of 99.996% and a relative standard deviation of $\pm 0.07\%$ ($n = 20$). The time needed for a typical titration was 10 to 13 min.

2. Flow-Injection Determination of Tetraphenylborate: Applications to Sequential Titrations of Metal Ions (Figs. 38.5 and 38.6).

Figure 38.5 Strip-chart recording of tetraphenylborate (TPB) standards obtained using a TPB-selective electrode in a flow-injection configuration. *(Reprinted from S. S. M. Hassan and I. H. Badr, "PVC Membrane Electrodes for Manual and Flow-Injection Determination of Tetraphenylborate: Applications to Separate and Sequential Titrations of Some Metal Ions," Talanta, 41(4), p. 526, Figure 1, copyright 1989 with kind permission of Elsevier Science—NL, Sara Burgerhartstraat 25, 1055 KV Amsterdam, The Netherlands.)*

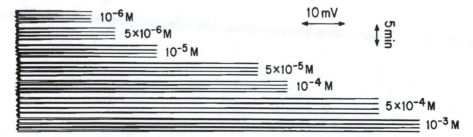

Figure 38.6 Potentiometric titration curves for two- and three-component mixtures of Ca²⁺, Sr²⁺, and Ba²⁺ using a TPB-selective electrode. *(Reprinted from S. S. M. Hassan and I. H. Badr, "PVC Membrane Electrodes for Manual and Flow-Injection Determination of Tetraphenylborate: Applications to Separate and Sequential Titrations of Some Metal Ions," Talanta, 41(4), p. 528, Figure 4, copyright 1989 with kind permission of Elsevier Science—NL, Sara Burgerhartstraat 25, 1055 KV Amsterdam, The Netherlands.)*

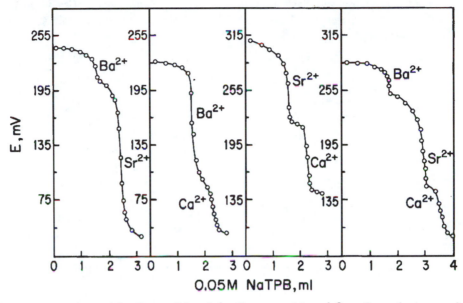

3. Combination Potassium/Carbon Dioxide Sensor Used for Continuous In Vivo Blood Measurements (Fig. 38.7).

Figure 38.7 Comparison of data obtained in vivo (solid line) using (a) a CO₂ gas sensor and (b) K⁺-selective electrode with data obtained in vitro (squares). *(Reprinted with permission from M. E. Collison et al., "Potentiometric Combination Ion/Carbon Dioxide Sensors for In Vitro and In Vivo Blood Measurements,"Anal. Chem., 1989, 61, p. 2368. Copyright 1989 American Chemical Society.)*

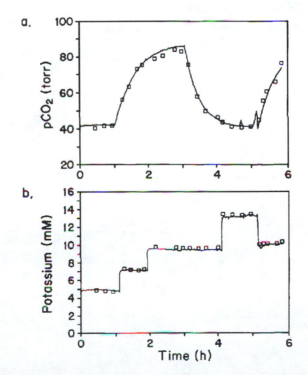

4. Disposable Bioselective Gas Sensors for Use in Urea Determination (Fig. 38.8).

Figure 38.8 Calibration curves for NH_4^+ and urea observed using an ammonium-based urease biosensor. *(Reprinted from Biosensors Bioelectron., 7, M. H. Gil et al., "Covalent Binding of Urease on Ammonium-Selective Potentiometric Membranes," p. 631, 1992, with kind permission from Elsevier Science S.A., P.O. Boc 564, 1001 Lausanne, Switzerland.)*

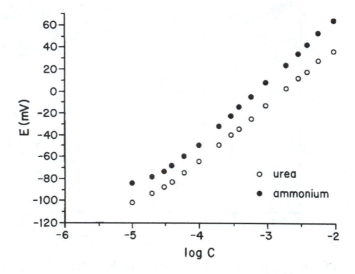

5. Flexible Microsensor Arrays for In Vivo Cardiovascular pH Determination (Fig. 38.9).

Figure 38.9 Two Kapton-based microelectrode array designs developed for in vivo use. *(Reprinted from J. Chem. Soc., Faraday Trans., 1993, 89(2), p. 365, Figure 3, with permission of The Royal Society of Chemistry.)*

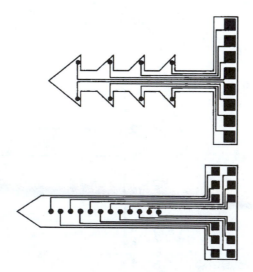

6. In Vitro Flow-Through Multianalyte Test System for the Determination of Blood Gases and Electrolytes (Fig. 38.10).

Figure 38.10 (a) Bedside multianalyte system that uses (b) a disposable cartridge for monitoring blood gases (CO_2 and O_2) and ions (H^+, K^+, Ca^{2+}, and Na^+). *(Reprinted with permission of the American Association for Clinical Chemistry, from* Clinical Chemistry, *40(9), 1994, p. 1571, Figure 4.)*

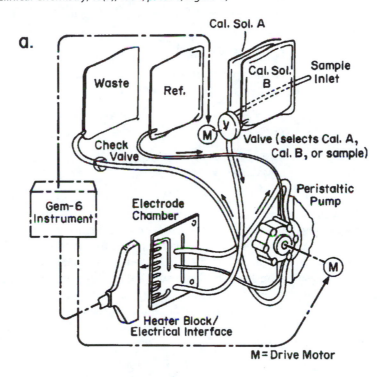

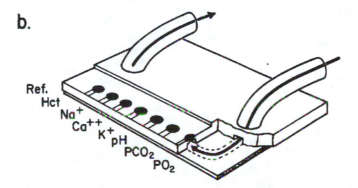

Nuts and Bolts

Relative Costs

pH meter	$$ to $$$$
Strip-chart recorder	$$ to $$$
ISEs	$ to $$
Reference electrodes	$ to $$
Fixed-temperature bath	$$$ to $$$$$
Peristaltic pump	$$$$
Supplies	$ to $$
Maintenance	$ to $$/year

$ = < $200, $$ = $200 to 500, $$$ = 0.5 to 1K, $$$$ = 1 to 2.5K,
$$$$$ = 2.5 to 5K.

The largest expense in potentiometry is typically the pIon/mV meter. An inexpensive meter may cost as little as a few hundred dollars but a more elaborate model with added accessories can exceed $2000. The application usually dictates which meter to buy. The remainder of the equipment is inexpensive. Specialty applications such as flow-injection analysis (requiring a peristaltic pump) can further increase the outlay costs. Low-end ($$$) fixed-temperature baths typically have a heating element but no means of cooling. Some of the needed supplies include cables and adapters, chart paper, stirrer bars, a magnetic stirrer, and jacketed beakers. Other accessories are available for the development of novel electrodes, such as ISE electrode bodies, gas-permeable membranes, plasticizers, and ionophores.

Vendors for Instruments and Accessories

Cole Palmer Instrument Co.
7425 N. Oak Park Ave.
Niles, IL 60714
phone: 800-323-4340
fax: 708-647-9660

Daigger Scientific
199 Carpenter Ave.
Wheeling, IL 60090
phone: 800-621-7193
fax: 800-320-7200

Diamond General Corp.
3965 Research Park Dr.
Ann Arbor, MI 48108
phone: 800-678-9856
fax: 313-747-9703

Fisher Scientific
711 Forbes Ave.
Pittsburgh, PA 15219-4785
phone: 800-766-7000
fax: 800-926-1166

Fluka Chemical Corp.
980 S. Second St.
Ronkonkoma, NY 11779
phone: 800-358-5287
fax: 800-441-8841

Hach Co.
P.O. Box 389
Loveland, CO 80539
phone: 800-227-4224
fax: 303-669-2932

Ingold Electrodes, Inc.
261 Ballardvale St.
Wilmington, MA 01887
phone: 800-352-8763
fax: 508-658-6973

Innovative Sensors, Inc.
4745 E. Bryson St.
Anaheim, CA 92807
phone: 714-779-8781
fax: 714-779-9315

Markson Science, Inc.
10201 S. 51st St., Ste. 100
Phoenix, AZ 85044
phone: 800-528-5114
fax: 602-496-8246

Microelectrodes, Inc.
298 Rockingham Rd.
Londonderry, NH 03053
phone: 603-668-0692
fax: 603-668-7926

Omega Engineering, Inc.
One Omega Dr.
Stamford, CT 06906
phone: 800-826-6342
fax: 203-359-7700

Orion Research Inc.
529 Main St.
Boston, MA 02129
phone: 800-225-1480
fax: 617-242-8594

PGC Scientifics Corp.
9161 Industrial Ct.
Gaithersburg, MD 20877
phone: 800-424-3300
fax: 301-990-0740

Radiometer America, Inc.
811 Sharon Dr.
Westlake, OH 44145
phone: 800-988-8110
fax: 216-899-1139

Rainin Instrument Co., Inc.
Mack Rd., P.O. Box 4026
Woburn, MA 01801-4628
phone: 800-472-4646
fax: 617-938-8157

TCI America
9211 N. Harborgate St.
Portland, OR 97203
phone: 800-423-8616
fax: 503-282-1987

Universal Sensors, Inc.
5258 Veterans Blvd., Ste. D
Metairie, LA 70006
phone: 504-885-8443

VWR Scientific
1310 Goshen Pkwy.
West Chester, PA 19380
phone: 800-932-5000
fax: 215-436-1761

World Precision Instruments, Inc.
375 Quinnipiac Ave.
New Haven, CT 06513
phone: 203-469-8281

Yellow Springs, Inc.
Yellow Springs, OH 45387
phone: 800-765-4974
fax: 513-767-9353

Required Level of Training

Basic potentiometry may be performed with a minimal amount of training by people with a high
school education. Analysis of the data, compensating for matrix effects, and preparing standard solutions requires a minimum of a college-level quantitative analytical chemistry course. Setting up
and using more elaborate electrode configurations requires a college degree in chemistry.

Service and Maintenance

All but the most basic pIon/mV meters come with a troubleshooting guide, and many of the more sophisticated electronic meters are designed to detect and indicate a problem when it occurs. Most pIon/mV meters come with a warranty, and it is rare that a problem requires professional attention. Electrodes themselves vary in the amount of maintenance required. Reference electrodes must have fresh solutions prepared at least weekly, and strip-chart recorders should be recalibrated frequently. Throw-away, no-maintenance electrodes are available that have a useful life of about one year. Commercial ISEs come with maintenance instructions (including storage conditions) and some, such as the pH electrode, can be occasionally serviced in-house to eliminate a sluggish, diminished response.

Suggested Readings

CAMPANELLA, L., AND M. TOMASSETTI, "The State-of-the-Art of Electrochemical Sensors," *Bull. Electrochem.*, 8 (1992), 229–38.

COSOFRET, V. V., AND R. P. BUCK, "Recent Advances in Pharmaceutical Analysis with Potentiometric Membrane Sensors," *Crit. Rev. Anal. Chem.*, 24 (1993), 1–58.

COVINGTON, A. K., ed., *Ion-Selective Electrode Methodology*, vols. I and II. Boca Raton, FL: CRC Press, 1979.

D'ORAZIO, P., M. F. BURRITT, AND S. F. SENA, eds., *Electrolytes, Blood Gases and Other Critical Analytes: The Patient, the Measurement and the Government.* Madison, WI: Omnipress, 1992.

JANATA, J., "Potentiometric Microsensors," *Chem. Rev.*, 90 (1990), 691–703.

KORYTA, J., "Theory and Applications of ISEs," *Anal. Chim. Acta*, 233 (1990), 1–30.

KORYTA, J., *Ions, Electrodes and Membranes*, 2nd ed., New York: Wiley, 1991.

LEWENSTAM, A., M. MAJ-ZURAWSKA, AND A. HULANICKI, "Application of Ion-Selective Electrodes in Clinical Analysis," *Electroanalysis*, 3 (1991), 727–34.

PRANITIS, D. M., M. T. DIAZ, AND M. E. MEYERHOFF, "Potentiometric Ion-, Gas-, and Bio-Selective Membrane Electrodes," *Crit. Rev. Anal. Chem.*, 23 (1992), 163–86.

THOMAS, J. D. R., "Ion-Selective Electrode and Enzyme Sensors for Flow-Type Environmental Analysis," *Collect. Czech. Chem. Commun.*, 56 (1991), 178–91.

UMEZAWA, Y., *Handbook of Ion-Selective Electrodes: Selectivity Coefficients.* Boca Raton, FL: CRC Press, 1990.

VADGAMA, P., "Membrane-Based Sensors: A Review," *J. Membrane Sci.*, 50 (1990), 141–52.

VADGAMA, P., M. DESAI, AND P. CRUMP, "Electrochemical Transducers for In Vivo Monitoring," *Electroanalysis*, 3 (1991), 597–606.

WOTRING, V. J., AND OTHERS, "Recent Advances in Polymer Membrane Anion-Selective Electrodes," in R. M. Nakamura, Y. Kasahara, and G. A. Rechnitz, eds., *Immunochemical Assays and Biosensor Technology for the 1990s.* Washington, D.C.: American Society for Microbiology, 1992.

References

1. Werner E. Morf, *The Principles of Ion-Selective Electrodes and of Membrane Transport* (Budapest: Elsevier Scientific and Akadémiai Kiadó, 1981), 60–61.

2. *Pure Appl. Chem.*, 48 (1976), 127–32.

3. B. D. Warner, G. Boehme, and K. H. Pool, *J. Chem. Ed.*, 59 (1982), 65–66.

4. H. A. Strobel and W. R. Heineman, *Chemical Instrumentation: A Systematic Approach* (New York: Wiley, 1989), 1016–17.

5. E. P. Sergeant, *Potentiometry and Potentiometric Titrations* (New York: Wiley, 1984), 129–32.

6. J. E. O'Reilly, *J. Chem. Ed.*, 56 (1979), 279.

7. A. Campiglio, *Analysis*, 118 (1993), 545-47.

8. S. S. M. Hassan and I. H. A. Badr, *Talanta*, 41 (1994), 523–30.

9. M. E. Collison and others, *Anal. Chem.*, 61 (1989), 2365–72.

10. M. H. Gil and others, *Biosensors Bioelectron.*, 7 (1992), 645–52.

11. E. Lindner and others, *J. Chem. Soc., Faraday Trans.*, 89 (1993), 361–7.

12. M. E. Meyerhoff, *Clin. Chem.*, 36 (1990), 1567–72.

Chapter 39

Electrolytic Conductivity

Paula Berezanski

National Institute of Standards and Technology

Summary

General Uses

- Measurement of conductivity of electrolytes such as salts, acids, and bases in aqueous solutions
- Measurement of conductivity of solutes in nonaqueous and mixed solvents

Common Applications

- Characterization of electrochemical behavior
- Determination of ionic dissociation constants
- Determination of ionic mobility
- Measurement of the ionic content of solutions
- Determination of endpoint of conductometric titrations
- Detection of elution of charged species in ion chromatography
- Monitoring of solution streams

Samples

State

Electrolytic conductivity of ionic substances is measured in solution in the liquid state.

Amount

The amount of sample depends on the volume of the conductivity cell. Pipette cells may require as little as 2 mL, whereas larger cells may require more than 50 mL.

Preparation

Preparation is minimal and usually involves drying samples and deionizing water or purifying solvents for use in the preparation of solutions. If the sample is a solution, there is no preparation.

Analysis Time

Instantaneous measurement may be performed if samples are not required to be equilibrated to desired temperatures, but rather are measured at other temperatures and corrected to the desired temperature. More accurate measurements require that a cell containing the sample be placed in a constant temperature bath at a temperature stability of within ±0.005 °C. A minimum of 30 min is usually required for the solution in the cell to come to thermal equilibrium.

Limitations

General

- Solvent must be of sufficient purity to give reproducible solvent conductivity.
- Strongly basic solutions or solutions with high concentrations of organic compounds tend to attack the conductivity cells and foul the electrodes.
- Very-low-conductivity measurements, that is, those in ultrapure water, must be performed without contact with atmospheric carbon dioxide.
- For relative uncertainties below 0.01%, temperature must be held stable (within ±0.002 °C).

Accuracy

- Relative standard uncertainty of hand-held meters is generally ±1%.
- Relative standard uncertainty of measurements performed with a Wheatstone bridge can be ±0.05% or less.
- As long as high-accuracy resistors are used and temperature is held constant to ±0.002 °C, relative standard uncertainty of ±0.02% may be achieved.

Sensitivity and Detection Limits

If the solvent conductivity and the conductivities of any interfering species are highly reproducible, a sensitivity better than 0.1 μS/cm may be achieved.

Complementary or Related Techniques

- pH and conductivity measurements are often coupled, especially if the only electrolyte in solution is an acid.

- Ion chromatography often uses conductimetric detection to determine the concentration of eluted species.

Introduction

Electrolytic conductivity is a measure of the ability of a solution to conduct electricity. If an ion is placed between two electrodes of opposite charge in an electric field, it migrates to the electrode of the opposite charge, thus conducting electricity. Because the ions in the solution conduct the electricity, the nature of the ions and the movement of the ions in a particular solvent determine the magnitude of the electrolytic conductivity. The nature of the ions can be described by properties such as size, charge, mobility, and concentration. The movement of the ions can be related to the temperature of the solution and the viscosity of the solvent. As temperature increases, ions move more quickly and conduct more electricity. An increase in viscosity slows the movement of the ions. The dielectric constant of the solvent also affects the electrolytic conductivity.

All ions in solution contribute to the electrolytic conductivity, and the contributions of individual ions cannot be distinguished from one another. Thus, electrolytic conductivity is nonspecific. Measurements of solutions of a single electrolyte are useful in determining the properties of that electrolyte. The electrolytic conductivity of solutions of single electrolytes and mixed electrolytes is useful in obtaining qualitative and quantitative information such as the purity of solvents and the relative ionic content of solutions.

How It Works

For a solution to conduct electricity, ions must be present. Ions are produced by the dissociation of electrolytes. An electrolyte $M_{v_+}A_{v_-}$ dissociates into cations M^{z+} and anions A^{z-} when placed in solution according to the general equation

$$M_{v_+}A_{v_-} \rightarrow v_+ M^{z+} + v_- A^{z-} \tag{39.1}$$

where v_+ is the number of cations produced by one molecule of electrolyte, z_+ is the charge number (valence) of the cation, v_- is the number of anions produced by one molecule of electrolyte, and z_- is the charge number of the anion.

For example, the electrolyte $La_2(SO_4)_3$ dissociates by

$$La_2(SO_4)_3 \rightarrow 2\ La^{3+} + 3\ SO_4^{2-} \tag{39.2}$$

where $v_+ = 2$, $z_+ = 3$, $v_- = 3$, and $z_- = -2$.

Strong electrolytes completely dissociate, whereas weak electrolytes exist as undissociated molecules as well as ions in solution. Ions of weak electrolytes may also form ion pairs. Strong electrolytes include many salts and inorganic acids. Weak electrolytes include many bases and acids (1).

The ability of an electrolyte solution (at a specific concentration and temperature) to conduct electricity is the electrolytic conductivity (sometimes called specific conductance). Electrolytic conductivity is usually measured by placing an electrolyte solution in a measurement cell of length

L and cross-sectional area A. The resistance R in this cell is then measured, and the electrolytic conductivity κ is calculated by

$$\kappa = \frac{L/A}{R} \qquad (39.3)$$

Because the dimensions of the cell are usually rigidly fixed, the quantity (L/A) is a constant for a given cell and is called the cell constant, K_{cell}. The electrolytic conductivity is generally given by

$$\kappa = \frac{K_{cell}}{R} \qquad (39.4)$$

In practice, the cell constant is determined experimentally by measurement of R for a solution of known κ. The SI unit of electrolytic conductivity is the S/m, where S is the symbol for the unit siemens, which is equal to an inverse ohm, Ω^{-1}. Most electrolytic conductivity measurements are given in $\mu S/cm$. Resistivity, ρ, sometimes given instead of electrolytic conductivity, is simply the inverse of the electrolytic conductivity,

$$\rho = 1/\kappa \qquad (39.5)$$

The SI unit of resistivity is the Ωm, but most measurements are given in $M\Omega cm$. A property of an electrolyte solution (at a specific concentration and temperature) in a specific cell is the conductance G (SI unit S), given by

$$G = 1/R \qquad (39.6)$$

Conductance should not be confused with electrolytic conductivity. Electrolytic conductivity is an intrinsic property of a solution, whereas conductance depends on the cell in which the solution is being measured.

The ability of individual electrolytes to conduct electric current is given by the molar conductivity. The molar conductivity Λ (Sm^2/mol) of an electrolyte is given by

$$\Lambda = \kappa/c \qquad (39.7)$$

where $\kappa = \kappa(\text{solution}) - \kappa(\text{solvent})$ and c is the amount-of-substance concentration (mol/m^3) of the electrolyte. The quantity equivalent conductivity, Λ_{eq} ($Scm^2/equiv$), is given by

$$\Lambda_{eq} = \frac{\kappa}{v_+z_+c} \qquad (39.8)$$

where v_+z_+c is the equivalent concentration, c_{eq} (equiv/L). Equivalent conductivity is often given instead of molar conductivity. The International Union of Pure and Applied Chemistry (IUPAC) recommends that the use of equivalent conductivity be discontinued and molar conductivity be used in its place because molar conductivity is experimentally defined, whereas the calculation of equivalent conductivity depends on an assumption about the chemical form of species present. Much theory has been developed in terms of equivalent conductivity and many current databases contain values for equivalent conductivity; therefore, extreme care should be practiced when working with these quantities. The molar conductivity is simply a multiple of the equivalent conductivity:

$$\Lambda = v_+z_+\Lambda_{eq} \qquad (39.9)$$

The molar conductivity of an electrolyte results from the contributions of all ions in the electrolyte such that

$$\Lambda = \Sigma v_i\lambda_i \qquad (39.10)$$

where λ is the ionic molar conductivity (Sm^2/mol). For example, the molar conductivity of $La_2(SO_4)_3$ is given by

$$\Lambda La_2(SO_4)_3 = 2\ \lambda La^{3+} + 3\ \lambda SO_4^{2-} \tag{39.11}$$

The ionic molar conductivity and ionic equivalent conductivity are related by $\lambda/\lambda_{eq} = |z|$. In dilute solutions, λ approaches λ^∞, the limiting ionic molar conductivity (λ at infinite dilution), and Λ approaches Λ^∞, the limiting molar conductivity (Λ at infinite dilution). The limiting ionic equivalent conductivities, and thus the limiting ionic molar conductivities, are known for many ions, some of which are listed in Table 39.1 (1-3).

Electrolytic conductivity depends on the concentration of electrolyte in solution. As the concentration of a strong electrolyte increases, the conductivity increases because more ions are present to conduct electricity. For weak electrolytes, the conductivity increases with increasing concentration, but the molar conductivity decreases with increasing concentration because of decreased dissociation. Thus the conductivity may go through a maximum value. At very high concentrations of ions, conductivity may decrease because of formation of ion pairs and increase in viscosity of the medium.

The conductivity of strong electrolytes in dilute solution may be estimated from the Kohlraush square root law (1-3),

$$\Lambda_{eq} = \Lambda_{eq}^\infty - Ac_{eq}^{1/2} \tag{39.12}$$

where A is a constant for the electrolyte and c_{eq} is the equivalent concentration, where $c_{eq} = v_+z_+c$.

Electrolytic conductivity depends on temperature. As thermal energy increases, Brownian motion increases, increasing the electrolytic conductivity. The uncertainty of the measurement of electrolytic conductivity thus depends on the accuracy and stability of the temperature of measurement. The temperature coefficient, α, of an electrolyte may be given as

Table 39.1 Limiting ionic molar conductivities and limiting ionic equivalent conductivities of selected ions in water at 25 °C.

Cation	$\lambda^\infty/(S{\cdot}cm^2{\cdot}mol^{-1})$	$\lambda_{eq}^\infty/(S{\cdot}cm^2{\cdot}equiv^{-1})$	Anion	$\lambda^\infty/(S{\cdot}cm^2{\cdot}mol^{-1})$	$\lambda_{eq}^\infty/(S{\cdot}cm^2{\cdot}equiv^{-1})$
H^+	349.81	349.81	OH^-	198.3	198.3
Li^+	38.68	38.68	F^-	55.4	55.4
Na^+	50.10	50.10	Cl^-	76.35	76.35
K^+	73.50	73.50	Br^-	78.14	78.14
Be^{2+}	90	45	$I-$	76.84	76.84
Mg^{2+}	106.10	53.05	NO_3^-	71.46	71.46
Ca^{2+}	119.00	59.50	SO_4^{2-}	160.04	80.02
Ba^{2+}	127.2	63.6	CH_3COO^-	40.90	40.90
Al^{3+}	183	61	$C_6H_5COO^-$	32.38	32.38
Cu^{2+}	107.2	53.6	HCO_3^-	44.50	44.50
Ag^+	61.90	61.90	CO_3^{2-}	138.6	69.3
Zn^{2+}	105.6	52.8	$Fe(CN)_6^{3-}$	302.7	100.9
Ce^{3+}	209.4	69.8	$Fe(CN)_6^{4-}$	442.0	110.5
NH_4^+	73.55	73.55	CH_3^-	54.59	54.59

$$\alpha = \frac{d\kappa/dt}{\kappa} \tag{39.13}$$

where t is temperature. The standard temperature at which electrolytic conductivity is usually measured is 25 °C. The temperature coefficients for many electrolytes are known. At 25 °C, $\alpha(\text{KCl, aq}) = 2\%/°C$. If the electrolytic conductivity at 25 °C, $\kappa_{25°C}$, and the temperature coefficient are known, the electrolytic conductivity at temperature t can be estimated by

$$\kappa_t = \kappa_{25\,°C}[1 + \alpha(t - 25\,°C)] \tag{39.14}$$

Some conductivity measurements are given in terms of total dissolved solids[1] (TDS, ppm), total ionized solids (TIS, ppm), or salinity (g/kg) and are meant to estimate the mass fraction of certain species in solution. Values of TDS and TIS depend on the closeness of the sample to the standard and thus yield reliably only relative values of TDS and TIS. Salinity measurements are used by many geochemists. In this case a specific standard (standard seawater with a salinity of 35 g/kg at 15 °C) and a defined salinity–conductivity relationship allows the calculation of the salinity of a sample. Accuracy of measurements of salinity by conductance is typically 0.0001% or better.

What It Does

The measurement of electrolytic conductivity requires some method for determining the electrical properties of a particular geometry of solution and electrodes. Electrolytic conductivity may be determined by two methods. The most common method uses electrodes immersed in a solution. The other method does not use electrodes and is called electrodeless conductivity measurement.

Immersed Electrode Measurements

The basic design of immersed electrode measurements is quite simple. An electrolyte solution is placed in a cell and current is passed through the solution. Two electrodes in contact with this solution allow measurement of resistance. The electrolytic conductivity is then determined using Eq. (39.4). The types of electrodes and methods for determining resistance vary with the type of current used.

AC Conductivity Determination

The most common method of electrolytic conductivity determination uses alternating current (AC) at frequencies below 5 kHz. Conductivity cells for AC measurements operate with two, three, or four electrodes. In the two-electrode cells, the electrodes are used for both the passage of current and the measurement of resistance. In three- and four-electrode cells, a pair of electrodes is used to measure resistance while one or two additional electrodes are used to pass current. The three- and four-electrode designs allow a lower current to be used, which often reduces the

---

1. Measurements of TDS and TIS are meant to estimate mass fraction. The common unit for TDS and TIS is parts per million (ppm), which is equivalent to the SI unit of μg/g.

common problems of polarization and fouling of electrodes. The electrodes in AC cells usually consist of a nonreactive material such as platinum, titanium, gold-plated nickel, tungsten, or graphite.

Conductivity cells for AC measurements can be divided into two types based on the method of determining the cell constant. Cells for absolute determination of electrolytic conductivity (Fig. 39.1) have a removable cylindrical tube of accurately known length and cross-sectional area. The resistance of the solution in this cell is measured with and without the tubing. The resistance is the difference between the two resistance values, $(R_{with} - R_{without})$. The electrolytic conductivity, κ, of the solution is then determined from Eq. (39.3). Because all of the quantities are determined by physical means, this measurement of conductivity is called absolute, that is, not dependent on other chemical standards or calibration. Recent efforts in AC methods of absolute conductivity determination have been in the determination of the electrolytic conductivity of primary standard solutions (4).

The high accuracy of absolute conductivity determinations is rarely required. Transfer cells are much less complicated in use. A transfer cell is one in which the cell constant is determined by calibration with standard solutions of accurately known electrolytic conductivity. The design of these cells varies with the level of electrolytic conductivity of the solution being measured and the application of the measurement. Much research has been performed for the purpose of minimizing effects that cause error in the resistance measurement (1). Some selected types of cells are described here.

Dip cells are the simplest and most commonly used conductivity cells, especially for field-work, because measurements may be performed in an open vessel. The dip cell consists of two electrodes, placed a fixed distance apart and surrounded by a glass or plastic sleeve to protect the electrodes (Fig. 39.2). Stirring may be required to prevent polarization of certain charged species at the electrodes.

Flow-through cells are commonly used in stream monitoring. Most flow-through cells consist of glass tubing in which disk electrodes are connected at fixed points. Flow-through cells

Figure 39.1 Absolute cell for AC measurement showing two compartments with electrodes and removable cylindrical connecting section.

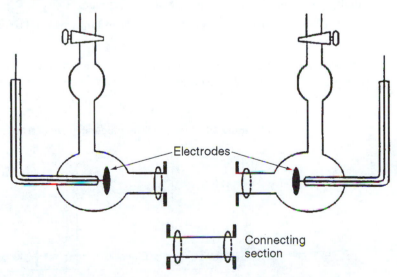

Figure 39.2 Dip cell.

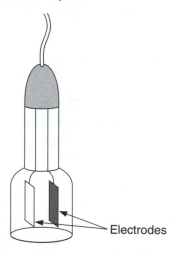

Electrodes

may be either the two- or the four-electrode type. Ion chromatographic detectors are commonly the flow-through type.

Pipette cells are used for extremely small sample volumes, sometimes as small as 2 mL. This cell is similar to the dip cell except that the electrodes are enclosed in a pipette instead of a sleeve.

Jones cells are used for more accurate measurements than dip and flow-through cells. For highest accuracy, these cells are usually made of borosilicate glass and use platinum electrodes. The Jones cell is usually composed of two separate glass bulbs containing platinum disk electrodes. The glass bulbs are then connected by a section of glass tubing of a specified diameter and length as to give the desired cell constant (Figs. 39.3 and 39.4).

Daggett cells, or Kraus cells (4), are used for accurate conductivity determination. The cell consists of an Erlenmeyer flask to which is attached a glass bulb containing the electrodes (Fig. 39.5). These cells are usually used for low conductivities because the cell constant is small, due to the size and juxtaposition of electrodes.

Single electrodes are commercially available and may be used to construct a cell of a desired or even variable cell constant.

The standard method of AC conductivity measurement uses a Wheatstone bridge to measure the resistance of the electrolyte solution in the cell. A frequency generator is used to supply an AC

Figure 39.3 Jones cell for moderate conductivity.

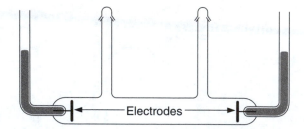

Electrodes

Figure 39.4 Jones cell for high conductivity.

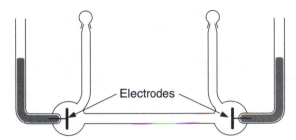

signal, normally in the range of 1 to 5 kHz in frequency, to the bridge. The cell is placed in one arm of the bridge and a variable resistor is placed in the other arm (Fig 39.6). A variable capacitor in parallel with the variable resistor is used to compensate for the impedance of the cell. An earth ground and a Wagner ground should be used to minimize the effects of stray capacitance on the accuracy of the null point. A differential input preamplifier, tuned amplifier, and oscilloscope may be used as a null detector. If the output of the tuned amplifier is connected to the vertical input and the reference signal from the signal generator is connected to the horizontal input of the oscilloscope, the displayed pattern, a Lissajous figure, can be used to indicate both capacitive and resistive balance of the bridge simultaneously. At the point where a null detector shows balance, the amplitudes of the potentials on each side of the null detector are equal and in phase and the impedance, Z (Ω), of the cell is equal to the impedance of the capacitor–resistor setup. The resistance of the cell, R, is determined by the resistance of the variable resistor. Because the resistance of the variable resistor equals the sum of the resistance of the electrolyte solution and the resistance of the leads, any resistance in the leads must be subtracted from the resistance given by the bridge. The resistance is measured at several frequencies and extrapolated to infinite frequency to eliminate any polarization effects (3).

Most commercial conductance bridges use the same basic circuit described above, but often measure at only one frequency. The fixed resistance arms of some bridges have a choice of fixed

Figure 39.5 Daggett cell for low conductivity.

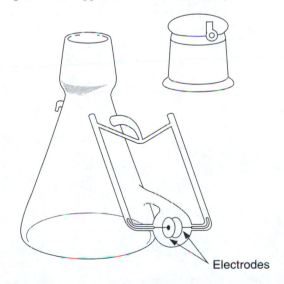

Figure 39.6 AC bridge for resistance measurement.

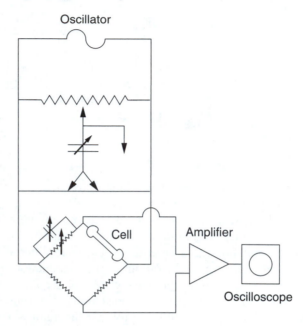

resistors in order to allow measurements over a wider range of conductivity. Some also have temperature probes, compensation, and displays with digital readouts.

DC Conductivity Determination

The measurement of high electrolytic conductivities can be accomplished using direct current. The DC method of absolute conductivity determination has been used to determine the electrolytic conductivity of primary standard solutions (5). The DC measurement is limited to cells of the four-electrode type (Figs. 39.7 and 39.8). Also, the DC measurement requires that all four electrodes be reversible to the electrolyte solution, such as silver/silver chloride electrodes for potassium chloride solutions. This is necessary to eliminate polarization effects.

The DC method of electrolytic conductivity measurement is much simpler in principle than the AC method and essentially operates on the principle of Ohm's law ($U = IR$, where U is the potential difference and I is the current). A standard resistor is placed in series with the conductivity cell. The current passing through the cell is equal to the current passing through the standard re-

Figure 39.7 Absolute cell for DC measurement.

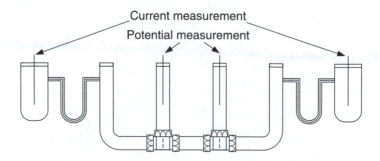

Figure 39.8 DC conductivity cell.

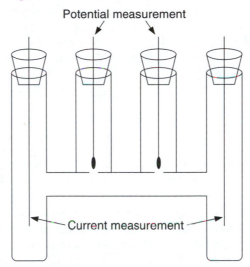

Potential measurement

Current measurement

sistor. A voltmeter is used to determine the potential difference across the standard resistor, and the current is then calculated by Ohm's law. The potential difference between the two electrodes in the solution of interest is then measured, and the resistance between the electrodes is then calculated by Ohm's law.

Electrodeless Conductivity Measurement Systems

There are two basic methods for determining electrolytic conductivity without placing electrodes in direct contact with the solution. One method uses inductive coupling and the other uses capacitive coupling.

Inductive measurements commonly use a pair of toroids to measure conductivity using frequencies around 20 kHz. An oscillator energizes the first toroid, which acts as a transformer and induces a current in a loop of solution. The solution loops through and energizes the second toroid. As the conductivity of the solution increases, the current induced in the second toroid increases. When the circuit is optimized, the current is linear with the conductivity of the solution. The toroids can be placed outside of nonconducting tubing for monitoring process streams or sealed in nonconducting material and immersed in caustic or corrosive solutions that would otherwise damage conventional electrodes.

Measurements using capacitive coupling are sometimes given the name "oscillometry" and require frequencies from 1 MHz to 10 MHz. Electrodes are placed outside of a glass cell. The current flowing through this cell is proportional to the resistance of the solution in the cell, thus allowing for the measurement of electrolytic conductivity.

Temperature Control

The most accurate conductivity measurements should be performed in an oil bath held to within ±0.005 °C. Constant-temperature water baths should be avoided because the water could induce capacitive effects across the cell walls in AC measurements and cause current leakage in DC measurements. It is not always practical to thermostat samples when measurements are performed in the field. As long as the temperature coefficient of the standard solution used to calibrate the cell is the same as that of the sample being measured, conductivity may still be determined to within a few percent. If maximum accuracy is not required, the conductivity may be measured at other temperatures and corrected back to 25 °C. This correction may be made manually or automatically by the use of a resistor that varies with temperature to about the same degree as the electrolyte solution being measured.

Standard Solutions for Electrolytic Conductivity

There are currently three primary standards for electrolytic conductivity that are recognized by the International Union of Pure and Applied Chemistry (IUPAC) and Organization of Legal Metrology (OIML). These solutions are defined on the demal scale (D), which is defined only at three specific concentrations of aqueous potassium chloride. Because the demal unit is not a customary unit of concentration, the National Institute of Standards and Technology (NIST), formerly the National Bureau of Standards (NBS), has recommended the adoption of primary standard solutions of aqueous potassium chloride for electrolytic conductivity based on molality. Molality (mol/kg) is defined as the amount of substance of solute in a solution divided by the mass of the solvent. These six primary standards are listed in Table 39.2. The electrolytic conductivities of these standards, measured by absolute methods, are listed in Table 39.3. The relative uncertainty of these electrolytic conductivities is ±0.02%. When using a primary standard, it is important to remember to include the electrolytic conductivity of the water used to prepare the standard in the total electrolytic conductivity used to calculate the cell constant.

Because it is not always practical to prepare primary standards, industrial standards are often used for calibration of cell constants. Standard reference materials (SRMs) are certified by NIST in the range of 5 μS/cm to 100,000 μS/cm for use as industrial standards. In addition, many companies provide standards that claim to be traceable to the NIST standards.

When calibrating conductivity cells with $K_{cell} \le 1 \text{ cm}^{-1}$, it is necessary to use solutions of very low electrolytic conductivity. Solutions commonly used are dilute hydrochloric acid and dilute aqueous potassium chloride. The equivalent conductivity of each electrolyte is related to its equivalent concentration by the following equations at 25 °C:

$$\Lambda_{eq}(\text{HCl}) = 426.16 - 158.68c_{eq}^{1/2} + 185.60c_{eq}\log c_{eq} + 500c_{eq} \tag{39.15}$$

Table 39.2 Preparation of primary standard KCl solutions.

KCl Solution	g KCl per kg Solution (in vacuum)	g KCl per kg Solution (in air)
1 demal	71.1352	71.1738
0.1 demal	7.41913	7.42343
0.01 demal	0.745263	0.745699
1 mol·kg^{-1}	69.379	69.416
0.1 mol·kg^{-1}	7.3999	7.4042
0.01 mol·kg^{-1}	0.74495	0.74539

and

$$\Lambda_{eq}(KCl) = 149.83 - 95.09c_{eq}^{1/2} + 38.66c_{eq}\log c_{eq} + 183.9c_{eq} \qquad (39.16)$$

where Λ_{eq} is given in S cm^2/equiv and c_{eq} is given in equiv L.

Analytical Information

Qualitative

The purity of a compound may be determined by the difference between the measured and theoretical values of conductivity. Process control measurements may be performed by monitoring the fluctuations in conductivity of streams.

Quantitative

If the specific nature of an impurity in a solvent is known, the concentration of that impurity may be calculated from the conductivity measurement. The dissociation constant and thermodynamic properties of electrolytes may be determined.

Table 39.3 Recommended electrolytic conductivity of primary standard KCl solutions from 0 °C to 50 °C (according to the International Temperature Standards (ITS-90) temperature scale).

$t/(°C)$	$\kappa/(\mu S \cdot cm^{-1})$					
	0.01 mol·kg^{-1} KCl*	0.1 mol·kg^{-1} KCl*	1.0 mol·kg^{-1} KCl[†]	0.01 demal KCl[‡]	0.1 demal KCl[‡]	1.0 demal KCl[†]
0	772.921	7,116.85	63,487	773.09	7,134.6	65,135
5	890.961	8,183.70	72,030	891.20	8,200.0	73,860
10	1,013.95	9,291.72	80,844	1,014.2	9,308.7	82,871
15	1,141.45	10,437.1	89,900	1,141.8	10,457	92,136
18	1,219.93	11,140.6	—	1,220.3	11,162	97,804
20	1,273.03	11,615.9	99,170	1,273.4	11,639	101,620
25	1,408.23	12,824.6	108,620	1,408.6	12,852	111,300
30	1,546.63	14,059.2	118,240	1,547.0	14,091	121,110
35	1,687.79	15,316.0	127,970	1,688.2	15,352	131,050
40	1,831.27	16,591.0	137,810	1,831.8	16,631	141,080
45	1,976.62	17,880.6	147,720	1,977.3	17,922	151,150
50	2,123.43	19,180.9	157,670	2,124.2	19,222	161,240

*Reference 6.
[†]Reference 5.
[‡]Reference 7.

Applications

1. Determination of the Purity of Water.

Highly pure water, free from carbon dioxide, has a theoretical conductivity of 0.055 µS/cm (resistivity of 18 MΩcm). Atmospheric carbon dioxide dissolved in water increases the conductivity to about 0.8 µS/cm to 1.2 µS/cm. Trace levels of mineral impurities such as calcium or sodium may increase the conductivity to several µS/cm. Conductivity detectors are often used on laboratory deionized water purifiers to indicate the purity of the water by its resistivity.

2. Determination of the Endpoint of a Conductimetric Titration.

In conductimetric titrations one ion is replaced with another of significantly different equivalent conductivity. The solution being titrated is often an acid or a base. Hydrogen and hydroxide ions have the greatest equivalent conductivity. In order to decrease the conductivity, an acid is used as the titrant to titrate a base, and vice versa. As the hydrogen or hydroxide ions become consumed in the formation of water and are replaced in solution by less conducting cations or anions, the conductivity decreases. After the endpoint has been reached the conductivity increases as excess acid or base is added. The endpoint is obtained by extrapolation of the measurements far from the endpoint. Conductimetric endpoints are also useful in precipitation and complex-formation titrations.

3. Monitoring of Solution Streams in Utilities, Semiconductor, Pulp and Paper, Chemical Processing, Biomedical, Biotechnical, and Pharmaceutical Industries to Monitor Operations of Ion Exchangers, Cooling Towers, Reverse Osmosis Systems, Boilers, and Rinse Tanks.

The purity of water is often essential knowledge in such systems as a boiler feedwater loop. In the textile industry, conductivity detectors are often used to monitor the concentration of bleach. In the food industry, conductivity detectors may be used to monitor cleanliness of pipelines and processing systems. In the pulp and paper industry, conductivity may be used to detect spills of costly chemicals. Conductivity sensors may be used for the control of concentration of alkaline solutions used to clean metals. Pure water is essential for the semiconductor wafer manufacturing industry.

4. Concentration Determination.

In a solution of only one species, concentration can be determined from known equations relating conductivity and concentration for that specific species. Because conductivity is nonspecific, the concentration of a given species in a mixed electrolyte cannot be measured. However, the total ionic content of a solution is often approximated as TDS, TIS, or salinity.

5. Ion Chromatography.

Conductivity detectors are often used in chromatographic equipment (Chapters 8, 9, 12) to determine the elution of charged analytes.

6. Salinity.

Many geochemists use salinity to determine the salt content in natural waters. Salinity is often found by determining the ratio of the conductivity of a sample of interest to the conductivity of a standard seawater sample and entering this ratio into a defined salinity–conductivity relationship.

Nuts and Bolts

Relative Costs

Conductivity bridge	$ to $$
Portable conductivity meter	$
Conductivity cell	$ to $$
Standard solutions	$
Precision temperature bath	$$$ to $$$$

$ = <1K, $$ = 1 to 5K, $$$ = 5 to 15K, $$$$ = 15 to 50K.

Vendors for Instruments and Accessories

Fischer Scientific Co.
711 Forbes Ave.
Pittsburgh, PA 15219-4785
phone: 412-562-8300, 800-766-7000
fax: 800-926-1166
email: info@fisher1.com
Internet: http://www.fisher1.com

Fischer Scientific UK
Bishop Meadow Rd.
Loughborough, Leics LE11 0RG
United Kingdom
phone: 44 1509 231166
fax: 44 1509 231893

Yellow Springs Instruments (YSI) Inc.
1725 Brannum Ln.
Yellow Springs, OH 45387
phone: 513-767-7241, 800-765-4974
fax: 513-767-9353

See the 1996 Lab Guide edition of *Analytical Chemistry* (*ACLG*), vol. 68, no. 16 (1996) for other listings under the headings "conductivity bridges" and "conductivity meters."

Required Level of Training

Use of commercial hand-held conductivity meters with dip cells requires minimal training. Measurements of this type may be performed at the technician level. Use of more accurate cells with resistance bridges requires more advanced training but still may be performed at the technician level.

Service and Maintenance

Bridges, meters, and temperature measuring equipment must be calibrated often to ensure maximum accuracy. Cell constraints must be checked periodically to ensure accurate measurements.

Suggested Readings

LIGHT, T. S., AND G. W. EWING, "Measurement of Electrolytic Conductance," in G. W. Ewing, ed., *Analytical Instrumentation Handbook*. New York: Marcel Dekker, 1990.

LOVELAND, J. W., "Conductance and Oscillometry," in G. D. Christian and J. E. O'Reilly, eds., *Instrumental Analysis*, 2nd ed. Boston: Allyn & Bacon, 1986.

PUNGOR, E., *Oscillometry and Conductometry*. Oxford: Pergamon Press, 1965.

REILLY, C. N., "High-Frequency Methods," in P. Delaney, ed., *New Instrumental Methods in Electrochemistry*. New York: Interscience, 1954.

STORK, J. T., "Two Centuries of Quantitative Electrolytic Conductivity," *Analytical Chemistry*, B56B (1984), 561A.

WU, Y. C., AND OTHERS, "Review of Electrolytic Conductance Standards," *Journal of Solution Chemistry*, 16, no. 12 (1987), 985.

References

1. R. A. Robinson and R. H. Stokes, *Electrolyte Solutions*, 3rd ed. (London: Butterworths, 1959).

2. H. S. Harned and B. B. Owen, *The Physical Chemistry of Electrolyte Solutions*, 3rd ed. (New York: Reinhold, 1958).

3. R. M. Fuoss and F. Accascina, *Electrolytic Conductance* (New York: Interscience, 1959).

4. Y. C. Wu, K. W. Pratt, and W. F. Koch, *Journal of Solution Chemistry*, 18, no. 6 (1989), 515.

5. Y. C. Wu, and others, *Journal of Research of the National Institute of Standard and Technology*, 99, no. 3 (1994), 241.

6. Y. C. Wu, W. F. Koch, and K. W. Pratt, *Journal of Research of the National Institute of Standard and Technology*, 96, no. 2 (1991), 191.

7. Y. C. Wu and W. F. Koch, *Journal of Solution Chemistry*, 20, no. 4 (1991), 391.

Microscopic and Surface Analysis

Chapter 40

Introduction

Gary E. McGuire

MCNC, Electronic Technologies Division

In this section, five techniques that are used for surface and interface analysis of solid specimens are discussed: atomic force microscopy (AFM), Auger electron spectroscopy (AES), scanning tunneling microscopy (STM), secondary ion mass spectrometry (SIMS), and X-ray photoelectron spectroscopy (XPS).

Each of these techniques provides valuable analytical information regarding the surface or interfacial composition of solid specimens, unlike other techniques for the analysis of bulk specimens. The techniques in this chapter may be grouped into two categories: imaging and compositional analysis techniques.

The imaging techniques AFM and STM provide an image of the outer surface. AFM images are generated by moving a stylus with a very sharp tip over the specimen surface and mapping out the topography on a near atomic scale. The topographical map is generated by monitoring the deflections of the stylus in the z-direction while scanning the x–y plane of the surface. The electron current that tunnels between a sharp metallic tip and a surface is imaged in STM. Atomic scale images are generated by maintaining a constant tunneling current while rastering the stylus over the surface. The height of the tip is adjusted as it scans over atomic scale features.

AES, SIMS, and XPS are elemental analysis techniques. They are used to determine the concentration of specific elements at surfaces or as a function of depth up to several micrometers within the specimen. AES and XPS are nondestructive techniques except (when using ion beam sputtering, as is required for SIMS) during in-depth analysis. Each of these techniques also provides chemical bonding information. In AES and XPS the chemical bonding information results from changes in the energy of electron transitions. The chemical bonding information in SIMS spectra results from the fragmentation patterns of the secondary ions, the molecular ions being subunits of the original specimen.

Comparison of Techniques

Like any analytical tool, each of the surface analysis techniques has strengths and weaknesses. In this section these strengths and weaknesses are reviewed and the special attributes of each are highlighted. The following table lists some of the attributes that are considered for the elemental analysis techniques (AES, SIMS, and XPS).

Summary of surface analytical technique capabilities and limitations.

Capability	AES	SIMS	XPS
Excitation source	Electron	Ion	X ray
Detected emission	Electron	Ion	Electron
Elements detected	$Z \geq 3$	$Z \geq 1$	$Z \geq 1$
Elemental identification	Excellent	Excellent	Excellent
Sensitivity variation	50	10,000	50
Detection limits	0.1%	.0001%	0.5%
Chemical information	Yes	Yes	Yes
Lateral resolution	50 nm	50 nm	150 µm
Depth resolution	5Å	20Å	5Å
Depth probed	Sputter depth	Sputter depth	Sputter depth
Depth analysis	Destructive	Destructive	Destructive
Beam damage	Yes	Low	Low
Sample charging	Yes	Yes	Minor
Standards required	Yes	Yes	Yes
Matrix effects	Minor	Yes	Minor
Special feature	Spatial resolution	High sensitivity	Chemical information

An electron beam is used for sample excitation in AES that provides an energetic probe with high current density that can be focused to a small beam diameter. SIMS relies on energetic ions that can be focused to a small probe diameter at high current density. An X-ray source is used to excite the photoelectrons in XPS. Energetic electron beams excite a variety of secondary electron transitions, among which is Auger electron emission resulting from the relaxation of excited atoms through filling of a hole in the core electron shells and emission of Auger electrons. The energetic ions used as probes in SIMS transfer momentum to the lattice in a collisional cascade, resulting in the ejection of secondary ions, which are detected with a mass spectrometer. Photoelectrons are emitted upon exposure of the specimen to X rays (preferably monochromatic), which, like Auger electrons, are detected using an electron spectrometer.

Auger electron emission requires three electrons: creation of the core hole during excitation, filling of the core hole during relaxation, and emission of the Auger electron. Because Auger emission is a three-electron process, only elements of $Z \geq 3$ can be detected. Most elements of $Z \geq 3$ exhibit one or more characteristic Auger transitions because any of the core electron energy levels in an atom can undergo excitation and Auger emission. As a result, AES provides excellent elemental identification. All mass spectrometers used in SIMS analysis can resolve better than an atomic mass unit, providing the ability to detect all elements. Mass interferences occur frequently due to molecular ions or ions with multiple charge. However, the occurrence of molecular ions also provides a means to detect elements when mass interference occurs. Photoelectrons are emitted from all electron energy levels when a photon of sufficient energy is used for excitation, allowing all elements $Z > 1$ to be detected. The low-energy photoelectrons from elements such as hydrogen and helium may overlap with those of the valence and conduction band of solids, making them difficult to uniquely identify. All other elements have multiple core-level electron transitions, which allow them to be identified when interfering transitions occur.

The probability of a primary electron resulting in Auger electron emission is 1 in 10^4. When coupled with the shallow escape depth of electrons, the resulting detection limit is 0.1% atomic. AES is sensitive to a fraction of a monolayer, with a sensitivity variation of about 50 because of the range of the Auger emission cross-section. SIMS, with detection limits of 10^{-4}% atomic, is the best surface analysis technique reviewed in this chapter. The sensitivity is greatest for elements that readily form positive or negative secondary ions. However, the sensitivity variation is about 10^4 due to the large variation in the probability of secondary ion generation. For most specimens only 2 to 4% of the sputtered material forms secondary ions. The detection limit for XPS is 0.5% atomic, with a sensitivity variation of about 50. Generation of an intense focused X-ray source required for XPS is more difficult than generation of an intense electron or ion beam required for AES or SIMS. This results in a poorer detection limit for XPS than for AES or SIMS, but XPS is still able to detect a fraction of a surface monolayer.

The spectra of all three of these techniques contain chemical information. XPS is most widely recognized for providing chemical information. The XPS spectra are rather simple, with narrow line widths and a low background. A change in the electron density surrounding the atom through chemical bonding results in a change in the core-level binding energies, called chemical shifts. AES spectra also exhibit chemical shifts but the spectra are broader and more complex, making it more difficult to resolve overlapping peaks. Chemical shifts almost twice those observed for photoelectrons are routinely observed for the Auger transitions for the same elements. The larger chemical shifts aid in chemical state identification even with the broader and more complex linewidths. The chemical information available from SIMS is usually obtained using static SIMS, where only a fraction of a monolayer of the specimen surface is removed during the analysis. The secondary molecular ions generated during sputtering represent fragments of the original molecular structure and may be used to identify the molecular structure of the specimen surface.

AES and SIMS have better spatial resolution than XPS because the primary beams for both consist of charged particles that can be focused to a small probe diameter. AES has better spatial resolution than SIMS because electron beams used for excitation in AES may be focused to a smaller probe diameter than the ions beams used in SIMS. The ultimate probe diameter is not a good criterion by which to determine the practical area of analysis. For any microprobe technique, the beam current decreases as the probe diameter is reduced and as the beam current is reduced the sensitivity decreases. Consequently, the practical spatial resolution is dictated by the signal-to-noise ratio required to analyze a specimen at a certain beam current. This may be a factor of 10 to 100 greater than the ultimate spatial resolution. The practical operating conditions may vary significantly for an imaging SIMS. X rays cannot be as easily focused but are focused by the diffracting crystal of the monochrometer to a spot size of less than 150 μm. Without a monochrometer, the spatial resolution of XPS is determined by a set of slits in front of the spectrometer or by a collection lens.

SIMS, AES, and XPS are all recognized surface analysis techniques. The depth resolution of static and dynamic SIMS is determined by the collisional volume created by the primary beam from which secondary ions originate. The collisional volume depends on the primary beam energy and current. Under high primary beam current conditions, the secondary ions originate from the top 30 to 50 Å of the surface. In static SIMS the objective is to analyze the outermost surface using low primary beam current conditions such that only a fraction of a monolayer is removed from the surface during the analysis. Using a parallel detection scheme to increase the sensitivity, the initial surface may be analyzed before it is removed by sputtering. In AES and XPS the depth resolution is determined by the escape depth of the excited Auger or photoelectron, not by the sample volume excited by the high-energy electron or X-ray source. An inelastic mean free path of 20 to 30 Å has been observed for electrons in the range of 0 to 2000 eV, the range of interest for most AES and XPS analyses. The inelastic mean free path is relatively independent of the atomic number of the matrix.

The depth resolutions for SIMS, AES, and XPS allow concentration versus depth profiles to be generated. By combining the surface sensitivity of these techniques with ion sputtering, the surface composition may be monitored while the specimen is gradually eroded by ion bombardment. The depth probed is limited only by the depth to which the specimen is sputtered. Practical times limit the depth probed to a few micrometers. The more focused ion beams used for AES and SIMS, which result in higher ion beam current densities, result in higher sputter rates and greater practical depths probed. The larger area of analysis in XPS results in the use of lower-current-density ion beams, which result in slower sputter rates and more shallow practical depths probed. In-depth analysis using any of the three techniques is inherently destructive due to the required use of ion sputtering.

The use of a focused electron beam, as in AES, may be destructive. High electron-current densities may result in electron-beam–induced desorption, decomposition, and evaporation. Beam-induced damage is less likely to occur for conductive specimens. Use of low beam currents or a rastered electron beam may reduce or eliminate beam damage. SIMS analysis is inherently destructive. The use of high ion beam energies may result in forward scattering and ion mixing. High ion current densities may produce localized heating, which induces diffusion or generates electric fields, which may cause mobile ion migration. X rays used in XPS are less destructive than electron or ion beams. The use of a nonmonochromized X-ray source is more likely to induce sample damage due to the Bremsstrahlung radiation.

Sample charging occurs frequently during charged particle excitation. During electron beam excitation, the surface potential of insulating samples is a function of the primary beam current and the secondary electron emission, including secondary, backscattered, and Auger electrons. Any change in surface potential may influence the measured electron energy in Auger spectroscopy. Lowering the primary electron beam energy and changing the angle of incidence are two methods of enhancing the secondary electron yield so that it is greater than the primary beam current. This generates a positive surface potential, which is a more stable situation during electron bombardment. Another method of reducing the surface potential is to place a grounded conductor close to the area of excitation. Ion bombardment of insulators may also generate surface potentials, which is established by the equilibrium between the primary ion beam current and the secondary ion and electron currents. Altering the primary ion beam energy and angle of incidence changes the secondary ion and electron yields and may be used to adjust the equilibrium surface potential. X-ray excitation presents only minor charging during XPS analysis of insulators. Photoemission generates a positive surface potential that may be neutralized by the Bremsstrahlung radiation excitation of secondary electrons from the foil used to protect the sample from high-energy electron bombardment of a nonmonochromized X-ray source or by a source of low-energy electrons when a monochromized X-ray source is used.

Quantitative analysis is difficult with all surface analysis techniques due to the difficulty of preparing standards. Because the depth of analysis is only 20 to 30A deep, the surface composition of a standard or specimen may be influenced by adsorbed gases or volatile hydrocarbons. Methods of removing these contaminants, such as ion sputtering, may alter the surface composition. As a result, all surface standards must be prepared carefully, especially because they are essential for quantitative analysis.

The Auger and photoelectron yield are influenced very little by the sample matrix, but the secondary ion yield during SIMS analysis may be strongly influenced by the matrix. The secondary ion yield may change by several orders of magnitude at surfaces or interfaces, especially if the matrix changes from an elemental one to a compound such as an oxide. Because very little can be done to eliminate this effect, it is important to recognize that it occurs and correct for it. Other techniques related to SIMS, such as resonance ionization spectroscopy (RIS) and sputtered neutral mass spectroscopy (SNMS), enhance the ion yield so that it does not change significantly as a function of the matrix.

Each technique has a special feature that makes its use attractive for analysis. For AES it is the high spatial resolution during surface analysis. SIMS is attractive because of its excellent sensitivity and large dynamic range. XPS provides the most chemical state information of all surface analysis techniques. These are the three most widely used surface analysis techniques for elemental and chemical state identification.

In addition to the elemental analysis techniques included in this section are the imaging techniques of scanning tunneling microscopy and atomic force microscopy. STM and AFM are surface imaging techniques with spatial resolutions of 1A and 2 to 5A, respectively. STM images are generated by the tunneling current between a biased metal tip and the substrate while rastering the tip in the x–y plane in order to generate a two-dimensional image. Individual atoms may be imaged, with best results obtained under ultra-high vacuum conditions. However, STM imaging is accomplished all the way from atmospheric pressure to 10^8 pascal. AFM imaging is usually conducted at atmospheric pressure because it does not rely on the conductivity of the specimen. In AFM the probe tip is an insulator that is scanned in the x–y plane in contact with the specimen surface. The tip deflections are monitored to provide a map of the surface topography. The imaging tools provide information regarding the microstructure of a specimen at the atomic level, which is complementary to the elemental and chemical state information available from the other surface analysis techniques.

Techniques based on infrared and Raman spectroscopy are also used for analysis of molecules on surfaces. These include attenuated total reflectance infrared spectrometry (ATRIS, pp. 261–262), diffuse reflectance infrared Fourier transform spectroscopy (DRIFTS, pp. 263–264), and surface enhanced Raman spectroscopy (SERS, p. 297).

Chapter 41

Atomic Force Microscopy and Scanning Tunneling Microscopy

Huub Salemink

IBM Zurich Research Laboratory

Summary

General Uses

Atomic force microscopy (AFM) and scanning tunneling microscopy (STM) are extreme surface sensitive techniques with high spatial resolution. AFM is usually done in ambient conditions:

- Surface topographic profiling in contact/noncontact mode
- In extreme cases, atomic arrangement of surface atoms (usually nanometer resolution)

STM is usually done in ultra-high vacuum (UHV):

- Real-space imaging of electronic local density of states on electrically conducting surfaces (metals, semiconductors) with atomic resolution
- Identification of electronic surface states near fermi level (spectroscopy)

Common Applications

AFM:

- Determination of roughness or flatness on oxidic and semiconductor surfaces
- Ordering in (an)organic and polymer films

STM:

- Mapping of reconstructed surface atomic structure
- Atomic adsorption and submonolayer nucleation
- Crystalline ordering within (differently) ordered nanometer-sized samples

State

- STM: Electrically conducting surfaces of metals and semiconductors
- AFM: Almost any insulating and conducting surface

Amount

- Surface area, depending on spatial resolution: atomic resolution (0.5 nm) over 50- × 50-nm areas, 100×100 nm maximum; 1 nm up to 1000×1000 nm depending on limits in image pixel density, digital-to-analog converters (DACs), and piezoelectric drivers
- Image acquisition rate from 1 frame/min to approximately 25 frames/sec.

Preparation

- STM: For highest quality, typical preparation as in surface science and technology (annealing, sputtering, cleaving, all in ultra-high vacuum, pressure < 1E–10 mbar).
- AFM: In ambient conditions, typically without surface preparation. Note that often unwanted features are reported due to surface contamination, notably adsorbed water layer.

Analysis Time

- Image acquisition rate is from 1 frame/min to 25 frames/sec, depending on resolution, number of pixels, and topographic accuracy.
- For UHV STM, extensive time may be used for preparation of samples and tunneling tips; most modern STM UHV systems are equipped so that samples and tips can be exchanged in high vacuum.

Limitations

- High-resolution STM in UHV and associated tunneling spectroscopy via acquisition of current–voltage (I–V) characteristics requires very careful treatment of tunneling tips, notably on GaAs-like surfaces; also requires very stable STM operation.
- Spatial resolution in STM (and AFM) in extreme (atomic) cases is very dependent on shape of tip-apex, hence on tip preparation.
- For variable-temperature operation, drift-free STM designs (symmetric or compensated) are required; generally small size is of advantage.
- For UHV operation of STMs, surface contamination usually limits observation of a particular sample surface, dependent on UHV pressure and adsorption rate or surface reactivity; similar applies to contamination of tunneling tips.

- Throughput of analysis in UHV systems is enhanced by efficient exchange of samples and tips, via attached preparation chambers and airlocks.
- Elemental analysis for specific chemical elements is generally difficult because surface reconstructions and associated surface states tend to obscure elemental properties; best possibilities for materials or compounds with low tendency for reconstruction.
- STM generally retrieves information from topmost one or two layers at the surface.

Complementary or Related Techniques

- Secondary ion mass spectrometry (SIMS)
- Auger electron spectroscopy (AES)
- High-resolution transmission electron microscopy (HRTEM)
- X-ray photoelectron spectroscopy (XPS)
- Photoelectron spectroscopy (PES)
- Low-energy electron diffraction (LEED)

Introduction

Scanning probe microscopy (SPM) (STM and AFM) and its derivatives have rapidly established themselves in the decade since their inception (1–3). Recently, reviews of these probe techniques have been made available that include basic overviews as well as examples of state-of-the-art work in each area (4–7). In particular, the NATO Special Program on Nanoscale Science has resulted in comprehensive references to areas in which SPM tools have delivered new contributions to existing fields or developed new fields (8). This chapter surveys the SPM techniques and provides an adequate directory to specialized reviews and papers; where appropriate, the SPM aspects are discussed in relation to well-established analysis techniques. In-depth discussions of analysis details and techniques can be found in the references cited.

How It Works

The SPM techniques form a class of extremely surface-sensitive tools, known mostly for their ability to deliver atomic resolution in real space. Equally important, however, are the various derivative techniques that have spun off the original STM developments, as these shed new light on many fields in science and technology. The SPM techniques bear strong resemblance to earlier work aimed at topographic profiling of metallic surfaces (9). The SPM techniques can roughly be divided into two application areas: one using STM, the other AFM. The STM is usually applied to electrically conductive materials, such as metallic or highly doped semiconductor surfaces, or to overlayers, such as metallic or conductive inorganic coverages. Many of the high-quality STM experiments are performed in a UHV environment with typical pressures less than 1×10^{-10} mbar, at the price of being more demanding in instrumental terms. Clearly, compared to work in ambient air, UHV conditions routinely offer the highest (atomic) spatial resolution, well-defined and clean surface conditions, and a superior STM operational stability. In addition, the UHV work allows a more quantitative analysis using STM spectroscopy (10). In contrast, AFM work is generally done under ambient laboratory conditions, mostly in air and, more importantly, does not require electrical surface conductivity. These two features allow a wide variety of oxidic and inorganic materials to be studied, with relatively simple and inexpensive AFM instruments.

The basic features of an STM are presented in Fig. 41.1. Figure 41.1(a) shows the geometric situation of a metallic tip (made of W, Pt, or PtIr) held at close proximity (typically 0.5 to 1.0 nm, a few atomic diameters) to a conductive surface. In the ideal case, the tip apex consists of a single metallic orbital, and the surface displays the localized metallic surface states or dangling bonds on a semiconductor surface (11). An applied voltage across the vacuum gap between tip and sample drives a tunneling current I_t, which to first order is described as $N \exp(-ks)$, where s is the tip–sample separation, k a decay constant, and N the convolution of the electron density of states (DOS) on both sides of the gap. Most semiconductor and metallic surfaces reconstruct their atomic positions at a free surface via the process of minimizing the surface free energy; this results in topographic superstructures with specific electronic states. Such surface states have already been extensively studied by photoemission spectroscopy (PES) or inverse photoemission spectroscopy (IPES) (12, 13), but STM can reveal such information on a much finer (nanometer) spatial scale.

Figure 41.1(b) shows the corresponding one-dimensional electronic scheme depicting the electron-tunneling process for electron transport from the tip through the vacuum barrier into the conductive (empty) metallic surface states. A negative tip (positive sample) voltage drives the small tunneling current through the vacuum barrier. The one-dimensional tunneling process has been extensively discussed in the literature (14–17), including refinements for image charges and tip-induced electrostatics. For closer proximity, the simple rectangular vacuum barrier must be cor-

Figure 41.1 STM. (a) Schematic of an STM tip at tunneling distance (approximately 1 nm) above the surface. (b) One-dimensional energy diagram for electron tunneling from a biased metal tip (right) via the vacuum barrier into the sample under study (left). The vertical axis represents electron energy, the horizontal one the spatial scale. Typical conditions for electron tunneling in UHV are tip–sample separation 1 nm, applied voltage (V_T) 50 mV to 2 V, tunneling current (I_t) 0.1 to 1 nA. (c) Schematic of STM operation for scanning: The vertical piezoelectric transducer (z) is used to maintain the tip–sample separation via feedback from the current-sensing circuit; the piezotransducers x and y are used for raster scanning. More recent STMs use compact tubular scanner arrangements.

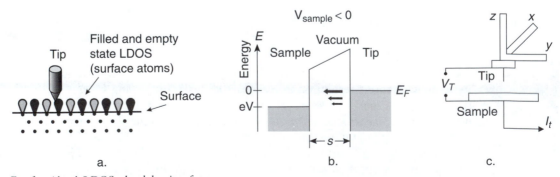

a.   b.   c.

E_F = fermi level. LDOS = local density of states.

rected for image charges, leading to an effective lowering of the barrier. The dominant result is an exponential dependence of the tunneling current I_t on the tip–sample separation, s, as is well known in quantum-mechanical tunneling, at a given voltage V, with an exponential decay constant given by $k \simeq 0.2$ nm^{-1}. As the electron transmission probability peaks near the injection energy, the I–V characteristic at a fixed separation, s, contains the electronic local density of states (10, 18, 19). The normalized derivative of the tunneling current to energy $(dI/dV) \cdot (V/I)$ compares well with photoemission data (12, 13), but now shows the exact spatial localization of the corresponding electronic states.

The surface sensitivity of SPMs is limited to the outermost (few) atomic layers of the surface under study. Depending on the actual tip conditions, the STM can deliver the surface topographic contours with nanometer resolution, in extreme (UHV) cases at the atomic scale. The STM uses a small tunneling current (0.1 to 1 nA) and a driving voltage typically in the range of 5 mV to 3 V. The electron scattering length is close to 0.5 nm, thus limiting the sensitivity to the topmost one or two atomic planes near the surface. Hence, the depth access is small compared with typical Auger electron spectroscopy operating at 0.5 to 1 keV, where the depth sensitivity is near 5.0 nm (20).

Critical Elements in SPM

Critical design and operation elements in SPM are the mechanical and thermal stability of the instrument (21). Important are a small size (compactness) and high stability in order to diminish the relative vibration and drift of tip relative to the sample. Moreover, the small size allows ease of integration into instruments, such as low-temperature stages (22) and electron microscopes (23), and can render additional vibration isolation redundant. Other features are multiaxis translator stages for the sample (24), allowing full three-dimensional movement of the surfaces under study and tip exchange within the UHV system (25). Most SPMs use piezoelectric drivers in the tip scanners for compact constructions and low-voltage operation. In early scanner versions, piezoelectric three-axis tripods were used. Now multielectrode tubular scanners (26) are used, which are more compact and have a higher response (displacement of typically 4.0 nm/V or more). Sample advancement and coarse positioning are mostly done with a slip-stick (sawtooth) operation of piezoelectric translation stages in two spatial directions.

Essential to all STMs is a (piezoelectric) feedback mechanism (Fig. 41.1(c)) that monitors the tunneling current (for fixed voltage) and regulates the tip–sample distance s. The tunnel current is amplified by typically a factor of 1×10^8 V/A using a closely placed current-to-voltage converter to minimize electrical interference. The piezoelectric transducers (PZT) with a displacement of 0.5 to 5.0 nm/V allow an ultimate spatial resolution of 0.005 nm (at 1 mV driving accuracy). The spatial scanning over the surface is performed by displacing the tip piezoelectrically in two planar (x and y) directions. All electrical signals from the feedback positions of the x and y scans, the tip voltage and tunneling current are stored in an image file on a PC or workstation. Graphical routines are then used to display the appropriate subsets of the data and for image processing. Such STM instruments are now readily available from several manufacturers for use in ambient as well as in UHV conditions. The UHV systems are usually equipped with exchangeable tips, allowing an uninterrupted vacuum environment; in some cases an in-UHV treatment of the tunneling tips is provided. The latter seems to be particularly important for the high-resolution analysis of direct band gap materials such as III–V(110) facets. The in-UHV tip preparation is usually done after the tip wires have been etched in the laboratory ambient by performing cycles of sputtering, annealing and field emission in UHV (25).

AFM

The basic concept of AFM is shown schematically in Fig. 41.2. Its implementation in ambient-air conditions has proven remarkably efficient. A cantilever with an attached sharp tip structure is contacted to the surface with a small loading force of typically 10^{-7} to 10^{-10} newtons. The lever can be made of micromachined silicon with the tip selectively grown and machined on the lever, or the tip can be mounted on the lever, allowing a choice of tip materials, shapes, and lever responsiveness (17). Readout of the deflection signal is mostly done optically by bouncing an incident laser beam onto the cantilever toward a quadrant detector or into an interferometer (27–29). The AFM can work in two modes of operation: in contact with the surface or in a noncontact mode. The contact mode essentially operates as a low-load, high-resolution surface profiler. In the noncontact case the attractive force near contact plays a significant role: As this mode is more difficult to achieve experimentally, only a few truly atom-resolved results are available on oxidic and semiconductor surfaces. This high-resolution noncontact work has so far been performed under elaborate UHV conditions, similar to those used for the best STM work. The design considerations for AFM are well documented (17, 21) and several types of instruments are commercially available. The contact-mode AFMs extend the use of conventional surface profilers into the nanometer range. Typical application areas are the determination of surface flatness or roughness on coated or oxidized materials, surface profiling of patterns modified (stamped) into the surface (compact disc bits) and the assembly of polymer or Langmuir–Blodgett (LB) films (30–32). Instruments are available with large stages suitable for nondestructive inspection of oxidic surfaces on semiconductor wafers in production lines. Note, however, that in many ambient conditions, no surface conditions as clear and clean as in UHV STM exist, with two major consequences: The interpretation might be more ambiguous because of ongoing surface oxidation and tip contamination, and it is more difficult to achieve high resolution and identification of species or molecular groups. Even with this restriction, the application areas are abundant and have led to new approaches in the analysis of small natural and artificial structures.

Figure 41.2 Diagram of basic AFM operation. A sharp tip on a cantilever is moved into proximity of or contact to the surface, and the lever displacement is sensed via an optical readout system (mainly quadrant photodetectors or interferometric designs).

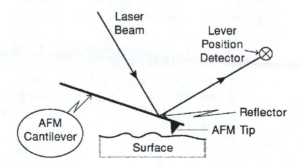

What It Does

STM Topography

A large part of the STM results in the literature deals with atomic-scale observations of reconstructed metallic and semiconductor surfaces, notably involving noble metals (33–35), alloys (36), and low-indexed semiconductor surfaces such as Si(111) (3, 37–39) and Si(001) (40–42). Furthermore, the (epitaxial) growth and metal deposition on these surfaces and the comparison with surface kinetics have been the subject of detailed studies (37, 42, 43), as well as the etching of such surfaces (29). An example of the imaging of Si(001) is given in Fig. 41.3, showing the terraced structure of Si(001). Note the alternating formation of the 2×1 dimer row orientations on subsequent terrace planes (42). On these semiconductors the STM topography actually traces the local DOS at the electron energy close to the fermi energy (E_F); thus, the image displays predominantly the local positions of the empty or filled states, depending on electron voltage and polarity (18, 44). Via this energy-selection various Si bond states on Si(111) have been resolved (38). Another recent example is the molecular adsorption onto metallic surfaces (45), work that is of interest regarding the natural and artificial configuration of molecular films and assemblies.

Figure 41.3 High-resolution STM topographic image of clean Si(001) in UHV. The Si dimers are reconstructed into 2×1 arrangements (joint orbitals). Adjacent dimers form rows. (a) Filled-state image of Si(001) at a sample voltage of −2.0 V. Crystal is 1.0° off-axis toward [110]. Image size is 53.8 nm × 53.8 nm. *(Image by B. Schwarzentruber, supplied courtesy of M. Lagally, University of Wisconsin, Madison.)*(b) Filled-state image of Si(001), covered with a 0.1-monolayer equivalent of Si atoms, at a sample voltage of −2.0 V. Surface is misoriented toward [110]. Image size is 38 nm × 38 nm. *(Image by Y. Wo, supplied courtesy of M. Lagally, University of Wisconsin, Madison.)*

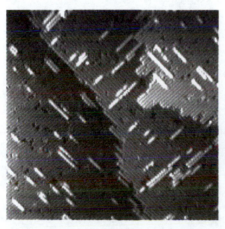

a. b.

STM Spectroscopy

The STM topography actually traces a field of constant local DOS at a particular, selected electron energy near the fermi energy E_F. By changing the electron injection energy this allows us to perform a type of local low-energy spectroscopy, typically in the energy window of $E_F \pm 3\,\text{eV}$; for this spectroscopy, the tunneling current–voltage (I–V) characteristic is measured under controlled conditions. The most important parameter to control is the tip–sample separation during the I–V sweep (lasting 0.01 to 1 sec). This requires a high STM device stability and low (thermal and mechanical) drift during the tip-holding phase. Two forms of spectroscopy are used: current imaging tunneling spectroscopy (CITS), which samples the I–V curves at discrete energy bins (typically 16 to 64) (38), and analog I–V, in which a full I–V curve is traced (some 256 to 1024 points) (10, 19), see Fig. 41.4. The latter technique has been refined for high-dynamic-range I–V spectroscopy on the (110) planes of III–V semiconductors (10). In contrast with the slightly metallic character of reconstructed semiconductors, no surface states exist within the (110) bandgap of III–V materials such as GaAs (46), and a large current range is required to define the valence and conduction band onsets accurately. This is achieved by a highly controlled tip approach during the I–V measurement. By using normalized I–V curves, $(dI/dV) \cdot (V/I)$, this spectroscopy mode allows a direct comparison with the electronic DOS (10) and photoemission experiments. Experimental difficulties arise from the requirement for stabilization of the tip–sample separation and from the relatively high field strength (several V/nm) that may occur during the voltage sweeps when large (± 3 eV) energy ranges are sampled.

AFM

AFM work on many different materials is abundant (5–8), and here only a few selected examples are discussed; the proceedings of recent STM conferences contain a large body of reference to the AFM work. There are many reports on the imaging of self-assembled molecules, showing the arrangement of molecular groups (32, 45) in thin deposited layers (Fig. 41.5). These experiments are performed in (near) ambient conditions, and the question of the exact surface conditions always

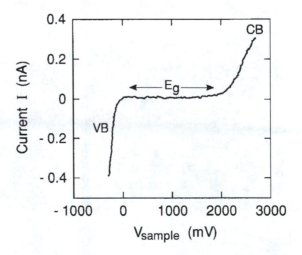

Figure 41.4 Tunneling current–voltage (I–V) characteristic on clean UHV-cleaved p-type GaAs(110) surface. The bandgap (E_g), valence (V_b), and conduction (C_b) band onsets are shown.

Figure 41.5 AFM topography of 8-cyanobiphenyl molecular arrangements MoS₂ substrate. Note orientation of the molecular chains. *(Reprinted from D. P. E. Smith, W. M. Heckl and H. A. Klagges, "Ordering of Alkylcyanobiphenyl Molecules at MoS₂ and Graphite Surfaces Studied by Tunneling Microscopy," Surface Science 278, 1992, pp. 166–174, copyright 1992, with kind permission from Elsevier Science—NL, Sara Burgerhartstraat 25, 1055 KV Amsterdam, The Netherlands.)*

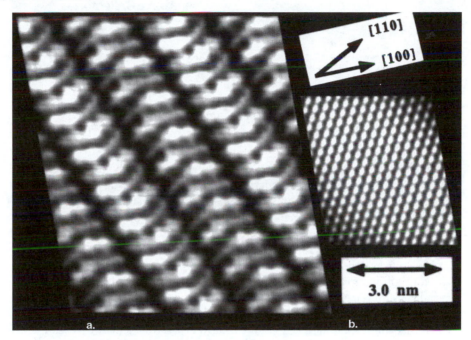

remains; in many cases an adsorbed (water) layer is suspected. Recently, realistic atomic resolution has been achieved with AFM at the price of working in UHV conditions (47); these experiments show a promising way to a high-resolution study of undoped semiconductors that are inaccessible by the electronic tunneling used in STM.

Derivative Techniques

Several interesting derivative techniques based on SPM have been invented. Voltage drops at material interfaces and *p–n* junctions have been analyzed by potentiometry (48); the STM tip is used as a third movable electrode between the fixed contact pads at the surface (in a miniature Wheatstone-bridge configuration). Operation of STM at higher electron energies (in the field-emission regimen) is used in a lithographic manner by locally depositing sufficient energy to break atomic bonds; several forms of this ultrafine lithography are being used (49), notably by writing into hydrogen-covered silicon (50). Typical line widths achieved are 5 to 20 nm at 4 to 20 eV. Issues of concern are the perfection of the material-removal process and methods to deposit material into the opened masks. In a complementary way, material has been deposited from metallized tips under field emission (51), forming small metallic bits on the oxidized silicon surface. Operating the STM at higher electron energies has led to the observation of local emission of light (photons) from the surface (52); the technique resembles cathode luminescence, but with a 0.5- to 5-nm spatial resolution that is mainly due to the low electron injection energy. Initially this led to the observation of surface plasmons (52), and in later work the bulk bandgap luminescence on cross-sections of quantum wells and wires has been observed (53). In the latter case, the light generation originates in a

Figure 41.6 Schematic arrangement for electron transmission in BEEM experiments. The tunnel current I_t is used for tip stabilization and the collector current I_c measures the current component through the buried interface.

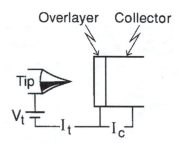

volume below the surface that is determined by electron confinement and by the carrier diffusion range, which can be in the micrometer range (54). Ballistic electron emission microscopy (BEEM) has been used to study tunneling transport across material interfaces (Fig. 41.6). A thin (5- to 10-nm) conductive layer is deposited onto a substrate, and an STM is operated on this layer; via a second current channel to the substrate, the current I_c transmitted through the interface is collected (55). Scanning the STM tip allows this second collector current to be mapped as a function of the position and thus allows any disturbances in electron transport at the buried interface (band gap discontinuities, growth defects) to be imaged. A large advantage is the unperturbed observation of realistic interfaces, including semiconductors, metals, and oxide (56). Issues of discussion are the conservation of energy and momentum (perpendicular and parallel) at the buried interface and the actual width (or cone angle) of the electron beam incident at this interface. Other remarkable techniques with AFM-derived tools are the fabrication stages for ultrasmall transistors (57) and the detection of small quantities of thermal energy in calorimetric sensors (58).

Using magnetically sensitive tips, AFM sensors have been used to image magnetic structures and domains, showing magnetic domain interfaces across 50 nm (59). Further related techniques, not discussed here, are the near-field tools for optical imaging below the diffraction limit (60) and the use of STM for the atomic configuration of nanometer structures (61).

Cross-Sectional STM

The atomic-scale roughness and the composition of epitaxially grown semiconductors are important for the performance of quantum devices (62). Using a planar observing STM, such information remains hidden because only the layer grown last (often reconstructed) is observed. During the growth of real multilayer devices, the thermodynamic and kinetic conditions can lead to atomic diffusion and subsequent modification of the initial interfaces. The III–V compound (110) cleavage plane (as in GaAs) has proven to be excellent for cross-sectional studies of epitaxially grown III–V multilayers, as shown in Fig. 41.7 (63). On these (110) planes, the geometry of the surface unit cell and the STM surface spectroscopy closely resemble the properties of the (truncated) bulk material because of the nonpolar character of this surface. This is in contrast with nearly all other semiconductor surfaces, where an excess charge density drives the reconstruction. With cross-sectional STM (XSTM), atomic-scale analyses of nanometer-sized quantum structures have been made (64), and in Fig. 41.8 the atomic-scale image of a quantum-well wire grown by molecular beam epitaxy (MBE) on the (110) plane is shown. A wealth of features can be derived from these data: absolute growth rates on high-index planes, identification of atomic species, the overgrowth of defects, the homogeneity of the

Figure 41.7 Schematic arrangement for STM on cross-sections of epitaxial multilayers in UHV environment. *(Reprinted from H. W. M. Salemink and O. Albrektsen, "Atomic Scale Survey of III-V Epitaxial Interfaces,"J. Vac. Sci. Technol. B 10(4), July/August 1992, 1799–1802. Copyright 1992 American Vacuum Society.)*

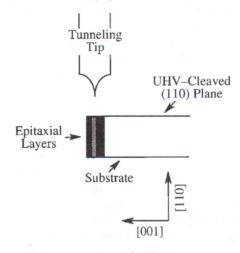

Figure 41.8 Filled-state image on (110) cross-sections of MBE-grown quantum wire structure. The As sublattice in GaAs/AlAs multilayer stack is shown. *(Adapted from Reference 67.)*

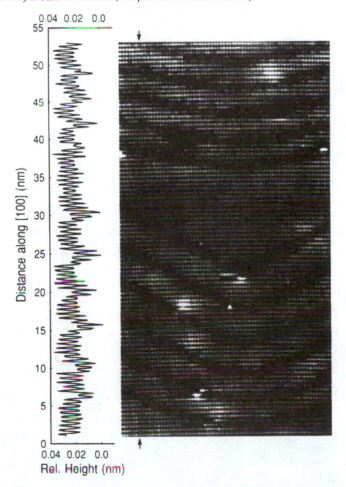

constituent thin layers, chemical intermixing at interfaces, and active doping concentrations (65). A comparison with high-resolution transmission-electron microscopy (HRTEM) can be made, but it should be kept in mind that STM essentially is sensitive to only one or two surface layers, whereas HRTEM relies on a projected depth profile across many layers (50 to 100) in the sample thickness (20 to 40 nm) (66). Because of this depth-averaging effect in HRTEM, the clustering and local ordering phenomena that can occur on a 5-nm scale are difficult to observe in TEM but become visible in an STM analysis. Such local-order and clustering phenomena are important in the description of the interfacial roughness (67) on the scale of the electron wavelength.

Outlook

In the domains of flatness analysis, of imaging small magnetic and replica-structures as well as polymers, AFM is well established. Similarly, in the fields of pure surface science, the STM is seen to complement photoemission and electron diffraction techniques because of its ease of operation, spatial resolution, and directness in interpretation. For the area of technology, complementary work by STM, HRTEM, SIMS, and AES analysis is foreseen, in which STM will explore the ultrasmall-scale structures; more quantitative results will have to be delivered by various forms of spectroscopy. It is expected that with the present state of SPM, a more thorough analysis of realistic quantum nanostructures can be undertaken with unprecedented detail; the goals are composition, atomic scale roughness, and dopant profiles.

Nuts and Bolts

Relative Costs

Ambient air operating AFM (estimated)

Probe head	$20K
Basic electronics	30K (feedback and interface)
Image processing	30K (including basic processing)
Basic total	80K

UHV STM (estimated)

Basic probe head	$50K
Basic electronics	30K
Image processing	30K
UHV enclosures	50K (dual chambers for STM and preparation)
Vacuum pumps	40K (ion and turbo pumps)
Preparation, airlock	50K (for sample and tip preparation)
Total	250K

Depending on final usage, additional costs for preparation of tips and samples may be incurred. Additional facilities for spectroscopy and advanced image processing will increase above amounts for both hardware and software by $10K to $100K. Running costs mainly include maintenance for turbo pumps ($5K/yr).

Vendors for Instruments and Accessories

AFM

Digital Instruments
520 E. Montecito St.
Santa Barbara, CA 93103
phone: 800-873-9750, 805-899-3380
fax: 805-899-3392
email: US_Sales@di.com
Internet: http://www.di.com

Topometrix
5403 Betsy Ross Dr.
Santa Clara, CA 95054-1162
phone: 408-982-9700
fax: 408-982-9751
Internet: http://www.topometrix.com

Veeco Instruments
Unit 8 Colne Way Court
Watford, Hertfordshire WD2 4NE, U.K.
phone: +44(0) 1923 235130
fax: +44(0) 1923 210044
email: sales@veeco.demon.co.uk
Internet: http://www.veeco.demon.co.uk

STM (UHV)

Omicron Vakuumphysik GmbH
Idsteinerstrasse 78
D-65232 Taunusstein, Germany
phone: +49 6128 987-0
fax: +49 6128 987-185
Internet: http://www.omicron-instruments.com

Oxford Instruments Inc
Old Station Way
Eynsham, Witney, Oxfordshire OX8 1TL, U.K.
phone: +44(0) 1865 882855
fax: +44(0) 1865 881567
email: info.ri@oxinst.co.uk
Internet: http://www.oxinst.com

Park Scientific
1171 Borregas Ave.
Sunnyvale, CA 94089
phone: 800-776-1602, 408-747-1600
fax: 408-747-1601
email: info@park.com
Internet: http://www.park.com

Required Level of Training

Scanning probe characterization of most materials can be taught by an experienced operator at the technician level; interpretation and operation on difficult and highly nonuniform material is an art, and mastery comes from experience. The operation of scanning probe instruments requires a good knowledge of computers and the controlling software, because data acquisition is highly automated. Design of experiments and data interpretation requires at least an undergraduate degree and preferably an advanced degree in physics, chemistry, materials science, or a related discipline. Although the basic principles of the techniques are easily described, the tip-surface interactions can be complex, and the surface being studied can be influenced by many factors. This makes data interpretation difficult and is aided by training in the field of surface science. In order to minimize atmospheric contaminants and maintain better control of the surface, many scanning probe instruments are operated in a vacuum. As a result, operators must have extensive experience with the principles of UHV equipment maintenance, operation, and control.

Service and Maintenance

Routine maintenance on air-based instruments is centered around the installation, characterization, and replacement of the scanning tips. Since tips frequently contact the surface, the probe tip may become contaminated or damaged. The condition of the probe tip may be determined by probing a well-characterized surface. In addition, the performance of the piezoelectric drivers used to position the tip may change over time. As a result, it may be necessary to calibrate the x, y, and z displacement as a function of voltage using reference materials with periodic surface features and topography. More extensive maintenance of the drive electronics may require the services of trained personnel. If the scanning probe is housed in a vacuum chamber, only vacuum compatible samples that do not out-gas significantly must be loaded for investigation. Routine maintenance of the vacuum pumping system must be conducted. The nature and frequency of the vacuum system maintenance depends on the vacuum pumping technology employed. Usually these are ion-pumped UHV systems, requiring maintenance with Cu gaskets, manipulators, and turbopump service costing about $5000 per year.

Suggested Readings

BONNELL, D. A., ED., *Scanning Tunneling Microscopy: Theory, Techniques and Applications*. New York: VCH, 1993.

CHEN, C. J., *Introduction to Scanning Tunneling Microscopy*. Oxford Series in Optical and Imaging Sciences. New York: Oxford University Press, 1993.

DiNARDO, N. J., *Nanoscale Characterization of Surfaces and Interfaces*. New York: VCH, 1994.

HERCULES, D. M., AND S. H. HERCULES, "Analytical Chemistry of Surfaces," *J. Chem. Ed.*, 61, no. 5 (1984), 402–09.

HORTON, O., AND M. AMREIN, EDS., *STM and SFM in Biology*. San Diego: Academic Press, 1993.

KELLY, M. J., "Adventures in the Micro World," *CHEMTECH*, 1 (1987), 31–33.

STROSCIO, J. A., AND W. J. KAISER, EDS., *Scanning Tunneling Microscopy*. Methods in Experimental Physics, vol. 27. San Diego: Academic Press, 1993.

WIESENDANGER, R., AND H. J. GUNTHERODT, EDS., *Scanning Tunneling Microscopy*, vols. I, II, and III, Springer Series in Surface Science, vols. 21, 28, and 29. Berlin: Springer Verlag, 1992, 1993, and 1994.

WIESENDANGER, R., *Scanning Probe Microscopy and Spectroscopy, Methods and Applications*. New York: University Press, 1994.

Proceedings of the NATO Advanced Research Workshops organized under the Special Program on Nanometer Scale Science; for example, R. Behm, N. Garcia, and H. Rohrer, eds., *Scanning Tunneling Microscopy*. NATO ASI Series E: Applied Sciences, vol. 184. Dordrecht, Netherlands: Kluwer Academic Publishers, 1990.

References

1. G. Binnig and others, *Appl. Phys. Lett.*, 40 (1982), 178.

2. G. Binnig and others, *Phys. Rev. Lett.*, 49 (1982), 57.

3. G. Binnig and others, *Phys. Rev. Lett.*, 50 (1983), 120.

4. IBM, *J. Res. Develop.*, 30, nos. 4 and 5 (1986).

5. R. Wiesendanger and H. J. Güntherodt, eds., *Scanning Tunneling Microscopy*, vols. I, II, and III, Springer Series in Surface Sciences, vols. 21, 28, and 29 (Berlin: Springer Verlag, 1992, 1994, and 1993).

6. Proceedings of the Conferences on Scanning Tunneling Microscopy (STM), 1991–1996.

7. R. Behm, N. Garcia, and H. Rohrer, eds., *Scanning Tunneling Microscopy*, NATO ASI Series E: Applied Sciences, vol. 184 (Dordrecht, Netherlands: Kluwer Academic Publishers, 1990).

8. Proceedings of the NATO Advanced Research Workshops organized under the Special Program on Nanoscale Science, *NATO ASI Series E: Applied Physics* (Dordrecht, Netherlands: Kluwer Academic Publishers).

9. R. Young, J. Ward, and F. Scire, *Rev. Sci. Instrum.*, 43 (1972), 999.

10. R. Feenstra, *Phys. Rev. B.*, 50 (1995), 4561.

11. A. Kahn, *Surf. Sci. Rep.*, 3 (1983), 193.

12. G. Hansson and R. Uhrberg, *Surf. Sci. Rep.*, 9 (1988), 197.

13. F. J. Himpsel, *Surf. Sci. Rep.*, 12 (1990), 1.

14. J. Tersoff and D. R. Hamann, *Phys. Rev. Lett.*, 50 (1983), 1998.

15. J. Tersoff and D. R. Hamann, *Phys. Rev. B.*, 31 (1985), 805.

16. C. J. Chen, *Introduction to Scanning Tunneling Microscopy* (Oxford: Oxford University Press, 1993).

17. R. Wiesendanger, *Scanning Probe Microscopy and Spectroscopy* (Cambridge: Cambridge University Press, 1994).

18. R. J. Hamers, *Ann. Rev. Phys. Chem.*, 40 (1989), 531.

19. J. A. Stroscio, R. Feenstra, and A. P. Fein, *Phys. Rev. Lett.*, 57 (1986), 2579.

20. M. P. Seah and W. Dench, *Surf. Interf. Anal.*, 1 (1979), 1.

21. D. Pohl, *IBM J. Res. Develop.*, 30 (1986), 417.

22. B. Reihl and others, *Physica B*, 197 (1993), 64.

23. C. Gerber and others, *Rev. Sci. Instrum.*, 57 (1986), 221; H. Salemink, O. Albrektsen, and M. Johnson, *J. Vac. Sci. Technol. B*, 12 (1994), 362.

24. R. R. Schlittler and J.K. Gimzewski, Presentation given at the 8th International Conference on Scanning Tunneling Microscopy/Spectroscopy and Related Techniques (STM '95), Snowmass, CO, July 23–28, 1995.

25. O. Albrektsen and others, *J. Vac. Sci. Technol. B*, 12 (1994), 3187.

26. G. Binnig and D. Smith, *Rev. Sci. Instrum.*, 57 (1986), 1688.

27. G. Meyer and N. Amer, *Appl. Phys. Lett.*, 53 (1988), 1045.

28. R. Erlandsson and others, *J. Vac. Sci. Technol. A*, 6 (1988), 266; A. den Boef, *Appl. Phys. Lett.*, 55 (1989), 439.

29. C. Schoenenberger and S. F. Alvarado, *Rev. Sci. Instrum.*, 60 (1989), 3131.

30. For a summary, see Chapter 6 of Reference 17.

31. J. K. Gimzewski, E. Stoll, and R. R. Schlittler, *Surf. Sci.*, 181 (1987), 267.

32. D. Smith, W. Heckl, and M. Klagges, *Surf. Sci.*, 278 (1992), 166.

33. V. Hallmark and others, *Phys. Rev. Lett.*, 59 (1987), 2879.

34. J. Wintterlin and others, *J. Microscopy*, 152 (1988), 423.

35. G. Binnig and others, *Surf. Sci.*, 131 (1983), L379.

36. D. Chambliss and S. Chiang, *Surf. Sci. Lett.*, 264 (1992), L187.

37. See Chapter 4.1 of Reference 17.

38. R. Hamers, R. Tromp, and J. Demuth, *Phys. Rev. Lett.*, 56 (1986), 1972.

39. R. Tromp, R. Hamers, and J. Demuth, *Phys. Rev. B*, 34 (1986), 1388.

40. R. Hamers, P. Avouris, and F. Boszo, *Phys. Rev. Lett.*, 59 (1987), 2071.

41. R. Tromp, R. Hamers, and J. Demuth, *Phys. Rev. Lett.*, 55 (1985), 1303.

42. R. Kariotis and M. Lagally, *Surf. Sci.*, 248 (1991), 295.

43. J. Stroscio and others, *J. Vac. Sci. Technol. A*, 8 (1990), 284.

44. J. Stroscio, R. Feenstra, and A. Fein, *Phys. Rev. Lett.*, 57 (1986), 2579.

45. T. A. Jung, J. K. Gimzewski, and R. R. Schlittler, Presentation at the European Conference on Surface Science (ECOSS-15), Lille, France, Sept. 4–8, 1995.

46. R. Feenstra and others, *Phys. Rev. Lett.*, 58 (1987), 1192.

47. H. Ueyama and others, Presentation at the 8th International Conference on Scanning Tunneling Microscopy/Spectroscopy and Related Techniques (STM '95), Snowmass, CO, July 23–28, 1995; *Jpn. J. Appl. Phys.*, 34 (1995), L1086.

48. P. Muralt and D. Pohl, *Appl. Phys. Lett.*, 48 (1986), 514; P. Muralt, *Appl. Phys. Lett.*, 49 (1986), 1441.

49. M. McCord and R. Pease, *J. Vac. Sci. Technol. B*, 3 (1985), 198; 4 (1986), 86; *Appl. Phys. Lett.*, 50 (1987), 569.

50. I. Lyo and P. Avouris, *Science*, 253 (1991), 1369; J. Lyding and others, *Appl. Phys. Lett.*, 64 (1994), 2010; G. Abeln and others, *Microelectron. Engin.*, 27 (1994), 23.

51. H. Mamin, P. Guethner, and D. Rugar, *Phys. Rev. Lett.*, 65 (1990), 2418; H. Mamin and others, *J. Vac. Sci. Technol. B*, 9 (1991), 1398.

52. J. Gimzewski and others, *Z. Phys. B*, 72 (1988), 497.

53. D. Abraham and others, *Appl. Phys. Lett.*, 56 (1990), 1564.

54. S. Alvarado and others, *J. Vac. Sci. Technol. B*, 9 (1991), 409.

55. W. Kaiser and L. Bell, *Phys. Rev. Lett.*, 60 (1988), 1406.

56. M. Prietsch and R. Ludeke, *Surf. Sci.*, 251/252 (1991), 413; see also Chapter 1.19 of Reference 17, and Abstracts of 148th WE Heraeus Seminar on STM-Related Spectroscopies of Semiconductor Interfaces, Bad Honnef, Germany, Aug. 30–Sept. 1, 1995.

57. S. Minne and others, *Appl. Phys. Lett.*, 66 (1995), 703.

58. J. Barnes and others, *Rev. Sci. Instrum.*, 65 (1994), 3793.

59. Y. Martin and H. Wickramasinghe, *Appl. Phys. Lett.*, 50 (1987), 1455; C. Schoenenberger and S. Alvarado, *Z. Phys. B*, 80 (1990), 373; D. Rugar and others, *J. Appl. Phys.*, 68 (1990), 1169.

60. D. Pohl, W. Denk, and M. Lanz, *Appl. Phys. Lett.*, 44 (1984), 651; E. Betzig, M. Isaacson, and A. Lewis, *Appl. Phys. Lett.*, 51 (1987), 2088.

61. D. Eigler and E. Schweitzer, *Nature*, 344 (1990), 524.

62. C. Weisbuch and B. Vinter, *Quantum Semiconductor Structures: Fundamentals and Applications* (New York: Academic Press, 1991).

63. H. Salemink, M. Johnson, and O. Albrektsen, *J. Vac. Sci. Technol. B*, 12 (1994), 362.

64. M. Pfister and others, *Appl. Phys. Lett.*, 67 (1995), 1459.

65. M. Johnson and others, *Appl. Phys. Lett.*, 63 (1993), 2923; M. Johnson and others, *Phys. Rev. Lett.*, 75 (1995), 1606.

66. A. Ourmazd and others, *Phys. Rev. Lett.*, 62 (1989), 933.

67. M. Pfister and others, *Appl. Phys. Lett.*, 65 (1994), 1168.

Chapter 42

Auger Electron Spectroscopy

A. R. Chourasia and D. R. Chopra

Texas A&M University–Commerce
Department of Physics

Summary

General Uses

- Identification of elements on surfaces of materials
- Quantitative determination of elements on surfaces
- Depth profiling by inert gas sputtering
- Phenomena such as adsorption, desorption, and surface segregation from the bulk
- Determination of chemical states of elements
- In situ analysis to determine the chemical reactivity at a surface
- Auger electron elemental map of the system

Common Applications

- Qualitative analysis through fingerprinting spectral analysis
- Identification of different chemical states of elements
- Determination of atomic concentration of elements
- Depth profiling
- Adsorption and chemisorption of gases on metal surfaces
- Interface analysis of materials deposited in situ on surfaces

Samples

State

Almost any solid can be analyzed.

Amount

Sample size depends on the instrument.

Preparation

Sample can be analyzed as it is. Because the analysis is done in high vacuum, some samples should be cleaned before loading in the chamber.

Analysis Time

Estimated time to obtain the survey spectrum from a sample varies from 1 to 5 min. High-resolution acquisition takes anywhere from 5 to 25 min per region of the spectrum and depends on the resolution required.

Limitations

General

- Analyzes conducting and semiconducting samples.
- Special procedures are required for nonconducting samples.
- Only solid specimens can be analyzed.
- Samples that decompose under electron beam irradiation cannot be studied.
- Quantification is not easy.

Accuracy

- The Auger spatial resolution common to most commercial instruments is of the order of 0.2 μm or less and is a function of analysis time.
- The sampling depth is about three monolayers.

Sensitivity and Detection Limits

The sensitivity is of the order of 0.3%.

Complementary or Related Techniques

- X-ray photoelectron spectroscopy.
- Atom probe field ion microscopy.

- Rutherford backscattering spectroscopy.
- Secondary ion mass spectrometry.
- Electron energy loss spectroscopy, X-ray absorption spectroscopy, Bremsstrahlung isochromat spectroscopy, inverse photoemission spectroscopy, and appearance potential spectroscopy. All provide complementary information on the unoccupied density of states.

Introduction

The Auger effect was discovered by Pierre Auger in 1925 while working with X rays and using a Wilson cloud chamber. Tracks corresponding to ejected electrons were observed along a beam of X rays (1).

Auger electron spectroscopy (AES) has now emerged as one of the most widely used analytical techniques for obtaining the chemical composition of solid surfaces. The basic advantages of this technique are its high sensitivity for chemical analysis in the 5- to 20-Å region near the surface, a rapid data acquisition speed, its ability to detect all elements above helium, and its capability of high-spatial resolution. The high-spatial resolution is achieved because the specimen is excited by an electron beam that can be focused into a fine probe. It was developed in the 1960s, when ultra-high vacuum (UHV) technology became commercially available (2, 3).

When an electron is ejected from an inner shell of an atom the resultant vacancy can be filled by either a radiative (X-ray) or nonradiative (Auger) process. In AES the atomic core levels are ionized by the incident electron beam and the resulting Auger electrons are detected with an electron spectrometer. These electrons form small peaks in the total energy distribution function, N(E) as shown in Fig. 42.1. The incident electrons entering a solid are scattered both elastically and inelastically. At the primary beam energy a sharp peak is observed, caused by electrons that have been elastically scattered back out of the specimen. For a crystalline specimen, these electrons carry the crystal structure information, which is exploited in techniques such as low-energy electron diffraction and reflection high-energy electron diffraction. At slightly lower energies there are smaller peaks due to electrons that have undergone characteristic energy losses. The information contained in this region is exploited in the technique of low-energy electron loss spectroscopy. At the other end of the spectrum (that is, on the low-energy side of the spectrum) there is a large peak corresponding to the secondary electrons. A few hundred eV above this peak is a loosely defined crossover energy. Above this point the distribution is dominated by backscattered primary electrons and below this point the secondary electrons form the major component. The crossover point depends on the primary beam energy and moves to higher energies with increasing primary beam energy. The peaks due to Auger electrons are superimposed on this distribution. The peaks become more pronounced by electronic spectral differentiation, which removes the large background.

The Auger process can be understood by considering the ionization process of an isolated atom under electron bombardment. The incident electron with sufficient primary energy, E_p, ionizes the core level, such as a K level. The vacancy thus produced is immediately filled by another electron from L_1. This process is shown in Fig. 42.2. The energy $(E_K - E_{L1})$ released from this transition can be transferred to another electron, as in the L_2 level. This electron is ejected from the atom as an Auger electron. The Auger electron will have energy given by

$$E = E_K - E_{L1} - E_{L2} \qquad (42.1)$$

Figure 42.1 The energy distribution for an elemental Si sample. The $N(E)$ and dN/dE spectra are displayed. The energy of primary electrons is 20 keV. The spectra are taken using PHI 670xi system. *(Reproduced with permission from Physical Electronics, Eden Prairie, MN.)*

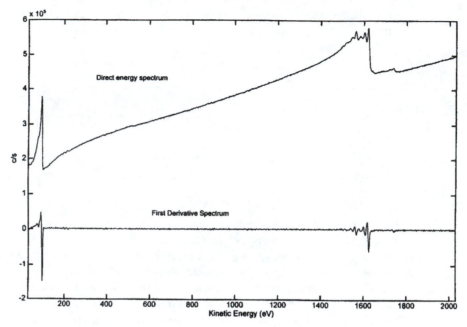

This excitation process is denoted as a KL_1L_2 Auger transition. It is obvious that at least two energy states and three electrons must take part in an Auger process. Therefore, H and He atoms cannot give rise to Auger electrons. Several transitions (KL_1L_1, KL_1L_2, LM_1M_2, etc.) can occur with various transition probabilities. The Auger electron energies are characteristic of the target material and independent of the incident beam energy.

Figure 42.2 Energy level diagram in an Auger process. Electron from L_1 drops into the K level with the emission of an L_2 electron.

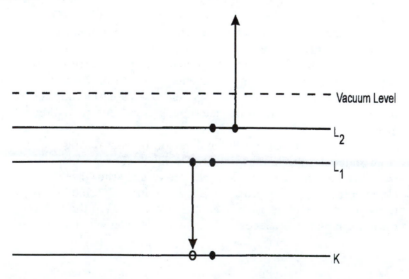

Isolated Li atoms having a single electron in the outermost level cannot give rise to Auger electrons. However, in a solid the valence electrons are shared and the Auger transitions of the type KVV occur involving the valence electrons of the solid. In general, the kinetic energy of Auger electrons originating from an ABC transition can be estimated from the empirical relation

$$E_{ABC} = E_A(Z) - E_B(Z) - E_C(Z + \Delta) - \Phi_A \tag{42.2}$$

where Φ_A represents the work function of the analyzer material and Z is the atomic number of the atom involved. The Δ term appears because the energy of the final doubly ionized state is somewhat larger than the sum of the energies for individual ionization of the same levels. Another expression for estimating Auger transition energies is

$$E_{ABC} = E_A(Z) - \frac{1}{2}[E_B(Z) + E_B(Z + 1)] - \frac{1}{2}[E_C(Z) + E_C(Z + 1)] - \Phi_A \tag{42.3}$$

The theoretical binding energies are discussed in the literature. A more detailed treatment of the Auger energies requires knowledge of a coupling in the final state that occurs between the two unfilled shells. For light elements the coupling scheme is pure L–S, for heavy elements j–j, and for elements in the middle of the periodic table intermediate coupling must be invoked.

The most pronounced Auger transitions observed in AES involve electrons of neighboring orbitals, such as KLL, LMM, MNN, NOO, MMM, and OOO families. The most prominent KLL transitions occur from elements with $Z = 3$ to 14, LMM transitions for elements with $Z = 14$ to 40, MNN transitions for elements with $Z = 40$ to 79, and NOO transitions for heavier elements. The Auger peak is commonly identified by the maximum negative peak in the $dN(E)/dE$ versus E spectrum.

How It Works

The schematic of the experimental arrangement for basic AES is shown in Fig. 42.3. The sample is irradiated with electrons from an electron gun. The emitted secondary electrons are analyzed for energy by an electron spectrometer. The experiment is carried out in a UHV environment because the AES technique is surface sensitive due to the limited mean free path of electrons in the kinetic energy range of 20 to 2500 eV. The essential components of an AES spectrometer are

- UHV environment
- Electron gun
- Electron energy analyzer
- Electron detector
- Data recording, processing, and output system

UHV Environment

The surface analysis necessitates the use of a UHV environment (4) because the equivalent of one monolayer of gas impinges on a surface every second in a vacuum of 10^{-6} torr. A monolayer is adsorbed on the surface of the specimen in about 1 second at 10^{-6} torr. Contamination of the specimen surface is critical for highly reactive surface materials, where the sticking coefficient for most residual gases is very high (near unity). The sticking coefficient for surfaces that are passivated through

Figure 42.3 Schematic arrangement of the basic elements of an Auger electron spectrometer.

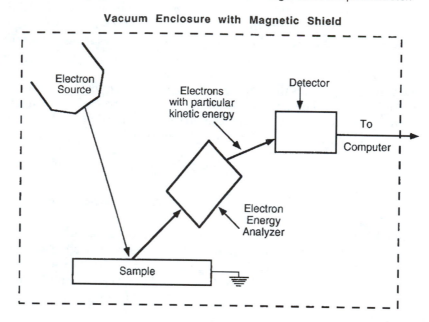

exposure to air is very low. The vacuum requirements are much less stringent for such samples. Generally, the background pressure is reduced to the low 10^{-10}-torr range in order to minimize the influence of residual gases in surface analysis measurements.

Until 1960 the advancements in surface analysis techniques were inhibited by two difficulties: constructing an apparatus suitable for operation in a UHV environment and production and measurement of UHV. In the early 1960s the glass enclosures were replaced by standardized, stainless steel hardware. The UHV environment could be easily achieved by pumping a stainless steel chamber with a suitable combination of ion, cryo, turbo molecular, or oil diffusion pumps (5). It is also possible to bake these systems (up to 200 °C) to achieve UHV conditions.

Electron Gun

The nature of the electron gun used for AES analysis depends on a number of factors:

- The speed of analysis (requires a high beam current)
- The desired spatial resolution (sets an upper limit on the beam current)
- Beam-induced changes to the sample surface (sets an upper limit to current density)

The range of beam currents normally used in AES is between 10^{-9} and 5×10^{-6} A. The lower current gives high spatial resolution whereas the higher current may be used to give speed and high sensitivity where spatial resolution is of little concern. In certain samples the high current used may induce surface damage to the specimen and should be avoided.

The electron gun optical system has two critical components: the electron source and the focusing forming lens. In most cases the electron source is thermionic but for the highest spatial resolution the brighter field emission source may be used. The field emission sources have problems of cost and stability and are therefore limited in their use. The commonly used thermionic sources are as follows:

- A tungsten hairpin filament. This filament has a life of about 100 hr due to repeated exposure to air. As a result of introduction of a load-lock system in the modern Auger electron spectrometers the average filament life exceeds 1000 hr.
- Lanthanum hexaboride (LaB_6). This source is brighter than tungsten but is more expensive and a little more complicated to operate.

The electron lenses used to focus the beam may be either magnetic or electrostatic. The magnetic lenses have low aberrations and therefore give the best performance. However, these lenses are complicated and expensive. The electrostatic lenses are easier to fit in a UHV system. For spatial resolution of the order of a micron, a 10 keV electrostatic gun could be easily used. For spatial resolution below 100 nm electromagnetic lenses are used.

Electron Energy Analyzer

The function of an electron energy analyzer is to disperse the secondary emitted electrons from the sample according to their energies. An analyzer may be either magnetic or electrostatic. Because electrons are influenced by stray magnetic fields (including the earth's magnetic field), it is essential to cancel these fields within the enclosed volume of the analyzer. The stray magnetic field cancellation is accomplished by using Mu metal shielding. Electrostatic analyzers are used in all commercial spectrometers today because of the relative ease of stray magnetic field cancellation (6).

The Cylindrical Mirror Analyzer (CMA)

The schematic of the CMA is shown in Fig. 42.4. The CMA consists of two coaxial cylinders with a negative potential (V) applied to the outer cylinder (with radius r_2) and ground potential applied to the inner cylinder (with radius r_1). The sample and the detector are located along the common axis of the cylinders. Electrons emitted from the sample at an angle α relative to the analyzer axis

Figure 42.4 Cross-sectional view of the single-pass CMA. *V* represents the voltage applied to the outer cylinder. The inner cylinder is grounded.

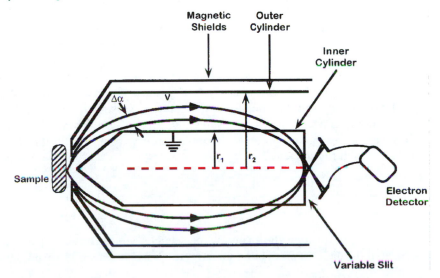

pass through the mesh-covered aperture in the inner cylinder. Only electrons that have a definite energy E_0 are deflected by the outer cylinder potential through the second mesh-covered aperture to a focus on the axis, where they pass into a detector. The focal condition is given by the relation

$$E_0 = \frac{KeV}{\ln(r_2/r_1)} \qquad (42.4)$$

where e is the electron charge and K a constant. A variety of entrance angles, α, can exist. However, in the special case where $\alpha = 42°18'$ the CMA becomes a second-order focusing instrument. At this focus the electrons from a very wide range of αs meet at the focal point. For this case the constant $K = 1.31$. Under this condition, the distance between the source and the focus is $6.1r_1$. All the CMA parameters scale with the radius r_1 of the inner cylinder. Once this is fixed, other parameters follow automatically. The energy resolution of the CMA is written in terms of the base resolution ΔE_B, defined as

$$\frac{\Delta E_B}{E} = \frac{0.36w}{r_1} + 5.55(\Delta\alpha)^3 \qquad (42.5)$$

where w is the slit width (equal at the entrance and exit) and $\Delta\alpha$ is the spread of the angle about α (Fig. 42.4). The CMA has a very large acceptance angle, comprising a full cone about the spectrometer axis. This gives a high sensitivity and reduces the dependency of the signal on specimen topography (6). The electron gun may be incorporated into the inner cylinder, making the design of the system a compact unit. Because of its superior signal-to-noise capability, the CMA is used almost exclusively for modern AES apparatus.

Concentric Hemispherical Analyzer (CHA)

The CHA consists of two hemispherical concentric shells of inner radius r_1 and outer radius r_2, as shown in Fig. 42.5. A potential difference, ΔV, is applied between the two surfaces such that the outer sphere is negative and the inner positive. Between the spheres there is an equipotential surface of radius r_0. The entrance and exit slits lie on a diameter and are centered at a distance r_0 from the center of curvature. The base resolution is given by

$$\frac{\Delta E}{E} = \frac{w}{r_0} + \alpha^2 \qquad (42.6)$$

where α is the entrance angle into the CHA and w is the width of the slit.

In the CHA, the sample must be placed at the slit of the analyzer. The working space between the sample and the analyzer is therefore increased by the use of an electron lens. The lens system takes electrons from the sample and injects them into the analyzer. The effective solid angle thus becomes comparable to that of a CMA. The main advantages the CHA has over the CMA are much better access to the sample and the ability to vary analyzer resolution electrostatically without changing physical apertures. Generally, a large entrance angle yields increased sensitivity. However, to reach a compromise between sensitivity and resolution it is a common practice to choose α such that $\alpha^2 = w/2r_0$.

Electron Detector

Having passed through the analyzer, the secondary electrons of a particular energy are spatially separated from electrons of different energies. Various detectors are used to detect these electrons.

Figure 42.5 Cross-sectional view of the CHA with input lens. The outer sphere has a negative voltage and the inner sphere has a positive voltage. The dashed lines indicate the trajectories followed by the emitted electrons. The central dashed line represents an equipotential surface. The entrance and the exit slits lie on a diameter and are centered at the mean radius from the center of curvature.

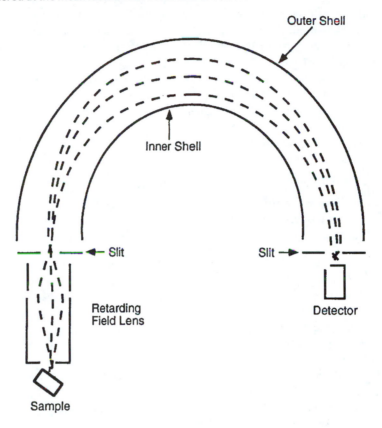

Single-Channel Detector (SCD)

The detector used in conventional instrumentation is a channel electron multiplier. It is an electrostatic device that uses a continuous dynode surface (a thin-film conductive layer on the inside of a tubular channel). It requires only two electrical connections to establish the conditions for electron multiplication. The output of this detector consists of a series of pulses that are fed into a pulse amplifier/discriminator and then into a computer. The advantage of such a detector is that it can be exposed to air for a long time without damage. It counts electrons with a high efficiency, even at essentially zero kinetic energy, and the background is 0.1 count/sec or lower. The only drawback is that a high count rate ($> 10^6$ counts/sec) causes a saturation effect.

Multichannel Detector (MCD)

A multiple detection system can be added at the output of the analyzer. The system may be in the form of a few multiple, parallel, equivalent detector chains or position-sensitive detectors spread across the whole of the analyzer output slit plane. Such an arrangement can be devised in a number of ways (7): using phosphor screens and TV cameras, phosphor screens and charge-coupled devices, resistive anode networks, or discrete anodes.

Data Recording, Processing, and Output System

The Auger electrons appear as peaks on a smooth background of secondary electrons. If the specimen surface is clean, the main peaks would be readily visible and identified. However, smaller peaks and those caused by trace elements present on the surface may be difficult to discern from the background. Because the background is usually sloping, even increasing the gain of the electron detection system and applying a zero offset is often not a great advantage. Therefore, the Auger spectra are usually recorded in a differential form. In the differential mode it is easy to increase the system gain to reveal detailed structure not directly visible in the undifferentiated spectrum (Fig. 42.1). The most distinctive characteristic of an Auger peak in the differential spectrum is the negative going feature at the high-energy side of the peak. The minimum of this feature represents the point of maximum slope of the parent peak. By convention, this feature is used to define the energy of the transition. For systems that record data in an analog mode, the differential Auger spectrum has the following advantages over the direct undifferentiated spectrum:

- The background is zero away from the peaks, so the amplifier gain can be easily changed.
- The differential peaks are sharper than the direct peaks. This provides a reproducible energy reference point.
- The peak-to-peak height provides a convenient measure of peak intensity.

At present, the data in commercial instruments are acquired digitally and can be presented in either analog or digital mode. The majority of AES instruments are controlled by computer. Major functions of the computer control system are to acquire and store data efficiently.

A conventional Auger electron spectrometer uses a lock-in amplifier. The oscillator in this amplifier superimposes a sinusoidal modulation (frequency 10 to 20 kHz) on the potential applied to the outer cylinder of the analyzer. The AC component of the signal is decoupled from the multiplier high voltage and detected in the lock-in amplifier. The output is then fed to an x–y recorder. The amplitude of the modulation is chosen to yield a compromise between sensitivity and resolution. A typical modern Auger electron spectrometer collects the data in the $N(E)$ versus E integral mode. The data are then mathematically differentiated using computer software to yield $dN(E)/dE$ versus E Auger spectra (Fig. 42.1).

The CMA or the CHA is used in the constant retardation ratio (CRR) mode in AES, which determines the resolution. Therefore, while scanning a spectrum the optimum modulation voltage will change. High-energy peaks generally require high modulation. Although a constant modulation voltage may be used throughout the scan when recording a spectrum, some instruments can make the modulation proportional to the analyzer energy so that the ratio of modulation to analyzer resolution is constant. This point should be considered while comparing the Auger spectra from different instruments.

What It Does

The high surface sensitivity of AES is due to the limited mean free path of electrons in the kinetic energy range 20 to 3000 eV. Auger electrons, which lose energy through plasma losses, core excitations, or interband transitions, are removed from the observed Auger peaks and contribute to the nearly uniform background on which the Auger peaks are superimposed. Because phonon losses are small compared with the natural width of Auger peaks, they do not affect the Auger escape depth. Hence the Auger yield is not dependent on the sample temperature.

Because the Auger transition probability and Auger electron escape depth are independent of the incident electron beam energy, E_p, the dependence of the Auger peak amplitude on E_p is governed completely by the ionization cross-section of the initial core level. Ionization occurs primarily by the incident electrons during their initial passage through the escape depth region (5 to 25Å thick). The backscattered primary electrons can also contribute to the Auger yield when the incident beam energy is substantially greater than the binding energy of the core level involved.

Variables Involved in the Production of Auger Electrons

An inner shell vacancy can be produced through a variety of methods, such as irradiation with electrons and X rays or bombardment with argon ions. Electron impact is usually used for producing Auger lines for analytical purposes. It provides an intense beam that can be brought to a fine focus. X-ray irradiation has its value in providing less radiation damage and better peak-to-background ratios.

High-Energy Satellite Lines

High-energy satellite structures have been observed in the Auger spectra of solids. The presence of such a structure has been interpreted as being due to plasmon gains. It is also believed that the high-energy lines arise from an initial multiple ionization or perhaps resonance absorption. The question of Auger satellites in solids is still under active consideration.

Characteristic Energy Losses

Electrons ejected from a solid can suffer characteristic energy losses, usually due to plasmon losses (see p. 820). Because Auger spectra are generally rather complex and often not well resolved and are spread over a considerable range of energies, peaks from characteristic energy losses are much more difficult to disentangle from the normal Auger spectrum than is usual in the case of photoelectron spectroscopy. Also, the surface contamination will alter the nature of the characteristic loss peaks considerably.

Charging in Nonconducting Samples

Charging as a result of an impinging beam of electrons on a nonconductor is a particularly serious problem in Auger spectroscopy. Often the charging and the resulting nonuniform surface potential prevent a meaningful Auger spectrum. However, this problem often can be overcome by choosing the proper angle of incidence and the energy of the primary electron beam. The important factor is the ratio δ (the number of secondary electrons leaving the target to the number impinging on the target). If $\delta = 1$, the charge is stabilized. If $\delta < 1$, the charge is negative, and if $\delta > 1$, it is positive. The choice of impact energy is also important. The factor δ becomes less than 1 if the energy of the impinging beam of electrons is either too large or too small. Generally, the primary beam energy lies between 1.5 and 3.0 keV depending on the application and the resolution required.

Scanning Auger Microscopy

With a finely focused electron beam for Auger excitation, AES can be used to perform two-dimensional surface elemental analysis (8). In this setup, the electron gun operation is similar to that used in conventional scanning electron microscopy (SEM). A set of deflection plates raster the electron beam on the sample. The scanning Auger system can be used to perform point Auger analysis with a spatial resolution on the order of 3 μm by using a minimum beam size of about 3 μm or to obtain a two-dimensional mapping of the concentration of a selected surface element. The low-energy secondary electron or absorbed current displays are used to monitor the surface topography and locate the areas of interest on the sample. To obtain an elemental map, the intensity of the display is controlled by the magnitude of the selected Auger peak. The most negative excursion in the differentiated Auger spectrum is taken as a measure of the Auger current. A two-dimensional elemental map of the surface is obtained by setting the pass energy of the electron spectrometer at the negative excursion of the Auger peak of interest, while the output of the lock-in amplifier is used to modulate the intensity of the record display as the electron beam is rastered across the sample. Three-dimensional analysis of the surface of a sample can be obtained by using a combination of scanning Auger microscopy and sputter etching.

Analytical Information

Qualitative

Identification of Elements

Elements of an unknown sample can be identified easily by recording the AES spectrum over a wide range (generally 0 to 2000 eV). The kinetic energies of the intense peaks present in the spectrum are then compared with the elemental values.

Chemical Effects

The AES peaks involving valence electrons carry significant information about the chemical state of the surface. The chemical effect may appear as a simple peak shift, a change in peak shape, or both. In effect the valence of molecular orbital structure is convoluted into the Auger peak structure. Usually X-ray photoelectron spectroscopy is the preferred technique for investigating the chemistry of the surface as photoelectron peaks are sharper and chemical effects are more easily interpreted than in the case of the corresponding Auger peaks. The interpretation is limited by the beam-induced effects. Many compounds, particularly oxides, are readily decomposed under electron irradiation. Therefore, the current density must be limited so that the specimen damage is acceptable in the time taken to make the measurement.

Quantitative

The incident electron beam, on striking the solid, penetrates with both elastic an inelastic scattering and ionizes atoms in the depth of 1 to 2 μm depending on the density of the material. Therefore, quantitative analysis from the observed AES signal depends on the average concentration of the element and how it is distributed within the first few atomic layers of the surface. The sensitivity to Auger electrons in the outer surface layer is greater than that to electrons originating from subsurface layers because of strong inelastic scattering. The Auger electron source volume generated by an incident electron beam of diameter d and energy E_p produces an Auger current of an ABC Auger transition in an element x, given by

$$I_x(ABC) = \int_\Omega \int_{E_A}^{E_p} \int_0^\infty I_p(E, Z)\sigma_x(E, E_A)N_x(Z)\gamma_x(ABC)\exp\left[\left(-\frac{Z}{\lambda}\right)d\Omega dEdZ\right] \quad (42.7)$$

where $I_p(E,Z)$ is the excitation electron flux density, $\sigma_x(E, E_A)$ is the ionization cross-section of the core level A, $N_x(Z)$ is the atomic density of the element x at a depth Z from the surface, $\exp(-Z/\lambda)$ is the Auger electron probability for escape, λ is mean free path of the electrons and is a function of depth or the matrix, and $\gamma_x(ABC)$ is the ABC Auger transition probability factor (9). In order to simplify this equation, it is assumed that the chemical composition is homogeneous over a depth of region for which the escape probability has a significant value. Also, the excitation flux density can be separated into two components:

$$I_p(E, A) = I_p + I_B(E, Z) \quad (42.8)$$

where $I_B(E,Z)$ is the excitation flux due to backscattered primary electrons and I_p is the primary electron current. With these assumptions, the detected Auger current can be expressed as

$$I_x(ABC) = I_p TN_x(Z)\gamma_x(ABC)\ \sigma_x(E_p, E_A)\ \lambda(1 + R_B) \quad (42.9)$$

where R_B is the backscattering factor and T is the energy-dependent transmission of the analyzer.

Thus, knowing the ionization cross-section, the Auger yield, and the backscattering factor, accurate quantitative analysis can be carried out using Eq. (42.9). In addition, the absolute Auger current must be accurately measured. A further complication is surface roughness, which generally reduces the Auger yield relative to flat surface. Because these requirements are generally not met for routine Auger analysis, quantitative analysis using first principles is not considered practical. Two methods are generally used for quantitative AES analysis.

Measurement with External Standards

In this method, the Auger spectra from the specimen of interest are compared with that of a standard with a known concentration of the element of interest. The concentration of element x in the unknown specimen, N_x^u is related to that in the standard, N_x^s using Eq. (42.9) and is given by

$$\frac{N_x^u}{N_x^s} = \frac{I_x^u}{I_x^s}\frac{\lambda^s}{\lambda^u}\frac{1 + R_B^s}{1 + R_B^u} \quad (42.10)$$

This method has an advantage in that ionization cross-section and Auger yield data are not required, and the Auger current is reduced to a relative measurement.

When the unknown sample composition is similar to that of the standard, the escape depth and backscattering factor are also eliminated from Eq. (42.10). The quantitative analysis thus reduces to the measurement of signal amplitude. The only requirement is that the measurements be made under identical experimental conditions. When the composition of the standard is not similar to that of the unknown sample, the influence of matrix on both the backscattering factor and the escape depth must be considered in detail.

The backscattering factor can be obtained by comparing Auger yield against E_p curves with theoretical ionization cross-section versus E_p curves or with Auger yield versus E_p data from gaseous specimens where the backscattering factor has negligible value.

Measurements with Elemental Sensitivity Factors

In quantitative AES analysis it is assumed that the composition of the sample in the near surface region is homogeneous. Quantitative analysis involving the use of elemental sensitivity factors is less accurate but is highly useful. The atomic concentration (C) of an element x in a sample is given by

$$C_x = \frac{I_x / S_x}{\sum (I_i / S_i)} \tag{42.11}$$

where I_x is the intensity of the Auger signal from the unknown specimen and S_i is the relative sensitivity of pure element i. The summation is for the corresponding ratios for all other elements present in the sample. Because it neglects variations in the backscattering factor and escape depth with material, this method is semiquantitative. The main advantages of this method are the elimination of standards and insensitivity to surface roughness. In order to avoid the need for a large number of pure elemental standards, the signal from the specimen is compared with that from a pure silver target. Thus, the elemental sensitivity factors relative to silver can be conveniently used in Eq. (42.11). Because the data are finally represented in the differential mode, it is necessary to base the sensitivity factors on the peak-to-peak height. This is valid only when the peak shape is invariant with the matrix. The atomic concentration of element x is then given by

$$C_x = \frac{I_x}{I_{Ag} S_x D_x} \tag{42.12}$$

where I_x is the peak-to-peak amplitude of the element x from the test specimen, I_{Ag} is the peak-to-peak amplitude from the Ag standard, and D_x is a relative scale factor between the spectra for the test specimen and silver. If the lock-in amplifier sensitivity (L_x), modulation energy ($E_{m,x}$), and primary beam current ($I_{p,x}$) settings used to obtain the test spectrum are different from those of the Ag spectrum (that is, L_{Ag}, $E_{m,Ag}$, and $I_{p,Ag}$) the relative scale factor is

$$D_x = \frac{L_x E_{m,x} I_{p,x}}{L_{Ag} E_{m,Ag} I_{p,Ag}} \tag{42.13}$$

Depth Profiling

Depth profiling is one of the most important applications of AES because it provides a convenient way of analyzing the composition of thin surface layers. It is a destructive technique. In this technique the sample is eroded by ion sputtering. The sample is bombarded with ions accelerated in an ion gun to an energy in the range 1 to 4 keV. As these energetic ions strike the sample a small amount of energy is transferred to the surface atoms, which causes them to leave the sample. The ion beam is rastered on the surface for a known time to remove a uniform layer of the sample. Under controlled conditions the layer removed can be calculated. The residual surface is then analyzed by AES, giving the depth distribution of different species in the sample. Inert gases (usually Ar) are commonly used as the ion sources (10).

Applications

Auger electron spectroscopy is a very powerful surface analytical technique that has found applications in many fields of solid-state physics and chemistry. AES is used to monitor the elemental composition of surfaces during physical property measurements. Several phenomena such as adsorption, desorption, surface segregation from the bulk, measurement of diffusion coefficients, and catalytic activity of surfaces have been investigated using AES. It has also been used to study the surface compositional changes in alloys during ion sputtering. Chemical properties such as corrosion, stress corrosion, oxidation, and catalytic activity and mechanical properties such as fatigue, wear, adhesion, resistance to deformation processes, and surface cracking depend on surface properties. Similarly, grain boundary chemistry influences mechanical properties such as low- and high-temperature ductility and fatigue, chemical properties such as intergranular corrosion and stress corrosion, and electrical properties. AES has been used to relate surface and grain boundary chemistry to properties of materials. AES has proved to be extremely valuable compared to most other techniques, which are limited by either large sampling depth or poor sensitivity. The main advantages of AES can be summarized as follows:

- Spatial resolution is high.
- Analysis is relatively rapid.
- Surface or subsurface analysis can be performed.
- It is sensitive to light elements (except H and He).
- It provides reliable semiquantitative analysis.
- Chemical information is available in some cases.

The disadvantages of this technique are as follows:

- Insulators are difficult to study due to surface charging.
- Surface may be damaged by the incident electron beam.
- Precise quantitative analysis may require extensive work.
- Sensitivity is modest (0.1 to 1 atom%).
- Depth profiling by ion sputtering or sectioning is destructive.

AES is expected to find increasing applications in many areas of science and technology requiring detailed information on elemental identification, surface composition, oxidation states, and chemical bonding.

Nuts and Bolts

Relative Costs

Vendors	Model and Description	Accessories	Cost
PHI	Model 680 Scanning Auger microprobe AES, SIMS, RGA, EDS	CMA, MCD 25-kV field emission Electron gun Multiple analysis points for simultaneous profiles Fracture stage	$$$$
PHI	Model 5600 Multitechnique system AES, XPS, ISS, SIMS	CHA, MCD, IG Sample manipulator (single or multiple) with x-, y-, and z-translations and tilt	$$$$
Comstock, Inc.	Model 951 Auger assembly	Retarding and focusing lens Electrostatic energy analyzer 50 eV to 5 kV electron source with focusing lens and x, y deflection Microchannel plate detector	$
VG Microtech	VG100AX Multitechnique AES, XPS	100-mm CHA Mu-metal analyzer housing Any orientation of sample possible Mounting flange: FC64 (114 mm/4.5-in. outside diameter)	$$
Kurt J. Lesker	CHA Electron sources Data acquisition software		$$, $$$ $ $

$ = 10,000 to 25,000, $$ = 25,000 to 50,000, $$$ = 50,000 to 100,000, $$$$ = 500 to 800 k$.
IG = ion gun (for sputter etching), RGA = residual gas analyzer, EDS = energy dispersive spectroscopy, ISS = ion scattering spectroscopy.

Vendors for Instruments and Accessories

Comstock, Inc.
1005 Alvin Weinberg Dr.
Oak Ridge, TN 37830
phone: 423-483-7690
fax: 423-481-3884
email: salesinfo@comstockinc.com
Internet: http://www.comstockinc.com

Kurt J. Lesker Company
1515 Worthington, Ave.
Clairton, PA 15025
phone: 800-245-1656
fax: 412-233-4275
Internet: http://www.lesker.com

Physical Electronics
6509 Flying Cloud Dr.
Eden Prairie, MN 55344
phone: 612-828-6100
fax: 612-828-6322

Vacuum Generators-Microtech
Bellbrook Business Park
Bolton Close, Uckfield
East Sussex, United Kingdom TN22IQZ
phone: 44 (0) 1825 761077
fax: 44 (0) 1825 768343
e-mail: sales-microtech@vacgen.fisons.co.uk

Required Level of Training

The operation of the instrument requires a good knowledge of vacuum pumps. An understanding of the computers and familiarity with the software are essential because the instruments are computer-controlled. Extensive training is required for each kind of instrument as these instruments are designed differently. The design of the experiment for the sample to be analyzed requires knowledge of the sample chemistry. A graduate-level knowledge regarding the spectroscopy, atomic physics, solid-state physics, and chemistry is required for understanding and interpreting the spectra.

Service and Maintenance

Routine adjustment and calibration of the energy analyzer are needed. The vacuum pumps need maintenance. Care must be taken to load only samples that are vacuum compatible, that is, ones that do not outgas in vacuum. The detection of malfunctioning electronic components is facilitated by plug-in cards. The detector is expensive and cumbersome to replace. The filament of the electron gun must be replaced often. Due to the design of the modules, most of these replacements must be carried out by trained personnel. The electron optics and the ion gun focusing components are not normally aligned on-site. They need regular attention to ensure proper operation of the instrument. Most companies offer service contracts but they are expensive.

Suggested Readings

BRIGGS, D., AND M. P. SEAH, EDS., *Practical Surface Analysis by Auger and X-ray Photoelectron Spectroscopy*. New York: Wiley, 1983.

CHILDS, K. D., AND OTHERS, in C. L. Hedberg, ed. *Handbook of Auger Electron Spectroscopy*. Eden Prairie, MN: Physical Electronics Publishing, 1995.

CZANDERNA, A. W., in S. P. Wolsky and A. W. Czanderna, eds. *Methods of Surface Analysis*. Amsterdam: Elsevier, 1988.

ERTL, G., AND J. KUPPERS, *Low Energy Electrons and Surface Chemistry*. Deerfield Beach, FL: VCH, 1985.

McGUIRE, G. E., AND OTHERS, "Surface Characterization," *Analytical Chemistry*, 65 (1995), 199R.

References

1. M. P. Auger, *Compt. Rend.*, 180 (1925), 65; *J. de Phys. Radium*, 6 (1925), 205; *Compt. Rend.*, 182 (1926), 773, 1215.

2. L. A. Harris, *J. Appl. Phys.*, 39 (1968), 1419.

3. L. A. Harris, *J. Appl. Phys.*, 39 (1968), 1428.

4. P. W. Palmberg, *J. Vac. Sci. Technol. A*, 12 (1994), 946.

5. M. H. Hablanian, *J. Vac. Sci. Technol. A*, 12 (1994), 897; P. A. Redhead, ibid., 904; K. M. Welch, ibid., 915.

6. M. P. Seah, in J. M. Walls, ed., *Methods of Surface Analysis* (Cambridge: Cambridge University Press, 1989), p. 57.

7. P. J. Hicks and others, *J. Phys. E: Scientific Instruments*, 13 (1980), 713.

8. N. C. MacDonald, *Appl. Phys. Lett.*, 16 (1970), 76; N. C. MacDonald and J. R. Waldrop, *Appl. Phys. Lett.*, 19 (1971), 315; D. J. Pocker and T. W. Haas, *J. Vac. Sci. Technol.*, 12 (1975), 370.

9. P. W. Palmberg, *Analytical Chemistry*, 45 (1973), 549A.

10. D. Briggs and M. P. Seah, eds., *Practical Surface Analysis by Auger and X-ray Photoelectron Spectroscopy* (New York: Wiley, 1983).

Chapter 43

X-Ray Photoelectron Spectroscopy

D. R. Chopra and A. R. Chourasia

Texas A&M University–Commerce
Department of Physics

Summary

General Uses

- Identification of elements on surfaces of materials
- Determination of binding energies of the core-level electrons in elements
- Quantitative determination of elements on surfaces
- Depth profiling by inert gas sputtering
- Depth profiling by angle-resolved technique (nondestructive)
- Determination of chemical states of elements
- In situ analysis to determine the chemical reactivity at a surface

Common Applications

- Identification of elements by matching spectrum with reference spectrum (fingerprinting)
- Identification of different chemical states of elements
- Determination of atomic concentration of elements
- Depth profiling
- Differentiation between conductors and insulators

- Chemical analysis of polymers and plastics
- Adsorption and chemisorption of gases on metal surfaces
- Interface analysis of materials deposited in situ on surfaces

Samples

State

Almost any solid can be analyzed.

Amount

Sample sizes range from 0.2 mm × 0.2 mm to 1 cm × 1 cm and depend on the instrument.

Preparation

Sample can be analyzed as it is. Because the analysis is done in high vacuum, the sample should be degreased before loading in the chamber.

Analysis Time

Estimated time to obtain the survey spectrum from a sample varies from 1 to 5 min. High-resolution acquisition takes anywhere from 5 to 25 min per region of the spectrum and depends on the resolution required.

Limitations

General

- Sample should be conducting.
- Special procedure is required for nonconducting samples.
- Only solid samples can be used.
- Samples that decompose under X-ray irradiation cannot be studied.

Accuracy

- The spatial resolution common to most commercial instruments is of the order of 0.2 mm.
- The sampling depth is about three monolayers.

Sensitivity and Detection Limits

The sensitivity is of the order of 0.3%.

Complementary or Related Techniques

- Auger electron spectroscopy provides additional information on elemental identification.
- Appearance potential spectroscopy, X-ray absorption spectroscopy, Bremsstrahlung isochromat spectroscopy, and inverse photoemission spectroscopy provide complementary information on the unoccupied density of states.

Introduction

The origin of X-ray photoelectron spectroscopy (XPS) can be traced back to the discovery of the photoelectric effect in which X rays were used as the exciting photon source. During the 1940s Siegbahn and his coworkers developed β-ray spectrometers with improved resolution and sensitivity. In these experiments the excitation source used was soft X rays. Precise measurement of binding energies of the core-level electrons led to the determination of chemical shifts. Identification of different oxidation states therefore became possible. Siegbahn's group named this technique electron spectroscopy for chemical analysis (1, 2). Because of his contribution, Siegbahn was honored by the awarding of the 1981 Nobel prize in physics. Surface scientists now prefer the term *X-ray photoelectron spectroscopy* for this technique.

The technique was intended for bulk analysis. In 1966 a publication (3) clearly explained that the signal obtained in this experiment did arise from the surface of the material. Interpretation of this paper led researchers to estimate the effective sampling depth of XPS to be 100 Å. Through the late 1960s and into the early 1970s, XPS was used primarily by chemists to investigate the electron density around atoms in molecules. With the advancements in vacuum technology in the early seventies, ultra-high vacuum (UHV) instruments became available, making analysis of clean surfaces possible. Furthermore, accurate data on the mean free path of slow electrons became available. As a result, the sampling depths were found to be in the range of 5 to 25 Å for metals and metal oxides, whereas these values ranged from 40 to 100 Å for organic and polymeric materials. It is the combination of high surface sensitivity and the ability to provide chemical information about species observed at surfaces that give XPS a unique position among surface analysis techniques.

A large number of reference spectra of the elements are available to assist in the identification of the specimen (fingerprinting). These are supplemented with high-resolution spectra of the most sensitive spectral region of the elements. Binding energy data on the most easily observable core levels are also provided for the elements and their compounds. An electronic database containing the spectra is also available (4).

In the basic XPS experiment, the sample surface is irradiated by a source of X rays under UHV conditions. Photoionization takes place in the sample surface and the resulting photoelectrons will have a kinetic energy, E_k, given by the Einstein relation

$$E_k = h\nu - E_b - \phi \tag{43.1}$$

where $h\nu$ is the energy of the incident beam, E_b is the binding energy of the photoelectron with respect to the fermi level, and ϕ is the work function of the spectrometer (Fig. 43.1). An XPS spectrum consists of a plot of the number of emitted electrons per energy interval versus their kinetic energy.

The X rays (photons) have path lengths on the order of a few microns. On the other hand, the path length of the electrons is on the order of tens of angstroms. Thus, whereas ionization occurs to a depth of a few microns, electrons originating within tens of angstroms below the surface and

Figure 43.1 Schematic of the physical basis of XPS process. The binding energy of the photoemitted electron is given by (*h*ν – kinetic energy of the photoelectron – work function of the spectrometer).

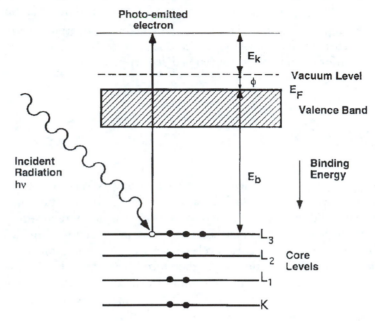

traversing without energy loss produce peaks in the spectra. The electrons that lose energy while emerging from the surface constitute the background on the high binding energy side of the peak.

The binding energy of core electrons is characteristic of the individual atom present. The excited core electrons have different kinetic energies depending on their binding energies. In addition, there is a different probability or cross-section for the excitation of the different core level electrons. The intensity of the peaks observed in an XPS spectrum depends on the excitation probability and the atomic concentration in the surface region. More than one type of element present on the surface may produce overlapping peaks. In such cases different spectral regions in which overlapping does not occur must be analyzed for reliable results.

In the photoelectric process, in addition to the emission of photoelectrons, Auger electrons are emitted as a result of relaxation of the energetic ions left after photoemission. In the Auger process, an outer electron falls into the inner orbital vacancy and a second electron is emitted, carrying the excess energy. The kinetic energy of these Auger electrons is the difference between the energy states of the singly charged initial ion and the doubly charged final ion, and is independent of the mode of the initial ionization. These Auger electrons also constitute peaks in an XPS spectrum.

How It Works

The schematic experimental arrangement for XPS is shown in Fig. 43.2. The sample is irradiated with soft X rays. The emitted electrons are analyzed for their energies. The analyzer acts as an energy window. It accepts those electrons that have an energy within the range of this fixed window, called the pass energy. The collection of these electrons is accomplished with the use of electron detectors. An XPS spectrometer is fabricated with the following components:

- UHV environment
- X-ray source
- Specimen holder
- Electron energy analyzer
- Electron detector
- Data recording, processing, and output system

Some of these are discussed below.

UHV Environment

The surface analysis necessitates the use of a UHV environment (5) because the equivalent of one monolayer of gas impinges on a surface every second in a vacuum of 10^{-6} torr. Moreover, the components of the spectrometer, such as X-ray source, electron and ion sources, and electron analyzers and detectors, operate only in a high vacuum or ultra-high vacuum environment. The background pressure must therefore be reduced to the low 10^{-10}-torr range in order to minimize the influence of residual gases in surface analysis measurements.

Until 1960, advances in surface analysis techniques were inhibited by two difficulties: constructing an apparatus suitable for operation in a UHV environment and producing and measuring UHV. In the early 1960s the glass enclosures were replaced by standardized, stainless steel

Figure 43.2 Schematic arrangement of the basic elements of an X-ray photoelectron spectrometer.

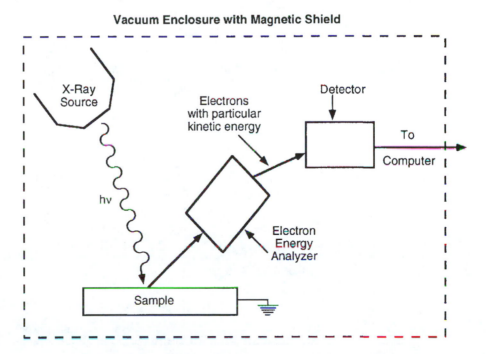

Vacuum Enclosure with Magnetic Shield

hardware. The UHV environment could be easily achieved by pumping a stainless steel chamber with a suitable combination of ion, cryo, turbo molecular, or oil diffusion pumps (6). It is also possible to bake these systems (up to 200 °C) to achieve UHV conditions.

X-Ray Sources

The X-ray source uses characteristic emission lines from an anode bombarded by high-energy electrons. Al-$K_{\alpha1,2}$ (hv = 1486.6 eV) and Mg-$K_{\alpha1,2}$ (hv = 1253.6 eV) are the most popular radiation sources. A thin (about 2-μm) aluminum or beryllium window separates the tube from the specimen. The Al and Mg sources may be combined as a switchable twin anode. Such a device helps in the separation of photoemitted electrons (whose kinetic energy depends on the incident radiation energy) and of Auger electrons (whose energy is fixed).

Unlike an electron beam, X rays cannot be easily focused on a small spot, limiting the lateral resolution of XPS. This difficulty is overcome by the use of scanning XPS. In this technique the sample is deposited on one side of a thin Al foil. A focused electron beam is then scanned over the other side of the foil. The characteristic Al-K_α X rays are thus produced in a small spot and, after passing through the foil, excite photoelectrons from a corresponding spot of the sample. The disadvantage of the technique is that it requires very thin samples (10^2 to 10^3 nm) to avoid excessive damping of the X rays before they reach the outer surface. Commercial instruments are available that provide small radiation spot a few millimeters wide. The minimum area that could be analyzed using the small radiation spot is about 15 μm × 15 μm.

The energy resolution of the spectrometer depends, among other factors, on the line width of the X-ray source. Therefore, for better resolution, the characteristic X-ray lines with sufficiently

Table 43.1 Energies and widths of some characteristic X-ray lines (eV).

	Line					
	K_α		L_α		M_ζ	
Anode Material	Energy	Width	Energy	Width	Energy	Width
Y			1922.6	1.5	132.3	0.47
Zr			2042.4	1.7	151.4	0.77
Nb					171.4	1.21
Mo					192.3	1.53
Ti	4510.0	2.0	395.3	3.0		
Cr	5417.0	2.1	572.8	3.0		
Ni			851.5	2.5		
Cu	8048.0	2.6	929.7	3.8		
Mg	1253.6	0.7				
Al	1486.6	0.85				
Si	1739.5	1.0				
Ag			2984.3	3.2		
Au					M_α:2122.9	2.5

small widths must be chosen. Table 43.1 lists X-ray energies and line widths for various characteristic lines from some materials. The X-ray spectrum from the source consists of the characteristic lines superimposed on a continuous background (Bremsstrahlung) radiation. The less intense lines therefore give rise to satellites and ghost peaks in the spectrum, causing assignment and interpretational problems. These problems can be eliminated by using a suitable X-ray monochromator. Monochromatization is achieved by back-diffraction of X rays from the face of a suitable crystal according to Bragg's law

$$n\lambda = 2d \sin\theta \tag{43.2}$$

where n is the order of diffraction, λ the X-ray wavelength, d the crystal spacing, and θ the Bragg angle. Such an arrangement has the additional advantage of a much narrower linewidth of the principal line, thereby increasing the spatial resolution. The problem, however, is a very low X-ray flux at the specimen as compared to conventional sources. The spatial resolution is limited by the finite energy-dispersion characteristics of the monochromator. For Al-K_α back-diffracted from a quartz crystal, the spot size (d in millimeters) is given by

$$d = 5 \times 10^{-4} R \tag{43.3}$$

where R is the monochromator Rowland circle diameter in millimeters. In conventional instruments, R is of the order of 200 mm, giving the best spatial resolution to be 100 μm.

Another way of improving the spatial resolution is the use of appropriate electron optical aperturing with conventional X-ray sources. In this method, the photoelectrons from a uniformly irradiated specimen are imaged through the analyzer input lens onto an area-defining aperture. This aperture transmits only electrons that originate from a small selected area on the specimen. This technique is often called small-spot XPS. The spatial resolution is limited only by spherical aberration in the input lens and is given by

$$d = 180\alpha_0^3 \tag{43.4}$$

where d is in millimeters and α_0 is the semiangle of acceptance of the input lens in radians. In a typical geometry, α_0 may be reduced to as little as 0.05 radians (3°), resulting in a spatial resolution of 22 μm.

The radiation obtained from a synchrotron is of continuous nature and is very intense as compared to conventional X-ray sources. It is very useful for studying low-energy (less than 500 eV) core levels and the valence band.

Electron Energy Analyzers

The function of an electron energy analyzer is to disperse the photoelectrons emitted from the sample according to their energies. An analyzer may be either magnetic or electrostatic. Because electrons are influenced by stray magnetic fields (including the earth's magnetic field), it is essential to cancel these fields within the enclosed volume of the analyzer. The stray magnetic field cancellation is accomplished by using Mu metal shielding. Electrostatic analyzers are used in all commercial spectrometers today because of the relative ease of stray magnetic field cancellation (7).

The Cylindrical Mirror Analyzer (CMA)

The schematic of the CMA is shown in Fig. 43.3. It consists of two coaxial cylinders with a negative potential (V) applied to the outer cylinder (with radius r_2) and ground potential applied to the inner cylinder (with radius r_1). The sample and the detector are located along the common axis of

Figure 43.3 Cross-sectional view of the double-pass CMA. The exit slit from the first stage forms the entrance slit of the second stage. *V* represents the voltage applied to the outer cylinder. The inner cylinder is grounded.

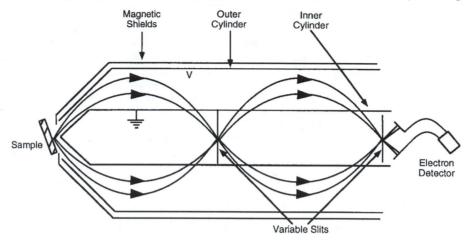

the cylinders. Electrons emitted from the sample at an angle α relative to the analyzer axis pass through the mesh-covered aperture in the inner cylinder. Only electrons that have a definite energy E_0 are deflected by the outer cylinder potential through the second mesh-covered aperture to a focus on the axis, where they pass into a detector. The focal condition is given by the relation

$$E_0 = \frac{KeV}{\ln(r_2/r_1)} \tag{43.5}$$

where e is the electron charge and K is a constant. A variety of entrance angles, α, can exist. However, in the special case where $\alpha = 42°18'$, the CMA becomes a second-order focusing instrument. At this focus the electrons from a very wide range of αs meet at the focal point. For this case the value of the constant $K = 1.31$. Under this condition, the distance between the source and the focus is 6.1 r_1. All the CMA parameters scale with the radius r_1 of the inner cylinder. Once this is fixed, other parameters follow automatically. The energy resolution of the CMA is written in terms of the base resolution ΔE_B, defined as

$$\frac{\Delta E_B}{E} = \frac{0.36w}{r_1} + 5.55(\Delta\alpha)^3 \tag{43.6}$$

where w is the slit width (equal at the entrance and exit) and $\Delta\alpha$ is the spread of the angle about α. The analyzer used for XPS consists of two CMAs in series, which is also known as a double-pass CMA.

Concentric Hemispherical Analyzer (CHA)

This is the most widely used analyzer for XPS. It consists of two hemispherical concentric shells of inner radius r_1 and outer radius r_2, as shown in Fig. 43.4. A potential difference, ΔV, is applied between the two surfaces such that the outer sphere is negative and the inner positive. Between the spheres there is an equipotential surface of radius r_0. The entrance and exit slits lie on a diameter and are centered on a distance r_0 from the center of curvature. If E is the kinetic energy of the electrons traveling the circular orbit of radius r_0, the energy difference, $e\Delta V$, between the hemispheres is given by

$$e\Delta V = E\left(\frac{r_2}{r_1} - \frac{r_1}{r_2}\right) \tag{43.7}$$

In the earlier form of spectrometers the midpotential of the analyzer was grounded. The scanning through the kinetic energy range was accomplished by varying the sample potential. In modern spectrometers the sample is grounded and the potential to the analyzer is varied for the scanning.

The base resolution is given by

$$\frac{\Delta E}{E} = \frac{w}{r_0} + \alpha^2 \tag{43.8}$$

where α is the entrance angle into the CHA and w is the width of the slit.

The CHA is used in two retarding modes. In one mode, the electrons are decelerated by a constant factor, or ratio from their initial kinetic energies. This mode is called the fixed retardation ratio (FRR), constant retard ratio (CRR), or constant $\Delta E/E$ mode. It has a constant relative resolution. In the other mode, the electrons are decelerated to a constant pass energy. This mode is called the fixed analyzer transmission (FAT), constant analyzer transmission (CAT), or constant

Figure 43.4 Cross-sectional view of the concentric hemispherical analyzer with input lens. The outer sphere has a negative voltage and the inner sphere has a positive voltage. The dashed lines indicate the trajectories followed by the photoelectrons. The central dashed line represents an equipotential surface. The entrance and the exit slits lie on a diameter and are centered at the mean radius from the center of curvature.

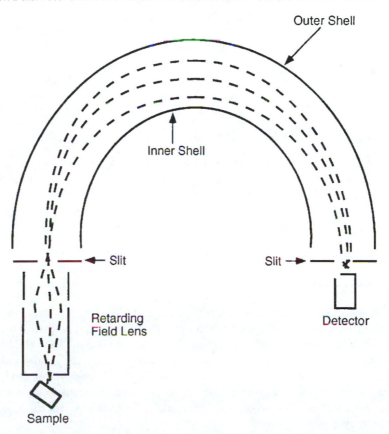

ΔE mode. It has constant absolute resolution. Because the absolute resolution in CAT mode is the same in all parts of the spectrum, the CAT mode is easier for quantification. The signal-to-noise ratio in the low-kinetic-energy part of the spectrum is poor, making it difficult for identification of the peaks in this region. On the other hand, in the CRR mode it is easier to identify small peaks in this part of the spectrum because of the better signal-to-noise ratio. However, it is difficult to quantify the results in this mode.

The sample must be placed at the slit of the analyzer. However, this is not convenient for the other treatments that might be required for the sample. The working space between the sample and the analyzer is therefore increased by the use of an electron lens. The lens system takes electrons from the sample and injects them into the analyzer. The working space thus created is advantageous because the X-ray source, the ion gun, and the electron gun can be brought closer to the sample.

Detector Systems

Having passed through the analyzer, the photoemitted electrons of a particular energy are spatially separated from electrons of different energies. Various detectors are used to detect these electrons.

Single-Channel Detector (SCD)

The detector commonly used in commercial instrumentation is a channel electron multiplier. It is an electrostatic device that uses a continuous dynode surface (a thin-film conductive layer on the inside of tubular channel). It requires only two electrical connections to establish the conditions for electron multiplication. The output of this detector consists of a series of pulses that are fed into a pulse amplifier/discriminator and then into a computer. The advantage of such a detector is that it can be exposed to air for a long time without damage. It counts electrons with a high efficiency, even at essentially zero kinetic energy, and the background is 0.1 count/sec or lower. The only drawback is that a high count rate (more than 10^6 counts/sec) will cause a saturation effect.

Multichannel Detector (MCD)

A multiple detection system can be added at the output of the analyzer. The system may be in the form of a few multiple, parallel, equivalent detector chains or position sensitive detectors spread across the whole of the analyzer output slit plane. Such an arrangement can be devised (8) in a number of ways: using phosphor screens and TV cameras, phosphor screens and charge-coupled devices, resistive anode networks, resistive filament, or discrete anodes.

Scanning the Energy Range

The energy range in the spectrum is scanned by altering the electrostatic or magnetic field and the counting rate is determined as a function of this field. The rate is proportional to the kinetic energy of the photoemitted electrons. The scanning is done with the help of a computer, which advances the voltage in discrete steps and stores the data, the total number of counts being recorded at each incremental step in a given channel. An alternative method is to sweep over the energy range with a sawtooth voltage in coincidence with the channel advance of a multichannel analyzer. In this method, the sweep is set on top of a bias voltage and the size of the sweep is adjusted for the region to be scanned.

Data Acquisition

The data in commercial instruments are acquired digitally and can be presented in either an analog or a digital mode. In the analog mode, the output of the pulse amplifier/discriminator is fed to a digital-to-analog converter, which drives the y-axis of an x–y recorder. The x-axis of the recorder is driven by a voltage that is proportional to the binding energy and synchronous with the scanning function of the analyzer. This results in the record of a spectrum of counting rate versus binding energy.

Digital systems use either a multichannel analyzer (MCA) or a computer to acquire and process the data. The data acquisition and processing in the MCA are hard-wired, whereas the computer uses software routines. In both configurations, it is possible to make multiple scans of a selected binding energy region to enhance the signal-to-noise ratio of the spectrum. Repetitive scanning is usually used to enhance weak signals. The software routines provided for data processing include smoothing, addition or subtraction of the spectra, background removal, normalization, presentation of derivative and integral curves, simultaneous display of spectra for comparison, peak deconvolution, and curve fitting.

What It Does

A typical spectrum of elemental manganese taken over a wide range of energy is shown in Fig. 43.5. This spectrum was recorded using a Mg anode and the analyzer was in the CAT mode with a pass energy of 89.45 eV. The origin of the energy scale is taken at the fermi level, E_F, of the solid. The separation of the peak from E_F is the binding energy, E_b, of that peak. The sample is electrically connected with the spectrometer so that their fermi energies are at the same level. In this figure, the spectrum is displayed as a plot of electron binding energy against the number of electrons in a fixed, small energy interval. The position on the kinetic energy scale equal to the photon energy minus the spectrometer work function corresponds to a binding energy of zero with reference to E_F. Therefore, the binding energy scale beginning at that point and increasing to the left is customarily used. As can be seen in the figure, the peaks ride on a background of photoelectrons that have lost energy while traversing to the surface. The various spectral features likely to be encountered in an XPS spectrum are discussed below.

Photoelectron Lines

These appear as the most intense, usually symmetrical, and typically the narrowest in the spectrum. However, lines obtained from metals are highly asymmetric due to the coupling with the conduction electrons. The binding energy values of all photoelectric lines for different elements are available in the literature.

Auger Lines

These lines appear as groups of lines. The commonly observed Auger series are *KLL*, *LMM*, *MNN*, and *NOO*. The nomenclature represents the initial and final vacancies in the Auger transition. Because Auger lines have kinetic energies independent of the ionizing radiation, they appear on the binding energy scale at different positions when the energy of the incident photons is changed. Tabulated values of the prominent Auger lines are available in the literature.

Figure 43.5 X-ray photoelectron spectrum of elemental manganese excited by Mg radiation. The spectrum has been recorded using a CHA in CAT mode with a pass energy of 89.45 eV.

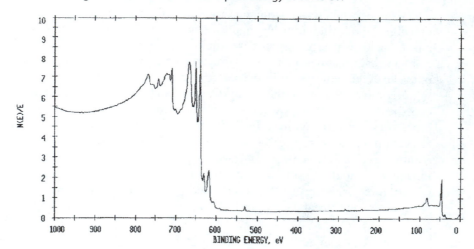

X-Ray Satellites and Ghosts

Because standard X-ray sources are not monochromatic, a series of lower-intensity lines, known as X-ray satellites, are also observed. For example, the principal line of Mg or Al anode is the un-resolved doublet $K_{\alpha1,2}$. Less-intense transitions (such as K_β or transitions in a multiply ionized atom) give rise to satellites.

The X-ray ghosts arise due to excitations from impurity elements in the X-ray source. These may be due to Mg impurity in the Al anode or vice versa, Cu from the anode base structure, or generation of Al-K_α X-rays when the secondary electrons strike the aluminum window. In dual-anode X-ray sources, misalignments inside the source can lead to crosstalk between filaments and anodes leading to the appearance of ghost peaks.

Shake-Up and Shake-Off Satellites

As the core electron leaves the atom due to photoemission, the valence electrons feel the increase of the nuclear charge. This perturbation gives rise to reorganization (called relaxation) of the valence electrons. This may involve excitation of one of the valence electrons to a higher unfilled level (shake-up). The energy required for this transition is taken from the primary photoelectron. As a result, a discrete structure is observed on the low-kinetic-energy (high-binding-energy) side of the photoelectron peaks. This structure is known as a shake-up satellite.

On the other hand, the valence electrons can be completely ionized, that is, excited to an un-bound continuum state. This process is called shake-off and leaves an ion with vacancies in both the core level and a valence level. In solids, this effect usually does not show up in the form of discrete peaks because they tend to fall into the energy region of inelastic secondary electrons.

Plasmon Loss Features

As the photoemitted electron passes through the solid it can excite one of the modes of collective oscillation of the sea of conduction electrons. These oscillations have frequencies characteristic of the material of the solid. Therefore, they need characteristic energies for excitation. The electron that has given up an amount of energy equal to one of these characteristic excitation energies is said to have undergone a plasmon loss.

Within the solid the loss is called bulk plasmon. If ω_b is the fundamental characteristic frequency of the bulk plasmon, then the plasmon energy loss is $\hbar\omega_b$. Because electrons that have suffered a plasmon loss in energy can suffer further losses of this kind in a sequential fashion, a series of losses, all equally spaced by $\hbar\omega_b$ but of decreasing intensity, will occur. Because the surface of a solid terminates abruptly, a rather localized type of collective oscillation can be excited, leading to surface plasmon. The fundamental frequency ω_s, characteristic of surface plasmon, is theoretically equal to $\omega_b/\sqrt{2}$ for free electron metals. The energy associated with this loss is called surface plasmon energy.

In the XPS spectrum from a solid a series of plasmon loss peaks is associated with the prominent peaks. The fundamental or first plasmon loss is always visible. Depending on the material and experimental conditions, several multiple plasmon losses of decreasing intensity may also become visible.

Analytical Information

Qualitative

Identification of Elements

Elements of an unknown sample can be identified easily by recording the XPS spectrum over a wide range (generally from 0 to 1000 eV). The binding energies of the intense lines present in the spectrum are then compared with the tabulated values.

Chemical Shift

The binding energies of the core levels of atoms are affected by their chemical environment. As a result, a shift in their positions is observed. This shift is called chemical shift and ranges from 0.1 to 10 eV. The shift arises from the variation of electrostatic screening experienced by core electrons as valence electrons are drawn toward or away from the atom of interest. The information of chemical states from the variations in binding energies of the photoelectron lines gives XPS a major advantage over other techniques. An extensive list of chemical shifts for various states of oxidation of many elements has been compiled (9, 10) and is available in the literature.

The higher resolution obtained by using synchrotron radiation has made it feasible to distinguish between the valence properties of the atoms at the surface and those in the bulk. These are known as surface core-level shifts. They are typically of the order of a few tenths of an eV and are associated with the reduced coordination of the atoms at the surface. The study of such effects with small catalyst particles will be an interesting field of future application.

If the sample is a nonconductor, then correction must be applied to measure the true binding energies. As the photoelectrons are removed from such a sample, positive charge is left behind. This charge tends to make the peaks appear at higher binding energy. There are various methods of correcting the charging effects on insulating samples.

The position of the C $1s$ line from adventitious hydrocarbon nearly always present in the sample is measured. This line appears at 284.6 eV. Any shift from this value can be taken as a measure of the static charge. A trace of gold is evaporated onto the sample after the spectra have been recorded. The Au 4*f* doublet is then recorded. The shift from the standard value of the binding energy then determines the static charge on the insulator. A low-energy electron flood gun is also used to stabilize the static charging of insulators. An electron beam of low energy (less than 1 eV) with respect to the vacuum chamber at ground potential is obtained from a source very close to the specimen.

Multiplet Splitting

Multiplet splitting (also called exchange or electrostatic splitting) of core-level peaks can occur when the system has unfilled shells containing unpaired electrons with a net spin $\neq 0$. In solids, there are transition metals with their unfilled *d* orbitals, and rare earths and actinides with their unfilled *f* orbitals. When an additional vacancy is created in the *s* level by photoionization, there can be a coupling between the unpaired electrons left behind following photoelectron ejection and the unpaired electrons in the originally incompletely filled shell. This leads to two different final states. Multiplet splitting can give a measurement of the bonding character of the unfilled valence shell.

Auger Chemical Shifts

The change in the chemical environment affects the Auger lines, causing a shift in them based on the same effects that result in chemical shifts for the XPS lines. For Auger transitions in which the final vacancies arise in the upper core levels, the chemical shift is of greater magnitude than the photoelectron chemical shift. The Auger chemical shift is very useful for identification of chemical states. If data for the various chemical states of an element are plotted, with the kinetic energy of the photoelectron line as the abscissa and that of the Auger line as the ordinate, a two-dimensional chemical state plot is obtained. A quantity called the Auger parameter, α, is defined as the difference between the kinetic energies of the Auger (A) and the photoelectron (P) lines:

$$\alpha = KE(A) - KE(P) = BE(P) - BE(A) \qquad (43.9)$$

which is also the difference in binding energy between the photoelectron and Auger lines. This difference can be accurately determined because static charge corrections cancel. When all the kinetic energies and binding energies are referenced to the Fermi level,

$$KE(P) = h\nu - BE(A) \qquad (43.10)$$

resulting in

$$KE(A) + BE(P) = \alpha + h\nu \qquad (43.11)$$

which indicates that the sum of the kinetic energy of the Auger line and the binding energy of the photoelectron line is equal to the Auger parameter plus the photon energy. Therefore, the plot of the Auger kinetic energy against the photoelectron binding energy then becomes independent of the energy of the photon. This plot is a powerful tool for identifying the chemical states of an element.

Auger Line Shape

During the Auger emission, if at least one final vacancy is in the valence levels, then the intensity distribution of the lines in the Auger series can vary greatly from one compound to another. The study of the Auger line shape helps in identification of the different chemical states.

Quantitative

XPS is being increasingly used to determine the chemical composition of the surface region of solids. For this it is necessary to quantify the XPS measurements. For a solid that is homogeneous to a depth of several electron mean free paths, the number of photoelectrons detected per second from an orbital of constituent atoms is given by

$$I = nf\sigma\phi\eta AT\lambda \qquad (43.12)$$

Here n is the number of atoms per cm^3 of the sample, f the flux of the X-ray photons impinging on the sample (in photons cm^{-2} s^{-1}), σ the photoelectric cross-section for the particular transition (in cm^2 per atom), ϕ the angular efficiency factor for the instrumental arrangement (angle between photon path and emitted photoelectron that is detected), η the efficiency of production of primary photoelectrons, A the area of the sample from which photoelectrons are detected, T the efficiency of detection of the photoelectrons emerging from the sample, and λ the mean free path of the photoelectrons in the sample. The values of σ are available in the literature (11).

For a given photoelectric transition in an atom, we define

$$S = \sigma\phi\eta AT\lambda \qquad (43.13)$$

as the atomic sensitivity factor. If transitions from two elements are considered, then

$$\frac{n_1}{n_2} = \frac{I_1/S_1}{I_2/S_2} \qquad (43.14)$$

This formula can be applied to a homogeneous material because the ratio S_1/S_2 is independent for all materials. This is valid because each of the ratios σ_1/σ_2 and λ_1/λ_2 remains nearly constant. Therefore, for any spectrometer a set of relative values of S can be developed for all of the elements.

To determine the atomic concentration of the elements on the surface, one must obtain a value for the peak intensity (peak area following background removal) of the most sensitive line of each element detected in the XPS spectrum. A fractional atomic concentration, C_A, of element A is then given by

$$C_A = \frac{I_A/S_A}{\sum (I_n/S_n)} \qquad (43.15)$$

where I_n is the measured peak intensity (peak area or peak height) for element n, S_n is the relative atomic sensitivity factor for that peak, and the summation is carried over all n. This equation is almost universally applicable. Values of S_n may be calculated from theory, or derived empirically by recording spectra from standard materials. In practice, the values of S_n used are from the published sets of experimentally derived relative atomic sensitivity factors.

The information on the depth of an element in the sample can be obtained either by destructive or nondestructive techniques.

Depth profiling is a destructive technique. In this technique the sample is eroded by ion sputtering. The sample is bombarded with ions accelerated in an ion gun to an energy in the range 1 to 4 keV. As these energetic ions strike the sample a small amount of energy is transferred to the surface atoms, which cause them to leave the sample. The ion beam is rastered on the surface for a known time to remove a uniform layer of the sample. Under controlled conditions the layer removed can be calculated. The residual surface is then analyzed by XPS, giving the depth distribution of different species in the sample. Inert gases (usually Ar) are commonly used as the ion sources.

Angle resolved XPS is a nondestructive technique that gives the distribution over the depth of the order of the mean free path of the electrons. The vertical depth sampled is given by

$$d = 3\lambda \sin\alpha \qquad\qquad (43.16)$$

where α is the take-off angle (that is, the angle between the sample surface and the axis of the analyzer). The information depth is a maximum for $\alpha = 90°$. If the sample consists of a uniform thin overlayer (o) on a substrate (s), then the angular variation of intensities is given by

$$I_s^d = I_s e^{-d/\lambda \sin\alpha} \qquad\qquad (43.17)$$

and for the overlayer

$$I_o^d = I_o(1 - e^{-d/\lambda \sin\alpha}) \qquad\qquad (43.18)$$

where λ is the appropriate inelastic mean free path for the observed photoelectron.

The experiment consists of tilting the sample at an angle α to the spectrometer. Because the instrument geometry also influences the spectra obtained at different α, it is customary to measure the relative values I_o/I_s to cancel the instrument effects. The technique is helpful in distinguishing the surface phenomena from the bulk properties, in the study of thin oxide layers on metals, and in polymer analysis.

Applications

Photoelectron spectroscopy is one of the exciting and rapidly growing branches in solid-state physics and chemistry. It is able to detect submonolayer surface concentrations. Chemical shifts give information about the atomic charge and the degree of ionicity. These phenomena are of particular interest with ions of transition metals, which have several oxidation states. XPS is also used to obtain information on the fermi energies and the density of states in the valence bands of metals, alloys, and semiconductors. The investigation of the electronic structure of binary alloys is at present an area of much experimental and theoretical activity. There are states that are localized at the surface and decay into the bulk and arise from the change of periodicity and the potential at the surface. These surface states have been discussed extensively for semiconductors but only recently have been studied by XPS. Intrinsic surface states of semiconductors are of great practical importance in semiconductor technology and therefore much experimental effort has been expended on this field. An extension of the important application of photoemission to surfaces lies in the detection of chemisorption levels. These are energy states that are created at the surface by the formation of chemisorption bonds between orbitals of the solids and the adsorbed particles. Knowledge of the energies involved is of vital importance in understanding the nature of chemisorption. XPS can be considered a nondestructive technique because the X-ray beam is relatively harmless to most materials. Therefore, the ability to obtain detailed chemical information from plastic and organic surfaces is an unparalleled capability of XPS.

One of the most important characteristics of XPS not shared by all other techniques is its ability to perform analysis at moderate vacuum levels. The chemical shifts observed can be used to gain inorganic and organic structure information. The presence of satellite structure can be used as a diagnostic tool to determine the magnetic characteristics of transition metal compounds. Through a repetitive sequential processing of XPS measurements followed by ion etching, depth-profiling information can be obtained. By tilting the sample at an angle to the spectrometer, it is possible to enhance the surface sensitivity of XPS and thus help distinguish surface phenomena from underlying bulk phenomena. This angular resolved spectroscopy is applicable to the study of very thin layers such as oxides on metals. XPS finds excellent industrial applications in the development of commercially important catalysts. XPS is expected to find increasing applications in many areas of science and technology requiring detailed information on elemental identification, surface composition, oxidation states, and chemical bonding.

Nuts and Bolts

Relative Costs

Vendors	Model and Description	Accessories	Costs
PHI	Model 1600 XPS system	NDA* or MS SCA SCD or MCD IG Input lens for the analyzer in both the constant retarding ratio and the fixed analyzer transmission mode Small area control option (minimum area analyzed: 75 μm diameter) Sample manipulator (single or multiple) with x-, y-, and z-translations, and tilt	$
PHI	Model 5600 Multitechnique system XPS, ISS, SIMS, AES	NDA* or MS SCA, MCD, IG Small area control option (minimum area analyzed: 75 μm diameter) Sample manipulator (single or multiple) with x-, y-, and z-translations, and tilt	$$
Scienta	ESCA 200	MS SCA (Mean diameter 200 mm) Position-sensitive detector Electrostatic lens system provides minimum area analysis region: 15 μm × 15 μm X–E and X–Y–E mapping Five-axis specimen manipulator (x, y, z, rotation, and tilt) Flood gun for charge neutralization	$$
	ESCA 310	Rotating anode as X-ray source with monochromator assembly SCA (mean diameter 300 mm) Position-sensitive detector Small-spot ESCA provides minimum area analysis region: 15 μm x 15 μm X–E and X–Y–E mapping Five-axis specimen manipulator (x, y, z, rotation, and tilt) Flood gun for charge neutralization	$$$
Fisons/VG Scientific	ESCALAB 220i/220i-XL Multitechnique instrument	NDA or MS SCA New lens system does away with sample scanning as well as slit scanning and provides a spatial resolution < 5 μm Four-axis specimen manipulator Flood gun for charge neutralization	$ $$ (with options)

*Options available are Mg/Al, Al/Al, Mg/Mg, Ag/Mg, Zr/Mg, Au/Mg, Si/Mg, and Cu/Mg.
$ = 300 to 500 k$, $$ = 500 to 800 k$, $$$ =>1,000 k$.
NDA = nonmonochromatic dual anode, MS = monochromated Al source, SCA = spherical capacitor analyzer, SCD = single-channel detector, MCD = multi-channel detector, CMA = cylindrical mirror analyzer, IG = ion gun (for sputter etching).

Vendors for Instruments and Accessories

Physical Electronics
6509 Flying Cloud Dr.
Eden Prairie, MN 55344
phone: 612-828-6100
fax: 612-828-6322

Scienta Instrument AB
Seminariegatan 33 H
Uppsala, Sweden 752 28
phone: 46 18 500160
fax: 46 18 543638
Internet: http://www.scienta.se

XELON Instrument Sales (U.S. sales representative for Scienta)
P.O. Box 311
Short Hills, NJ 07078
phone: 201-564-8833
fax: 201-564-8657

VG Scientific
The Birches Industrial Estate
Imberhorne Lane
East Grinstead
West Sussex, United Kingdom RH19 1UB
phone: 44(0) 1342 327211
fax: 44(0) 1342 324613
Internet: http://www.fisonsurf.co.uk/scinew/welcome.htm

Required Level of Training

The operation of the instrument requires a good knowledge of vacuum pumps. An understanding of the computers and familiarity with the software are essential because the instruments are computer-controlled. Extensive training is required for each kind of instrument because these instruments are designed differently. The design of the experiment for the sample to be analyzed requires knowledge of the sample chemistry. A graduate-level knowledge of spectroscopy, atomic physics, solid-state physics, and chemistry is required for understanding and interpreting the spectra.

Service and Maintenance

Routine adjustment and calibration of the energy analyzer are needed. The vacuum pumps need maintenance. Care must be taken to load only samples that are vacuum-compatible, that is, ones that do not outgas in vacuum. The detection of malfunctioning electronic components is facilitated by the plug-in cards. The detector is expensive and cumbersome to replace. The X-ray filaments and the Al window must be replaced often. Due to the design of the modules, most of these replacements must be done by trained personnel. The electrooptics and the ion gun focusing components are not normally aligned on-site. They need regular attention to ensure proper working of the instrument. Most companies offer service contracts but they are expensive.

Suggested Readings

BRIGGS, D., AND M. P. SEAH, EDS., *Practical Surface Analysis by Auger and X-ray Photoelectron Spectroscopy*. New York: Wiley, 1983.

CZANDERNA, A. W., in S. P. Wolsky and A. W. Czanderna, eds., *Methods of Surface Analysis*. Amsterdam: Elsevier, 1988.

ERTL, G., AND J. KUPPERS, *Low Energy Electrons and Surface Chemistry*. Deerfield Beach, FL: VCH, 1985.

McGUIRE, G. E., AND OTHERS, "Surface Characterization," *Analytical Chemistry*, 65 (1995), 199R.

MOULDER, J. F., AND OTHERS, in J. Chastian, ed., *Handbook of X-Ray Photoelectron Spectroscopy*. Eden Prairie, MN: Physical Electronics Publishing, 1992.

References

1. K. Siegbahn and others, *ESCA-Atomic, Molecular and Solid State Structure Studied by Means of Electron Spectroscopy* (Uppsala, Sweden: Almquist and Wicksell, 1967).

2. K. Siegbahn and others, *ESCA Applied to Free Molecules* (Amsterdam: North Holland, 1969).

3. K. Larsson and others, *Acta Chem. Scand.*, 20 (1966), 2880.

4. *Surface Science Spectra*, published by American Institute of Physics for the American Vacuum Society. (MCNC, Center for Microelectronics, 3021 Cornwallis Road, Caller Box 13994, Research Triangle Park, N.C., 27709 USA.)

5. P. W. Palmberg, *J. Vac. Sci. Technol. A*, 12 (1994), 946.

6. M. H. Hablanian, *J. Vac. Sci. Technol. A*, 12 (1994), 897; P. A. Redhead, *ibid.*, 904; K. M. Welch, *ibid.*, 915.

7. M. P. Seah, in J. M. Walls, ed., *Methods of Surface Analysis* (Cambridge: Cambridge University Press, 1989), p. 57.

8. P. J. Hicks and others, *J. Phys. E: Scientific Instruments*, 13 (1980), 713.

9. J. F. Moulder and others, in J. Chastian, ed., *Handbook of X-Ray Photoelectron Spectroscopy* (Eden Prairie, MN: Perkin-Elmer Corporation, Physical Electronics Division, 1992).

10. D. Briggs and M. P. Seah, eds., *Practical Surface Analysis by Auger and X-Ray Photoelectron Spectroscopy* (New York: Wiley, 1983).

11. J. H. Scofield, *J. Elect. Spec. Rel. Phenom.*, 8 (1976), 129.

Chapter 44

Secondary Ion Mass Spectrometry

Mark A. Ray, Edward A. Hirsch,
and Gary E. McGuire

MCNC
Electronic Technologies Division

Summary

General Uses

- Identification of elements on surfaces of materials
- Elemental surface distribution maps
- Identification of molecular species on surfaces
- Determination of the depth distribution of elements in specimens
- Trace analysis of elements in materials
- Determination of isotope ratios

Common Applications

- Identification of surface contaminants such as adsorbed or chemisorbed gases, cleaning residues, and surface segregated residues
- Determination of the distribution of elements associated with corrosion, erosion, or non-uniform growth of materials
- Identification of organic residues or polymer surfaces through molecular fragmentation

- In-depth distribution analysis of ion implanted or diffused dopants in semiconductors
- Analysis of impurities that affect the electrical properties of semiconductor devices at surfaces or buried interfaces or in solids

Samples

State

Almost any solid can be analyzed.

Amount

The total sample size (assuming a flat surface) typically varies from 1 mm^2 to 4 cm^2. The analyzed area is typically in the range of 4 to 60,000 μm^2. The typical sputter etched sample depth (per second) ranges from 1 to 30 Å.

Preparation

Most samples can be analyzed with minimal preparation. Because the analysis is surface-sensitive and is done in ultra-high vacuum (UHV), the sample should not be volatile and requires precaution in handling to prevent surface residues. Some samples may require cleaning in a solvent before loading into the analysis chamber.

Analysis Time

Estimated time to acquire a survey spectrum of a surface varies typically from 0.1 to 5 min for dynamic secondary ion mass spectrometry (SIMS) analyses. In time-of-flight (TOF) instruments the time required for a high-resolution (M/δM of about 10,000) static SIMS surface survey is typically 0.5 to 2 min depending on the mass range scanned. Static SIMS typically requires only a few minutes per specimen. Depth profiles of typical semiconductor junctions range from 5 to 20 min each, depending on the junction depth and erosion rate.

Limitations

General

- Sample should be solids or liquids or vapors condensed on a cold surface.
- Samples that out-gas or are volatile in high vacuum may require extra preparation.
- Insulating samples require charge neutralization during analysis.

Accuracy

The depth resolution of dynamic SIMS can be as low as 20 to 50 Å but increases with depth into the sample. This is due primarily to bombardment-induced mixing and surface topographical changes. Sputtered atoms originate from the top two or three atomic layers, thereby giving static

SIMS sensitivity to fractions of a monolayer. The ultimate spatial resolution is about 1 μm and depends heavily on the beam size. However, typical dynamic SIMS analysis is in the hundreds of microns range.

Sensitivity and Detection Limits

The dynamic SIMS sensitivity can range from 1 ppb (5E13 cm^{-3}) to 1 ppm (5E16 cm^{-3}) for bulk specimens and varies significantly depending on the element analyzed, the matrix, and the analysis conditions. The surface sensitivity is approximately 1 ppm.

Complementary or Related Techniques

- Auger electron spectroscopy (AES) provides additional information on elemental composition of surfaces and higher spatial distribution than SIMS. AES also provides some chemical oxidation state information and is more quantitative for surfaces or interfaces.
- X-ray photoelectron spectroscopy (XPS) identifies chemical oxidation states of surfaces. Like AES, XPS is less sensitive than SIMS but is more quantitative for surface and interface analysis.
- Rutherford backscattering spectrometry (RBS) provides elemental concentration data as a function of depth to a sensitivity limit of about 0.01 atomic %.

Introduction

Secondary ion mass spectrometry (SIMS) is the most sensitive and challenging of all surface analysis techniques. The intricacies of the technique are completely revealed in the listed review articles (1–10). SIMS is the mass analysis of positive and negative ions sputtered from the surface of a solid during ion bombardment. A spectrum of the intensity of the secondary ions versus mass is generated. The ion beam used to sputter the sample is called the primary ion beam and typically has a kinetic energy of 0.5 to 25 kilovolts. The sputter erosion process removes atomic and molecular ions and neutrals at a rate that depends primarily on the intensity, mass, energy, and angle of incidence of the primary species and the physical and chemical nature of the specimen. Typically, ions are produced at a rate of 1 ion per 10^3 to 10^6 sputtered atoms. The ions produced at the surface for analysis make up the secondary ion beam. The detected secondary ions originate in the top few monolayers of the surface.

When energetic ions impact a solid sample, a collision cascade is set up during which energy is transferred to the atoms in the solid via direct or indirect collisions, resulting in the rupture of chemical bonds. This results in ejection of both neutral particles and ions and is called sputtering. The sputter yield is defined as the total number of ions produced for each primary ion while the sputter rate is the erosion rate of the specimen. The sputtering rate increases if the primary beam current density (beam current divided by the rastered area) increases, the ion energy increases, or the angle of incidence increases. The yield of sputtered atoms increases until reaching a peak near 60 to 70 degrees relative to the surface normal (1). The sputter rate also is dependent on the mass of the primary ion, the mass of the sample, the specific composition of the sample, crystal orientation, and surface binding energy. Sputtered atoms and ions originate from the top two to three monolayers of a sample surface, making SIMS one of the most surface-sensitive analysis techniques (2).

Because only secondary ions are detected during SIMS analysis, it is important to look at the factors that contribute to ionization. The ionization efficiency can vary by orders of magnitude (10^4) for different chemical species and is a complex function of the electronic and vibrational states of the sputtered species and the solid surface (3). A number of parameters are operative during analysis and influence the secondary ion yield, some of which are used to enhance the SIMS sensitivity and others that are complicating factors. The two most widely used primary ion beam species, O_2^+ and Cs^+, were chosen to enhance the secondary ion yield and to increase the sensitivity of the technique. The use of any electropositive primary ion species such as Cs^+ enhances the negative secondary ion yield, whereas the use of any electronegative primary ion species such as O_2^+ enhances the positive secondary ion yield. In addition, changes in the composition of the specimen matrix can affect the ion yield. The presence of electronegative elements such as O, N, Cl, or F at the surface increases the positive secondary ion yield; conversely, the presence of electropositive elements at the surface increases the negative ion yield. The matrix effect on ion yield complicates SIMS analysis of specimens with changing composition at surfaces or interfaces where ion yields may change by several orders of magnitude. The most widely encountered matrix effect is the increased ion yield at the surface of specimens as a result of a native oxide. The surface oxide causes an increase in the ion yield for many species during a depth profile through the oxide region. A similar effect can be observed for a clean surface until equilibrium conditions are established between the implantation of the primary beam species, such as O_2^+, and sputter removal of the surface (4).

Dynamic SIMS is a surface analysis technique with a depth resolution of 20 to 50Å, determined by the depth from which secondary ions originate. The in-depth distribution of elements in the specimen is altered by the sputtering process through recoil and cascade mixing, radiation-enhanced diffusion, and surface topographical changes. The depth of the ion-bombardment–induced mixing zone is a function of the primary ion's mass, energy, and angle of incidence and the nature of the specimen. This determines the depth resolution. Depth resolution can be characterized by interface width, delta-layer broadening, or decay length. Interface width is defined as the depth interval over which the signal intensity drops from 84% to 16% of the maximum. When assuming normal statistics (which they rarely are), this interval is equal to two standard deviations. Broadening of a delta layer is the direct measure of the change in width of a thin layer that has been determined precisely by another technique such as transmission electron microscopy (TEM). Because depth resolution depends on the zone of mixing induced by the primary beam, factors that reduce its penetration depth result in sharper interfaces. For example, lower-energy (2 keV) Cs^+ or Xe^+ primary ions provide sharper interfaces than O_2^+.

How It Works

SIMS analyses are performed at either relatively high (dynamic SIMS) or very low (static SIMS) primary ion intensities. High-intensity primary ion beams result in high sputter rates, faster erosion of the sample surface, and consequently high secondary ion production. High secondary ion production provides the best detection limits. The enhanced detection limits result from the high sputter rate, not from a change in the ion yield. Some samples are too thin or shallow to allow the high sputter rates necessary to achieve high detection sensitivities. For example, low-energy ion implantation to form shallow junction depths in semiconductors may result in a maximum in the distribution of the ion-implanted species below 1000 Å. Rapid erosion of the specimen during SIMS analysis results in too few data points to accurately determine the implant profile. A slower sputter rate results in more data points but a lower ion production rate and consequently a lower detection limit.

Whereas dynamic SIMS is inherently destructive, static SIMS results in the removal of only the top few monolayers of the sample. Low-intensity primary ion beams are used to sputter the constituents of the surface of a solid. This results in low secondary ion production, with detection limits ranging from 0.1% to a few parts per million. Static SIMS supplies information on all elements, including hydrogen and, most importantly, molecular ion species. Static SIMS can be applied to all kinds of materials, but in comparison to dynamic SIMS it is most widely recognized for its enhanced ability to analyze organic and polymer surfaces. The molecular ions produced during ion sputtering are fragments of the original molecular structure and represent a fingerprint to identify the specimen's surface composition.

The basic components of a SIMS instrument are the following:

- UHV chamber
- Primary ion source
- Specimen holder
- Secondary ion extraction optics
- Mass spectrometer
- Secondary ion detector
- Data recording, processing, and output system

UHV Environment

Surface analysis necessitates the use of a UHV environment because the equivalent of one monolayer of gas impinges on a surface every second in a vacuum of 10^{-4} pascal. Because static SIMS analyzes only the outer few monolayers of the surface, it is generally performed under UHV conditions of 10^{-6} pascal or less to minimize contamination of the surface. The vacuum requirements for dynamic SIMS are less stringent because typical sample erosion rates are 10 to 50 monolayers per second.

Primary Ion Source

Most SIMS systems are equipped with two ion sources: one to enhance positive secondary ion yields (typically a duoplasmatron used with O_2) and a second to enhance negative secondary ion yields (typically a Cs ion source). Positive or negative ions of oxygen or inert gases are extracted from the plasma created in the duoplasmatron. The ions are accelerated from 1 to 13 keV and then focused onto the specimens by means of electrostatic lenses in the ion column. Duoplasmatron ion sources have been developed with spot sizes down to 0.5 μm, but most analytical data are taken with beam diameters ranging from 30 to 75 μm.

To enhance the yield of negative secondary ions, a Cs^+ ion gun is typically used. A reservoir of Cs is heated to generate Cs atoms. The atoms diffuse through a glass frit and a porous W plug that is kept at a high voltage. At a critical temperature (for a given bias and evaporation rate), some of the Cs atoms undergo a surface ionization process at the W surface. These ions are extracted, accelerated, and focused down the ion column. The ions may be focused to a spot as small as 0.15 μm, but more typically beam diameters from 30 to 75 μm are used. The ion energies used for the Cs ion source are the same as for the duoplasmatron.

A primary ion mass filter is sometimes used on both the duoplasmatron and the Cs ion sources. It removes the interfering primary ion species (such as doubly charged ions), neutrals, and impurities. In systems without a primary ion filter, a 2° bend is placed in the ion path through the column to remove neutrals from the beam.

Occasionally other ion sources, such as liquid metal ion sources, are used. The metal used in a liquid metal ion gun (LMIG) is most often Ga. Ions are produced via a field emission from a tip that is formed from a liquid Ga surface under a high electrostatic field. Some instruments use a solid tip coated with Ga during the tip formation process. The ion beams are operated typically from 10 to 30 keV.

Specimen Holder

Ion microprobes traditionally maintain the sample stage at ground potential, so the specimen holders are shaped to accommodate the geometry of the samples of interest. Typically, sample pieces up to 1 cm^3 may be examined with holders provided by the instrument manufacturer. There is actually no limit to the sample size for an ion microprobe. However, because microelectronic applications dominate the SIMS field, instruments usually accommodate 1.5 cm × 1.5 cm cleaved pieces of a Si wafer.

In an ion microscope, the sample surface is an integral part of the ion extraction and focusing. It is therefore elevated to a high potential and must be kept coplanar with the immersion lens. The problem is that a sample that is not flat must be made to fit an extraction system that works best for flat samples. Therefore, custom alterations are typically made for irregularly shaped samples to fit a regularly shaped sample holder.

Secondary Ion Extraction Optics

The purpose of the secondary ion optics is to transport as many ions from the sample surface to the spectrometer as possible. This is usually accomplished by a series of electrostatic lenses and apertures. The lenses may use either slit apertures, round apertures, or a combination of both. A round single-aperture lens geometry produces a focal length, f, given by

$$f = \frac{V}{E_2 - E_1} \qquad (44.1)$$

where V is the potential at the circular electrode, and E_1 and E_2 are the axial field strengths before and after the circular lens aperture (5). When $E_2 < E_1$ the lens is diverging and when $E_2 > E_1$ the lens is converging. Double (or multiple) immersion lenses are also used that provide electrodes at increasing potentials to provide a constantly converging lens system. The einzel lens may also be used in the secondary ion column. This is a three-electrode lens with the entrance and exit sides at the same potential. The relationship for the focal length of the immersion and einzel lenses is more complicated than for the simple-single aperture lens, but analytical relations do exist (5).

An energy filter is also located in the secondary ion column before the entrance to the mass spectrometer. The energy filter is a spherical capacitor and the voltages are fixed ratios of the desired acceleration voltage. Ions with greater than desired energies collide with the outer plate while ions with lower energies collide with the inner plate. For quadrupole-based SIMS instruments, the energy window of the energy filter is typically 4% of the acceleration voltage. Thus, the energy window can vary reproducibly from 2 to 16 eV. For ion microscopes, the energy filter is used in combination with a continuously adjustable energy slit that provides a 130-eV energy window. The difficulty is that it is not always easy to know the exact energy window given the combination of an energy analyzer with a manually operated energy slit. The kinetic energy distribution of molecular ions is narrower than for atomic ions. For example, Si^+ ion energy distributions are at least twice as broad (measured by the full width at half maximum intensity) as the Si_2^+ molecular ion energy distribution.

Mass Spectrometer

Three types of mass analyzers dominate in SIMS analysis: double-focusing magnetic sector instruments, quadrupole mass filters, and TOF mass spectrometers. Each of these mass spectrometers has unique features that make them attractive for SIMS analyses. The choice of mass analyzer type depends on such factors as mass resolution and range, transmission efficiency, imaging capability, resolution, and cost. It also depends on the type of SIMS analysis needed (dynamic or static SIMS).

Quadrupole secondary ion analysis systems contain an extraction lens (or lenses), beam-defining apertures, an energy filter, a quadrupole mass filter, and an electron multiplier. The sputtered ions are accelerated and focused into the energy filter. The energy filter allows the transmission of only a narrow band of the extracted ion energy distribution. For example, at an acceleration potential of 200 V, the energy window is typically ±4 V. Ions that make it through are refocused into the quadrupole mass filter. The quadrupole mass filter is made up of four evenly spaced rods equal in length. The diameter of the rods determines the ultimate resolution of the mass filter ($M/\delta M$). Only ions with a particular mass-to-charge (M/Z) ratio pass through the quadrupole; all other ions collide with the rods of the quadrupole. Mass filtering is accomplished by applying and varying the ratio of radiofrequency (RF) and direct current (DC) voltages to the rods. The opposite quadrupole rods are electrically connected. One pair is biased positive and the other is biased negative. The RF voltages are applied between the sets of rods. The result is that positive electrodes act as a high-pass mass filter, the negative electrodes act as a low-pass mass filter, and the combination of the two acts as a band pass filter. The ions that pass through the mass filter now strike an electron multiplier. The surface of the electron multiplier is coated with a material that has a high secondary electron yield. When ions strike the front of the electron multiplier, secondary electrons are produced that undergo a series of collisions with the walls of the multiplier, each collision increasing the number of electrons because of the high potential across the multiplier. These electrons are counted and sent to the computer system for further tabulation and analysis. No direct ion-imaging system is possible but ion mapping of the surfaces can be achieved. Three-dimensional image depth profiles can also be attained by mapping during depth profiling. The mass range of most quadrupoles is 255 amu, with some going as high as 1024 amu. The mass resolving power (MRP, defined as $M/\delta M$) is typically no better than 250. The transmission efficiency is 0.5 to 10%. The major advantages to these systems are as follows:

- Rapid switching between masses and from positive to negative ion detection
- More flexibility to establish the appropriate conditions for the analysis of difficult samples such as insulating or polymeric specimens
- Simpler and cheaper to operate (good cost-to-performance ratio)

The double-focusing magnetic sector instruments use a magnetic field to separate the charge-to-mass ratio and achieve mass analysis. These instruments are typically ion microscopes that keep the sample holder at a high potential. The sputtered ions gain approximately an electron volt per micron as they are accelerated toward the immersion lens. Thus, the ions are focused rapidly and their relative ejection point is maintained throughout the travel in the secondary ion column. The secondary ion column contains extraction (immersion) lenses, beam and field defining apertures, an energy filter, a magnetic sector mass filter, an electron multiplier or a faraday cup (for ion detection), and a dual-channel plate or resistive anode for image detection. A drawback of magnetic sector instruments has been the slow acquisition speeds required because of the magnetic field saturation. However, recent instruments have demonstrated excellent throughput and this no longer seems to be of any concern. Direct ion imaging of the surface is possible with a lateral resolution of 5000 Å. Currently, ion microprobes achieve the highest lateral resolution, which approaches 1500 Å. High spatial resolution is achieved in the ion microprobe by compromising the mass resolution and image area or analysis time. Imaging in the ion microscope is fast and performed at high mass resolution but is adversely affected by sample topography of a few micrometers. Three-dimensional image depth profiles are also possible. The mass range for this system extends to 280 amu at 10 keV and 560 amu at 5 keV. The MRP extends to 25,000 but a more practical limit is 10,000. The transmission efficiency is 35% at 300 MRP but drops to 1% at 8000 MRP. The major advantages of these systems are as follows:

- Direct stigmatic ion imaging capability
- High sensitivity over large mass range
- High mass resolution, but a simultaneous loss of sensitivity

In both the magnetic sector and quadrupole mass spectrometers, secondary ions are generated continuously by constant primary ion beam bombardment. Because only a single mass is analyzed at a time, most of the secondary ions that are generated go undetected. In a TOF instrument, the sample is bombarded by a pulsed primary ion beam (typically Cs^+ or Ga^+ ions in the nanosecond range). The secondary ions are accelerated and mass separated in a TOF analyzer. The TOF analyzer contains lenses and apertures near the sample surface, followed by the drift tube. The drift tube most often used is a reflectron with a two-stage electrostatic field gradient. The ions enter the tube and traverse a fixed path (typically 1 m) toward the back of the drift tube. The first gradient is a decelerating stage between the first and second grids, followed by a retarding stage between the second and third grids. The trajectory of the ions is reversed in the second field gradient and focused back onto the channel plate detector. Ions leave the tube at approximately uniform energies, thus allowing lighter ions to arrive at the detector before heavier ions based on the simple relation $KE = 1/2 \, mV^2$, where KE is the kinetic energy, m is mass, and V is velocity. The time of arrival of the different mass ions gives rise to a complete spectrum. All the secondary ion species can be analyzed very rapidly. The mass resolution is essentially constant (MRP approximately 10,000) as a function of mass but does decrease slightly for lower masses (masses ≤ 25 amu). The mass range for this system is almost infinite. Large-molecular-weight ions with masses greater than 10,000 amu have been analyzed. Analysis of all the mass fragments gives a powerful measure of the surface composition. The transmission is in the 10 to 20% range and is essentially the same for any mass resolution. Depth profiling may be accomplished with these systems by alternating between two ion beams: one for profiling and another for TOF analysis. Although two-beam systems have been designed for dynamic SIMS analyses, TOF instruments are most often used as static SIMS tools. Imaging is possible using the TOF SIMS with a dual-channel plate detector. The major advantages of TOF SIMS are as follows:

- High and nearly constant sensitivity for all masses
- High mass resolution without loss of sensitivity
- Almost simultaneous collection and analysis of all ions (very efficient for static SIMS analysis)

A summary of some of these SIMS parameters is given in Table 44.1. Some of the most commonly available instruments are listed.

Table 44.1 Operational parameters available on commercial SIMS instruments.

Mass Spectrometer	Mass Range (AMU)	Mass Resolution (M/δM)	Transmission Efficiency (%)	Imaging Mode	Image Resolution (Å)	Manufacturers
Double-focusing magnetic sector	0–500	10,000	0.5–35 (from high MRP to low MRP)	Ion microscope	5,000–10,000	Cameca SA
Quadrupole	0–1,000	~250	0.5–10 (from high MRP to low MRP)	Ion microprobe	500 (LMIG)	Physical Electronics VG Microtech
Time of flight (TOF)	0–10,000	10,000		Ion microscope	10,000	Physical Electronics
Time of flight (TOF)	0–10,000	10,000	10–20	Ion microprobe	~1,000 (LMIG)	Physical Electronics Cameca Ion TOF

LMIG = Liquid metal ion gun.

What It Does

Mass Spectra

A typical mass spectrum of elemental silicon taken over a range of 0 to 100 amu is shown in Fig. 44.1. This spectrum was recorded using a 5-keV O_2^+ primary beam and a quadrupole mass spectrometer on a Perkin-Elmer PHI 6300 instrument. The spectral intensities are recorded on a log scale due to the large dynamic range of the data. The relative intensities of the many secondary ions that contain silicon may be observed. These include species made up of singly charged matrix ions, molecular ions consisting of matrix atoms, and matrix atoms plus primary ion beam species. This abundance of species in the spectrum adds to the information content available but may result in frequent mass interferences. At this juncture, isotopic abundance or high mass resolution is necessary to determine the exact species present.

Static SIMS analysis of the surface produces a mass spectrum of the topmost surface layer. When a primary ion strikes the sample, it creates a damaged region. The cross-sectional area of the damaged region can be obtained from compendia of ion damage and projected ranges or from ion bombardment simulations, such as transport of ions in materials (TRIM) (11). The probability of the damage regions overlapping can be calculated by the following relation:

$$P = \frac{\Sigma \sigma}{A} \tag{44.2}$$

where P is the probability of cross-sectional overlap, A is the ion-irradiated area, and σ is the damage cross-sectional area for an individual ion. If $P \ll 1$, then the mass spectral information comes from a surface that is relatively unaffected by ion bombardment damage. The static limit is typically 10^{10}

Figure 44.1 A dynamic SIMS survey of a clean Si surface. The Si atomic and molecular species are evident along with the species formed from primary ion bombardment with a 5-keV O_2^+ ion beam.

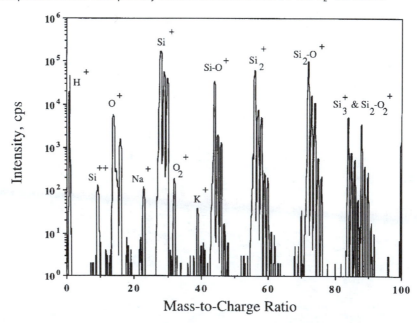

to 10^{12} ions/cm^2. This is most important for monolayer and submonolayer analyses, surface reaction and catalysis studies, and identification and analysis of organic surface species. TOF analysis is especially suited for, and most often used in, static SIMS analyses. An example of a high-mass-resolution static SIMS analysis of a Si surface is shown in Fig. 44.2. This high-mass-resolution, high-sensitivity survey was obtained in less than 1 min of acquisition time. The results also point out the extensive coverage of adventitious carbon on the surface, as evidenced by the $C_2H_5^+$ intensity.

Bulk Analysis

When a sample is sputter etched rapidly, the number of secondary ions generated is higher. A mass spectrum under these conditions provides the most sensitive analysis (10^{14} to 10^{16} atoms/cm^3) for bulk specimens. The mass spectrum may be used to identify unknown or suspected components. The volume of material removed is relatively small. However, if the sample is a thin layer, the sample thickness, the sputter rate, and the time required to obtain the data must be considered. This analysis is especially suited for trace analysis and isotopic abundance using dynamic SIMS instruments.

Depth Profile

Because the sputter process is inherently destructive, the specimen surface is gradually eroded away under continuing ion bombardment. The secondary ions detected at any time are those coming from the top few angstroms of the exposed surface. If one or more masses are monitored as a

Figure 44.2 Static SIMS survey of Si surface using a TOF SIMS system. The survey was acquired at high MRP ($M/\delta M = 9060$) using a Ga$^+$ primary ion beam.

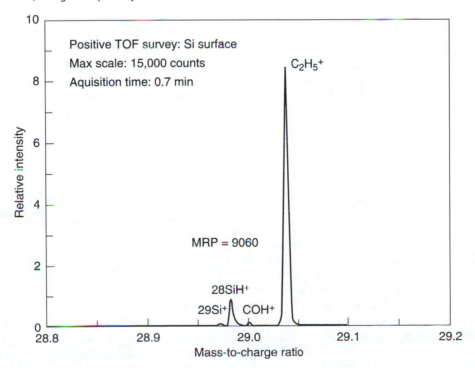

function of sputtering time, a depth distribution profile of each element is generated. More than one mass is monitored during the depth profile by sequentially switching the mass filter so that each mass is analyzed repetitively.

Accurate depth profiles require uniform bombardment of the sampled area. Extraneous signals may arise if the walls of the crater, generated as a result of the sputtering process, are sampled. This is avoided by gating the signal so that it is monitored only when it originates from the flat portion at the crater bottom. This is accomplished by rastering the primary beam over an area larger than the analyzed area. The area of analysis is defined by the use of an aperture or by electronic gating. Occasionally, extraneous secondary ions originate due to collisions with the secondary ion extraction lens, sample mount, or other surfaces in proximity to the specimen.

Because the sputter rate depends on the material, it is necessary to calibrate the sputter rate by measuring the depth of the crater following SIMS analysis or through the use of an acceptable standard. For multilayer samples, the individual sputter rates for each layer must be determined before an accurate measure of the true depth can be obtained. Measurement of the crater depth is often accomplished using a stylus technique such as a profilometer. For thin samples, the etch rate must be tailored to the layer thickness.

Once the depth is calibrated, a suitable standard for the unknown species must also be analyzed. SIMS is most often used as a tool for providing the impurity concentration as a function of sample depth. Ion implanted samples of the same matrix are typically used for concentration calibration standards. A sample of a constant known impurity level can also be used for concentration calibration.

The true in-depth distribution of elements in a matrix can be altered by the sputtering process. Ion beam mixing due to recoil and cascade effects and radiation-enhanced diffusion result in measured in-depth distributions greater than the true values. Surface topographical changes as a result of ion bombardment also alter the depth distribution of the analyzed species.

Ion Images

SIMS can also be used in an imaging mode analogous to scanning electron microscopy (SEM), where the secondary ions are imaged as opposed to secondary electrons. By mass analyzing the secondary ion (element, isotope, or molecular species) species-specific images can be generated. Several species-specific images can be generated from the same specimen and then compared to identify any nonuniform distribution of constituents associated with corrosion, segregation, growth, or defects. Ion images can be obtained during depth profiling to provide a three-dimensional distribution of elements within a solid.

Analytical Information

Qualitative

SIMS analysis is used to identify the elemental composition and distribution of molecular species present at surfaces, thin films, or bulk specimens. The conditions used for the analysis depend on the type of specimen and instrument used. Analysis of surfaces requires low-intensity ion beams so that the surface layer is not removed during the analysis. Surface analysis is accomplished in all

SIMS systems, but the TOF SIMS system is best suited, with its high transmission and parallel detection scheme. High-intensity ion beams are used to gain the highest sensitivity or best detection limits. These conditions are used for bulk specimens where the amount of material removed during the analysis is not critical. Thin films are often analyzed to determine the matrix composition as well as to determine the presence of trace impurities. With the large dynamic range of SIMS this is possible, however, the exact conditions for analysis may be determined by the thickness of the layer and the required detection limits. Very thin films may require a lower primary ion beam current for slower sputter rate, resulting in a loss of sensitivity for trace elements, yet the matrix composition will still be determined. Thin films of sufficient thickness, so that high sputter rates may be used, may be analyzed under the same conditions as bulk specimens.

Quantitative

Because of its extreme sensitivity, SIMS is most often used to discriminate between parts-per-million or parts-per-billion levels of atomic or molecular species in solids. SIMS is not often used to discriminate between a few atomic percent of those same species. A calibration curve method could be used to determine the concentration of a species A in a matrix B if sufficient samples of species A in matrix B could be found. If the concentration of species A in matrix C were needed, a new calibration curve would have to be determined. Calibration curves obtained by SIMS are generally continuous curves that increase monotonically from several atomic percent to 100 atomic percent. Nonetheless, specific effects such as isotopic effects (variations in the ion yield for specific isotopes), bombardment-induced effects (radiation-induced segregation or radiation-enhanced diffusion), or phase transitions can all lead to discontinuous calibration curves. This does not account for the errors induced by primary and secondary ion current measurements. The intensive work involved in calibrating a materials system or compound makes it an impractical task. A simple analysis by XPS would provide as accurate a result, with little or no need for calibration in many instances.

Most often, SIMS is used to determine the concentration of species A in matrix B as a function of depth into the sample. The ion intensity versus depth is converted to concentration versus depth using a sensitivity factor. The sensitivity factor is commonly obtained from an implanted standard where the implanted concentration varies by several decades as a function of depth. An example of a depth profile converted to atomic concentration is shown in Fig. 44.3. In this case, the raw data were obtained as a function of counts versus sputtering time. In its simplest form, the analysis follows these steps:

1. The sputtering time was converted to depth by measuring the crater depth (assuming a constant sputtering rate).

2. The concentration scale was obtained by determining the sensitivity factor, using the following relation:

$$\text{Sensitivity factor} = \frac{\text{Implanted dose (atoms/cm}^2) \times \text{Total matrix counts}}{\text{Total implant counts} \times \text{Depth (cm)}} \qquad (44.3)$$

3. Once the sensitivity factor was obtained, it was used to determine the concentration scale given the following relation:

$$\text{Atomic concentration} = \text{Sensitivity factor} \times \frac{\text{Implant counts}}{\text{Matrix counts}} \qquad (44.4)$$

A standard of a known constant impurity level can also be used to determine the sensitivity factor.

Figure 44.3 A SIMS depth profile of a $^{11}B^+$ implant (dose = 10^{14} atoms/cm^2) in Si converted to atomic concentration.

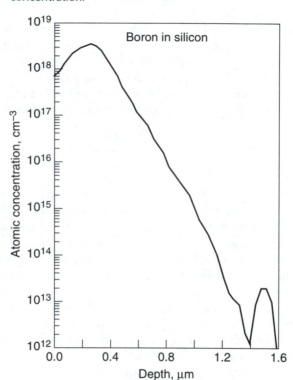

When using sensitivity factors, the assumption is made that the sensitivity factor for a species is independent of the concentration of that species in the desired matrix. For most practical analyses, this is of no concern. The standard used to calculate the sensitivity factor should be identical in composition and structure to the unknown being analyzed. Deviations in composition, for example, give rise to changes in secondary ion yields, which affect the measurement accuracy. It is unclear at what point compositional changes become dominant, but small changes in composition (1% or less) are, for the most part, irrelevant. The condition of analysis must also be the same for the standard and the unknown sample. For example, the angle of incidence of an O_2^+ ion beam must be maintained during analysis of the standard and the unknown because oxygen incorporation can vary with angle of incidence. Increased oxygen incorporation leads to oxygen enhancement of the secondary ion yield, which affects the sensitivity factor measurement (4). Also, sensitivity factors are not transferable between different SIMS instruments. The relative sensitivity factor (RSF) is defined more critically as

$$\text{RSF} = \frac{\phi C I_m t}{d\Sigma I_i - dI_b C} \tag{44.5}$$

where ϕ is the ion implantation fluence (atoms/cm^2), C is the number of data cycles, I_m is the matrix isotope ion intensity (counts/data cycle), d is the crater depth (cm), ΣI_i is the sum of isotope impurity ion counts, I_b is the background ion intensity (counts/data cycle), and t is the analysis time (sec/data cycle).

The RSF is unique to the instrument make and model. Comparisons from instrument to instrument show that there is considerable chance of error in the analysis (10). As a practical matter, all quantitative sensitivity factor analyses should be considered unique to each instrument and sample. This again points to the challenging nature of SIMS analysis. A complete and very pragmatic treatment of quantitative analysis by SIMS can be found in Reference 4.

Quantitative analysis of static SIMS or TOF SIMS data is of considerable interest. However, suitable surface standards must be formulated and tested before quantitative analyses are obtained. At the present time, few standards have been properly characterized.

Nuts and Bolts

Vendors for Instruments and Accessories

Cameca SA
103 Boulevard Saint Denis, BP6
92403 Courebevoie Cedex, France
phone: 33143 3430 60

Physical Electronics
6509 Flying Cloud Dr.
Eden Prairie, MN 55344
phone: 612-828-6100

VG Microtech
Bellbrook Business Park
Bolton Close, Uckfield
East Sussex TN22 1QZ, U.K.
phone: 44 1825 761 077

Required Level of Training

To competently operate a SIMS instrument requires knowledge of vacuum equipment, ion beams and ion/surface interactions, mass spectrometry, and computer-controlled instruments. Understanding the individual nuances of each sample and specific analytical technique is also required in order to coax the best possible data from each sample. Typically, SIMS analysts have taken graduate-level courses in analytical chemistry, physics, or materials science. Graduate-level courses in solid-state physics, analytical chemistry, spectroscopy, and materials science will prepare the analyst for many eventualities. However, the specific knowledge required for analyzing samples using a SIMS instrument is not taught in most graduate schools. It is most often acquired by patient examination of samples under the tutelage of someone skilled in the technique and by familiarity with the large body of SIMS literature.

The analyst/operator must know how far to push the SIMS instrument and what performance can be obtained. The knowledge required to optimize the instrument is acquired by experience. Each SIMS instrument is unique and requires special attention to various aspects of the analysis, such as charge neutralization or beam selection. Manufacturers offer training on the instruments they produce; this training is highly recommended for any potential analyst, regardless of skill level. However, optimization must still be done on each instrument to ascertain the limits of that instrument. Analysts must also keep up with the SIMS literature to keep abreast of changes in analysis techniques for their respective instruments and samples.

Service and Maintenance

There are several components of a SIMS system that require frequent service from routine use, such as ion guns, vacuum pumps, spectrometers, and electron sources. Servicing tips for these items are given below. There are also aspects of sample selection and SIMS instrument operation that influence service and maintenance. For example, sample selection is important for SIMS operation since samples may not be vacuum-compatible. The quality of the data from vacuum-sensitive samples may be degraded, or, in the extreme case, damage to the instrument may occur. Out-gassing of certain constituents from a vacuum-sensitive sample may result in damage to ion or electron guns and vacuum pumps. As a result, care must be taken during operation to ensure that the vacuum is acceptable and that the electron and ion guns are still operating properly. Oxygen flooding is a normal instrument operation that increases the required frequency of service. Oxygen flooding of the specimen during SIMS analysis (at levels $\geq 5 \times 10^{-7}$ torr) is a common procedure that yields beneficial results. However, the high pressure of a reactive gas affects the system components.

The state of the instrument can be monitored by maintaining a log book of operating conditions. Maximum ion current and ion current stability should be monitored daily; significant changes in the operating parameters of the ion gun may serve as early warning signs of potential malfunction. In a similar fashion, total electron current to the specimen and emission current fluctuations should be noted if an electron beam is used on a SIMS system. Pumps should be inspected on a weekly basis and pump fluids replenished when low or replaced when they become dirty. Calibration of the mass spectrometer should be monitored on a weekly basis and optimization performed if necessary (most TOF SIMS systems calibrate each spectrum using software).

Many SIMS instruments are operated so frequently that monitoring their operation without taxing an overburdened schedule is difficult. When this occurs, if mass resolution is not critical, a quick way to evaluate the operating conditions is to perform a depth profile analysis. This is accomplished by obtaining two ion-implanted Si samples as depth-profile references, one for positive-ion analysis and one for negative-ion analysis. These samples are routinely profiled under identical conditions and the profiles examined for similarity. The analysis takes only a few minutes. Deviations in the depth profiles from previously acquired spectra highlight problems in the instrument operation. Implant profiles taken years apart from properly maintained SIMS instruments look identical.

Ion Beams

Duoplasmatron ion guns used to generate oxygen ions are one of the highest maintenance items on SIMS systems. This type of ion gun requires maintenance every 3 to 6 months, depending on the hours of use. Maintenance requires cleaning or replacing the hollow cathode and intermediate electrode, as well as cleaning the anode. Most duoplasmatron ion sources can be maintained by the user after proper training. The ion column can become contaminated with sputtered metals from the ion source. The manufacturer's representative typically cleans the column.

Cesium (Cs) ion sources need replenishment of the Cs every couple of years, as well as cleaning of metals deposited on column parts similar to that required for duoplasmatron ion sources.

Characteristically, the Cs ion source consists of a hot filament at high voltage that causes the Cs metal to become molten, which when placed in a high extraction field becomes ionized. The filament needs replacement periodically, depending on use. Like the duoplasmatron ion column, the column of the Cs ion gun needs maintenance depending on the frequency of use. The Cs source should be maintained by qualified personnel, because Cs is a highly reactive metal that undergoes spontaneous combustion if exposed to air.

Electron bombardment ion guns use a hot tungsten (W) filament that ionizes an inert gas by passing the gas between a hot filament and an extraction field; this causes the electrons emitted from the hot filament to collide with the gas atoms and ionize them. Replacement of the filament assembly is required when it burns out or if the ceramic insulators become coated, causing electrical leakage to occur. As with all ion guns, maintenance of the column is required, although less frequently than the columns of duoplasmatron and Cs ion guns, because the electron bombardment ion guns are simpler.

Gallium (Ga) liquid metal ion guns (LMIG) are another option for SIMS systems and are used in imaging applications because of their very high spatial resolution (for ion sources). Maintenance involves coating the tip with fresh Ga approximately every 40 hours of operation. This is accomplished by heating the Ga reservoir located in the beam and allowing the molten Ga to wet the tip, following the manufacturer's recommended procedure. The Ga reservoir becomes depleted approximately every 500 to 2000 hours, depending on the size of the reservoir. The Ga reservoir is usually attached to the filament, so the tip and reservoir are replaced together once the Ga is depleted. The process is not difficult, but the proper procedure should be obtained from the manufacturer.

Vacuum Pumps

Ion pumps are excellent UHV pumps because they have no moving parts or lubricants, making them very clean and virtually maintenance free. Ion pumps last 5–10 years, depending on the base pressure at which they operate and the pressure at which they are first started on pumpdown from atmospheric pressure. Ion pumps fail due to coating of the ceramic insulators, shorts between the anode and cathode, and the release of excess gas back into the vacuum chamber. Signs that service is needed include excessive current being drawn and the inability to pump inert gases in order to maintain a stable pressure. Replacement of the pumping elements requires disassembly of the vacuum chamber.

Titanium (Ti) sublimation pumps are commonly used in conjunction with ion pumps. The principle component is a Ti filament, which is heated by passing current (45–55 Å) through it, causing it to sublime, coating the chamber walls and gettering the active gases present in the system. The filaments must be replaced often, depending on their frequency of use and the pressure at which they are operated. Guidelines for use of the Ti sublimation pump should be included with the system and typically recommend one cycle of the pump every 2 to 3 weeks when the base pressure of the system is below 1×10^{-9} Torr.

Turbomolecular pumps are commonly used in sample loading chambers, differentially-pumped ion guns, and on some main chambers when the system is frequently brought to atmospheric pressure. This type of pump requires higher maintenance than other types of pumps. A mechanical pump is required when a turbomolecular pump is employed. One risk is contamination of the system with oils that may backstream from the mechanical pump. The use of oil-free mechanical pumps to support the turbomolecular pump is one means of eliminating backstreaming; however, oil-free pumps are more expensive. The bearings of turbomolecular pumps require oiling twice a year for trouble-free maintenance. When they are employed on imaging systems, turbomolecular pumps induce vibrations that may degrade image quality. The oil in the mechanical pumps used should be changed twice per year, following the manufacturer's directions. Oil-free mechanical pumps use a diaphragm that needs replacement once or twice per year. When choosing a mechanical pump, it is important to understand the application required since the ultimate pressure achievable and the capacity of the oil-free pump greatly differs from its counterpart.

Cryo-pumps have various parts that require maintenance including a compressor, a cold head, and a charcoal trap. Cryo-pumps operate at very low temperature (usually liquid He) to condense the gases present in the system. Cryo-pumps are an excellent pump choice if excessive gas volumes are being pumped. They required low maintenance but have extensive vibration associated with a mechanical piston located in the cold head that compresses the gas. As with any mechanical item, the piston in the cryo-pump wears out. Every couple of years, the piston will need new rings and associated parts. Cryo-pumps need occasional regeneration to allow the frozen gases to escape from the charcoal trap. The frequency depends on the operating pressure; the higher the pressure the more frequent the regeneration interval. The regeneration procedure should be detailed in the pump manual. An adsorber associated with the compressor needs to be changed in accordance with the manufacturer's recommendations.

Spectrometers

There are two basic types of dynamic SIMS analyzers: magnetic sector and quadrupole. There are also two types of TOF systems: reflectron and magnetic sector. Each type of analyzer has advantages and disadvantages regarding maintenance. Problems for each type vary in their complexity to diagnose and repair. Usually, the most difficult challenge is diagnosing the problem. All analyzers use some type of multiplier that required periodic replacement. Some analyzers make use of grids and apertures that require replacement. Memory effects (the detection of a signal of a material that has been analyzed recently) occur on some SIMS systems and may be minimized by taking precautions. The proper operating procedures and system limitations are usually documented in the literature provided by the vendor.

General calibration of dynamic SIMS systems includes adjusting the spectrometer for proper mass calibration. Quadrupole spectrometers use an RF supply that needs to be adjusted for proper linearity. Magnetic sector systems are adjusted before each analysis, because of their sensitivity to current changes in the environment and the main power supply. The high mass resolution of the magnetic analyzers makes them even more sensitive. TOF spectrometers are usually software-calibrated for each spectrum, so no hardware calibration is required. Because of the high mass resolution of TOF systems, at least three peaks in the spectrum must be identified for proper calibration.

Electron Sources

On most modern SIMS systems, an electron source is present to help charge-compensate insulating samples. Electrons are typically extracted from a hot W filament and accelerated to the sample. The W filament usually can be replaced in the field, which may require the assistance of a manufacturer's representative. LaB_6 filaments are more expensive than W filaments but provide higher spatial resolution and longer life.

There are tradeoffs between the different types of SIMS systems. Maintenance and reliability are key features in the selection of the system. Determine if the manufacturer offers a warranty and service contract. Be sure to inquire about the nearest location of the service support provided by the vendor. How many engineers are located within an acceptable travel distance to your site? What spare parts are available and where are they located? Is there a guaranteed response time to your phone call, and how long will it take to send an engineer if the problem cannot be resolved over the phone?

References

1. K. Wittmaack, *Nuclear Instruments and Methods*, B2 (1984), 674.

2. J. W. Burnett and others, *Journal of Vacuum Science and Technology A*, 6 (1988), 2064.

3. J. A. McHugh, "Secondary Ion Mass Spectrometry," in A. W. Czanderna, ed., *Methods of Surface Analysis* (Amsterdam: Elsevier, 1975), p. 223.

4. R. G. Wilson, F. A. Stevie, and C. W. Magee, *Secondary Ion Mass Spectrometry* (New York: Wiley, 1989).

5. A. Benninghoven, F. G. Rudenauer, and H. W. Werner, *Secondary Ion Mass Spectrometry: Basic Concepts, Instrumental Aspects, Applications, and Trends* (New York: Wiley, 1987).

6. R. W. Odom, *Applied Spectroscopy Reviews*, 29, no. 1 (1994), 67–116.

7. D. Briggs, "Characterization of Surfaces," in C. Booth and C. Price, eds., *Comprehensive Polymer Science*, vol. 1 (Oxford: Pergamon, 1989).

8. A. Benninghoven, *Surface Science*, 299/300 (1994), 246–60.

9. P. K. Chu, *Materials Chemistry and Physics*, 38 (1994), 203–23.

10. D. S. Simons and others, in A. Benninghoven and others, eds., *Secondary Ion Mass Spectrometry: SIMS VII* (New York: Wiley, 1989), p. 111.

11. J . F. Ziegler, J. P. Biersack, and U. Littmark, *The Stopping and Range of Ions in Matter* (New York: Pergamon, 1985).

Polymer Analysis

Chapter 45

Introduction

Charles Smith
C & J Associates

Analysis of polymers and other large molecules offers unique challenges that are not experienced with analysis of smaller molecules. For one thing, commercially useful polymers often are not made up solely of the macromolecule itself; these samples also contain additives, fillers, residual monomers, or catalysts that affect the performance of the polymer. Also, chromatographic and spectroscopic techniques described in earlier chapters of this volume often necessitate solubilizing the sample. The size of polymeric molecules and reactions such as grafting and crosslinking tend to limit the solubility of many macromolecules. Some polymers are soluble only at elevated temperatures, so we must understand the effects of temperature on the stability of these macromolecules to ensure that any solutions prepared at elevated temperatures truly represent the original sample.

Unlike inorganic compounds and small organic molecules, polymers are generally not characterized by a single, unique molecular weight. Macromolecules generally have a molecular weight distribution or they are characterized by terms such as *weight-average molecular weight* or *number-average molecular weight*. Numerical values that define the molecular weight characteristics of a polymeric sample are obtained from a number of techniques, including size exclusion chromatography or low-angle laser light scattering experiments, which are described in this section.

Techniques such as viscometry, ultracentrifugation, osmometry, end-group analysis, and precipitation-liquid chromatography are also used to characterize the molecular weight of polymeric systems. Polymers, particularly those prepared using emulsion technology, also contain particles of different sizes. Consequently, particle size techniques such as light obstruction, discussed in this section, are an important part of polymer analysis.

In addition to the previously mentioned additives and fillers, the composition of macromolecules includes homopolymers prepared from a single monomer, copolymers prepared from two or more monomers, or blends of polymers. The conventional spectroscopic techniques (such as infrared, nuclear magnetic resonance, X ray, and mass spectrometry) detailed in earlier chapters of this book are used to define the composition of complex polymeric systems. Pyrolysis techniques discussed in this section offer another technology for analysis of polymers that is not normally used or needed for analysis of smaller molecules. These pyrolysis techniques are beneficial for polymer analysis because they are applicable to solid insoluble samples, which eliminates the need to solubilize the macromolecules, and they are based on the controlled degradation of the large molecules to smaller molecules, which are amenable to conventional chromatographic and spectroscopic analyses. Although not extensively discussed in this section, pyrolysis techniques are also applicable to characterization of biological and environmental systems.

In addition to pyrolysis technology, other thermal techniques such as differential scanning colorimetry (DSC), thermogravimetric analysis (TGA), and dynamic mechanical methods are invaluable for comprehensive characterization of polymeric systems. These and related thermal analysis methods are used to define the state of a polymer sample, including properties such as the glass transition temperature, melting point, and degree of crystallinity. In addition, thermal techniques are also used to study the thermal and oxidative stability of polymers.

Finally, the commercial success of polymeric materials depends on the performance of molded or extruded parts in an end-use application. Testing mechanical properties such as impact, tensile, creep, and elongation gives information related to the ultimate performance of a polymeric system in a particular application. The end-use performance of a polymeric system also depends on the molecular weight, composition, particle size, and thermal properties of the polymer. In combination, the analysis techniques presented in this section provide comprehensive characterization of many synthetic polymers. Each technique may measure a specific property of the polymer, but data from all, or at least many, of these techniques are required to fully characterize any polymeric system.

Chapter 46

Molecular Weight Determinations

David M. Meunier

The Dow Chemical Company

Summary

General Uses

- Determination of absolute and apparent molecular weight distribution (MWD) for polymers and biopolymers
- Determination of polymer concentration in a matrix
- Determination of copolymer and blend composition (multidetector methods)
- Determination of branching, in conjunction with low-angle laser light scattering or viscometry
- Sample cleanup
- Chromatographic cross-fractionation of copolymers (in conjunction with precipitation or adsorption chromatography)
- Separation of small molecules
- Preparative size exclusion chromatography (SEC) for isolating relatively large quantities of particular components

Common Applications

- Determination of MWD of organic and aqueous-soluble polymers
- Establishing molecular weight/property relationships
- Quality control
- Sample cleanup for determination of additives or oligomer distributions

Samples

State

Solid or liquid polymer samples can be analyzed.

Amount

For polydisperse samples, 25 to 100 mg is desirable. For monodisperse samples, 1 to 20 mg is desirable.

Preparation

Sample is dissolved in SEC eluent and filtered before analysis. Care is taken not to fractionate the sample on the filter membrane.

Analysis Time

Sample preparation consists of weighing the sample, adding a specific volume of solvent, and using light agitation to achieve dissolution. Dissolution can be the rate-limiting step for SEC analysis time, requiring on the order of 0.5 to 24 hr or more, depending on the sample. Because alteration of the property to be measured may occur, devices that induce high shear on the polymer sample (such as ultrasonic baths) should not be used to expedite sample dissolution. Additionally, crystalline polymers often require heating to achieve dissolution, but one should determine whether heating changes the sample MWD before applying heat.

A chromatogram can be obtained in 10 to 60 min depending on the number of columns used and the eluent flow rate. Data analysis (with a computer) requires only a few minutes per chromatogram.

Calibration with narrow MWD standards requires 1 to 2 hr depending on the number of standards used to establish the calibration. Typically, up to four narrow fraction standards can be analyzed in a single injection, provided there are sufficient molecular weight differences to achieve baseline resolution.

Limitations

General

- Only apparent molecular weights are determined unless molecular weight standards exist of the same composition and topology as the unknown, LALLS or viscometry is used, or the Mark–Houwink parameters of standards and sample are known.
- Suitable or compatible solvents must be found for characterization of difficult-to-dissolve polymers (such as nylon).
- SEC is an inherently low-resolution technique, but this is typically not a real limitation for broad-MWD synthetic polymers.

Accuracy

For samples where absolute molecular weight calibration is possible, we can readily achieve accuracy within ±5%.

Complementary or Related Techniques

For Molecular Weight Distributions

- Thermal field flow fractionation: For a given composition, polymer molecules are separated according to differences in the ordinary (Fick's) diffusion coefficient. Polymers of different composition are separated according to differences in the thermal diffusion coefficient.
- Mass spectrometry: Matrix-assisted laser desorption time-of-flight mass spectrometry has been applied to synthetic polymers and proteins of relatively low molecular weight (less than 150,000 Da).
- Ultracentrifugation: This tedious, older technique relies on the measurement of sedimentation coefficients, which can be used to calculate molecular weights. Although still used sparingly, this technique has been made obsolete by more current polymer characterization techniques.

For Molecular Weight Averages Only

- Static low-angle laser light scattering for absolute weight-average molecular weight
- Colligative properties for absolute number-average molecular weight (such as vapor phase osmometry)
- End-group analysis for absolute number-average molecular weight (such as titrations or nuclear magnetic resonance)
- Solution viscosity for measurement of viscosity-average molecular weight, provided one knows or can measure the Mark–Houwink coefficients
- Quasielastic light scattering (or photon correlation spectroscopy) to determine diffusion coefficients of polymers, which are related to polymer molecular weight

Introduction

Size exclusion chromatography (SEC) is an extremely important technique for characterizing macromolecules. SEC is a high-performance liquid chromatographic (HPLC) technique in which molecules are separated according to differences in hydrodynamic volume. This separation mechanism is made possible by the packing material in the column. The packing material is made of porous (rigid or semirigid) spherical particles (3 to 20 μm). The retention in SEC is governed by the partitioning (or exchange) of the macromolecular solute molecules between the mobile phase (the eluent flowing through the column) and the stagnant liquid phase that occupies the interior of the pores. The range of macromolecular sizes that can be separated with a given column depends on the size (or size distribution) of the pores. The resulting chromatogram in an SEC experiment thus represents a molecular size distribution.

The relationship between molecular size and molecular weight depends on the conformation of the dissolved solute molecules. However, for any solute conformation (such as random coil, rigid rod, or hard sphere), the molecular size increases with molecular weight. The rate at which the molecular size increases with molecular weight varies for the different possible solute conformations. Most synthetic polymer samples can be categorized as random coil solutes. As such, the relationship between molecular size (radius of gyration, R_g) and molecular weight is $R_g \propto M^{\alpha}$, where the exponent is a constant dependent on the solute composition, the temperature, and the solvent. Given that the molecular size is related to the molecular weight of the polymer, the molecular size distribution obtained in the SEC experiment can be converted into a molecular weight distribution (MWD) if we can establish a molecular weight versus retention volume calibration.

How It Works

The diagram in Fig. 46.1 illustrates the sequence of events that occur during an SEC experiment. Larger solutes elute from the column first; smaller solutes elute from the column later. Consider the analysis of a hypothetical mixture of large and small polymer solutes by SEC. The first frame in Fig. 46.1 shows the injection of the mixture as a narrow band at the head of the column. In the second frame, the development of size separation is illustrated. In this frame, the larger polymer solutes are clearly migrating through the column at a faster rate than the smaller solutes. The larger solutes are more excluded from the pores in the packing, and thus spend more time in the mobile phase than in the stagnant mobile phase contained within the pores. The smaller solutes can enter more of the pores containing the stagnant mobile phase. As a result, the smaller solutes spend more time in the column. In the third frame, the larger molecules elute from the column and are detected by a suitable detector, and in the last frame, the small molecules elute.

The sequence of events illustrated in Fig. 46.1 serves only to describe the mechanism of separation in SEC. Real samples of synthetic polymers usually contain a broad distribution of molecular sizes that cannot be separated into individual SEC peaks. Rather, an SEC chromatogram of a given polymer sample is typically a broad peak, with the earlier-eluting species representing the high-molecular-weight species and the late-eluting species representing the low-molecular-weight species.

In SEC, retention depends on the continuous exchange of solute molecules between the mobile phase and the stagnant mobile phase within the pores of the packing. This exchange is an equilibrium, so entropy-controlled processes and enthalpic processes such as adsorption are undesirable in SEC. Thus, the SEC retention volume (V_r) is expressed as shown in Eq. (46.1).

$$V_r = V_o + V_p K_{SEC} + V_s K_{LC} \qquad (46.1)$$

where V_o = the exclusion volume or interstitial volume between the packing particles (defined in Fig. 46.2), V_p = the pore volume (defined in Fig. 46.2), V_s = the stationary phase volume, K_{SEC} = the SEC solute distribution coefficient, and K_{LC} = the liquid chromatography solute distribution coefficient.

Because ideal SEC retention is governed only by entropic contributions, the column packing material/eluent combination is chosen such that K_{LC} is minimized or ideally $K_{LC} = 0$.

The dependence of V_r on solute molecular weight represents the SEC calibration curve. An example calibration curve, generated from an injection of four narrow-fraction molecular weight

Figure 46.1 Illustrative description of separation in SEC. *(From* Introduction to Modern Liquid Chromatography, *2nd edition by L. Snyder and J. J. Kirkland, © 1979 by John Wiley & Sons, Inc. Reprinted by permission of John Wiley & Sons, Inc.)*

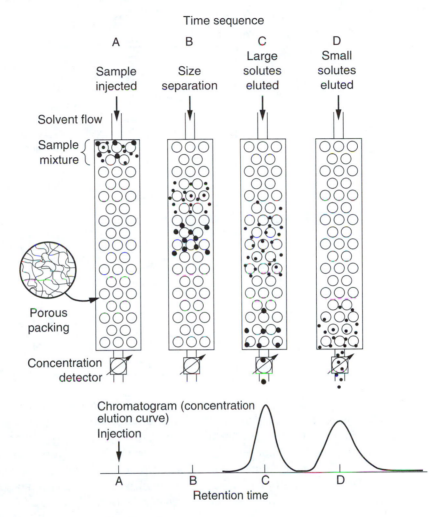

standards, is shown in Fig. 46.2. In this example, the highest-molecular-weight species is entirely excluded from the pores in the packing, so its retention volume is equal to the interstitial or exclusion volume, in which case K_{SEC} is zero. The lowest-molecular-weight standard permeates all of the pores within the SEC column, so its retention volume equals the sum of the interstitial volume and the pore volume. This volume is called the total volume or total permeation limit. The value of K_{SEC} for species eluting at the total permeation limit is 1. The two intermediate-molecular-weight standards can permeate the pores to some extent, so they are separated according to their respective hydrodynamic volumes. The value of K_{SEC} for peaks eluting in this region is $0 < K_{SEC} < 1$.

The linear region in Fig. 46.2 defines the useful region for SEC separation of macromolecules. The molecular weight range covered by the linear region of the SEC calibration curve depends on the size of the pores in the packing material. For single pore size columns, the useful region for SEC covers approximately 2 decades in molecular weight. (Gels with pore sizes ranging from 106 to 50 Å are typically available in a wide variety of compositions.) Most synthetic polymer

Figure 46.2 Log molecular weight versus retention volume plot for a typical SEC experiment with four narrow-fraction molecular weight standards. *(From* Introduction to Modern Liquid Chromatography, *2nd edition by L. Snyder and J. J. Kirkland, © 1979 by John Wiley & Sons, Inc. Reprinted by permission of John Wiley & Sons, Inc.)*

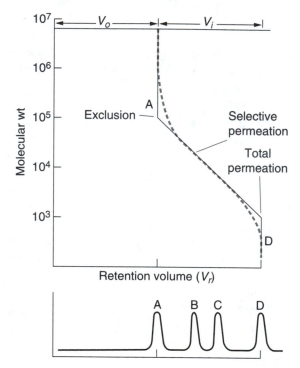

samples have broad MWDs that span more than 2 decades in molecular weight. Individual pore size columns can be coupled to provide the required separation range, or mixed-bed columns (columns containing a mixture of pore sizes) can be used. Mixed-bed columns are designed to provide several orders of magnitude (4 to 5) in molecular weight separation range. The choice of column depends on the sample composition, solvent options, and molecular weight range. The goals are to have ideal SEC retention (that is, separation by size only) and optimal resolution. Resolution in SEC depends on the slope (and column dispersion) of the linear calibration region. The column should be chosen such that the linear range encompasses the entire MWD of the sample with the smallest (absolute value) slope available. A detailed discussion of resolution in SEC and its impact on molecular weight accuracy is beyond the scope of this chapter. The interested reader is encouraged to consult the Suggested Readings.

What It Does

Size exclusion chromatography is widely used for determination of polymer molecular weight and MWD. For polymer characterization, molecular weight and MWD are important because they can have significant impact on physical properties. Because SEC is useful for determination of molecular weight and MWD, it is an important technique for aiding in the establishment of structure/property relationships for polymers. Once molecular weight/property relationships are established, SEC can be used in quality control fashion to ensure the production of polymers having desirable physical properties. SEC can also be used to aid in optimization of process conditions for production of materials with desirable molecular weight. SEC may also be used for comparing the MWD of samples that may perform differently in end-use applications.

Synthetic polymers differ from small molecules in that they cannot be characterized by a single molecular weight. In a synthetic polymerization, a distribution of chain lengths (that is, molecular weights) is produced. This distribution can be described by any number of molecular weight averages. The most common molecular weight averages used in establishing molecular weight/property relationships are the number-average (M_n), weight-average (M_w), and z-average (M_z) molecular weights. These averages are defined by the following expressions:

$$M_n = \frac{\sum\limits_i N_i M_i}{\sum\limits_i N_i} = \frac{\sum\limits_i W_i}{\sum\limits_i W_i/M_i} = \frac{\sum\limits_i h_i}{\sum\limits_i h_i/M_i} \tag{46.2}$$

$$M_w = \frac{\sum\limits_i N_i M_i^2}{\sum\limits_i N_i M_i} = \frac{\sum\limits_i W_i M_i}{\sum\limits_i M_i} = \frac{\sum\limits_i h_i M_i}{\sum\limits_i M_i} \tag{46.3}$$

$$M_z = \frac{\sum\limits_i N_i M_i^3}{\sum\limits_i N_i M_i^2} = \frac{\sum\limits_i W_i M_i^2}{\sum\limits_i W_i M_i} = \frac{\sum\limits_i h_i M_i^2}{\sum\limits_i h_i M_i} \tag{46.4}$$

Here N_i and W_i are the number and weight, respectively, of molecules having molecular weight M_i. The subscript i is an index representing all molecular weights present in the ensemble of chains. Note that each of the above equations contains three representations of the particular molecular weight average. The third representation in each case (farthest right) defines how one obtains these averages from SEC chromatograms. h_i is the height (from baseline) of the SEC curve at the ith elution increment and M_i is the molecular weight of species eluting at this increment. M_i is obtained via calibration with appropriate standards.

Conversion of the retention volume axis in SEC to a molecular weight axis (that is, calibration) can be accomplished in a number of ways including peak position, universal calibration, broad standard (integral and linear), and determination of actual molecular weight. In the peak position method, a series of well-characterized, narrow-fraction molecular weight standards of known peak molecular weight (M_p) are injected onto the SEC system and the retention volumes determined. A plot of log M_p versus retention volume is constructed as shown in Fig. 46.3. Depending on the type of SEC columns used, the calibration data points can be fit with either a linear or a third-order polynomial function. For determination of absolute molecular weights, the peak position calibration is of limited utility because only a few commercially available standards exist, and in order for the molecular weight calibration to be valid, the composition and topology of the standards and unknown must be the same.

We could prepare and characterize calibration standards via preparative fractionation schemes and absolute molecular weight techniques (such as LALLS), but these efforts can be costly and time-consuming. In some instances, the accuracy of the calibration is not important because the researcher may be interested only in comparing two samples to determine whether there is a relative difference in molecular weight. In these cases, only apparent molecular weight information is obtained, and for comparison of sample MWDs, calibration may not even be necessary. However, in other cases, absolute molecular weight data may be necessary. In these cases, alternative calibration schemes exist.

Figure 46.3 SEC calibration curve for 17 narrow-fraction polystyrene molecular weight standards analyzed using two mixed-gel SEC columns and tetrahydrofuran as the mobile phase. The line is a third-order polynomial fit of the data points.

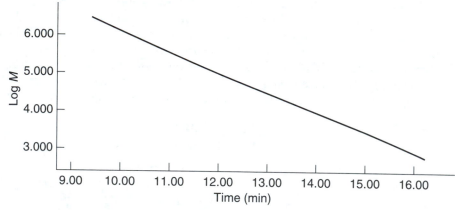

Polynomial coefficients: Log $M = A + BT + CT^2 + DT^3$
$A = 1.737168e + 001$, $B = -1.911779e + 000$, $C = 1.043279e - 001$,
$D = -2.578582e - 003$

One such scheme is the universal calibration principle shown in Fig. 46.4. The product of intrinsic viscosity, [η], and molecular weight is proportional to hydrodynamic volume, and in ideal SEC, molecules are separated according to hydrodynamic volume. In 1967, Benoit showed that the calibration for polymers of different types can be merged into a single line when plotted as log [η]M versus retention volume, as opposed to the typical log M versus retention volume shown in Fig. 46.3. There are two ways to use the universal calibration principle to glean absolute molecular weight data. One way is to use an on-line viscometer to determine [η] at each SEC retention increment. Well-characterized narrow-fraction molecular weight standards can be used to generate the universal calibration curve. The plot of log [η]M versus retention volume and the measured [η]s for the sample allow one to calculate M at each retention volume increment.

Alternatively, one may combine the universal calibration principle and the Mark–Houwink equation to determine absolute molecular weights. From the universal calibration principle,

$$[\eta]_{i,A} M_{i,A} = [\eta]_{i,B} M_{i,B} \tag{46.5}$$

where the subscript i refers to the particular SEC retention time increment and the subscripts A and B refer to polymers of different composition. The Mark–Houwink equation is given below.

$$[\eta] = kM^\alpha \tag{46.6}$$

where k and α are constants that depend on polymer composition, temperature, and solvent. Combining Eqs. (46.5) and (46.6) yields

$$k_A M_{i,A}^{1+\alpha A} = k_B M_{i,B}^{1+\alpha B} \tag{46.7}$$

Rearrangement of Eq. (46.7) gives

$$M_{i,A} = (k_B/k_A)^{1/1+\alpha A} M_{i,B}^R \tag{46.8}$$

where $R = (1 + \alpha_B)/(1 + \alpha_A)$.

Thus, with knowledge of the Mark–Houwink parameters for the standards and the sample (in the SEC eluent), one can determine absolute sample molecular weights. Published values of

Figure 46.4 Universal calibration plot. *(From H. Benoit et al., Journal of Polymer Science, Part B, 5, 753, copyright © 1967. Reprinted by permission of John Wiley & Sons, Inc.)*

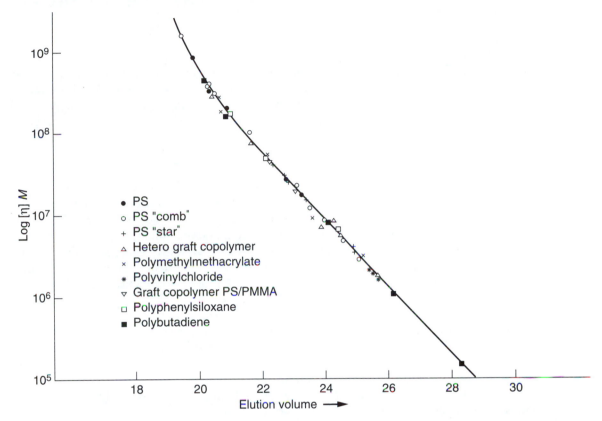

Mark–Houwink parameters in common SEC solvents are available for many polymers. When published constants are not available, they may be measured via Eq. (46.6). However, the measurement of k and α require standards of known molecular weight.

A third type of calibration in SEC is called broad-standard-calibration. Many types of broad-standard calibration schemes are used. The integral calibration scheme requires the complete MWD of a broad standard be known. Accurate characterization of a sample for the purposes of establishing it as a broad standard can be tedious and time-consuming. In many instances, a model such as the Flory most-probable model is used with the measured M_n or M_w to predict the complete MWD of a broad standard. Use of this approach requires that the polymer standard closely follow the theoretical model, which limits the applicability of this approach.

A second, more widely used method for broad-standard calibration is the linear method. This approach requires a broad standard of known M_n and M_w. The method consists of an iterative search for the coefficients of a linear calibration equation (log M versus retention volume), such that the equation yields the known values of M_w and M_n for the broad standard.

Finally, one may determine the absolute calibration by using an absolute molar mass sensitive detector, such as a LALLS detector. In this approach, the absolute M_w is determined at each SEC elution volume increment. The plot of the determined log M_w versus retention volume represents the calibration curve. The calibration curve is then applied to the concentration detector chromatogram for determination of molecular weight averages and MWD.

Analytical Information

Qualitative

SEC is typically not used for qualitative analyses, as separation is governed by size, not composition. However, simple comparison of SEC chromatograms can provide qualitative information about differences in MWDs for unknowns having the same composition and topology.

Quantitative

SEC is used for determination of MWDs and molecular weight averages. MWDs are typically calculated and presented in one of two ways. The cumulative weight fraction MWD of a typical polystyrene sample is shown in Fig. 46.5. The right-hand y-axis is the cumulative weight percentage of chains having molecular weight less than M_i. A second, more popular, way of presenting MWD data is the differential weight (dw) fraction MWD, also shown for the same polystyrene sample in Fig. 46.5. Most commercially available SEC software packages provide these options for plotting MWD data. Consult the Suggested Readings section for more information regarding the details of converting a SEC chromatogram to a MWD plot.

Integrated SEC peak areas may also be used for determining the concentration of polymer in a matrix. For example, the concentration of polymer in a monomer sample may be determined via SEC. Typically, in these types of experiments, MWD information may not be desired. In these cases, a small pore size column is chosen such that the polymer is excluded (and thus well separated from the monomer peak). The excluded peak area is then used to calculate the concentration of polymer in the monomer sample. Determination of the concentration of polymer in other small-molecule matrices may be accomplished in a similar fashion.

Figure 46.5 Cumulative and differential weight fraction log MWDs for the same polystyrene sample. $M_p = 204685$, $M_n = 100988$, $M_w = 255949$, and $M_z = 482307$; polydispersity (M_w/M_n) = 2.534 with peak area = 906374.

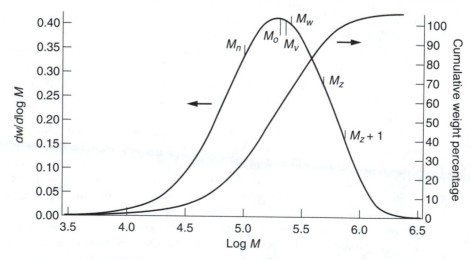

Applications

The number of applications of SEC to the characterization of synthetic polymers and other macromolecules is enormous. The scope of this handbook does not allow for discussion of even a small fraction of the total SEC applications. A number of review articles have been written that adequately cover the recent applications of SEC to macromolecular characterization, and are cited in the Suggested Readings section. Two examples are presented here to give the reader a small sampling of typical SEC applications.

1. Determination of the MWD and Molecular Weight Averages of a Polystyrene Sample.

The absolute MWD of a polystyrene sample was determined by SEC using a peak position calibration curve generated from narrow-fraction polystyrene molecular weight standards. A solution of the polymer was prepared in tetrahydrofuran (THF) at a concentration of 2.0 mg/mL. The solution was filtered through a 0.2 μm filter before injection. Separation was performed on two mixed-gel SEC columns designed to provide a linear separation range for polystyrene from approximately 3×10^6 Da to 500 Da. THF at a flowrate of 1 mL/min was used as the eluent. A differential refractive index detector was used to detect the eluting peaks. A total of 17 narrow fraction polystyrene molecular weight standards were injected to determine the $\log M_w$ versus retention time calibration via the peak position method. The calibration curve was fit to a third-order polynomial as shown in Fig. 46.3. The resulting SEC chromatogram of the polystyrene sample is shown in Fig. 46.6, and the MWD is shown in Fig. 46.5 along with the calculated molecular weight averages.

Figure 46.6 SEC chromatogram of a polystyrene sample.

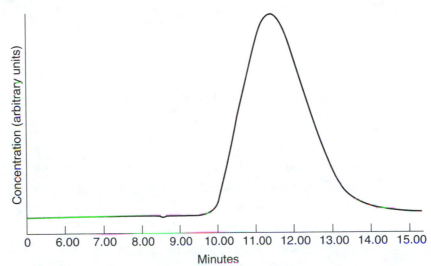

Figure 46.7 PEO apparent MWDs of polyurethane samples after various thermal processing steps. Virgin = solid curve, compounded = dashed curve, injection-molded = dot-dashed curve.

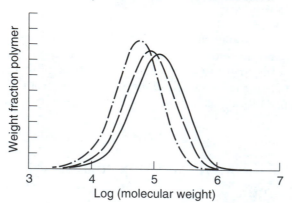

2. Comparison of the Apparent MWDs of Polyurethane Samples After Various Processing Steps.

Three polyurethane samples that had undergone various thermal cycles were analyzed by SEC. The samples were virgin pellets, compounded pellets, and an injection-molded part. SEC was performed on the same column set and detector described in the previous example. The eluent (and solvent for sample dissolution) was N,N,dimethylformamide containing 0.4% (*w:v*) lithium nitrate. Narrow-fraction polyethylene oxide (PEO) molecular weight standards were used for calibration. Thus, the results generated in this case are PEO apparent data. The resulting MWDs are overlaid in Fig. 46.7. Although the data are based on PEO standards, clearly there is a systematic decrease in molecular weight after each thermal processing step.

Nuts and Bolts

Relative Costs

Instruments for high-temperature SEC can be obtained for $$$$ to $$$$$.

Pumps	$$
Columns	$
Autosamplers	$$
Concentration detectors	$ to $$
Data system	$$ to $$$
Viscosity detector	$$$

$ = 1 to 5 K, $$ = 5 to 15 K, $$$ = 15 to 50K, $$$$ = 50 to 100K, $$$$$ = >100K.

Vendors for Instruments and Accessories

Alltech Associates (pumps, detectors, columns)
2051 Waukegan Rd.
Deerfield, IL 60015-1899
phone: 847-948-8600
fax: 847-948-1078
email: 73554.3372@compuserve.com
Internet: http://www.alltechweb.com

Hewlett-Packard (LC systems, data systems)
2850 Centerville Rd.
Wilmington, DE 19808
phone: 302-633-8504, 800-227-9770
fax: 302-633-8902
Internet: http://www.hp.com/go/chem

Micra Scientific Inc. (columns)
1955 Techny Rd., Suite 1
Northbrook, IL 60062
phone: 847-272-7877
fax: 847-272-7893
Internet: http://www.micrasci.com

Perkin-Elmer (LC systems, data systems)
761 Main Ave.
Norwalk, CT 06859-0001
phone: 800-762-4000
email: info@perkin-elmer.com
Internet: http://www.perkin-elmer.com

Polymer Laboratories (columns, detectors, high-temperature SEC instruments,
data systems, molecular weight standards)
160 Old Farm Rd.
Amherst, MA 01002
phone: 413-253-9554
fax: 413-253-2476

TosoHaas (columns)
156 Keystone Dr.
Montgomeryville, PA 18936
phone: 215-283-9385, 800-366-4875
fax: 215-283-5035
Internet: http://www.rohmhaas.com/tosohaas

Varian Associates (LC systems, data systems, columns)
2700 Mitchell Dr.
Walnut Creek, CA 94598
phone: 510-939-2400, 800-926-3000
fax: 510-945-2102
Internet: http://www.varian.com

Waters Corp. (columns, high-temperature SEC instruments,
LC components, data systems)
34 Maple St.
Milford, MA 01757
phone: 508-478-2000, 800-252-4752
fax: 508-872-1990
email: info@waters.com
Internet: http://www.waters.com

Required Level of Training

The skill levels required for performing SEC experiments can vary greatly depending on the particular application. For example, a trained chemical technician can perform most routine SEC experiments where conditions for ideal SEC have been defined. On the other hand, development of ideal SEC conditions for complex polymer systems requires a far greater skill level. In these cases, training in analytical chemistry, separations science, and polymer chemistry and materials characterization is most useful.

Service and Maintenance

Service and maintenance requirements for SEC are no different from those required for conventional HPLC. Column performance can be evaluated on a daily basis through the calculation of $D\sigma$ (where D is the slope of the linear region of the SEC calibration curve and σ is the peak variance for a monodisperse polymer peak), peak asymmetry, and number of theoretical plates.

Suggested Readings

BARTH, H. G., AND B. E. BOYES, "Size Exclusion Chromatography," *Analytical Chemistry*, 64 (1992), 428R–42R.

BARTH, H. G., AND J. W. MAYS, EDS., *Modern Methods of Polymer Characterization*. New York: Wiley, 1991.

BRANDRUP, J., AND E. H. IMMERGUT, EDS., *Polymer Handbook*. New York: Wiley, 1989.

PROVDER, T., ED., *Chromatography of Polymers: Characterization by SEC and FFF, ACS Symposium Series 521*. Washington, D.C.: American Chemical Society, 1993.

SMITH, C. G., AND OTHERS, "Analysis of Synthetic Polymers and Rubbers," *Analytical Chemistry*, 65 (1993), 217R–43R.

YAU, W. W., J. J. KIRKLAND, AND D. D. BLY, *Modern Size-Exclusion Liquid Chromatography*. New York: Wiley, 1979.

Chapter 47

Low-Angle Laser Light Scattering

Deidre Strand

The Dow Chemical Company

Summary

General Uses

- Determination of absolute weight average molecular weight of homopolymers
- Determination of absolute molecular weight distributions of homopolymers
- Determination of absolute molecular weights of a restricted set of copolymers

Common Applications

- Absolute molecular weight measurements for branched polymers
- Absolute molecular weight measurements for polymers for which molecular weight standards do not exist
- Molecular weight determinations for proteins and biopolymers
- Characterization of intrinsic viscosity/molecular weight relationships

Samples

State

Soluble polymers in liquid or solid form can be analyzed.

Amount

- Bulk low-angle laser light scattering; 1 g is desirable.
- Size-exclusion chromatography coupled to low-angle laser light scattering; 20 mg is desirable.
- Measurement of specific refractive index increment; 1 g is desirable.

Preparation

- Polymer must be dissolved to accurately known concentration.
- Clarification of solution to remove dust particulates is critical. This may be accomplished by filtration or ultracentrifugation. Choice of solvent must yield favorable specific refractive index increment.

Analysis Time

Bulk low-angle laser light scattering can take up to several hours once solutions are prepared.

When size-exclusion chromatography (SEC) coupled to low-angle laser light scattering is used as an SEC detector, experimental time is governed by the chromatographic time scale.

Limitations

General

- Determination of molecular weights below about 10,000 g/mol are difficult due to low light scattering signals.
- Appropriate solvent must be found such that a favorable specific refractive index increment exists.
- Sample must be relatively pure for accurate determinations of concentration and specific refractive index increment.

Accuracy

Accuracy depends on molecular weight of sample and specific refractive index increment of sample/solvent combination. In favorable conditions, weight average molecular weight can be determined within a few percent.

Complementary or Related Techniques

- SEC provides relative molecular weights typically. It can provide absolute molecular weights provided that appropriate standards are available.
- Colligative methods of molecular weight measurements provide absolute molecular weights.

- Dynamic light scattering (photon correlation spectroscopy) is used to obtain rotational diffusion coefficients of polymers.
- Neutron scattering is analogous to light scattering; it is used for analysis of polymers in the solid state.

Introduction

Low-angle laser light scattering (LALLS) is used primarily to determine the absolute weight average molecular weight (M_w) of homopolymers. A LALLS experiment may be performed on a bulk solution in which only M_w is obtained. A LALLS detector may also be used in conjunction with an SEC experiment in which a measure of the absolute molecular weight distribution may be obtained.

Although SEC is the most common method used to determine molecular weight, LALLS offers several advantages. SEC separates on the basis of hydrodynamic size. A calibration is necessary to correlate size with retention time. Unless appropriate molecular weight standards are available, the experiment yields molecular weight averages relative only to the particular standards used for calibration. Ideally, the standards should match the unknowns in terms of composition and topology (linear or branched, for example). Unfortunately, relevant standards are not typically available for most polymers. The use of SEC coupled to LALLS eliminates the need for calibration, as the LALLS detector provides absolute molecular weight data. Thus, the use of LALLS allows for accurate molecular weight measurements for homopolymers where suitable SEC standards are not available. In addition, LALLS may be used to distinguish between linear and branched samples which may have the same molecular size as determined by SEC, yet differ in absolute molecular weight.

The static or bulk LALLS experiment is useful for determining absolute weight average molecular weights of homopolymers, which are not well suited for SEC experiments.

How It Works

Determination of absolute molecular weight by light scattering involves measurement of the ratio of scattered light intensity at a particular angle, θ, to the intensity of incident light, per unit volume. This quantity is called the Rayleigh ratio, R_θ, defined by the following equation:

$$R_\theta = \frac{\mathcal{J}_\theta}{I_O V} \tag{47.1}$$

where $\mathcal{J}_\theta$ = intensity of scattered light at angle θ, I_o = incident intensity, and V = scattering volume.

Polymer solutions are considered to be optically heterogeneous due to local random fluctuations in density and refractive index. The fluctuation theory of light scattering (1) can be used to relate the Rayleigh ratio to physical characteristics of the polymer molecule causing the scattering by the following equations:

$$\frac{kc}{R_\theta} = \frac{1}{M_w P_\theta} + \frac{2A_2 c}{P_\theta} + \frac{3A_2 c^2}{P_\theta} + \cdots \qquad (47.2)$$

$$k = \left(\frac{2\pi^2 n^2}{\lambda^4 N}\right)\left(\frac{dn}{dc^2}\right)(1 + \cos^2\theta) \qquad (47.3)$$

where c = concentration of scattering species in g/mL, M_w = absolute weight average molecular weight of the scattering species, n = solution refractive index, λ = wavelength of incident light in vacuo, N = Avagadro's number, $A_2, A_3, \ldots$ = virial coefficients, P_θ = form factor, which depends on size and shape of scattering species, dn/dc = specific refractive index increment, $R_\theta = R_{\theta,\text{solution}} - R_{\theta,\text{solvent}}$ where $R_{\theta,\text{solution}}$ and $R_{\theta,\text{solvent}}$ represent the Rayleigh factors for the solution and solvent, respectively.

The specific refractive index increment is measured in an independent experiment using a differential refractometer or it is measured on-line using some commercial instruments.

The form factor, P_θ, arises when the scattering molecules are large with respect to the wavelength of incident light, as is typically the case for polymers. For these larger particles, different parts of the molecule are exposed to incident light of different phase and magnitude. Thus, the scattered light will be of various phases and magnitudes, which results in interference of the scattered light coming from different parts of the polymer molecule. The net result is an angularly dependent scattered light intensity. Particle scattering functions have been derived for various particle shapes, as shown in Fig. 47.1 (2). In all cases, P_θ approaches unity as the size parameter (which is directly proportional to the ratio of the particle diameter to the incident wavelength in solution) becomes small.

In general, P_θ is related to the square of the molecular radius of gyration (R_g). In the limiting case of small angle ($\theta \to 0$), the relationship may be expanded in a power series, keeping only the first-order terms as follows:

$$\lim_{\theta \to 0} P_\theta = \frac{1 - 16\pi^2}{3\lambda^2 \sin^2(\theta/2) R_g^2} \qquad (47.4)$$

In this expression, P_θ reduces to unity as θ approaches zero.

Figure 47.1 Variation in the particle scattering function (P_θ) with the size parameter. *(From* Textbook of Polymer Science *by F. W. Billmeyer, copyright © 1984. Reprinted by permission of John Wiley & Sons, Inc.)*

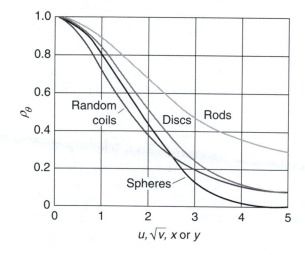

What It Does

In a classic multiangle laser light scattering experiment, the excess Rayleigh ratio, R_θ, is determined at a number of angles for a series of polymer concentrations. A double extrapolation to zero concentration and zero angle is then performed in what is typically called a Zimm plot, as shown in Fig. 47.2 (2). The extrapolation to zero angle corrects for the effect of finite particle size. The slope of the θ-dependence plot in this form can be shown to be equal to the quantity $16\pi^2 R_g^2/3\lambda^2$. Thus, an average radius of gyration is obtained in this type of experiment.

The second extrapolation, to zero concentration, corrects for the effect of finite concentration. The intercepts of both the zero concentration extrapolation and the zero angle extrapolation can be shown to be equal to $1/M_w$.

A LALLS experiment takes advantage of the fact that at small angle θ, the form factor P_θ approaches unity regardless of the molecular size or shape. Equation (47.5) reduces to the following in the limit of small angle:

$$\frac{kc}{R_\theta} = \frac{1}{M_w} + 2A_2 c + \cdots \tag{47.5}$$

If R_θ for a series of sufficiently low concentrations is measured such that higher-order concentration terms may be neglected, a plot of kc/R_θ versus concentration will be linear with an intercept of $1/M_w$ and a slope of $2A_2$.

In the SEC/LALLS experiment, the LALLS detector is used to measure R_θ and a differential refractive index (DRI) detector is typically used to measure concentration, c, at each retention time increment ($2A_2 c$ is assumed to be negligible). The data are then plotted as log (M_w) versus retention volume over the range of the chromatogram. This curve or a fit to the curve may then be used as the SEC calibration curve and applied to the DRI chromatogram to generate an absolute molecular weight distribution.

Figure 47.2 Zimm plot from light scattering data for polystyrene in butanone. *(From Textbook of Polymer Science by F. W. Billmeyer, copyright © 1984. Reprinted by permission of John Wiley & Sons, Inc.)*

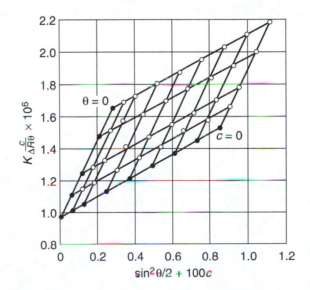

A number of low-angle and multiangle light scattering instruments are commercially available with a variety of features. Some of the instruments may be used both in the static (or bulk) mode as well as an SEC detector. Others may be limited to one application.

Common components of light scattering instrumentation include a source, incident optics, sample cell, output collection optics, and a detector for the scattered light. Instrument design is crucial because the scattered light intensity is inherently small. Therefore, reflections and scattering from all interfaces must be minimized for optimum sensitivity.

Laser sources have typically replaced mercury arc sources in currently available instruments. The increased available intensity of the lasers results in increased scattering intensities and, hence, increased sensitivity. Incident optics may include focusing lenses, polarizing lenses, and filters depending on the type of source used. Neutral density filters may also be necessary for measurement of the incident light intensity. The sample cell design varies greatly among commercial instrumentation. Cells designed for use in multiangle instruments are typically cylindrical cuvettes, which allow detection of scattered light at multiple angles from incident. LALLS cells may be rectangular, from which scattered light is collected via an annulus at a fixed angle. These cells are often designed as flow-through cells, which may be coupled with an SEC experiment. Output collection optics may include lenses, polarizers, and apertures. Finally, a photomultiplier tube, photodiode, or photodiode array may be used for detection of the scattered light.

Other features available in commercial instruments include temperature control and simultaneous differential refractive index (DRI) detection. Advantages to the DRI detection capability are that *dn/dc* may be measured on-line at the wavelength of the light scattering instrument and that no interdetector delay exists between the LALLS and the DRI signals in the coupled SEC/LALLS experiment.

Most currently available light scattering instrumentation is easily interfaced to data acquisition systems for data collection. Sophisticated data handling software is also available from a variety of sources.

Analytical Information

Qualitative

LALLS is not typically used for qualitative analytical purposes.

Quantitative

The primary use of LALLS is quantitative determination of absolute molecular weights of homopolymers, as previously described. LALLS is extremely sensitive to high-molecular-weight fractions of polymers and, therefore, is most accurate for high-molecular-weight polymers. Accuracy decreases for low-molecular-weight polymers (less than 10,000 g/mol), but may still be acceptable if the experiments can be performed in a solvent where *dn/dc* is large.

Applications

1. Determination of Absolute Weight Average Molecular Weight, M_w, of Polystyrene by LALLS.

A determination of absolute M_w was performed on a polystyrene sample without the use of molecular weight standards. The specific refractive index increment of the polystyrene sample in tetrahydrofuran (THF) was measured using a differential refractometer. Solutions of the sample were prepared in THF at concentrations ranging from 0.2 mg/mL to 2.5 mg/mL using volumetric glassware. The excess Rayleigh ratios of the solutions were measured using a LALLS photometer at an angle of 4.6°.

A plot of kc/R_θ versus concentration is shown in Fig. 47.3. As described earlier, the intercept of the plot yields $1/M_w$ and the slope of the plot is equal to $2A_2$.

2. Determination of Absolute Molecular Weight Distribution of Polystyrene by SEC Coupled to LALLS.

The absolute molecular weight distribution of a polystyrene sample was determined without the use of molecular weight standards using SEC coupled to LALLS. The specific refractive index increment of the polystyrene sample in tetrahydrofuran was measured using a differential refractometer. A solution of the sample in THF was prepared at 2 mg/mL using volumetric glassware. A known quantity of the sample was injected onto the SEC system using DRI and LALLS detection with a known interdetector delay.

Figure 47.4 gives the DRI and LALLS chromatograms for the polystyrene sample. A plot of the logarithm of the absolute M_w of the eluting species as a function of retention time may be

Figure 47.3 Plot of Rayleigh ratio (kc/R_θ) versus concentration. *(Reprinted with the permission of the American Chemical Society.)*

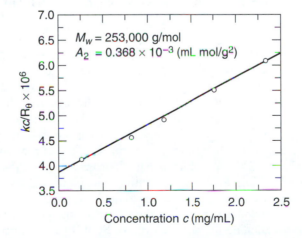

Figure 47.4 Comparison of LALLS and DRI chromatograms for a polystyrene sample.

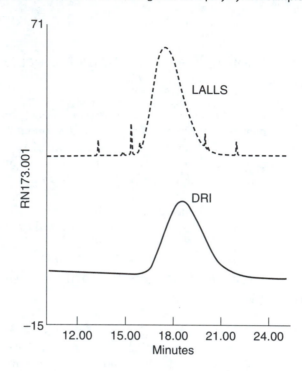

generated knowing the optical constants of the detector and the flow rate. The DRI response is used to measure the concentration as a function of retention time, which is necessary for the molecular weight calculation. A fit of a line to the plot was used as an SEC calibration curve to calculate the absolute molecular weight distribution of the polystyrene sample from the DRI chromatogram.

3. Determination of the Amount of Long-Chain Branching in Polystyrene by SEC Coupled to LALLS.

SEC coupled to LALLS was used to quantify the amount of branching in branched polystyrene samples. The branched polystyrene was prepared using benzocyclobutenoyl peroxide (BCBPO) as a branching agent (3). The experimental procedure used was identical to that described in the previous application. Curvature in the plot of the logarithm of the absolute M_w versus retention time is attributed to branching in the sample. This is an effect of the SEC, which separates polymer chains on the basis of hydrodynamic size. For a given absolute molecular weight, a branched sample will have a smaller hydrodynamic radius than a linear sample. Therefore, the branched sample elutes from the SEC column later than a linear sample of the same molecular weight, resulting in curvature of the log (M_w) versus retention time plot.

The use of models developed by Zimm and Stockmayer may then be used to calculate the long-chain branching per 1000 carbon units of the polymer. Figure 47.5 shows the calculated amounts of branching for various levels of BCBPO added during the polymerization.

Figure 47.5 Long-chain branching versus concentration of BCBPO. *(Reprinted with permission from Macromolecules 27 (1994), 1311, Figure 7. Reprinted with permission of the American Chemical Society.)*

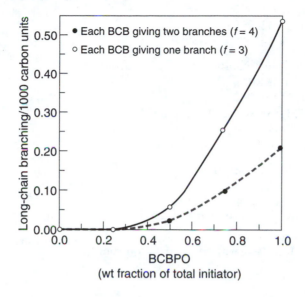

Nuts and Bolts

Relative Costs

Multiangle laser light scattering photometer (flow-through capabilities)	$$$$ to $$$$$
LALLS photometer (static)	$$$
LALLS photometer (flow-through and static)	$$$ to $$$$
Differential refractometer	$$ to $$$
Maintenance	5 to 10% of purchase price/year

$ = 1 to 5 K, $$ = 5 to 15 K, $$$ = 15 to 50K, $$$$ = 50 to 100K, $$$$$ = >100K.

Vendors for Instruments and Accessories

Brookhaven Instruments
750 Blue Point Rd.
Holtsville, NY 11742
phone: 516-758-3200
fax: 516-758-3255
email: sales@bic.com
Internet: http://www.bic.com

Polymer Laboratories
160 Old Farm Rd.
Amherst, MA 01002
phone: 413-253-9554
fax: 413-253-2476

Precision Detectors, Inc.
160 Old Farm Rd.
Amherst, MA 01002
phone: 413-256-0516, 800-472-6934
fax: 413-256-0507
Internet: http://www.lightscatter.com

Thermo-Separations, Inc.
355 River Oaks Pkwy.
San Jose, CA 95134
phone: 408-526-1100, 800-532-4752
fax: 408-526-1074
Internet: http://www.thermoseparation.com

Wyatt Technology Corp.
802 E. Cota St.
Santa Barbara, CA 93103
phone: 805-963-5904
fax: 805-965-4898
email: wyatt@wyatt.com
Internet: http://www.wyatt.com

Required Level of Training

Instrument operation varies considerably depending on type of experiment being performed and type of instrument being used. Routine analyses could be performed by a trained chemical technologist. However, data handling and data interpretation require a great deal of experience. Prior experience and knowledge of polymer chemistry and polymer characterization are necessary.

Service and Maintenance

Service primarily involves optic alignment and source replacement. Realignment may be done by a trained user, but often may require a service technician. Primary maintenance consists of periodic cleaning of sample cells and exposed optics, which may be done by a trained user.

Suggested Readings

BILLMEYER, F. W., JR., *Textbook of Polymer Science*, 3rd ed. New York: Wiley, 1984.

CHU, B., *Laser Light Scattering*, 2nd ed. San Diego, CA: Academic Press, 1991.

HUGLIN, M. B., *Light Scattering from Polymer Solutions*. New York: Academic Press, 1972.

References

1. B. Chu, *Laser Light Scattering*, 2nd ed. (San Diego, CA: Academic Press, 1991), 21.

2. F. W. Billmeyer, Jr., *Textbook of Polymer Science*, 3rd ed. (New York: Wiley, 1984), 201, 202.

3. S. L. DeLassus and others, *Macromolecules*, 27 (1994), 1307–12.

Chapter 48

Particle Size Measurements

Stewart P. Wood and Grant A. Von Wald

The Dow Chemical Company

Summary

General Uses

- Determination of particle size distributions
- Determination of mean particle size
- Determination of particle concentration

Common Applications

- Determination of particle size distributions of inorganic, organic, and polymeric materials
- Measurement of droplet size in some emulsions and in suspension
- Quantitation of particulate contamination in liquids, gases, and solids
- Determination of filter efficiencies
- Determination of the rate of swelling or dissolution of a particle
- Quality control

Samples

State

- Airborne solids and liquids
- Liquid-borne solids, liquids, and gases

Amount

- Typically less than 500 mg of a solid is needed.

Preparation

- Surface of dry particles is wet using a surfactant.
- When necessary, sample is diluted in an appropriate fluid. The fluid used for dilution must be free of particulates and it must not dissolve or swell the particles of interest.
- Materials containing particles are dissolved, liberating the particles of interest.

Analysis Time

Estimated time to obtain a particle size distribution can range from 10 to 300 sec depending on the particle size range. Sample preparation time is between 1 and 15 min depending on the nature of the sample matrix.

Limitations

General

- Analysis is limited to low particle concentrations typically less than 50,000 particles per mL.
- Size range is between approximately 1 and 2500 μm and any single sensor has an approximately 50-fold dynamic range.
- Limited in some instances to low-viscosity (< 50 cPs) suspending fluids.

Accuracy

- Accuracy is approximately 3% relative using median particle diameter assignments.
- Accuracy of size assignment can depend on the particle and suspending medium refractive index.

Complementary or Related Techniques

- Sedimentation, a density-dependent technique that provides particle size distributions between approximately 0.1 and 50 μm
- Capillary hydrodynamic chromatography (HDC), which provides particle size distributions between approximately 0.05 and 0.7 μm
- Fraunhofer diffraction, which provides particle size distributions
- Field flow fractionation, which provides particle size distributions

- Electrical conductance, in which analysis in a conducting medium provides particle size distributions
- Phase Doppler anemometry, which provides particle size and velocity information for spherical particles
- Ultrasonic spectrometry

Introduction

There are numerous analytical techniques available that focus on the measurement of particle size. As a result of the large number of available particle size analysis techniques, the first step in performing a particle size analysis is deciding which technique to use. To make such a decision, one must first determine what type of size information is needed and what are the physical and chemical properties of the sample of interest. The type of size information obtained can be as detailed as a complete size distribution or as simple as a mean size. Contained in a size distribution is information pertaining to the breadth of the distribution, the number of individual populations in the sample, and the percentage of particles greater than or less than a particular diameter. A mean diameter, on the other hand, is a way of describing particle size using a single number. Some commercially available particle size analyzers report only a mean diameter. Such analyzers may be capable of identifying differences in the particle size, but one must realize that any two samples having the same mean diameter may have completely different size distributions.

Physical properties such as particle refractive index and density may be needed for some particle size techniques. The refractive index of the particle and suspending medium is often required for size measurements based on light scattering, whereas densities are needed for techniques based on sedimentation and acoustics.

Examining the particles of interest under a microscope allows one to determine the general range of sizes and shapes present. Because most size measurement techniques have a finite range of applicability with respect to particle size, a preliminary examination under the microscope is very useful. Nearly all particle size analysis techniques are sensitive to particle shape. In general, the more the shapes of the particles of interest deviate from spherical, the greater the error in the size assignment. Particles whose shapes deviate significantly from spherical, such as needles or rods, are best analyzed using microscopy.

Knowing the chemical properties of the particulate also helps one determine what size analysis technique can be used. Particles are often suspended in a fluid to facilitate the analysis. The fluid dilutes the particles so that they can be delivered to the analyzer at the appropriate concentration. When chosen properly, the suspending fluid helps dissipate any cohesive forces that tend to lead to particle agglomeration. Moreover, when chosen properly, the suspending fluid helps produce a suspension that is homogeneous in particle concentration and size, which can be delivered to the particle size instrumentation. The suspending fluid chosen must not dissolve the particles of interest and must be compatible with the instrument's materials of construction.

Light obscuration (LO) counting is a versatile light scattering technique because of the broad size range, approximately 1 to 2500 μm, and because of the flexibility in the choice of suspending fluid. Light obscuration is based on the fact that particles whose refractive index differs from that of the medium in which they are suspended scatter light away from a detector, as shown in Fig.

Figure 48.1 Ray trace for a light obscuration particle size experiment.

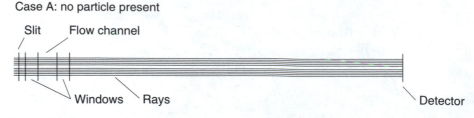

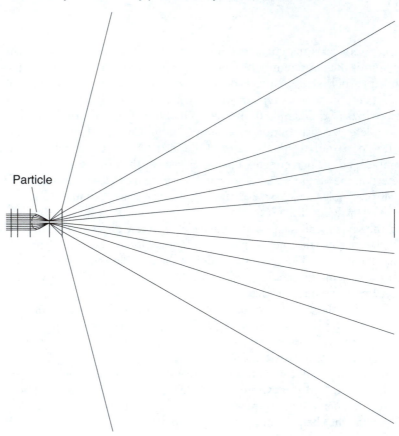

48.1. When a single particle is in the measurement volume at any given time, the amount of light scattered is proportional to the projected cross-sectional area of the particle.

Another common technique used in the particle size range of 1 to 2500 μm, called Fraunhofer diffraction, suffers from a number of deficiencies. Fraunhofer diffraction has much lower size resolution than does light obscuration. The measurement technique for Fraunhofer diffraction is insensitive to systematic errors such as solubility of the sample in the suspending fluid or high particulate background in the fluid. Finally, Fraunhofer diffraction has been found to be unable to weigh accurately the contributions of particles of widely differing sizes to the size distribution. The widespread use of Fraunhofer diffraction stems from the apparent ease of use and the reproducibility of the measurement.

How It Works

The light obscuration instrument consists of a sensor where the light scattering experiment takes place, a pulse height analyzer, and generally a computer that handles all data storage and analysis. Some type of sample handling system is required to pump the sample suspension through the sensor. The design of the sample handling system varies with each instrument, but can be as simple as a vacuum source with regulator, a large-volume vacuum flask to collect the fluid after it passes through the sensor, and a beaker for the sample suspension. Some attention must be given to the stirring of the sample suspension for particles that are dense or larger than 500 μm in diameter to ensure good mixing and representative sampling.

The pulses from the sensor are fed into a multichannel analyzer or similar electronic device that sorts the pulses based on the total change in light intensity as a consequence of scatter from the particles. In order to relate the pulse sizes to particle diameter, calibration using particles of known diameter is required. The number of counts in each pulse size range is then communicated to a computer, where average diameters are calculated, graphic results are presented, and the data are stored.

For best results, analysis parameters should be optimized as highlighted in the following paragraphs. Sensors are chosen according to the size of the material to be analyzed. If the range of particle sizes in the material is not known, it can be determined by examining the material using a light microscope. The two principal factors to balance include avoiding a plugged sensor if particles are too large and having a small enough lower size limit to detect the majority of the particles. Common sensor ranges are 40 to 2000 μm, 20 to 1000 μm, 10 to 500 μm, 5 to 250 μm, and 2 to 100 μm. Manufacturers' recommendations for solvent compatibility should be consulted when choosing a sensor.

The choice of a suspending fluid is crucial. The fluid should wet the particles of interest, allowing formation of a stable dispersion without causing dissolution. To check for dissolution of the material, the sample is prepared for analysis and a portion of the suspension is analyzed. After 3 to 5 min of stirring, another portion is analyzed. Dissolution of the particles is indicated if the total particle count has decreased by more than 10%. The particle count may also be decreased by particle agglomeration or deposition on the sample suspension beaker.

When changing suspending fluids, care should be taken to wash out the old fluid. If the two fluids are not miscible, an intermediate solvent should be found that is soluble in both. Isopropanol is generally a good intermediate between most organic fluids and water.

It is imperative that the suspending fluid be free of contamination (particles) and bubbles. Bubbles will be detected by the sensor and included in the particle size analysis, so care should be taken to avoid creating bubbles when transferring fluid.

A 2- or 4-L sample suspension beaker should be used with the 500-μm or larger sensors. The beaker should contain a stirrer and baffle. The addition of a baffle has been found to be effective in eliminating the formation of a vortex. Bubbles will be formed from entrapped air whenever a vortex is present. The baffle also promotes mixing and homogeneity of the suspension. For smaller sensors, a 250-mL beaker will suffice and baffles are usually not necessary unless the particles are dense (greater than 3 g/mL).

The flowrate should be the same as that used to calibrate the sensor. The flowrate can be measured easily using a glass cylinder and a stopwatch or by measuring the weight loss per time. It is good practice to measure the flowrate frequently.

The background particulate level should be measured with a clean sample of fluid before sample analysis. The background level should be measured down to the lower size limit of the sensor. The background count rate should be no more than 2% of the expected count rate for the sample.

If the background count rate is too high, an effort should be made to clean the beaker, baffle, and stir bar or the fluid should be allowed to stand for several minutes to allow bubbles to purge. If this does not reduce the background, the fluid must be filtered.

All glassware, stir bars, baffles, and other equipment must be kept free of particulate contamination. Multiple rinses with clean fluid are often necessary after cleaning the sample vessel. If a non-aqueous fluid is used, water must be removed from the sample handling system by drying or rinsing with a fluid that is compatible with the nonaqueous fluid and water. In addition, care should be taken to obtain a representative sample using techniques described in the text by T. Allen (1).

To be wetted by the suspending fluid (particularly water), some powders require a surfactant. It is important to ensure that the particles are wetted by the surfactant before adding them to a large volume of suspending fluid.

Dispersion of particles may be assisted by sonication. A stable dispersion will have a size distribution that is invariant to sonication. However, it is important to bear in mind the goal of the analysis when considering sonication. If the goal of the analysis is to determine the size of the ultimate particles, which are held strongly together, then sonication is appropriate. On the other hand, if the size distribution of the particles as they are, even if held loosely together, is needed, sonication is obviously inappropriate.

Dilute the sample such that the concentration is appropriate for the sensor. If the concentration is out of range, dilute or add material as appropriate. It often facilitates the measurement if the dilution steps are recorded so that they can be repeated.

The concentration of the sample should be high enough to count a statistically significant number of particles, but not so high as to create too much double counting or coincidence counting from having two or more particles in the sensing volume simultaneously. The coincidence level can be calculated using the following equation:

$$Vdc = \left(\frac{C^* Sv^* \sqrt{2}}{1 + C^* Sv^* \sqrt{2}} \right) 100 \qquad (48.1)$$

where Vdc = the volume percentage of the measured size distribution due to coincidence counting, C = the number of particles per volume, and Sv = the volume of the sensing zone.

For samples with broad size distributions, the coincidence level should not exceed 10%. For samples with uniform size distributions, the effects of double counting will be observed above 2% coincidence level as a population of concentration-dependent particles in the size distribution at roughly $\sqrt{2}$ times greater diameter than the main mode of the size distribution.

A standard of known size should be measured with each series of samples. If the measured average diameter differs by more than 3% from the expected result, consider recalibration of the light obscuration system.

Calibration of sensors requires a series of standards which span the range of the sensor. Five standards, generally narrow-size-distribution styrene–divinylbenzene beads, that are supplied by a number of vendors, are generally adequate. Calibrations are generally stable for periods up to several months.

Each standard is analyzed as a sample. The method by which the standard data are converted to a calibration depends on the instrumentation and software provided by the vendor. One procedure involves plotting cumulative counts versus pulse height in millivolts. The number median of the standard peak is chosen as the point halfway through the counts. The calibration is constructed by plotting the voltage that corresponds to the middle of the standards size distribution versus the diameter of the standard on a log–log plot, as shown in Fig. 48.2. The voltage and diameter pairs for the standards are then fit to the following expression:

$$\text{Diameter} = \left(\frac{\text{Voltage} - \text{Asymptote}}{\text{Intercept}} \right)^{1/\text{Slope}} \qquad (48.2)$$

Figure 48.2 Calibration plot (particle size versus pulse height) for a light obscuration particle size instrument.

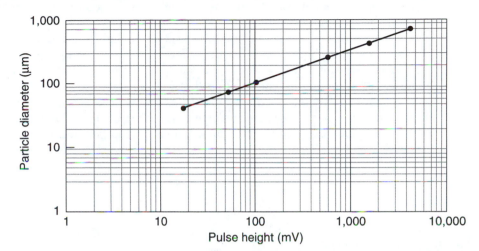

where the asymptote, intercept, and slope are calculated using a least-squares fit of the calibration data. The slope is expected to have a value within 5% of 2 because the sensor measures the projected area of the particles. For spheres, the projected area is proportional to the square of the diameter.

It is also necessary to choose a lower size limit for each sensor. This lower limit must be greater than the noise limit for the sensor. The noise limit can be determined by having the sensor count without pumping fluid through the sensor. The voltage level at which the count rate is roughly one count per second can be used as a noise level. However, the calibration graph of voltage versus diameter should also be examined to ensure that the lower size limit is in the range where the logarithmic plot of voltage versus diameter is linear, as this plot tends to deviate from the expected slope of 2 as the lower limit of the sensor is approached.

An additional effect that should be considered when calibrating light obscuration sensors is the dependence of the light scattering on particle composition when the index of refraction of the particle is within 5% of that of the suspending fluid. If the contrast between the index of refraction of the particle and that of the fluid is not sufficiently large, the size of the particles will be underestimated. In that case, calibration with standards and suspending fluids having the same index of refraction ratio as the actual sample and suspending medium may suffice, provided that the sample has uniform composition.

What It Does

The final result of particle size analysis using light obscuration counting instrumentation is a size distribution that consists of number of particles versus diameter. From this result, a volume-weighted size distribution can be calculated assuming spherical shape. Computer software can easily calculate a variety of averages from the counts versus diameter data. The text by T. Allen (1) describes a number of the more common averages used to reflect the central tendency of size distribution. The volume or number median is often emphasized instead of the mean diameters because the median is a more robust measure of average size and it is not as strongly affected by a few outlying data points.

For spherical samples, the result measured by the light obscuration sensor will be accurate. However, for nonspherical particles, the accuracy of the size assignment depends on how closely the projected area is approximated by the square of the diameter. No useful size information can be measured from any instrumentation that uses an equivalent sphere assumption for particles such as rods and needles. Useful measurements can be made using light obscuration for particles that are irregular, but whose general shape does not deviate too strongly from that of a sphere. No quantitative guidelines are available, but in general if the aspect ratio of any two dimensions of the particle is less than 2, useful size information can be obtained by this instrumentation, which makes an equivalent sphere assumption.

Analytical Information

Qualitative

Light obscuration is typically used for quantitative particle size and concentration measurements. However, the output of a light obscuration experiment could be sent to a simple pulse height analyzer, allowing qualitative sample comparisons. Qualitative particle concentrations are determined from the total number of pulses registered by the pulse height analyzer, whereas qualitative information about the size distribution is obtained from the registered distribution of pulse heights.

Quantitative

Quantitative particle size analyses are obtained when there is a known relationship between particle size and pulse height. As described previously, such a relationship is obtained by measuring the distribution of pulse heights for several uniform particle size standards. Once a pulse height versus size relationship is obtained, the pulse height information is converted directly to a particle size. The signal processor places the individual size measurements into discrete size classes, thus forming a size distribution. The breadth and number of the size classes vary from system to system. Particle size distributions can typically be displayed in terms of particle number, area, and volume.

A particle size distribution for a blend of uniform polystyrene standards is given in Fig. 48.3. The solid line shown on the distribution is the cumulative volume percent of the analyzed particles. The individual bins represent the volume fraction within each individual size class. If the particles are spheres and the densities of all the analyzed particles are the same, the volume fraction and weight fraction will be equivalent.

For accurate size distribution measurements, one particle must be present in the measurement volume at any given time. As the concentration of particles in the suspending fluid increases, the probability of having more than one particle present in the measurement volume also increases. When multiple particles are present in the measurement volume, they appear as though they were

Figure 48.3 Particle size distribution of a blend of uniform polystyrene standards.

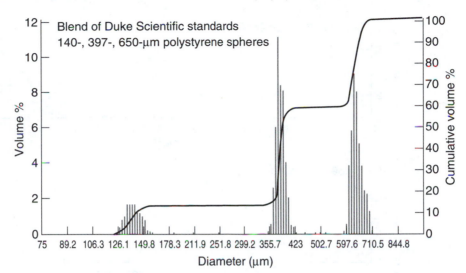

one particle with a cross-sectional area approximately equal to the sum of the individual cross-sectional areas. This double counting or coincident measurement generally leads to size distributions that are skewed to larger sizes. In general, the degree of coincidence should be kept below 5 to 10%, as discussed earlier in this chapter. The effects of coincidence can be readily observed with samples having uniform or narrow particle size distributions. At coincidence levels below approximately 1%, size distributions have a single population. At a coincidence level greater than approximately 2%, a distinct second population is observed in the measured size distribution. If further analysis of the new population is performed, one will see that the average size corresponds to a particle with a cross-sectional area that is twice as large as the population observed at low particle concentrations.

Applications

1. Characterizing the Effectiveness of a Screening Operation.

Particle size measurements of a synthetic polymer resin were performed to determine the effectiveness of a sample fractionation procedure. Approximately 1 mL of a 1% aqueous solution of sodium lauryl sulfate (SLS) was added to 100 mg of a synthetic polymer resin. The SLS was used to facilitate dispersion of the resin into water by wetting the surface of the resin particles. The wet resin was then added to 4 L of distilled water, forming a dilute dispersion. Before the formation of a dispersion, the distilled water was filtered through a 0.1-μm filter. To create homogeneity, the dispersion was stirred using a magnetic stirrer. Baffles were added to the vessel that contained the dispersion to eliminate the vortex present and to promote mixing. A dip tube was added to the entrance of the LO sensor and a vacuum was applied to the exit of the sensor. The vacuum was

Figure 48.4 Particle size distributions of a series of fractionated ion exchange resins.

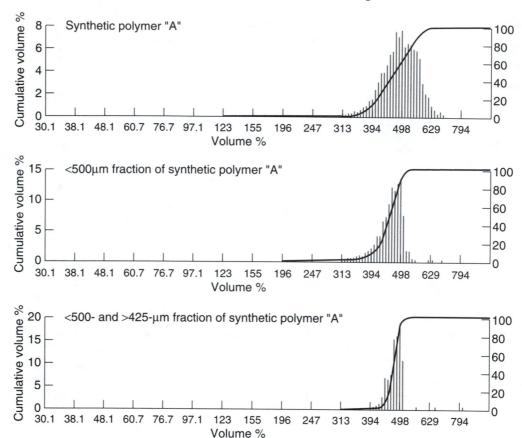

adjusted so that the flowrate through the sensor was approximately 500 mL per minute. The sensor was calibrated with six different samples of uniform polystyrene spheres before the analysis. The analysis time was 3 min. Figure 48.4 gives the size distribution of the unfractionated resin, the original resin with the less-than-250-μm particles removed, and the original resin with the less-than-250-μm and greater-than-500-μm particles removed. Analysis of the various fractions was consistent with the analysis procedure previously described.

2. Size Distribution Analysis of a Sample of Poly(Methyl Methacrylate) Beads.

With an analysis procedure similar to the one described in the previous application, a sample of poly(methyl methacrylate) beads suspended in water were analyzed using a light obscuration particle size analyzer. The particle size distribution of the beads is shown in Fig. 48.5.

Figure 48.5 Particle size distribution of a sample of poly(methyl methacrylate) beads.

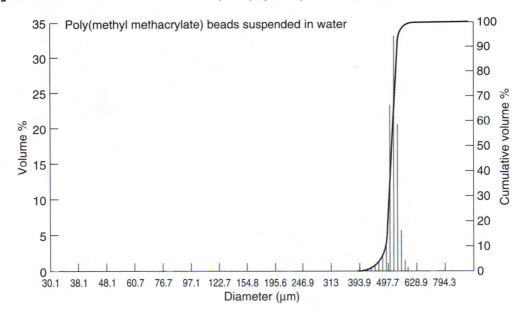

Nuts and Bolts

Relative Costs

Signal processor	$$ to $$$
Sensors	$
Sample delivery systems	$ to $$
Calibration standards	<$
Computer	$
Supplies	<$

$ = 1 to 5 K, $$ = 5 to 15 K, $$$ = 15 to 50K, $$$$ = 50 to 100K, $$$$$ = >100K.

Vendors for Instruments and Accessories

Aerometrics, Inc.
755 N. Mary Ave.
Sunnyvale, CA 94086
phone: 408-738-6688
fax: 408-738-6871
email: sales@aeromatrics.com
Internet: http://www.ccnet.com/~aero

Climet Instruments Co.
P.O. Box 1760
Redlands, CA 92373
phone: 714-793-2788

Duke Scientific Corp. (calibration standards)
2463 Faber Place
Palo Alto, CA 94303
phone: 800-334-3883
fax: 415-424-1158
email: info@dukesci.com
Internet: http://www.dukesci.com

HIAC/RoyCo
11801 Tech Rd.
Silver Spring, MD 20904
phone: 800-638-2790, 301-680-7000
fax: 301-622-0714

Malvern Instruments Inc.
10 Southville Rd.
Southborough, MA 01772
phone: 508-480-0200
fax: 508-460-9692
Internet:http://www.malvern-particle.com

Met One, Inc.
481 California Ave.
Grant Pass, OR 97526
phone: 503-479-1248
fax: 541-479-3057
Internet: http://www.metoneinc.com

Particle Measurement Systems Inc.
5475 Airport Blvd.
Boulder CO 80301
phone: 303-443-7100
fax: 303-449-6870

Particle Sizing Systems, Inc.
75 Aero Camino, Ste. B
Santa Barbara, CA 93117
phone: 805-968-1497
fax: 805-968-0361

PPM, Inc.
11428 Kingston Pike
Knoxville, TN 37922
phone: 615-966-8796

Required Level of Training

Light obscuration instrumentation has been successfully operated by high school students and plant operators. Ensuring the accuracy of the measurements and maintaining the instrumentation, however, requires a trained analytical chemist or technician. The chemist need not be present for every analysis, but an expert is needed to calibrate the instrument, to choose the analysis parameters for the initial measurement of a new sample, to periodically review data, and to troubleshoot the instrument.

Service and Maintenance

Some light obscuration sensors use tungsten filament lamps as the light source. Although such lamps are very stable, they have a finite lifetime and will need replacement. The flow channel in the sensor can become plugged and may require service. Most plugging problems can be overcome by simply reversing the flow direction through the sensor. The sensor calibration will last for extended periods of time (more than 4 mo). One should check the calibration by periodically analyzing a standard or by analyzing a standard before each analysis.

Suggested Readings

ALLEN, T., *Particle Size Measurement*, 3rd ed. New York: Chapman & Hall, 1981.

BARTH, H., *Modern Methods of Particle Size Analysis*. New York: Wiley, 1984.

KAYE, B., *Direct Characterization of Fine Particles*. New York: Wiley, 1981.

References

1. T. Allen, *Particle Size Measurement*, 3rd ed. (New York: Chapman & Hall, 1981).

Chapter 49

Pyrolysis Measurements

Charles Smith
C & J Associates

Summary

General Uses

- Qualitative identification of components in a copolymer or polymer blend
- Identification of low-level polymer contaminants
- Characterization of copolymer sequencing
- Differentiation between copolymers and physical blends of homopolymers
- Determination of monomer ratios in copolymers
- Studying polymer kinetics and degradation mechanisms

Common Applications

- Fingerprint comparison of pyrograms with standard pyrograms to identify major components of a copolymer or polymer blend
- Quantitative determination of the composition of copolymers and polymer blends
- Triad sequencing in vinyl chloride–vinylidene chloride copolymers
- Forensic identification of paints, fibers, adhesives, and plastics
- Studying polymer kinetics and degradation mechanisms

Samples

State

Any solid or cross-linked polymer and polymers soluble in volatile solvents (such as methylene chloride, tetrahydrofuran, or H_2O) can be analyzed.

Amount

Solids The amount ranges from 10 to 200 μg. For low-level detection of contaminants, use up to 1000 μg, but there is a risk that the sample may not completely pyrolyze.

Solutions Generally, 1 to 3 μL of a 1 to 3% solution is needed.

Preparation

For soluble polymers, a 1 to 3% (w/v) solution is generally prepared in a volatile solvent such as methylene chloride, tetrahydrofuran (THF), or H_2O. Curie point wires are dip-coated or aliquots (1 to 3 μL) are deposited on the pyrolyzer filament. In the latter case, the solvent is removed (flashed) by heating the filament to 100 °C in air for a few seconds.

For solids (insoluble polymers), 100 to 300 μg are weighed into a quartz tube or boat, which is placed into a coiled platinum filament. Curie point wires are available in spiral and tube configurations for accommodating solid samples. Other systems use foils, prepared from Curie point alloys, that can be folded around the sample.

Analysis Time

Pyrolysis generally takes place within a matter of milliseconds. For pyrolysis gas chromatography experiments, the limiting factor is the chromatographic time needed to resolve all of the pyrolysis products. Interpretation time for comparing pyrograms or interpreting mass spectra varies from several minutes to several hours depending on the complexity of the pyrolysis data.

Limitations

General

- Polymers containing inorganic fillers may pyrolyze differently from unfilled polymers because of catalytic effects of the filler.
- Changes in the thermal characteristics of a filament affect the quality of subsequent pyrolysis data. Deterioration of Pt filaments by HCl from the pyrolysis of vinyl chloride polymers is an example of an application that may result in changes in the thermal characteristics if a filament is used too long.
- To avoid changes in composition, thermally unstable polymers cannot be left too long in a heated interface before pyrolysis. If a sample begins to degrade or react in a heated inter-

face, products observed after the planned pyrolysis of the sample are not really representative of the original sample.

Accuracy

Accuracy depends on the quality of calibration standards. Relative precision (2σ) of $\pm 5\%$ is readily achieved for calculation of copolymer composition.

Sensitivity and Detection Limits

With sufficiently large samples, polymer contaminants can be determined at concentrations as low as 0.01% (by weight).

Complementary or Related Techniques

- Infrared spectroscopy and nuclear magnetic resonance spectroscopy are often used with data from pyrolysis experiments to define the composition of complex polymeric systems.
- Thermal techniques such as differential scanning calorimetry and thermogravimetric analysis (TGA) are also used to study degradation mechanisms for polymers and the combined TGA–mass spectrometry technique gives valuable data.

Introduction

Pyrolysis, defined as the degradation of macromolecules by thermal energy alone, is one of several commercially available degradative techniques used routinely for the characterization of synthetic polymers and other complex macromolecules. The development and historical perspectives on pyrolysis as an analytical tool are reported in the literature (1, 2). In the broadest sense, pyrolysis involves the application of thermal energy to a sample in the absence of oxygen. This results in breaking of molecular bonds and fragmentation into small molecular species that are related to the original sample composition. These small, characteristic molecules are used to qualitatively identify the structure of the original macromolecule and, with proper standardization, the characteristic small molecules can be used to determine quantitative information on the composition of the macromolecule.

Mechanisms for pyrolysis include elimination of small neutral molecules, chain scission or depolymerization, or random cleavage. The mechanism of pyrolysis or thermal degradation depends on the type of polymer or copolymer, whereas the types and relative yields of various pyrolysis products depend on variables including the pyrolyzer type and the pyrolysis parameters. To obtain reproducible pyrolysis data, these parameters must be optimized for the particular polymeric system being investigated and they must be carefully controlled. Continuous-mode and pulse-mode pyrolyzers are commercially available to control the necessary parameters needed to give reproducible pyrolysis.

Because thermal degradation of complex macromolecules often results in a complex mixture of smaller pyrolysis product molecules, gas chromatography provides easy separation of individual species. This coupling of the thermal degradation event (pyrolysis) with readily available separation science (gas chromatography) results in the technique called pyrolysis gas chromatography (PyGC). The reader is referred to other publications for detailed information on gas chromatographic separations (Chapter 8) (3–5). When the separation and detection system includes a gas chromatograph mass spectrometer (GC-MS, Chapter 31), the technique is called pyrolysis gas chromatography mass spectrometry (PyGC-MS). Thermal degradation of some macromolecules results in undesirable secondary reactions and some PyGC systems may result in discrimination or loss of degradation products during transfer from the pyrolyzer to the chromatographic system. When these problems occur or pyrolysis product separation is undesirable, a useful alternative involves pyrolysis within the mass spectrometer, giving rise to the technique known as pyrolysis mass spectrometry (PyMS).

How It Works

The polymer sample is rapidly heated in an oxygen-free atmosphere to a high temperature called the final pyrolysis temperature (FT). Helium is often the preferred atmosphere because the high thermal conductivity facilitates heat transfer from the sample to minimize secondary degradation reactions and helium is a common GC carrier gas. The time required to raise the temperature of the sample from ambient (or another initial temperature) to this final pyrolysis temperature is the temperature rise time (TRT) and the total time required to raise the sample temperature and pyrolyze it at the final temperature is called the pyrolysis interval or total heating time (THT). Andersson and Ericsson (6) discuss the determination of the temperature–time profile for a sample in PyGC in considerable detail.

The relationship between these various temperatures and times depends on the type of pyrolyzer. Historically, pyrolyzers are classified as either continuous-mode or pulse-mode systems. Furnace and microfurnace pyrolyzers, which are preheated to the desired final pyrolysis temperature before introduction of the sample, fall into the category of continuous-mode pyrolyzers. Figure 49.1 shows a schematic for a commercial microfurnace pyrolyzer.

Pulse-mode pyrolyzers include systems using resistively heated filaments, shown schematically in Fig. 49.2, or ferromagnetic metal wires inductively heated with radiofrequencies. The latter, shown schematically in Fig. 49.3, are called Curie point (CP) pyrolyzers. Although not routinely used for pyrolysis of polymers, laser pyrolysis systems are another class of pyrolyzer.

The reproducible yield of characteristic small molecule fragments from a polymer, called pyrolysis products, is also affected by the cleanliness of the sample holder and other sampling considerations. Pyrolysis using a contaminated surface can result in catalytic effects that can drastically alter the type and amount of pyrolysis products. Heating Pt filaments in air at 1000 °C is adequate to remove most organic residues, although some inorganic materials (such as fillers) may remain behind even under these conditions. Cleaning Curie point wires or filaments in a flame, however, may form metal oxides that will affect subsequent pyrolysis results. For Curie point systems, the position of the wire in the induction coil also greatly affects the nature and the amount of pyrolysis products.

Techniques of sample deposition can also affect the quality of pyrolysis results. For filament and Curie point pyrolyzers, use of uniform, thin polymer coatings gives the best results. Samples are normally deposited as a few microliters of a polymer solution or, in the case of Curie point wires, simply dip-coated from a polymer solution. With microfurnace pyrolyzers or filament sys-

Figure 49.1 Schematic of a two-stage microfurnace pyrolyzer. *(Reprinted by permission of Huthig-Fachveriage.)*

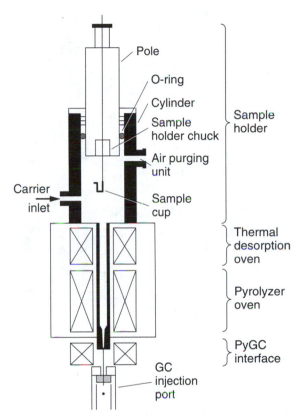

tems using coiled filaments with quartz tubes, the tubes should be carefully washed and heated before use. Sample sizes should be kept small to facilitate good heat transfer from the pyrolyzer to the sample. Andersson and Ericsson (7) discuss the effects of sample size on the reproducibility of pyrolysis results, and Wampler and Levy (8) discuss the effects of sample size, sample geometry, contamination, and other variables on the reproducibility of pyrolysis data.

Figure 49.2 Schematic of a resistively heated filament pyrolyzer. *(Reprinted from S. Liebman and T. P. Wampler, Chromatogr. Sci. 29, Marcel Dekker Inc., 1985, p. 64, courtesy of Marcel Dekker, Inc.)*

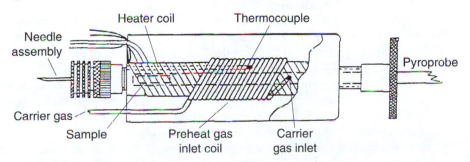

Figure 49.3 Schematic of an inductively heated Curie point pyrolyzer. *(From W. Irwin,* Analytical Pyrolysis: A Comprehensive Guide. *Marcel Dekker Inc., New York, 1982, p. 64, courtesy of Marcel Dekker, Inc.)*

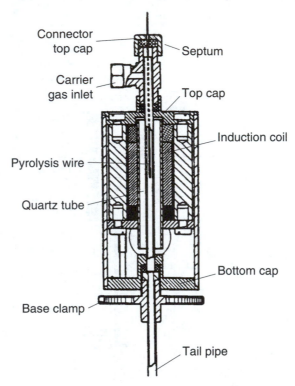

What It Does

Pyrolysis gas chromatographic systems require interfacing the pyrolyzer with the injection port of the gas chromatograph. This is often achieved using a simple needle through the septum of any commercial chromatograph. Other pyrolyzers require a separate interface or pyrolysis chamber, which is connected to the chromatographic injection port by a simple needle assembly. For any of these PyGC systems, this interface is critical to successful, reproducible pyrolysis experiments. Cold spots (places where pyrolysis products can condense before injection into the chromatograph) or overheated zones that facilitate secondary thermal degradations will result in loss or undesirable differentiation of significant pyrolysis products. Consequently, care must be given to adequately heat needle assemblies and any transfer lines from the pyrolyzer to the gas chromatographic inlet. Obviously, direct PyGC-MS experiments eliminate the need for these interfaces and the potential loss of reactive or polar pyrolysis products.

The quality of the PyGC data depends on the quality of the chromatographic system. Generally, current PyGC systems use fused-silica capillary columns, which give a high degree of resolution to pyrolysis products (9). Conventional thermal conductivity detectors (TCDs) and flame ionization detectors (FIDs) are widely used to detect the separated pyrolysis products, although electron capture or flame photometric detectors can be used to obtain specificity for specific elements (such as Cl, S, and N) in the pyrolysis products.

Pyrolysis mass spectrometry (PyMS) systems eliminate some of the problems associated with the transfer of pyrolysis products from an external pyrolyzer to a gas chromatograph. Curie point systems are commercially available that can be interfaced with mass spectrometers so that pyrolysis occurs near the ion source. Transfer of temperature profiles from PyGC to PyMS may not be totally valid because the cooling processes are slower in the vacuum environment of a mass spectrometer. The direct insertion probe option for many spectrometers permits heating samples at temperatures up to 500 °C directly in the ion source. Some instrument suppliers also provide pyrolysis systems that are used directly in the ion source as a replacement for the conventional direct insertion probe.

Tuning of the pyrolysis temperature is important to obtain reproducible pyrolysis data using filament or pulse-type pyrolyzers. This standardization or tuning of the pyrolysis temperature is extremely important for kinetic studies or attempts to differentiate between homopolymer blends and copolymers based on changes in the pyrograms. Curie point wire temperature is a well-defined physical parameter based on the metal alloy composition of the wire. The temperature repeatability for these Curie point wires is dictated by the quality of the metals used to manufacture the wires. For pulse-mode pyrolyzers from CDS Analytical, Inc., each filament probe is accompanied by a calibration table that correlates the controller set point with the resulting filament temperature as measured with an optical pyrometer. Assuming a clean filament probe, this calibration table serves as a guide to the user to define the controller settings needed to obtain a desired probe temperature. Obviously, contaminations that alter the thermal characteristics of a probe will invalidate the calibration table.

To compensate for changes in filament probe calibration (that is, actual temperature relative to the set temperature), Levy (10) proposed the model molecular thermometer approach for tuning pyrolyzers. This approach was based on the pyrolysis of a styrene–isoprene (86/14%) copolymer whose fragmentation was highly sensitive to the actual pyrolysis temperature. Under a strict set of experimental conditions, the ratio of major pyrolysis products (ratio of isoprene:dipentene) varied linearly with the pyrolysis temperature. By pyrolyzing this temperature-sensitive copolymer at different temperatures and examining the relative yields of isoprene and dipentene, an analyst can evaluate variations in the actual temperatures achieved by the pyrolyzer. In 1984, ASTM Committee E-19 reported results of correlation trials attempting to standardize parameters for PyGC data collection based on this tuning concept (11).

Automation of the pyrolysis process is achieved using systems such as the Fischer AP-12 pyrolyzer, which is shown schematically in Fig. 49.4. This unit, readily adapted to the injection port of a gas chromatograph, accommodates 12 Curie point wires, which are placed into separate induction fields. Using a series of valves and an accurate timing sequence, this unit permits unattended pyrolysis of the 12 samples with a high degree of accuracy and precision.

Older automated systems were based on commercially available solid sample injectors (12), but CDS Analytical now markets a special coiled filament pyrolysis insert to replace conventional gas chromatographic injection port liners. This coiled filament accommodates automated injection of polymer solutions using commercial autosampler technology.

Analytical Information

Qualitative

Each type of polymeric material gives a unique set of pyrolysis products based on the various mechanisms of thermal degradation. Polymer degradation mechanisms are generally classified as depolymerization, or unzipping; elimination, which features loss of a small, neutral molecule possibly followed by kick-back fission or cyclization; and random cleavage, which occurs when bond energies in a polymer are similar and rearrangement reactions are not possible. Table 49.1 presents examples

Figure 49.4 Cross-section schematic of the Fischer automated Curie point pyrolyzer. *(Reprinted by permission from Fischer America, Inc. and Expotech U.S.A., Inc.)*

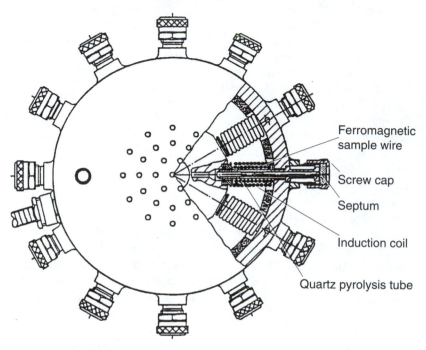

Ferromagnetic sample wire

Screw cap

Septum

Induction coil

Quartz pyrolysis tube

that demonstrate these various degradation mechanisms for vinyl polymers and some condensation polymers. Gas chromatographic separation of these pyrolysis products results in a fingerprint or pyrogram that is considered to be unique for each type of polymer. For PyMS experiments, the mass spectrum is a composite of the mass spectra of the individual pyrolysis products.

Qualitative identification of an unknown polymer by PyGC is achieved by comparison of the sample pyrogram with reference chromatograms from pyrolysis of standard polymers. Tsuge and

Table 49.1 Examples of typical polymer degradation mechanisms.

Mechanism	Polymer	Major Pyrolysis Products
Depolymerization	Polystyrene	Styrene, dimers, trimers
	Poly(methyl methacrylate)	Methyl methacrylate
	Styrene–butadiene rubbers	Styrene, 1,3-butadiene, 4-vinylcyclohexene, oligomers
	Polyamide (Nylon 6)	Monomer (ε-caprolactam)
Elimination	Poly(vinyl chloride)	HCl, benzene
	Poly(vinyl acetate)	Acetic acid
Random cleavage	Polyethylene	Homologous series of alkenes, alkanes, and dienes
	Polyurethanes	Diisocyanates, mono-ols, diols, and cyclic compounds

Table 49.2 Individual pyrolysis products useful for identification of synthetic polymers.

Polymer	Pyrolysis Products
Polyisoprene	Isoprene, dipentene
Polybutadiene	1,3-Butadiene, 4-vinylcyclohexene
Butyl rubbers	Isobutylene
Nitrile rubbers	1,3-Butadiene, acrylonitrile
Polystyrene	Styrene, oligomers
Polyacrylates	Acrylate monomer
Polychloroprene	Chloroprene
Styrene–butadiene	Styrene, 1,3-butadiene, 4-vinylcyclohexene

Ohtani (13) prepared a compilation of pyrograms for polymers using methyl silicone fused silica capillary column technology. This and other publications are sources for reference pyrograms. If available, standard reference polymers can be pyrolyzed and characterized using the same system used for the unknown polymer.

Rather than use the entire pyrogram, some workers recommend use of several specific pyrolysis products as the basis for identification of unknowns by PyGC. Table 49.2 shows some of the useful comparisons achieved by Alekseeva (14).

Stepwise pyrolysis followed by gas chromatography is a useful approach to differentiate between copolymers and homopolymer blends. Pyrolysis of a sample at a series of different temperatures results in different relative yields of unique products from a blend and a copolymer.

Quantitative

Quantitative aspects of pyrolysis experiments include the calculation of copolymer composition, microstructure measurements, and the determination of kinetic and degradation parameters. The following paragraphs detail some of the important quantitative aspects of pyrolysis gas chromatography.

Composition calculations based on PyGC data require identification of one specific and unique pyrolysis product peak that is related to each monomeric component of a copolymer or homopolymer blend. In the example of an acrylonitrile–butadiene–styrene (ABS) copolymer, pyrolysis yields characteristic peaks for the three monomers (acrylonitrile, 1,3-butadiene, and styrene). To quantify the normalized composition of a sample such as an ABS copolymer, a series of equations are written to define the relationship between the unique monomer peaks in the pyrogram and the concentration of these various moieties in the copolymer backbone.

$$K1 = \frac{P_s}{P_a} \div \frac{A_s}{A_a} \qquad K2 = \frac{P_b}{P_a} \div \frac{A_b}{A_a} \qquad (49.1)$$

$$P_s = K1 P_a \frac{A_s}{A_a} \qquad P_b = K2 P_a \frac{A_b}{A_a} \qquad (49.2)$$

$$P_a + P_b + P_S = 100\% \qquad (49.3)$$

where P_a, P_b, P_s = weight % monomer units in a copolymer of acrylonitrile, butadiene, and styrene and A_a, A_b, A_s = monomer peak area responses from the pyrogram for the respective monomers.

Copolymer standards of known composition are pyrolyzed. The area responses for the monomer peaks and the known composition data are substituted into Eqs. (49.1) and these equations are solved to determine the values of constants $K1$ and $K2$. Once the values for these constants are known, these same equations, rearranged as shown in Eqs. (49.2) and (49.3), and the respective peak areas from the unknown sample pyrogram are used to calculate concentrations P_a, P_b, and P_s for the unknown.

Standardization for quantitative analyses based on PyGC data requires careful choice of reference polymers. Pyrolysis of homopolymer blends results in product peaks that are unique for each component of the blend. Pyrolysis of a copolymer containing the same monomeric moieties, however, results in a pyrogram containing peaks for hybrid oligomers in addition to these unique monomeric species. Yields of unique monomeric pyrolysis products can be significantly different for blends and copolymers that have the same nominal composition because monomer yield from a homopolymer blend may be independent of polymer composition. As a result, use of blended standards to calculate the composition of copolymers may lead to erroneous results.

Figure 49.5 compares 1,3-butadiene yields from pyrolysis of a series of polystyrene–polybutadiene blends and rubber-modified impact polystyrene samples with the same nominal compositions and similar rubber microstructure. Note that the difference in 1,3-butadiene yield between the blend and the true copolymer increases with increasing nominal concentration of polybutadiene. The slopes of these correlation curves are considerably different for samples containing higher levels of polybutadiene. Based on a 100-μg sample, a 1,3-butadiene peak height was equivalent to 6% polybutadiene in a blend, but the same peak height was equivalent to 6.7% polymerized butadiene in a styrene–butadiene copolymer.

Using an internal standard may improve the precision when PyGC data are used to calculate the concentration of a single polymeric component of a blend or copolymer (such as polybutadiene in a polystyrene sample). To be useful as an internal standard for PyGC experiments, a polymer must degrade (fragment) quantitatively into a single pyrolysis product and the peak for this product must occur at a retention time free of interference by peaks for products from pyrolysis of the sample. For example, poly(ethyl methacrylate) pyrolyzes to give predominantly ethyl methacrylate

Figure 49.5 Comparison of 1,3-butadiene yields from pyrolysis of copolymers and homopolymer blends with polystyrene.

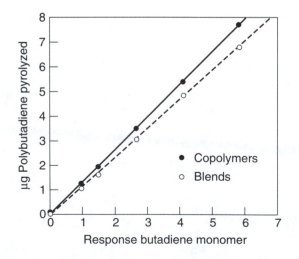

Table 49.3 Triad sources for products from pyrolysis of vinyl chloride (V)–vinylidene chloride (D) copolymers.

Pyrolysis Product	Sequence Designation	Contributing Triad Sequences
Benzene	V3	VVV
Chlorobenzene	V2	VDV, DVV, VVD
m-Dichlorobenzene	V1	DVD, DDV, VDD
1,3,5-Trichlorobenzene		DDD
Vinylidene chloride		DDD

monomer, so this polymer was used by Esposito (15) as an internal standard in the PyGC examination of other acrylic polymers.

Microstructure measurements from pyrolysis data take advantage of the dependence of product yields from pyrolysis of copolymers on the effects of boundaries between the different monomeric units. For example, the pyrolysis of vinyl chloride (V)–vinylidene chloride (D) copolymers follows a dehydrochlorination–cyclization mechanism. Copolymer molecules initially lose HCl, then the resulting six-membered alkene chains cyclize to form chlorinated benzenes (16). The yields of monochloro benzene, m-dichlorobenzene, and 1,3,5-trichlorobenzene are related to the relative number of various VVD, VDV, VDD triads in the copolymer, as shown in Table 49.3.

In another application, Nakagawa and Tsuge (17) report the use of pyrolysis followed by high-resolution capillary gas chromatography to study cross-linking in styrene–divinylbenzene (DVB) copolymers. Linear correlations were obtained between the normalized yields of DVB isomers (counts per microgram) and the known DVB content of styrene–DVB copolymers. Similar correlations were observed using the isomers of ethylvinylbenzene.

Kinetic calculations are made from data obtained by pyrolyzing samples over a range of temperatures using different periods of time. From these pyrolysis data, one calculates the isothermal pyrolysis rate for the polymer at different temperatures, then the Arrhenius equation is used to determine the activation energy for the degradation. With pulse-mode pyrolyzers, a sequential pyrolysis approach was used by Andersson and Ericsson (6) to obtain these data. The same polymer sample was pyrolyzed for a fixed time at some preset temperature. The product yield decreased for each subsequent pyrolysis and the process was repeated until no additional product peaks were observed. Data from these experiments were used to calculate the rate constant for formation of the various pyrolysis products. Repeating the experiment at other temperatures gave data useful in calculation of Arrhenius parameters.

Applications

1. Identification of Components in a Polymer Blend.

Figure 49.6 compares the pyrograms using a 60-m fused silica capillary column (J&W DBWAX; 0.25 μm film) from pyrolysis (CDS Pyroprobe set temperature = 700 °C) of 300 μg of an unknown polymer blend and comparable portions of two reference polymers. Highlighted peaks in the ABS copolymer pyrogram include acrylonitrile (AN), 1,3-butadiene (B), and styrene (S), whereas the major peak from pyrolysis of nylon 6 polyamide was ε-caprolactam (C). Examination of the

Figure 49.6 Pyrograms of a polymer blend (ABS–nylon 6) and the respective polymer standards.

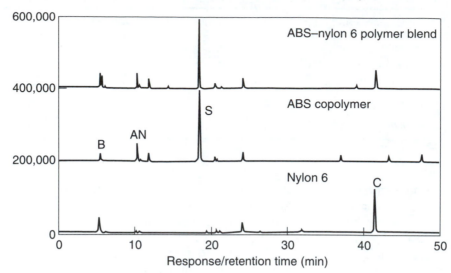

unknown blend pyrogram shows the presence of these peaks attributable to the pyrolysis of the ABS copolymer and nylon 6 components of the blend.

2. Identification of Poly(methyl methacrylate) Contamination in a Styrene Copolymer.

The increased selectivity and sensitivity of the PyGC-MS technique are demonstrated in this example. Figure 49.7 compares the partial (0- to 5-min region) ion current pyrograms for a contaminated copolymer sample and a sample of the virgin copolymer. From mass spectral data, the peak at 2.5 min in the contaminated sample pyrogram (M/Z = 41, 39, 69, 100, 99) was identified as methyl methacrylate (MMA) monomer from the depolymerization or unzipping of poly(methyl methacrylate). Pyrolysis of a series of blends containing poly(methyl methacrylate) and the virgin copolymer resulted in a curve correlating the yield of MMA monomer with the concentration of poly(methyl methacrylate) in the blend. The level of PMMA contamination in the sample was estimated using this correlation curve and the MMA peak response in the contaminated sample pyrogram.

3. Determination of the Vinyl Chloride Triad Sequence Distribution of Vinyl Chloride–Vinylidene Chloride Copolymers.

The pyrogram (450 °C) of a vinyl chloride–vinylidene chloride copolymer, presented in Fig. 49.8, shows resolution of vinylidene chloride, benzene, and chlorinated benzene isomers, which are related to the various triad sequences as summarized in Table 49.3. Chromatography of a synthetic mixture containing benzene, vinylidene chloride, chlorobenzene, m-dichlorobenzene, and 1,3,5-trichlorobenzene allows calculation of molar response values for these pyrolysis products relative to benzene. Division of the peak areas in a sample pyrogram by these relative molar response val-

Figure 49.7 Ion current pyrograms of a copolymer sample contaminated with poly(methyl methacrylate) and a virgin copolymer standard.

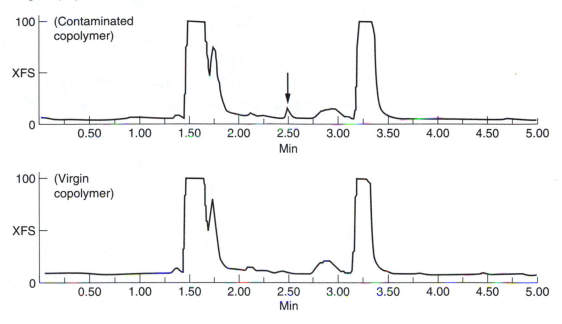

ues gives the respective molar areas which are used in Eq. (49.4) to calculate the mole fraction vinyl chloride (V) in the copolymer (18,19).

$$V = \frac{B + (2/3)C + (1/3)D}{\sum A + B + C + D + E}$$

(49.4)

Figure 49.8 Pyrogram for a vinyl chloride–vinylidene chloride copolymer.

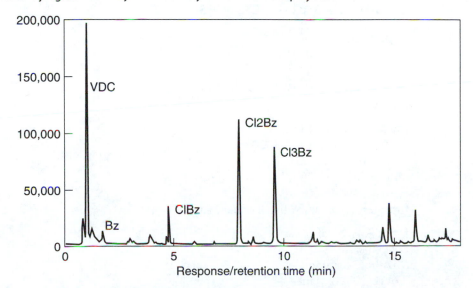

In Eq. (49.4), A = 1/3 molar area of vinylidene chloride peak, B = molar area of benzene peak, C = molar area of chlorobenzene peak, D = molar area of m-dichlorobenzene peak, and E = molar area of 1,3,5-trichlorobenzene peak.

Equation (49.5) and the respective molecular weights for vinyl chloride (62.5) and vinylidene chloride (96.95) are used to convert from mole fraction vinyl chloride to the corresponding weight percent vinyl chloride in the copolymer sample.

$$\text{Wt\% Vinyl chloride} = 6250V/62.5V + 96.95(1 - V) \tag{49.5}$$

Finally, the relative vinyl chloride (VC) percentages in the triads described in Table 49.3 are calculated using the molar area responses, the total VC concentration, and Eqs. (49.6), (49.7), and (49.8).

$$V3 = (B)\%VC/B + 2/3C + 1/3D \tag{49.6}$$

$$V2 = (2/3C)\%VC/B + 2/3C + 1/3D \tag{49.7}$$

$$V1 = (1/3D)\%VC/B + 2/3C + 1/3D \tag{49.8}$$

Nuts and Bolts

Relative Costs

Filament (CDS Pyroprobe 1000)	$$ to $$$
Curie point (Fischer)	$$$
Microfurnace (Frontier Lab)	$$ to $$$

$ = 1 to 5 K, $$ = 5 to 15 K, $$$ = 15 to 50K, $$$$ = 50 to 100K, $$$$$ = >100K.

Vendors for Instruments and Accessories

CDS Analytical, Inc.
7000 Limestone Rd.
Oxford, PA 19363
phone: 800-541-6593, 610-932-3636
fax: 610-932-4158

Fischer America Inc.
1880 Dairy Ashford, Ste. 205
Houston, TX 77077
phone: 281-597-0001
fax: 281-496-0400
Internet: http://www.fischeramerica.com

Japan Analytical Industry Co., Ltd.
208 Musashi
Mizuho, Nishitama
Tokyo 190-12, Japan
phone: 0425 572331

Microscience Inc.
412 Accord Park Dr.
Norwell, MA 02061
phone: 617-871-0308

Pyrolab
P.O. Box 766
S-220 07 Lund, Sweden
phone: 46 46 139797
fax: 46 46 139698
email: pyrolab@algonct.se

Scientific Glass Engineering, Inc.
2007 Kramer Ln.
Austin, TX 78758
phone: 512-837-7190
fax: 512-836-9159

Required Level of Training

Most commercially available pyrolyzers are simple add-on modifications to the injection systems of standard gas chromatographs or mass spectrometers. Pyrolyzer operation does not require special training, and the techniques for efficient loading of samples onto pyrolysis probes are an easily acquired skill. To get the most from the PyGC technique, users must be proficient in gas chromatographic techniques. Techniques for qualitative comparison of unknown and reference pyrograms are easily learned, but interpretation of mass spectral data from PyMS experiments requires more extensive training.

Service and Maintenance

Commercial pyrolyzers use relatively simple electronic circuits, which can be maintained by anyone with an elementary knowledge of electronics. Many commercial suppliers of filament-type pyrolyzers provide service including replacement and recalibration of these probes. The maintenance is restricted to periodically replacing O-rings and seals on interface connections and cleaning quartz tubes, filaments, and needles.

Suggested Readings

IRWIN, W. J., *Analytical Pyrolysis: A Comprehensive Guide*. New York: Marcel Dekker, 1982, 1–578.

SMITH, C. G., AND OTHERS, *CRC Handbook of Chromatography: Polymers.*, vol. 1. Boca Raton, FL: CRC Press, 1982, 1–183.

VOORHEES, K. J., *Analytical Pyrolysis: Techniques and Applications*. London: Butterworth, 1984, 1–485.

References

1. W. J. Irwin, *Analytical Pyrolysis: A Comprehensive Guide*, Chromatographic Science Series, vol. 22 (New York: Marcel Dekker, 1982), 3–44.

2. R. L. Levy, *Chromatographic Reviews*, 8 (1966), 48–89.

3. G. Zweig and J. Sherma, *CRC Handbook Series in Chromatography*, vol. 11, Section I.II (Boca Raton, FL: CRC Press, 1972), 7–23.

4. C. G. Smith and others, "Gas Chromatography," in J. Sherma, ed., *CRC Handbook of Chromatography: Polymers*, vol. I (Boca Raton, FL: CRC Press, 1982), 3–4.

5. C. G. Smith and others, "Gas Chromatography," in J. Sherma, ed., *CRC Handbook of Chromatography: Polymers*, vol. II (Boca Raton, FL: CRC Press, 1994), 3–9.

6. E. M. Andersson and I. Ericsson, *Journal of Analytical Applied Pyrolysis*, 1 (1979), 27–38.

7. E. M. Andersson and I. Ericsson, *Journal of Analytical Applied Pyrolysis*, 2 (1980), 97–107.

8. T. P. Wampler and E. J. Levy, *Journal of Analytical Applied Pyrolysis*, 12 (1987), 75–82.

9. C. G. Smith and others, "Gas Chromatography," in J. Sherma, ed., *CRC Handbook of Chromatography: Polymers*, vol. II (Boca Raton, FL: CRC Press, 1994), 3–9, 23–24.

10. E. J. Levy, "Concept of a Model Molecular Thermometer," Conference on Analytical Chemistry and Spectroscopy, Pittsburgh, 1977.

11. E. J. Levy and J. Q. Walker, *Journal of Chromatographic Science*, 22 (1984), 49–54.

12. G. L. Coulter and W. C. Thompson, "Automated Analysis of Tyre Rubber Blends by Computer-Linked Pyrolysis–Gas Chromatography," in C. E. R. Jones and C. A. Cramers, eds., *Analytical Pyrolysis* (Amsterdam: Elsevier, 1977), 1–15.

13. S. Tsuge and H. Ohtani, *Pyrolysis–Gas Chromatography of High Polymers: Fundamentals and Data Compilation* (Tokyo: Techno-Systems, 1989), 1–371 (in Japanese).

14. K. V. Alekseeva, *Journal of Analytical Applied Pyrolysis*, 2 (1980), 19–34.

15. G. G. Esposito, *Analytical Chemistry*, 36 (1964), 2183–85.

16. S. Tsuge, T. Tadaoki, and T. Takeuchi, *Die Makromolekulare Chemie*, 123 (1969), 123–29.

17. H. Nakagawa and S. Tsuge, *Macromolecules*, 18 (1985), 2068–72.

18. H. J. Harwood and W. M. Ritchey, *Journal of Polymer Science*, B-2 (1964), 601–7.

19. W. A. Dietz, *Journal of Gas Chromatography*, 5 (1967), 68–71.

Chapter 50

Thermal Analysis Techniques

Andrew J. Pasztor
The Dow Chemical Company

Summary

General Uses and Common Applications

- Determination of glass transition temperature and melting point of polymer systems
- Determination of degradation temperatures and rates
- Determination of linear and volumetric expansion
- Determination of modulus and mechanical properties
- Determination of dielectric properties
- Determination of the crystallinity level of a polymer system
- Determination of relative levels of components in blends
- Screening for level and effectiveness of additives and stabilizers
- Determination of the rate of crystallization

Samples

State

Usually solids, although liquids can also be studied. Allows for the study of materials as-made for investigation of the effects of various processing conditions.

Amount

1 mg to several grams, depending on the test. With nonhomogeneous samples, these small sample size requirements sometimes create problems in obtaining a representative sample.

Preparation

Depending on the purpose of the study, sample preparation can range from simple to difficult. To study material properties, obtaining a representative sample and then removing thermal history requires some time, but this process is usually relatively easy and straightforward. To study the properties of an as-made part, sampling is difficult but little or no sample preparation is required after obtaining the sample.

Analysis Time

Simple differential scanning calorimetry (DSC) scans are 20 to 40 min, whereas thermogravimetric analysis (TGA) scans can take 30 min to several hours. Thermomechanical analysis (TMA), dynamic mechanical spectroscopy (DMS), and dielectric measurements are usually on the order of hours and kinetic studies can require hours.

Limitations

General

Thermal techniques often are not specific for the analysis of a polymer system. With supporting methods, these techniques often aid in identification of polymer systems.

Accuracy and Sensitivity

For quantitative analysis, both accuracy and sensitivity are at best about ±2%. Other factors generally complicate the system and can decrease both the accuracy and sensitivity.

Complementary or Related Techniques

- Infrared (IR) provides additional structural information and helps identify types of materials present.
- Mass spectrometry is used to identify fragments in TGA experiments and help define the degradation mechanism.
- Pyrolysis techniques help identify the various decomposition products.
- Microscopy helps to visualize presence of different phases present in the sample.

Introduction

Thermal analysis for polymers is a series of techniques, including calorimetry, TGA, TMA, dynamic mechanical analysis (DMA), and dielectric analysis, which are used to characterize different aspects of polymer systems. Thermal analysis methods often cannot be used for identification purposes by themselves without data from some complementary techniques. For example, a polymer with a melting point of about 220 °C could be either nylon 6 or poly(butylene terephthalate). Appearance of an amine functionality in the IR spectrum of the same material would identify the material as a nylon, but may not be specific as to which type. Using both pieces of data, thermal and IR, the material could be identified as nylon 6, demonstrating the complementary nature of thermal analysis.

Thermal analysis often is most useful for quantitative analysis and studying the state of a polymer. Thermal methods are good qualitative tools because they are used to rapidly determine whether a polymer is the same or different from a standard, or to determine the quantity of a component after it has already been established that the component is present in the system. Characterizing the state of a system can include determining physical properties, such as heat capacities and linear and volumetric expansion coefficients, as well as reaction, including degradation reactions and kinetics.

Following a brief description on the theory of each technique, several examples of the more usual applications are given to show how these techniques can be applied to polymer characterization.

How It Works

Differential scanning calorimetry (DSC) is the most often used calorimetric method because it is rapid, easy to operate, and readily available. In DSC, a sample contained in an appropriate holder and an empty reference holder are placed in the DSC cell in the sample and reference positions. Power is applied to heaters in the system to either ramp the temperature of the DSC at a specified rate (such as 10 °C/min) or to hold the DSC isothermally at a given temperature. The DSC instrument measures the difference in the heat flow between the sample and the reference.

Differential scanning calorimeters are classified as either a power-compensated DSC or a heat-flux DSC. In the power-compensated DSC, the temperature of the sample and reference are measured and compared as shown in Fig. 50.1. Depending on the difference in observed temperature, more or less power is applied to the sample to adjust the sample temperature so that it is the same as the reference temperature, and the instrument records this change in power. In a heat-flux DSC, a single heating element is used to heat both the sample and the reference, which are positioned on a thermocouple as shown in Fig. 50.2. The difference in temperature is then measured

Figure 50.1 Schematic of a power-compensated DSC. The computer generates the program temperature (T_p) signal and the platinum resistance thermometer (PRT) sensor detects a small error signal between T_p and the sample temperature. The resistance heater provides power to each heater to keep both sample holders on program. The difference in power required between the sample and reference is amplified and recorded. *(Reprinted by permission of Perkin-Elmer Corporation.)*

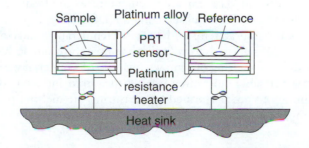

Figure 50.2 Schematic of a heat-flux DSC. *(Courtesy of TA Instruments, Inc.)*

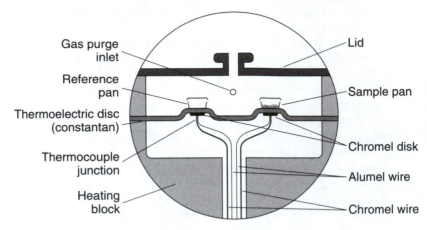

and used to determine the heat flow. A more extensive discussion of DSC instrument types is presented by Turi (1).

A typical DSC experiment involves encapsulating a small portion (0.5 to 100 mg, usually 2 to 10 mg) of the desired sample into a sample pan. Because small samples are used, obtaining a representative sample must be a matter of concern. Sample pans are most often made from aluminum, although pans made from gold, stainless steel, and in some cases even sealed glass ampoules can be used for DSC. A gas purge is usually applied when the sample is placed into the DSC cell. In different experiments, performing the study in an inert atmosphere such as nitrogen is desirable; in other cases the effects of oxidation may be studied by using an air or oxygen atmosphere. Most commercial DSCs are able to run at low temperatures, ranging to about −150 °C, although in practice special care is required to obtain good data below about −100 °C. Most commercial DSCs have an upper temperature limit of about 700 °C, which is more than adequate for most polymer studies. There are instruments that will go to 1700 °C, but these are used primarily in the study of inorganic materials.

Once the sample is in the DSC cell with the desired purge, the temperature is adjusted to 10 to 40 °C below where the first data point is desired because it can take 1 to 2 min for the DSC to come to an equilibrium heating rate. Until this equilibrium rate is achieved, heat flow data represent equilibration of the cell, not the sample.

Most commercial DSCs now use a computer interface to set up temperature programs, control the DSC during the heating program, and collect data. When the sample is at starting temperature, the analyst simply begins the temperature profile and the computer carries out the experiment.

As part of the computer control package, most instrument manufacturers also provide software packages for the analysis of the collected data. This software allows the analyst to determine melting points and glass transition temperatures and to integrate peaks to quantify the heat from exotherms or endotherms observed in the DSC scan.

DSCs are calibrated for both heat and temperature and American Society for Testing and Materials (ASTM) standards exist for both temperature and enthalpy calibrations (2,3). For enthalpic calibrations, the melting of a known quantity of a material with a well-characterized heat of fusion (indium is probably most often used) is measured and a correction factor determined. Temperature calibrations are performed by measuring the melting point for well-characterized materials and comparing the observed melting point to the literature value. Temperature calibrations should be performed over a range wide enough to include the temperature range being studied for a particular sample.

TGA is used to follow the change in mass of a sample as it is heated or held isothermally at a specific temperature. For TGA, the balance mechanism is usually housed outside the furnace area to maintain good mass sensitivity. Various manufacturers use either a horizontal or vertical configuration for the relationship of the balance to the furnace, with wires or rods connecting the sample and the balance, to permit mass determination. A typical TGA schematic is given in Fig. 50.3. Both configurations are subject to drift with increasing temperature. While the balance is outside of the furnace area, temperature effects remain on the ability to measure weight. One common problem is buoyancy effects as the sample is heated, particularly for instruments where the sample hangs vertically from the balance. The horizontal furnace design was developed to overcome this problem; however, in this case the material used to connect the sample to the balance, usually a quartz rod, expands upon heating and thus the force (mass times distance) measured by the balance changes with changes in the length of the rod. The TGA instrument interprets this as a change in mass.

Operating temperatures for TGA typically range from about room temperature to 1000 °C, although at least one manufacturer has an instrument that has a maximum temperature of 2400 °C. The 1000 °C temperature range allows the use of quartz tubes for containing the experiment. TGA must use high-temperature inert sample holders, and quartz, ceramic materials, and platinum are the most commonly used materials of construction. In a TGA experiment, a gas is purged through the system to provide a suitable atmosphere for the measurement and remove decomposition

Figure 50.3 Schematic of a typical TGA balance and heater assembly. As the sample weight changes, the balance tips, resulting in a change in the light hitting the sensor. The sensor output is amplified and provides current to the coil, which restores the beam position. This restoring current is the ordinate (y) and the thermocouple signal is the abscissa (x) of a conventional TGA plot. (Reprinted by permission of Perkin-Elmer Corporation.)

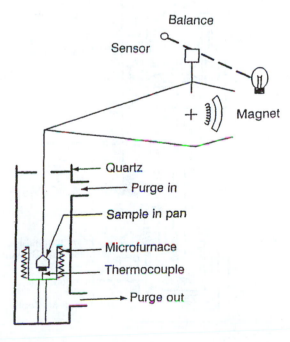

products from the sample chamber. Purge gases for TGA experiments are usually nitrogen (inert) or air (oxidizing) although other atmospheres are also used. Caution must be used when a reducing atmosphere is desired because the use of hydrogen as a purge gas can create a serious hazard. Decomposition products for polymer systems often are condensable tars, which must be trapped or flushed from the system. Coating parts of the sample holder system with these tars results in recording an inaccurate weight loss. Consequently, TGA experiments must be performed in well-ventilated areas, such as a hood, in order to prevent buildup of degradation gases in the area.

For a TGA experiment, sample mass ranges from about 2 mg to as much as 3 g in some systems, with typical sample masses being 5 to 150 mg. Many typical TGA system balances are sensitive to better than 5 μm (not taking into account buoyancy problems), giving TGA a sensitivity that can reach 10 ppm.

Temperature calibration in TGA experiments is performed at a series of temperatures using either melting point standards or the Curie temperature of ferromagnetic materials. ASTM is currently developing a standard for temperature calibration of TGAs.

Thermomechanical analysis (TMA) and thermodilitometry are the methods used to characterize volumetric changes in polymer systems. These changes are often used to characterize the changes in the physical state of a polymer and are critical for many applications of polymer systems. A commercial TMA, shown in Fig. 50.4, usually consists of a dimensionally stable sample

Figure 50.4 Schematic of a typical TMA assembly. Current is supplied to the force coil to support the weight of the core rod and probe. Additional current is provided to apply the desired force on the sample. The LVDT position transducer provides a voltage proportional to the deflection from center null. As the furnace heats, the sample expands or softens and the analyzer output is the sample height versus sample temperature. *(Reprinted by permission of Perkin–Elmer Corporation.)*

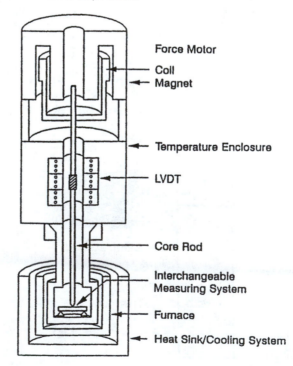

Force Motor

Coil

Magnet

Temperature Enclosure

LVDT

Core Rod

Interchangeable Measuring System

Furnace

Heat Sink/Cooling System

holder (quartz is typical), a measuring probe (again with good dimensional stability), a device to measure the deflection of the probe, a linear variable differential transformer (LVDT), and a temperature control unit. TMA experiments can be run from about –100 °C to 700 °C; however, the upper use temperature is usually much lower and is dictated by the thermal stability of the polymer sample. Because it is critical to maintain good contact between the probe and the sample, a load is usually applied to the system. The load is typically a force of a few grams, although newer TMAs have the ability to electronically set the force to range from 0.1 to about 100 Newtons. Probes for TMA instruments can range from a fine-tipped quartz rod for measuring penetration of the sample to various large, flat surfaces intended to measure the expansion of the sample.

Although the instrument is designed and constructed with materials of low thermal expansion, obtaining accurate results requires calibrating the instrument for temperature and expansion. This calibration must be performed under the load to be used for testing the polymer. Temperature calibration is performed using various metals as melting point standards, with indium, tin, and zinc being typical metals (4). Aluminum is often used if a sample of the melting standard is sandwiched between two higher-melting metal discs. In this case, a deflection of the TMA probe is observed when the melting standard melts and the temperature at which this occurs corresponds to the melting point of the standard.

Dimensional calibration of the LVDT is performed by running a sample (such as aluminum) of known expansivity. In this method, the initial length of the sample is determined at room temperature and then the change in length of the sample is determined with changes in the temperature (5). As the temperature is changed, a finite time is required for the sample to come to the equilibrium length for that new temperature. This requires using slow temperature ramp rates (less than 5 °C/min) or temperature step isothermal methods. The linear coefficient of thermal expansion (LCTE) is defined by Eq. (50.1).

$$LCTE = (1/L)(\delta L/\delta T) \qquad\qquad (50.1)$$

where L = length of the sample and T = temperature (°K). The calculated LCTE is then compared to the standard value for the standard and a correction factor is applied to the data.

Once the instrument is calibrated, a TMA experiment is run in a manner similar to the calibration experiment. The sample is placed onto the sample holder, the temperature is adjusted to the starting temperature, and the temperature is ramped or stepped over the desired temperature range. As with other modern thermal analysis instruments, the entire experiment, sometimes including the initial measurement of the sample dimension, is computer controlled and data analysis software allows for easy calculation of the LCTE or deflection temperature.

Dynamic mechanical spectroscopy (DMS) applies a physical force to one portion of a test specimen and measures the response of the sample at some distance from the specimen (1,6,7). A test specimen mounted in the dynamic mechanical spectrometer is subjected to a sinusoidal mechanical deformation of a preset force. Commonly, the frequency (typically 0.001 to 10 Hz) is varied but the applied force remains constant during any given experiment. Various types of deformation can be used, including torsional, elongational, and bending motions. DMS experiments are often performed over a range of temperatures (typically –100 °C to 500 °C). The frequency shift and the force transmitted through the sample are measured and used to determine the storage and loss moduli (8). The ratio of the loss modulus to storage modulus remains nearly constant, or changes only slightly in a regular fashion, over temperature ranges that permit no new relaxations. At temperatures where new molecular motions such as glass transition temperature (T_g) begin, there is an increase in the loss modulus and often a decrease in the storage modulus so that there is a maximum in their ratio. This maximum shows the temperature range over which the molecular motion begins to occur and it is used to assign the temperature of the transition.

Figure 50.5 The glass transition temperature as observed by changes in volume and expansion (V and α), heat flow and heat capacity changes (H and C_p), and the shear modulus and loss modulus (G' and G^*). *(From "Transitions and Relaxations in Amorphous and Semicrystalline Organic Polymers and Copolymers," by R. F. Boyer, in* Encyclopedia of Polymer Science and Technology, *supplement vol. II, p. 747, copyright © 1977. Reprinted by permission of John Wiley & Sons, Inc.)*

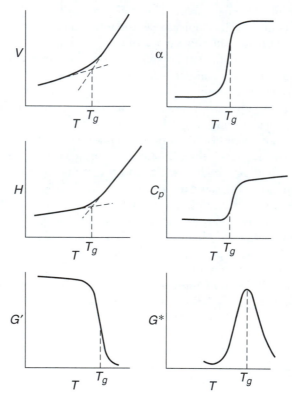

The loss modulus curve in Fig. 50.5 illustrates another point important in polymer analysis. If you are asked to determine the melting point of a pure inorganic material or small organic molecule (such as indium metal or benzoic acid), a single sharp melting peak is observed usually over a small temperature range of 1 to 3 °C. In characterizing polymer systems, however, that is usually not the case. Most polymers consist of a molecular weight distribution and many polymers also have a variety of defects (such as branches). Therefore, a thermal transition, such as T_g, does not occur at a single temperature but over a range of temperatures. For DMS testing, the use of a variety of frequencies helps to determine the distribution of properties in a sample because a response to a deformation is observed only when the relaxation time for the motion under study is the same as the deformation frequency. If DMS were used to study the T_g relaxation in a polymer, a different value for the loss modulus peak maximum would be observed for the different frequencies used in the test.

In a dielectric experiment, a test specimen is fixed between capacitor electrodes (plates). A sinusoidal voltage oscillation (ranging from about 0.001 to 10^7 Hz) at a specified voltage is applied across a pair of electrodes. The shift in phase and amplitude of the electric current, relative to the applied field, is measured as the field passes through the polymer. As in the case of mechanical testing, the measured values are used to determine the stored energy (capacitance) and lost energy (resistance) for the sample (7). These values are commonly reported as the dielectric constant or permitivity and the dielectric loss value or dissipation factor.

Figure 50.6 The dielectric tan δ (top) and mechanical tan δ (bottom) for poly(methyl methacrylate). *(From Introduction to Polymer Viscoelasticity, by J. J. Aklonis and W. J. MacKnight, copyright © 1983. Reprinted by permission of John Wiley & Sons, Inc.)*

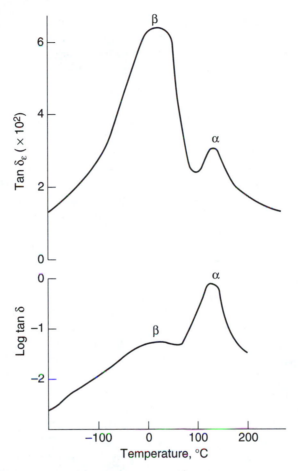

With the similarities between DMS and dielectric spectroscopy, why choose one method over the other? One answer to this question depends on the type of transition being studied because some transitions are more sensitive to mechanical deformation and others to electrical probing (8). Fig. 50.6 illustrates this point by comparing the dielectric and mechanical tan δ (dielectric loss factor/dielectric constant, loss modulus/storage modulus) for poly(methyl methacrylate) (PMMA) (8). For PMMA, the α transition is T_g and the β transition is the rotation of the methacrylate group around the main polymer chain. Because T_g is a main chain motion that affects modulus and polymer stiffness, this motion is most intense for a mechanical deformation. The rotation of the methacrylate group, on the other hand, is not a main chain motion and it only weakly affects the stiffness of the polymer, and is therefore a weak mechanical transition. The methacrylate group is a highly polar segment and it gives a more intense dielectric response. This demonstrates that comparison of mechanical and dielectric responses, combined with a knowledge of the polymer structure, will give added insight to the relaxation temperatures and what motions are occurring at these temperatures.

What It Does

Calorimetry is one of the most commonly applied techniques for polymer characterization. In polymer systems it is possible, using different techniques, to observe a number of transitions. However, two transitions that involve main chain motions, and therefore are of most importance, are the glass transition and the melting of semicrystalline polymers.

At the glass transition temperature (T_g), the physical properties of a polymer change dramatically. At this temperature, where large-scale molecular motions involving more than 20 atoms of the main chain are allowed, the polymer goes from a glassy state to a rubbery state. The physical changes are represented in Fig. 50.5. At the glass transition, there is a change in the rate of volume expansion and a change in the total heat content. For the expansion coefficient and heat capacity, this change at T_g leads to step changes in the linear coefficient of thermal expansion (α) and the heat capacity (c_p). This change in heat capacity is readily measured calorimetrically. This step in heat capacity can be directly related to the difference in heat capacity for the glassy and liquid (rubbery) states of the polymer. For a pure high-molecular-weight polymer, both the temperature where a glass transition occurs and the specific heat capacity change (ΔC_p/gram of polymer) are constants; however, variations occur when studying low-molecular-weight polymers or polymers with additives. Calorimetric methods are often the easiest method for determining T_g.

The melting point (T_m) and the extent of crystallinity are two other important thermal transitions of polymers. Not all polymers can crystallize and a variety of factors affect the ability of a polymer to crystallize. A detailed explanation of these factors is beyond the scope of this text, but probably the most important factor affecting whether a polymer can crystallize is polymer tacticity. Atactic polystyrene cannot crystallize, but both isotactic and syndiotactic polystyrene can crystallize, although they have different melting points and heats of fusion. This being the case, the melting point and the heat of fusion can often be used to determine the presence and the quantity of a polymer present in blend. The heat of fusion can be used to determine the extent of crystallinity in a polymer sample, and because the extent of crystallinity can have a dramatic effect on polymer properties, this can be a critical measurement for characterizing the polymer.

Thermogravimetric analysis (TGA) is a method used to study the weight loss characteristics of a polymer when the sample is heated over a temperature range or held isothermally at a temperature. TGA is often used to determine the temperature for the onset of degradation, which becomes the upper use temperature for a polymer. TGA is also used to characterize polymer systems and the TGA decomposition curve is used for identification purposes (9). Additionally, TGA is used to quantify the presence of volatile components (such as adsorbed water in a hygroscopic polymer) that may be present in a sample.

When studying degradation, the analyst must be certain to understand the meaning of the term *degradation* and know that degradation as measured by different techniques can mean different things. For degradation phenomena to be observed by a TGA experiment requires the generation of a molecular fragment that is small enough to be volatile at the experimental temperature. If a polymer chain with 1000 carbon atoms in the backbone "degrades" into two polymer chains, each with 500 carbon atoms in the backbone, it is likely that no product that would be volatile at 300 °C would be made. However, because the molecular weight was reduced by a factor of two, changes in other important properties (such as modulus) may have occurred and these are best examined using other techniques.

Thermogravimetric analysis requires a sensitive balance and a well-controlled furnace or other suitable heating system to maintain a specified temperature or temperature ramp program. Data collected during a TGA experiment includes time, temperature, and sample weight. Commercially available TGAs are interfaced with a computer that can be used to set up the experiment, control the instrument during the experiment, collect data, and analyze the data after the experiment.

Thermogravimetric analysis can be linked to other methods to allow for even greater characterization of a sample. Methods such as TGA Fourier transform infrared, TGA mass spectrometry, or TGA gas chromotography mass spectroscopy are used to characterize the degradation products given off by the sample. In a multicomponent sample, these data are used to determine which part of the sample is degrading over a given temperature range, resulting in some evaluation of the degradation mechanism.

Thermomechanical analysis (TMA) and thermodilitometry are the methods used to characterize volumetric changes in polymer systems. As noted in Fig. 50.5, the glass transition temperature can be characterized as a change in volumetric expansion. One author described TMA as the measurement of the temperature of deformation under a static load, whereas thermodilitometry determines the change in dimensions under a static load (1). The same instrument is used for both measurements and is usually called TMA.

Dynamic mechanical spectroscopy (DMS or DMA) and dielectric spectroscopy are used to determine changes in modulus, the last example defining T_g. Of the three different manifestations of the glass transition temperature shown in Fig. 50.5, changes in enthalpy/heat capacity and volume/expansion were already discussed. Polymers, being long-chain molecules, have a variety of possible molecular motions that, upon heating from a low enough temperature, are allowed at specific temperatures (10). For some motions, such as T_g or the melting point, there are major changes in bulk polymer properties, such as modulus or toughness. The temperature at which these and other molecular motions are allowed is also useful in identifying and defining a polymer. In some cases, this temperature information helps to define the occurrence of less dramatic, although important, changes in polymer properties, such as the change from brittle to ductile fracture (10, 11). These sub-T_g relaxations usually involve the motion of molecular fragments that are only a few atoms in length, and are significantly weaker in intensity than the T_g motion. This often makes it difficult to detect these transitions, and in fact they often cannot be detected using DSC or TMA but are detectable using DMS or dielectric spectroscopy.

Dielectric spectroscopy is much like DMS in that a sinusoidal oscillation is applied to a test specimen, resulting in a measurable response. In the case of dielectric spectroscopy, however, an electrical force is used instead of a mechanical force. Polymer chains that contain dipoles, such as acrylates or halogenated polymers, are studied by dielectric spectroscopy. Although we would not expect a nonpolar polymer such as polyethylene to have a significant dielectric response, the presence of small levels of impurities, such as oxidized polymer fragments, gives enough of a dielectric response that dielectric measurements on these polymers often are possible.

Analytical Information

Qualitative

Although glass transition temperature and melting point are useful for classifying the polymers present in a sample, thermal methods generally cannot be used for the absolute identification of materials present in a polymer system. Thermal methods, particularly DSC and TGA, are often used in quality control applications to verify that a given lot of material is the same as a control sample of that chemical.

Quantitative

Once the presence of a given material has been demonstrated by other analytical techniques, thermal methods are used for quantitative analysis (such as the determination of vinyl acetate content of poly(ethylene-co-vinyl acetate). The use of DSC to determine heats of fusion and the extent of crystallization for crystalline materials are discussed in the Applications section of this chapter. Levels of inorganic fillers are also determined by thermogravimetric analysis.

Applications

1. Determination of Glass Transition.

From a thermodynamic and mechanical standpoint, the glass transition temperature is one of the most important parameters for characterizing a polymer system (1, 2, 10, 11). Therefore, the determination of the glass transition temperature is often one of the first analyses to be performed on a polymer system. The glass transition temperature can be determined by a variety of techniques, and examples of three techniques, TMA, DMA, and DSC, are summarized in a paper by Cassel and Twombly (12). In this paper, these three methods are used to characterize a cured epoxy system. In addition to describing the experimental methodology used for determining T_g by each of the methods, the authors discussed the differences in T_g as determined by the different methods.

Figure 50.7 illustrates the type of data obtained for each of the three methods for an epoxy/woven glass composite. At T_g, the TMA curve drops sharply as the material softens. The extrapolated onset on this curve is used for defining the glass transition temperature as 122.9 °C. In the DSC curve (solid line), a step change in the heat capacity is observed in the same region. The mid-

Figure 50.7 The glass transition temperature for an epoxy/woven glass composite as determined by TMA, DMA, and DSC. *(Copyright ASTM. Reprinted with permission.)*

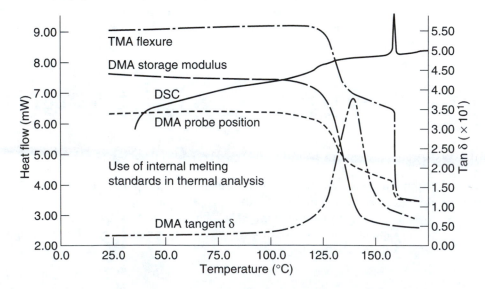

point of the step transition, 121.3 °C for this sample, is taken as T_g. The other three curves in the figure are various representations of the DMA data for this sample.

As previously mentioned, glass transitions occur over a temperature range, and there are a variety of options for where to choose the glass transition temperature. In this work, the extrapolated onset of the storage modulus at 0.1 Hz gave a value of 122.1 °C which was in good agreement with values obtained by TMA and DSC. The authors noted that, for this sample, these DMA conditions give the best agreement with other methods. For other materials, some other temperature, possibly corresponding to the loss modulus, tangent delta extrapolated onset temperature, or peak temperature might give better agreement with other methods. The effects of shifts in frequency (from 0.1 to 10 Hz) were also discussed. It is important to remember that it is not critical that all methods give the same value of T_g. Because the different methods can represent different deformation conditions, it is often valuable to use a test method that most closely represents the conditions under which the polymer will be used. For strictly analytical purposes, the determination of T_g by DSC using published parameters (13,14) is usually the method of choice, because this test is more rapid than TMA or DMA.

2. Heat Capacity Determinations.

The basis for all calorimetric measurements is the determination of heat capacities. In the DSC scan of a polymer system, in the absence of any other transition, the DSC curve represents the change in heat capacity of the sample over the experimental temperature range (9). Even at the glass transition temperature of a polymer, the step observed in the DSC curve is related to the change in heat capacity when the polymer goes through the transition from the glassy to the liquid state (10,15,16).

The determination of heat capacities is often performed using DSC methods (1,9,17). In this method, a DSC scan of an empty pan is recorded followed by the DSC scan of a well-characterized heat capacity standard, such as sapphire. Finally, the sample of interest is run in the same DSC pan. The data for the empty pan is subtracted from the curve data for the reference standard, and the DSC is calibrated for heat capacity. (Calibration of the DSC for heat flow, with indium for example, is not the same as calibrating for heat capacity determinations.) The baseline is then subtracted from the sample run and the heat capacity of the unknown is determined using the heat capacity calibration.

The reported determination of the heat capacity of various nylon polymers is a good example of how to determine heat capacities (17). Of particular interest in this work was the relationship between the molecular structure of the nylons studied, that is, the number of methylene groups and the number of amide groups. An equation was developed to calculate the heat capacity at any temperature between the polymer T_g and about 600 °K from the chemical structure. To be analytically useful, this type of data also helps to understand the relative molecular motions in the polymer liquid.

3. Characterization of Polymer Blends.

As seen in Fig. 50.7, a shift in the heat capacity curve appears at T_g in a DSC experiment, a peak in the loss modulus or tan δ appears in DMA experiments, and a change in the volumetric expansion occurs for TMA experiments. When more than one polymer is present, such as in a blend, we observe more than one T_g (15). One of the most commonly used materials is rubber modified high-impact polystyrene (HIPS). This material contains two polymeric phases, often a polybutadiene phase with a T_g near –85 °C and a polystyrene phase with a T_g near 104 °C. When scanned from a low temperature (about –120 °C) to 200 °C, two steps in the heat capacity are observed for DSC measurements and two maxima in tan δ are observed for DMA experiments (18).

This can be useful for studying blends of similar polymers such as the characterization example of styrene–butadiene latex blends. Because both components are composed of styrene and butadiene, a blend of a 75% styrene and 25% butadiene latex with one containing 25% styrene and 75% butadiene would look like a single polymer based on data from spectroscopic methods such as infrared (IR). In a DSC scan, however, different T_gs would be observed for each of the styrene–butadiene compositions, confirming the existence of two different compositions in the blend.

4. Compositional Effects on the Glass Transition.

Homopolymers and blends of homopolymers, such as the high-impact polystyrene example discussed above, show a T_g for each of the polymers present in the blend. In the case of copolymers, where two or more monomers are blended together and reacted in an essentially random fashion, the T_g is intermediate between the T_gs of the homopolymers for each of the monomers used. Various relationships for calculating the change in T_g with composition have been developed and a simple relationship is given in Eq. (50.2).

$$1/T_g = W^a/T_g^a + W^b/T_g^b \qquad (50.2)$$

where W = the weight fraction of monomers a and b and T_g = respective T_g for homopolymers of monomers a and b.

Figure 50.8 shows the change in T_g for various copolymers of styrene with other monomer units (11). Therefore, for any copolymer, at any mole fraction of styrene, only a single T_g value exists. If the T_g of a copolymer is determined and the T_g composition curve exists, then the composition of the copolymer can be approximated from the T_g. This holds only when the copolymers being studied are prepared in a random fashion.

As polymers are being developed to have more specialized properties, it is becoming more common to see terpolymer systems containing three or more monomer types. Although the equation for calculating T_g can be expanded, the relationship between T_g and composition becomes more complicated, and therefore of less analytical value.

5. Additive Analysis.

Differential scanning calorimetry (DSC) is used to study additives added to polymeric systems to increase flowability in the melt, to reinforce the toughness of samples, or to improve the oxidative stability or reduce the flammability of polymer samples.

One example of DSC use for studying oxidative stability and the effect of adding antioxidants involves characterization of the oxidative stability of polyethylene (19). The oxidation of nearly any polymer is an exothermic event, so when the polymer begins to oxidize an exotherm is observed. The presence of a material to inhibit the oxidation of the polymer, an antioxidant, significantly reduces the rate of oxidation of the polymer until the antioxidant is consumed, and the exotherm is then observed.

A scanning method and an isothermal method are used to study the oxidation of polymers. In the scanning method, the sample is heated at a specified rate in an air atmosphere to a high enough temperature that the exotherm of oxidation is observed. The extrapolated onset temperature is then determined. By using this method with polymers containing different antioxidants or different levels of the same antioxidant, it is possible to determine the relative oxidative stability of polymers in this manner.

Figure 50.8 Glass transition temperature at various styrene mole fractions for copolymers of styrene with acrylic acid (AS), acrylamide (AA), t-butyl acrylate (BA), and butadiene (BU). *(Reprinted with permission from H. G. Elias, ed.,* Macromolecules: Structure and Properties, *1977. Reprinted with permission from Plenum Press.)*

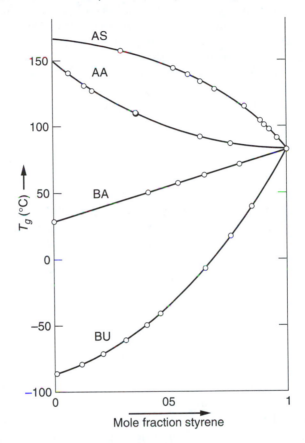

In the second method, an isothermal test (19, 20), the sample is rapidly heated to the temperature of interest, which was determined from previous range-finding experiments. The time the sample is held in the oxidative atmosphere at temperature is used to determine the time the polymer system can spend at that temperature in air before the onset of oxidation. This points out the difference between the scanning and the isothermal test. The scanning test is a faster test for screening and is good for relative information (determining whether antioxidant 1 is better than antioxidant 2). The isothermal test takes longer but gives a more accurate picture of the lifetime of a polymer under use conditions. The time to the onset of the oxidation peak is a kinetic parameter, so if a series of tests at different temperatures are run on the same sample, the Arrhenius kinetics can be determined, and the time to oxidation at lower temperatures, where longer times are observed, can be calculated if there is no change in mechanism over the range where the extrapolation is made.

Thermogravimetric analysis (TGA) is used to determine inert (inorganic) additives and organic additives such as pigments (19). Glass fibers are sometimes added to polymer to increase their rigidity above T_g. Glass fabric is sometimes used as the support for epoxy resins in larger cured parts such as aircraft parts and computer boards. In an oxidative atmosphere, the polymer will degrade but the glass will not. When a sample of a glass-filled polymer is heated to about 900 °C in an air atmosphere, the polymer will be totally oxidized and the weight remaining will be that of the glass fiber from the composite formulation.

With inert atmospheres (such as nitrogen and helium), a polymer will unzip or depolymerize (that is, return to monomer form) when heated to a sufficiently high temperature. Polyethylene and polystyrene are examples of this type of polymer. Some organic additives such as carbon black are of a more complex structure and they do not undergo any weight loss when heated in the same inert atmosphere to temperatures as high as 600 °C. However, these materials will oxidize in air at these temperatures. TGA methods for the determination of carbon black in polyethylene (1,19) involve rapidly heating the polymer sample in an inert atmosphere (over 100 °C/min) to about 600 °C and holding the sample isothermally at that temperature for some period of time. The atmosphere is then changed to air and the sample is heated to about 1000 °C. The weight loss observed after switching from inert to oxidative atmosphere gives the level of carbon black in the system.

6. Crystallinity and Crystallization Rate.

In crystalline polymers, the level of crystallinity can be one of the most important factors for determining properties of the polymer systems including strength, ultimate elongation, and barrier properties. Thus, a method for determining the level of crystallinity is important for studying polymer systems. There are a variety of methods, including IR spectroscopy, density, X-ray diffraction, and calorimetric methods, including some ASTM standard methods for the calorimetric techniques (21–23). In many cases, calorimetric methods are the easiest method for determining crystallinity. Additionally, calorimetric methods are not sensitive to orientation of the polymer crystal structure, and this can be an advantage for looking at total crystallinity. This is a disadvantage, however, if we are interested in studying the effects of orientation, in which case IR spectroscopy or X-ray diffraction would be preferable methods.

Use of calorimetric methods for determining the level of crystallinity is based on a knowledge of the heat of fusion for the sample. Inorganics and small molecule organic materials have well-characterized heats of fusion, which can be used to determine the quantity of the material present in a system. Polymers also can have well-characterized heats of fusion, and by measuring the heat of fusion the level of crystallinity can be determined. Dole gives a derivation of the expression in Eq. (50.3) for determination of the level of crystallinity in a polymer (24).

$$\text{Percent crystallinity} = \Delta H^{\text{sample}} / \Delta H^{0} \tag{50.3}$$

where ΔH^{sample} = measured heat of fusion for the sample of interest and ΔH^{0} = heat of fusion for the fully crystalline sample.

It is important to distinguish the two meanings of ΔH^{0}. In one case, ΔH^{0} can be the heat of fusion for a sample that is 100% crystalline. It is virtually impossible to make a sample that is 100% crystalline, so it can be difficult to determine the value for the 100% crystalline sample. Values of the heat of fusion for polymers are tabulated or can be determined (25). Two methods that are used to determine the heat of fusion include X ray and heat capacity measurements. In the X-ray method, the level of crystallinity is determined by X ray for several samples and the heat of fusion is determined for the same samples by DSC. The heat measured is then divided by the fraction crystallinity, as determined by X ray and ΔH^{0} is then determined.

A second method involves the use of both heat of fusion measurements and the heat capacity change at T_g. As previously discussed, the heat capacity change at T_g for a polymer is used for quantitation of the amount of polymer present. In the present application this can be related to the amount of polymer present in the glassy phase. To determine the heat of fusion for a 100% crystalline polymer, both the heat of fusion and change in heat capacity at T_g are measured for samples of the polymer that have different levels of crystallinity (that is, different ΔH values). A plot of ΔH versus ΔC_p should give a straight line, and the intercept where ΔC_p is zero is the heat of fusion for the 100%-crystalline polymer (16,26).

Various authors report using DSC measurements to characterize the crystallinity of polymer systems, including measuring the level of crystallinity and the relationship to branching (27). In this work, the change in crystallinity level and melting point were studied and found to decrease as a function of the number of branch points in the polymer chain. Another study also reported the effects of branching on the DSC trace and the extent of branching characterized by DSC traces (28).

Another aspect of crystallinity involves studying the rate of crystallization, or crystal growth, in a polymer system. Calorimetry is also used for studying crystallization kinetics. Crystallization is an exothermic event, as shown in Fig. 50.9, and calorimetry is used to follow the rate of heat release from the system (1,9). Differential scanning calorimetry was applied to the study of the crystallization of polymer blends as well as homopolymers (29,30).

7. Reaction Kinetic Studies

Like the crystallization process, many chemical reactions are exothermic, and as such are readily followed by calorimetric methods (1,9). In using calorimetry to study reaction kinetics, the important parameters include the determination of the amount and the rate ($\delta H/\delta_{\text{time}}$) of heat released

Figure 50.9 DSC scan for a sample of poly(ethylene terephthalate) that was quenched from the melt. T_g is observed at 78.89 °C. The onset of crystallization is observed at about 137 °C and the onset of melting is observed at about 232 °C. The melting endotherm (40.13 joules/gram) is slightly larger than the crystallization exotherm (36.36 joules/gram) indicating that the sample was not fully amorphous when quenched. (*Courtesy of TA Instruments, Inc.*)

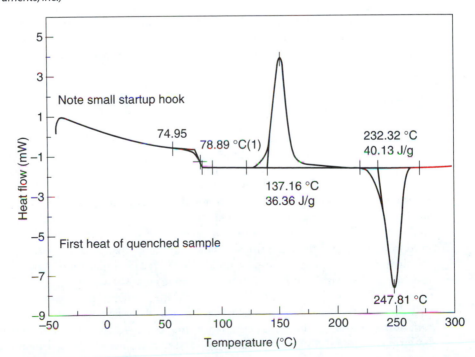

during the course of the reaction. From this knowledge the extent of conversion versus time can be determined.

In studying the kinetics of polymerization, various complicating factors must be considered. When studying the polymerization of monomers, the volatility of the monomer often creates a problem in determining accurate heats because the mass can change during the course of the experiment. Some instrument manufacturers market a high-pressure DSC pan to overcome this volatility problem and a sealed glass capillary method is described in the literature (31). When studying curing reactions for epoxy systems, volatility is not normally a concern.

In the study of polymerization reactions, the physical changes that occur during the reaction can also affect the measured kinetics. A material below its glass transition temperature is a glass, and as such has limited mobility. If reaction kinetics for a system are studied at temperatures below the maximum T_g for the final polymer, which is often the case for thermoset systems, then the reaction will occur only to an overall extent of reaction such that the T_g of the product is essentially the same as the reaction temperature. If the sample is then heated to a higher temperature (above the current T_g of the sample), the reaction will continue until the reaction completes or the T_g of the sample reaches the current reaction temperature. As a reaction nears completion, the viscosity of the system can limit the kinetics, so the rate of reaction becomes diffusion controlled rather than kinetically controlled. This phenomenon is often observed in the study of cure kinetics for thermoset systems.

Polymerization kinetics are studied both in a scanning and an isothermal mode. Reaction kinetics are often determined from a scanning DSC curve, using the method of Borchardt and Daniels (32). A good discussion of the overall method of the study of cure kinetics is given by Prime (1). The kinetics of cure for phenol-formaldehyde resins was reported by Chow and Steiner (33). In this application, scanning DSC was used to distinguish the reaction of two different resin systems. Differences in the reaction peak temperatures were used to distinguish the reaction kinetics.

Isothermal DSC experiments were reported for studying the cure of various epoxy systems (34,35). In these works, the extent of conversion was followed in two ways. The heat of reaction (or the rate of heat release) was determined at various times, and these measurements were used to determine the kinetics. Additionally, kinetics are determined by holding the sample isothermally at the reaction temperature for some time interval, and then cooling the sample and determining the remaining heat (extent) of reaction. In this case, the extent of reaction is 1 minus the unreacted fraction. The rate of conversion is determined by doing this at several isothermal times.

8. Polymer Degradation.

The degradation of polymers can be important for determining the maximum use temperature of polymer systems. As previously discussed, the term *degradation* can be somewhat ambiguous, so methods for measuring degradation can lead to different results. When a polymer chain breaks, some endothermic or exothermic energy is related with the process. Measuring the energy release gives a measure of the extent of degradation. Studying the same process by TGA may not give the same results because TGA requires that the degradation product formed be volatile at the test temperature. In deciding what test to use to follow degradation, we must determine which method will most closely represent the end-use conditions and the conditions that would constitute material failure.

In addition to being used to study the oxidative degradation of polymers, calorimetry is used to study the oxidation of polymers without additives (36).

Thermogravimetric analysis is often used for studying degradation. Two aspects of degradation that are studied are thermal degradation and oxidative degradation of polymers. Thermal degradation is important because degradation that occurs during processing (such as extrusion) may be largely thermal and not oxidative. A classic example of thermal degradation is the study of halogenated aliphatic polymers. In studying the degradation of halogenated polymers, such as poly(vinylidene chloride), TGA is often the preferred method. These polymers often degrade by the elimination of a hydrohalogen molecule, so the weight loss is directly related to the rate of polymer degradation (37).

Oxidative degradation is also studied by TGA. In these studies, the weight loss profiles are different from those for thermal degradation studies, but they may be more applicable to performance at high temperatures encountered during the use of the part or in some processing applications (such as thermoforming).

For either the thermal or oxidative degradation of polymers, understanding the degradation pathway is important for designing end-use applications for the polymer and this information may enhance the ability to stabilize the polymer. In addition to pyrolysis methods, various combined techniques such as TGA infrared, TGA mass spectrometry, or TGA gas chromotography mass spectrometry have been developed for studying polymer degradation (38,39). These methods allow real-time analysis of the degradation products of the sample to be characterized. This information aids in building a model for the degradation mechanism for polymer systems under a variety of end-use applications.

In some instances, TGA degradation studies are useful for quantitative analysis. Poly(ethylene-co-vinyl acetate) is made by reacting ethylene and vinyl acetate. When heated, a polymer containing vinyl acetate eliminates acetic acid, producing an ethylene linkage in the polymer. Because this is a stoichiometric elimination, each mole of acetic acid lost corresponds to one mole of vinyl acetate in the copolymer. Fig. 50.10 shows the weight loss for several copolymers with different vinyl acetate levels, and demonstrates how TGA is used for quantitative analysis.

Figure 50.10 Determination of vinyl acetate levels in poly(ethylene-co-vinyl acetate) by TGA. As the vinyl acetate level increases, the weight loss corresponding to the elimination of acetic acid increases, giving a quantitative measure of the vinyl acetate level. *(Courtesy of TA Instruments, Inc.)*

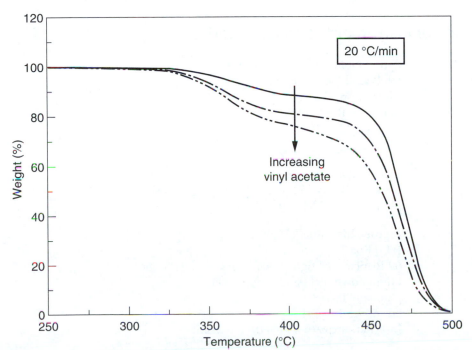

Nuts and Bolts

Relative Costs

DSC	$$ to $$$$
TGA	$$$
TMA	$$$ to $$$$
DMA	$$$ to $$$$
Dielectric	$$$ to $$$$
Maintenance	$
Supplies	$

$ = 1 to 5 K, $$ = 5 to 15 K, $$$ = 15 to 50K, $$$$ = 50 to 100K, $$$$$ = >100K.

For thermal analysis instrumentation, a wide range of options are available, so we encounter a wide range of prices. Simple DSC systems useful in an academic laboratory are fairly inexpensive, but systems with autosamplers and cooling capabilities to −150 °C are significantly more expensive ($$$$). Addition of IR or MS to TGA substantially increases the cost.

For the first use of thermal analysis equipment, work with several instrument manufacturers to help determine the needs for your specific application and how a particular instrument meets these needs. This will help to define the type of instrument that will best suit your needs and avoid buying more instrument than necessary.

Vendors for Instruments and Accessories

Astra Scientific International, Inc.
6900 Koll Center Pkwy., Ste. 417
Pleasanton, CA 94556
phone: 510-426 6900
fax: 510-426-6990

Giangarlo Scientific Co., Inc.
162 Steuben St.
Pittsburgh, PA 15220
phone: 412-922-8850

Mettler Instruments Corp.
P.O. Box 71
69 Princeton-Hightstown Rd.
Hightstown, NJ 08520-0071
phone: 800-638-8537
fax: 609-586 5451
Internet: http://www.mettler-ta.com

Perkin-Elmer Corp.
761 Main Ave.
Norwalk, CT 06859-0001
phone: 800-762-4000
fax: 203-762-4228
email: info@perkin-elmer.com
Internet: http://www.perkin-elmer.com

Shimadzu Scientific Instruments, Inc.
7102 Riverwood Dr.
Columbia, MD 21046
phone: 800-477-1227
fax: 410-381-1222
Internet: http://www.shimadzu.com

TA Instruments, Inc.
109 Lukens Dr.
New Castle, DE 19720
phone: 302-427-4000
fax: 302-427-4001
Internet: http://www.tainst.com

Required Levels of Training

Operation of Instrument

Routine analysis can be performed by people with a broad range of academic background including students, technicians, engineers, and research personnel. The setup of analysis methods often requires a person with a knowledge of the systems involved and some practical experience in thermal analysis.

Data Interpretation

For simple systems (inorganics, small-molecule organics), the experience gained in a typical undergraduate chemistry curriculum is adequate training. Analysis of more complicated samples, such as polymers, requires a greater level of training or experience. Several courses are taught in the field of thermal analysis of polymers. Many current users of thermal analysis techniques have developed expertise through years of practical experience.

Service and Maintenance

Service and maintenance problems are classified as instrument or computer problems. The mechanical components of most thermal analysis instruments are fairly rugged and fairly easily replaced when worn or damaged. Many instruments include a significant number of electronic components; more often than not, when a problem occurs it is in the system electronics. The same is true of the computer systems. Many manufacturers use computers that they have modified for their needs so the purchase of a personal computer will not suffice for operating thermal analysis instruments. When a problem develops with the computer it is often necessary to have the manufacturer service agent repair the instrument.

Suggested Readings

ASTM Standard E 967-92, "Standard Practice for Temperature Calibration of Differential Scanning Calorimeters and Differential Thermal Analyzers." 1992.

ASTM Standard E 968-83, "Standard Practice for Heat Flow Calibration of Differential Scanning Calorimeters." 1983.

ASTM Standard E 1363-90, "Standard Test Method for Temperature Calibration of Thermomechanical Analyzers." 1990.

ASTM Standard E 831-86, "Standard Test Method for Linear Thermal Expansion of Solid Materials by Thermomechanical Analysis." 1986.

ASTM Method D 3418-82, "Standard Test Method for Transition Temperatures of Polymers by Thermal Analysis." 1982.

ASTM Method E 1356-91, "Standard Test Method for Glass Transition Temperatures by Differential Scanning Calorimetry or Differential Thermal Analysis." 1991.

ASTM Method E 793-85, "Standard Test Method for Heats of Fusion and Crystallinity by Differential Scanning Calorimetry." 1985.

ASTM Method E 794-85, "Standard Test Method for Melting and Crystallization Temperatures by Thermal Analysis." 1985.

ASTM Method D 3417-83, "Standard Test Method for Heats of Fusion and Crystallization of Polymers by Thermal Analysis." 1983.

References

1. E. A. Turi, ed., *Thermal Characterization of Polymeric Materials* (New York: Academic Press, 1981).

2. ASTM Standard E 967-92, "Standard Practice for Temperature Calibration of Differential Scanning Calorimeters and Differential Thermal Analyzers." 1992.

3. ASTM Standard E 968-83, "Standard Practice for Heat Flow Calibration of Differential Scanning Calorimeters." 1983.

4. ASTM Standard E 1363-90, "Standard Test Method for Temperature Calibration of Thermomechanical Analyzers." 1990.

5. ASTM Standard E 831-86, "Standard Test Method for Linear Thermal Expansion of Solid Materials by Thermomechanical Analysis." 1986.

6. C. A. Berglund, *Styrene Polymers, Dynamic Mechanical Properties, Encyclopedia of Polymer Science and Engineering*, vol. 16 (New York: Wiley, 1989), 142–48.

7. N. G. McCrum, B. E. Read, and G. Williams, *Anaelastic and Dielectric Effects in Polymer Solids* (New York: Dover Publications, 1967).

8. J. J. Aklonis and W. J. MacKnight, *Introduction to Polymer Viscoelasticity*, 2nd ed. (New York: Wiley, 1983).

9. B. Wunderlich, *Thermal Analysis* (Boston: Academic Press, 1990).

10. R. F. Boyer, "Transitions and Relaxations in Amorphous and Semicrystalline Organic Polymers and Copolymers," in *Encyclopedia of Polymer Science and Technology*, supplement vol. II (New York: Wiley, 1977), 745–839.

11. H. G. Elias, *Macromolecules*, vol. 1, translated by J. W. Stafford (New York: Plenum Press, 1977), Ch. 10.

12. B. Cassel and B. Twombly, "Glass Transition Determination by Thermomechanical Analysis, A Dynamic Mechanical Analyzer, and a Differential Scanning Calorimeter," in A. T. Riga and C. M. Neag, eds., *Materials Characterization by Thermomechanical Analysis, ASTM STP 1136* (Philadelphia: American Society for Testing and Materials, 1991), 108–119.

13. ASTM Method D 3418-82, "Standard Test Method for Transition Temperatures of Polymers by Thermal Analysis." 1982.

14. ASTM Method E 1356-91, "Standard Test Method for Glass Transition Temperatures by Differential Scanning Calorimetry or Differential Thermal Analysis." 1991.

15. H. E. Bair, "Thermoanalytic Measurements of Impact Modified Blends," in R. S. Porter and J. F. Johnson, eds., *Analytical Calorimetry* (New York: Plenum Press, 1970), 2, 51.

16. F. E. Karasz, H. E. Bair, and J. M. O'Reilly, *J. Phys. Chem.*, 69, no. 8 (1965), 2657–67.

17. A. Xenopoulos, and B. Wunderlich, *J. Polymer Science: Part B: Polymer Physics*, 28 (1990), 2271–90.

18. S. G. Turley, *J. Polymer Science: Part C*, *1*, 106, 101–16.

19. W. P. Brennan, *Thermochimica Acta*, 18 (1977), 101–11.

20. T. Schwarz, G. Steiner, and J. Koppelmann, *J. Applied Polymer Science*, 38 (1989), 1–7.

21. ASTM Method E 793-85, "Standard Test Method for Heats of Fusion and Crystallinity by Differential Scanning Calorimetry." 1985.

22. ASTM Method E 794-85, "Standard Test Method for Melting and Crystallization Temperatures by Thermal Analysis." 1985.

23. ASTM Method D 3417-83, "Standard Test Method for Heats of Fusion and Crystallization of Polymers by Thermal Analysis." 1983.

24. M. Dole, *J. Polymer Science: Part C*, 18 (1967), 57–68.

25. R. A. Miller, "Crystallographic Data for Various Polymers," in J. Brandrup and E. H. Immergut, eds., *Polymer Handbook*, 2nd ed. (New York: Wiley, 1975), 1–138.

26. A. J. Pasztor, Jr., B. G. Landes, and P. J. Karjala, *Thermochimica Acta*, 177 (1991), 187–95.

27. S. Hosoda, *Polymer Journal*, 20, no. 5 (1988), 383–97.

28. E. Karbashewski and others, *J. Applied Polymer Science*, 44 (1992), 425–34.

29. J. Quintana and others, *Polymer*, 32 no. 15 (1991), 2793–98.

30. G. M. Kerch, and L. Irgen, *J. Thermal Analysis*, 36 (1990), 129–35.

31. L. Whiting, M. LaBean, and S. Eadie, *Thermochimica Acta*, 136 (1988), 231–45.

32. H. J. Borchardt and F. Daniels, *J. Amer. Chem. Soc.*, 79 (1957), 41–46.

33. S. Chow and P. R. Steiner, *J. Applied Polymer Science*, 23 (1979), 1973–85.

34. S. Sourour and M. R. Kamal, *Thermochimica Acta*, 14 (1979), 41–59.

35. K. Horie and others, *J. Polymer Science: Part A-1*, 3 (1970), 1357–72.

36. ASTM Method E 537-86, "Standard Test Method for Assessing the Thermal Stability of Chemicals by Methods of Differential Thermal Analysis." 1986.

37. B. A. Howell, *Thermochimica Acta*, 148 (1989), 375–80.

38. M. R. Holdeiness, *Thermochimica Acta*, 75 (1984), 361–99.

39. J. Chiu and C. S. McLaren, *Thermochimica Acta*, 101 (1986), 231–44.

Chapter 51

Mechanical Testing Techniques

Barbara Furches

The Dow Chemical Company

Summary

General Uses

- Determine material response to ranges of stress in uniaxial and biaxial modes
- Evaluate viscoelastic properties of plastics
- Measure response of material to high-speed impact
- Estimate performance of material in bending mode
- Evaluate qualitatively fundamental properties of materials
- Meet specifications for material applications
- Evaluate material response to stress over time and temperature ranges
- Measure response of material to cyclic stress
- Measure response of material under tensile or compressive stress

Common Applications

- Basis for prediction of material performance in use
- Criteria in quality control
- Comparison of materials
- Data for design calculations and concept proofs
- Monitor of manufacturing process
- Verification of changes in material formulations
- Criteria in failure analysis

Samples

State

For most mechanical testing of plastics a solid sample of very specific geometry is required. Dynamic mechanical testing is relatively independent of geometry, but uses solid samples.

Amount

The number and size of specimen are method dependent. Although some qualitative data could be generated on one specimen, the variability of material preparation and geometry dictates multiple (usually 5 as a minimum and for some tests up to 40) specimens.

Preparation

Because of the viscoelastic nature of thermoplastics, sample preparation has a significant effect on their response to stress loadings. Samples may be prepared by injection molding, extrusion (sheet), or compressive molding, or by milling samples from parts prepared by any of these techniques. Depending on the conditions used to prepare the sample and the geometry of the part being prepared, different levels of anisotropy or crystallinity may be present. These effects may be relieved in some materials by annealing techniques. However, in testing parts the effect of strength orientation from anisotropy is important in determining suitability for some applications. In creep studies the aging of samples can have serious effects on the data generated. For thermosetting materials, the degree of cure has a significant effect on their measured properties.

Analysis Time

Mechanical tests for plastics are often short-term, single-point tests that require less than one-half hour for the typical 5 to 10 specimens to be tested. Cyclic tests or creep tests may require thousands of hours for completion under each set of conditions.

Dynamic mechanical testing may require from one to a few hours for the test depending on range of frequency or temperature sweeps being performed.

Limitations

General

Short-term quality control tests are best used for comparison of properties of similar materials prepared under the same conditions.

Geometry constraints limit many tests to qualitative information regarding any fundamental property of the material.

The viscoelastic nature of plastics makes their properties dependent on both temperature and rate of loading. Many tests are performed under very limited ranges of temperature or strain rate.

Accuracy

Quality control tests may be reproducible within ±5% and some tests may have variance of up to ±25%. Dynamic mechanical testing of peak tan d is reproducible within ±5%.

Sensitivity and Detection Limits

This attribute is dependent on instrumentation and sample geometry. Load cells with measuring capacity down to 1 g are available.

Complementary or Related Techniques

- The wide range of mechanical tests are complementary to each other; each provides additional data on the performance of a material under different types of stress.
- Some of the data generated by dynamic mechanical testing may also be determined by thermal techniques such as differential scanning calorimetry or dynamic electrical analysis, which uses dielectric constants for monitoring the change in ionic conductivity for materials where the movement of dipoles or ionic species can be electrically stimulated.

Introduction

Many of the test methods used to characterize plastic materials were adapted from metals and other materials testing. Because these techniques did not adequately reflect the differences in plastics, they were modified by each user. The inability to compare data from the multitude of these adapted tests eventually led to establishment of standard specifications for test methods through the American Society for Testing and Materials (ASTM) and other national standards organizations and, more recently, the International Standards Organization (ISO). These test methods are used to characterize the performance of plastic materials under set conditions of loading and deformation.

Short-term mechanical tests typically used to determine compliance to specifications are tensile, flexural, shear, compression, and impact. These tests do not give fundamental property data for a plastic material because of sample preparation effects (anisotropy, degree of crystallinity, or degree of cure), specimen geometry constraints, and the faulty techniques and errors shown in Table 51.1. However, they are widely used in quality control testing and in early stages of material selection by designers.

Stress–strain curves (in some cases recalculated from load–deformation curves) are generated in performing most of these short-term tests. Although the data for any specimen are not the same for the various methods of applying stress, the curves generated are similar in shape and are useful in qualitative evaluation of the material's properties.

Plastics are viscoelastic materials, so their properties depend on stress type, temperature, strain rate, and geometric effects in addition to the normal variations of testing. In evaluating the data generated by either short- or long-term tests, the history of the material, its preparation method, internal stresses, geometrically induced stresses, and conditions of testing must be known for correct interpretation of results.

Long-term tests, such as creep and stress relaxation, are able to give more fundamental property data for a plastic material. The large amount of time (greater than 1000 hr) required for completion of these tests prohibits their usefulness for quality control or for generating data for product development.

Dynamic mechanical analysis was originally developed as a rheological testing technique to measure the viscoelastic properties of materials. This type of testing approaches a good compromise

Table 51.1 Imperfections and errors in deriving raw data and property values.

Imperfections in Raw Data	Errors in Raw Data	Errors in Translation of Raw Data
Unregistered or distorted events on force deformation curve, natural vibrations registered as noise	Extraneous forces or displacements included in detected response	Difficulty in conversion of force–deformation to stress–strain
Loss of precision and accuracy, reading errors, or guillotined signal	Sensor movement relative to specimen, erroneous register of displacement	Naive interpretation of force–displacement curves
	Constraint or modification of local displacements in specimen	
Caused by:		Caused by:
Unsuitable data processing response time	Caused by:	Reduction in cross-sectional area
Unsuitable sensitivity of chosen sensor	Incorrect positioning of strain sensor	Heterogeneous strains
	Insecure or constraining attachment of strain sensor	

Reprinted by permission of ASM INTERNATIONAL.

in allowing generation of more fundamental data on the nature of a plastic material than found in short-term quality control tests without the excessive time required to generate data for the long-term tests. The major drawbacks to use in quality control are the cost of the equipment and the sophistication required for interpretation of the data.

How It Works

Mechanical testing is performed using a load frame with a movable crosshead (pictured in Fig. 51.1) with various accessories to apply stress in a uniaxial or a biaxial mode. Universal load frames, developed in the late 1800s, have been used commercially for both short- and long-term test data generation.

A specimen of very specific geometry is supported in a fixture attached to the load frame. The load frame has sufficient stiffness to allow application of a stress (load), usually by movement of the crosshead, that forces a deformation (strain) in the specimen. (In some tests strain is applied.) The deformation is detected by specific instrumentation, including extensometers for tensile and deflectometers or crosshead movement for flexural deformations; the load required to cause the deformation is monitored through load cells or proof rings. The imbalance of the strain gauge bridge circuit within a load cell sends an electrical signal that is amplified and converted to an output signal proportional to the applied force. The applied load versus deformation is recorded via oscilloscope or microprocessors and transferred to hard-copy printouts that are generally translated as stress–strain curves.

The response of an ideally elastic material to stress is a proportional amount of strain (Hooke's law), but plastics rarely show this exact response. Several typical stress–strain curves are shown in

Figure 51.1 Servo hydraulic test frame in tensile testing mode with contact linear variable differential transformer extensometer.

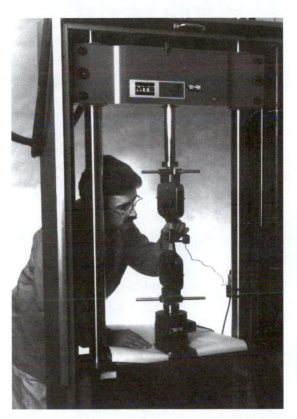

Fig. 51.2. Although these curves were generated via tensile testing, similar curves are generated for other mechanical tests. The curves are related to the general properties of the materials by the amount of load required to make the material yield and the degree of deformation the material undergoes before complete failure. Strength is related to the load and ductility or toughness is related to deformation or elongation before break.

Insulated chambers capable of both cooling and heating may be used to vary the test temperature with either of the generic types of load frames. Electromechanical load frames use multiple drive screws and nuts to move a crosshead with a specimen fixture attached. Another portion of the fixture is attached to the nonmoving base of the load frame. Closed loop servo drive systems ensure that the crosshead moves at a constant speed.

Servo hydraulic load frames, on the other hand, use forces from hydraulic actuation systems to move the crosshead. These units can generate both higher speed (greater than 100 m/sec) and larger loads than the electromechanical units. Servo hydraulic equipment can also be used for testing in load, strain, or displacement rate control.

Grips or chucks for holding samples fall into several categories. Wedge-action grips are the most commonly used grips for tensile, creep, stress relaxation, and fatigue testing of plastics. The degree of friction between the specimen and gripping face is controlled by use of different surface patterns on faces in contact with the specimen.

Hydraulic or pneumatic grips allow remote actuation of the clamping of specimen. Strip chucks are used for very strong laminates and sheet materials. With these grips, holes are drilled

Figure 51.2 Typical stress–strain curves with stress to yield indicating strength and amount of strain indicating ductility. *(Courtesy of Ellis Horwood Ltd.)*

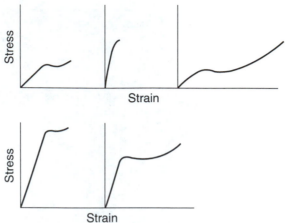

in the tab ends of the specimen and bolted to the chucks. Vice chucks are used for thin, brittle film specimens to reduce stress and self-tightening jaw chucks are useful in gripping soft elastomeric and plastic sheeting.

The force required to accomplish the deformation of the specimen is measured by means of a load cell. Load cell capacity can range from a low of 1 g of force to 50 kN force. Strain gauge load cells are constructed with a bridge circuit that measures the electrical resistance changes during mechanical deformation. These, in combination with amplifier circuits and converters, give an output signal that is proportional to the applied force. Proof ring cells, on the other hand, use transducers to sense the movement of the sides of a slit in a steel block that widens in response to the applied load. The transducer produces an electrical signal that is proportional to the force. Amplification and conversion are similar to those of strain gauge systems.

Output signals from the instruments are transmitted to transient signal recorders or directly to microprocessor system controllers for analysis. The data are then sent to storage media or directly to a printer or plotter.

With dynamic mechanical analysis (DMA), a sinusoidal stress (or strain) to a perfectly elastic material would result in a response proportional to and in phase with the applied stress. Application of the same sinusoidal stress (or strain) to an ideal viscous liquid would be a response 90° out of phase with the applied stress. As has been already discussed, plastics are a hybrid of elastic and viscous behavior. Therefore, when a sinusoidal stress (or strain) is applied to a plastic specimen, it responds in a manner between in phase and 90° out of phase to the applied stress. The degree the response (normally strain) is out of phase with applied stress is called the phase angle δ. This measurement is an indication of the degree of elasticity or viscosity of the sample, as illustrated in Fig. 51.3 for a poly(phenylene sulfide) (PPS) sample.

Both fixed frequency and resonance modes of operation are possible. Fixed frequency scans, the preferred technique, allow finer resolution of a material response (1).

In addition to the highly automated DMA instruments, equipment allowing free vibrational motion is available, although this instrumentation is no longer commonly used. The torsional pendulum used for these tests clamps a specimen at one end with a rigid fixed clamp. The movable clamp is attached to an inertial member of known moment of inertia. A differential transformer and recorder are used to track angular displacement versus time. Insulated chambers with both

Figure 51.3 DMA curve of tan δ versus temperature for PPS showing the effect of heat aging on the peak associated with T_g. *(Reprinted by permission of Dicten & Masch Manufacturing Company.)*

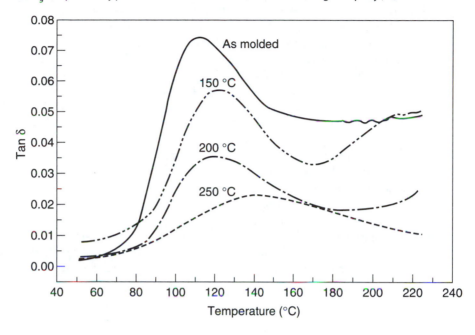

Figure 51.3 DMA curve of tan δ versus temperature for PPS showing the effect of heat aging on the peak associated with T_g. *(Reprinted by permission of Dicten & Masch Manufacturing Company.)*

heating and cooling elements are used for either technique to allow a wide range of temperatures used for dynamic testing.

What It Does

This section briefly reviews each mode of mechanical testing, the specimen used, equipment and accessories required for the tests, and the advantages and limitations of each test. Where standards exist, the associated ASTM and ISO standard designation is listed for reference on the exact techniques for performing these tests. Other national standards such as the German Institute of Standards (DIN) or the British Standards Institute (BSI) may have similar tests, but they are not listed here.

In tensile testing (rigid and semi-rigid polymers, ASTM D638, ISO 527-1; elastomers, ASTM D412; composites, ASTM D3039 and ISO 3268), dumbbell-shaped, flat specimens are subjected to a uniaxial stress, stretching the sample. The dumbbell geometry limits the area for most probable failure to the narrow parallel section between the tabs that are placed in the grips. A portion of this narrow section becomes the gauge length and it is the basis for monitoring and calculating any deformation (elongation) of the specimen during yield or postyield break. Although in many cases the specimen deforms sufficiently during yield or postyield to change the cross-sectional area of the part, all calculations are made on the original specimen dimensions, primarily to simplify the needed mathematics and permit reference back to a constant condition.

For some situations, crosshead travel is an adequate measure for elongation of the specimen. However, this includes deformation outside of the gauge length, any slippage of the specimen in the grips, and any slack in the setup of the load frame–grip system. Most polymers require careful measurement of deformation, both near and postyield. Contact and noncontact extensometers are accessories used to measure gauge length deformation. Contact extensometers may use strain gauges or linear variable differential transformers (LVDT), which are electromechanical devices with output voltage proportional to displacement to measure the change in length of the gauge area. Noncontact extensometers are optical (camera) or laser-based systems that use monitoring of targets' movement on the specimen to measure elongation. These are particularly useful for films or lightweight specimens, such as foams, where the weight of the contact extensometer would affect the generated data. These are also useful in insulated chambers where contact units may be affected by temperature extremes. Figure 51.4 shows a typical laser extensometer (noncontact) used to monitor tensile testing of foams.

Special types of extensometers are used for measuring samples in which anisotropy may be high enough in the specimen to significantly affect results (averaging extensometer) or for monitoring lateral contraction during elongation (biaxial extensometer). This latter unit is valuable in the determination of Poisson's ratio, often used in design calculations with plastics.

Yield point strength from tensile data is typically listed as a physical property of polymers on supplier data sheets, although this test, like other short-term mechanical tests, is restricted by sample preparation, geometry, and measuring limitations that interfere with the data as a fundamental polymer property. Stress–strain curves can be produced that indicate general polymer perfor-

Figure 51.4 Laser extensometer used to monitor tensile testing of foam.

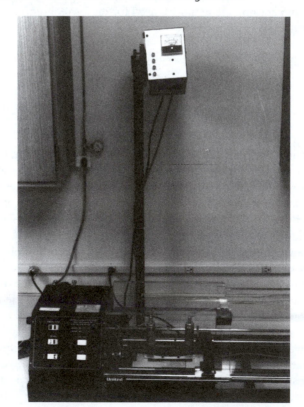

mance under short-term ramped loading. Typical tensile test curves are shown in Fig. 51.2. For most rigid polymers, the curve includes a relatively linear portion wherein the material is elastic. If stress is removed during this area of the curve, there is no apparent change in the specimen. Near yield point, defined in ASTM D638 as the point where additional strain occurs without any increase in stress, the polymer may begin to deviate from linearity. This point typically appears as the first inflection in the stress–strain curve. Deformation beyond yield point is not reversible by removal of stress.

A modulus value may be measured in the linear section of the curve and is the ratio of stress to strain within this region. For materials showing a very limited or no linear region, a tangent to the curve at a specific level of strain may be used to calculate the modulus. Modulus (of elasticity) or Young's modulus is used in product design calculations. Tensile properties, including modulus values, are rate-of-strain dependent. At high extension rates, polymers show higher strength at yield, higher modulus, and lower elongation beyond yield.

In flexural testing (ASTM D790, ISO 178), a simple beam specimen is supported at two points, near each end of the beam, and the load is applied at the center, either by a single nose (three-point bend) or with two loading noses (four-point bend), each an equal distance from the adjacent support. The load is applied perpendicular to the longitudinal axis of the specimen. Four-point bend is often used for composite materials, whereas three-point bend is the commonly used quality control test for rigid plastics. Both compressive and tensile forces are induced in flexural tests. The load is applied until the specimen breaks or until strain in the outer fiber of the specimen reaches 5%. These tests are dependent on strain rate as well as the thickness of the specimen.

Flexural testing is simpler than tensile testing because less orientation (anisotropy) is found in a simple beam specimen than in the dumbbell specimen of tensile. There are fewer specimen alignment problems and fewer clamping problems. The deformation is large enough to improve the accuracy of measurements with either a deflectometer or crosshead movement.

Like tensile tests, flexural modulus is also determined from the stress–strain curve. Flexural modulus is a measure of stiffness and gives similar, but generally not the same, values as the tensile modulus of elasticity. Although many applications for plastics are modeled by bending or flexural stress, flexural modulus is not used in calculations for product design. This is because it is not a pure stress, but a combination of compressive and tensile elements.

Compressive tests (ASTM D695, ISO 604) use flat plate fixtures to crush the specimen. Because of the tendency of materials to buckle or bend, the specimen, usually a cube or cylinder of material, must truly be parallel (especially the end faces) and the stress must be applied in a truly axial manner. Although this is a relatively simple test, frictional forces between the plates and material may constrain lateral expansion and cause barreling of the specimen. A deflectometer or compressometer is used to measure deformation. Again, a stress–strain curve can be generated and a compressive modulus determined. Compressive properties are rarely used in most plastics design.

Shear testing (ASTM D732, ISO 4585) includes a simple form called punch shear, which approximates true shear. In contrast to tensile or compressive testing, the forces are parallel to the plane on which they act. Punch shear is usually performed on sheet or film samples and may allow excessive stretch of the specimen before failure when these thin samples are used.

Iosipescu shear testing was developed for isotropic materials, but this test has been used primarily on composites. With polymers, a double-notched specimen is placed in a fixture such that the force applied in tension or compression results in pure shear across the area between notches, as shown in Fig. 51.5. When used with polymers and composites, an increased notch angle (beyond the original 90°) and a radius on the notch versus a sharp notch relieves stress concentration at the root of the notch (2). Highly brittle materials have local cracking at the notch, which relieves stress concentration before reaching the ultimate applied shear stress (3). The main advantage of this shear test procedure is that it allows capture of the complete shear stress–strain curve to failure.

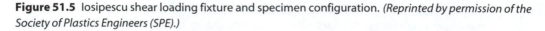

Figure 51.5 Iosipescu shear loading fixture and specimen configuration. *(Reprinted by permission of the Society of Plastics Engineers (SPE).)*

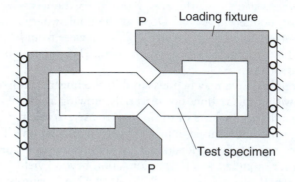

Impact tests, classified as pendulum (ASTM D256, ISO 179 and 180 Izod) or drop weight (ASTM D3029, ISO 6603-2; High-speed puncture, ASTM D3763; ISO 6603-3) can be performed manually or can be readily instrumented. The pendulum impact tests (such as notched Izod/Charpy) rely on a notch in the test specimen to act as a stress concentrator. The notch dimensions, radius, and smoothness are all important in obtaining reproducible results. Notched Izod uses a weighted hammer (pendulum) to strike above the notch in the specimen that has been clamped in the base perpendicular to the base. A total energy value is obtained that includes the energy to propagate a crack from the base of the notch through the specimen and toss the broken end of the specimen. For Charpy impact tests, the specimen is notched similarly to the notched Izod specimen; however, it is not clamped into the equipment, but positioned and supported horizontally. The weighted hammer then strikes the unnotched edge of the specimens, as shown in Fig. 51.6.

The results of pendulum testing are dependent on sample preparation, particularly of the notch in each specimen. A slight change in radius or depth of notch may significantly affect the results obtained. With some polymers, even high cutting speed of the notched may affect the results (4). The high variability of this test and the limited total energy value obtained prevent it from being useful in design data calculations. Notched Izod does measure the notch sensitivity of a material. Because it is a simple test to run and the equipment is inexpensive, it is often used in quality control testing of materials.

Gardner and manual drop weight impact tests are similar in procedure and analysis. Variations include the hammer configuration (tup) and the energy available during the test. In both systems, a flat plaque or disk of material is used as a specimen and a weight or weighted tup is raised to different heights in a supported channel and dropped on the surface of the normally unclamped specimen. The specimen is visually inspected for damage on the reverse side from that hit. Height and weights are varied to change the energy available in the system. The data from these tests are analyzed by step method or probit analysis and, therefore, they require testing of multiple (up to 40) specimens to give a statistical 50% failure energy. Reproducibility is generally poor based on the subjective evaluation of failures, sample specimen variation, and variability of impact velocity. However, these tests are more reflective of actual application impact because they are multiaxial in nature, as opposed to the uniaxial pendulum impact.

Instrumented dart impact is a variation of drop weight using either a servo hydraulic load frame or a gravity-type free-fall instrument in which the dart or tup has been instrumented. These tests are designed to force a failure of every impact event and, with the instrumentation, record the total event from contact through failure of the clamped specimen. Instrumented dart impacts generate

Figure 51.6 Manual instrument configured for Charpy Izod test.

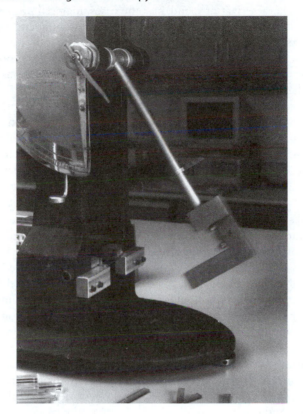

stress–strain curves giving significantly greater information regarding the failure event than the manual systems. This also requires fewer samples, normally five specimens, to complete the test.

More recently developed impact testing equipment is designed for high-speed impact to better simulate actual impact failure in plastics applications. These tests are limited to dart or ball impact procedures, with impact velocity capability of over 42 m/sec with hydraulically powered machines (4). These procedures are typically used to mimic automotive crash test velocities of 12 m/sec or less. The equipment uses an oscilloscope or other transient recorder for electronic storage of data transmitted from the load cell. These data can be analyzed by microprocessors to give an applied load–displacement curve for the complete impact event as well as record energy absorbed over time. Frequency filters may be used to reduce noise in curves of brittle materials, but these must be applied after the data are generated, not to the signal from the cell. Adequate energy to force the specimen to failure or to not lose more than 20% of velocity through the impact is important in obtaining reproducible data for peak energy values (5).

The versatility of instrumented impact equipment also allows reproducible partial penetration or only indentation of a specimen. The specimen may also vary from thin film to complete parts designed for final application. Use of different strain rates can also be used to evaluate the ductile-brittle failure transition of a material.

Additional short-term impact tests include the pendulum method for tensile impact, a high-speed tension procedure that correlates with drop weight, the Underwriters Laboratories swinging ball impact test for electronic enclosures, the ASTM D2463 drop impact resistance test for blow-molded thermoplastic containers, and the air cannon impact test to determine impact resistance of exterior building components to projectiles. These short-term tests are all used for quality control, material characterizations, and, in some cases, material properties. Few of the results of these tests can be used in part design calculations for determining end-use part properties.

In contrast, long-term property tests are more indicative of fundamental material properties. Creep, stress relaxation, fatigue, and fracture mechanics properties may have similar test configurations as used in short-term tests, but tests may require thousands of hours per condition to obtain useful data.

In creep testing (tensile: ASTM D2990, ISO 899; flexural: ISO 6602), a fixed load is applied to a specimen and the deformation is recorded as a function of time. Creep testing may be in tensile, flexural, or compressive modes, with shear and torsional modes used less often. The specimen and fixtures are essentially the same as the short-term static tests. Temperature and usually humidity (especially where moisture may affect a material's properties, as with polyamides) are held constant. Careful preparation of samples is more important for these long-term tests, where variation from anisotropy or degree of crystallinity or cure will affect the data to a greater degree than in the short-term tests. Applying the load in a true uniaxial mode for tensile and compression testing is also more important where the small strains that are typical for these tests may be affected by lateral forces from deviations from axial loading.

Measurement of deformation is more exacting because low levels of strain are expected even at the end of these long-term tests. Extensometers, primarily LVDT (transducer) or optical non-contact types, are recommended to meet accuracy requirements of less than ±1% of total strain during the test (6). Careful attachment of extensometers is important to ensure that no additional stress is applied to the specimen where knife edges or screw tips contact the plastic specimen, which can lead to localized yielding.

Testing is usually conducted under several different loadings and at a range of temperatures because creep modulus is not independent of either factor. To avoid an initial increase in stress during the test and vibration that could affect reproducibility, it is important to apply the load smoothly and quickly (usually within 1 sec) to the specimen. Measurement of strain is not started until 1 min after loading (ISO procedure) to minimize loading effects on data.

Because deformation versus time curves are difficult to compare for various stresses for different materials, isochronous (constant time) stress–strain curves are calculated from the strain–time curves at different stresses. From these or deformation versus time curves, creep modulus (initial applied stress–creep strain at a point in time) may be calculated and plotted against time on a log scale to give a straight line. This type of plot, shown in Fig. 51.7, is often provided by material manufacturers for comparison at different stress levels or a single stress and different temperatures. Care must still be taken in comparing data from different suppliers to check on sample preparation techniques and mode of creep testing, which will affect data. Flexural creep may be more practical in terms of more realistic loading of parts. However, design calculations are more straightforward using tensile creep data.

Stress relaxation (ASTM D2991) testing is conducted by introducing and maintaining a constant strain and monitoring the load over the time required to maintain the strain. Stress decreases over time as the polymer chains relax. This technique is more difficult to use, so data are less available. The testing is performed on equipment for tensile measurement and requires very tight monitoring of deformation, usually by transducer.

Fatigue testing (ASTM D671, ISO 338) subjects a specimen to repeated cyclic loading, generally to failure. Mode of loading may be in tensile, flex, compression, or torsion and may be constant amplitude of stress or strain. Data are plotted as stress versus the log number of cycles to failure, known as S–N curves, as presented in Fig. 51.8. An endurance limit is extrapolated from these curves and defined as the stress level below which the material can endure an infinite number of stress cycles without failure. For many polymers this limit is approximately 30% of static tensile strength (7).

The ASTM method uses a cantilever beam in flexural constant strain cycles and a fixed frequency of 30 Hz. For most polymers, this frequency would generate unacceptable heat

Figure 51.7 Creep Modulus versus Time curve for a polymer blend at different loading with an ASTM D2990 injection molded specimen, flexural load (simple beam bending with load at the center of 2-in. span), and the strain measurement equaling the deflection at the center of the beam.

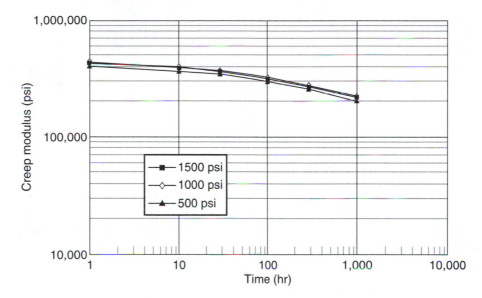

levels that would cause thermal failure, so much lower frequencies (3 Hz or lower) are preferred.

Fatigue data are not reproducible, so these data are not recommended for use in design of parts. Previously indicated thermal effects, as well as local temperature variations during testing, effects of specimen thickness, and the changes that occur in stress levels of the test once fracture or cracking has begun, contribute to the poor reproducibility.

Fracture mechanics (ASTM D5045) has recently been used to improve understanding of the failure modes of plastics as a crack grows. Determination of K_{ic}, the critical stress intensity factor, has recently become a standard test procedure in ASTM and development of an ISO standard is in progress. The procedure uses fatigue testing of an infinite plane specimen with a razor-sharp crack to monitor crack growth rate. These data are valuable in understanding fundamental properties of plastic materials.

Dynamic mechanical analysis (ASTM D4065, ISO 6721; bending: ASTM D5023, compression: ASTM D5024, tensile: ASTM D5026, and elastomers: ISO 2856) applies a sinusoidal deformation or stress to the sample with variation of frequency, temperature, amount of strain, or time during ramped changes. Frequency sweeps over wide temperature ranges are commonly used to determine fundamental polymer properties. In combination with time–temperature sweeps, the data can be used to generate time–temperature superposition curves that are used to derive a master curve using the Williams, Landel, and Ferrys (WLF) equation. Further discussion of master curves is beyond the scope of this chapter. Chapter 7 of I. M. Ward's book is recommended for further explanation of the WLF equation and master curves (8).

Various geometries are possible for testing with solid or fiber samples: three-point bending (flexural), fiber–film parallel plate (compression), dual cantilever, monofilament fiber, and shear sandwich. The tests are typically run at low strains, within the linear portion of the viscoelastic curve. Temperature ranges vary from the boiling point of liquid nitrogen to several hundred degrees Celsius.

During the application of the sinusoidal stress (or strain), the material responds by deformation that is between the response of a completely elastic material that would be in phase with the

Figure 51.8 *S–N* curve showing tensile fatigue with marking of the extrapolated endurance limit. Tensile load = 3 Hz, *R* ratio = 0.1; specimen = injection molded T-bar; test at 23 °C with 50% relative humidity.

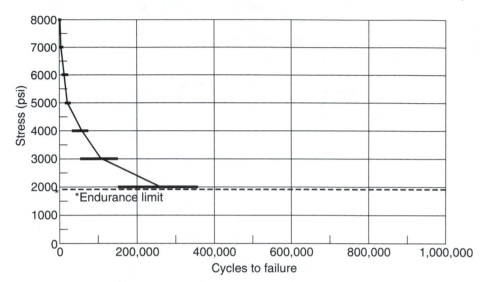

applied stress and that of a completely viscous liquid that would be 90° out of phase with the applied stress. The amount that the strain is out of phase with applied stress is called the phase angle δ and indicates the degree of elasticity or viscosity of the sample. This complex response is resolved into the elastic (or storage) modulus, G', and the viscous (or loss) modulus, G''.

Instruments are available to input stress or strain with a fixed or variable amplitude and frequency. All can be correlated when the mode of deformation, operating frequency, material orientation and conformation, and sensitivity of the individual instrument to external factors are accounted in making measurements (9).

The most useful data obtained from DMA or sinusoidal testing is the determination of transitions that can be correlated to changes in molecular structure. Beyond glass-transition and melting points, other types of transitions such as motion of chain as a unit, chain segment motion, side group motion, and interaction between amorphous and crystalline areas can be detected through this type of test.

Analytical Information

Qualitative

Short-term mechanical testing techniques for plastics give an indication of the material response under relatively instantaneous loading. These are quick quality control tests for relative comparison of materials. The data do not reflect molecular properties of the test materials. The data can show trends of material performance, but should not be used for plastic part design calculations.

Long-term tests, such as creep, stress relaxation, and fatigue tests are used to determine material response over time and, with appropriate controls, temperature. The modulus data obtained from these tests can be used in design calculations. Fatigue studies are of limited value unless performed under low frequency and with tight control of specimen preparation and temperature.

Dynamic mechanical analysis can be used to determine transitions in polymer structure that can then be related to property performance. Alpha (α) peaks (transition) are related to glass transition temperature, beta (β) peaks are related to impact performance (in polyethylene the intensity of β peaks varies with the degree of chain branching (10)), and gamma (γ) peaks are related to low temperature impact.

The data must be generated at low levels of stress or strain to keep within the elastic limits of the material being tested. Therefore, the applications of plastics wherein high stresses and deformations may occur cannot be accurately mapped.

Quantitative

Eqs. (51.1) to (51.13) are useful for property calculations from data obtained from each of the previously described short-term and long-term tests. Modulus is always measured in the elastic portion of the generated curves where a straight line portion exists. For some materials (particularly elastomers), a chord or secant modulus is used. Again, before comparing properties for different materials or grades of the same material, we must determine the mode of preparation for each specimen.

Tensile strength at yield (T_y, in MPa) is defined by Eq. (51.1) and tensile strength at break (T_b) is similar in form to T_y.

$$T_y = \frac{F}{A} \qquad (51.1)$$

where T_y = tensile stress at yield, F = tensile force at yield, and A = cross-sectional area of initial specimen.

Elongation (E, in percent) is calculated from Eq. (51.2).

$$E = \frac{100(l_1 - l_0)}{l_0} \qquad (51.2)$$

where l_1 = length of specimen at break or yield and l_0 = length of original specimen.

Young's modulus or the elastic modulus (E, in MPa) is defined by Eq. (51.3).

$$E = \frac{\sigma L}{\varepsilon L} \qquad (51.3)$$

where σL = stress within the elastic limit and εL = strain within the elastic limit.

Flexural stress (σF) using the three-point bend procedure is defined by Eq. (51.4) and the expression for the four-point bend appears as Eq. (51.6).

$$\sigma F = \frac{3FL}{2bh^2} \qquad (51.4)$$

where F = force at midpoint, L = length of span, b = width of specimen, and h = thickness of specimen. This equation is modified as shown in Eq. (51.5) to account for the horizontal component of the flexural moment.

$$\sigma F = \frac{3FL}{2bh^2} \frac{1 + 4d^2}{L^2} \qquad (51.5)$$

where d = deflection of the specimen.

$$\sigma F = \frac{6FL_1}{bh^2} \qquad (51.6)$$

where L = length between support and the adjacent loading nose.

Modulus of elasticity in flexure (E) is defined by Eq. (51.7).

$$E = \frac{L^3}{4bb^3}\frac{F}{Y} \tag{51.7}$$

where F/Y = slope of the initial force deformation curve or linear section.

Shear modulus (G) is defined by Eq. (51.8) and shear strength by the punch shear test is defined by Eq. (51.9).

$$G = \frac{T}{\gamma a} \tag{51.8}$$

where T = shear stress, γ = shear strain ($\delta a/a$, where δa is the displacement), and a = specimen length.

$$\text{Shear strength} = \frac{F}{\pi DT} \tag{51.9}$$

where F = force at fracture, D = diameter of the punch, and T = mean thickness of the test specimen.

For creep testing, the apparent modulus (M_c) at time t equals the initial applied stress divided by the creep strain.

With fracture mechanics testing, the stress intensity factor (K_i) is defined by Eq. (51.10) where correction factors are necessary to account for geometry constraints.

$$K_i = p\sqrt{\pi a} \tag{51.10}$$

where p = stress over an infinite uniformly loaded plate and a = 1/2 the crack length.

With dynamic mechanical testing, complex modulus (G^* or E^*), elastic modulus (G' or E'), and viscous modulus (G'' or E'') are defined by Eqs. (51.11) to (51.13), respectively. The loss tangent (tan δ) equals G''/G'.

$$G^* \quad \text{or} \quad E^* = \frac{T}{\gamma} \tag{51.11}$$

where T = stress and γ = strain.

$$G' \quad \text{or} \quad E' = \frac{T'}{\gamma} = \frac{T^*\cos\delta}{\gamma} \tag{51.12}$$

where δ = phase angle between the applied stress and the resultant strain.

$$G'' \quad \text{or} \quad E'' = \frac{T''}{\gamma} = \frac{T^*\sin\delta}{\gamma} \tag{51.13}$$

Applications

1. Short-term Mechanical Tests.

Short-term usage of mechanical testing data includes comparison of properties either to standards or specifications for quality control. Another important usage involves comparison of properties of similar products from different manufacturers' technical data sheets to choose a material for use in an application. The latter use is valid only if other data have been used to ensure that the material type is appropriate for the application and that the data sheets are based on similarly prepared specimens tested under the same conditions.

Some data from tensile, compressive, or in some cases flexural testing may be indicative of actual material physical properties. Tensile modulus, primarily, is indicative of the stiffness of a material and these data are useful in designing parts for instantaneous loading. Other short-term static tests give useful data when carried out under a range of temperatures and strain rates. Short-term static tests may also provide useful data on fabricated products when the final manufactured product can be tested. The most common testing of this type is impact; either a drop test of the part or a drop test of a weight onto the part can be valuable in determining ductility of the material and part energy absorption characteristics if tested under conditions reflective of the actual use environment. Hydraulically controlled instrumented impact equipment is used to test parts to mimic indentations or energy absorption before permanent yielding or break of the part.

These short-term tests are also useful for correlation with one of the long-term tests. Ireland discusses this in the example of an automotive flex fan in which instrumented impact data were correlated to the fatigue failure of the flex fan blades (11). In this case, the time to fracture in an instrumented Charpy pendulum impact was found to correlate with the number of cycles to failure of the fan in an engine simulation test.

2. Long-term Mechanical Tests.

Longer-term tests such as creep, stress relaxation, fatigue, and fracture mechanics are used to determine material responses to stress or strain over time or combined static and dynamic (cyclic) loading responses. These data are used in design calculations for application of plastic materials. Brown (12) suggests the following criteria for minimum testing required to generate adequate data for use in engineering design:

- Determine isochronous stress–strain curve at one temperature close to standard room temperature (23 °C) for a specified short time (60 or 100 sec).

- For the same temperature and material state as in the first criterion, determine at least four creep curves at different stress levels for at least 1 yr.

- Determine creep recovery for at least 10% of the period under stress at the end of each creep test in the second criterion.

- Determine the effect of temperature using isochronous stress–strain curves at various temperatures and creep curves at some selected stress levels and temperatures.

- Determine effects of fabrication variable, environmental, and humidity factors using isochronous stress–strain tests.

Such extensive test data for a material are rarely available. In another application of long-term data, Bonmin, Dunn, and Turner report using the ratio of the tensile creep modulus to shear creep modulus as an index to the degree of anisotropy in molding a material (13).

Bennett and Quick estimate that over 80% of all service failures of equipment are caused by fatigue (14). However, static (creep rupture) or dynamic fatigue testing have limited utility without agreed-upon standard tests. Only recently have national and international standards been developed and accepted for fracture mechanics testing that is not geometry specific. Better prediction of fatigue performance should be a result of adaptation of this testing mode. Typical S–N curves published on cyclic fatigue using 30-Hz frequency cannot reliably be used to predict material performance because of both geometric and temperature effects that cause wide variations in the data.

3. Dynamic Mechanical Testing.

Dynamic mechanical testing provides fundamental data on characterizing a material. Several applications were developed, from refining data normally obtained from thermal tests such as differential scanning calorimetry (DSC) and thermogravimetric analysis (TGA) to correlations with impact performance of a material. DMA tests were used to study the molded-in strain effects in injection-molded parts via correlation with broadening of transitions and increased degree of strain (15). With appropriate computer software, DMA scans were used to calculate Arrhenius and WLF functions and produce master curves by superposition for creep studies (16). Strain sweeps by DMA were used in monitoring both the quality of additives and the effects of types of compounding equipment on the uniformity of the additive in polypropylene compounds (17). In the developing area of blends and plastic alloys, DMA is often used to characterize these materials through resolution of T_g. Immiscible blends give two T_gs that match values for the individual polymers, but miscible blends give a single T_g that shifts to higher temperatures as the higher T_g blend component increases, as shown in Fig. 51.9.

Driscoll showed the use of DMA to evaluate effects of hardener chemistry on the temperature behavior of cured epoxy resins. The storage modulus declined while the tan δ peaks associated with the T_g were significantly altered by the type of hardener (18).

The effects of polymer aging can also be monitored using DMA because structural changes normally occur even well below the deflection temperature. In Fig. 51.10, the storage modulus is plotted against temperature for polyphenylene sulfide (PPS) aged for 500 hr at different thermal aging temperatures and compared to as-molded samples. Aging at temperatures below 250 °C results in an increase in crystallinity, as shown by the increase in the plateau modulus. Aging

Figure 51.9 Plot of E'' versus temperature for neat poly(butylene terephthalate) (PBT) and polycarbonate (PC) and a PC/PBT alloy. *(Reprinted with permission from N. P. Cheremisinoff, ed.,* Elastomer Technology Handbook. *Copyright CRC Press, Boca Raton, FL.)*

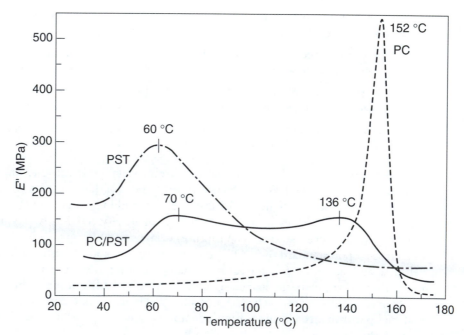

Figure 51.10 Plot of E' versus Temperature for a 40% glass fiber reinforced PPS heated 500 hr in air at various temperatures. *(Reprinted by permission of Dicten & Masch Manufacturing Company.)*

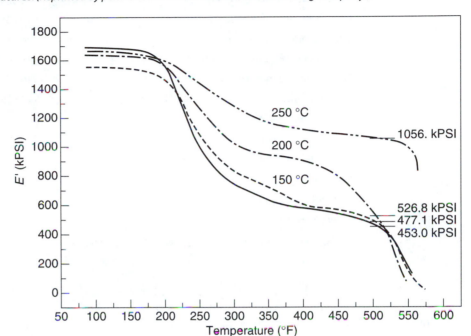

at 250 °C adds the effect of cross-linking to further increase the plateau modulus. A plot of tan δ versus temperature for the same materials shows a loss in peak height indicative of lost ductility (19).

Finally, DMA is used to characterize curing and cross-linking of thermosetting materials. As an incompletely cured material is heated during a temperature scan, the storage modulus will show a recovery as additional curing takes place. Landi and Merserau used DMA with tan δ being monitored to plot sets of curves for the onset of T_g versus time for mold temperatures. Construction of master curves for any set reference temperature allowed them to predict the resultant T_g for a given time–temperature profile inside the mold (20).

Nuts and Bolts

Relative Costs

Electromechanical universal test frame	$$ to $$$
Hydraulic activated universal test frame	$$$ to $$$$
Fixtures: grips, flex supports (cage), and plates	$ to $$
Load cells and proof rings	$ to $$

Extensometers and deflectometers (LVDT)	$ to $$
Extensometers (optical/laser)	$ to $$$
Extensometers (biaxial/special)	$$ to $$$
Oscilloscopes and microprocessors	$ to $$
High-speed instrumented impact	$$$$ to $$$$$
Pendulum and manual drop weight	$ to $$
Environmental chambers, liquid N_2, CO_2	$$ to $$$
Environmental chambers, mechanical	$$$
Dynamic mechanical analyzer	$$$$ to $$$$$
Maintenance (yearly)	$ to $$
Computer hardware	$ to $$
Computer software	$$ to $$$
Reference data	$

$ = 1 to 5 K, $$ = 5 to 15 K, $$$ = 15 to 50K, $$$$ = 50 to 100K, $$$$$ = >100K.

The major factors affecting the price of load frames are the required stiffness (usually determined by the maximum load required for testing) and whether the equipment is screw-driven (electromechanical) or hydraulic actuated. Load cells can range from a gram to 50 kN, but the cost is similar regardless of size.

Because of the auxiliary equipment requirements, noncontact extensometers or measuring devices are more expensive than contact devices. The higher-resolution units are normally found in research and development labs for characterizing materials.

For quality control laboratories, pendulum impact equipment, manual drop weight, or gravity-based high-speed instrumented impact and moderately priced universal test frames are adequate for performing many of the short-term tests. Computer software and hardware, although not required for quality control testing, significantly improve productivity and data precision and reduce levels of work requiring trained interpretation of data.

Vendors for Instruments and Accessories

Applied Test Systems, Inc.
348 New Castle Rd.
Butler, PA 16003
phone: 412-762-8623, 800-441-0215
fax: 412-283-6570

C.W. Brabender Instruments
50 East Wesley St.
S. Hackensack, NJ 07606
phone: 201-343-8425
fax: 201-343-0608
email: CWBI@ix.netcom.com
Internet: http://www.plasticsnet.com/cwbrabender

CEAST,USA, Inc.
377 Carowinds Blvd., Ste. 207
Ft. Mill, SC 29715
phone: 803-548-6093
fax: 803-548-1954
Internet: http://www.ceast.com

Custom Scientific Instruments
13 Wing Dr.
Cedar Knolls, NJ 07929
phone: 201-538-8500
fax: 201-984-6793

GRC Instruments
5383 Hollister Ave.
Santa Barbara, CA 93111
phone: 805-681-8825
fax: 805-964-2914
Internet: http://www/grcinstrument.com

Instron Corp.
100 Royall St.
Canton, MA 02021-1089
phone: 617-828-2500, 800-564-8378
fax: 617-575-5751

Kayeness
115 Thousand Oaks Blvd.
Morgantown, PA 19543
phone: 610-286-7555
fax: 610-286-7555

MTS Systems Corp.
14000 Technology Dr.
Eden Prairie, MN 55344-2290
phone: 800-944-1687
fax: 612-537-4515
Internet: http://www.mts.com

Rheometric Scientific, Inc.
One Possumtown Rd.
Piscataway, NJ 08854
phone: 908-560-8550
fax: 908-560-7451
Internet: http://www.rheosci.com

SATEC Systems, Inc.
900 Liberty St.
Grove City, PA 16127
phone: 412-458-9610
fax: 412-458-9614
Internet: http://www.satec.com

Shimadzu Scientific Instruments, Inc.
7102 Riverwood Dr.
Columbia, MD 21046
phone: 800-477-1227
fax: 410-381-1222
Internet: http://www.shimadzu.com

Testing Machines, Inc.
400 Bayview Ave.
Amityville, NY 11701
phone: 516-842-5400, 800-678-3221
fax: 516-842-5220

Tinius Olsen Testing Machine Co., Inc.
P.O. Box 429
Easton Rd.
Willow Grove, PA 19090
phone: 215-675-7100
fax: 215-441-0899
email: Info@TiniusOlsen.com
Internet: http://www.TiniusOlsen.com

United Testing Systems Inc.
5171 Exchange Dr.
Flint, MI 48507
phone: 810-732-2800
fax: 810-732-2872

Required Level of Training

Operation of equipment required for performing most plastics testing can be done by people with a high school education or associate degree. Some knowledge of chemistry or physics or mechanical aptitude is helpful. Experience is required to apply principles to new materials.

Interpretation of data requires a minimum of a college physics or mechanical engineering course for mechanical short- or long-term testing. Interpretation of dynamic mechanical analysis data requires both analytical chemistry and introductory polymer courses. Specific short courses are offered by ASTM and several equipment vendors in both mechanical tests and DMA.

Service and Maintenance

Modular units are used to simplify servicing of equipment. Routine service can be completed by the operator with some minimal training. Troubleshooting should be handled by the operator in conjunction with a trained instrument repair technician. Equipment manufacturers supply on-site service personnel for repair at relatively high hourly costs and often with charges for travel time and expenses. Maintenance contracts are also available for some equipment that give preference in service scheduling.

Replacement of accessory equipment is straightforward and part of standard operator training. Repair of broken extensometers and load cells requires special training and they should be returned to the manufacturer for repair. These are the most common replacements and cost more than $1,000 to repair or replace.

Calibration to National Institute of Standards and Technology traceable standards for any load cells used should be completed on at least a yearly basis. Testing of a reference material commonly used in the laboratory on a regular basis and tracking test results assists in ensuring that the equipment is operating properly and these data can assist in troubleshooting.

Suggested Readings

ADAMS, D. F., "The Iosipescu Shear Test Method as used for Testing Polymers and Composite Materials," *Polym. Compos.*, 11, no. 5 (1990), 286–90.

ASTM Volumes 8.01, 8.02, 8.03, and 8.04 /ISO Volumes 1 and 2 contain specific procedures for over 4000 plastic tests.

BROWN, R. P., ed., *Handbook of Plastics Test Methods*, 3rd ed. New York: Longman Scientific & Technical co-published with Wiley, 1988.

GALLI, E., "Properties Testing: Dynamic Mechanical Testing," *Plast. Cmpd.*, 7, no. 4, 15–25.

HAN, P., ed., *Tensile Testing*. Materials Park, OH: ASM International, 1992.

HAWLEY, S. W., "Physical Testing of Thermoplastics," *RAPRA, Rev. Rep.*, 5, no. 12 (1992).

HOROWITZ, E., "Plastics Testing," *Kirk-Othmer Encyclopedia of Chemical Technology*. New York: Wiley, 1985, 207–28.

MANLEY, T. R., "Thermal Analysis of Rubber and Plastics," *Proj. Rubber Plast. Technol.*, 5, no. 4 (1989), 253–88.

POURNOOR, K., AND J. C. SEFERIS, "Dynamic Mechanical Measurements with Instruments Having Distinct Operating Principles," *SPE ANTEC*, 1989.

SHAH, V., *Handbook of Plastics Testing Technology*. New York: Wiley, 1984.

References

1. M. Sepe, "The Uses of Thermal Analysis in Polymer Characterization," in N. Cheremisinoff, ed. *Elastomer Technology Handbook* (Boca Raton, FL: CRC Press, 1994), 175.

2. D. F. Adams and D. E. Walrath, *J. Compos. Mater.*, 21, no. 6 (1987), 494–507.

3. D. F. Adams, *Polym. Compos.*, 11, no. 5 (1990), 286–90.

4. V. Shah, *Handbook of Plastics Testing Technology* (New York: Wiley, 1984), 171.

5. D3763 Standard Test Method for Puncture Properties of Plastics, ASTM vol. 8.02. Philadelphia: ASTM.

6. R. P. Brown, *Handbook of Plastics Test Methods*, 3rd ed. (New York: Wiley, 1988), 186.

7. B. J. Lazan and A. Yorgiodis, *ASTM STP*, 59 (1944), 66.

8. I. M. Ward, *Mechanical Properties of Solid Polymers*, 2nd ed. (New York: Wiley, 1984).

9. K. Pournoor and J. C. Seferis, *SPE ANTEC* (1989), 1103.

10. D. M. Jeddy and E. Galli, *Plast. Cmpd.*, 7, no. 4, 17.

11. D. R. Ireland, "Instrumented Impact Testing for Evaluating End-Use Performance," *Physical Testing of Plastics* (Baltimore, MD: ASTM, 1984), 50–52.

12. R. P. Brown, ed., *Handbook of Plastic Test Methods*, 3rd ed. (New York: Wiley, 1988), 188.

13. M. J. Bonmin, C. M. R. Dunn, and S. Turner, *Plastics and Polymers*, 37 (1969), 517.

14. J. Bennett and G. W. Quick, *NBS Circular*, 550 (Sept. 1954).

15. C. L. Rohn and K. W. Herb, *Plast. Engng.*, 44, no. 10 (Oct. 1988), 33–6.

16. T. R. Manley, *Prog. Rubber Plast. Technol.*, 5, no. 4 (1989), 253–88.

17. S. B. Driscoll, "Using ASTM D4065 for Predicting Processability and Properties," *ASTM Symposium*, (Nov. 1983).

18. S. B. Driscoll, "Using ASTM D4065-82 for Predicting Processability and Properties in High Modulus Composites in Ground Transportation and High Volume Applications," in D.W. Wilson, ed., *ASTM STP 873* (Philadelphia: ASTM, 1985), 144.

19. M. P. Sepe, "The Use of Thermal Analysis in Polymer Characterization," in N. Cheremisinoff, ed., *Elastomer Technology Handbook* (Boca Raton, FL: CRC Press, 1994), 221.

20. V. R. Landi and J. M. Merserau, *SPE ANTEC*, 32 (1986), 1369.

Glossary

The numbers in parentheses at the ends of glossary entries refer to chapters in which these terms appear.

[M – H]⁻: an even electron ion derived by deprotonating a molecule. (28–34)

Φ_{CL}: quantum efficiency for chemiluminescence. (27)

Φ_{EM}: quantum efficiency for emission from an excited state. (27)

Φ_{EX}: quantum efficiency for chemical production of excited states. (27)

AAS: atomic absorption spectrometry. (19–27)

absorbance: the negative log of the transmittance. (25)

absorptiometer: an instrument designed to measure transmittance or absorbance of samples, usually using a prism or grating monochromator as the wavelength isolation device. (25)

absorption: attenuation of the radiant power of a beam of electromagnetic radiation as it traverses a sample of matter. The absorption process proceeds by raising one or more constituents of the sample to excited states. (19–27)

absorptivity: a proportionality constant used in the Beer–Lambert–Bouguer law. (25)

accuracy: the degree of agreement of a measured value with the true or expected value of the quantity of concern. (1–6)

activity: a thermodynamic function that relates changes in the chemical potential with changes in experimentally measurable quantities such as voltages, partial pressures, etc.

ADC: see **analog-to-digital converter.**

additive interference: a sample constituent (other than the analyte) that produces a signal that adds to the signal produced by the analyte. (19–27)

adduct ions: ions formed by addition of cations (usually Na^+, K^+, NH_4^+) or matrix ions (in matrix-assisted laser desorption/ionization applications). (28–34)

adsorptive stripping voltammetry (AdSV): technique in which material is accumulated on an electrode by adsorption and determined by the voltammetric signal for its oxidation or reduction. (37)

AED: atomic emissions detector.

AES: atomic emission spectrometry. (19–27)

AES: see **Auger electron spectroscopy.**

AFM: see **atomic force microscopy.**

AFS: atomic fluorescence spectrometry. (19–27)

aliquot: the portion of a sample resulting from dividing it into an equal number of parts leaving no remainder. (1–6)

alpha cleavage: cleavage of a bond adjacent to a heteroatom in an organic compound. (28–34)

AMD: automated multiple development. (13)

amperometric detection: measurement of the current or charge resulting from an oxidation or reduction at the surface of the working electrode. (9, 10, 12, 36)

AMU: see **atomic mass unit.**

analog-to-digital converter (ADC): an electronic component that converts an analog signal to a digital signal. (6, 41)

analyte: the sample constituent (compound, element, or ion) whose concentration (or amount) is to be determined in a quantitative analytical procedure.

anodic stripping voltammetry (ASV): technique for determination of metals or other species that can be deposited cathodically in or on an electrode and then determined from the anodic voltammetric signal. (37)

API: atmospheric pressure ionization. (28–34)

array detector: any of a number of solid-state devices that behave as a spatial array of individual detector elements. Common array detectors include diode arrays and charge-coupled devices. (19–27)

ASE: accelerated solvent extraction. (2)

assay: quantitative determination of the composition of a material.

atom cell: in atomic absorption, emission, or fluorescence spectrometry, the region containing the cloud of gaseous atoms that produce the analytical signal. Also called atom reservoir. (19–27)

atomic force microscopy (AFM): a surface imaging technique with atomic resolution capability based on the scanning of a sharp probe tip over a specimen while monitoring the tip displacement. (40, 41)

atomic mass unit (AMU): a unit of measurement that expresses the relative mass of specific isotopes of an element.

atomizer: in atomic spectroscopy, a device for converting a liquid or solid sample to a cloud of gas-phase atoms. Plasmas, furnaces, and flames are the most widely-used atomizers in atomic absorption, emission, and fluorescence spectrometry. (19–27)

ATR: see **attenuated total reflectance.**

attenuated total reflectance (ATR): a reflection with reduced (attenuated) radiant power, due to a coupling absorbing mechanism, during a process of total internal reflection. (15)

Auger electron spectroscopy (AES): a surface analysis technique based on the energy analysis of Auger electrons typically excited by a focused electron beam. (40–44)

autogain: the process of changing the gain of a spectrophotometer amplifier so as to compensate for changes in source intensity and detector response as wavelength is scanned. (25)

autoslit: the process of changing the slit width of a spectrophotometer so as to compensate for changes in source intensity and detector response as wavelength is scanned. (25)

auxochromes: molecular groups attached to chromophores that tend to alter the chromophores' absorption characteristics. (25)

background corrector: in atomic absorption or fluorescence spectrometry, any of several devices that attempt to correct for errors produced by scattering of source radiation and atomic and molecular absorption or emission. (19–27)

ballistic electron emission microscopy (BEEM): an interface-sensitive microscopy/spectroscopy based on the analysis of electrons injected with a scanning tunneling microscope. (41)

bathochromic shift: a shift in the wavelength to longer wavelengths (red shift). (25)

BEEM: see **ballistic electron emission microscopy.**

Beer–Lambert–Bouguer law: in atomic or molecular absorption spectrometry, the fundamental equation that relates the absorbance of the sample to the analyte concentration. Usually called simply Beer's law. (19, 20, 25, 27)

beta cleavage: cleavage of a bond once-removed from a heteroatom, following a hydrogen migration. (28–34)

bias: a systematic error inherent in a method or one caused by an artifact or idiosyncrasy of the measurement system. (4)

bioluminescence (BL): chemiluminescence produced by certain living organisms such as fireflies, jellyfish, and some bacteria. (27)

Bjerrum's method: a procedure for determining the stability constant of a complex ion. (25)

BL: see **bioluminescence.**

blank: the measured value obtained in the absence of the analyte. In such cases the apparent measured value for the analyte is due to artifacts; hence it should be subtracted from the measured value to give a net value for the analyte. A blank measurement must be made to validate the determination of the analyte. (See also **calibration blank** and **reagent blank.**)

buffer: a solution that resists changes in pH. Usually composed of a weak acid and its salt or a weak base and its salt.

bulk property detector: a detector, such as a refractive index detector, that measures a property of both the solute and the mobile phase. (7–13)

calibration (1): comparison of a measurement standard or instrument with another standard or instrument to report or eliminate by adjustment any variation in the accuracy of the item being compared. (4)

calibration (2): the act of ascertaining the relationship between a measured signal and the concentration of analyte producing the signal.

calibration blank: a solution, as free of the analyte as possible, used to give the null reading for the measuring instrument. (4)

calibration curve: a plot of the measured signal as a function of the analyte concentration. (19–27)

capillary electrophoresis (CE): separation of analytes by means of electric field in a solution flowing through a capillary. (10, 36)

capillary gel electrophoresis (CGE): a type of capillary electrophoresis in which the capillary is filled with a gel and the separation is on the basis of differences in sizes of the sample molecules. Often used for analysis of large biomolecules. (10)

capillary isoelectric focusing (CIEF): a type of capillary electrophoresis in which sample ions are separated on the basis of differences in their isoelectric points. Often used for determination of isoelectric points of proteins. (10)

capillary isotachophoresis (CITP): a type of capillary electrophoresis in which the sample is sandwiched between leading and terminating buffers. Separation is on the basis of differences in mobilities of the sample ions. (10)

capillary zone electrophoresis (CZE): also called free-solution capillary electrophoresis (FSCE). A type of capillary electrophoresis in which the capillary contains only an electrolyte. Sample ions are separated on the basis of differences in their charge-to-size ratios. (10)

carbon dioxide: the most common mobile phase used in supercritical fluid extraction and chromatography. (10)

CCD: see **charge-coupled device.**

CE: see **capillary electrophorsis.**

cell constant (K_{cell}) (SI unit m^{-1}, commonly expressed in cm^{-1}): constant that is specific for a given cell at a given temperature. (39)

CGE: see **capillary gel electrophoresis.**

CHA: see **concentric hemispherical analyzer.**

charge transfer devices: generally divided into two types: the charge-coupled device (CCD) and the charge-injection device (CID). (14)

charge-coupled device (CCD): a type of array detector (q.v.) of electromagnetic radiation. (19–27)

charge-injection device (CID): a charge transfer device used to detect photons.

check standard: in physical calibration, any property measured periodically. The results are plotted on a control chart to evaluate the measurement process.

chelating agent: an organic compound that forms multiple coordinate bonds with metal ions in solution. These agents can be used to remove interfering ions and thus enhance the specificities of an analytical method.

chemical interference: in general, a substance that prevents accurate determination of the analyte. In atomic absorption, emission, or fluorescence spectrometry, a sample constituent that acts as a multiplicative interference (q.v.) by decreasing the efficiency with which the analyte element is converted to gas-phase atoms. (4, 19–27)

chemical ionization (CI): ionization produced by reaction with chemical reagents. (28–34)

chemical shift: the difference between the resonant frequency of a nucleus in one type of chemical environment and the frequency of a reference, divided by the spectrometer frequency. It is expressed in ppm. (17)

chemical shift anisotropy (CSA): in nuclear magnetic resonance, the directional dependence of electronic shielding in a molecule. (17)

chemiluminescence (CL): light emission resulting from a chemical reaction that causes production of species in an electronically excited state. (27)

chiral molecules: molecules that are not superimposable with their mirror images. (9)

chromatogram: the output from a detector in most chromatographic techniques. A plot of detector response versus time. In thin-layer chromatography, the chromatogram plate with the separated substances is often called the chromatogram. (7–13, 31, 33)

chromatography: a technique used to separate components in a sample mixture by the differential distribution of the components between a mobile phase and a stationary phase. (7–13, 31, 33)

chromophores: aggregates of atoms within a molecule capable of absorbing UV/VIS electromagnetic radiation. (25)

CI: see **chemical ionization.**

CID: see **charge-injection device.**

CID: collision-induced dissociation. (28–34)

CIDNIP: chemically induced dynamic nuclear polarization. (17)

CIEF: see **capillary isoelectric focusing.**

CITP: see **capillary isotachophoresis.**

CITS: see **current imaging tunneling spectroscopy.**

CL: see **chemiluminescence.**

classically ruled grating: a diffraction grating that has interference lines scribed by a mechanical device called a ruling engine. (25)

cluster ion: ion formed by cumulative additions of matrix molecules to a protonated or cationized molecular ion. (28–34)

CMA: see **cylindrical mirror analyzer.**

coated wire electrodes (CWE): potentiometric electrodes prepared by coating a wire with the sensing material. (37, 38)

COLOC: see **correlation by long-range coupling.**

color comparators: devices for comparing and determining relative color intensities. (25)

colorimeter: an instrument used to measure color in terms of three primary colors; also erroneously described as an instrument for measuring optical absorption of colored samples. (25)

colorimetry: the science of quantitatively measuring color in terms of three primary colors; also used erroneously to describe the process of measuring optical absorption of colored solutions. (25)

column: the part of the instrument in which the chromatographic separation occurs. May be capillary type of very small diameter and open tubular with the stationary phase coated on the inner walls, or may be packed type of larger diameter and packed with a solid adsorbent or solid support material coated with a liquid phase. (3, 7–13, 31, 46)

compressive test: a mechanical test utilizing a flat plate fixture to crush the specimen. (45–51)

concentric hemispherical analyzer (CHA): an electron energy analyzer that uses two electrostatically biased concentric hemispheres to create an electric field to disperse electrons according to their energy through a set of slits onto an electron multiplier. (42, 43)

conductometric detection: measurement of the current resulting from the mobility of ions in an applied electric field. (7–13)

control sample: a material of known composition similar to the test sample, which is analyzed concurrently with test samples to evaluate a method or measurement process.

corrected spectrum: excitation or emission spectrum that has been corrected for variation of such parameters as source power, detector sensitivity, and monochromator throughput with wavelength. (26)

correlation by long-range coupling (COLOC): in nuclear magnetic resonance, a two-dimensional experiment that produces long-range correlations between protons and carbons. (17)

correlation spectroscopy (COSY): a general term for any two-dimensional NMR experiment that produces signals that correlate two nuclei by their chemical shifts. It has also come to mean the proton–proton chemical shift correlation experiment. (17)

COSY: see **correlation spectroscopy.**

CPMAS: see **cross-polarization magic angle spinning.**

creep test: a mechanical test where a fixed load is applied to a specimen and the deformation is recorded as a function of time. (51)

cross-polarization magic angle spinning (CPMAS): a combination of two techniques used in solid-state NMR for sensitivity enhancement and line narrowing. (17)

cross-sectional scanning tunneling microscopy (XSTM): the use of scanning tunneling microscopy to characterize the cross-section of a sample. (41)

cryostat: a system of concentric dewars enclosed in a vacuum vessel for maintaining a liquid helium bath in which a superconducting solenoid magnet is immersed. (17)

CSA: see **chemical shift anisotropy.**

CTD: see **charge transfer devices.**

Curie point pyrolysis: a pyrolysis system based on the use of ferromagnetic metal wires that are inductively heated with radio frequencies. (49)

current imaging tunneling spectroscopy (CITS): generation of an image of a surface based on the tunneling current between a conductive probe tip and the surface. The probe tip is rastered in a two-dimensional plane within the electron tunneling distance of the surface to produce the image. Simultaneously with the topography, a two-dimensional set of discrete I-V spectra are obtained at a given mesh. (41)

current-voltage (I-V): the current generated in a semiconductor junction at a specific voltage. (41)

CV: see **cyclic voltammetry.**

CWE: see **coated wire electrodes.**

cyclic voltammetry (CV): technique in which potential is varied at a constant rate from an initial potential to a final potential and back. (37)

cylindrical mirror analyzer (CMA): an electron energy analyzer that uses two electrostatically biased cylinders to create an electric field to disperse electrons according to their energy through a set of slits onto an electron multiplier. (17, 42, 43)

CZE: see **capillary zone electrophoresis.**

Da: see **dalton.**

DAC: see **digital-to-analog converter.**

dalton (Da): unit of atomic mass, based on the definition of carbon-12 as exactly 12 mass units. (28–34)

dark current: a current found in light-sensitive detectors arising primarily from thermal motion of electrons in the unilluminated detector. (25)

DCI: desorption chemical ionization. (28–34)

decay time: the time required for an emission signal (fluorescence or phosphorescence) to decay to 1/e of its initial value. (26)

degassing: in liquid chromatography, the process of removing dissolved gases from the mobile phase by sparging or applying a vacuum. (9)

demal: term used specifically for three primary standard potassium chloride solutions (1.0 demal, 0.1 demal, and 0.01 demal), where 1.0 demal is defined as 1 g•mol of KCl dissolved in 1000 cm^3 of solution. (39)

density of states (DOS): the density of electronic states at a semiconductor surface. Also referred to as the local DOS (LDOS) when used to describe the density of electronic states on the atomic or nanometer scale. (41)

depolymerization: a degradation mechanism resulting in the unzipping of polymer chains to yield monomers and oligomers. (45–51)

DEPT: see **distortionless enhancement by polarization transfer.**

derivatization: see **indirect determination.**

detection limit: the smallest concentration or amount of the analyte that can be determined by a single measurement with a stated level of confidence. (4)

detector: the transducer that transforms chemical or physical properties of an analyte into an electrical signal.

differential pulse voltammetry (DPV): technique in which electrode potential is pulsed and the current signal is the difference of currents before and near the end of the pulse. (37)

differential scanning calorimetry (DSC): a thermal technique that measures the difference in heat flow between the sample and a reference material. (50)

differential themogravimetric analysis (DGA): technique in which change in mass of sample is measured as a function of temperature. (50)

diffraction grating: a common dispersive element for the monochromator. It consists of equally spaced grooves positioned on a metal surface and can disperse the radiation into component energy bands of different wavelengths by rotating the grating relative to the radiation source. (15)

digital-to-analog converter (DAC): an electronic component that converts a digital signal to an analog signal. (6, 41)

DIL: direct liquid introduction.

diode array: a common type of array detector (q.v.) of electromagnetic radiation. (19–27)

direct determinations: a photometric or spectrophotometric method of analysis where the analyte or a derivative of it possesses the chromophore being measured. (25)

distortionless enhancement by polarization transfer (DEPT): a very sensitive ^{13}C experiment that identifies the number of protons attached to each carbon in a molecule. (17, 41)

DLI: direct liquid introduction. (28–34)

DMA: see **dynamic mechanical analysis.**

DMS: dynamic mechanical spectroscopy.

DOS: see **density of states.**

DPV: see **differential pulse voltammetry.**

DRI: differential refractive index. (47)

dropping mercury electrode (DME): a continuously renewed electrode formed by hydrostatic flow of mercury through a fine capillary. (37)

DSC: see **differential scanning calorimetry.**

duplicate: repeated analysis of the same sample with a given procedure to assist in the evaluation of variance.

dynamic mechanical analysis (DMA): a thermal technique that applies a physical force to one portion of a test specimen and measures the response of the sample at some difference from the specimen. (51)

dynamic range: sometimes known as linear dynamic range, the analyte concentration range over which response is a well-defined (usually linear) function of the analyte concentration.

dynode: a component of a photomultiplier tube with a photoelectron-sensitive surface geometrically arranged with other dynodes to focus electrons to succeeding dynodes for the purpose of photoelectron gain. (25)

ECD: electron capture detector. (8)

ECL: see **electrogenerated chemiluminescence.**

ECNI: electron capture negative ionization. (28–34)

EDL: electrodeless discharge lamp. (19–27)

EDTA: ethylenediaminetetraacetic acid, a complexing agent for metal ions. (38)

EDXRF: energy-dispersive X-ray fluorescence; see **X-ray fluorescence.** (24)

EELS: see **electron energy loss spectroscopy.**

efficiency: in chromatography, measurement of the sharpness of an eluting band or peak. High efficiencies indicate low diffusion and dispersion, which are major contributors to peak broadening. Can be calculated as theoretical plates. (7–13)

effluent: the liquid or mobile phase after it has passed through the analytical column. (7–13)

EI: see **electron ionization.**

ELCD: electrolytic conductivity detector. (8)

electrogenerated chemiluminescence or electrochemiluminescence (ECL): initiation of chemiluminescence by direct electrochemical triggering of the CL reaction or by electrochemical production of a key reagent for the CL reaction. (27)

electrokinetic injection: a sample introduction technique used in capillary electrophoresis in which the sample is forced into the capillary by electroosmosis. (10)

electrolyte: also called the run buffer. The conductive solution in the capillaries and reservoirs in capillary electrophoresis. (10)

electrolytic conductivity (κ) (SI unit $S{\bullet}m^{-1}$, commonly expressed in $\Lambda S/cm$): ability of a solution to carry electric current. (39)

electron energy loss spectroscopy (EELS): the energy analysis of electrons to determine the energy loss of electrons backscattered from a surface. (42)

electron ionization (EI): ionization produced by an electron beam. (28–34)

electroosmotic flow: the flow of the electrolyte in capillary electrophoresis caused by electroosmosis. (10)

electropherogram: the output from a detector in capillary electrophoresis. A plot of detector response versus time. (10)

electropherograph: the instrument used to perform capillary electrophoresis. (10)

electrophoresis: the movement of charged particles in a conductive solution due to application of an electric field. (10)

electrophoretic mobility: in capillary electrophoresis, the mobility of an ion in a conductive medium, under the influence of an electric field. Directly related to the charge-to-size ratios of sample ions. (10)

electrothermal atomization: in atomic absorption or fluorescence spectrometry, a technique for sample atomization in which a small volume of sample is inserted into a small furnace (usually made of graphite) and then subjected to rapid resistive heating. (20)

elimination: a degradation mechanism resulting in the loss of small neutral molecules. (49)

eluent: the liquid phase, sometimes called the mobile phase, in a chromatographic separation. (7–13)

elution order: the order in which the components pass out of the column after separation. (7–13)

emission spectrum: spectrum obtained when the excitation wavelength in a fluorescence spectrometer is held constant and the emission wavelength is scanned. Also known as fluorescence spectrum. (26)

emission: in atomic spectroscopy, emission of light by an atom that had previously been raised to an excited state by collisional processes in an atom reservoir (usually a flame or plasma). (19–27)

end absorption: a strong absorption resulting from the existence of a very strong absorption band lying beyond the instrument's lower wavelength limit. (25)

endcapping: the process of covering the exposed silanol groups on the stationary phase of a column. (7–13)

enhanced chemiluminescence: the increased emission obtained with the luminol/horseradish peroxidase/ H_2O_2 system when certain phenolic compounds are also present. (27)

EOF: see **electroosmotic flow.**

equivalent conductivity (Λ_{eq}) (unit $S\bullet cm^2\bullet equiv^{-1}$): ability of individual electrolytes in a solution to conduct electric current, equal to the electrolytic conductivity divided by equivalent concentration, related to molar conductivity by $\Lambda_{eq}=\Lambda/(v_+|z_+|)$. (39)

ESI-MS: electrospray ionization mass spectrometry. (33)

ESI: electrospray ionization. (33)

ETA: see **electrothermal atomization.**

excitation spectrum: spectrum obtained when the emission wavelength in a fluorescence spectrometer is held constant and the excitation wavelength is scanned. (26)

excitation–emission matrix: mathematical compilation of all data in excitation and fluorescence spectra of a sample in the form of a matrix; useful in analysis of complex samples. (26)

external standard: a standard used to calibrate an instrument in a different measurement than that used for the analyte. (28–34)

FAAS: flame atomic absorption spectrometry. (20)

FAB: see **fast atom bombardment.**

fast atom bombardment (FAB): a technique by which high energy ions bombard a solid surface producing secondary ions for analysis by mass spectrometry. (32)

fast Fourier transform (FFT): a computer algorithm developed for the purpose of transforming time domain data (an interferogram) to frequency domain data (a spectrum). (25)

FAT: see **fixed (constant) analyzer transmission.**

fatigue testing: a mechanical test that subjects the specimen to repeated cyclic loading, generally to failure of the specimen. (51)

FD: field desorption. (28–34)

FFT: see **fast Fourier transform.**

FIA: see **flow injection analysis.**

FID: see **flame ionization detector.**

FID: see **free induction decay.**

field ion microscopy (FIM): the imaging of a surface using ions extracted by the application of high electrostatic fields. (42)

FIM: see **field ion microscopy.**

final temperature (FT): the final temperature for a pyrolysis experiment. (49)

fingerprint: a pyrolysis chromatogram or pattern of pyrolysis products that is generally considered unique for each polymer type. (49)

fingerprinting: in general, the matching of data from a sample with data from a reference material. (15, 49)

finite slit width effect: anomalous absorbance readings, such as decreased absorbance peaks or unresolved peaks, resulting from the fact that the spectral bandwidth of the instrument is not sufficiently small relative to the compound's natural bandwidth. (25)

fixed (constant) analyzer transmission (FAT or CAT): operation of an electron energy analyzer in the fixed or constant analyzer transmission mode. (43)

fixed (constant) retardation ratio (FRR or CRR): operation of an electron energy analyzer in the fixed or constant retardation ratio mode. (43)

fixed signal method: a method of measuring the rate of a chemical reaction by measuring the time necessary for a predetermined change of the signal (absorbance) to occur. (25)

fixed time method: a method of measuring the rate of a chemical reaction by measuring the change of the signal (absorbance) over a predetermined time interval. (25)

fixed wavelength detector: an absorbance detector that can be used to measure absorbance at several discrete wavelengths, but only one at a time.

flame ionization detector (FID): a common type of detector used in gas chromatography and supercritical fluid chromatography. (8)

flexural testing: a mechanical test involving application of a load to the center of a beam specimen that is supported at two points near each end of the beam. (51)

flow injection analysis (FIA): technique for treating samples and presenting them to a detector as a plug in a flowing stream of carrier liquid. (37)

flowrate: in chromatography, the volumetric flowrate of the mobile phase. (7–13)

fluorescence: emission of electromagnetic radiation by an excited atom or molecule, the excited species having been produced by absorption of electromagnetic radiation by the atom or molecule in question. (19–27)

fluorescence advantage: the combination of factors that cause fluorescence spectrometry to exhibit much lower limits of detection than absorption spectrometry if the analyte exhibits a reasonably large fluorescence quantum yield and the blank produced by scattering and spurious fluorescence is not large. (26)

fluorescence spectrum: see **emission spectrum.** (26)

Fourier transform (FT): a mathematical operation that can be used to convert an intensity-versus-time spectrum to an intensity-versus-frequency (wavenumber) spectrum. (15)

Fourier transform ion cyclotron resonance (FTICR): synonymous with Fourier transform mass spectrometry (FTMS). (28–34)

Fourier transform mass spectrometry (FTMS): synonymous with Fourier transform ion cyclotron resonance (FTICR). (28–34)

FPD: flame photometric detector. (8)

free induction decay (FID): the NMR signal induced in the coil by the freely precessing net magnetization vector. The signal decays with time, giving rise to the linewidth of the NMR lines. (17)

free-solution capillary electrophoresis (FSCE): see **capillary zone electrophoresis.** (10)

front-surface geometry: optical arrangement in a fluorescence spectrometer when fluorescence is viewed from the surface of the sample container on which the incident radiation impinges. Used primarily for solid and very strongly absorbing or turbid liquid samples. (26)

FSCE: see **capillary zone electrophoresis.**

FTICR: see **Fourier transform ion cyclotron resonance.**

FTIR: Fourier transform infrared spectroscopy.

FTMS: see **Fourier transform mass spectrometry.**

full width at half maximum (fwhm): the width of a peak in wavelength units at one-half of its peak intensity. (25)

furnace pyrolyzer: a pyrolysis system that is preheated to the final pyrolysis temperature prior to the introduction of the sample. (49)

GC-HRMS: gas chromatography high-resolution mass spectrometry. (31)

GC-MS: gas chromatography mass spectrometry. (31)

GC: gas chromatography. (8)

geometric dispersion term: a term used to describe the dispersion of a prism that results from the geometric design of the prism and its configuration. (25)

GFAAS: graphite furnace atomic absorption spectrometry. (20)

GFLEAFS: graphite furnace laser-excited atomic fluorescence spectrometry. (20)

glass transition temperature (T_g): the temperature at which the physical properties of a polymer change from a glassy state to a rubbery state. (45–51)

GLPC: gas–liquid partition chromatography (synonymous with gas chromatography). (8, 31)

gradient elution: in liquid chromatography, the elution of solutes using a variable mixture of solvents. (9)

gradients: magnetic field strength gradients applied to the NMR sample for many different applications, including magnetic resonance imaging. (17)

Harney–Miller cell: a sample cell that holds a sample in a glass capillary that is bathed in a stream of nitrogen whose temperature is controlled and monitored. (14)

HCL: hollow cathode lamp. (19–27)

helium: an inert gas commonly used as the carrier gas in gas chromatography. (8, 31)

heteronuclear multibond correlation (HMBC): two-dimensional experiments performed using inverse detection for long-range and directly bonded heteronuclear correlations, respectively. (17)

high-resolution transmission electron microscopy (HRTEM): the characterization of the microstructure of materials by imaging high-energy electrons transmitted through a thinned specimen. By operating the microscope in the high resolution mode, individual atomic planes may be imaged. (41)

HMBC: see **heteronuclear multibond correlation.**

holographically ruled grating: a diffraction grating that has interference lines made by laser etching a holographic image on a photosensitive optical surface. (25)

HPLC: high performance liquid chromatography. (9, 33, 37)

hydrodynamic injection: a sample introduction technique used in capillary electrophoresis in which the sample is forced into the capillary by pressure or siphoning. (10)

hydrogen overvoltage: an overvoltage occurring at an electrode as a result of the liberation of hydrogen gas. (36–39)

hyperchromic: refers to an increase in the intensity of absorbance of radiation in the ultraviolet/visible region of the spectrum. (25)

hyphenated techniques: analytical techniques employing one or more instrumental separation or measurement systems functioning in a sequential or simultaneous mode. Gas chromatography mass spectrometry (GC-MS) is an example. (25)

hypochromic: refers to a decrease in the intensity of absorbance of radiation in the ultraviolet/visible region of the spectrum. (25)

hypsochromic: refers to a shift in the wavelength to shorter wavelengths (blue shift). (25)

I_{CL}: intensity of chemiluminescence (CL) emission. (27)

ICP-AES: inductively coupled plasma atomic emission spectrometry. (21)

ICP-MS: inductively coupled plasma mass spectrometry. (22)

ICP: see **inductively coupled plasma.**

ICR: ion cyclotron resonance.

immunoassay: an assay based on immune response; for example, a specifically treated antigen fluoresces in ultraviolet light and can thus be used to detect homologous antigens.

in situ: in the original location. Analysis is performed with a probe, and the sample is not removed from its original location.

indirect determination: a photometric or spectrophotometric method of analysis where the analyte does not possess the chromophore but is made to quantitatively react with a reagent that possesses a chromophore either before or after the reaction; also known as derivatization. (25)

indirect fluorescence: any of several procedures in which the amount or concentration of a nonfluorescent molecule is determined by measuring the extent to which it displaces a fluorescent species from the observation region or causes the concentration of a fluorescent species to be altered. (26)

inductively coupled plasma (ICP): widely used atom reservoir in atomic emission (ICP-AES) and mass spectrometry (ICP-MS). (21, 22)

inner-filter effect: type of interference in which a concomitant decreases the fluorescence signal for an analyte by absorbing source radiation or fluorescence emitted by the analyte. (26)

integrator: an electrical device commonly used to collect, perform calculations on, and display the detector output signal or chromatogram. (8)

interference or interferant: any sample constituent (other than the analyte) that complicates the process of detecting or quantifying the analyte. Can be subdivided into additive and multiplicative interferences (q.v.). (4)

interferogram: record of the detector response for the interference signals that are produced by an interferometer. (15)

interferometer: a device used to split a radiation beam into two or more beams, create differences in the optical paths between the split beams, and recombine them in order to generate interference patterns with repetitive, sinusoidal waves for each modulated component. (15)

intermolecular energy transfer: any of several processes by which an excited molecule transfers energy to a second molecule, causing fluorescence of the donor molecule to be quenched and that of the acceptor molecule to be sensitized. (26)

internal laboratory spike: a sample prepared from an independent standard that falls within the instrument calibration range used to serve as an independent check of technique, methodology, and standards.

internal standard method: a calibration technique in which known amounts of some species other than the analyte are added to samples to assist in quantifying the analyte. Also known as internal reference technique. (19–27)

internal standard: a compound added in known amounts to a sample before analysis to assist in quantifying the analyte. (28–34)

inverse detection: in nuclear magnetic resonance, proton-detected one- and two-dimensional experiments especially useful for dilute samples or insensitive nuclei. (17)

inverse photoemission spectroscopy (IPES): a surface analysis technique based on the photon spectra emitted from a sample under electron bombardment. (41)

ion chromatography: the chromatographic separation and measurement of ionic species. (12)

ion-exchange affinity: the attraction an ion has for an ion-exchange site on the stationary phase. (12)

ion-exchange capacity: the number of available ion-exchange sites, usually expressed as equivalents per column. (7–13)

ion-selective electrodes (ISE): potentiometric electrodes that are highly specific for one or a small group of ions. (38)

ionic equivalent conductivity (λ_{eq}) (unit $S \bullet cm^2 \bullet equiv^{-1}$): ability of individual ions in a solution to conduct electric current, related to the ionic molar conductivity by $\lambda_{eq} = \lambda/|z|$. (39)

ionic molar conductivity (λ) (SI unit $S \bullet m^2 mol^{-1}$, commonly expressed in $S \bullet cm^2 \bullet mol^{-1}$) ability of individual ions in a solution to conduct electric current. (39)

IPES: see **inverse photoemission spectroscopy.**

ISE: see **ion-selective electrodes.**

isocratic elution: elution of solutes using a single solvent system throughout the separation. (9, 33)

isothermal mode: elution of solutes using a single column oven temperature throughout the separation. (8)

isotope dilution: an internal standard technique used in mass spectrometry and ICP-MS in which the added "standard" is another isotopic form of the analyte element. (22)

J-coupling: the interaction of two or more spins through the bonding electrons that connect them. J-coupling produces the multiplet structure often found in NMR lines. (17)

jet separator: a type of gas chromatography mass spectrometry interface. (31)

LALLS: see **low-angle laser light scattering.**

Langmuir–Blodgett (LB) film: a film produced by adsorbing monolayers of fatty acid molecules one after the other to a diffraction element. (41)

LC-CFFAB-MS: liquid chromatography continuous-flow fast atom bombardment mass spectrometry. (32)

LC-ESI-MS: liquid chromatography electrospray ionization mass spectrometry. (33)

LC-MS-MS: liquid chromatography tandem mass spectrometry. (33)

LC-MS: liquid chromatography mass spectrometry. (33)

LCEC: see **liquid chromatography/electrochemical detection.**

LD: laser desorption. (34)

LEAFS: laser-excited atomic fluorescence spectrometry. (23)

LEED: see **low-energy electron diffraction.**

LIDAR: see **light detection and ranging.**

LIF: laser-induced fluorescence. (23)

light detection and ranging (LIDAR): a spectroscopic method whereby the Rayleigh scattering from an intense laser beam is collected by a telescope. (14)

light obscuration (LO): a technique for characterization of polymer particle size. (45–51)

limit of detection (LOD): the smallest concentration or amount of analyte that can be established at a reasonable statistical confidence level as being different from a blank; the concentration of analyte that produces a signal that exceeds the signal from a blank by an amount three times the standard deviation of the blank signal. (4)

limit of quantification (LOQ): the smallest concentration or amount of analyte for which reliable quantitative determination can realistically be achieved. (4)

limiting equivalent conductivity (Λ_{eq}^{∞}) (unit $S \bullet cm^2 \bullet equiv^{-1}$): equivalent conductivity extrapolated to infinite dilution. (39)

limiting molar conductivity (Λ^{∞}) (SI unit $S \bullet m^2 \bullet mol^{-1}$, commonly expressed in $S \bullet cm^2 \bullet mol^{-1}$): molar conductivity extrapolated to infinite dilution. (39)

LIMS: laboratory information management system. (5)

liquid chromatography/electrochemical detection (LCEC): the combined technique of separation by chromatography and detection on-line by amperometry. (36)

liquid metal ion gun (LMIG): an ion gun based on the field ionization of a sharp metal tip, usually gallium, which is molten at a low temperature. (44)

liquid secondary ion mass spectrometry (LSIMS): a technique similar to fast atom bombardment, in which a beam of primary ions bombards a liquid matrix to produce secondary ions for analysis by mass spectrometry. (32)

LLE: liquid–liquid extraction. (2)

LMIG: see **liquid metal ion gun.**

LO: see **light obscuration.**

LOD: see **limit of detection.**

LOQ: see **limit of quantification.**

low-angle laser light scattering (LALLS): a technique used to determine absolute polymer molecular weights. (45–51)

low-energy electron diffraction (LEED): the characterization of the crystal structure of a surface based upon the diffraction of low-energy electrons. (41)

LSIMS: see **liquid secondary ion mass spectrometry.**

luminescence: light emission by substances which have absorbed energy; photoluminescence and chemiluminescence are two of the major subdivisions of luminescence. (27)

m/z ratio: mass-to-charge ratio of ions in mass spectrometry (often called mass). (28–34)

magnetogyric ratio: the ratio of a nucleus' magnetic moment to its angular momentum. This ratio is unique for each magnetically active isotope. (17)

MALDI-TOFMS: matrix-assisted laser desorption/ionization time-of-flight mass spectrometry. (28–34)

MALDI: matrix-assisted laser desorption/ionization. (28–34)

mass chromatogram: plot of abundance of one selected mass as a function of time. (31)

mass resolving power (MRP): a unit that represents a mass spectrometer's ability to discriminate between adjacent masses of detected ions. (28–34)

matrix: the general environment of the analyte; for example, sodium ion analyte determined in a blood matrix.

matrix effect: see **multiplicative interference.**

matrix isolation: low-temperature sampling technique in which the analyte is sublimed and diluted in an inert diluent (matrix) in the gas phase. (26)

matrix spike: a predetermined quantity of analyte added to a sample matrix prior to sample preparation and measurement to provide a measure of the accuracy of the methods for the analyte in a given matrix. The concentration of the spike should be at the regulatory standard level or the practical quantitation level for the method.

MBE: see **molecular beam epitaxy.**

MCA: see **multichannel analyzer.**

McLafferty rearrangement: in mass spectrometry, beta cleavage rearrangement of organic ions. (28–34)

MECC: see **micellar electrokinetic chromatography.**

method of continuous variations: sometimes called Job's method, a method for determining the metal-to-ligand ratio of complex ions. (25)

method: the specific application of one or more techniques to provide the desired information. Includes all steps from sampling to reporting.

MFS: molecular fluorescence spectrometry. (26)

MH⁺: in mass spectrometry, an even electron ion derived by protonating a molecule. (28–34)

micellar electrokinetic chromatography (MECC): a type of capillary electrophoresis in which the sample molecules are separated on the basis of their water solubility. Micelles are added to the electrolyte and analyte molecules partition between the micelles and the electrolyte. Used for the determination of nonionic molecules. (10)

migration time (t_m): the time it takes a sample ion to migrate through a capillary in capillary electrophoresis. Usually expressed in minutes. (10)

mobile phase: the liquid or gas phase that passes through the column carrying the sample through the instrument. In gas chromatography, also called carrier gas. Also called the solvent or eluent in other types of chromatography. (7–13, 31, 33)

mobile-phase distance: in thin-layer chromatography, the distance traveled by the mobile phase traveling along the medium from the line of sample application to the mobile phase front. (13)

mobile-phase front: the leading edge of the mobile phase as it traverses the planar media in thin-layer chromatography. In all forms of development except circular and anticircular, the mobile phase front is essentially a straight line parallel to the mobile phase surface (also called liquid front or solvent front). (13)

modifiers: additives (such as methanol or organic acids) that are introduced to the supercritical fluid mobile phase in order to improve the peak shape and quality of separation. (11)

molality (m) (SI unit mol•kg⁻¹): amount of solute divided by the mass of solvent. (39)

molar absorptivity: the absorptivity expressed in L/mole⁻¹/cm⁻¹. (25)

molar conductivity (Λ) (SI unit S•m²•mol⁻¹): ability of individual ions in a solution to conduct electric current, equal to the electrolytic conductivity divided by amount-of-substance concentration. (39)

molecular beam epitaxy (MBE): the growth of epitaxial films using molecular beams. (41)

molecular ion: an odd electron ion derived by loss of an electron to form M^+ or the addition of an electron to form M^-. (28–34)

monochromator: a device that can be used to disperse a broad spectrum of electromagnetic radiation into a continuously calibrated series of monochromatic component bands. Common dispersive elements are prisms and gratings. (15, 25)

MRP: see **mass resolving power.**

MS: mass spectrometry. (28–34)

MS/MS: see **tandem mass spectrometry.**

multichannel analyzer (MCA): an analyzer with multiple channels for separating and storing data. (43)

multiple reflection path: the phenomenon of light being multiply internally reflected inside a sample cell. (25)

multiplicative interference: a sample constituent (other than the analyte) that alters the expected signal magnitude produced by a given concentration of analyte. Often called matrix effect, especially in atomic spectroscopy and X-ray fluorescence. (4, 19–27)

MWD: molecular weight determinations.

natural bandwidth (NBW): the full width at half maximum (fwhm) of a completely resolved absorption band. (25)

NBW: see **natural bandwidth.**

NCI: see **negative chemical ionization.**

negative chemical ionization (NCI): in mass spectrometry, the production of negative ions. The common modes of ionization usually produce positive ions. (28–34)

NIMS: negative ion mass spectrometry. (28–34)

NMR: nuclear magnetic resonance. (17)

NOE: see **nuclear Overhauser effect.**

NOESY: see **nuclear Overhauser effect spectroscopy.**

nominal mass: the isotopic species of lowest mass. (28–34)

normal phase: a type of partition chromatography that uses a polar stationary phase and a relatively nonpolar mobile phase. (7–13, 31, 33)

normal pulse voltammetry (NPV): technique in which electrode potential is pulsed and the current signal is measured near the end of the pulse. (37)

NPD: nitrogen-phosphorous detector. (8)

NPV: see **normal pulse voltammetry.**

nuclear Overhauser effect (NOE): an effect that occurs between nuclei that are very close in space to one another. The NOE is used for determining internuclear distances and for the enhancement of signals from insensitive nuclei. (17)

nuclear Overhauser effect spectroscopy (NOESY): a two-dimensional experiment using the NOE for generating correlation peaks. (17)

OBW: observed bandwidth. (19–27)

optical dispersion: a term used to describe the dispersion of a prism that results from variations of the refractive index of the prism with respect to wavelength. (25)

optical null balance: the process of balancing reference and sample signals in a spectrophotometer by attenuating one of the beams, usually the reference beam. (25)

p–n junction: the junction of two types of semiconductors used in the operation of solid-state diodes and transistors. (25)

PAD: pulsed amperometric detection. (12)

PAH: polycyclic aromatic hydrocarbon. (19–27)

PCB: polychlorinated biphenyl.

peak: the graphic part of the chromatogram that represents the plot of detector signal versus time for a separated component (also called band or zone). (7–13)

peptide mapping: combination of chemical cleavage with mass spectrometry analysis of fragments to determine peptide sequence. (28–34)

perfluorokerosene (PFK): a compound used as a calibration standard in mass spectrometry. (28–34)

PES: see **photoemission spectroscopy.**

PFK: see **perfluorokerosene.**

phosphorescence: emission process in which the upper and lower states have different spin multiplicities. In organic molecules, the upper state is usually the lowest triplet state and the lower state is the ground state. (26)

photodiode array: an array of photodiodes used in the simultaneous detection of various wavelengths of optical energy.

photoemission spectroscopy (PES): the analysis of a surface using photoemission of electrons with ultraviolet or x-ray radiation. (41)

photoluminescence: light emission resulting from a molecule, ion, or atom absorbing a photon to result in an electronically excited state. Fluorescence and phosphorescence are the most common photoluminescence processes. (27)

photometer: an instrument designed to measure transmittance or absorbance of samples, usually using a filter as the wavelength isolation device. (25)

photomultiplier tube (PMT): widely used detector for spectroscopy in the ultraviolet and visible spectral regions. (19–27)

piezoelectric transducer (PZT): a transducer based on the piezoelectric effect of a multicomponent oxide under high electric bias. (41)

PIXE: particle induced X-ray emission. (24)

plasma: a very hot gas in which a substantial fraction of the constituent atoms are ionized. Widely used as an atom reservoir for atomic emission and mass spectrometry. The inductively coupled plasma is the most widely used plasma system in atomic spectroscopy. (21, 22)

PMT: see **photomultiplier tube.**

polarity index: a measure of the relative polarity of a solvent (used to select solvents in partition chromatography). (7–13)

PQL: see **practical quantitation limit.**

practical quantitation limit (PQL): the lowest level of the analyte that can be reliably achieved within specified limits of accuracy and precision using routine laboratory operating conditions.

precision: the agreement among a set of replicate measurements without assumption or knowledge of the true value. The precision determines the standard deviation and hence the repeatability or reproducibility of the method.

preconcentration: any of several procedures used to increase the analyte concentration in a sample before measuring the concentration. (2, 3, 19–27)

primary standard: a substance whose value can be accepted (within specific limits) without question when used to establish the value of the same or related property of another material. The primary standard of one user may be the secondary standard of another. (4)

prism: a common dispersive element for the monochromator. The refractive index of a prism changes as the frequency of radiation changes. (15)

procedure: a set of general, systematic instructions for using a method of sampling, sample preparation, measurement, or data manipulation and storage. (1)

protocol: a specified, detailed procedure to be used when performing a measurement or related operations to obtain results acceptable to the specifier. (1)

pulse mode pyrolyzer: a pyrolysis system that generally uses resistively heated metal filaments. (49)

pyrolysis gas chromatography (PyGC): coupling of the pyrolysis event with the separation capabilities of gas chromatography. (49)

pyrolysis gas chromatography mass spectrometry (PyGC/MS): coupling of the pyrolysis event with the separation and identification capabilities of gas chromatography and mass spectrometry. (49)

pyrolysis: the degradation of macromolecules by thermal activity alone. (49)

PZT: see **piezoelectric transducer.**

QA: quality assurance.

QC: quality control.

quadrature phase detection: the technique of splitting the NMR signal into two channels with 90° phase difference. This results in an improved signal-to-noise ratio. (17)

qualitative analysis: determination of the identity of the analyte(s) in a sample. (7–13)

quality: an estimation of acceptability or suitability of a product or service to satisfy stated or implied needs. (1–6)

quantitative analysis: determination of the amount or concentration of the analyte(s) in a sample. (7–13)

quantum efficiency: the efficiency (expressed as a fraction) of a chemical or physical process initiated by electromagnetic radiation. Most often applied to the efficiencies of molecular or atomic fluorescence phenomena. Also known as quantum yield. (23, 26)

quantum yield: fraction of excited molecules (produced by absorption of light) that decay by a particular process; for example, the fluorescence quantum yield is the fraction of excited molecules that decay by fluorescence. (26)

quenching: any process in which the fluorescence quantum yield of an analyte is caused to decrease as a result of interactions between excited analyte molecules and another species (called the quencher). (23, 26)

racemic mixture: a compound that is a mixture of equal quantities of dextorotatory and levoratory isomers and is therefore optically inactive. (17)

random cleavage: a degradation mechanism that occurs when bond energies in a polymer are similar and rearrangements are impossible. (45–51)

Rayleigh ratio (R_τ): the ratio of the intensity of scattered light at angle τ ($\mathcal{J}_\tau$) to the product of the incident light intensity (I_0) and the scattering volume (V). (45–51)

Rayleigh scattering: the simplest type of scattering, involving very small, spherical, and optically isotropic particles. (14)

RBS: see **Rutherford backscattering spectrometry.**

reaction rate methods: methods of analysis based on determining the analyte when it is the rate determining species by following the rate of the reaction. (25)

reagent blank: a blank containing the same concentrations of reagents (solvents, etc.) as the samples used to determine the analyte concentrations. Reagent blanks undergo the same sample preparation and processing procedures as samples and are thus used to correct for possible contamination resulting from sample preparation or processing.

recovery: a means of ascertaining whether a methodology measures all of the analyte in the sample. The recovery is stated as the percentage of analyte measured with respect to the amount of analyte known to be in the sample. It is best evaluated by determining the analyte in reference materials or other samples of known composition. If this is not possible, spikes or surrogates may be added to the sample matrix.

redox: reduction-oxidation, a type of chemical reaction involving the transfer of electrons; forms the basis for many electrochemical techniques. (35–38)

reference material: a material, one or more of whose properties have been established and certified, that is used to calibrate instruments, assess methods, or assign values to other materials. Both government laboratories, such as the National Institute of Standards and Technology, and private companies provide standard reference materials.

relative sensitivity factor (RSF): a measure of the relative detection sensitivity for specific secondary ions given a specific set of instrumental conditions. (44)

relaxation: the return of nuclear spins to their equilibrium energy state after excitation. (17)

relayed coherence transfer: in nuclear magnetic resonance, the relay experiment generation of correlation peaks between nuclei within a spin system even though the nuclei may not share a J-coupling. (17)

resolution: a measurement of how well two signal peaks are separated from each other. Term applies to data presentation from many techniques.

resonance ionization spectroscopy (RIS): a surface analysis technique based on the analysis of secondary ions from resonant laser ionization of neutral species produced by ion sputtering from a solid surface or introduced into the laser beam by other means. (40)

restrictors: devices used to control the mass flow of the mobile phase under different pressure conditions through a supercritical fluid chromatography or supercritical fluid extraction system. (11)

retention time: in chromatography, the elapsed time from the point of injection of the sample to the measurement of the maximum signal detected by the detector for a particular component. (7–13)

reversed phase chromatography: a type of partition chromatography that uses a nonpolar stationary phase and a polar mobile phase. (9, 33)

right-angle geometry: optical arrangement in a fluorescence or Raman spectrometer wherein the emitted radiation from the sample is viewed at 90° to the direction of illumination of the sample. (14, 26)

RIS: see **resonance ionization spectroscopy.**

RP-HPLC: reversed phase high-performance liquid chromatography. (28–34)

RSF: see **relative sensitivity factor.**

Rutherford backscattering spectroscopy (RBS): a surface analysis technique based on the energy analysis of backscattered ions. (42, 44)

S(H)MDE: see **static (hanging) mercury drop electrode.**

S/N: signal-to-noise ratio.

sample stacking: in capillary electrophoresis, a technique whereby the sample is concentrated by dissolving it in a solution that is more dilute than the electrolyte. (10)

saturated calomel electrode (SCE): reference electrode based on the Hg/Hg_2Cl_2 electrode reaction. (37, 38)

SBW: see **spectral bandwidth.**

scanning probe microscopy (SPM): surface imaging techniques based on the interaction of a sharp probe scanned in proximity to or in contact with the surface of a sample. (41)

scanning tunneling microscopy (STM): a surface imaging technique with atomic resolution capability based on the tunneling current produced when an electric field is established between a conductive probe tip and the specimen surface. (40, 41)

SCE: see **saturated calomel electrode.**

SEC calibration curve: a plot defining the dependence of the specific retention volume (V_R) on the solute molecular weight. (45–51)

SEC: see **size exclusion chromatography.**

secondary ion mass spectrometry (SIMS): a surface analysis technique based on the analysis of secondary ions produced by ion bombardment of the sample surface. (40–42, 44)

secondary standard: a standard whose value is based on comparison with a primary standard. Note that a secondary standard, once its value is established, can become a primary standard for other users.

selected ion monitoring (SIM): in mass spectrometry, measurement of ions of selected masses, rather than a complete mass scan of all ions. (28–34)

selectivity: the ability of a methodology or instrumentation to respond to a specific analyte and not to other species in a given sample matrix. Also the extent to which a particular quantitative method is free from additive interference. In chromatographgy, the relative affinities of a pair of analytes for a stationary phase under a given set of chromatographic conditions. For detectors, the types of compounds that will cause a detector response.

sensitivity: the ability of a methodology or instrumentation to discriminate among samples containing different amounts of analyte; the change in analyte signal with a corresponding change in analyte concentration.

sensitized fluorescence: fluorescence in which the emitting species is not directly excited by absorption of light from the source. Generally, excitation of the emitting species occurs via intermolecular energy transfer. (26)

SERS: see **surface-enhanced Raman spectroscopy.**

SFC: see **supercritical fluid chromatography.**

SFE: see **supercritical fluid extraction.**

shim: in NMR, a small electromagnet used to apply small corrective magnetic fields to the sample volume for improved field homogeneity. (17)

Shpol'skii effect: the tendency for absorption and fluorescence spectra to sharpen dramatically when certain analytes (for example, aromatic hydrocarbons) are incorporated into organic solvents (for example, n-alkanes) at low temperature. (26)

SID: surface-induced dissociation. (28–34)

SIM: see **selected ion monitoring.**

SIMS: see **secondary ion mass spectrometry.**

simultaneous multicomponent analysis: a procedure where the concentrations of a mixture of analytes are determined by establishing and solving simultaneous equations. (25)

singlet state: electronic state in a molecule in which the total spin quantum number is 0 (that is, all spins are paired). (26)

size exclusion chromatography (SEC): a high-performance liquid chromatographic (HPLC) technique where molecules are separated according to differences in hydrodynamic volume. (46)

slope method: a method of measuring the rate of a chemical reaction by observing the rate of change of the signal (absorbance) with time. (25)

slope ratio method: a method for determining the metal-to-ligand ratio of complex ions. (25)

SNMS: see **sputtered neutral mass spectroscopy.**

solute distance: in thin-layer chromatography, the distance traveled by the solute along the medium from the starting (application) point or line to the center of the solute spot. (10)

solute property detector: a detector, such as a UV detector, that measures a property of the solute only. (7–13)

solvent strength index: a measure of the relative solvent strength (used to select solvents in adsorption chromatography. (7–13)

SPE: solid phase extraction. (2)

speciation: the determination of the presence, amount, or concentration of an element in a specific oxidation or chelation state as opposed to total analysis of the element in all forms.

specific conductance (κ): term used for electrolytic conductivity (defined as the conductance of a 1 cm^3 cube of electrolyte solution given in the same units as electrolytic conductivity). (39)

specificity: the sensitivity that a detector or method has for an analyte compared to its sensitivity for another, potentially interfering species.

spectral bandwidth (SBW): the width of the band of light emerging from a wavelength-isolation device that makes up 75% of the total emergent optical energy (25)

spectral interference: in spectrometry, any sample constituent other than the analyte that absorbs or emits in the same wavelength region as the analyte. (19–27)

spectral resolution: a term describing how well-separated two adjacent spectral bands are. It can be defined as the ratio $\lambda/\delta\lambda$, where λ is the wavelength of radiation being examined and $\delta\lambda$ is the spectral bandwidth (the width in wavelength units at half-height of the spectral band). (15)

spectrophotometer: an instrument designed to measure transmittance or absorbance of samples, usually using a prism or grating monochromator as the wavelength-isolation device. (25)

spike: addition of a known quantity of the analyte or some other species to a sample for calibration purposes; see **internal standard method** and **standard addition**. (19–27)

split ratio: in capillary gas chromatography, the ratio of the amount of the injected sample "split" or "thrown away" versus the amount that goes into the column for analysis. (8, 31)

SPM: see **scanning probe microscopy.**

spot: a zone in paper and thin-layer chromatography of approximately circular appearance. (10)

sputtered neutral mass spectroscopy (SNMS): a surface analysis technique based on the analysis of secondary ions produced by nonresonant photoionization of neutral species produced by ion sputtering of a solid surface. (40)

square wave voltammetry (SWV): technique in which electrode potential modulated by a square wave and the current signal is the difference of currents on the two half-cycles of the square wave. (37)

SRM: see **standard reference material.**

SRM: selected reaction monitoring (used in tandem mass spectrometry). (28–34)

standard addition: a method in which small, measured increments of the analyte are added to a sample to determine the amount of analyte originally present in the sample. This method has the advantage of minimizing the effect of the sample matrix on the determination of the analyte. (4)

standard reference material (SRM): a sample whose composition is certified by the National Institute of Standards and Technology or other authoritative organization.

standard: a substance or material whose properties are known with sufficient accuracy to permit it to be used in the evaluation of the same properties in other substances. In chemical analysis, it often describes a solution or other substance prepared by the analyst to establish a calibration curve or response function for an instrument.

starting point or line: in thin-layer (planar) chromatography, the point or line on a chromatographic paper or layer where the substance to be chromatographed is applied. (10)

static (hanging) mercury drop electrode (S(H)MDE): renewable mercury electrode in which drops are formed at the end of a capillary by a computer-controlled mechanical device. (37)

stationary phase: the material inside the column which may be a solid or viscous liquid phase. Also called the solvent in gas chromatography when the phase is a liquid. (7–13)

Stern–Volmer equation: equation relating the extent to which a particular species quenches the fluorescence of an analyte to the concentration of the quencher. (26)

STM: see **scanning tunneling microscopy.**

Stokes shift: difference (expressed in wavelength or wavenumber units) between positions of the fluorescence and absorption maxima for a compound. (26)

stray light: radiation falling outside the nominal spectral bandpass of a wavelength selector (filter or monochromator) that impinges on the sample. Of concern in any form of absorption spectrometric measurement. (19–27)

stray light peak: an anomalous absorbance peak resulting from an excessive amount of stray light falling on the sample. (25)

stress relaxation: a mechanical test conducted by introducing and maintaining a constant strain and monitoring the load over the time required to maintain the strain. (51)

supercritical fluid: any substance that is above its critical temperature (the highest temperature at which a gas can be converted to a liquid by an increase in pressure) and critical pressure (the highest pressure at which a liquid can be converted to a traditional gas by an increase in the liquid temperature). The fluid possesses properties of both gas and liquid. (11)

supercritical fluid chromatography (SFC): chromatographic process whereby the mobile phase is a supercritical fluid. (11)

supercritical fluid extraction (SFE): extraction process using supercritical fluids that have high solvating power, rather than traditional organic solvents, for sample preparation of substances to be analyzed by a subsequent instrumental technique. (11)

suppressor: in ion chromatography, a device used to reduce the background conductance of an eluent by selective removal of either anions or cations. (12)

surface-enhanced Raman spectroscopy: samples adsorbed onto microscopically roughened surfaces or colloids that give stronger than expected Raman spectra. (14)

SWV: see **square wave voltammetry.**

synchronous spectrum: spectrum obtained by scanning the excitation and emission wavelengths simultaneously; widely used in mixture analysis. (26)

T_1: an exponential time constant for the longitudinal (z-axis) relaxation of nuclear spins toward equilibrium. (17)

T_2: an exponential time constant for the transverse (x–y plane) relaxation of nuclear spins toward equilibrium. (17)

T_2^*: the observed exponential time constant for the transverse (x–y plane) relaxation of nuclear spins toward equilibrium. (17)

tandem mass spectrometry (MS/MS): the use of two sequential analyzers in mass spectrometry. (28–34)

TCD: thermal conductivity detector. (8)

TDS: see **total dissolved solids.**

technique: a physical or chemical principle used separately or in combination with other techniques to determine the composition or structure of materials. (1)

temperature programming: elution of solutes using a temperature gradient of the column or oven temperature throughout the separation. (7–13)

temperature rise time (TRT): the time required to raise the sample temperature from ambient (or some other initial temperature) to the final pyrolysis temperature. (49)

tensile testing: mechanical tests where dumbbell-shaped, flat specimens are subjected to a uniaxial stress that stretches the sample. (51)

tetramethyl silane (TMS): a compound used to calibrate nuclear magnetic resonance spectrometers. (17)

TGA: see **thermogravimetric analysis.**

thermogravimetric analysis (TGA): a thermal technique used to monitor the change in the mass of a sample as it is heated or held isothermally at a specific temperature. (50)

thermomechanical analysis (TMA): a thermal technique that monitors volumetric changes in a polymeric system. (50)

THT: see **total heating time.**

TIC: see **total ion current chromatogram.**

time-of-flight (TOF): a type of mass spectrometer that separates ions of different mass by the flight time of the ions from the sample surface to the detector. (44)

TIS: see **total ionized solids.**

TLC: thin layer (planar) chromatography. (13).

TMA: see **thermomechanical analysis.**

TMS: see **tetramethyl silane.**

TOCSY: see **total correlation spectroscopy.**

TOF: see **time of flight.**

TOFMS: time-of-flight mass spectrometry. (28–34)

total correlation spectroscopy (TOCSY): a two-dimensional experiment that correlates all of the spins in a spin system. (17)

total dissolved solids (TDS): term used to approximate the total mass fraction of dissolved solids in a sample from electrolytic conductivity measurement, where the TDS value is dependent on the standard solution used for calibration. Usually expressed in $\mu g/g$ or parts per million (ppm) by mass. (39)

total heating time (THT): the total time required to raise the sample temperature and pyrolyze the sample at the final temperature. (45–51)

total ion current (TIC) chromatogram: a plot of mass spectrometry total ion current versus time in a gas chromatography-mass spectrometry analysis. (31, 33)

total ionized solids (TIS): term used to approximate the total ionic mass fraction in a sample from electrolytic conductivity measurement, where the TIS value is dependent on the standard solution used for calibration. Usually expressed in $\mu g/g$ or parts per million (ppm) by mass. (39)

transmittance: the mathematical ratio of the optical power of the transmitted beam to the optical power of the incident beam. (25)

transport of ions in materials (TRIM): a Monte Carlo simulation of ion-materials interactions that occur as a result of ion implantation or bombardment. (44)

TRIM: see **transport of ions in materials.**

triplet state: electronic state in a molecule in which the total spin quantum number is 1 (that is, two unpaired electron spins). (26)

TRT: see **temperature rise time.**

two-dimensional nuclear magnetic resonance: experiments using the observation of two time domains and therefore two frequency domains. An evolution time is observed during which J-coupling, chemical exchange, or NOE produce correlation peaks. (17)

u: mass unit, based on carbon-12 standard, units of dalton. (28–34)

UHV: see **ultra-high vacuum.**

ultra-high vacuum (UHV): a base pressure in a vacuum chamber of 1×10^{-8} mbar or lower. (41–44)

UV/Vis: ultraviolet/visible molecular absorption spectrometry.

UV: ultraviolet. (19–27)

validation: the process by which a sample, a measurement method, data, or information is deemed useful for a specified purpose. The process includes (1) initial specifications and implementation, (2) testing to see if specifications are met, and (3) documentation of entire process and results.

variable-wavelength detector: an absorbance detector that can be used to sense a range of wavelengths, but only one at a time.

VOC: volatile organic compound. (8)

V_R **(retention volume for a SEC experiment):** a term relating the exclusion volume or interstitial volume between the packing particles, the particle pore volume, the stationary phase volume, and the distribution constants for the SEC solute (K_{SEC}) and the LC solute (K_{LC}). (46)

WDXRF: wavelenghth-dispersive X-ray fluorescence. (24)

X-ray fluorescence spectrometry (XFS): an atomic spectrometric method based on the detection of emitted X-ray radiation from exicited atoms. (24)

X-ray photoelectron spectroscopy (XPS): a surface analysis technique based on the analysis of photoelectrons excited by x-ray radiation. (43)

XFS: see **X-ray fluorescence spectrometry.**

XHCORR: X (any other nucleus) H (hydrogen) CORR (correlation), a two-dimensional NMR experiment that produces heteronuclear (different isotopes) chemical shift correlations. This typically refers to proton–carbon chemical shift correlation. (17)

XPS: see **X-ray photoelectron spectroscopy.**

XRF: X-ray fluorescence. (19–27)

XRMF: X-ray microfluorescence. (24)

XSTM: see **cross-sectional scanning tunneling microscopy.**

Index

About the Editor

Frank A. Settle is professor emeritus of chemistry at the Virginia Military Institute where he taught analytical chemistry for 28 years. He is currently a consultant in analytical chemistry focusing on laboratory automation and information management. He is a co-author of *Instrumental Methods of Analysis* (Wadsworth, 1987) and numerous papers.

Technique Selection Software to accompany
"Handbook of Instrumental Techniques for Analytical Chemistry"
by Frank Settle

Minimum System Requirements

486 SX, 33 Mhz or higher
Windows 3.1, 95 or NT
8 MB RAM
5 MB available hard-disk space
Color Monitor (256 or more colors)
2x CD-ROM Drive

Installation

Insert the CD into your CD-ROM drive.

If you are using Windows 3.1:
-- from the Windows Program Manager select Run from the File Menu
-- from the Command Line box, type "d:setup" (where "d" is the letter of your CD-ROM drive, otherwise insert the proper drive letter).

If you are using Windows 95 or NT:
-- click on the Start Button in the Taskbar then select Run
-- in the text box, type "d:setup" (or your CD ROM drive letter).

Once the installation process begins, the main installer screen will appear. You will be asked where you want to install the technique selection software. You may accept the suggested location, which will place the program on your C: drive in a directory called "HANDBOOK", or you may designate another location.

When the installation is complete, a new Program Group will appear called "Technique Selection Software"

How to use the Technique Selection Software

To begin using the software double-click the Technique Selection Software icon. A screen will appear prompting you for a Sample ID and User Name. Type in the appropriate information then click the Start button.

An opening screen called Desired Output will offer two options.

USING PARAMETERS

Click the Parameters button. A screen with eight parameters will come up. Each box has a pull down menu describing the attributes of your sample.

**NOTE: You must select an attribute for each box. Once you have done this click the Continue button.

Another screen will appear with your desired output. It will list a technique you can use to evaluate your sample. In this screen you will see three buttons.
-- Change a Parameter will take you back to the previous screen, where parameters may be altered.
-- Next Record will display another technique you can use with that sample. Keep in mind that there will not always be more than one technique available.
-- Previous Record lets you scroll back through all the techniques appropriate for your sample.
Depending on the combination of parameters you select, there may be no suitable technique. A prompt will appear stating this.

To go back to the opening Desired Output screen, select File -> Restart from the menu which appears at the top of the screen.

USING TECHNIQUES

Click the Technique button. In the next screen, You will see an empty box called Technique. Scroll down the list until you find the technique you would like more information about. After a few seconds the screen will display all the information regarding that technique and the parameters your sample must meet in order to use it. To select a new technique simply choose one from the menu.

At all times you may exit the program by selecting File -> Exit from the menu at the top of the screen.

Technique Selection Software
Prentice-Hall, Inc.

YOU SHOULD CAREFULLY READ THE TERMS AND CONDITIONS BEFORE USING THE DISKETTE PACKAGE. USING THIS DISKETTE PACKAGE INDICATES YOUR ACCEPTANCE OF THESE TERMS AND CONDITIONS.

Prentice-Hall, Inc. provides this program and licenses its use. You assume responsibility for the selection of the program to achieve your intended results, and for the installation, use, and results obtained from the program. This license extends only to use of the program in the United States or countries in which the program is marketed by authorized distributors.

LICENSE GRANT

You hereby accept a nonexclusive, nontransferable, permanent license to install and use the program ON A SINGLE COMPUTER at any given time. You may copy the program solely for backup or archival purposes in support of your use of the program on the single computer. You may not modify, translate, disassemble, decompile, or reverse engineer the program, in whole or in part.

TERM

The License is effective until terminated. Prentice-Hall, Inc. reserves the right to terminate this License automatically if any provision of the License is violated. You may terminate the License at any time. To terminate this License, you must return the program, including documentation, along with a written warranty stating that all copies in your possession have been returned or destroyed.

LIMITED WARRANTY

THE PROGRAM IS PROVIDED "AS IS" WITHOUT WARRANTY OF ANY KIND, EITHER EXPRESSED OR IMPLIED, INCLUDING, BUT NOT LIMITED TO, THE IMPLIED WARRANTIES OR MERCHANTABILITY AND FITNESS FOR A PARTICULAR PURPOSE. THE ENTIRE RISK AS TO THE QUALITY AND PERFORMANCE OF THE PROGRAM IS WITH YOU. SHOULD THE PROGRAM PROVE DEFECTIVE, YOU (AND NOT PRENTICE-HALL, INC. OR ANY AUTHORIZED DEALER) ASSUME THE ENTIRE COST OF ALL NECESSARY SERVICING, REPAIR, OR CORRECTION. NO ORAL OR WRITTEN INFORMATION OR ADVICE GIVEN BY PRENTICE-HALL, INC., ITS DEALERS, DISTRIBUTORS, OR AGENTS SHALL CREATE A WARRANTY OR INCREASE THE SCOPE OF THIS WARRANTY.

SOME STATES DO NOT ALLOW THE EXCLUSION OF IMPLIED WARRANTIES, SO THE ABOVE EXCLUSION MAY NOT APPLY TO YOU. THIS WARRANTY GIVES YOU SPECIFIC LEGAL RIGHTS AND YOU MAY ALSO HAVE OTHER LEGAL RIGHTS THAT VARY FROM STATE TO STATE.

Prentice-Hall, Inc. does not warrant that the functions contained in the program will meet your requirements or that the operation of the program will be uninterrupted or error-free.

However, Prentice-Hall, Inc. warrants the diskette(s) on which the program is furnished to be free from defects in material and workmanship under normal use for a period of ninety (90) days from the date of delivery to you as evidenced by a copy of your receipt.

The program should not be relied on as the sole basis to solve a problem whose incorrect solution could result in injury to person or property. If the program is employed in such a manner, it is at the user's own risk and Prentice-Hall, Inc. explicitly disclaims all liability for such misuse.

LIMITATION OF REMEDIES

Prentice-Hall, Inc.'s entire liability and your exclusive remedy shall be:

1. the replacement of any diskette not meeting Prentice-Hall, Inc.'s "LIMITED WARRANTY" and that is returned to Prentice-Hall, or
2. if Prentice-Hall is unable to deliver a replacement diskette that is free of defects in materials or workmanship, you may terminate this agreement by returning the program.

IN NO EVENT WILL PRENTICE-HALL, INC. BE LIABLE TO YOU FOR ANY DAMAGES, INCLUDING ANY LOST PROFITS, LOST SAVINGS, OR OTHER INCIDENTAL OR CONSEQUENTIAL DAMAGES ARISING OUT OF THE USE OR INABILITY TO USE SUCH PROGRAM EVEN IF PRENTICE-HALL, INC. OR AN AUTHORIZED DISTRIBUTOR HAS BEEN ADVISED OF THE POSSIBILITY OF SUCH DAMAGES, OR FOR ANY CLAIM BY ANY OTHER PARTY.

SOME STATES DO NOT ALLOW FOR THE LIMITATION OR EXCLUSION OF LIABILITY FOR INCIDENTAL OR CONSEQUENTIAL DAMAGES, SO THE ABOVE LIMITATION OR EXCLUSION MAY NOT APPLY TO YOU.

GENERAL

You may not sublicense, assign, or transfer the license of the program. Any attempt to sublicense, assign or transfer any of the rights, duties, or obligations hereunder is void.

This Agreement will be governed by the laws of the State of New York.

Should you have any questions concerning this Agreement, you may contact Prentice-Hall, Inc. by writing to:
Director of New Media
Higher Education Division
Prentice-Hall, Inc.
1 Lake Street
Upper Saddle River, NJ 07458

Should you have any questions concerning technical support, you may write to:
New Media Production
Higher Education Division
Prentice-Hall, Inc.
1 Lake Street
Upper Saddle River, NJ 07458

YOU ACKNOWLEDGE THAT YOU HAVE READ THIS AGREEMENT, UNDERSTAND IT, AND AGREE TO BE BOUND BY ITS TERMS AND CONDITIONS. YOU FURTHER AGREE THAT IT IS THE COMPLETE AND EXCLUSIVE STATEMENT OF THE AGREEMENT BETWEEN US THAT SUPERSEDES ANY PROPOSAL OR PRIOR AGREEMENT, ORAL OR WRITTEN, AND ANY OTHER COMMUNICATIONS BETWEEN US RELATING TO THE SUBJECT MATTER OF THIS AGREEMENT.